A Second Course in Linear Algebra

Linear algebra is a fundamental tool in many fields, including mathematics and statistics, computer science, economics, and the physical and biological sciences. This undergraduate textbook offers a complete second course in linear algebra, tailored to help students transition from basic theory to advanced topics and applications. Concise chapters promote a focused progression through essential ideas, and contain many examples and illustrative graphics. In addition, each chapter contains a bullet list summarizing important concepts, and the book includes over 600 exercises to aid the reader's understanding.

Topics are derived and discussed in detail, including the singular value decomposition, the Jordan canonical form, the spectral theorem, the QR factorization, normal matrices, Hermitian matrices (of interest to physics students), and positive definite matrices (of interest to statistics students).

Stephan Ramon Garcia is Associate Professor of Mathematics at Pomona College. He is the author of two books and over seventy research articles in operator theory, complex analysis, matrix analysis, number theory, discrete geometry, and other fields. He is on the editorial boards of the *Proceedings of the American Mathematical Society* (2016–), *Involve* (2011–), and *The American Mathematical Monthly* (2017–). He received three NSF research grants as principal investigator, five teaching awards from three different institutions, and was twice nominated by Pomona College for the CASE US Professors of the Year award.

Roger A. Horn was Research Professor of Mathematics at the University of Utah. His publications include the Cambridge University Press books Matrix Analysis and Topics in Matrix Analysis as well as more than 100 research articles in matrix analysis, statistics, health services research, complex variables, probability, differential geometry, and analytic number theory. He was Editor of *The American Mathematical Monthly* (1996–2001), the MAA *Spectrum* book series (1992–1995), and the MAA *Carus Mathematical Monographs* (2002–2005). He has also served on the editorial boards of the *SIAM Journal of Matrix Analysis*, *Linear Algebra and its Applications*, and the *Electronic Journal of Linear Algebra*.

A Second Course in Linear Algebra

Stephan Ramon Garcia
Pomona College, California

Roger A. Horn
University of Utah

CAMBRIDGE
UNIVERSITY PRESS

CAMBRIDGE
UNIVERSITY PRESS

University Printing House, Cambridge CB2 8BS, United Kingdom

One Liberty Plaza, 20th Floor, New York, NY 10006, USA

477 Williamstown Road, Port Melbourne, VIC 3207, Australia

4843/24, 2nd Floor, Ansari Road, Daryaganj, Delhi – 110002, India

79 Anson Road, #06–04/06, Singapore 079906

Cambridge University Press is part of the University of Cambridge.

It furthers the University's mission by disseminating knowledge in the pursuit of education, learning, and research at the highest international levels of excellence.

www.cambridge.org
Information on this title: www.cambridge.org/9781107103818

First published 2017

Printed in the United Kingdom by TJ International Ltd. Padstow

A catalogue record for this publication is available from the British Library.

Library of Congress Cataloging-in-Publication data
Names: Garcia, Stephan Ramon. | Horn, Roger A.
Title: A second course in linear algebra / Stephan Ramon Garcia,
Pomona College, California, Roger A. Horn, University of Utah.
Description: Cambridge : Cambridge University Press, 2017. |
Includes bibliographical references and index.
Identifiers: LCCN 2016033754 | ISBN 9781107103818 (Hardback : alk. paper)
Subjects: LCSH: Algebras, Linear–Textbooks.
Classification: LCC QA184.2 .G37 2017 | DDC 512/.5–dc23
LC record available at https://lccn.loc.gov/2016033754

ISBN 978-1-107-10381-8 Hardback

To our families:

Gizem, Reyhan, and Altay
Susan;
Craig, Cori, Cole, and Carson;
Howard, Heidi, Archer, and Ella Ceres

Contents

Preface

Linear algebra and matrix methods are increasingly relevant in a world focused on the acquisition and analysis of data. Consequently, this book is intended for students of pure and applied mathematics, computer science, economics, engineering, mathematical biology, operations research, physics, and statistics. We assume that the reader has completed a lower-division calculus sequence and a first course in linear algebra.

Noteworthy features of this book include the following:

- Block matrices are employed systematically.

- Matrices and matrix factorizations are emphasized.

- Transformations that involve unitary matrices are emphasized because they are associated with feasible and stable algorithms.

- Numerous examples appear throughout the text.

- Figures illustrate the geometric foundations of linear algebra.

- Topics for a one-semester course are arranged in a sequence of short chapters.

- Many chapters conclude with sections devoted to special topics.

- Each chapter includes a problem section (more than 600 problems in total).

- Notes sections provide references to sources of additional information.

- Each chapter concludes with a bullet list of important concepts introduced in the chapter.

- Symbols used in the book are listed in a table of notation, with page references.

- An index with more than 1700 entries helps locate concepts and definitions, and enhances the utility of the book as a reference.

Matrices and vector spaces in the book are over the complex field. The use of complex scalars facilitates the study of eigenvalues and is consistent with modern numerical linear algebra software. Moreover, it is aligned with applications in physics (complex wave functions and Hermitian matrices in quantum mechanics), electrical engineering (analysis of circuits and signals in which both phase and amplitude are important), statistics (time series and characteristic functions), and computer science (fast Fourier transforms, convergent matrices in iterative algorithms, and quantum computing).

While studying linear algebra with this book, students can observe and practice good mathematical communication skills. These skills include how to state (and read) a theorem carefully; how to choose (and use) hypotheses; how to prove a statement by induction, by contradiction, or by proving its contrapositive; how to improve a theorem by weakening its hypotheses or strengthening its conclusions; how to use counterexamples; and how to write a cogent solution to a problem.

Many topics that are useful in applications of linear algebra fall outside the realm of linear transformations and similarity, so they may be absent from textbooks that adopt an abstract operator approach. These include:

- Geršgorin's theorem

- Householder matrices

- The QR factorization

- Block matrices

- Discrete Fourier transforms

- Circulant matrices

- Matrices with nonnegative entries (Markov matrices)

- The singular value and compact singular value decompositions

- Low-rank approximations to a data matrix

- Generalized inverses (Moore–Penrose inverses)

- Positive semidefinite matrices

- Hadamard (entrywise) and Kronecker (tensor) products

- Matrix norms

- Least squares and minimum norm solutions

- Complex symmetric matrices

- Inertia of normal matrices

- Eigenvalue and singular value interlacing

- Inequalities involving eigenvalues, singular values, and diagonal entries

The book is organized as follows:

Chapter 0 is a review of definitions and results from elementary linear algebra.

Chapters 1 and 2 review complex and real vector spaces, including linear independence, bases, dimension, rank, and matrix representations of linear transformations.

The "second course" topics begin in Chapter 3, which establishes the block-matrix paradigm used throughout the book.

Chapters 4 and 5 review geometry in the Euclidean plane and use it to motivate axioms for inner product and normed linear spaces. Topics include orthogonal vectors, orthogonal projections, orthonormal bases, orthogonalization, the Riesz representation theorem, adjoints, and applications of the theory to Fourier series.

Chapter 6 introduces unitary matrices, which are used in constructions throughout the rest of the book. Householder matrices are used to construct the QR factorization, which is employed in many numerical algorithms.

Chapter 7 discusses orthogonal projections, best approximations, least squares/minimum norm solutions of linear systems, and use of the QR factorization to solve the normal equations.

Chapter 8 introduces eigenvalues, eigenvectors, and geometric multiplicity. We show that an $n \times n$ complex matrix has between one and n distinct eigenvalues, and use Geršgorin's theorem to identify a region in the complex plane that contains them.

Chapter 9 deals with the characteristic polynomial and algebraic multiplicity. We develop criteria for diagonalizability and define primary matrix functions of a diagonalizable matrix. Topics include Fibonacci numbers, the eigenvalues of AB and BA, commutants, and simultaneous diagonalization.

Chapter 10 contains Schur's remarkable theorem that every square matrix is unitarily similar to an upper triangular matrix (with a related result for a commuting family). Schur's theorem is used to show that every square matrix is annihilated by its characteristic polynomial. The latter result motivates introduction of the minimal polynomial and a study of its properties. Sylvester's theorem on linear matrix equations is proved and used to show that every square matrix is similar to a block diagonal matrix with unispectral diagonal blocks.

Chapter 11 builds on the preceding chapter to show that every square matrix is similar to a special block diagonal upper bidiagonal matrix (its Jordan canonical form) that is unique up to permutation of its direct summands. Applications of the Jordan canonical form include initial value problems for linear systems of differential equations, an analysis of the Jordan structures of AB and BA, characterizations of convergent and power-bounded matrices, and a limit theorem for Markov matrices that have positive entries.

Chapter 12 is about normal matrices: matrices that commute with their conjugate transpose. The spectral theorem says that a matrix is normal if and only if it is unitarily diagonalizable; many other equivalent characterizations are known. Hermitian, skew-Hermitian, unitary, real orthogonal, real symmetric, and circulant matrices are all normal.

Positive semidefinite matrices are the subject of Chapter 13. These matrices arise in statistics (correlation matrices and the normal equations), mechanics (kinetic and potential energy in a vibrating system), and geometry (ellipsoids). Topics include the square root function, Cholesky factorization, and the Hadamard and Kronecker products.

The principal result in Chapter 14 is the singular value decomposition, which is at the heart of many modern numerical algorithms in statistics, control theory, approximation, image compression, and data analysis. Topics include the compact singular value decomposition and polar decompositions, with special attention to uniqueness of these factorizations.

In Chapter 15 the singular value decomposition is used to compress an image or data matrix. Other applications of the singular value decomposition discussed are the generalized inverse (Moore–Penrose inverse) of a matrix; inequalities between singular values and eigenvalues; the spectral norm of a matrix; complex symmetric matrices; and idempotent matrices.

Chapter 16 investigates eigenvalue interlacing phenomena for Hermitian matrices that are bordered or are subjected to an additive perturbation. Related results include an interlacing theorem for singular values, a determinant criterion for positive definiteness, and inequalities that characterize eigenvalues and diagonal entries of a Hermitian matrix. We prove Sylvester's inertia theorem for Hermitian matrices and a generalized inertia theorem for normal matrices.

A comprehensive list of symbols and notation (with page references) follows the Preface. A review of complex numbers and a list of references follow Chapter 16. A detailed index is at the end of the book.

The cover art is an image of a 2002 oil painting "Summer Again" (72 × 52 inches) by Lun-Yi Tsai, a New York City artist whose work has often been inspired by mathematical themes.

We thank Zachary Glassman for producing many of our illustrations and for answering our LATEX questions.

We thank Dennis Merino, Russ Merris, and Zhongshan Li for carefully reading a preliminary draft.

We thank the students who attended the first author's advanced linear algebra courses at Pomona College during Fall 2014 and Fall 2015. In particular, we thank Ahmed Al Fares, Andreas Biekert, Andi Chen, Wanning Chen, Alex Cloud, Bill DeRose, Jacob Fiksel, Logan Gilbert, Sheridan Grant, Adam He, David Khatami, Cheng Wai Koo, Bo Li, Shiyue Li, Samantha Morrison, Nathanael Roy, Michael Someck, Sallie Walecka, and Wentao Yuan for catching several mistakes in a preliminary version of this text.

Special thanks to Ciaran Evans, Elizabeth Sarapata, Adam Starr, and Adam Waterbury for their eagle-eyed reading of the text.

S.R.G.
R.A.H.

Notation

$\in, \notin$	is / is not an element of
$\subseteq$	is a subset of
$\varnothing$	the empty set
$\times$	Cartesian product
$f : X \to Y$	f is a function from X into Y
$\implies$	implies
$\iff$	if and only if
$x \mapsto y$	implicit definition of a function that maps x to y
$\mathbb{N} = \{1, 2, 3, \ldots\}$	the set of all natural numbers
$\mathbb{Z} = \{\ldots, -2, -1, 0, 1, 2, \ldots\}$	the set of all integers
$\mathbb{R}$	the set of real numbers
$\mathbb{C}$	the set of complex numbers
$\mathbb{F}$	field of scalars ($\mathbb{F} = \mathbb{R}$ or $\mathbb{C}$)
$[a, b]$	a real interval that includes its endpoints a, b
$\mathcal{U}, \mathcal{V}, \mathcal{W}$	vector spaces
$\mathscr{U}, \mathscr{V}$	subsets of vector spaces
$a, b, c, \ldots$	scalars
$\mathbf{a}, \mathbf{b}, \mathbf{c}, \ldots$	(column) vectors
$A, B, C, \ldots$	matrices
δ_{ij}	Kronecker delta (p. 3)
I_n	$n \times n$ identity matrix (p. 3)
I	identity matrix (size inferred from context) (p. 3)
$\operatorname{diag}(\cdot)$	diagonal matrix with specified entries (p. 4)
$A^0 = I$	convention for zeroth power of a matrix (p. 4)
A^T	transpose of A (p. 5)
$A^{-\mathsf{T}}$	inverse of A^T (p. 5)
$\overline{A}$	conjugate of A (p. 5)
A^*	conjugate transpose (adjoint) of A (p. 5)
A^{-*}	inverse of A^* (p. 5)
$\operatorname{tr} A$	trace of A (p. 6)
$\det A$	determinant of A (p. 8)
$\operatorname{adj} A$	adjugate of A (p. 9)
$\operatorname{sgn} \sigma$	sign of a permutation σ (p. 10)
$\deg p$	degree of a polynomial p (p. 12)
$\mathcal{P}_n$	set of complex polynomials of degree at most n (p. 21)

$\mathcal{P}_n(\mathbb{R})$	set of real polynomials of degree at most n (p. 21)
$\mathcal{P}$	set of all complex polynomials (p. 22)
$\mathcal{P}(\mathbb{R})$	set of all real polynomials (p. 22)
$C_{\mathbb{F}}[a,b]$	set of continuous $\mathbb{F}$-valued functions on $[a,b]$, $\mathbb{F} = \mathbb{C}$ or $\mathbb{R}$ (p. 22)
$C[a,b]$	set of continuous $\mathbb{C}$-valued functions on $[a,b]$ (p. 22)
null A	null space of a matrix A (p. 23)
col A	column space of a matrix A (p. 23)
$\mathcal{P}_{\text{even}}$	set of even complex polynomials (p. 23)
$\mathcal{P}_{\text{odd}}$	set of odd complex polynomials (p. 23)
$A\mathcal{U}$	A acting on a subspace $\mathcal{U}$ (p. 23)
span $\mathcal{S}$	span of a subset $\mathcal{S}$ of a vector space (p. 24)
$\mathbf{e}$	all-ones vector (p. 26)
$\mathcal{U} \cap \mathcal{W}$	intersection of subspaces $\mathcal{U}$ and $\mathcal{W}$ (p. 26)
$\mathcal{U} + \mathcal{W}$	sum of subspaces $\mathcal{U}$ and $\mathcal{W}$ (p. 27)
$\mathcal{U} \oplus \mathcal{W}$	direct sum of subspaces $\mathcal{U}$ and $\mathcal{W}$ (p. 27)
$\mathbf{v}_1, \mathbf{v}_2, \ldots, \widehat{\mathbf{v}}_j, \ldots, \mathbf{v}_r$	list of vectors with $\mathbf{v}_j$ omitted (p. 30)
$\mathbf{e}_1, \mathbf{e}_2, \ldots, \mathbf{e}_n$	standard basis of $\mathbb{F}^n$ (p. 35)
E_{ij}	matrix with (i,j) entry 1 and all others 0 (p. 35)
dim $\mathcal{V}$	dimension of $\mathcal{V}$ (p. 35)
$[\mathbf{v}]_\beta$	coordinate vector of $\mathbf{v}$ with respect to a basis β (p. 40)
$\mathcal{L}(\mathcal{V}, \mathcal{W})$	set of linear transformations from $\mathcal{V}$ to $\mathcal{W}$ (p. 41)
$\mathcal{L}(\mathcal{V})$	set of linear transformations from $\mathcal{V}$ to itself (p. 41)
ker T	kernel of T (p. 42)
ran T	range of T (p. 42)
I	identity linear transformation (p. 44)
row A	row space of a matrix A (p. 59)
rank A	rank of a matrix A (p. 60)
$\star$	unspecified matrix entry (p. 65)
$A \oplus B$	direct sum of matrices A and B (p. 66)
$[A, B]$	commutator of A and B (p. 71)
$A \otimes B$	Kronecker product of matrices A and B (p. 74)
vec A	vec of A (p. 75)
$\langle \cdot, \cdot \rangle$	inner product (p. 87)
$\perp$	orthogonal (p. 90)
$\| \cdot \|$	norm (p. 90)
$\| \cdot \|_2$	Euclidean norm (p. 91)
$\| \cdot \|_1$	ℓ^1 norm (absolute sum norm) (p. 97)
$\| \cdot \|_\infty$	ℓ^∞ norm (max norm) (p. 97)
$_\gamma[T]_\beta$	matrix representation of $T \in \mathcal{L}(\mathcal{V}, \mathcal{W})$ with respect to bases β and γ (p. 110)
F_n	$n \times n$ Fourier matrix (p. 129)
$\mathcal{U}^\perp$	orthogonal complement of a set $\mathcal{U}$ (p. 149)
$P_\mathcal{U}$	orthogonal projection onto $\mathcal{U}$ (p. 155)

$d(\mathbf{v}, \mathcal{U})$	distance from $\mathbf{v}$ to $\mathcal{U}$ (p. 160)
$G(\mathbf{u}_1, \mathbf{u}_2, \ldots, \mathbf{u}_n)$	Gram matrix (p. 164)
$g(\mathbf{u}_1, \mathbf{u}_2, \ldots, \mathbf{u}_n)$	Gram determinant (p. 164)
$\mathrm{spec}\, A$	spectrum of A (p. 183)
$\mathcal{E}_\lambda(A)$	eigenspace of A for eigenvalue λ (p. 186)
$p_A(\cdot)$	characteristic polynomial of A (p. 201)
$\mathcal{F}'$	commutant of a set of matrices $\mathcal{F}$ (p. 213)
e^A	matrix exponential (p. 212)
$m_A(\cdot)$	minimal polynomial of A (p. 229)
C_p	companion matrix of the polynomial p (p. 230)
$J_k(\lambda)$	$k \times k$ Jordan block with eigenvalue λ (p. 244)
J_k	$k \times k$ nilpotent Jordan block (p. 245)
$w_1, w_1, \ldots, w_q$	Weyr characteristic of a matrix (p. 252)
$\rho(A)$	spectral radius of A (p. 260)
$p(n)$	number of partitions of n (p. 271)
$\Delta(A)$	defect from normality of A (p. 285)
$A \circ B$	Hadamard product of A and B (p. 319)
$|A|$	modulus of A (p. 336)
$\sigma_{\max}(A)$	maximum singular value (p. 348)
$\sigma_{\min}(A)$	minimum singular value (p. 350)
$\sigma_1(A), \sigma_2(A), \ldots$	singular values of A (p. 350)
$A^\dagger$	pseudoinverse of A (p. 356)
$\kappa_2(A)$	spectral condition number of A (p. 359)
$\mathrm{Re}\, z$	real part of the complex number z (p. 398)
$\mathrm{Im}\, z$	imaginary part of the complex number z (p. 398)
$|z|$	modulus of the complex number z (p. 401)
$\arg z$	argument of the complex number z (p. 401)

0 Preliminaries

In this chapter, we review some concepts from elementary linear algebra and discuss mathematical induction. We document some facts about complex polynomials (including the Fundamental Theorem of Algebra, the division algorithm, and Lagrange interpolation) and introduce polynomial functions of a matrix.

0.1 Functions and Sets

Let $\mathscr{X}$ and $\mathscr{Y}$ be sets. The notation $f : \mathscr{X} \to \mathscr{Y}$ indicates that f is a *function* whose *domain* is $\mathscr{X}$ and *codomain* is $\mathscr{Y}$. That is, f assigns a definite value $f(x) \in \mathscr{Y}$ to each $x \in \mathscr{X}$. A function may assign the same value to two different elements in its domain, that is, $x_1 \neq x_2$ and $f(x_1) = f(x_2)$ is possible. But $x_1 = x_2$ and $f(x_1) \neq f(x_2)$ is not possible.

The *range* of $f : \mathscr{X} \to \mathscr{Y}$ is

$$\operatorname{ran} f = \{f(x) : x \in \mathscr{X}\} = \{y \in \mathscr{Y} : y = f(x) \text{ for some } x \in \mathscr{X}\},$$

which is a subset of $\mathscr{Y}$. A function $f : \mathscr{X} \to \mathscr{Y}$ is *onto* if $\operatorname{ran} f = \mathscr{Y}$, that is, if the range and codomain of f are equal. A function $f : \mathscr{X} \to \mathscr{Y}$ is *one to one* if $f(x_1) = f(x_2)$ implies that $x_1 = x_2$. Equivalently, f is one to one if $x_1 \neq x_2$ implies that $f(x_1) \neq f(x_2)$; see Figure 0.1.

We say that elements $x_1, x_2, \ldots, x_k$ of a set are *distinct* if $x_i \neq x_j$ whenever $i, j \in \{1, 2, \ldots, k\}$ and $i \neq j$.

0.2 Scalars

We denote the real numbers by $\mathbb{R}$ and the complex numbers by $\mathbb{C}$. Real or complex numbers are called *scalars*. The only scalars that we consider are complex numbers, which we sometimes restrict to being real. See Appendix A for a discussion of complex numbers.

0.3 Matrices

An $m \times n$ *matrix* is a rectangular array

$$A = [a_{ij}] = \begin{bmatrix} a_{11} & a_{12} & \cdots & a_{1n} \\ a_{21} & a_{22} & \cdots & a_{2n} \\ \vdots & \vdots & \ddots & \vdots \\ a_{m1} & a_{m2} & \cdots & a_{mn} \end{bmatrix} \tag{0.3.1}$$

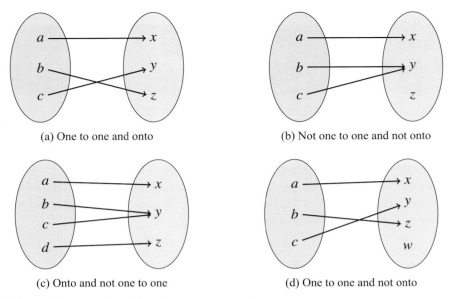

<table>
<tr><td>(a) One to one and onto</td><td>(b) Not one to one and not onto</td></tr>
<tr><td>(c) Onto and not one to one</td><td>(d) One to one and not onto</td></tr>
</table>

Figure 0.1 Properties of functions: one to one and onto.

of real or complex numbers. The (i, j) *entry* of A is a_{ij}. Two matrices are *equal* if they have the same size (the same number of rows and columns) and if their corresponding entries are equal. An $n \times n$ matrix is a *square matrix*. The set of all $m \times n$ matrices with complex entries is denoted by $\mathsf{M}_{m \times n}(\mathbb{C})$, or by $\mathsf{M}_n(\mathbb{C})$ if $m = n$. For convenience, we write $\mathsf{M}_n(\mathbb{C}) = \mathsf{M}_n$ and $\mathsf{M}_{m \times n}(\mathbb{C}) = \mathsf{M}_{m \times n}$. The set of $m \times n$ matrices with real entries is denoted by $\mathsf{M}_{m \times n}(\mathbb{R})$, or by $\mathsf{M}_n(\mathbb{R})$ if $m = n$. In this book, we consider only matrices with real or complex entries.

Rows and Columns For each $i = 1, 2, \ldots, m$, the ith *row* of the matrix A in (0.3.1) is the $1 \times n$ matrix

$$[a_{i1} \ a_{i2} \ \ldots \ a_{in}].$$

For each $j = 1, 2, \ldots, n$, the jth *column* of A is the $m \times 1$ matrix

$$\mathbf{a}_j = \begin{bmatrix} a_{1j} \\ a_{2j} \\ \vdots \\ a_{mj} \end{bmatrix}.$$

It is often convenient to write the matrix (0.3.1) as a $1 \times n$ array of columns

$$A = [\mathbf{a}_1 \ \mathbf{a}_2 \ \ldots \ \mathbf{a}_n].$$

Addition and Scalar Multiplication If $A = [a_{ij}]$ and $B = [b_{ij}]$ are $m \times n$ matrices, then $A + B$ is the $m \times n$ matrix whose (i, j) entry is $a_{ij} + b_{ij}$. If $A \in \mathsf{M}_{m \times n}$ and c is a scalar, then $cA = [ca_{ij}]$ is the $m \times n$ matrix obtained by multiplying each entry of A by c. A *zero matrix* is an $m \times n$ matrix whose entries are all zero. Such a matrix is denoted by 0, although subscripts can be attached to indicate its size. Let $A, B \in \mathsf{M}_{m \times n}$ and let c, d be scalars.

(a) $A + B = B + A$.

(b) $A + (B + C) = (A + B) + C$.

(c) $A + 0 = A = 0 + A$.

(d) $c(A + B) = cA + cB$.

(e) $c(dA) = (cd)A = d(cA)$.

(f) $(c + d)A = cA + dA$.

Multiplication If $A = [a_{ij}] \in \mathsf{M}_{m \times r}$ and $B = [b_{ij}] \in \mathsf{M}_{r \times n}$, then the (i, j) entry of the product $AB = [c_{ij}] \in \mathsf{M}_{m \times n}$ is

$$c_{ij} = \sum_{k=1}^{r} a_{ik} b_{kj}. \tag{0.3.2}$$

This sum involves entries in the ith row of A and the jth column of B. The number of columns of A must be equal to the number of rows of B. If we write $B = [\mathbf{b}_1\ \mathbf{b}_2\ \ldots\ \mathbf{b}_n]$ as a $1 \times n$ array of its columns, then (0.3.2) says that

$$AB = [A\mathbf{b}_1\ A\mathbf{b}_2\ \ldots\ A\mathbf{b}_n].$$

See Chapter 3 for other interpretations of matrix multiplication.

We say that $A, B \in \mathsf{M}_n$ *commute* if $AB = BA$. Some pairs of matrices in M_n do not commute. Moreover, $AB = AC$ does not imply that $B = C$. Let A, B, and C be matrices of appropriate sizes and let c be a scalar.

(a) $A(BC) = (AB)C$.

(b) $A(B + C) = AB + AC$.

(c) $(A + B)C = AC + BC$.

(d) $(cA)B = c(AB) = A(cB)$.

Identity Matrices The matrix

$$I_n = \begin{bmatrix} 1 & 0 & 0 & \cdots & 0 \\ 0 & 1 & 0 & \cdots & 0 \\ 0 & 0 & 1 & \cdots & 0 \\ \vdots & \vdots & \vdots & \ddots & \vdots \\ 0 & 0 & 0 & \cdots & 1 \end{bmatrix} \in \mathsf{M}_n$$

is the $n \times n$ *identity matrix*. That is, $I_n = [\delta_{ij}]$, in which

$$\delta_{ij} = \begin{cases} 1 & \text{if } i = j, \\ 0 & \text{if } i \neq j, \end{cases}$$

is the *Kronecker delta*. If the size is clear from context, we write I in place of I_n. For every $A \in \mathsf{M}_{m \times n}$,

$$AI_n = A = I_m A.$$

Triangular Matrices Let $A = [a_{ij}] \in M_n$. We say that A is *upper triangular* if $a_{ij} = 0$ whenever $i > j$; *lower triangular* if $a_{ij} = 0$ whenever $i < j$; *strictly upper triangular* if $a_{ij} = 0$ whenever $i \geq j$; and *strictly lower triangular* if $a_{ij} = 0$ whenever $i \leq j$. We say that A is *triangular* if it is either upper triangular or lower triangular.

Diagonal Matrices We say that $A = [a_{ij}] \in M_n$ is *diagonal* if $a_{ij} = 0$ whenever $i \neq j$. That is, any nonzero entry of A must lie on the *main diagonal* of A, which consists of the *diagonal entries* $a_{11}, a_{22}, \ldots, a_{nn}$; the entries a_{ij} with $i \neq j$ are the *off-diagonal entries* of A. The notation $\mathrm{diag}(\lambda_1, \lambda_2, \ldots, \lambda_n)$ is used to denote the $n \times n$ diagonal matrix whose diagonal entries are $\lambda_1, \lambda_2, \ldots, \lambda_n$, in that order. A *scalar matrix* is a diagonal matrix of the form $\mathrm{diag}(c, c, \ldots, c) = cI$ for some scalar c. Any two diagonal matrices of the same size commute.

Superdiagonals and Subdiagonals The (first) *superdiagonal* of $A = [a_{ij}] \in M_n$ contains the entries $a_{12}, a_{23}, \ldots, a_{n-1,n}$. The kth superdiagonal contains the entries $a_{1,k+1}, a_{2,k+2}, \ldots, a_{n-k,n}$. The kth *subdiagonal* contains the entries $a_{k+1,1}, a_{k+2,2}, \ldots, a_{n,n-k}$.

Tridiagonal and Bidiagonal Matrices A matrix $A = [a_{ij}]$ is *tridiagonal* if $a_{ij} = 0$ whenever $|i - j| \geq 2$. A tridiagonal matrix is *bidiagonal* if either its subdiagonal or its superdiagonal contains only zero entries.

Submatrices A *submatrix* of $A \in M_{m \times n}$ is a matrix whose entries lie in the intersections of specified rows and columns of A. A $k \times k$ *principal submatrix* of A is a submatrix whose entries lie in the intersections of rows $i_1, i_2, \ldots, i_k$ and columns $i_1, i_2, \ldots, i_k$ of A, for some indices $i_1 < i_2 < \cdots < i_k$. A $k \times k$ *leading principal submatrix* of A is a submatrix whose entries lie in the intersections of rows $1, 2, \ldots, k$ and columns $1, 2, \ldots, k$. A $k \times k$ *trailing principal submatrix* of A is a submatrix whose entries lie in the intersections of rows $n - k + 1, n - k + 2, \ldots, n$ and columns $n - k + 1, n - k + 2, \ldots, n$.

Inverses We say that $A \in M_n$ is *invertible* if there exists a $B \in M_n$ such that

$$AB = I_n = BA. \tag{0.3.3}$$

Such a matrix B is an *inverse* of A. If A has no inverse, then A is *noninvertible*. Either of the equalities in (0.3.3) implies the other. That is, if $A, B \in M_n$, then $AB = I$ if and only if $BA = I$; see Theorem 2.2.19 and Example 3.1.8.

Not every square matrix has an inverse. However, a matrix has at most one inverse. As a consequence, if A is invertible, we speak of *the* inverse of A, rather than *an* inverse of A. If A is invertible, then the inverse of A is denoted by A^{-1}. It satisfies

$$AA^{-1} = I = A^{-1}A.$$

If $ad - bc \neq 0$, then

$$\begin{bmatrix} a & b \\ c & d \end{bmatrix}^{-1} = \frac{1}{ad - bc} \begin{bmatrix} d & -b \\ -c & a \end{bmatrix}. \tag{0.3.4}$$

For $A \in M_n$, define

$$A^0 = I \quad \text{and} \quad A^k = \underbrace{AA \cdots A}_{k \text{ times}}.$$

If A is invertible, we define $A^{-k} = (A^{-1})^k$ for $k = 1, 2, \ldots$. Let A and B be matrices of appropriate sizes, let j, k be integers, and let c be a scalar.

(a) $A^j A^k = A^{j+k} = A^k A^j$.

(b) $(A^{-1})^{-1} = A$.

(c) $(A^j)^{-1} = A^{-j}$.

(d) If $c \neq 0$, then $(cA)^{-1} = c^{-1} A^{-1}$.

(e) $(AB)^{-1} = B^{-1} A^{-1}$.

Transpose The *transpose* of $A = [a_{ij}] \in \mathsf{M}_{m \times n}$ is the matrix $A^\mathsf{T} \in \mathsf{M}_{n \times m}$ whose (i, j) entry is a_{ji}. Let A and B be matrices of appropriate sizes and let c be a scalar.

(a) $(A^\mathsf{T})^\mathsf{T} = A$.

(b) $(A \pm B)^\mathsf{T} = A^\mathsf{T} \pm B^\mathsf{T}$.

(c) $(cA)^\mathsf{T} = cA^\mathsf{T}$.

(d) $(AB)^\mathsf{T} = B^\mathsf{T} A^\mathsf{T}$.

(e) If A is invertible, then $(A^\mathsf{T})^{-1} = (A^{-1})^\mathsf{T}$. We write $(A^{-1})^\mathsf{T} = A^{-\mathsf{T}}$.

Conjugate The *conjugate* of $A \in \mathsf{M}_{m \times n}$ is the matrix $\overline{A} \in \mathsf{M}_{m \times n}$ whose (i, j) entry is $\overline{a_{ij}}$, the complex conjugate of a_{ij}. Thus,

$$\overline{(\overline{A})} = A, \qquad \overline{A + B} = \overline{A} + \overline{B}, \qquad \text{and} \qquad \overline{AB} = \overline{A}\, \overline{B}.$$

If A has only real entries, then $A = \overline{A}$.

Conjugate Transpose The *conjugate transpose* of $A \in \mathsf{M}_{m \times n}$ is the matrix $A^* = \overline{A^\mathsf{T}} = (\overline{A})^\mathsf{T} \in \mathsf{M}_{n \times m}$ whose (i, j) entry is $\overline{a_{ji}}$. If A has only real entries, then $A^* = A^\mathsf{T}$. The conjugate transpose of a matrix is also known as its *adjoint*. Let A and B be matrices of appropriate sizes and let c be a scalar.

(a) $I_n^* = I_n$.

(b) $0_{m \times n}^* = 0_{n \times m}$.

(c) $(A^*)^* = A$.

(d) $(A \pm B)^* = A^* \pm B^*$.

(e) $(cA)^* = \overline{c} A^*$.

(f) $(AB)^* = B^* A^*$.

(g) If A is invertible, then $(A^*)^{-1} = (A^{-1})^*$. We write $(A^{-1})^* = A^{-*}$.

Special Types of Matrices Let $A \in \mathsf{M}_n$.

(a) If $A^* = A$, then A is *Hermitian*; if $A^* = -A$, then A is *skew Hermitian*.

(b) If $A^\mathsf{T} = A$, then A is *symmetric*; if $A^\mathsf{T} = -A$, then A is *skew symmetric*.

(c) If $A^*A = I$, then A is *unitary*; if A is real and $A^TA = I$, then A is *real orthogonal*.

(d) If $A^*A = AA^*$, then A is *normal*.

(e) If $A^2 = I$, then A is an *involution*.

(f) If $A^2 = A$, then A is *idempotent*.

(g) If $A^k = 0$ for some positive integer k, then A is *nilpotent*.

Trace The *trace* of $A = [a_{ij}] \in M_n$ is the sum of the diagonal entries of A:

$$\operatorname{tr} A = \sum_{i=1}^n a_{ii}.$$

Let A and B be matrices of appropriate sizes and let c be a scalar.

(a) $\operatorname{tr}(cA \pm B) = c \operatorname{tr} A \pm \operatorname{tr} B$.

(b) $\operatorname{tr} A^T = \operatorname{tr} A$.

(c) $\operatorname{tr} \overline{A} = \overline{\operatorname{tr} A}$.

(d) $\operatorname{tr} A^* = \overline{\operatorname{tr} A}$.

If $A = [a_{ij}] \in M_{m \times n}$ and $B = [b_{ij}] \in M_{n \times m}$, let $AB = [c_{ij}] \in M_m$ and $BA = [d_{ij}] \in M_n$. Then

$$\operatorname{tr} AB = \sum_{i=1}^m c_{ii} = \sum_{i=1}^m \sum_{j=1}^n a_{ij} b_{ji} = \sum_{j=1}^n \sum_{i=1}^m b_{ji} a_{ij} = \sum_{j=1}^n d_{jj} = \operatorname{tr} BA. \qquad (0.3.5)$$

Be careful: $\operatorname{tr} ABC$ need not equal $\operatorname{tr} CBA$ or $\operatorname{tr} ACB$. However, (0.3.5) ensures that

$$\operatorname{tr} ABC = \operatorname{tr} CAB = \operatorname{tr} BCA.$$

0.4 Systems of Linear Equations

An $m \times n$ *system of linear equations* (a *linear system*) is a list of linear equations of the form

$$
\begin{aligned}
a_{11}x_1 + a_{12}x_2 + \cdots + a_{1n}x_n &= b_1, \\
a_{21}x_1 + a_{22}x_2 + \cdots + a_{2n}x_n &= b_2, \\
\vdots \quad \vdots \quad \vdots \quad \vdots \quad \ddots \quad \vdots \quad \vdots \quad &\vdots \\
a_{m1}x_1 + a_{m2}x_2 + \cdots + a_{mn}x_n &= b_m.
\end{aligned}
\qquad (0.4.1)
$$

It involves m *linear equations* in the n *variables* (or *unknowns*) $x_1, x_2, \ldots, x_n$. The scalars a_{ij} are the *coefficients* of the system (0.4.1); the scalars b_i are the *constant terms*.

By a *solution* to (0.4.1) we mean a list of scalars $x_1, x_2, \ldots, x_n$ that satisfy the m equations in (0.4.1). A system of equations that has no solution is *inconsistent*. If a system has at least one solution, it is *consistent*. There are exactly three possibilities for a system of linear equations: it has no solution, exactly one solution, or infinitely many solutions.

Homogeneous Systems The system (0.4.1) is *homogeneous* if $b_1 = b_2 = \cdots = b_m = 0$. Every homogeneous system has the *trivial solution* $x_1 = x_2 = \cdots x_n = 0$. If there are other solutions, they are called *nontrivial solutions*. There are only two possibilities for a homogeneous system: it has infinitely many solutions, or it has only the trivial solution. A homogeneous linear system with more unknowns than equations has infinitely many solutions.

Matrix Representation of a Linear System The linear system (0.4.1) is often written as

$$A\mathbf{x} = \mathbf{b}, \tag{0.4.2}$$

in which

$$A = \begin{bmatrix} a_{11} & a_{12} & \cdots & a_{1n} \\ a_{21} & a_{22} & \cdots & a_{2n} \\ \vdots & \vdots & \ddots & \vdots \\ a_{m1} & a_{m2} & \cdots & a_{mn} \end{bmatrix}, \quad \mathbf{x} = \begin{bmatrix} x_1 \\ x_2 \\ \vdots \\ x_n \end{bmatrix}, \quad \text{and} \quad \mathbf{b} = \begin{bmatrix} b_1 \\ b_2 \\ \vdots \\ b_m \end{bmatrix}. \tag{0.4.3}$$

The *coefficient matrix* $A = [a_{ij}] \in \mathsf{M}_{m \times n}$ of the system has m rows and n columns if the corresponding system of equations (0.4.1) has m equations in n unknowns. The matrices $\mathbf{x}$ and $\mathbf{b}$ are $n \times 1$ and $m \times 1$, respectively. Matrices such as $\mathbf{x}$ and $\mathbf{b}$ are *column vectors*. We sometimes denote $\mathsf{M}_{n \times 1}(\mathbb{C})$ by $\mathbb{C}^n$ and $\mathsf{M}_{n \times 1}(\mathbb{R})$ by $\mathbb{R}^n$. When we need to identify the entries of a column vector in a line of text, we often write $\mathbf{x} = [x_1 \ x_2 \ \ldots \ x_n]^\mathsf{T}$ instead of the tall vertical matrix in (0.4.3).

An $m \times n$ homogeneous linear system can be written in the form $A\mathbf{x} = \mathbf{0}_m$, in which $A \in \mathsf{M}_{m \times n}$ and $\mathbf{0}_m$ is the $m \times 1$ column vector whose entries are all zero. We say that $\mathbf{0}_m$ is a *zero vector* and write $\mathbf{0}$ if the size is clear from context. Since $A\mathbf{0}_n = \mathbf{0}_m$, a homogeneous system always has the trivial solution.

If $A \in \mathsf{M}_n$ is invertible, then (0.4.2) has the unique solution $\mathbf{x} = A^{-1}\mathbf{b}$.

Reduced Row Echelon Form Three elementary operations can be used to solve a system (0.4.1) of linear equations:

(I) Multiply an equation by a nonzero constant.

(II) Interchange two equations.

(III) Add a multiple of one equation to another.

One can represent the system (0.4.1) as an *augmented matrix*

$$[A \ \mathbf{b}] = \begin{bmatrix} a_{11} & a_{12} & \cdots & a_{1n} & b_1 \\ a_{21} & a_{22} & \cdots & a_{2n} & b_2 \\ \vdots & \vdots & \ddots & \vdots & \vdots \\ a_{m1} & a_{m2} & \cdots & a_{mn} & b_m \end{bmatrix} \tag{0.4.4}$$

and perform *elementary row operations* on (0.4.4) that correspond to the three permissible algebraic operations on the system (0.4.1):

(I) Multiply a row by a nonzero constant.

(II) Interchange two rows.

(III) Add a multiple of one row to another.

Each of these operations is reversible.

The three types of elementary row operations can be used to *row reduce* the augmented matrix (0.4.4) to a simple form from which the solutions to (0.4.1) can be obtained by inspection. A matrix is in *reduced row echelon form* if it satisfies the following:

(a) Rows that consist entirely of zero entries are grouped together at the bottom of the matrix.

(b) If a row does not consist entirely of zero entries, then the first nonzero entry in that row is a one (a *leading one*).

(c) A leading one in a higher row must occur further to the left than a leading one in a lower row.

(d) Every column that contains a leading one must have zero entries everywhere else.

Each matrix has a unique reduced row echelon form.

The number of leading ones in the reduced row echelon form of a matrix is equal to its rank; see Definition 2.2.6. Other characterizations of the rank are discussed in Section 3.2. It is always the case that rank $A = \text{rank}\, A^{\mathsf{T}}$; see Theorem 3.2.1.

Elementary Matrices An $n \times n$ matrix is an *elementary matrix* if it can be obtained from I_n by performing a single elementary row operation. Every elementary matrix is invertible; the inverse is the elementary matrix that corresponds to reversing the original row operation. Multiplication of a matrix on the left by an elementary matrix performs an elementary row operation on that matrix. Here are some examples:

(I) Multiply a row by a nonzero constant:

$$\begin{bmatrix} k & 0 \\ 0 & 1 \end{bmatrix} \begin{bmatrix} a_{11} & a_{12} \\ a_{21} & a_{22} \end{bmatrix} = \begin{bmatrix} ka_{11} & ka_{12} \\ a_{21} & a_{22} \end{bmatrix}.$$

(II) Interchange two rows:

$$\begin{bmatrix} 0 & 1 \\ 1 & 0 \end{bmatrix} \begin{bmatrix} a_{11} & a_{12} \\ a_{21} & a_{22} \end{bmatrix} = \begin{bmatrix} a_{21} & a_{22} \\ a_{11} & a_{12} \end{bmatrix}.$$

(III) Add a nonzero multiple of one row to another:

$$\begin{bmatrix} 1 & k \\ 0 & 1 \end{bmatrix} \begin{bmatrix} a_{11} & a_{12} \\ a_{21} & a_{22} \end{bmatrix} = \begin{bmatrix} a_{11} + ka_{21} & a_{12} + ka_{22} \\ a_{21} & a_{22} \end{bmatrix}.$$

Multiplication of a matrix on the right by an elementary matrix corresponds to performing column operations. An invertible matrix can be expressed as a product of elementary matrices.

0.5 Determinants

The determinant function $\det : \mathsf{M}_n(\mathbb{C}) \to \mathbb{C}$ is of great theoretical importance, but of limited numerical use. Computation of determinants of large matrices should be avoided in applications.

Laplace Expansion We can compute the determinant of an $n \times n$ matrix as a certain sum of determinants of $(n-1) \times (n-1)$ matrices. Let $\det[a_{11}] = a_{11}$, let $n \geq 2$, let $A \in M_n$, and let $A_{ij} \in M_{n-1}$ denote the $(n-1) \times (n-1)$ matrix obtained by deleting row i and column j of A. Then for any $i, j \in \{1, 2, \ldots, n\}$, we have

$$\det A = \sum_{k=1}^{n} (-1)^{i+k} a_{ik} \det A_{ik} = \sum_{k=1}^{n} (-1)^{k+j} a_{kj} \det A_{kj}. \tag{0.5.1}$$

The first sum is the *Laplace expansion by minors along row i* and the second is the *Laplace expansion by minors along column j*. The quantity $\det A_{ij}$ is the (i, j) *minor of* A; $(-1)^{i+j} \det A_{ij}$ is the (i, j) *cofactor* of A.

Using Laplace expansions, we compute

$$\det \begin{bmatrix} a_{11} & a_{12} \\ a_{21} & a_{22} \end{bmatrix} = a_{11} \det[a_{22}] - a_{12} \det[a_{21}]$$

$$= a_{11}a_{22} - a_{12}a_{21}$$

and

$$\det \begin{bmatrix} a_{11} & a_{12} & a_{13} \\ a_{21} & a_{22} & a_{23} \\ a_{31} & a_{32} & a_{33} \end{bmatrix} = a_{11} \det \begin{bmatrix} a_{22} & a_{23} \\ a_{32} & a_{33} \end{bmatrix} - a_{12} \det \begin{bmatrix} a_{21} & a_{23} \\ a_{31} & a_{33} \end{bmatrix} + a_{13} \det \begin{bmatrix} a_{21} & a_{22} \\ a_{31} & a_{32} \end{bmatrix}$$

$$= a_{11}a_{22}a_{33} + a_{12}a_{23}a_{31} + a_{13}a_{32}a_{21}$$

$$- a_{11}a_{23}a_{32} - a_{22}a_{13}a_{31} - a_{33}a_{12}a_{21}.$$

Following the same rule, the determinant of a 4×4 matrix can be written as a sum of four terms, each involving the determinant of a 3×3 matrix.

Determinants and the Inverse If $A \in M_n$, the *adjugate* of A is the $n \times n$ matrix

$$\operatorname{adj} A = [(-1)^{i+j} \det A_{ji}],$$

which is the transpose of the matrix of cofactors of A. The matrices A and $\operatorname{adj} A$ satisfy

$$A \operatorname{adj} A = (\operatorname{adj} A)A = (\det A)I. \tag{0.5.2}$$

If A is invertible, then

$$A^{-1} = (\det A)^{-1} \operatorname{adj} A. \tag{0.5.3}$$

Properties of Determinants Let $A, B \in M_n$ and let c be a scalar.

(a) $\det I = 1$.

(b) $\det A \neq 0$ if and only if A is invertible.

(c) $\det AB = (\det A)(\det B)$.

(d) $\det AB = \det BA$.

(e) $\det(cA) = c^n \det A$.

(f) $\det \overline{A} = \overline{\det A}$.

(g) $\det A^{\mathsf{T}} = \det A$.

(h) $\det A^* = \overline{\det A}$.

(i) If A is invertible, then $\det(A^{-1}) = (\det A)^{-1}$.

(j) If $A = [a_{ij}] \in \mathsf{M}_n$ is upper or lower triangular, then $\det A = a_{11}a_{22}\cdots a_{nn}$.

(k) $\det A \in \mathbb{R}$ if $A \in \mathsf{M}_n(\mathbb{R})$.

Be careful: $\det(A + B)$ need not equal $\det A + \det B$. Property (c) is the *product rule* for determinants.

Determinants and Row Reduction The determinant of an $n \times n$ matrix A can be computed with row reduction and the following properties:

(I) If A' is obtained by multiplying each entry of a row of A by a scalar c, then $\det A' = c \det A$.

(II) If A' is obtained by interchanging two different rows of A, then $\det A' = -\det A$.

(III) If A' is obtained from A by adding a scalar multiple of a row to a different row, then $\det A' = \det A$.

Because $\det A = \det A^{\mathsf{T}}$, column operations have analogous properties.

Permutations and Determinants A *permutation* of the list $1, 2, \ldots, n$ is a one-to-one function $\sigma : \{1, 2, \ldots, n\} \to \{1, 2, \ldots, n\}$. A permutation induces a reordering of $1, 2, \ldots, n$. For example, $\sigma(1) = 2$, $\sigma(2) = 1$, and $\sigma(3) = 3$ defines a permutation of $1, 2, 3$. There are $n!$ distinct permutations of the list $1, 2, \ldots, n$.

A permutation $\tau : \{1, 2, \ldots, n\} \to \{1, 2, \ldots, n\}$ that interchanges precisely two elements of $1, 2, \ldots, n$ and leaves all others fixed is a *transposition*. Each permutation of $1, 2, \ldots, n$ can be written as a composition of transpositions in many different ways. However, the parity (even or odd) of the number of transpositions involved depends only upon the permutation. We say that a permutation σ is *even* or *odd* depending upon whether an even or odd number of transpositions is required to represent σ. The *sign* of σ is

$$\operatorname{sgn}\sigma = \begin{cases} 1 & \text{if } \sigma \text{ is even,} \\ -1 & \text{if } \sigma \text{ is odd.} \end{cases}$$

The determinant of $A = [a_{ij}] \in \mathsf{M}_n$ can be written

$$\det A = \sum_{\sigma} \left(\operatorname{sgn}\sigma \prod_{i=1}^{n} a_{i\sigma(i)} \right),$$

in which the sum is over all $n!$ permutations of $1, 2, \ldots, n$.

Determinants, Area, and Volume If $A = [\mathbf{a}_1 \ \mathbf{a}_2] \in \mathsf{M}_2(\mathbb{R})$, then $|\det A|$ is the area of the parallelogram determined by $\mathbf{a}_1$ and $\mathbf{a}_2$ (its vertices are at $\mathbf{0}$, $\mathbf{a}_1$, $\mathbf{a}_2$, and $\mathbf{a}_1 + \mathbf{a}_2$). If $B = [\mathbf{b}_1 \ \mathbf{b}_2 \ \mathbf{b}_3] \in \mathsf{M}_3(\mathbb{R})$, then $|\det B|$ is the volume of the parallelepiped determined by $\mathbf{b}_1$, $\mathbf{b}_2$, and $\mathbf{b}_3$.

(a) After you prove the base case, you know that S_1 is true.

(b) After you prove the inductive step, you know that S_1 implies S_2, so S_2 is true ...

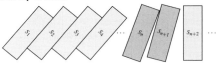

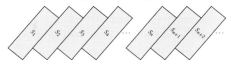

(c) ... and you know that S_n implies S_{n+1}.

(d) Mathematical induction ensures that S_n is true for all $n = 1, 2, \ldots$.

Figure 0.2 Domino analogy for induction.

0.6 Mathematical Induction

Suppose that $S_1, S_2, \ldots$ are mathematical statements. The *principle of mathematical induction* asserts that if the statements:

(a) "S_1 is true," and

(b) "If S_n is true, then S_{n+1} is true,"

are true, then S_n is true for all $n \geq 1$. The statement "S_n is true" is the *induction hypothesis*.

The principle of mathematical induction is plausible. If we have proved the *base case* (a), then S_1 is true. If we have also proved the *inductive step* (b), then the truth of S_1 implies the truth of S_2, which implies the truth of S_3, and so forth; see Figure 0.2.

For example, let S_n be the statement

$$1 + 2 + \cdots + n = \frac{n(n+1)}{2}.$$

We use mathematical induction to prove that S_n is true for $n = 1, 2, \ldots$. Since $1 = \frac{1 \cdot 2}{2}$, S_1 is true. This establishes the base case. To show that S_n implies S_{n+1} we must show that the induction hypothesis

$$1 + 2 + \cdots + n = \frac{n(n+1)}{2} \tag{0.6.1}$$

implies that

$$1 + 2 + \cdots + n + (n+1) = \frac{(n+1)((n+1)+1)}{2}. \tag{0.6.2}$$

Add $n + 1$ to both sides of (0.6.1) and obtain

$$(1 + 2 + \cdots + n) + (n+1) = \frac{n(n+1)}{2} + (n+1)$$
$$= \frac{n(n+1) + 2(n+1)}{2}$$

$$= \frac{(n+1)(n+2)}{2}$$
$$= \frac{(n+1)((n+1)+1)}{2},$$

which is (0.6.2). Therefore, S_{n+1} is true if S_n is true. The principle of mathematical induction ensures that S_n is true for all $n = 1, 2, \ldots$.

A statement S_n can be true for many initial values of n without being true for all n. For example, the polynomial $p(n) = n^2 + n + 41$ has the property that $p(n)$ is prime for $n = 1$, $2, \ldots, 39$. However, the streak ends with $n = 40$ since $p(40) = 41^2$ is not prime. The polynomial $p(n) = n^6 + 1091$ has the property that $p(n)$ is not prime for $n = 1, 2, \ldots, 3905$, but $p(3906)$ is prime.

Complete Induction An equivalent variant of mathematical induction is *complete induction*, in which one replaces the inductive step (b) with the seemingly stronger statement:

(b′) If S_m is true for $1 \leq m \leq n$, then S_{n+1} is true.

The statement "S_m is true for $1 \leq m \leq n$" is the *induction hypothesis* for complete induction. Anything that can be proved with mathematical induction can be proved with complete induction, and vice-versa. However, (b′) is sometimes more convenient to use in practice.

To prove that every natural number greater than 1 is a product of (one or more) prime numbers, we can use complete induction. The base case $n = 2$ is true, since 2 is a prime number. Suppose that every natural number less than or equal to n is the product of prime numbers. Then either $n + 1$ is prime itself, or $n + 1 = n_1 n_2$, in which n_1, n_2 are natural numbers between 2 and n. By the induction hypothesis, n_1 and n_2 are products of prime numbers and hence $n + 1$ is as well.

0.7 Polynomials

Let $c_0, c_1, \ldots, c_k$ be complex numbers. A *complex polynomial of degree* $k \geq 0$ is a function of the form

$$p(z) = c_k z^k + c_{k-1} z^{k-1} + \cdots + c_1 z + c_0, \qquad c_k \neq 0. \qquad (0.7.1)$$

The scalars $c_0, c_1, \ldots, c_k$ are the *coefficients* of p. We denote the degree of p by $\deg p$. If all of its coefficients are real, then p is a *real polynomial*. The *zero polynomial* is the function $p(z) = 0$; by convention, the degree of the zero polynomial is $-\infty$. If $c_k = 1$ in (0.7.1), then p is a *monic* polynomial. Polynomials of degree 1 or greater are *nonconstant polynomials*; polynomials of degree less than one are *constant polynomials*.

A real polynomial is a special type of complex polynomial. The word "polynomial," when used without qualification, means "complex polynomial." For p and c_k as in (0.7.1), the polynomial $p(z)/c_k$ is monic.

If p and q are polynomials and c is a scalar, then cp, $p + q$, and pq are polynomials. For example, if $c = 5, p(z) = z^2 + 1$, and $q(z) = z - 3$, then $cp(z) = 5z^2 + 5, p(z) + q(z) = z^2 + z - 2$, and $p(z)q(z) = z^3 - 3z^2 + z - 3$.

Zeros and Roots Let λ be a complex number and let p be a polynomial. Then λ is a *zero of p* (alternatively, λ is a *root of the equation* $p(z) = 0$) if $p(\lambda) = 0$.

A real polynomial might have no real zeros. For example, $p(z) = z^2 + 1$ has no real zeros, but $\pm i$ are non-real complex zeros of p. One of the reasons for using complex scalars in linear algebra is the following result, which can be proved with methods developed in complex analysis or topology.

Theorem 0.7.2 (Fundamental Theorem of Algebra) *Every nonconstant polynomial has a zero in* $\mathbb{C}$.

The Division Algorithm The following polynomial version of long division with remainder is known as the *division algorithm*. If f and g are polynomials such that $1 \leq \deg g \leq \deg f$, then there are unique polynomials q and r such that $f = gq + r$ and $\deg r < \deg g$. The polynomials f, g, q, r are the *dividend*, *divisor*, *quotient*, and *remainder*, respectively. For example, if

$$f(z) = 2z^4 + z^3 - z^2 + 1 \quad \text{and} \quad g(z) = z^2 - 1,$$

then

$$q(z) = 2z^2 + z + 1 \quad \text{and} \quad r(z) = z + 2.$$

Factoring and Multiplicities of Zeros If $\deg p \geq 1$ and $\lambda \in \mathbb{C}$, then the division algorithm ensures that $p(z) = (z - \lambda)q(z) + r$, in which r is a constant polynomial. If λ is a zero of p, then $0 = p(\lambda) = 0 + r$, so $r = 0$ and $p(z) = (z - \lambda)q(z)$, in which case we have *factored out* the zero λ from p.

This process can be repeated. For the polynomial (0.7.1), we obtain

$$p(z) = c_k(z - \lambda_1)(z - \lambda_2) \cdots (z - \lambda_k), \tag{0.7.3}$$

in which the (not necessarily distinct) complex numbers

$$\lambda_1, \lambda_2, \ldots, \lambda_k \tag{0.7.4}$$

are the zeros of p. If $\mu_1, \mu_2, \ldots, \mu_d$ are the *distinct* elements of the list (0.7.4), we can write (0.7.3) as

$$p(z) = c_k(z - \mu_1)^{n_1}(z - \mu_2)^{n_2} \cdots (z - \mu_d)^{n_d}, \tag{0.7.5}$$

in which $\mu_i \neq \mu_j$ whenever $i \neq j$. The exponent n_i in (0.7.5), the number of times that μ_i appears in the list (0.7.4), is the *multiplicity* of the zero μ_i of p. Thus,

$$n_1 + n_2 + \cdots + n_d = n = \deg p.$$

For example, the polynomial $p(z) = z^3 - 2z^2 + z = z(z - 1)^2$ has $d = 2$, $\mu_1 = 0$, $n_1 = 1$, $\mu_2 = 1$, and $n_2 = 2$.

Identity Theorems for Polynomials Let f and g be polynomials and suppose that $\deg f = n \geq 1$.

(a) The sum of the multiplicities of the zeros of f is n.

(b) f has at most n distinct zeros.

(c) If $\deg g \le n$ and g has at least $n+1$ distinct zeros, then g is the zero polynomial.

(d) If $\deg g \le n$, if $z_1, z_2, \ldots, z_{n+1}$ are distinct complex numbers, and if $f(z_i) = g(z_i)$ for each $i = 1, 2, \ldots, n+1$, then $f = g$.

(e) If $g(z) = 0$ for infinitely many distinct values of z, then g is the zero polynomial.

(f) If fg is the zero polynomial, then g is the zero polynomial.

The final assertion in the preceding list follows from observing that if g is not the zero polynomial, then fg is polynomial of degree 1 or more, so it has only finitely many zeros.

Lagrange Interpolation The following polynomial interpolation theorem has many important applications in linear algebra; see P.0.15 for a different proof.

Theorem 0.7.6 (Lagrange Interpolation) *Let $n \ge 1$, let $z_1, z_2, \ldots, z_n$ be distinct complex numbers, and let $w_1, w_2, \ldots, w_n \in \mathbb{C}$. There is a unique polynomial p of degree at most $n-1$ such that $p(z_i) = w_i$ for $i = 1, 2, \ldots, n$. If the data $z_1, z_2, \ldots, z_n, w_1, w_2, \ldots, w_n$ are real, then p is a real polynomial.*

Proof If $n = 1$, let $p(z) = w_1$. Now suppose that $n \ge 2$. For $j = 1, 2, \ldots, n$, define the polynomials

$$\ell_j(z) = \prod_{\substack{1 \le k \le n \\ k \ne j}} \frac{z - z_k}{z_j - z_k},$$

each of degree $n-1$, and observe that

$$\ell_j(z_k) = \begin{cases} 0 & \text{if } j \ne k, \\ 1 & \text{if } j = k. \end{cases}$$

Therefore, the polynomial $p(z) = \sum_{j=1}^{n} w_j \ell_j(z)$ has degree at most $n-1$ and satisfies $p(z_k) = \sum_{j=1}^{n} w_j \ell_j(z_k) = w_k$. The uniqueness of p among polynomials of degree at most $n-1$ follows from (d) of the preceding section. If the interpolation data are real, then each polynomial $\ell_j(z)$ is real and each coefficient w_j is real, so p is real. $\square$

0.8 Polynomials and Matrices

For a polynomial

$$p(z) = c_k z^k + c_{k-1} z^{k-1} + \cdots + c_1 z + c_0$$

and $A \in \mathsf{M}_n$, we define

$$p(A) = c_k A^k + c_{k-1} A^{k-1} + \cdots + c_1 A + c_0 I.$$

Consider

$$A = \begin{bmatrix} 1 & 2 \\ 3 & 4 \end{bmatrix}$$

and $p(z) = z^2 - 2z + 1$. Then

$$p(A) = \begin{bmatrix} 1 & 2 \\ 3 & 4 \end{bmatrix} \begin{bmatrix} 1 & 2 \\ 3 & 4 \end{bmatrix} - 2 \begin{bmatrix} 1 & 2 \\ 3 & 4 \end{bmatrix} + \begin{bmatrix} 1 & 0 \\ 0 & 1 \end{bmatrix}$$

$$= \begin{bmatrix} 7 & 10 \\ 15 & 22 \end{bmatrix} - \begin{bmatrix} 2 & 4 \\ 6 & 8 \end{bmatrix} + \begin{bmatrix} 1 & 0 \\ 0 & 1 \end{bmatrix} = \begin{bmatrix} 6 & 6 \\ 9 & 15 \end{bmatrix}.$$

The same result is obtained using the factorization $p(z) = (z - 1)^2$:

$$p(A) = (A - I)^2 = \begin{bmatrix} 0 & 2 \\ 3 & 3 \end{bmatrix} \begin{bmatrix} 0 & 2 \\ 3 & 3 \end{bmatrix} = \begin{bmatrix} 6 & 6 \\ 9 & 15 \end{bmatrix}.$$

If p and q are polynomials and $A \in \mathsf{M}_n$, then

$$p(A) + q(A) = (p + q)(A)$$

and

$$p(A)q(A) = (pq)(A) = (qp)(A) = q(A)p(A).$$

For a diagonal matrix $\Lambda = \mathrm{diag}(\lambda_1, \lambda_2, \ldots, \lambda_n)$ we have

$$p(\Lambda) = \mathrm{diag}(p(\lambda_1), p(\lambda_2), \ldots, p(\lambda_n)).$$

Intertwining Let $A \in \mathsf{M}_m$, $B \in \mathsf{M}_n$, and $X \in \mathsf{M}_{m \times n}$. Then X *intertwines* A and B if $AX = XB$.

If X is square and invertible, then X intertwines A and B if and only if $A = XBX^{-1}$, that is, A and B are similar; see Definition 2.4.16. If $A = B$, then intertwining is *commuting*: $AX = XA$. Thus, intertwining is a generalization of both similarity and commuting. The following theorem says that if X intertwines A and B, then X intertwines $p(A)$ and $p(B)$.

Theorem 0.8.1 *Let $A \in \mathsf{M}_m$, $B \in \mathsf{M}_n$, and $X \in \mathsf{M}_{m \times n}$. If $AX = XB$, then $p(A)X = Xp(B)$ for any polynomial p. Moreover, A commutes with $p(A)$.*

Proof We first use induction to prove that $A^j X = XB^j$ for $j = 0, 1, 2, \ldots$. The base case $j = 0$ is $IX = XI$, which is true. For the induction step, suppose that $A^j X = XB^j$ for some j. Then $A^{j+1} X = AA^j X = AXB^j = XBB^j = XB^{j+1}$. This completes the induction.

Let $p(z) = c_k z^k + \cdots + c_1 z + c_0$. Then

$$p(A)X = (c_k A^k + \cdots + c_1 A + c_0 I)X$$
$$= c_k(A^k X) + \cdots + c_1(AX) + c_0 X$$
$$= c_k(XB^k) + \cdots + c_1(XB) + c_0 X$$
$$= X(c_k B^k + \cdots + c_1 B + c_0 I)$$
$$= Xp(B).$$

See P.0.22 for the second assertion. $\qquad\qquad\qquad\qquad\qquad\qquad\qquad\qquad\quad\square$

Polynomials and Similarity Let $A, B, X \in M_n$, let X be invertible, and let p be a polynomial. If $A = XBX^{-1}$, then $AX = XB$, so the preceding theorem ensures that $p(A)X = Xp(B)$, which implies that

$$p(A) = Xp(B)X^{-1}. \tag{0.8.2}$$

This identity has a special form if $B = \Lambda = \text{diag}(\lambda_1, \lambda_2, \ldots, \lambda_n)$ is diagonal:

$$p(A) = Xp(\Lambda)X^{-1} = X \, \text{diag} \left(p(\lambda_1), p(\lambda_2), \ldots, p(\lambda_n) \right) X^{-1}. \tag{0.8.3}$$

If $p(z) = z + c$, then (0.8.2) reveals the *shift property* of similarity: $A = XBX^{-1}$ implies that

$$A + cI = X(B + cI)X^{-1}. \tag{0.8.4}$$

0.9 Problems

P.0.1 Let $f : \{1, 2, \ldots, n\} \to \{1, 2, \ldots, n\}$ be a function. Show that the following are equivalent: (a) f is one to one. (b) f is onto. (c) f is a permutation of $1, 2, \ldots, n$.

P.0.2 Show that (a) the diagonal entries of a Hermitian matrix are real; (b) the diagonal entries of a skew-Hermitian matrix are purely imaginary; (c) the diagonal entries of a skew-symmetric matrix are zero.

P.0.3 Use mathematical induction to prove that $1^2 + 2^2 + \cdots + n^2 = \frac{n(n+1)(2n+1)}{6}$ for $n = 1, 2, \ldots$.

P.0.4 Use mathematical induction to prove that $1^3 + 2^3 + \cdots + n^3 = \left(\frac{n(n+1)}{2} \right)^2$ for $n = 1, 2, \ldots$.

P.0.5 Let $A \in M_n$ be invertible. Use mathematical induction to prove that $(A^{-1})^k = (A^k)^{-1}$ for all integers k.

P.0.6 Let $A \in M_n$. Use mathematical induction to prove that $A^{j+k} = A^j A^k$ for all integers j, k.

P.0.7 Use mathematical induction to prove Binet's formula (9.5.5) for the Fibonacci numbers.

P.0.8 Use mathematical induction to prove that $1 + z + z^2 + \cdots + z^{n-1} = \frac{1 - z^n}{1 - z}$ for complex $z \neq 1$ and all positive integers n.

P.0.9 (a) Compute the determinants of the matrices

$$V_2 = \begin{bmatrix} 1 & z_1 \\ 1 & z_2 \end{bmatrix}, \qquad V_3 = \begin{bmatrix} 1 & z_1 & z_1^2 \\ 1 & z_2 & z_2^2 \\ 1 & z_3 & z_3^2 \end{bmatrix}, \qquad V_4 = \begin{bmatrix} 1 & z_1 & z_1^2 & z_1^3 \\ 1 & z_2 & z_2^2 & z_2^3 \\ 1 & z_3 & z_3^2 & z_3^3 \\ 1 & z_4 & z_4^2 & z_4^3 \end{bmatrix},$$

and simplify your answers as much as possible. (b) Use mathematical induction to evaluate the determinant of the $n \times n$ *Vandermonde matrix*

$$V_n = \begin{bmatrix} 1 & z_1 & z_1^2 & \cdots & z_1^{n-1} \\ 1 & z_2 & z_2^2 & \cdots & z_2^{n-1} \\ \vdots & \vdots & \vdots & \ddots & \vdots \\ 1 & z_n & z_n^2 & \cdots & z_n^{n-1} \end{bmatrix}. \tag{0.9.1}$$

(c) Find conditions on $z_1, z_2, \ldots, z_n$ that are necessary and sufficient for V_n to be invertible.

P.0.10 Consider the polynomial $p(z) = c_k z^k + c_{k-1} z^{k-1} + \cdots + c_1 z + c_0$, in which $k \geq 1$, each coefficient c_i is a nonnegative integer, and $c_k \geq 1$. Prove the following statements: (a) $p(t + 2) = c_k t^k + d_{k-1} t^{k-1} + \cdots + d_1 t + d_0$, in which each d_i is a nonnegative integer and $d_0 \geq 2^k$. (b) $p(nd_0 + 2)$ is divisible by d_0 for each $n = 1, 2, \ldots$. (c) $p(n)$ is not a prime for infinitely many positive integers n. This was proved by C. Goldbach in 1752.

P.0.11 If p is a real polynomial, show that $p(\lambda) = 0$ if and only if $p(\bar{\lambda}) = 0$.

P.0.12 Show that a real polynomial can be factored into real linear factors and real quadratic factors that have no real zeros.

P.0.13 Show that every real polynomial of odd degree has a real zero. *Hint*: Use the Intermediate Value Theorem.

P.0.14 Let $h(z)$ be a polynomial and suppose that $z(z - 1)h(z) = 0$ for all $z \in [0, 1]$. Prove that h is the zero polynomial.

P.0.15 (a) Prove that the $n \times n$ Vandermonde matrix (0.9.1) is invertible if and only if the n complex numbers $z_1, z_2, \ldots, z_n$ are distinct. *Hint*: Consider the system $V_n \mathbf{c} = \mathbf{0}$, in which $\mathbf{c} = [c_0 \ c_1 \ \cdots \ c_{n-1}]^T$, and the polynomial $p(z) = c_{n-1} z^{n-1} + \cdots + c_1 z + c_0$. (b) Use (a) to prove the Lagrange Interpolation Theorem (Theorem 0.7.6).

P.0.16 If c is a nonzero scalar and p, q are nonzero polynomials, show that (a) $\deg(cp) = \deg p$, (b) $\deg(p + q) \leq \max\{\deg p, \deg q\}$, and (c) $\deg(pq) = \deg p + \deg q$. What happens if p is the zero polynomial?

P.0.17 Prove the uniqueness assertion of the division algorithm. That is, if f and g are polynomials such that $1 \leq \deg g \leq \deg f$ and if q_1, q_2, r_1 and r_2 are polynomials such that $\deg r_1 < \deg g$, $\deg r_2 < \deg g$, and $f = gq_1 + r_1 = gq_2 + r_2$, then $q_1 = q_2$ and $r_1 = r_2$.

P.0.18 Give an example of a nonconstant function $f : \mathbb{R} \to \mathbb{R}$ such that $f(t) = 0$ for infinitely many distinct values of t. Is f a polynomial?

P.0.19 Let $A = \mathrm{diag}(1, 2)$ and $B = \mathrm{diag}(3, 4)$. If $X \in \mathsf{M}_2$ intertwines A and B, what can you say about X? For a generalization, see Theorem 10.4.1.

P.0.20 Verify the identity (0.5.2) for a 2×2 matrix, and show that the identity (0.3.4) is (0.5.3).

P.0.21 Deduce (0.5.3) from the identity (0.5.2).

P.0.22 Deduce the second assertion in Theorem 0.8.1 from the first.

P.0.23 Let $A = \begin{bmatrix} 0 & 1 \\ 0 & 0 \end{bmatrix}$, $B = \begin{bmatrix} 4 & 3 \\ 1 & 2 \end{bmatrix}$, and $C = \begin{bmatrix} 3 & 4 \\ 1 & 2 \end{bmatrix}$. Show that $AB = AC$ even though $B \neq C$.

P.0.24 Let $A \in \mathsf{M}_n$. Show that A is idempotent if and only if $I - A$ is idempotent.

P.0.25   Let $A \in \mathsf{M}_n$ be idempotent. Show that A is invertible if and only if $A = I$.

P.0.26   Let $A, B \in \mathsf{M}_n$ be idempotent. Show that $\operatorname{tr}((A - B)^3) = \operatorname{tr}(A - B)$.

0.10 Some Important Concepts

- Mathematical induction.

- The Fundamental Theorem of Algebra.

- The division algorithm for polynomials.

- Lagrange interpolation.

- Polynomial functions of a matrix.

- Intertwining.

1   Vector Spaces

Many types of mathematical objects can be added and scaled: vectors in the plane, real-valued functions on a given real interval, polynomials, and real or complex matrices. Through long experience with these and other examples, mathematicians have identified a short list of essential features (axioms) that define a consistent and inclusive mathematical framework known as a vector space.

The theory of vector spaces and linear transformations provides a conceptual framework and vocabulary for linear mathematical models of diverse phenomena. Even inherently nonlinear physical theories may be well approximated for a broad range of applications by linear theories, whose natural setting is in real or complex vector spaces.

Examples of vector spaces include the two-dimensional real plane (the setting for plane analytic geometry and two-dimensional Newtonian mechanics) and three-dimensional real Euclidean space (the setting for solid analytic geometry, classical electromagnetism, and analytical dynamics). Other kinds of vector spaces abound in science and engineering. For example, standard mathematical models in quantum mechanics, electrical circuits, and signal processing use complex vector spaces. Many scientific theories exploit the formalism of vector spaces, which supplies powerful mathematical tools that are based only on the axioms for a vector space and their logical consequences, not on the details of a particular application.

In this chapter we provide a formal definition (and many examples) of a real or complex vector space. Among the important concepts introduced are linear combinations, span, linear independence, and linear dependence.

1.1 What is a Vector Space?

A vector space comprises four things that work together in harmony:

(a) A field $\mathbb{F}$ of *scalars*, which in this book is either the complex numbers $\mathbb{C}$ or the real numbers $\mathbb{R}$.

(b) A set $\mathcal{V}$ of objects called *vectors*.

(c) An operation of *vector addition* that takes any pair of vectors $\mathbf{u}, \mathbf{v} \in \mathcal{V}$ and assigns to them a vector in $\mathcal{V}$ denoted by $\mathbf{u} + \mathbf{v}$ (their *sum*).

(d) An operation of *scalar multiplication* that takes any scalar $c \in \mathbb{F}$ and any vector $\mathbf{u} \in \mathcal{V}$ and assigns to them a vector in $\mathcal{V}$ denoted by $c\mathbf{u}$.

Definition 1.1.1 Let $\mathbb{F} = \mathbb{R}$ or $\mathbb{C}$. Then $\mathcal{V}$ is a *vector space over the field* $\mathbb{F}$ (alternatively, $\mathcal{V}$ is an $\mathbb{F}$-*vector space*) if the scalars $\mathbb{F}$, the vectors $\mathcal{V}$, and the operations of vector addition and scalar multiplication satisfy the following axioms:

(i)   There is a unique additive identity element $\mathbf{0} \in \mathcal{V}$ such that $\mathbf{0} + \mathbf{u} = \mathbf{u}$ for all $\mathbf{u} \in \mathcal{V}$. The vector $\mathbf{0}$ is called the *zero vector*.

(ii)   Vector addition is commutative: $\mathbf{u} + \mathbf{v} = \mathbf{v} + \mathbf{u}$ for all $\mathbf{u}, \mathbf{v} \in \mathcal{V}$.

(iii)   Vector addition is associative: $\mathbf{u} + (\mathbf{v} + \mathbf{w}) = (\mathbf{u} + \mathbf{v}) + \mathbf{w}$ for all $\mathbf{u}, \mathbf{v}, \mathbf{w} \in \mathcal{V}$.

(iv)   Additive inverses exist and are unique: for each $\mathbf{u} \in \mathcal{V}$ there is a unique vector $-\mathbf{u} \in \mathcal{V}$ such that $\mathbf{u} + (-\mathbf{u}) = \mathbf{0}$.

(v)   The number 1 is the scalar multiplication identity element: $1\mathbf{u} = \mathbf{u}$ for all $\mathbf{u} \in \mathcal{V}$.

(vi)   Multiplication in $\mathbb{F}$ and scalar multiplication are compatible: $a(b\mathbf{u}) = (ab)\mathbf{u}$ for all $a, b \in \mathbb{F}$ and all $\mathbf{u} \in \mathcal{V}$.

(vii)   Scalar multiplication distributes over vector addition: $c(\mathbf{u} + \mathbf{v}) = c\mathbf{u} + c\mathbf{v}$ for all $c \in \mathbb{F}$ and all $\mathbf{u}, \mathbf{v} \in \mathcal{V}$.

(viii)   Addition in $\mathbb{F}$ distributes over scalar multiplication: $(a+b)\mathbf{u} = a\mathbf{u} + b\mathbf{u}$ for all $a, b \in \mathbb{F}$ and all $\mathbf{u} \in \mathcal{V}$.

A vector space over $\mathbb{R}$ is a *real vector space*; a vector space over $\mathbb{C}$ is a *complex vector space*. To help distinguish vectors from scalars, we often denote vectors (elements of the set $\mathcal{V}$) by boldface lowercase letters. In particular, it is important to distinguish the scalar 0 from the vector $\mathbf{0}$.

We often need to derive a conclusion from the fact that a vector $c\mathbf{u}$ is the zero vector, so we should look carefully at how that can happen.

Theorem 1.1.2 *Let $\mathcal{V}$ be an $\mathbb{F}$-vector space, let $c \in \mathbb{F}$, and let $\mathbf{u} \in \mathcal{V}$. The following statements are equivalent: (a) either $c = 0$ or $\mathbf{u} = \mathbf{0}$; (b) $c\mathbf{u} = \mathbf{0}$.*

Proof   (a) $\Rightarrow$ (b) Axioms (i) and (vii) ensure that $c\mathbf{0} = c(\mathbf{0} + \mathbf{0}) = c\mathbf{0} + c\mathbf{0}$. Use axiom (iv) to add $-(c\mathbf{0})$ to both sides of this identity and conclude that

$$c\mathbf{0} = \mathbf{0} \text{ for any } c \in \mathbb{F}. \tag{1.1.3}$$

Now invoke axiom (viii) and consider the vector $0\mathbf{u} = (0 + 0)\mathbf{u} = 0\mathbf{u} + 0\mathbf{u}$. Use axioms (iv) and (i) to add $-(0\mathbf{u})$ to both sides of this identity and conclude that

$$0\mathbf{u} = \mathbf{0} \text{ for any } \mathbf{u} \in \mathcal{V}.$$

(b) $\Rightarrow$ (a) Suppose that $c \neq 0$ and $c\mathbf{u} = \mathbf{0}$. Then (1.1.3) ensures that $c^{-1}(c\mathbf{u}) = c^{-1}\mathbf{0} = \mathbf{0}$, and axioms (vi) and (v) ensure that $\mathbf{0} = c^{-1}(c\mathbf{u}) = (c^{-1}c)\mathbf{u} = 1\mathbf{u} = \mathbf{u}$.   □

Corollary 1.1.4 *Let $\mathcal{V}$ be an $\mathbb{F}$-vector space. Then $(-1)\mathbf{u} = -\mathbf{u}$ for every $\mathbf{u} \in \mathcal{V}$.*

Proof   Let $\mathbf{u} \in \mathcal{V}$. We must show that $(-1)\mathbf{u} + \mathbf{u} = \mathbf{0}$. Use the vector space axioms (v) and (viii), together with the preceding theorem, to compute

$$(-1)\mathbf{u} + \mathbf{u} = (-1)\mathbf{u} + 1\mathbf{u} \qquad \text{Axiom (v)}$$
$$= (-1+1)\mathbf{u} \qquad \text{Axiom (viii)}$$
$$= 0\mathbf{u}$$
$$= \mathbf{0}. \qquad \text{Theorem 1.1.2} \qquad \square$$

1.2 Examples of Vector Spaces

Every vector space contains a zero vector (axiom (i)), so a vector space cannot be empty. However, the axioms for a vector space permit $\mathcal{V}$ to contain only the zero vector. Such a vector space is not very interesting, and we often need to exclude this possibility in formulating our theorems.

Definition 1.2.1 Let $\mathcal{V}$ be an $\mathbb{F}$-vector space. If $\mathcal{V} = \{\mathbf{0}\}$, then $\mathcal{V}$ is a *zero vector space*; if $\mathcal{V} \neq \{\mathbf{0}\}$, then $\mathcal{V}$ is a *nonzero vector space*.

In each of the following examples, we describe the elements of the set $\mathcal{V}$ (the *vectors*), the zero vector, and the operations of scalar multiplication and vector addition. The field $\mathbb{F}$ is always either $\mathbb{C}$ or $\mathbb{R}$.

Example 1.2.2 Let $\mathcal{V} = \mathbb{F}^n$, the set of $n \times 1$ matrices (column vectors) with entries from $\mathbb{F}$. For typographical convenience, we often write $\mathbf{u} = [u_i]$ or $\mathbf{u} = [u_1 \ u_2 \ \ldots \ u_n]^{\mathsf{T}}$ instead of

$$\mathbf{u} = \begin{bmatrix} u_1 \\ u_2 \\ \vdots \\ u_n \end{bmatrix} \in \mathbb{F}^n, \quad u_1, u_2, \ldots, u_n \in \mathbb{F}.$$

Vector addition of $\mathbf{u} = [u_i]$ and $\mathbf{v} = [v_i]$ is defined by $\mathbf{u} + \mathbf{v} = [u_i + v_i]$, and scalar multiplication by elements of $\mathbb{F}$ is defined by $c\mathbf{u} = [cu_i]$; we refer to these as *entrywise operations*. The zero vector in $\mathbb{F}^n$ is $\mathbf{0}_n = [0 \ 0 \ldots 0]^{\mathsf{T}}$. We often omit the subscript from a zero vector when its size can be inferred from the context.

Example 1.2.3 Let $\mathcal{V} = \mathsf{M}_{m \times n}(\mathbb{F})$, the set of $m \times n$ matrices with entries from $\mathbb{F}$. Vector addition and scalar multiplication are defined entrywise, as in the preceding example. The zero vector in $\mathsf{M}_{m \times n}(\mathbb{F})$ is the matrix $0_{m \times n} \in \mathsf{M}_{m \times n}(\mathbb{F})$, all entries of which are zero. We often omit the subscripts from a zero matrix when its size can be inferred from the context.

Example 1.2.4 Let $\mathcal{V} = \mathcal{P}_n$, the set of polynomials of degree at most n with complex coefficients. If we wish to emphasize that complex coefficients are permitted, we write $\mathcal{V} = \mathcal{P}_n(\mathbb{C})$. The set of polynomials of degree at most n with real coefficients is denoted by $\mathcal{P}_n(\mathbb{R})$. Addition of polynomials is defined by adding the coefficients of corresponding monomials. For example, with $p(z) = iz^2 + 11z - 5$ and $q(z) = -7z^2 + 3z + 2$ in $\mathcal{P}_2$, we have $(p+q)(z) = (i-7)z^2 + 14z - 3$. Scalar multiplication of a polynomial by a scalar c is defined by multiplying each coefficient by c. For example, $(4p)(z) = 4iz^2 + 44z - 20$. The zero vector in $\mathcal{P}_n$ is the zero polynomial; see Definition 0.7.1.

Example 1.2.5 Let $V = \mathcal{P}$ (sometimes we write $\mathcal{P}(\mathbb{C})$), the set of all polynomials with complex coefficients; $V = \mathcal{P}(\mathbb{R})$ denotes the set of all polynomials with real coefficients. The operations of vector addition and scalar multiplication are the same as in the preceding example, and the zero vector in $\mathcal{P}$ is again the zero polynomial.

Example 1.2.6 Let $V = C_{\mathbb{F}}[a, b]$, the set of continuous $\mathbb{F}$-valued functions on an interval $[a, b] \subseteq \mathbb{R}$ with $a < b$. If the field designator is absent, it is understood that $\mathbb{F} = \mathbb{C}$, that is, $C[0, 1]$ is $C_{\mathbb{C}}[0, 1]$. The operations of vector addition and scalar multiplication are defined pointwise. That is, if $f, g \in C_{\mathbb{F}}[a, b]$, then $f + g$ is the $\mathbb{F}$-valued function on $[a, b]$ defined by $(f + g)(t) = f(t) + g(t)$ for each $t \in [a, b]$. If $c \in \mathbb{F}$, the $\mathbb{F}$-valued function cf is defined by $(cf)(t) = cf(t)$ for each $t \in [a, b]$. It is a theorem from calculus that $f+g$ and cf are continuous if f and g are continuous, so sums and scalar multiples of elements of $C_{\mathbb{F}}[a, b]$ are in $C_{\mathbb{F}}[a, b]$. The zero vector in $C_{\mathbb{F}}[a, b]$ is the *zero function*, which takes the value zero at every point in $[a, b]$.

Example 1.2.7 Let V be the set of all infinite sequences $\mathbf{u} = (u_1, u_2, \ldots)$, in which each $u_i \in \mathbb{F}$ and $u_i \neq 0$ for only finitely many values of the index i. The operations of vector addition and scalar multiplication are defined entrywise. The zero vector in V is the *zero infinite sequence* $\mathbf{0} = (0, 0, \ldots)$. We say that V is the $\mathbb{F}$-vector space of *finitely nonzero sequences*.

1.3 Subspaces

Definition 1.3.1 A *subspace* of an $\mathbb{F}$-vector space V is a subset U of V that is an $\mathbb{F}$-vector space with the same vector addition and scalar multiplication operations as in V.

Example 1.3.2 If V is an $\mathbb{F}$-vector space, then $\{\mathbf{0}\}$ and V itself are subspaces of V.

A subspace is nonempty; it is a vector space, so it contains a zero vector.

To show that a subset U of an $\mathbb{F}$-vector space V is a subspace, we do not need to verify the vector space axioms (ii)–(iii) and (v)–(viii) because they are automatically satisfied; we say that U *inherits* these properties from V. However, we must show the following:

(a) Sums and scalar multiples of elements of U are in U (that is, U is *closed under vector addition and scalar multiplication*).

(b) U contains the zero vector of V.

(c) U contains an additive inverse for each of its elements.

The following theorem describes a streamlined way to verify these three conditions.

Theorem 1.3.3 *Let V be an $\mathbb{F}$-vector space and let U be a nonempty subset of V. Then U is a subspace of V if and only if $c\mathbf{u} + \mathbf{v} \in U$ whenever $\mathbf{u}, \mathbf{v} \in U$ and $c \in \mathbb{F}$.*

Proof If U is a subspace, then $c\mathbf{u} + \mathbf{v} \in U$ whenever $\mathbf{u}, \mathbf{v} \in U$ and $c \in \mathbb{F}$ because a subspace is closed under scalar multiplication and vector addition.

Conversely, suppose that $c\mathbf{u} + \mathbf{v} \in \mathcal{U}$ whenever $\mathbf{u}, \mathbf{v} \in \mathcal{U}$ and $c \in \mathbb{F}$. We must verify the properties (a), (b), and (c) in the preceding list. Let $\mathbf{u} \in \mathcal{U}$. Corollary 1.1.4 ensures that $(-1)\mathbf{u}$ is the additive inverse of $\mathbf{u}$, so $\mathbf{0} = (-1)\mathbf{u}+\mathbf{u} \in \mathcal{U}$. This verifies (b). Since $(-1)\mathbf{u} = (-1)\mathbf{u}+\mathbf{0}$, it follows that the additive inverse of $\mathbf{u}$ is in $\mathcal{U}$. This verifies (c). We have $c\mathbf{u} = c\mathbf{u} + \mathbf{0} \in \mathcal{U}$ and $\mathbf{u} + \mathbf{v} = 1\mathbf{u} + \mathbf{v} \in \mathcal{U}$ for all $c \in \mathbb{F}$ and all $\mathbf{u}, \mathbf{v} \in \mathcal{U}$. This verifies (a). $\qquad\square$

Example 1.3.4 Let $A \in \mathsf{M}_{m \times n}(\mathbb{F})$. The *null space* of A is

$$\text{null } A = \{\mathbf{x} \in \mathbb{F}^n : A\mathbf{x} = \mathbf{0}\} \subseteq \mathbb{F}^n. \tag{1.3.5}$$

Since $A\mathbf{0}_n = \mathbf{0}_m$, the zero vector of $\mathbb{F}^n$ is in null A, which is therefore not empty. If $\mathbf{x}, \mathbf{y} \in \mathbb{F}^n$, $A\mathbf{x} = \mathbf{0}$, $A\mathbf{y} = \mathbf{0}$, and $c \in \mathbb{F}$, then $A(c\mathbf{x} + \mathbf{y}) = cA\mathbf{x} + A\mathbf{y} = c\mathbf{0} + \mathbf{0} = \mathbf{0}$, so $c\mathbf{x} + \mathbf{y} \in \text{null } A$. The preceding theorem ensures that null A is a subspace of $\mathbb{F}^n$.

Example 1.3.6 $\mathcal{U} = \{[x_1\ x_2\ x_3]^\mathsf{T} \in \mathbb{R}^3 : x_1 + 2x_2 + 3x_3 = 0\}$ is the plane in $\mathbb{R}^3$ that contains the zero vector and has normal vector $[1\ 2\ 3]^\mathsf{T}$. If $A = [1\ 2\ 3] \in \mathsf{M}_{1 \times 3}(\mathbb{R})$, then $\mathcal{U} = \text{null } A$, so it is a subspace of $\mathbb{R}^3$.

Example 1.3.7 Let $A \in \mathsf{M}_{m \times n}(\mathbb{F})$. The *column space* of A is

$$\text{col } A = \{A\mathbf{x} : \mathbf{x} \in \mathbb{F}^n\} \subseteq \mathbb{F}^m. \tag{1.3.8}$$

Since $A\mathbf{0} = \mathbf{0}$, the zero vector of $\mathbb{F}^m$ is in col A, which is therefore not empty. If $\mathbf{u}, \mathbf{v} \in \text{col } A$ then there are $\mathbf{x}, \mathbf{y} \in \mathbb{F}^n$ such that $\mathbf{u} = A\mathbf{x}$ and $\mathbf{v} = A\mathbf{y}$. For any $c \in \mathbb{F}$, we have $c\mathbf{u} + \mathbf{v} = cA\mathbf{x} + A\mathbf{y} = A(c\mathbf{x} + \mathbf{y})$, so $c\mathbf{u} + \mathbf{v} \in \text{col } A$. The preceding theorem ensures that col A is a subspace of $\mathbb{F}^m$.

Example 1.3.9 $\mathcal{P}_5$ is a subspace of $\mathcal{P}$; see Examples 1.2.4 and 1.2.5. Sums and scalar multiples of polynomials of degree 5 or less are in $\mathcal{P}_5$.

Example 1.3.10 $\mathcal{P}_5(\mathbb{R})$ is a subset of $\mathcal{P}_5(\mathbb{C})$, but it is not a subspace. For example, the scalar 1 is in $\mathcal{P}_5(\mathbb{R})$ but $i1 = i \notin \mathcal{P}_5(\mathbb{R})$. The issue here is that the scalars for the vector space $\mathcal{P}_5(\mathbb{R})$ are the real numbers and the scalars for the vector space $\mathcal{P}_5(\mathbb{C})$ are the complex numbers. A subspace and the vector space that contains it must have the same field of scalars.

Example 1.3.11 A polynomial p is *even* if $p(-z) = p(z)$ for all z. We denote the set of even polynomials by $\mathcal{P}_{\text{even}}$. A polynomial p is *odd* if $p(-z) = -p(z)$ for all z. We denote the set of odd polynomials by $\mathcal{P}_{\text{odd}}$. For example, $p(z) = 2 + 3z^2$ is even and $p(z) = 5z + 4z^3$ is odd. Constant polynomials are even; the zero polynomial is both odd and even. Each of $\mathcal{P}_{\text{even}}$ and $\mathcal{P}_{\text{odd}}$ is a subspace of $\mathcal{P}$.

Example 1.3.12 $\mathcal{P}$ is a subspace of the complex vector space $C_{\mathbb{C}}[a, b]$. Certainly, $\mathcal{P}$ is nonempty. Moreover, every polynomial is a continuous function, and $cp + q \in \mathcal{P}$ whenever $p, q \in \mathcal{P}$ and $c \in \mathbb{C}$. Theorem 1.3.3 ensures that $\mathcal{P}$ is a subspace of $C_{\mathbb{C}}[a, b]$.

Example 1.3.13 Let $A \in \mathsf{M}_m(\mathbb{F})$ and let $\mathcal{U}$ be a subspace of $\mathsf{M}_{m \times n}(\mathbb{F})$. We claim that

$$A\mathcal{U} = \{AX : X \in \mathcal{U}\}$$

is a subspace of $\mathsf{M}_{m \times n}(\mathbb{F})$. Since $0 \in \mathcal{U}$, we have $0 = A0 \in A\mathcal{U}$, which is therefore not empty. Moreover, $cAX + AY = A(cX + Y) \in A\mathcal{U}$ for any scalar c and any $X, Y \in \mathcal{U}$. Theorem 1.3.3 ensures that $A\mathcal{U}$ is a subspace of $\mathsf{M}_{m \times n}(\mathbb{F})$. For example, if we take $\mathcal{U} = \mathsf{M}_{m \times n}(\mathbb{F})$, then it follows that $A\mathsf{M}_{m \times n}(\mathbb{F})$ is a subspace of $\mathsf{M}_{m \times n}(\mathbb{F})$.

1.4 Linear Combinations and Span

Definition 1.4.1 Let $\mathcal{U}$ be a nonempty subset of an $\mathbb{F}$-vector space $\mathcal{V}$. A *linear combination* of elements of $\mathcal{U}$ is an expression of the form

$$c_1 \mathbf{v}_1 + c_2 \mathbf{v}_2 + \cdots + c_r \mathbf{v}_r, \tag{1.4.2}$$

in which r is a positive integer, $\mathbf{v}_1, \mathbf{v}_2, \ldots, \mathbf{v}_r \in \mathcal{U}$, and $c_1, c_2, \ldots, c_r \in \mathbb{F}$. A linear combination (1.4.2) is *trivial* if $c_1 = c_2 = \cdots = c_r = 0$; otherwise, it is *nontrivial*.

A linear combination is, by definition, a sum of finitely many scalar multiples of vectors.

Example 1.4.3 Every element of $\mathcal{P}$ is a linear combination of the elements $1, z, z^2, \ldots$.

Definition 1.4.4 A *list* of vectors in an $\mathbb{F}$-vector space $\mathcal{V}$ is a nonempty, finite, ordered sequence $\mathbf{v}_1, \mathbf{v}_2, \ldots, \mathbf{v}_r$ of vectors in $\mathcal{V}$. We often denote a list by a Greek letter, for example, $\beta = \mathbf{v}_1, \mathbf{v}_2, \ldots, \mathbf{v}_r$.

A subtle, but important, point is that a given vector can appear more than once in a list. For example, $\beta = z, z^2, z^2, z^2, z^3$ is a *list* of five vectors in $\mathcal{P}_3$. However, the *set* of vectors in the list β is $\{z, z^2, z^3\}$.

Definition 1.4.5 Let $\mathcal{U}$ be a subset of an $\mathbb{F}$-vector space $\mathcal{V}$. If $\mathcal{U} \neq \varnothing$, then span $\mathcal{U}$ is the set of linear combinations of elements of $\mathcal{U}$; we define span $\varnothing = \{\mathbf{0}\}$. If $\beta = \mathbf{v}_1, \mathbf{v}_2, \ldots, \mathbf{v}_r$ is a list of $r \geq 1$ vectors in $\mathcal{V}$, we define span $\beta = \text{span}\{\mathbf{v}_1, \mathbf{v}_2, \ldots, \mathbf{v}_r\}$, that is, the span of a list is the span of the set of vectors in the list.

Example 1.4.6 If $\mathbf{u} \in \mathcal{V}$, then $\text{span}\{\mathbf{u}\} = \{c\mathbf{u} : c \in \mathbb{F}\}$ is a subspace of $\mathcal{V}$. In particular, $\text{span}\{\mathbf{0}\} = \{\mathbf{0}\}$.

Example 1.4.7 Let $A = [\mathbf{a}_1 \ \mathbf{a}_2 \ \ldots \ \mathbf{a}_n] \in \mathsf{M}_{m \times n}(\mathbb{F})$ and consider the list $\beta = \mathbf{a}_1, \mathbf{a}_2, \ldots, \mathbf{a}_n$ in the $\mathbb{F}$-vector space $\mathbb{F}^m$. Then

$$\text{span } \beta = \{x_1 \mathbf{a}_1 + x_2 \mathbf{a}_2 + \cdots + x_n \mathbf{a}_n : x_1, x_2, \ldots, x_n \in \mathbb{F}\}$$
$$= \{A\mathbf{x} : \mathbf{x} \in \mathbb{F}^n\} = \text{col } A,$$

that is, the span of the columns of a matrix is its column space.

Example 1.4.8 Consider the list $\beta = \{z, z^2, z^3\}$ of elements of $\mathcal{P}_3$. Then span $\beta = \{c_3 z^3 + c_2 z^2 + c_1 z : c_1, c_2, c_3 \in \mathbb{C}\}$ is a subspace of $\mathcal{P}_3$ because it is nonempty and

$$c(a_3z^3 + a_2z^2 + a_1z) + (b_3z^3 + b_2z^2 + b_1z)$$
$$= (ca_3 + b_3)z^3 + (ca_2 + b_2)z^2 + (ca_1 + b_1)z$$

is a linear combination of vectors in the list β for all $c, a_1, a_2, a_3, b_1, b_2, b_3 \in \mathbb{C}$.

The span of a subset of a vector space is always a subspace.

Theorem 1.4.9 *Let $\mathcal{U}$ be a subset of an $\mathbb{F}$-vector space $\mathcal{V}$.*

(a) $\operatorname{span} \mathcal{U}$ *is a subspace of $\mathcal{V}$.*

(b) $\mathcal{U} \subseteq \operatorname{span} \mathcal{U}$.

(c) $\mathcal{U} = \operatorname{span} \mathcal{U}$ *if and only if $\mathcal{U}$ is a subspace of $\mathcal{V}$.*

(d) $\operatorname{span}(\operatorname{span} \mathcal{U}) = \operatorname{span} \mathcal{U}$.

Proof First suppose that $\mathcal{U} = \varnothing$. Then by definition, $\operatorname{span} \mathcal{U} = \{\mathbf{0}\}$, which is a subspace of $\mathcal{V}$. The empty set is a subset of every set, so it is contained in $\{\mathbf{0}\}$. Both implications in (c) are vacuous. For the assertion in (d), see Example 1.4.6.

Now suppose that $\mathcal{U} \neq \varnothing$. If $\mathbf{u}, \mathbf{v} \in \operatorname{span} \mathcal{U}$ and $c \in \mathbb{F}$, then each of $\mathbf{u}, \mathbf{v}, c\mathbf{u}$, and $c\mathbf{u} + \mathbf{v}$ is a linear combination of elements of $\mathcal{U}$, so each is in $\operatorname{span} \mathcal{U}$. Theorem 1.3.3 ensures that $\operatorname{span} \mathcal{U}$ is a subspace. The assertion in (b) follows from the fact that $1\mathbf{u} = \mathbf{u}$ is an element of $\operatorname{span} \mathcal{U}$ for each $\mathbf{u} \in \mathcal{U}$. To prove the two implications in (c), first suppose that $\mathcal{U} = \operatorname{span} \mathcal{U}$. Then (a) ensures that $\mathcal{U}$ is a subspace of $\mathcal{V}$. Conversely, if $\mathcal{U}$ is a subspace of $\mathcal{V}$, then it is closed under vector addition and scalar multiplication, so $\operatorname{span} \mathcal{U} \subseteq \mathcal{U}$. The containment $\mathcal{U} \subseteq \operatorname{span} \mathcal{U}$ in (b) ensures that $\mathcal{U} = \operatorname{span} \mathcal{U}$. The assertion in (d) follows from (a) and (c). $\qquad \square$

Theorem 1.4.10 *Let $\mathcal{U}$ and $\mathcal{W}$ be subsets of an $\mathbb{F}$-vector space $\mathcal{V}$. If $\mathcal{U} \subseteq \mathcal{W}$, then $\operatorname{span} \mathcal{U} \subseteq \operatorname{span} \mathcal{W}$.*

Proof If $\mathcal{U} = \varnothing$, then $\operatorname{span} \mathcal{U} = \{\mathbf{0}\} \subseteq \operatorname{span} \mathcal{W}$. If $\mathcal{U} \neq \varnothing$, then every linear combination of elements of $\mathcal{U}$ is a linear combination of elements of $\mathcal{W}$. $\qquad \square$

Example 1.4.11 Let $\mathcal{U} = \{1, z - 2z^2, z^2 + 5z^3, z^3, 1 + 4z^2\}$. We claim that $\operatorname{span} \mathcal{U} = \mathcal{P}_3$. To verify this, observe that

$$1 = 1,$$
$$z = (z - 2z^2) + 2(z^2 + 5z^3) - 10z^3,$$
$$z^2 = (z^2 + 5z^3) - 5z^3, \text{ and}$$
$$z^3 = z^3.$$

Thus, $\{1, z, z^2, z^3\} \subseteq \operatorname{span} \mathcal{U} \subseteq \mathcal{P}_3$. Now invoke the two preceding theorems to compute

$$\mathcal{P}_3 = \operatorname{span}\{1, z, z^2, z^3\} \subseteq \operatorname{span}(\operatorname{span} \mathcal{U}) = \operatorname{span} \mathcal{U} \subseteq \mathcal{P}_3.$$

Definition 1.4.12 Let $\mathcal{V}$ be an $\mathbb{F}$-vector space. Let $\mathcal{U}$ be a subset of $\mathcal{V}$ and let β be a list of vectors in $\mathcal{V}$. Then $\mathcal{U}$ *spans* $\mathcal{V}$ ($\mathcal{U}$ is a *spanning set*) if $\operatorname{span} \mathcal{U} = \mathcal{V}$. We say that β *spans* $\mathcal{V}$ (β is a *spanning list*) if $\operatorname{span} \beta = \mathcal{V}$.

Example 1.4.13 Each of the sets $\{1, z, z^2, z^3\}$ and $\{1, z - 2z^2, z^2 + 5z^3, z^3, 1 + 4z^2\}$ spans $\mathcal{P}_3$.

Example 1.4.14 Let $A = [\mathbf{a}_1 \ \mathbf{a}_2 \ \ldots \ \mathbf{a}_n] \in M_n(\mathbb{F})$ be invertible, let $\mathbf{y} \in \mathbb{F}^n$, and let $A^{-1}\mathbf{y} = [x_i]_{i=1}^n$. Then $\mathbf{y} = A(A^{-1}\mathbf{y}) = x_1\mathbf{a}_1 + x_2\mathbf{a}_2 + \cdots + x_n\mathbf{a}_n$ is a linear combination of the columns of A. We conclude that $\mathbb{F}^n$ is spanned by the columns of any $n \times n$ invertible matrix.

Example 1.4.15 The identity matrix I_n is invertible and its columns are

$$\mathbf{e}_1 = [1 \ 0 \ \ldots \ 0]^\mathsf{T}, \quad \mathbf{e}_2 = [0 \ 1 \ \ldots \ 0]^\mathsf{T}, \ldots, \quad \mathbf{e}_n = [0 \ 0 \ \ldots \ 1]^\mathsf{T}. \tag{1.4.16}$$

Consequently, $\mathbf{e}_1, \mathbf{e}_2, \ldots, \mathbf{e}_n$ span $\mathbb{F}^n$. One can see this directly by observing that any $\mathbf{u} = [u_i] \in \mathbb{F}^n$ can be expressed as

$$\mathbf{u} = u_1\mathbf{e}_1 + u_2\mathbf{e}_2 + \cdots + u_n\mathbf{e}_n.$$

For example, the *all-ones vector* in $\mathbb{F}^n$ can be expressed as

$$\mathbf{e} = \mathbf{e}_1 + \mathbf{e}_2 + \cdots + \mathbf{e}_n = [1 \ 1 \ \ldots \ 1]^\mathsf{T}.$$

1.5 Intersections, Sums, and Direct Sums of Subspaces

Theorem 1.5.1 *Let $\mathcal{U}$ and $\mathcal{W}$ be subspaces of an $\mathbb{F}$-vector space $\mathcal{V}$. Their intersection*

$$\mathcal{U} \cap \mathcal{W} = \{\mathbf{v} : \mathbf{v} \in \mathcal{U} \text{ and } \mathbf{v} \in \mathcal{W}\}$$

is a subspace of $\mathcal{V}$.

Proof The zero vector is in both $\mathcal{U}$ and $\mathcal{W}$, so it is in $\mathcal{U} \cap \mathcal{W}$. Thus, $\mathcal{U} \cap \mathcal{W}$ is nonempty. If $\mathbf{u}, \mathbf{v} \in \mathcal{U}$ and $\mathbf{u}, \mathbf{v} \in \mathcal{W}$, then for any $c \in \mathbb{F}$ the vector $c\mathbf{u} + \mathbf{v}$ is in both $\mathcal{U}$ and $\mathcal{W}$ since they are subspaces. Consequently, $c\mathbf{u} + \mathbf{v} \in \mathcal{U} \cap \mathcal{W}$, so Theorem 1.3.3 ensures that $\mathcal{U} \cap \mathcal{W}$ is a subspace of $\mathcal{V}$. $\qquad\square$

The union of subspaces need not be a subspace.

Example 1.5.2 In the real vector space $\mathbb{R}^2$ (thought of as the xy-plane) the x-axis $\mathcal{X} = \{[x \ 0]^\mathsf{T} : x \in \mathbb{R}\}$ and y-axis $\mathcal{Y} = \{[0 \ y]^\mathsf{T} : y \in \mathbb{R}\}$ are subspaces, but their union is not a subspace of $\mathbb{R}^2$ since $[1 \ 0]^\mathsf{T} + [0 \ 1]^\mathsf{T} = [1 \ 1]^\mathsf{T} \notin \mathcal{X} \cup \mathcal{Y}$.

Example 1.5.3 In $\mathcal{P}$, the union $\mathcal{P}_{\text{even}} \cup \mathcal{P}_{\text{odd}}$ is the set of polynomials that are either even or odd. However, $z^2 \in \mathcal{P}_{\text{even}}$ and $z \in \mathcal{P}_{\text{odd}}$, but $z + z^2 \notin \mathcal{P}_{\text{even}} \cup \mathcal{P}_{\text{odd}}$, so $\mathcal{P}_{\text{even}} \cup \mathcal{P}_{\text{odd}}$ is not a subspace of $\mathcal{P}$.

The span of the union of subspaces is a subspace since the span of *any* set is a subspace.

Definition 1.5.4 Let $\mathcal{U}$ and $\mathcal{W}$ be subspaces of an $\mathbb{F}$-vector space $\mathcal{V}$. The *sum* of $\mathcal{U}$ and $\mathcal{W}$ is the subspace $\text{span}(\mathcal{U} \cup \mathcal{W})$. It is denoted by $\text{span}(\mathcal{U} \cup \mathcal{W}) = \mathcal{U} + \mathcal{W}$.

In the preceding definition, $\text{span}(\mathcal{U} \cup \mathcal{W})$ consists of all linear combinations of vectors that are either in $\mathcal{U}$ or in $\mathcal{W}$. Since both $\mathcal{U}$ and $\mathcal{W}$ are closed under vector addition, it follows that

$$\mathcal{U} + \mathcal{W} = \text{span}(\mathcal{U} \cup \mathcal{W}) = \{\mathbf{u} + \mathbf{w} : \mathbf{u} \in \mathcal{U} \text{ and } \mathbf{w} \in \mathcal{W}\}.$$

Example 1.5.5 In the real vector space $\mathbb{R}^2$, we have $\mathcal{X} + \mathcal{Y} = \mathbb{R}^2$ since $[x \ y]^\mathsf{T} = [x \ 0]^\mathsf{T} + [0 \ y]^\mathsf{T}$ for all $x, y \in \mathbb{R}$; see Example 1.5.2.

Example 1.5.6 In the complex vector space $\mathcal{P}$, we have $\mathcal{P}_{\text{even}} + \mathcal{P}_{\text{odd}} = \mathcal{P}$. For example,

$$z^5 + iz^4 - \pi z^3 - 5z^2 + z - 2 = (iz^4 - 5z^2 - 2) + (z^5 - \pi z^3 + z)$$

is a sum of a vector in $\mathcal{P}_{\text{even}}$ and a vector in $\mathcal{P}_{\text{odd}}$.

In the two preceding examples, the respective pairs of subspaces have an important special property that we identify in the following definition.

Definition 1.5.7 Let $\mathcal{U}$ and $\mathcal{W}$ be subspaces of an $\mathbb{F}$-vector space $\mathcal{V}$. If $\mathcal{U} \cap \mathcal{W} = \{\mathbf{0}\}$, then the sum of $\mathcal{U}$ and $\mathcal{W}$ is a *direct sum*. It is denoted by $\mathcal{U} \oplus \mathcal{W}$.

Example 1.5.8 Let

$$\mathcal{U} = \left\{ \begin{bmatrix} a & 0 \\ b & c \end{bmatrix} \in \mathsf{M}_2 : a, b, c \in \mathbb{C} \right\} \text{ and } \mathcal{W} = \left\{ \begin{bmatrix} x & y \\ 0 & z \end{bmatrix} \in \mathsf{M}_2 : x, y, z \in \mathbb{C} \right\}$$

be the subspaces of M_2 consisting of the lower triangular and upper triangular matrices, respectively. Then $\mathcal{U} + \mathcal{W} = \mathsf{M}_2$, but the sum is not a direct sum because $\mathcal{U} \cap \mathcal{W} \neq \{\mathbf{0}\}$; it is the subspace of all diagonal matrices in M_2. Every matrix can be expressed as a sum of a lower triangular matrix and an upper triangular matrix, but the summands need not be unique. For example,

$$\begin{bmatrix} 1 & 2 \\ 3 & 4 \end{bmatrix} = \begin{bmatrix} 1 & 0 \\ 3 & 4 \end{bmatrix} + \begin{bmatrix} 0 & 2 \\ 0 & 0 \end{bmatrix} = \begin{bmatrix} -1 & 0 \\ 3 & 3 \end{bmatrix} + \begin{bmatrix} 2 & 2 \\ 0 & 1 \end{bmatrix}.$$

Direct sums are important because any vector in a direct sum has a unique representation with respect to the direct summands.

Theorem 1.5.9 *Let $\mathcal{U}$ and $\mathcal{W}$ be subspaces of an $\mathbb{F}$-vector space $\mathcal{V}$ and suppose that $\mathcal{U} \cap \mathcal{W} = \{\mathbf{0}\}$. Then each vector in $\mathcal{U} \oplus \mathcal{W}$ is uniquely expressible as a sum of a vector in $\mathcal{U}$ and a vector in $\mathcal{W}$.*

Proof Suppose that $\mathbf{u}_1, \mathbf{u}_2 \in \mathcal{U}$, $\mathbf{w}_1, \mathbf{w}_2 \in \mathcal{W}$, and $\mathbf{u}_1 + \mathbf{w}_1 = \mathbf{u}_2 + \mathbf{w}_2$. Then $\mathbf{u}_1 - \mathbf{u}_2 = \mathbf{w}_2 - \mathbf{w}_1 \in \mathcal{U} \cap \mathcal{W}$. But $\mathcal{U} \cap \mathcal{W} = \{\mathbf{0}\}$, so $\mathbf{u}_1 - \mathbf{u}_2 = \mathbf{0}$ and $\mathbf{w}_2 - \mathbf{w}_1 = \mathbf{0}$. Thus, $\mathbf{u}_1 = \mathbf{u}_2$ and $\mathbf{w}_1 = \mathbf{w}_2$. $\quad\square$

Example 1.5.10 Although $\mathcal{P}_4 = \text{span}\{1, z, z^2, z^3\} + \text{span}\{z^3, z^4\}$, this sum is not a direct sum since $\text{span}\{1, z, z^2, z^3\} \cap \text{span}\{z^3, z^4\} = \{cz^3 : c \in \mathbb{C}\} \neq \{\mathbf{0}\}$.

Example 1.5.11 We have $\mathcal{P} = \mathcal{P}_{\text{even}} \oplus \mathcal{P}_{\text{odd}}$ since the only polynomial that is both even and odd is the zero polynomial.

Example 1.5.12 M_2 is the direct sum of the subspace of strictly lower triangular matrices and the subspace of upper triangular matrices.

1.6 Linear Dependence and Linear Independence

Definition 1.6.1 A list of vectors $\beta = v_1, v_2, \ldots, v_r$ in an $\mathbb{F}$-vector space $\mathcal{V}$ is *linearly dependent* if there are scalars $c_1, c_2, \ldots, c_r \in \mathbb{F}$, not all zero, such that $c_1 v_1 + c_2 v_2 + \cdots + c_r v_r = \mathbf{0}$.

Here are some facts about linear dependence:

- If $\beta = v_1, v_2, \ldots, v_r$ is linearly dependent, then the commutativity of vector addition ensures that any list of r vectors obtained by rearranging the vectors in β is also linearly dependent. For example, the list $v_r, v_{r-1}, \ldots, v_1$ is linearly dependent.

- Theorem 1.1.2 ensures that a list consisting of a single vector v is linearly dependent if and only if v is the zero vector.

- A list of two vectors is linearly dependent if and only if one of the vectors is a scalar multiple of the other. If $c_1 v_1 + c_2 v_2 = \mathbf{0}$ and $c_1 \neq 0$, then $v_1 = -c_1^{-1} c_2 v_2$; if $c_2 \neq 0$, then $v_2 = -c_2^{-1} c_1 v_1$. Conversely, if $v_1 = c v_2$, then $1 v_1 + (-c) v_2 = \mathbf{0}$ is a nontrivial linear combination.

- A list of three or more vectors is linearly dependent if and only if one of the vectors is a linear combination of the others. If $c_1 v_1 + c_2 v_2 + \cdots + c_r v_r = \mathbf{0}$ and $c_j \neq 0$, then $v_j = -c_j^{-1} \sum_{i \neq j} c_i v_i$. Conversely, if $v_j = \sum_{i \neq j} c_i v_i$, then $1 v_j + \sum_{i \neq j} (-c_i) v_i = \mathbf{0}$ is a nontrivial linear combination.

- Any list of vectors that includes the zero vector is linearly dependent. For example, if the list is $v_1, v_2, \ldots, v_r$ and $v_r = \mathbf{0}$, then consider the nontrivial linear combination $0 v_1 + 0 v_2 + \cdots + 0 v_{r-1} + 1 v_r = \mathbf{0}$.

- Any list of vectors that includes the same vector twice is linearly dependent. For example, if the list is $v_1, v_2, \ldots, v_{r-1}, v_r$ and $v_{r-1} = v_r$, then consider the nontrivial linear combination $0 v_1 + 0 v_2 + \cdots + 0 v_{r-2} + 1 v_{r-1} + (-1) v_r = \mathbf{0}$.

Example 1.6.2 Let

$$v_1 = \begin{bmatrix} 1 \\ 1 \end{bmatrix}, \qquad v_2 = \begin{bmatrix} 1 \\ -1 \end{bmatrix}, \quad \text{and} \quad v_3 = \begin{bmatrix} 1 \\ 2 \end{bmatrix}.$$

Then $-3 v_1 + 1 v_2 + 2 v_3 = \mathbf{0}$, so the list $\beta = v_1, v_2, v_3$ is linearly dependent.

Theorem 1.6.3 *If $r \geq 1$ and a list of vectors $v_1, v_2, \ldots, v_r$ in an $\mathbb{F}$-vector space $\mathcal{V}$ is linearly dependent, then the list $v_1, v_2, \ldots, v_r, v$ is linearly dependent for any $v \in \mathcal{V}$.*

Proof If $\sum_{i=1}^{r} c_i v_i = \mathbf{0}$ and some $c_j \neq 0$, then $\sum_{i=1}^{r} c_i v_i + 0 v = \mathbf{0}$ and not all of the scalars $c_1, c_2, \ldots, c_r, 0$ are zero. $\qquad \square$

The opposite of linear dependence is linear independence.

Definition 1.6.4 A list of vectors $\beta = \mathbf{v}_1, \mathbf{v}_2, \ldots, \mathbf{v}_r$ in an $\mathbb{F}$-vector space $\mathcal{V}$ is *linearly independent* if it is not linearly dependent. That is, β is linearly independent if the only scalars $c_1, c_2, \ldots, c_r \in \mathbb{F}$ such that $c_1\mathbf{v}_1 + c_2\mathbf{v}_2 + \cdots + c_r\mathbf{v}_r = \mathbf{0}$ are $c_1 = c_2 = \cdots = c_r = 0$.

It is convenient to say that "$\mathbf{v}_1, \mathbf{v}_2, \ldots, \mathbf{v}_r$ are linearly independent (respectively, linearly dependent)" rather than the more formal "the list of vectors $\mathbf{v}_1, \mathbf{v}_2, \ldots, \mathbf{v}_r$ is linearly independent (respectively, linearly dependent)." Unlike lists, which are finite by definition, sets can contain infinitely many distinct vectors. We define linear independence and linear dependence for sets of vectors as follows:

Definition 1.6.5 A subset $\mathscr{S}$ of a vector space $\mathcal{V}$ is *linearly independent* if every list of distinct vectors in $\mathscr{S}$ is linearly independent; $\mathscr{S}$ is *linearly dependent* if some list of distinct vectors in $\mathscr{S}$ is linearly dependent.

Here are some facts about linear independence:

- Whether a list is linearly independent does not depend on how it is ordered.

- A list consisting of a single vector $\mathbf{v}$ is linearly independent if and only if $\mathbf{v}$ is not the zero vector.

- A list of two vectors is linearly independent if and only if neither vector in the list is a scalar multiple of the other.

- A list of three or more vectors is linearly independent if and only if no vector in the list is a linear combination of the others.

Linear independence of a list of n vectors in $\mathbb{F}^m$ can be formulated as a statement about the null space of a matrix in $\mathsf{M}_{m \times n}(\mathbb{F})$.

Example 1.6.6 Let $A = [\mathbf{a}_1 \ \mathbf{a}_2 \ \ldots \ \mathbf{a}_n] \in \mathsf{M}_{m \times n}(\mathbb{F})$. The list $\beta = \mathbf{a}_1, \mathbf{a}_2, \ldots, \mathbf{a}_n$ is linearly independent if and only if the only $\mathbf{x} = [x_1 \ x_2 \ \ldots \ x_n]^\mathsf{T} \in \mathbb{F}^n$ such that

$$\underbrace{x_1\mathbf{a}_1 + x_2\mathbf{a}_2 + \cdots + x_n\mathbf{a}_n}_{A\mathbf{x}} = \mathbf{0}$$

is the zero vector. That is, β is linearly independent if and only if $\text{null } A = \{\mathbf{0}\}$.

Example 1.6.7 In $\mathcal{P}$, the vectors $1, z, z^2, \ldots, z^n$ are linearly independent. A linear combination $c_0 + c_1 z + \cdots + c_n z^n$ is the zero polynomial if and only if $c_0 = c_1 = \cdots = c_n = 0$.

Example 1.6.8 In $C[-\pi, \pi]$, the vectors $1, e^{it}, e^{2it}, \ldots, e^{nit}$ are linearly independent for each $n = 1, 2, \ldots$. This follows from P.5.9 and Theorem 5.1.10.

Example 1.6.9 Let $A = [\mathbf{a}_1 \ \mathbf{a}_2 \ \ldots \ \mathbf{a}_n] \in \mathsf{M}_n(\mathbb{F})$ be invertible and suppose that $x_1, x_2, \ldots, x_n \in \mathbb{F}$ are scalars such that $x_1\mathbf{a}_1 + x_2\mathbf{a}_2 + \cdots + x_n\mathbf{a}_n = \mathbf{0}$. Let $\mathbf{x} = [x_i] \in \mathbb{F}^n$. Then

$$A\mathbf{x} = x_1\mathbf{a}_1 + x_2\mathbf{a}_2 + \cdots + x_n\mathbf{a}_n = \mathbf{0}$$

and hence

$$\mathbf{x} = A^{-1}(A\mathbf{x}) = A^{-1}\mathbf{0} = \mathbf{0}.$$

It follows that $x_1 = x_2 = \cdots = x_n = 0$, so $\mathbf{a}_1, \mathbf{a}_2, \ldots, \mathbf{a}_n$ are linearly independent. We conclude that the columns of any invertible matrix are linearly independent.

Example 1.6.10 The identity matrix I_n is invertible. Consequently, its columns $\mathbf{e}_1, \mathbf{e}_2, \ldots, \mathbf{e}_n$ (see (1.4.16)) are linearly independent in $\mathbb{F}^n$. One can see this directly by observing that

$$[u_i] = u_1 \mathbf{e}_1 + u_2 \mathbf{e}_2 + \cdots + u_n \mathbf{e}_n = \mathbf{0}$$

if and only if $u_1 = u_2 = \cdots = u_n = 0$.

The most important property of a linearly independent list of vectors is that it provides a unique representation of each vector in its span.

Theorem 1.6.11 *Let* $\mathbf{v}_1, \mathbf{v}_2, \ldots, \mathbf{v}_r$ *be a list of linearly independent vectors in an* $\mathbb{F}$*-vector space* $\mathcal{V}$*. Then*

$$a_1 \mathbf{v}_1 + a_2 \mathbf{v}_2 + \cdots + a_r \mathbf{v}_r = b_1 \mathbf{v}_1 + b_2 \mathbf{v}_2 + \cdots + b_r \mathbf{v}_r \tag{1.6.12}$$

if and only if $a_j = b_j$ *for each* $j = 1, 2, \ldots, r$.

Proof The identity (1.6.12) is equivalent to

$$(a_1 - b_1)\mathbf{v}_1 + (a_2 - b_2)\mathbf{v}_2 + \cdots + (a_r - b_r)\mathbf{v}_r = \mathbf{0}, \tag{1.6.13}$$

which is satisfied if $a_j = b_j$ for each $j = 1, 2, \ldots, r$. Conversely, if (1.6.13) is satisfied, then the linear independence of $\mathbf{v}_1, \mathbf{v}_2, \ldots, \mathbf{v}_r$ implies that $a_j - b_j = 0$ for each $j = 1, 2, \ldots, r$. $\square$

Under certain circumstances, a linearly independent list can be extended to a longer linearly independent list. Under other circumstances, some elements of a list that spans $\mathcal{V}$ can be omitted to obtain a shorter list that still spans $\mathcal{V}$. It is convenient to have a notation for the list obtained by omitting a vector from a given list.

Definition 1.6.14 Let $r \geq 2$ and let $\beta = \mathbf{v}_1, \mathbf{v}_2, \ldots, \mathbf{v}_r$ be a list of vectors in an $\mathbb{F}$-vector space $\mathcal{V}$. If $j \in \{1, 2, \ldots, r\}$, the list of $r - 1$ vectors obtained by omitting $\mathbf{v}_j$ from β is denoted by $\mathbf{v}_1, \mathbf{v}_2, \ldots, \widehat{\mathbf{v}_j}, \ldots, \mathbf{v}_r$.

Example 1.6.15 If $\beta = \mathbf{v}_1, \mathbf{v}_2, \mathbf{v}_3, \mathbf{v}_4$ and $j = 3$, then $\mathbf{v}_1, \mathbf{v}_2, \widehat{\mathbf{v}_3}, \mathbf{v}_4$ is the list $\mathbf{v}_1, \mathbf{v}_2, \mathbf{v}_4$.

If we remove any vector from a linearly independent list of two or more vectors, the list that remains is still linearly independent.

Theorem 1.6.16 *Let* $r \geq 2$ *and suppose that a list* $\mathbf{v}_1, \mathbf{v}_2, \ldots, \mathbf{v}_r$ *of vectors in an* $\mathbb{F}$*-vector space* $\mathcal{V}$ *is linearly independent. Then the list* $\mathbf{v}_1, \mathbf{v}_2, \ldots, \widehat{\mathbf{v}_j}, \ldots, \mathbf{v}_r$ *is linearly independent for any* $j \in \{1, 2, \ldots, r\}$.

Proof The claim is equivalent to the assertion that if $\mathbf{v}_1, \mathbf{v}_2, \ldots, \widehat{\mathbf{v}_j}, \ldots, \mathbf{v}_r$ is linearly dependent, then $\mathbf{v}_1, \mathbf{v}_2, \ldots, \mathbf{v}_r$ is linearly dependent. Theorem 1.6.3 ensures that this is the case. $\square$

Theorem 1.6.17 *Let $\beta = \mathbf{v}_1, \mathbf{v}_2, \ldots, \mathbf{v}_r$ be a list of vectors in a nonzero $\mathbb{F}$-vector space $\mathcal{V}$.*

(a) *Suppose that β is linearly independent and does not span $\mathcal{V}$. If $\mathbf{v} \in \mathcal{V}$ and $\mathbf{v} \notin \operatorname{span} \beta$, then the list $\mathbf{v}_1, \mathbf{v}_2, \ldots, \mathbf{v}_r, \mathbf{v}$ is linearly independent.*

(b) *Suppose that β is linearly dependent and $\operatorname{span} \beta = \mathcal{V}$. If $c_1\mathbf{v}_1 + c_2\mathbf{v}_2 + \cdots + c_r\mathbf{v}_r = \mathbf{0}$ is a nontrivial linear combination and $j \in \{1, 2, \ldots, r\}$ is any index such that $c_j \neq 0$, then $\mathbf{v}_1, \mathbf{v}_2, \ldots, \widehat{\mathbf{v}}_j, \ldots, \mathbf{v}_r$ spans $\mathcal{V}$.*

Proof (a) Suppose that $c_1\mathbf{v}_1 + c_2\mathbf{v}_2 + \cdots + c_r\mathbf{v}_r + c\mathbf{v} = \mathbf{0}$. If $c \neq 0$, then $\mathbf{v} = -c^{-1} \sum_{i=1}^{r} c_i\mathbf{v}_i \in \operatorname{span}\{\mathbf{v}_1, \mathbf{v}_2, \ldots, \mathbf{v}_r\}$, which is a contradiction. Thus, $c = 0$, and hence $c_1\mathbf{v}_1 + c_2\mathbf{v}_2 + \cdots + c_r\mathbf{v}_r = \mathbf{0}$. The linear independence of β implies that $c_1 = c_2 = \cdots = c_r = 0$. We conclude that the list $\mathbf{v}_1, \mathbf{v}_2, \ldots, \mathbf{v}_r, \mathbf{v}$ is linearly independent.

(b) If $r = 1$, then the linear dependence of the list $\beta = \mathbf{v}_1$ implies that $\mathbf{v}_1 = \mathbf{0}$, so $\mathcal{V} = \operatorname{span} \beta = \{\mathbf{0}\}$, which is a contradiction. Thus, $r \geq 2$. Since $c_j \neq 0$, we have $\mathbf{v}_j = -c_j^{-1} \sum_{i \neq j} c_i\mathbf{v}_i$. This identity can be used to eliminate $\mathbf{v}_j$ from any linear combination in which it appears, so any vector that is a linear combination of the r vectors in the list β (namely, every vector in $\mathcal{V}$) is also a linear combination of the $r-1$ vectors in the list $\mathbf{v}_1, \mathbf{v}_2, \ldots, \widehat{\mathbf{v}}_j, \ldots, \mathbf{v}_r$. $\qquad\square$

1.7 Problems

P.1.1 In the spirit of the examples in Section 1.2, explain how $\mathcal{V} = \mathbb{C}^n$ can be thought of as a vector space over $\mathbb{R}$. Is $\mathcal{V} = \mathbb{R}^n$ a vector space over $\mathbb{C}$?

P.1.2 Let $\mathcal{V}$ be the set of real 2×2 matrices of the form $\mathbf{v} = \begin{bmatrix} 1 & v \\ 0 & 1 \end{bmatrix}$. Define $\mathbf{v} + \mathbf{w} = \begin{bmatrix} 1 & v \\ 0 & 1 \end{bmatrix}\begin{bmatrix} 1 & w \\ 0 & 1 \end{bmatrix}$ (ordinary matrix multiplication) and $c\mathbf{v} = \begin{bmatrix} 1 & cv \\ 0 & 1 \end{bmatrix}$. Show that $\mathcal{V}$ together with these two operations is a real vector space. What is the zero vector in $\mathcal{V}$?

P.1.3 Show that the intersection of any (possibly infinite) collection of subspaces of an $\mathbb{F}$-vector space is a subspace.

P.1.4 Let $\mathcal{U}$ be a subset of an $\mathbb{F}$-vector space $\mathcal{V}$. Show that $\operatorname{span} \mathcal{U}$ is the intersection of all the subspaces of $\mathcal{V}$ that contain $\mathcal{U}$. What does this say if $\mathcal{U} = \varnothing$?

P.1.5 Let $\mathcal{U}$ and $\mathcal{W}$ be subspaces of an $\mathbb{F}$-vector space $\mathcal{V}$. Prove that $\mathcal{U} \cup \mathcal{W}$ is a subspace of $\mathcal{V}$ if and only if either $\mathcal{U} \subseteq \mathcal{W}$ or $\mathcal{W} \subseteq \mathcal{U}$.

P.1.6 Give an example of a linearly dependent list of three vectors in $\mathbb{F}^3$ such that any two of them are linearly independent.

P.1.7 Let $n \geq 2$ be a positive integer and let $A \in \mathsf{M}_n(\mathbb{C})$. Which of the following subsets of $\mathsf{M}_n(\mathbb{C})$ is a subspace of the complex vector space $\mathsf{M}_n(\mathbb{C})$? Why? (a) All invertible matrices; (b) All noninvertible matrices; (c) All A such that $A^2 = 0$; (d) All matrices whose first column are zero; (e) All lower triangular matrices; (f) All $X \in \mathsf{M}_n(\mathbb{C})$ such that $AX + X^{\mathsf{T}}A = 0$.

P.1.8 Let $\mathcal{V}$ be a real vector space and suppose that the list $\beta = \mathbf{u}, \mathbf{v}, \mathbf{w}$ of vectors in $\mathcal{V}$ is linearly independent. Show that the list $\gamma = \mathbf{u}-\mathbf{v}, \mathbf{v}-\mathbf{w}, \mathbf{w}+\mathbf{u}$ is linearly independent. What about the list $\delta = \mathbf{u} + \mathbf{v}, \mathbf{v} + \mathbf{w}, \mathbf{w} + \mathbf{u}$?

P.1.9 Let $\mathcal{V}$ be an $\mathbb{F}$-vector space, let $\mathbf{w}_1, \mathbf{w}_2, \ldots, \mathbf{w}_r \in \mathcal{V}$, and suppose that at least one $\mathbf{w}_j$ is nonzero. Explain why $\text{span}\{\mathbf{w}_1, \mathbf{w}_2, \ldots, \mathbf{w}_r\} = \text{span}\{\mathbf{w}_i : i = 1, 2, \ldots, r$ and $\mathbf{w}_i \neq \mathbf{0}\}$.

P.1.10 Review Example 1.4.8. Prove that $\mathcal{U} = \{p \in \mathcal{P}_3 : p(0) = 0\}$ is a subspace of $\mathcal{P}_3$ and show that $\mathcal{U} = \text{span}\{z, z^2, z^3\}$.

P.1.11 State the converse of Theorem 1.6.3. Is it true or false? Give a proof or a counterexample.

P.1.12 In $\mathsf{M}_n(\mathbb{C})$, let $\mathcal{U}$ denote the set of strictly lower triangular matrices and let $\mathcal{W}$ denote the set of strictly upper triangular matrices. (a) Show that $\mathcal{U}$ and $\mathcal{W}$ are subspaces of $\mathsf{M}_n(\mathbb{C})$. (b) What is $\mathcal{U} + \mathcal{W}$? Is this sum a direct sum? Why?

P.1.13 Let $a, b, c \in \mathbb{R}$, with $a < c < b$. Show that $\{f \in C_\mathbb{R}[a, b] : f(c) = 0\}$ is a subspace of $C_\mathbb{R}[a, b]$.

P.1.14 Let $\mathbf{v}_1, \mathbf{v}_2, \ldots, \mathbf{v}_r \in \mathbb{C}^n$ and let $A \in \mathsf{M}_n$ be invertible. Prove that $A\mathbf{v}_1, A\mathbf{v}_2, \ldots, A\mathbf{v}_r$ are linearly independent if and only if $\mathbf{v}_1, \mathbf{v}_2, \ldots, \mathbf{v}_r$ are linearly independent.

P.1.15 Show that $\{\sin 2n\pi t : n = 1, 2, \ldots,\}$ and $\{t^k : k = 0, 1, 2, \ldots\}$ are linearly independent sets in $C_\mathbb{R}[-\pi, \pi]$.

1.8 Notes

A *field* is a mathematical construct whose axioms capture the essential features of ordinary arithmetic operations with real or complex numbers. Examples of other fields include the real rational numbers (ratios of integers), complex algebraic numbers (zeros of polynomials with integer coefficients), real rational functions, and the integers modulo a prime number. The only fields we deal with in this book are the real and complex numbers. For information about general fields, see [DF04].

1.9 Some Important Concepts

- Subspaces of a vector space.

- Column space and null space of a matrix.

- Linear combinations and span.

- Linear independence and linear dependence of a list of vectors.

Bases and Similarity

Linearly independent lists of vectors that span a vector space are of special importance. They provide a bridge between the abstract world of vector spaces and the concrete world of matrices. They permit us to define the dimension of a vector space and motivate the concept of matrix similarity. They are at the heart of the relationship between the rank and nullity of a matrix.

2.1 What is a Basis?

Definition 2.1.1 Let $\mathcal{V}$ be an $\mathbb{F}$-vector space and let n be a positive integer. A list of vectors $\beta = \mathbf{v}_1, \mathbf{v}_2, \ldots, \mathbf{v}_n$ in $\mathcal{V}$ is a *basis* for $\mathcal{V}$ if span $\beta = \mathcal{V}$ and β is linearly independent.

By definition, a basis is a finite list of vectors.

Example 2.1.2 Let $\mathcal{V}$ be the real vector space $\mathbb{R}^2$ and consider the list $\beta = [2\ 1]^\mathsf{T}, [1\ 1]^\mathsf{T}$. The span of β consists of all vectors of the form

$$x_1 \begin{bmatrix} 2 \\ 1 \end{bmatrix} + x_2 \begin{bmatrix} 1 \\ 1 \end{bmatrix} = \begin{bmatrix} 2 & 1 \\ 1 & 1 \end{bmatrix} \begin{bmatrix} x_1 \\ x_2 \end{bmatrix} = A\mathbf{x},$$

in which

$$A = \begin{bmatrix} 2 & 1 \\ 1 & 1 \end{bmatrix} \quad \text{and} \quad \mathbf{x} = \begin{bmatrix} x_1 \\ x_2 \end{bmatrix} \in \mathbb{R}^2.$$

The columns of A are the vectors in the ordered list β. A calculation reveals that

$$A^{-1} = \begin{bmatrix} 1 & -1 \\ -1 & 2 \end{bmatrix}.$$

Thus, any $\mathbf{y} = [y_1\ y_2]^\mathsf{T} \in \mathbb{R}^2$ can be written uniquely as $\mathbf{y} = A\mathbf{x}$, in which

$$\mathbf{x} = A^{-1}\mathbf{y} = \begin{bmatrix} y_1 - y_2 \\ -y_1 + 2y_2 \end{bmatrix}.$$

This shows that span $\beta = \mathbb{R}^2$ and that β is linearly independent. Thus, β is a basis for $\mathbb{R}^2$.

The preceding example is a special case of the following theorem. For its converse, see Corollary 2.4.11.

Theorem 2.1.3 *Let* $A = [\mathbf{a}_1 \ \mathbf{a}_2 \ \ldots \ \mathbf{a}_n] \in \mathsf{M}_n(\mathbb{F})$ *be invertible and let* $\beta = \mathbf{a}_1, \mathbf{a}_2, \ldots, \mathbf{a}_n$. *Then* β *is a basis for* $\mathbb{F}^n$.

Proof The columns of any invertible matrix in $\mathsf{M}_n(\mathbb{F})$ span $\mathbb{F}^n$ and are linearly independent; see Examples 1.4.14 and 1.6.9. $\square$

Let β be a basis of an $\mathbb{F}$-vector space $\mathcal{V}$. Each vector in $\mathcal{V}$ is a linear combination of the elements of β since span $\beta = \mathcal{V}$. That linear combination is unique since β is linearly independent; see Theorem 1.6.11. However, a vector space can have many different bases. Our next task is to investigate how they are related.

Lemma 2.1.4 (Replacement Lemma) *Let* $\mathcal{V}$ *be a nonzero* $\mathbb{F}$-vector space and let r be a positive integer. Suppose that $\beta = \mathbf{u}_1, \mathbf{u}_2, \ldots, \mathbf{u}_r$ spans $\mathcal{V}$. Let $\mathbf{v} \in \mathcal{V}$ be nonzero and let

$$\mathbf{v} = \sum_{i=1}^{r} c_i \mathbf{u}_i. \tag{2.1.5}$$

(a) $c_j \neq 0$ for some $j \in \{1, 2, \ldots, r\}$.

(b) If $c_j \neq 0$, then

$$\mathbf{v}, \mathbf{u}_1, \mathbf{u}_2 \ldots, \widehat{\mathbf{u}}_j, \ldots, \mathbf{u}_r \tag{2.1.6}$$

     spans $\mathcal{V}$.

(c) If β is a basis for $\mathcal{V}$ and $c_j \neq 0$, then the list (2.1.6) is a basis for $\mathcal{V}$.

(d) If $r \geq 2$, β is a basis for $\mathcal{V}$, and $\mathbf{v} \notin \mathrm{span}\{\mathbf{u}_1, \mathbf{u}_2, \ldots, \mathbf{u}_k\}$ for some $k \in \{1, 2, \ldots, r-1\}$, then there is an index $j \in \{k+1, k+2, \ldots, r\}$ such that

$$\mathbf{v}, \mathbf{u}_1, \ldots, \mathbf{u}_k, \mathbf{u}_{k+1}, \ldots, \widehat{\mathbf{u}}_j, \ldots, \mathbf{u}_r \tag{2.1.7}$$

     is a basis for $\mathcal{V}$.

Proof (a) If all $c_i = 0$, then $\mathbf{v} = \sum_{i=1}^{r} c_i \mathbf{u}_i = \mathbf{0}$, which is a contradiction.

(b) The list $\mathbf{v}, \mathbf{u}_1, \mathbf{u}_2, \ldots, \mathbf{u}_r$ is linearly dependent because $\mathbf{v}$ is a linear combination of $\mathbf{u}_1, \mathbf{u}_2, \ldots, \mathbf{u}_r$. Therefore, the assertion follows from Theorem 1.6.17.b.

(c) We must show that the list (2.1.6) is linearly independent. Suppose that $c\mathbf{v} + \sum_{i \neq j} b_i \mathbf{u}_i = \mathbf{0}$. If $c \neq 0$, then

$$\mathbf{v} = -c^{-1} \sum_{i \neq j} b_i \mathbf{u}_i, \tag{2.1.8}$$

which is different from the representation (2.1.5), in which $c_j \neq 0$. Having two different representations for $\mathbf{v}$ as a linear combination of elements of β would contradict Theorem 1.6.11, so we must have $c = 0$. Consequently, $\sum_{i \neq j} b_i \mathbf{u}_i = \mathbf{0}$ and the linear independence of β ensures that each $b_i = 0$. Thus, the list (2.1.6) is linearly independent.

(d) Because $\mathbf{v} \notin \mathrm{span}\{\mathbf{u}_1, \mathbf{u}_2, \ldots, \mathbf{u}_k\}$, in the representation (2.1.5) we must have $c_j \neq 0$ for some index $j \in \{k+1, k+2, \ldots, r\}$. The assertion now follows from (c). $\square$

The next theorem shows that the number of elements in a basis for $\mathcal{V}$ is an upper bound for the number of elements in any linearly independent list of vectors in $\mathcal{V}$.

Theorem 2.1.9 *Let V be an $\mathbb{F}$-vector space and let r and n be positive integers. If $\beta = \mathbf{u}_1, \mathbf{u}_2, \ldots, \mathbf{u}_n$ is a basis for V and $\gamma = \mathbf{v}_1, \mathbf{v}_2, \ldots, \mathbf{v}_r$ is linearly independent, then $r \leq n$. If $r = n$, then γ is a basis for V.*

Proof There is nothing to prove if $r < n$, so assume that $r \geq n$. For each $k = 1, 2, \ldots, n$, we claim that there are indices $i_1, i_2, \ldots, i_{n-k} \in \{1, 2, \ldots, n\}$ such that the list of vectors $\gamma_k = \mathbf{v}_k, \mathbf{v}_{k-1}, \ldots, \mathbf{v}_1, \mathbf{u}_{i_1}, \mathbf{u}_{i_2}, \ldots, \mathbf{u}_{i_{n-k}}$ is a basis for V. We proceed by induction. The base case $k = 1$ follows from (a) and (c) of the preceding lemma. The induction step follows from (d) of the preceding lemma because the linear independence of γ ensures that $\mathbf{v}_{k+1} \notin \mathrm{span}\{\mathbf{v}_1, \mathbf{v}_2, \ldots, \mathbf{v}_k\}$. The case $k = n$ tells us that the list $\gamma_n = \mathbf{v}_n, \mathbf{v}_{n-1}, \ldots, \mathbf{v}_1$ is a basis for V. If $r > n$, then $\mathbf{v}_{n+1}$ would be in the span of $\mathbf{v}_n, \mathbf{v}_{n-1}, \ldots, \mathbf{v}_1$, which would contradict the linear independence of γ. $\qquad\square$

Corollary 2.1.10 *Let r and n be positive integers. If $\mathbf{v}_1, \mathbf{v}_2, \ldots, \mathbf{v}_n$ and $\mathbf{w}_1, \mathbf{w}_2, \ldots, \mathbf{w}_r$ are bases of an $\mathbb{F}$-vector space V, then $r = n$.*

Proof The preceding theorem ensures that $r \leq n$ and $n \leq r$. $\qquad\square$

2.2 Dimension

The preceding corollary is of fundamental importance, and it permits us to define the dimension of a vector space.

Definition 2.2.1 Let V be an $\mathbb{F}$-vector space and let n be a positive integer. If there is a list $\mathbf{v}_1, \mathbf{v}_2, \ldots, \mathbf{v}_n$ of vectors that is a basis for V, then V is *n-dimensional* (or, V has *dimension n*). The zero vector space has *dimension zero*. If V has dimension n for some nonnegative integer n, then V is *finite dimensional*; otherwise, V is *infinite dimensional*. If V is finite dimensional, its dimension is denoted by $\dim V$.

Example 2.2.2 The vectors $\mathbf{e}_1, \mathbf{e}_2, \ldots, \mathbf{e}_n$ in (1.4.16) are linearly independent and their span is $\mathbb{F}^n$. They comprise the *standard basis* for $\mathbb{F}^n$. There are n vectors in this basis, so $\dim \mathbb{F}^n = n$.

Example 2.2.3 In the $\mathbb{F}$-vector space $\mathsf{M}_{m \times n}(\mathbb{F})$, consider the matrices E_{pq}, for $1 \leq p \leq m$ and $1 \leq q \leq n$, defined as follows: the (i, j) entry of E_{pq} is 1 if $(i, j) = (p, q)$; it is 0 otherwise. The mn matrices E_{pq} (arranged in a list in any desired order) comprise a basis, so $\dim \mathsf{M}_{m \times n}(\mathbb{F}) = mn$. For example, if $m = 2$ and $n = 3$,

$$E_{11} = \begin{bmatrix} 1 & 0 & 0 \\ 0 & 0 & 0 \end{bmatrix}, \qquad E_{12} = \begin{bmatrix} 0 & 1 & 0 \\ 0 & 0 & 0 \end{bmatrix}, \qquad E_{13} = \begin{bmatrix} 0 & 0 & 1 \\ 0 & 0 & 0 \end{bmatrix},$$

$$E_{21} = \begin{bmatrix} 0 & 0 & 0 \\ 1 & 0 & 0 \end{bmatrix}, \qquad E_{22} = \begin{bmatrix} 0 & 0 & 0 \\ 0 & 1 & 0 \end{bmatrix}, \qquad E_{23} = \begin{bmatrix} 0 & 0 & 0 \\ 0 & 0 & 1 \end{bmatrix},$$

and $\dim \mathsf{M}_{2 \times 3}(\mathbb{F}) = 6$.

Example 2.2.4 In Example 1.6.7 we saw that the vectors $1, z, z^2, \ldots, z^n$ are linearly independent in the vector space $\mathcal{P}$ for each $n = 1, 2, \ldots$. Theorem 2.1.9 says that if $\mathcal{P}$ is finite dimensional, then $\dim \mathcal{P} \geq n$ for each $n = 1, 2, \ldots$. Since this is impossible, $\mathcal{P}$ is infinite dimensional.

Example 2.2.5 In the vector space $\mathcal{V}$ of finitely nonzero sequences (see Example 1.2.7), consider the vectors $\mathbf{v}_k$ that have a 1 in position k and zero entries elsewhere. For each $n = 1, 2, \ldots$, the vectors $\mathbf{v}_1, \mathbf{v}_2, \ldots, \mathbf{v}_n$ are linearly independent, so $\mathcal{V}$ is infinite dimensional.

Definition 2.2.6 Let $A = [\mathbf{a}_1 \ \mathbf{a}_2 \ \ldots \ \mathbf{a}_n] \in \mathsf{M}_{m \times n}(\mathbb{F})$ and let $\beta = \mathbf{a}_1, \mathbf{a}_2, \ldots, \mathbf{a}_n$, a list of vectors in the $\mathbb{F}$-vector space $\mathbb{F}^m$. Then $\dim \operatorname{span} \beta = \dim \operatorname{col} A$ is the *rank* of A.

The following theorem says two things about a nonzero $\mathbb{F}$-vector space $\mathcal{V}$: (a) any finite set that spans $\mathcal{V}$ contains a subset whose elements comprise a basis; and (b) if $\mathcal{V}$ is finite dimensional, then any linearly independent list of vectors can be extended to a basis.

Theorem 2.2.7 *Let $\mathcal{V}$ be a nonzero $\mathbb{F}$-vector space, let r be a positive integer, and let $\mathbf{v}_1, \mathbf{v}_2, \ldots, \mathbf{v}_r \in \mathcal{V}$.*

(a) *If $\operatorname{span}\{\mathbf{v}_1, \mathbf{v}_2, \ldots, \mathbf{v}_r\} = \mathcal{V}$, then $\mathcal{V}$ is finite dimensional, $n = \dim \mathcal{V} \leq r$, and there are indices $i_1, i_2, \ldots, i_n \in \{1, 2, \ldots, r\}$ such that the list $\mathbf{v}_{i_1}, \mathbf{v}_{i_2}, \ldots, \mathbf{v}_{i_n}$ is a basis for $\mathcal{V}$.*

(b) *Suppose that $\mathcal{V}$ is finite dimensional and $\dim \mathcal{V} = n > r$. If $\mathbf{v}_1, \mathbf{v}_2, \ldots, \mathbf{v}_r$ are linearly independent, then there are $n - r$ vectors $\mathbf{w}_1, \mathbf{w}_2, \ldots, \mathbf{w}_{n-r} \in \mathcal{V}$ such that $\mathbf{v}_1, \mathbf{v}_2, \ldots, \mathbf{v}_r, \mathbf{w}_1, \mathbf{w}_2, \ldots, \mathbf{w}_{n-r}$ is a basis for $\mathcal{V}$.*

Proof (a) The hypothesis is that $\operatorname{span}\{\mathbf{v}_1, \mathbf{v}_2, \ldots, \mathbf{v}_r\} = \mathcal{V}$. We may assume that each $\mathbf{v}_i \neq \mathbf{0}$ since zero vectors contribute nothing to the span; see P.1.9. If $\mathbf{v}_1, \mathbf{v}_2, \ldots, \mathbf{v}_r$ are linearly independent, then they comprise a basis. If they are linearly dependent, consider the following algorithm. Theorem 1.6.17.b ensures that some vector can be omitted from the list and the remaining $r - 1$ vectors still span $\mathcal{V}$. If this shorter list is linearly independent, then stop. If not, invoke Theorem 1.6.17.b again and obtain a shorter list of vectors that still spans $\mathcal{V}$; repeat until a linearly independent list is obtained. At most $r - 1$ repetitions are required since each $\mathbf{v}_i$ is nonzero and hence constitutes a one-element linearly independent list.

(b) The hypothesis is that $\mathbf{v}_1, \mathbf{v}_2, \ldots, \mathbf{v}_r \in \mathcal{V}$ are linearly independent and $r < n = \dim \mathcal{V}$. Since $\mathbf{v}_1, \mathbf{v}_2, \ldots, \mathbf{v}_r$ do not span $\mathcal{V}$ (Corollary 2.1.10), Theorem 1.6.17.a ensures that some vector can be appended to the list and the augmented list is still linearly independent. If this longer list spans $\mathcal{V}$, we have a basis. If not, invoke Theorem 1.6.17.a again. Theorem 2.1.9 ensures that this process terminates in $n - r$ steps. $\qquad\square$

One sometimes says that a basis is simultaneously a *maximal linearly independent list* and a *minimal spanning list*.

Corollary 2.2.8 *Let n be a positive integer and let $\mathcal{V}$ be an n-dimensional $\mathbb{F}$-vector space. Let $\beta = \mathbf{v}_1, \mathbf{v}_2, \ldots, \mathbf{v}_n \in \mathcal{V}$.*

(a) *If β spans V, then it is a basis.*

(b) *If β is linearly independent, then it is a basis.*

Proof (a) If β is not linearly independent, then the preceding theorem ensures that a strictly shorter list comprises a basis. This contradicts the assumption that $\dim V = n$.

(b) If β does not span V, then the preceding theorem ensures that a strictly longer list comprises a basis, which contradicts the assumption that $\dim V = n$.   □

Theorem 2.2.9 *Let U be a subspace of an n-dimensional $\mathbb{F}$-vector space V. Then U is finite dimensional and $\dim U \leq n$, with equality if and only if $U = V$.*

Proof If $U = \{0\}$, then $\dim U = 0$ and there is nothing to prove, so we may assume that $U \neq \{0\}$. Let $\mathbf{v}_1 \in U$ be nonzero. If $\operatorname{span}\{\mathbf{v}_1\} = U$, then $\dim U = 1$. If $\operatorname{span}\{\mathbf{v}_1\} \neq U$, Theorem 1.6.17.a ensures that there is a $\mathbf{v}_2 \in U$ such that the list $\mathbf{v}_1, \mathbf{v}_2$ is linearly independent. If $\operatorname{span}\{\mathbf{v}_1, \mathbf{v}_2\} = U$, then $\dim U = 2$; if $\operatorname{span}\{\mathbf{v}_1, \mathbf{v}_2\} \neq U$, Theorem 1.6.17.a ensures that there is a $\mathbf{v}_3 \in U$ such that the list $\mathbf{v}_1, \mathbf{v}_2, \mathbf{v}_3$ is linearly independent. Repeat until a linearly independent spanning list is obtained. Since no linearly independent list of vectors in V contains more than n elements (Theorem 2.1.9), this process terminates in $r \leq n$ steps with a linearly independent list of vectors $\mathbf{v}_1, \mathbf{v}_2, \ldots, \mathbf{v}_r$ whose span is U. Thus, $r = \dim U \leq n$ with equality only if $\mathbf{v}_1, \mathbf{v}_2, \ldots, \mathbf{v}_n$ is a basis for V (Theorem 2.1.9 again), in which case $U = V$.   □

The preceding theorem ensures that any pair of subspaces U and W of a finite-dimensional vector space V has a finite-dimensional sum $U + W$ and a finite-dimensional intersection $U \cap W$, since each is a subspace of V. Moreover, $U \cap W$ is a subspace of U and of W, so any basis for $U \cap W$ can be extended to a basis for U; it can also be extended to a basis for W. Careful consideration of how these bases interact leads to the identity in the following theorem; see Figure 2.1.

Theorem 2.2.10 *Let U and W be subspaces of a finite-dimensional $\mathbb{F}$-vector space V. Then*

$$\dim(U \cap W) + \dim(U + W) = \dim U + \dim W. \tag{2.2.11}$$

Proof Let $k = \dim(U \cap W)$. Since $U \cap W$ is a subspace of both U and W, the preceding theorem ensures that $k \leq \dim U$ and $k \leq \dim W$. Let $p = \dim U - k$ and $q = \dim W - k$. Let $\mathbf{v}_1, \mathbf{v}_2, \ldots, \mathbf{v}_k$ be a basis for $U \cap W$. Theorem 2.2.7.b ensures that there are $\mathbf{u}_1, \mathbf{u}_2, \ldots, \mathbf{u}_p$ and $\mathbf{w}_1, \mathbf{w}_2, \ldots, \mathbf{w}_q$ such that

$$\mathbf{v}_1, \mathbf{v}_2, \ldots, \mathbf{v}_k, \mathbf{u}_1, \mathbf{u}_2, \ldots, \mathbf{u}_p \tag{2.2.12}$$

is a basis for U and

$$\mathbf{v}_1, \mathbf{v}_2, \ldots, \mathbf{v}_k, \mathbf{w}_1, \mathbf{w}_2, \ldots, \mathbf{w}_q \tag{2.2.13}$$

is a basis for W. We must show that

$$\dim(U + W) = (p + k) + (q + k) - k = k + p + q.$$

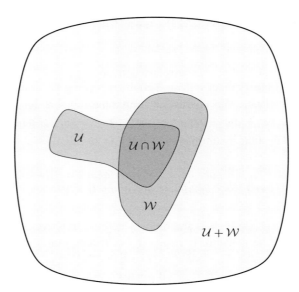

Figure 2.1 Illustration of intersection and sum of subspaces.

Since every vector in $\mathcal{U} + \mathcal{W}$ is the sum of a vector in $\mathcal{U}$ and a vector in $\mathcal{W}$, the span of

$$\mathbf{v}_1, \mathbf{v}_2, \ldots, \mathbf{v}_k, \mathbf{u}_1, \mathbf{u}_2, \ldots, \mathbf{u}_p, \mathbf{w}_1, \mathbf{w}_2, \ldots, \mathbf{w}_q \qquad (2.2.14)$$

is $\mathcal{U} + \mathcal{W}$. It suffices to show that the list (2.2.14) is linearly independent. Suppose that

$$\sum_{i=1}^{k} a_i \mathbf{v}_i + \sum_{i=1}^{p} b_i \mathbf{u}_i + \sum_{i=1}^{q} c_i \mathbf{w}_i = \mathbf{0}. \qquad (2.2.15)$$

Then

$$\sum_{i=1}^{k} a_i \mathbf{v}_i + \sum_{i=1}^{q} c_i \mathbf{w}_i = \sum_{i=1}^{p} (-b_i) \mathbf{u}_i. \qquad (2.2.16)$$

The right-hand side of (2.2.16) is in $\mathcal{U}$ and its left-hand side is in $\mathcal{W}$, so both sides are in $\mathcal{U} \cap \mathcal{W}$. Thus, there are scalars $d_1, d_2, \ldots, d_k$ such that

$$\sum_{i=1}^{k} a_i \mathbf{v}_i + \sum_{i=1}^{q} c_i \mathbf{w}_i = \sum_{i=1}^{k} d_i \mathbf{v}_i.$$

Consequently,

$$\sum_{i=1}^{k} (a_i - d_i) \mathbf{v}_i + \sum_{i=1}^{q} c_i \mathbf{w}_i = \mathbf{0}.$$

The linear independence of the vectors (2.2.13) ensures that

$$c_1 = c_2 = \cdots = c_q = 0$$

and it follows from (2.2.15) that

$$\sum_{i=1}^{k} a_i \mathbf{v}_i + \sum_{i=1}^{p} b_i \mathbf{u}_i = \mathbf{0}.$$

The linear independence of the list (2.2.12) implies that

$$b_1 = b_2 = \cdots = b_p = a_1 = a_2 = \cdots = a_k = 0.$$

We conclude that the list (2.2.14) is linearly independent. $\square$

The identity (2.2.11) leads to a criterion for the intersection of two subspaces to contain a nonzero vector. The key to understanding this criterion is the inequality in Theorem 2.2.9. For subspaces $\mathcal{U}$ and $\mathcal{W}$ of a finite-dimensional vector space $\mathcal{V}$, their sum is a subspace and therefore $\dim(\mathcal{U} + \mathcal{W}) \le \dim \mathcal{V}$.

Corollary 2.2.17 *Let $\mathcal{U}$ and $\mathcal{W}$ be subspaces of a finite-dimensional $\mathbb{F}$-vector space $\mathcal{V}$ and let k be a positive integer.*

(a) *If $\dim \mathcal{U} + \dim \mathcal{W} > \dim \mathcal{V}$, then $\mathcal{U} \cap \mathcal{W}$ contains a nonzero vector.*

(b) *If $\dim \mathcal{U} + \dim \mathcal{W} \ge \dim \mathcal{V} + k$, then $\mathcal{U} \cap \mathcal{W}$ contains k linearly independent vectors.*

Proof The assertion (a) is the case $k = 1$ of the assertion (b). Under the hypothesis in (b),

$$\dim(\mathcal{U} \cap \mathcal{W}) = \dim \mathcal{U} + \dim \mathcal{W} - \dim(\mathcal{U} + \mathcal{W})$$
$$\ge \dim \mathcal{U} + \dim \mathcal{W} - \dim \mathcal{V} \ge k,$$

so $\mathcal{U} \cap \mathcal{W}$ has a basis comprising at least k vectors. $\square$

Another application of Theorem 2.2.9 is to a result about left and right matrix inverses that is typically approached via determinants. A matrix $B \in \mathsf{M}_n(\mathbb{F})$ is a *left inverse* (respectively, *right inverse*) of $A \in \mathsf{M}_n(\mathbb{F})$ if $BA = I$ (respectively, $AB = I$). A square matrix need not have a left inverse, but if it does, then that left inverse is also a right inverse. We now show how this remarkable fact (Theorem 2.2.19) follows from the finite dimensionality of $\mathsf{M}_n(\mathbb{F})$.

Lemma 2.2.18 *Let $A, B, C \in \mathsf{M}_n(\mathbb{F})$ and suppose that $AB = I = BC$. Then $A = C$.*

Proof If $AB = BC = I$, then $A = AI = A(BC) = (AB)C = IC = C$. $\square$

For any $A \in \mathsf{M}_n(\mathbb{F})$, define

$$A\mathsf{M}_n(\mathbb{F}) = \{AX : X \in \mathsf{M}_n(\mathbb{F})\}.$$

Example 1.3.13 shows that $A\mathsf{M}_n(\mathbb{F})$ is a subspace of $\mathsf{M}_n(\mathbb{F})$.

Theorem 2.2.19 *Let $A, B \in \mathsf{M}_n(\mathbb{F})$. Then $AB = I$ if and only if $BA = I$.*

Proof It suffices to consider the case $AB = I$. Since $B^{k+1}X = B^k(BX)$ for all $X \in \mathsf{M}_n(\mathbb{F})$, we have $B^{k+1}\mathsf{M}_n(\mathbb{F}) \subseteq B^k\mathsf{M}_n(\mathbb{F})$ for $k \ge 1$. Consider the descending sequence

$$\mathsf{M}_n(\mathbb{F}) \supseteq B\mathsf{M}_n(\mathbb{F}) \supseteq B^2\mathsf{M}_n(\mathbb{F}) \supseteq B^3\mathsf{M}_n(\mathbb{F}) \supseteq \cdots$$

of subspaces of $\mathsf{M}_n(\mathbb{F})$. Theorem 2.2.9 ensures that

$$n^2 = \dim \mathsf{M}_n(\mathbb{F}) \ge \dim B\mathsf{M}_n(\mathbb{F}) \ge \dim B^2\mathsf{M}_n(\mathbb{F}) \ge \cdots \ge 0. \qquad (2.2.20)$$

Since only finitely many (in fact, at most n^2) of the inequalities in (2.2.20) can be strict inequalities, there is a positive integer k such that $\dim B^k \mathsf{M}_n(\mathbb{F}) = \dim B^{k+1} \mathsf{M}_n(\mathbb{F})$, in which case, Theorem 2.2.9 ensures that $B^k \mathsf{M}_n(\mathbb{F}) = B^{k+1} \mathsf{M}_n(\mathbb{F})$. Since $B^k = B^k I \in B^k \mathsf{M}_n(\mathbb{F}) = B^{k+1} \mathsf{M}_n(\mathbb{F})$, there is a $C \in \mathsf{M}_n(\mathbb{F})$ such that $B^k = B^{k+1} C$. Then

$$A^k B^k = A^k B^{k+1} C = (A^k B^k) BC. \tag{2.2.21}$$

We use induction to prove that $A^r B^r = I$ for $r = 1, 2, \ldots$. The base case $r = 1$ is our hypothesis. If $r \geq 1$ and $A^r B^r = I$, then $A^{r+1} B^{r+1} = A(A^r B^r)B = AIB = AB = I$. Therefore, (2.2.21) says that $I = A^k B^k = IBC = BC$. Lemma 2.2.18 ensures that $C = A$, so $BA = I$. $\square$

Finite dimensionality is an essential assumption in the preceding theorem; see P.2.7.

2.3 Basis Representations and Linear Transformations

Definition 2.3.1 Let $\beta = \mathbf{v}_1, \mathbf{v}_2, \ldots, \mathbf{v}_n$ be a basis for a finite-dimensional $\mathbb{F}$-vector space $\mathcal{V}$. Write any vector $\mathbf{u} \in \mathcal{V}$ as a (unique) linear combination

$$\mathbf{u} = c_1 \mathbf{v}_1 + c_2 \mathbf{v}_2 + \cdots + c_n \mathbf{v}_n. \tag{2.3.2}$$

The function $[\cdot]_\beta : \mathcal{V} \to \mathbb{F}^n$ defined by

$$[\mathbf{u}]_\beta = \begin{bmatrix} c_1 \\ c_2 \\ \vdots \\ c_n \end{bmatrix} \tag{2.3.3}$$

is the *β-basis representation function*. For a given $\mathbf{u} \in \mathcal{V}$, the scalars $c_1, c_2, \ldots, c_n$ in (2.3.2) are the *coordinates of* $\mathbf{u}$ with respect to the basis β; $[\mathbf{u}]_\beta$ is the *β-coordinate vector of* $\mathbf{u}$.

Example 2.3.4 Let $\mathcal{V} = \mathbb{R}^2$ and consider the basis $\beta = [2\ 1]^\mathsf{T}, [1\ 1]^\mathsf{T}$. In Example 2.1.2 we found that if $\mathbf{y} = [y_1\ y_2]^\mathsf{T}$, then

$$[\mathbf{y}]_\beta = \begin{bmatrix} y_1 - y_2 \\ -y_1 + 2y_2 \end{bmatrix}.$$

Example 2.3.5 Let $\mathcal{V} = \mathcal{P}_2$ and consider the basis $\beta = f_1, f_2, f_3$, in which

$$f_1 = 1, \qquad f_2 = 2z - 1, \quad \text{and} \quad f_3 = 6z^2 - 6z + 1;$$

this basis plays a role in Example 5.1.5. A calculation reveals that

$$1 = f_1, \qquad z = \tfrac{1}{2}f_1 + \tfrac{1}{2}f_2, \quad \text{and} \quad z^2 = \tfrac{1}{3}f_1 + \tfrac{1}{2}f_2 + \tfrac{1}{6}f_3,$$

so the representation of a polynomial $p(z) = c_0 1 + c_1 z + c_2 z^2$ with respect to β is

$$[p]_\beta = \frac{1}{6} \begin{bmatrix} 6c_0 + 3c_1 + 2c_2 \\ 3c_1 + 3c_2 \\ c_2 \end{bmatrix}.$$

The β-basis representation function (2.3.3) provides a one-to-one correspondence between vectors in the $\mathbb{F}$-vector space $\mathcal{V}$ and vectors in $\mathbb{F}^n$. Theorem 1.6.11 ensures that it is one to one. It is onto because, for any given column vector on the right-hand side of (2.3.3), the vector $\mathbf{u}$ defined by (2.3.2) satisfies the identity (2.3.3). In addition, the β-basis representation function has the following important property. If $\mathbf{u}, \mathbf{w} \in \mathcal{V}$,

$$\mathbf{u} = a_1\mathbf{v}_1 + a_2\mathbf{v}_2 + \cdots + a_n\mathbf{v}_n,$$
$$\mathbf{w} = b_1\mathbf{v}_1 + b_2\mathbf{v}_2 + \cdots + b_n\mathbf{v}_n,$$

and $c \in \mathbb{F}$, then

$$c\mathbf{u} + \mathbf{w} = (ca_1 + b_1)\mathbf{v}_1 + (ca_2 + b_2)\mathbf{v}_2 + \cdots + (ca_n + b_n)\mathbf{v}_n.$$

Consequently,

$$[c\mathbf{u} + \mathbf{w}]_\beta = \begin{bmatrix} ca_1 + b_1 \\ ca_2 + b_1 \\ \vdots \\ ca_n + b_n \end{bmatrix} = c\begin{bmatrix} a_1 \\ a_2 \\ \vdots \\ a_n \end{bmatrix} + \begin{bmatrix} b_1 \\ b_2 \\ \vdots \\ b_n \end{bmatrix} = c[\mathbf{u}]_\beta + [\mathbf{w}]_\beta.$$

This identity is deceptively obvious, but it says something subtle and important. The addition and scalar multiplication operations on its left-hand side are operations in the $\mathbb{F}$-vector space $\mathcal{V}$; the addition and scalar multiplication operations on its right-hand side are operations in the $\mathbb{F}$-vector space $\mathbb{F}^n$. The β-basis representation function links these two pairs of operations and there is a one-to-one correspondence between linear algebraic operations in $\mathcal{V}$ and in $\mathbb{F}^n$. Informally, we conclude that any n-dimensional $\mathbb{F}$-vector space $\mathcal{V}$ is fundamentally the same as $\mathbb{F}^n$; formally, we say that any two n-dimensional $\mathbb{F}$-vector spaces are *isomorphic*.

Definition 2.3.6 Let $\mathcal{V}$ and $\mathcal{W}$ be vector spaces over the same field $\mathbb{F}$. A function $T : \mathcal{V} \to \mathcal{W}$ is a *linear transformation* if

$$T(c\mathbf{u} + \mathbf{v}) = cT\mathbf{u} + T\mathbf{v}$$

for all $\mathbf{u}, \mathbf{v} \in \mathcal{V}$ and all $c \in \mathbb{F}$. If $\mathcal{V} = \mathcal{W}$, a linear transformation $T : \mathcal{V} \to \mathcal{V}$ is sometimes called a *linear operator* (or just an *operator*). The set of linear transformations from $\mathcal{V}$ to $\mathcal{W}$ is denoted by $\mathcal{L}(\mathcal{V}, \mathcal{W})$. If $\mathcal{V} = \mathcal{W}$, this is abbreviated to $\mathcal{L}(\mathcal{V}, \mathcal{V}) = \mathcal{L}(\mathcal{V})$.

For notational convenience (and by analogy with the conventional notation for matrix-vector products), $T(\mathbf{v})$ is usually written $T\mathbf{v}$.

Example 2.3.7 For a given basis β of an n-dimensional $\mathbb{F}$-vector space $\mathcal{V}$, the function $T\mathbf{v} = [\mathbf{v}]_\beta$ is a linear transformation from $\mathcal{V}$ to the $\mathbb{F}$-vector space $\mathbb{F}^n$.

Example 2.3.8 For a given matrix $A \in \mathsf{M}_{m \times n}(\mathbb{F})$, properties of matrix arithmetic ensure that the function $T_A : \mathbb{F}^n \to \mathbb{F}^m$ defined by $T_A\mathbf{x} = A\mathbf{x}$ is a linear transformation.

Definition 2.3.9 The linear transformation T_A defined in the preceding example is the *linear transformation induced by A*.

Example 2.3.10 On the complex vector space $\mathcal{P}$, the function $T : \mathcal{P} \to \mathcal{P}$ defined by $Tp = p'$ (differentiation) is a linear operator. This is because the derivative of a polynomial is a polynomial and

$$(cp + q)' = cp' + q'$$

for any $c \in \mathbb{C}$ and any $p, q \in \mathcal{P}$.

Example 2.3.11 The function $T : C_{\mathbb{R}}[0, 1] \to C_{\mathbb{R}}[0, 1]$ defined by

$$(Tf)(t) = \int_0^t f(s)\, ds$$

is a linear operator. This is because the indicated integral of a continuous function is continuous (even better, it is differentiable) and

$$\int_0^t (cf(s) + g(s))\, ds = c \int_0^t f(s)\, ds + \int_0^t g(s)\, ds$$

for any $c \in \mathbb{R}$ and any $f, g \in C_{\mathbb{R}}[0, 1]$.

Example 2.3.12 On the complex vector space $\mathcal{V}$ of finitely nonzero sequences (see Examples 1.2.7 and 2.2.5), define the *right shift* $T(x_1, x_2, \ldots) = (0, x_1, x_2, \ldots)$ and the *left shift* $S(x_1, x_2, x_3, \ldots) = (x_2, x_3, \ldots)$. A computation reveals that both T and S are linear operators. See P.2.7 for other properties of these operators.

Let $\mathcal{V}$ and $\mathcal{W}$ be vector spaces over the same field $\mathbb{F}$ and let $T \in \mathcal{L}(\mathcal{V}, \mathcal{W})$. The *kernel* and *range* of T are

$$\ker T = \{\mathbf{v} \in \mathcal{V} : T\mathbf{v} = \mathbf{0}\} \quad \text{and} \quad \operatorname{ran} T = \{T\mathbf{v} : \mathbf{v} \in \mathcal{V}\}.$$

The same arguments used in Example 1.3.4 and Example 1.3.7 to show that the null space and column space of a matrix are subspaces also show that $\ker T$ is a subspace of $\mathcal{V}$ and $\operatorname{ran} T$ is a subspace of $\mathcal{W}$. A convenient way to show that a subset of a vector space is a subspace is to identify it as the kernel or range of a linear transformation; see P.2.2.

Theorem 2.3.13 *Let $\mathcal{V}$ and $\mathcal{W}$ be vector spaces over $\mathbb{F}$. Then $T \in \mathcal{L}(\mathcal{V}, \mathcal{W})$ is one to one if and only if $\ker T = \{\mathbf{0}\}$.*

Proof Suppose that T is one to one. Since T is linear, $T\mathbf{0} = \mathbf{0}$. Thus, if $T\mathbf{x} = \mathbf{0}$, then $\mathbf{x} = \mathbf{0}$ and hence $\ker T = \{\mathbf{0}\}$. Now suppose that $\ker T = \{\mathbf{0}\}$. If $T\mathbf{x} = T\mathbf{y}$, then $\mathbf{0} = T\mathbf{x} - T\mathbf{y} = T(\mathbf{x} - \mathbf{y})$, which says that $\mathbf{x} - \mathbf{y} \in \ker T$. Consequently, $\mathbf{x} - \mathbf{y} = \mathbf{0}$ and $\mathbf{x} = \mathbf{y}$. $\square$

The most important fact about a linear transformation on a finite-dimensional vector space is that if its action on a basis is known, then its action on every vector is determined. The following example illustrates the principle, which is formalized in a theorem.

Example 2.3.14 Consider the basis $\beta = 1, z, z^2$ of $\mathcal{P}_2$ and the linear transformation $T : \mathcal{P}_2 \to \mathcal{P}_1$ defined by $Tp = p'$ (differentiation). Then $T1 = 0, Tz = 1$, and $Tz^2 = 2z$. Consequently, for any $p(z) = c_2 z^2 + c_1 z + c_0$,

$$Tp = c_2 Tz^2 + c_1 Tz + c_0 T1 = c_2(2z) + c_1.$$

Theorem 2.3.15 *Let V and W be vector spaces over the same field $\mathbb{F}$ and suppose that V is finite dimensional and nonzero. Let $\beta = \mathbf{v}_1, \mathbf{v}_2, \ldots, \mathbf{v}_n$ be a basis for V and let $T \in \mathcal{L}(V, W)$. If $\mathbf{v} = c_1 \mathbf{v}_1 + c_2 \mathbf{v}_2 + \cdots + c_n \mathbf{v}_n$, then $T\mathbf{v} = c_1 T\mathbf{v}_1 + c_2 T\mathbf{v}_2 + \cdots + c_n T\mathbf{v}_n$, so $\operatorname{ran} T = \operatorname{span}\{T\mathbf{v}_1, T\mathbf{v}_2, \ldots, T\mathbf{v}_n\}$. In particular, $\operatorname{ran} T$ is finite dimensional and $\dim \operatorname{ran} T \le n$.*

Proof Compute

$$\begin{aligned}
\operatorname{ran} T &= \{T\mathbf{v} : \mathbf{v} \in V\} \\
&= \{T(c_1 \mathbf{v}_1 + c_2 \mathbf{v}_2 + \cdots + c_n \mathbf{v}_n) : c_1, c_2, \ldots, c_n \in \mathbb{F}\} \\
&= \{c_1 T\mathbf{v}_1 + c_2 T\mathbf{v}_2 + \cdots + c_n T\mathbf{v}_n : c_1, c_2, \ldots, c_n \in \mathbb{F}\} \\
&= \operatorname{span}\{T\mathbf{v}_1, T\mathbf{v}_2, \ldots, T\mathbf{v}_n\}.
\end{aligned}$$

Theorem 2.2.7.a ensures that $\operatorname{ran} T$ is finite dimensional and $\dim \operatorname{ran} T \le n$. □

If V and W are nonzero finite-dimensional vector spaces over the same field $\mathbb{F}$, the preceding theorem can be further refined. Let $\beta = \mathbf{v}_1, \mathbf{v}_2, \ldots, \mathbf{v}_n$ be a basis for V, let $\gamma = \mathbf{w}_1, \mathbf{w}_2, \ldots, \mathbf{w}_m$ be a basis for W, and let $T \in \mathcal{L}(V, W)$. Express $\mathbf{v} \in V$ as

$$\mathbf{v} = c_1 \mathbf{v}_1 + c_2 \mathbf{v}_2 + \cdots + c_n \mathbf{v}_n,$$

that is, $[\mathbf{v}]_\beta = [c_i]$. Then

$$T\mathbf{v} = c_1 T\mathbf{v}_1 + c_2 T\mathbf{v}_2 + \cdots + c_n T\mathbf{v}_n,$$

so

$$[T\mathbf{v}]_\gamma = c_1 [T\mathbf{v}_1]_\gamma + c_2 [T\mathbf{v}_2]_\gamma + \cdots + c_n [T\mathbf{v}_n]_\gamma. \tag{2.3.16}$$

For each $j = 1, 2, \ldots, n$, write $T\mathbf{v}_j = a_{1j} \mathbf{w}_1 + a_{2j} \mathbf{w}_2 + \cdots + a_{mj} \mathbf{w}_m$ as a linear combination of the vectors in the basis γ and define

$$\gamma[T]_\beta = \big[[T\mathbf{v}_1]_\gamma \ [T\mathbf{v}_2]_\gamma \ \ldots \ [T\mathbf{v}_n]_\gamma \big] = [a_{ij}] \in \mathsf{M}_{m \times n}(\mathbb{F}), \tag{2.3.17}$$

whose respective columns are the γ-coordinate vectors of the $T\mathbf{v}_j$. We can rewrite (2.3.16) as

$$[T\mathbf{v}]_\gamma = \gamma[T]_\beta [\mathbf{v}]_\beta, \tag{2.3.18}$$

in which

$$\gamma[T]_\beta = \begin{bmatrix} a_{11} & a_{12} & \cdots & a_{1n} \\ a_{21} & a_{22} & \cdots & a_{2n} \\ \vdots & \vdots & \ddots & \vdots \\ a_{m1} & a_{m2} & \cdots & a_{mn} \end{bmatrix}.$$

Once we have fixed a basis β of V and a basis γ of W, the process of determining $T\mathbf{v}$ splits into two parts. First, compute the β-γ *matrix representation* $\gamma[T]_\beta$. This must be done only once and it can be used in all subsequent computations. Then, for each $\mathbf{v}$ of interest, compute its β-coordinate vector $[\mathbf{v}]_\beta$ and calculate the product in (2.3.18) to determine the γ-coordinate vector $[T\mathbf{v}]_\gamma$, from which one can recover $T\mathbf{v}$ as a linear combination of the vectors in γ.

Example 2.3.19 In Example 2.3.14, consider the basis $\gamma = 1, z$ of $\mathcal{P}_1$. Then

$$[T1]_\gamma = \begin{bmatrix} 0 \\ 0 \end{bmatrix}, \qquad [Tz]_\gamma = \begin{bmatrix} 1 \\ 0 \end{bmatrix}, \qquad [Tz^2]_\gamma = \begin{bmatrix} 0 \\ 2 \end{bmatrix},$$

and hence

$$_\gamma[T]_\beta = \begin{bmatrix} 0 & 1 & 0 \\ 0 & 0 & 2 \end{bmatrix} \quad \text{and} \quad [p]_\beta = \begin{bmatrix} c_0 \\ c_1 \\ c_2 \end{bmatrix}.$$

We have

$$[Tp]_\gamma = {}_\gamma[T]_\beta[p]_\beta = \begin{bmatrix} 0 & 1 & 0 \\ 0 & 0 & 2 \end{bmatrix} \begin{bmatrix} c_0 \\ c_1 \\ c_2 \end{bmatrix} = \begin{bmatrix} c_1 \\ 2c_2 \end{bmatrix},$$

so $Tp = c_1 1 + (2c_2)z = 2c_2 z + c_1$.

2.4 Change of Basis and Similarity

The identity (2.3.18) contains a wealth of information. Consider the special case in which $\mathcal{W} = \mathcal{V}$ is n-dimensional and $n \geq 1$. Suppose that $\beta = \mathbf{v}_1, \mathbf{v}_2, \ldots, \mathbf{v}_n$ and $\gamma = \mathbf{w}_1, \mathbf{w}_2, \ldots, \mathbf{w}_n$ are bases for $\mathcal{V}$. The *identity linear transformation* $I \in \mathcal{L}(\mathcal{V})$ is the function defined by $I\mathbf{v} = \mathbf{v}$ for all $\mathbf{v} \in \mathcal{V}$. The β-β basis representation of I is

$$\begin{aligned}
_\beta[I]_\beta &= \begin{bmatrix} [I\mathbf{v}_1]_\beta & [I\mathbf{v}_2]_\beta & \cdots & [I\mathbf{v}_n]_\beta \end{bmatrix} \\
&= \begin{bmatrix} [\mathbf{v}_1]_\beta & [\mathbf{v}_2]_\beta & \cdots & [\mathbf{v}_n]_\beta \end{bmatrix} \\
&= \begin{bmatrix} \mathbf{e}_1 & \mathbf{e}_2 & \cdots & \mathbf{e}_n \end{bmatrix} \\
&= I_n \in \mathsf{M}_n(\mathbb{F}),
\end{aligned}$$

the $n \times n$ identity matrix. What can we say about

$$_\gamma[I]_\beta = \begin{bmatrix} [\mathbf{v}_1]_\gamma & [\mathbf{v}_2]_\gamma & \cdots & [\mathbf{v}_n]_\gamma \end{bmatrix} \tag{2.4.1}$$

and

$$_\beta[I]_\gamma = \begin{bmatrix} [\mathbf{w}_1]_\beta & [\mathbf{w}_2]_\beta & \cdots & [\mathbf{w}_n]_\beta \end{bmatrix}? \tag{2.4.2}$$

For any $\mathbf{v} \in \mathcal{V}$ use (2.3.18) to compute

$$I_n[\mathbf{v}]_\gamma = [\mathbf{v}]_\gamma = [I\mathbf{v}]_\gamma = {}_\gamma[I]_\beta[\mathbf{v}]_\beta = {}_\gamma[I]_\beta[I\mathbf{v}]_\beta = {}_\gamma[I]_\beta {}_\beta[I]_\gamma[\mathbf{v}]_\gamma,$$

so $I_n\mathbf{x} = {}_\gamma[I]_\beta {}_\beta[I]_\gamma \mathbf{x}$ for all $\mathbf{x} \in \mathbb{F}^n$. It follows that

$$I_n = {}_\gamma[I]_\beta {}_\beta[I]_\gamma. \tag{2.4.3}$$

The calculation

$$I_n[\mathbf{v}]_\beta = [\mathbf{v}]_\beta = [I\mathbf{v}]_\beta = {}_\beta[I]_\gamma[\mathbf{v}]_\gamma = {}_\beta[I]_\gamma[I\mathbf{v}]_\gamma = {}_\beta[I]_\gamma {}_\gamma[I]_\beta[\mathbf{v}]_\beta \tag{2.4.4}$$

leads in the same way to

$$I_n = {}_\beta[I]_\gamma {}_\gamma[I]_\beta. \tag{2.4.5}$$

The identities (2.4.3) and (2.4.5) tell us that the matrix $_\gamma[I]_\beta$ is invertible and that $_\beta[I]_\gamma$ is its inverse. For another approach to this conclusion, see P.2.8.

Definition 2.4.6 The matrix $_\gamma[I]_\beta$ defined in (2.4.1) is the β-γ *change of basis* matrix.

This matrix describes how to represent each vector in the basis β as a linear combination of the vectors in the basis γ.

Example 2.4.7 Figure 2.2 illustrates the standard basis $\beta = \mathbf{e}_1, \mathbf{e}_2$ in $\mathbb{R}^2$ and the basis $\gamma = \mathbf{w}_1, \mathbf{w}_2$, in which $[\mathbf{w}_1]_\beta = [1\ 2]^\mathsf{T}$ and $[\mathbf{w}_2]_\beta = [3\ 1]^\mathsf{T}$. A computation reveals that $[\mathbf{e}_1]_\gamma = [-\frac{1}{5}\ \frac{2}{5}]^\mathsf{T}$ and $[\mathbf{e}_2]_\gamma = [\frac{3}{5}\ -\frac{1}{5}]^\mathsf{T}$. We have

$$_\beta[I]_\gamma = \big[[\mathbf{w}_1]_\beta\ [\mathbf{w}_2]_\beta\big] = \begin{bmatrix} 1 & 3 \\ 2 & 1 \end{bmatrix}$$

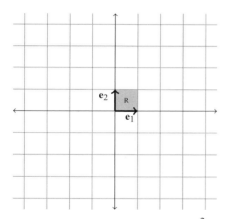

(a) The standard basis $\beta = \mathbf{e}_1, \mathbf{e}_2$ in $\mathbb{R}^2$.

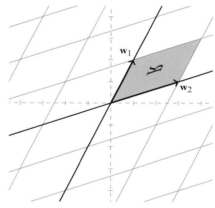

(b) The basis $\gamma = \mathbf{w}_1, \mathbf{w}_2$ in $\mathbb{R}^2$, in which $\mathbf{w}_1 = [1\ 2]^\mathsf{T}$ and $\mathbf{w}_2 = [3\ 1]^\mathsf{T}$.

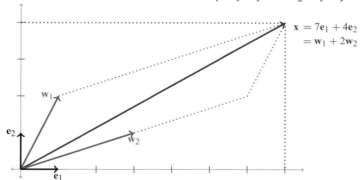

(c) Change of basis from β to γ. The coordinate vectors of $\mathbf{x}$ with respect to β and γ are $[\mathbf{x}]_\beta = [7\ 4]^\mathsf{T}$ and $[\mathbf{x}]_\gamma = [2\ 1]^\mathsf{T}$.

Figure 2.2 Relationship between two bases in $\mathbb{R}^2$.

and

$$\beta[I]_\gamma^{-1} = \begin{bmatrix} -\frac{1}{5} & \frac{3}{5} \\ \frac{2}{5} & -\frac{1}{5} \end{bmatrix},$$

which is consistent with the identity

$$\gamma[I]_\beta = \begin{bmatrix} [e_1]_\gamma & [e_2]_\gamma \end{bmatrix} = \begin{bmatrix} -\frac{1}{5} & \frac{3}{5} \\ \frac{2}{5} & -\frac{1}{5} \end{bmatrix}.$$

Figure 2.2(c) illustrates how a fixed vector $\mathbf{x}$ is represented as a linear combination of vectors in the respective bases β and γ. We have

$$\mathbf{x} = 7\mathbf{e}_1 + 4\mathbf{e}_2 = 7\left(-\tfrac{1}{5}\mathbf{w}_1 + \tfrac{2}{5}\mathbf{w}_2\right) + 4\left(\tfrac{3}{5}\mathbf{w}_1 - \tfrac{1}{5}\mathbf{w}_2\right) = \mathbf{w}_1 + 2\mathbf{w}_2.$$

Example 2.4.8 Each of the lists $\beta = \mathbf{x}_1, \mathbf{x}_2$ and $\gamma = \mathbf{y}_1, \mathbf{y}_2$, with

$$\mathbf{x}_1 = \begin{bmatrix} 1 \\ 1 \end{bmatrix}, \quad \mathbf{x}_2 = \begin{bmatrix} 1 \\ -1 \end{bmatrix} \quad \text{and} \quad \mathbf{y}_1 = \begin{bmatrix} 1 \\ 2 \end{bmatrix}, \quad \mathbf{y}_2 = \begin{bmatrix} 1 \\ 3 \end{bmatrix},$$

is a basis of $\mathbb{R}^2$. To compute the columns of $\gamma[I]_\beta = [[\mathbf{x}_1]_\gamma \ [\mathbf{x}_2]_\gamma]$, we must solve some linear equations. For example, the entries of $[\mathbf{x}_1]_\gamma = [a \ b]^\mathsf{T}$ are the coefficients in the representation

$$\mathbf{x}_1 = \begin{bmatrix} 1 \\ 1 \end{bmatrix} = a\begin{bmatrix} 1 \\ 2 \end{bmatrix} + b\begin{bmatrix} 1 \\ 3 \end{bmatrix} = \begin{bmatrix} a+b \\ 2a+3b \end{bmatrix}$$

of $\mathbf{x}_1$ as a linear combination of $\mathbf{y}_1$ and $\mathbf{y}_2$. The solution is $a = 2$ and $b = -1$, so $[\mathbf{x}_1]_\gamma = [2 \ -1]^\mathsf{T}$. Solving the linear equation $\mathbf{x}_2 = a\mathbf{y}_1 + b\mathbf{y}_2$ provides the second column of $\gamma[I]_\beta$, which is $[\mathbf{x}_2]_\gamma = [4 \ -3]^\mathsf{T}$. Thus,

$$\gamma[I]_\beta = \begin{bmatrix} 2 & 4 \\ -1 & -3 \end{bmatrix} \quad \text{and} \quad \beta[I]_\gamma = \gamma[I]_\beta^{-1} = \begin{bmatrix} \frac{3}{2} & 2 \\ -\frac{1}{2} & -1 \end{bmatrix}.$$

To check that $\beta[I]_\gamma$ is the inverse of $\gamma[I]_\beta$, we compute the entries of $[\mathbf{y}_1]_\beta = [a \ b]^\mathsf{T}$, which are the coefficients in the representation

$$\mathbf{y}_1 = \begin{bmatrix} 1 \\ 2 \end{bmatrix} = a\begin{bmatrix} 1 \\ 1 \end{bmatrix} + b\begin{bmatrix} 1 \\ -1 \end{bmatrix} = \begin{bmatrix} a+b \\ a-b \end{bmatrix}$$

of $\mathbf{y}_1$ as a linear combination of $\mathbf{x}_1$ and $\mathbf{x}_2$. As expected, the solution is $a = 3/2$ and $b = -1/2$. This example suggests a general algorithm to compute any change of basis matrix in $\mathbb{F}^n$; see P.2.6.

Theorem 2.4.9 *Let n be a positive integer, let $\mathcal{V}$ be an n-dimensional $\mathbb{F}$-vector space, and let $\beta = \mathbf{v}_1, \mathbf{v}_2, \ldots, \mathbf{v}_n$ be a basis for $\mathcal{V}$.*

(a) *Let $\gamma = \mathbf{w}_1, \mathbf{w}_2, \ldots, \mathbf{w}_n$ be a basis for $\mathcal{V}$. The change of basis matrix $\gamma[I]_\beta \in \mathsf{M}_n(\mathbb{F})$ is invertible and its inverse is $\beta[I]_\gamma$.*

(b) *If $S \in \mathsf{M}_n(\mathbb{F})$ is invertible, then there is a basis γ for $\mathcal{V}$ such that $S = \beta[I]_\gamma$.*

Proof We proved the first assertion in the preceding discussion. Let $S = [s_{ij}]$ and define

$$\mathbf{w}_j = s_{1j}\mathbf{v}_1 + s_{2j}\mathbf{v}_2 + \cdots + s_{nj}\mathbf{v}_n, \qquad j = 1, 2, \ldots, n. \tag{2.4.10}$$

We claim that $\gamma = \mathbf{w}_1, \mathbf{w}_2, \ldots, \mathbf{w}_n$ is a basis for $\mathcal{V}$. It suffices to show that $\operatorname{span} \gamma = \mathcal{V}$ (see Corollary 2.2.8). Let $S^{-1} = [\sigma_{ij}]$. For each $k = 1, 2, \ldots, n$,

$$\sum_{j=1}^{n} \sigma_{jk}\mathbf{w}_j = \sum_{j=1}^{n} \sum_{i=1}^{n} \sigma_{jk} s_{ij}\mathbf{v}_i = \sum_{i=1}^{n} \sum_{j=1}^{n} s_{ij}\sigma_{jk}\mathbf{v}_i = \sum_{i=1}^{n} \delta_{ik}\mathbf{v}_i = \mathbf{v}_k.$$

Thus, every vector in the basis β is in $\operatorname{span} \gamma$, so $\operatorname{span} \gamma = \mathcal{V}$. Then (2.4.10) ensures that $S = {}_\beta[I]_\gamma$. $\qquad\square$

The following result provides a converse to Theorem 2.1.3.

Corollary 2.4.11 *If $\beta = \mathbf{a}_1, \mathbf{a}_2, \ldots, \mathbf{a}_n$ is a basis of $\mathbb{F}^n$, then $A = [\mathbf{a}_1 \ \mathbf{a}_2 \ \ldots \ \mathbf{a}_n] \in M_n(\mathbb{F})$ is invertible.*

Proof Let $\mathcal{V} = \mathbb{F}^n$ and let $\gamma = \mathbf{e}_1, \mathbf{e}_2, \ldots, \mathbf{e}_n$ be the standard basis. Then part (a) of the preceding theorem says that $A = {}_\gamma[I]_\beta = [\mathbf{a}_1 \ \mathbf{a}_2 \ \ldots \ \mathbf{a}_n]$ is invertible. $\qquad\square$

Corollary 2.4.12 *Let $A \in M_n(\mathbb{F})$. Then A is invertible if and only if $\operatorname{rank} A = n$.*

Proof By definition, $\operatorname{rank} A = \dim \operatorname{col} A$. If A is invertible, Theorem 2.1.3 says that the columns of A comprise a basis for $\mathbb{F}^n$. Therefore, $\operatorname{col} A = \mathbb{F}^n$ and $\dim \operatorname{col} A = n$.

Conversely, suppose that $\dim \operatorname{col} A = n$. Since $\operatorname{col} A$ is a subspace of $\mathbb{F}^n$, Theorem 2.2.9 ensures that $\operatorname{col} A = \mathbb{F}^n$. Therefore, the columns of A span $\mathbb{F}^n$ and Corollary 2.2.8.a says that they comprise a basis for $\mathbb{F}^n$. Finally, Corollary 2.4.11 ensures that A is invertible. $\qquad\square$

Now return to the identity (2.3.18), which gives us the tools to understand how two different basis representations of a linear transformation are related.

Theorem 2.4.13 *Let n be a positive integer, let $\mathcal{V}$ be an n-dimensional $\mathbb{F}$-vector space, and let $T \in \mathcal{L}(\mathcal{V})$.*

(a) *Let β and γ be bases for $\mathcal{V}$, and let $S = {}_\gamma[I]_\beta$. Then S is invertible and*

$$_\gamma[T]_\gamma = {}_\gamma[I]_\beta \, {}_\beta[T]_\beta \, {}_\beta[I]_\gamma = S \, {}_\beta[T]_\beta \, S^{-1}. \tag{2.4.14}$$

(b) *Let $S \in M_n(\mathbb{F})$ be invertible and let β be a basis for $\mathcal{V}$. Then there is a basis γ for $\mathcal{V}$ such that $_\gamma[T]_\gamma = S \, {}_\beta[T]_\beta \, S^{-1}$.*

Proof (a) The preceding theorem ensures that S is invertible. Let $\mathbf{v} \in \mathcal{V}$ and compute

$$_\gamma[T]_\gamma[\mathbf{v}]_\gamma = [T\mathbf{v}]_\gamma = [I(T\mathbf{v})]_\gamma = {}_\gamma[I]_\beta \, [T\mathbf{v}]_\beta$$
$$= {}_\gamma[I]_\beta \, {}_\beta[T]_\beta[\mathbf{v}]_\beta = {}_\gamma[I]_\beta \, {}_\beta[T]_\beta \, [I\mathbf{v}]_\beta$$
$$= {}_\gamma[I]_\beta \, {}_\beta[T]_\beta \, {}_\beta[I]_\gamma \, [\mathbf{v}]_\gamma.$$

Consequently, $_\gamma[T]_\gamma \, [\mathbf{v}]_\gamma = {}_\gamma[I]_\beta \, {}_\beta[T]_\beta \, {}_\beta[I]_\gamma \, [\mathbf{v}]_\gamma$ for all $\mathbf{v} \in \mathcal{V}$, which implies (2.4.14).

(b) The preceding theorem ensures that there is a basis γ for $\mathcal{V}$ such that $S = {}_\gamma[I]_\beta$, so $S^{-1} = {}_\beta[I]_\gamma$ and the assertion follows from (2.4.14). $\qquad\qquad\square$

Example 2.4.15 Let $\beta = \mathbf{e}_1, \mathbf{e}_2$ be the standard basis of $\mathbb{R}^2$ and let the basis $\gamma = \mathbf{y}_1, \mathbf{y}_2$ be as in Example 2.4.8. Then

$$_\beta[I]_\gamma = \begin{bmatrix} [\mathbf{y}_1]_\beta & [\mathbf{y}_2]_\beta \end{bmatrix} = \begin{bmatrix} 1 & 1 \\ 2 & 3 \end{bmatrix} \quad \text{and} \quad _\gamma[I]_\beta = {}_\beta[I]_\gamma^{-1} = \begin{bmatrix} 3 & -1 \\ -2 & 1 \end{bmatrix}.$$

Theorem 2.3.15 says that a linear transformation on $\mathbb{R}^2$ is uniquely determined by its action on a basis. Let $T : \mathbb{R}^2 \to \mathbb{R}^2$ be the linear transformation such that $T\mathbf{e}_1 = 2\mathbf{e}_1$ and $T\mathbf{e}_2 = 3\mathbf{e}_2$. Then

$$T\mathbf{y}_1 = T(\mathbf{e}_1 + 2\mathbf{e}_2) = T\mathbf{e}_1 + 2T\mathbf{e}_2 = 2\mathbf{e}_1 + 6\mathbf{e}_2,$$

$$T\mathbf{y}_2 = T(\mathbf{e}_1 + 3\mathbf{e}_2) = T\mathbf{e}_1 + 3T\mathbf{e}_2 = 2\mathbf{e}_1 + 9\mathbf{e}_2,$$

so

$$_\beta[T]_\beta = \begin{bmatrix} 2 & 0 \\ 0 & 3 \end{bmatrix} \quad \text{and} \quad _\beta[T]_\gamma = \begin{bmatrix} [T\mathbf{y}_1]_\beta & [T\mathbf{y}_2]_\beta \end{bmatrix} = \begin{bmatrix} 2 & 2 \\ 6 & 9 \end{bmatrix}.$$

The preceding theorem ensures that

$$_\gamma[T]_\gamma = {}_\gamma[I]_\beta \, {}_\beta[T]_\beta \, {}_\beta[I]_\gamma = \begin{bmatrix} 3 & -1 \\ -2 & 1 \end{bmatrix} \begin{bmatrix} 2 & 0 \\ 0 & 3 \end{bmatrix} \begin{bmatrix} 1 & 1 \\ 2 & 3 \end{bmatrix} = \begin{bmatrix} 0 & -3 \\ 2 & 5 \end{bmatrix}.$$

Definition 2.4.16 Let $A, B \in \mathsf{M}_n(\mathbb{F})$. Then A and B are *similar over* $\mathbb{F}$ if there is an invertible $S \in \mathsf{M}_n(\mathbb{F})$ such that $A = SBS^{-1}$.

If $A = SBS^{-1}$ and there is a need to emphasize the role of S, we say that *A is similar to B via the similarity matrix S (or S^{-1}).*

Corollary 2.4.17 *Let $A, B \in \mathsf{M}_n(\mathbb{F})$. The following are equivalent:*

(a) *A and B are similar over $\mathbb{F}$.*

(b) *There is an n-dimensional $\mathbb{F}$-vector space $\mathcal{V}$, bases β and γ for $\mathcal{V}$, and a linear operator $T \in \mathcal{L}(\mathcal{V})$ such that $A = {}_\beta[T]_\beta$ and $B = {}_\gamma[T]_\gamma$.*

Proof (a) $\Rightarrow$ (b) Let $S \in \mathsf{M}_n(\mathbb{F})$ be an invertible matrix such that $A = SBS^{-1}$. Let $\mathcal{V} = \mathbb{F}^n$ and let $T_A : \mathbb{F}^n \to \mathbb{F}^n$ be the linear transformation induced by A (see Definition 2.3.9). Let β be the standard basis of $\mathbb{F}^n$ and let γ be the ordered list of columns of S; Theorem 2.1.3 ensures that γ is a basis. Then $_\beta[T_A]_\beta = A$ and $_\beta[I]_\gamma = S$, so

$$SBS^{-1} = A = {}_\beta[T_A]_\beta = {}_\beta[I]_\gamma \, {}_\gamma[T_A]_\gamma \, {}_\gamma[I]_\beta = S_\gamma[T_A]_\gamma S^{-1}.$$

Consequently, $SBS^{-1} = S_\gamma[T_A]_\gamma S^{-1}$, which implies that $B = {}_\gamma[T_A]_\gamma$.

(b) $\Rightarrow$ (a) This implication is part (a) of the preceding theorem. $\qquad\qquad\square$

The following theorem identifies an important fact that is also noted in (0.8.4): shifting by a scalar matrix preserves similarity.

Theorem 2.4.18 *Let $A, B \in M_n(\mathbb{F})$.*

(a) *If A is similar to B, then $A - \lambda I$ is similar to $B - \lambda I$ for every $\lambda \in \mathbb{F}$.*

(b) *If there is a $\lambda \in \mathbb{F}$ such that $(A - \lambda I)$ is similar to $(B - \lambda I)$, then A is similar to B.*

Proof Let $S \in M_n$ be invertible. If $A = SBS^{-1}$, then $S(B-\lambda I)S^{-1} = SBS^{-1} - \lambda SS^{-1} = A - \lambda I$ for all $\lambda \in \mathbb{F}$. If there is a $\lambda \in \mathbb{F}$ such that $A - \lambda I = S(B - \lambda I)S^{-1}$, then $A - \lambda I = SBS^{-1} - \lambda SS^{-1} = SBS^{-1} - \lambda I$, so $A = SBS^{-1}$.

Alternatively, we can use (b) in the preceding corollary. If A and B represent the same linear operator T, then $A - \lambda I$ and $B - \lambda I$ both represent $T - \lambda I$. Conversely, if $A - \lambda I$ and $B - \lambda I$ both represent the same linear operator T, then A and B represent $T + \lambda I$. $\square$

The trace and determinant are also preserved by similarity.

Theorem 2.4.19 *Let $A, B \in M_n$ be similar. Then $\operatorname{tr} A = \operatorname{tr} B$ and $\det A = \det B$.*

Proof Let $S \in M_n$ be invertible and such that $A = SBS^{-1}$. Then (0.3.5) ensures that

$$\operatorname{tr} A = \operatorname{tr} S(BS^{-1}) = \operatorname{tr}(BS^{-1})S = \operatorname{tr} B.$$

The product rule for determinants ensures that

$$\det A = \det SBS^{-1} = (\det S)(\det B)(\det S^{-1})$$
$$= (\det S)(\det S)^{-1}(\det B) = \det B. \qquad \square$$

Since similarity of $n \times n$ matrices A and B means that each represents the same linear operator T, the following properties of similarity are self-evident:

Reflexive A is similar to A for all $A \in M_n(\mathbb{F})$.

Symmetric A is similar to B if and only if B is similar to A.

Transitive If A is similar to B and B is similar to C, then A is similar to C.

For example, transitivity follows from the fact that if A and B represent T, and if B and C represent T, then A, B, and C all represent T. See P.2.10 for another approach.

Definition 2.4.20 A relation between pairs of matrices is an *equivalence relation* if it is reflexive, symmetric, and transitive.

It has taken many steps to arrive at this destination, but it has been worth the journey to learn one of the most important facts in linear algebra. Similar matrices represent (with respect to possibly different bases) the same linear operator, so they can be expected to share many important properties. Some of these shared properties are: rank, determinant, trace, eigenvalues, characteristic polynomial, minimal polynomial, and Jordan canonical form. We have a lot to look forward to as we study these properties in the following chapters.

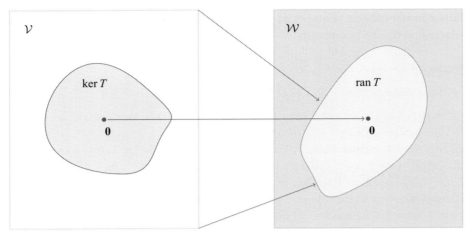

Figure 2.3 For a linear transformation $T : \mathcal{V} \to \mathcal{W}$, the range of T need not equal $\mathcal{W}$; this occurs if T is onto. The kernel of T need not be $\{\mathbf{0}\}$; this occurs if T is one to one.

2.5 The Dimension Theorem

Figure 2.3 illustrates the kernel and range of a linear transformation $T : \mathcal{V} \to \mathcal{W}$. If $\mathcal{V}$ is finite dimensional, an important relationship between the dimensions of $\ker T$ and $\operatorname{ran} T$ follows from Theorem 2.2.7.b.

Theorem 2.5.1 (The Dimension Theorem for Linear Transformations) *Let $\mathcal{V}$ and $\mathcal{W}$ be vector spaces over the same field $\mathbb{F}$. Suppose that $\mathcal{V}$ is finite dimensional and let $T \in \mathcal{L}(\mathcal{V}, \mathcal{W})$. Then*

$$\dim \ker T + \dim \operatorname{ran} T = \dim \mathcal{V}. \tag{2.5.2}$$

Proof Let $n = \dim \mathcal{V}$ and let $k = \dim \ker T$, so $0 \le k \le n$. If $n = 0$ or if $k = n$, there is nothing to prove, so we may assume that $0 \le k < n$.

If $k = 0$, let $\beta = \mathbf{w}_1, \mathbf{w}_2, \ldots, \mathbf{w}_n$ be a basis for $\mathcal{V}$. Theorem 2.3.15 ensures that

$$\operatorname{ran} T = \operatorname{span}\{T\mathbf{w}_1, T\mathbf{w}_2, \ldots, T\mathbf{w}_n\},$$

so it suffices to show that the list $\gamma = T\mathbf{w}_1, T\mathbf{w}_2, \ldots, T\mathbf{w}_n$ is linearly independent. If

$$\mathbf{0} = c_1 T\mathbf{w}_1 + c_2 T\mathbf{w}_2 + \cdots + c_n T\mathbf{w}_n$$
$$= T(c_1 \mathbf{w}_1 + c_2 \mathbf{w}_2 + \cdots + c_n \mathbf{w}_n),$$

then $c_1 \mathbf{w}_1 + c_2 \mathbf{w}_2 + \cdots + c_n \mathbf{w}_n \in \ker T = \{\mathbf{0}\}$, so the linear independence of β implies that $c_1 = c_2 = \cdots = c_n = 0$ and hence γ is linearly independent.

If $k \ge 1$, let $\mathbf{v}_1, \mathbf{v}_2, \ldots, \mathbf{v}_k$ be a basis for $\ker T$ and extend it to a basis

$$\beta = \mathbf{v}_1, \mathbf{v}_2, \ldots, \mathbf{v}_k, \mathbf{w}_1, \mathbf{w}_2, \ldots, \mathbf{w}_{n-k}$$

for $\mathcal{V}$. Since $T\mathbf{v}_1 = T\mathbf{v}_2 = \cdots = T\mathbf{v}_k = \mathbf{0}$, Theorem 2.3.15 again ensures that

$$\operatorname{ran} T = \operatorname{span}\{T\mathbf{v}_1, T\mathbf{v}_2, \ldots, T\mathbf{v}_k, T\mathbf{w}_1, T\mathbf{w}_2, \ldots, T\mathbf{w}_{n-k}\}$$
$$= \operatorname{span}\{T\mathbf{w}_1, T\mathbf{w}_2, \ldots, T\mathbf{w}_{n-k}\}.$$

It suffices to show that $\gamma = T\mathbf{w}_1, T\mathbf{w}_2, \ldots, T\mathbf{w}_{n-k}$ is linearly independent. If

$$c_1 T\mathbf{w}_1 + c_2 T\mathbf{w}_2 + \cdots + c_{n-k} T\mathbf{w}_{n-k} = \mathbf{0},$$

then

$$T(c_1\mathbf{w}_1 + c_2\mathbf{w}_2 + \cdots + c_{n-k}\mathbf{w}_{n-k}) = \mathbf{0}.$$

Therefore, $c_1\mathbf{w}_1 + c_2\mathbf{w}_2 + \cdots + c_{n-k}\mathbf{w}_{n-k} \in \ker T$ and there are scalars $a_1, a_2, \ldots, a_k$ such that

$$c_1\mathbf{w}_1 + c_2\mathbf{w}_2 + \cdots + c_{n-k}\mathbf{w}_{n-k} = a_1\mathbf{v}_1 + a_2\mathbf{v}_2 + \cdots + a_k\mathbf{v}_k.$$

Then

$$c_1\mathbf{w}_1 + c_2\mathbf{w}_2 + \cdots + c_{n-k}\mathbf{w}_{n-k} - a_1\mathbf{v}_1 - a_2\mathbf{v}_2 - \cdots - a_k\mathbf{v}_k = \mathbf{0},$$

so the linear independence of β implies that $c_1 = c_2 = \cdots = c_{n-k} = a_1 = a_2 = \cdots = a_k = 0$. We conclude that γ is linearly independent. $\square$

Perhaps the most important special case of the dimension theorem is for a linear operator T on a finite-dimensional vector space: T is one to one if and only if it is onto. The following corollary is a formal statement of a slight generalization of this observation.

Corollary 2.5.3 *Let $\mathcal{V}$ and $\mathcal{W}$ be finite-dimensional vector spaces over the same field $\mathbb{F}$. Suppose that $\dim \mathcal{V} = \dim \mathcal{W}$ and let $T \in \mathcal{L}(\mathcal{V}, \mathcal{W})$. Then $\ker T = \{\mathbf{0}\}$ if and only if $\operatorname{ran} T = \mathcal{W}$.*

Proof If $\ker T = \{\mathbf{0}\}$, then $\dim \ker T = 0$ and (2.5.2) ensures that $\dim \operatorname{ran} T = \dim \mathcal{V} = \dim \mathcal{W}$. But $\operatorname{ran} T$ is a subspace of $\mathcal{W}$, so Theorem 2.2.9 tells us that $\operatorname{ran} T = \mathcal{W}$.

Conversely, if $\operatorname{ran} T = \mathcal{W}$, then $\dim \operatorname{ran} T = \dim \mathcal{W} = \dim \mathcal{V}$ and (2.5.2) ensures that $\dim \ker T = 0$, so $\ker T = \{\mathbf{0}\}$. $\square$

Here is a matrix version of these results.

Corollary 2.5.4 (The Dimension Theorem for Matrices) *Let $A \in \mathsf{M}_{m \times n}(\mathbb{F})$. Then*

$$\dim \operatorname{null} A + \dim \operatorname{col} A = n. \tag{2.5.5}$$

If $m = n$, then $\operatorname{null} A = \{\mathbf{0}\}$ if and only if $\operatorname{col} A = \mathbb{F}^n$.

Proof Apply the preceding theorem to the linear transformation $T_A : \mathbb{F}^n \to \mathbb{F}^m$ induced by A. $\square$

The dimension of the null space of a matrix A is the *nullity* of A. The preceding corollary says that the nullity of A plus its rank is equal to the number of its columns.

2.6 Problems

P.2.1 Let $\beta = \mathbf{u}_1, \mathbf{u}_2, \ldots, \mathbf{u}_n$ be a basis for a nonzero $\mathbb{F}$-vector space $\mathcal{V}$. (a) If any vector is appended to β, explain why the resulting list still spans $\mathcal{V}$ but is not linearly independent. (b) If any vector in β is omitted, explain why the resulting list is still linearly independent but no longer spans $\mathcal{V}$.

P.2.2 Review Example 1.3.11. Show that the sets $\mathcal{P}_{\text{even}}$ and $\mathcal{P}_{\text{odd}}$ are subspaces because they are kernels of linear operators on the vector space $\mathcal{P}$.

P.2.3 Let $\mathcal{V}$ be an n-dimensional $\mathbb{F}$-vector space with $n \geq 2$ and let $\beta = \mathbf{v}_1, \mathbf{v}_2, \ldots, \mathbf{v}_r$ be a list of vectors in $\mathcal{V}$ with $1 \leq r < n$. Show that β does not span $\mathcal{V}$.

P.2.4 Let $\mathcal{U}, \mathcal{V}$, and $\mathcal{W}$ be vector spaces over the same field $\mathbb{F}$. Let $S \in \mathcal{L}(\mathcal{U}, \mathcal{V})$ and $T \in \mathcal{L}(\mathcal{V}, \mathcal{W})$. Define the function $T \circ S : \mathcal{U} \to \mathcal{W}$ by $(T \circ S)\mathbf{u} = T(S\mathbf{u})$. Show that $T \circ S \in \mathcal{L}(\mathcal{U}, \mathcal{W})$.

P.2.5 Let $\mathcal{V}$ be a nonzero n-dimensional $\mathbb{F}$-vector space and let $T \in \mathcal{L}(\mathcal{V}, \mathbb{F})$ be a nonzero linear transformation. Explain why $\dim \ker T = n - 1$.

P.2.6 Let $n \geq 1$, and let $\beta = \mathbf{v}_1, \mathbf{v}_2, \ldots, \mathbf{v}_n$ and $\gamma = \mathbf{w}_1, \mathbf{w}_2, \ldots, \mathbf{w}_n$ be bases of $\mathbb{F}^n$. Define the $n \times n$ matrices $B = [\mathbf{v}_1 \ \mathbf{v}_2 \ \ldots \ \mathbf{v}_n]$ and $C = [\mathbf{w}_1 \ \mathbf{w}_2 \ \ldots \ \mathbf{w}_n]$. Let $S = {}_\gamma[I]_\beta$ be the β-γ change of basis matrix. (a) Explain why $B = CS$ and deduce that $S = C^{-1}B$. (b) Compute B, C, and $S = C^{-1}B$ for the bases β and γ in Example 2.4.8. Discuss. (c) Why is ${}_\beta[I]_\gamma = B^{-1}C$?

P.2.7 Let $\mathcal{V}$ be the complex vector space of finitely nonzero sequences. Let T and S be the right-shift and left-shift operators defined in Example 2.3.12. Show that $ST = I$ but $TS \neq I$. Does this contradict Theorem 2.2.19? Discuss.

P.2.8 Use Theorem 2.2.19 to deduce (2.4.5) from (2.4.3) without the computation (2.4.4).

P.2.9 Let $A \in \mathsf{M}_n$. (a) Show that A is similar to I if and only if $A = I$. (b) Show that A is similar to 0 if and only if $A = 0$.

P.2.10 Use Definition 2.4.16 to prove that similarity is an equivalence relation on $\mathsf{M}_n(\mathbb{F})$.

P.2.11 Consider the matrix

$$A = \begin{bmatrix} 1 & 1 & 1 \\ 1 & -1 & 2 \end{bmatrix}.$$

(a) Why is $\dim \operatorname{col} A \leq 2$? (b) Why is $\dim \operatorname{col} A = 2$? (c) Use Corollary 2.5.4 to determine $\dim \operatorname{null} A$. (d) Find a basis for $\operatorname{null} A$ and explain what this has to do with Example 1.6.2.

P.2.12 Let $A \in \mathsf{M}_n(\mathbb{R})$ and $B \in \mathsf{M}_n(\mathbb{C})$. Let $B = X + iY$, in which $X, Y \in \mathsf{M}_n(\mathbb{R})$. If $AB = I$, show that $AX = XA = I$ and $Y = 0$. Explain why a real square matrix has a complex inverse if and only if it has a real inverse.

P.2.13 Let $A \in \mathsf{M}_n(\mathbb{F})$. Use Corollary 2.5.4 to show that the linear system $A\mathbf{x} = \mathbf{y}$ has a unique solution for some $\mathbf{y} \in \mathbb{F}^n$ if and only if it has a solution for each $\mathbf{y} \in \mathbb{F}^n$.

P.2.14 Let $A \in \mathsf{M}_n(\mathbb{F})$. Use Corollary 2.5.4 to show that the linear system $A\mathbf{x} = \mathbf{y}$ has a solution for every $\mathbf{y} \in \mathbb{F}^n$ if and only if $\mathbf{x} = \mathbf{0}$ is the only solution to $A\mathbf{x} = \mathbf{0}$.

P.2.15 Let $\mathcal{V} = \mathbb{C}^n$ and $\mathbb{F} = \mathbb{R}$. Describe a basis for $\mathcal{V}$ and explain why $\dim \mathcal{V} = 2n$.

P.2.16 Let $\mathcal{V} = \operatorname{span}\{AB - BA : A, B \in \mathsf{M}_n\}$. (a) Show that the function $\operatorname{tr} : \mathsf{M}_n \to \mathbb{C}$ is a linear transformation. (b) Use the dimension theorem to prove that $\dim \ker \operatorname{tr} = n^2 - 1$. (c) Prove that $\dim \mathcal{V} = n^2 - 1$. (d) Let $E_{ij} = \mathbf{e}_i \mathbf{e}_j^\mathsf{T}$, every entry of which is zero except for a 1 in the (i,j) position. Show that $E_{ij}E_{k\ell} = \delta_{jk}E_{i\ell}$ for $1 \leq i,j,k,\ell \leq n$. (e) Find a basis for $\mathcal{V}$. *Hint*: Work out the case $n = 2$ first.

P.2.17 Let $\mathcal{U}$ and $\mathcal{W}$ be subspaces of a finite-dimensional vector space. Show that $\mathcal{U} + \mathcal{W}$ is a direct sum if and only if $\dim(\mathcal{U} + \mathcal{W}) = \dim \mathcal{U} + \dim \mathcal{W}$.

P.2.18 Let $A \in \mathsf{M}_{m \times k}$ and $B \in \mathsf{M}_{k \times n}$. Let $\mathcal{W} = \operatorname{col} B$, $\mathcal{U} = \operatorname{null} A \cap \mathcal{W}$, and $\mathcal{Z} = \operatorname{col} AB$. Let $\mathbf{u}_1, \mathbf{u}_2, \ldots, \mathbf{u}_p$ be a basis for $\mathcal{U}$. (a) Why are there vectors $\mathbf{w}_1, \mathbf{w}_2, \ldots, \mathbf{w}_q$ such that $\mathbf{u}_1, \mathbf{u}_2, \ldots, \mathbf{u}_p, \mathbf{w}_1, \mathbf{w}_2, \ldots, \mathbf{w}_q$ is a basis for $\mathcal{W}$? (b) Show that $\operatorname{span}\{A\mathbf{w}_1, A\mathbf{w}_2, \ldots, A\mathbf{w}_q\} = \mathcal{Z}$. (c) Show that $A\mathbf{w}_1, A\mathbf{w}_2, \ldots, A\mathbf{w}_q$ are linearly independent. (d) Conclude that

$$\dim \operatorname{col} AB = \dim \operatorname{col} B - \dim(\operatorname{null} A \cap \operatorname{col} B). \tag{2.6.1}$$

This identity is used in P.3.25.

P.2.19 Let $\mathcal{U}_1, \mathcal{U}_2$, and $\mathcal{U}_3$ be subspaces of a finite-dimensional $\mathbb{F}$-vector space $\mathcal{V}$. Show that

$$\dim(\mathcal{U}_1 \cap \mathcal{U}_2 \cap \mathcal{U}_3) \geq \dim \mathcal{U}_1 + \dim \mathcal{U}_2 + \dim \mathcal{U}_3 - 2 \dim \mathcal{V}.$$

2.7 Some Important Concepts

- Basis.

- Dimension of a vector space.

- Functions that are linear transformations.

- The action of a linear transformation on a basis determines its action on all vectors.

- Basis representations of a linear transformation.

- The connection between change of basis and matrix similarity.

- The dimension theorem for linear transformations and for matrices.

3   Block Matrices

A matrix is not just an array of scalars. It can be thought of as an array of submatrices (a block matrix) in many different ways. We exploit this concept to explain why the row and column ranks of a matrix are equal, and to discover several inequalities involving rank. We discuss determinants of block matrices, and derive Cramer's rule as well as Cauchy's formula for the determinant of a bordered matrix. As an illustration of how block matrices can be used in a proof by induction, we characterize the square matrices that have zero trace (Shoda's theorem). The Kronecker product provides a way to construct block matrices that have many interesting properties; we discuss this in the final section of this chapter.

3.1   Row and Column Partitions

Let $\mathbf{b}_1, \mathbf{b}_2, \ldots, \mathbf{b}_n$ be the columns of $B \in \mathsf{M}_{r \times n}(\mathbb{F})$. Then the presentation

$$B = [\mathbf{b}_1 \ \mathbf{b}_2 \ \ldots \ \mathbf{b}_n] \tag{3.1.1}$$

is *partitioned according to its columns*. For any $\mathbf{x} = [x_1 \ x_2 \ \ldots \ x_n]^\mathsf{T} \in \mathbb{F}^n$, a computation reveals that

$$B\mathbf{x} = [\mathbf{b}_1 \ \mathbf{b}_2 \ \ldots \ \mathbf{b}_n] \begin{bmatrix} x_1 \\ \vdots \\ x_n \end{bmatrix} = x_1 \mathbf{b}_1 + x_2 \mathbf{b}_2 + \cdots + x_n \mathbf{b}_n \tag{3.1.2}$$

is a linear combination of the columns of B. The coefficients are the entries of $\mathbf{x}$.

Let $A \in \mathsf{M}_{m \times r}(\mathbb{F})$. Use the same column partition of B, to write $AB \in \mathsf{M}_{m \times n}(\mathbb{F})$ as

$$AB = [A\mathbf{b}_1 \ A\mathbf{b}_2 \ \ldots \ A\mathbf{b}_n]. \tag{3.1.3}$$

This presentation partitions AB according to its columns, each of which is a linear combination of the columns of A. The coefficients are the entries of the corresponding column of B.

The identity (3.1.2) is plausible; it is a formal product of a row object and a column object. To prove that it is correct, we must verify that corresponding entries of the left-hand and right-hand sides are the same. Let $B = [b_{ij}] \in \mathsf{M}_{r \times n}(\mathbb{F})$, so the vectors that comprise the columns of B are

$$\mathbf{b}_j = \begin{bmatrix} b_{1j} \\ b_{2j} \\ \vdots \\ b_{rj} \end{bmatrix}, \qquad j = 1, 2, \ldots, n.$$

For any $i \in \{1, 2, \ldots, r\}$, the ith entry of $B\mathbf{x}$ is $\sum_{k=1}^{n} b_{ik}x_k$. This equals the ith entry of

$$x_1\mathbf{b}_1 + x_2\mathbf{b}_2 + \cdots + x_n\mathbf{b}_n,$$

which is

$$x_1 b_{i1} + x_2 b_{i2} + \cdots + x_n b_{in}.$$

Example 3.1.4 Let

$$A = \begin{bmatrix} 1 & 2 \\ 3 & 4 \end{bmatrix}, \qquad B = \begin{bmatrix} 4 & 5 & 2 \\ 6 & 7 & 1 \end{bmatrix}, \tag{3.1.5}$$

$$\mathbf{b}_1 = \begin{bmatrix} 4 \\ 6 \end{bmatrix}, \qquad \mathbf{b}_2 = \begin{bmatrix} 5 \\ 7 \end{bmatrix}, \quad \text{and} \quad \mathbf{b}_3 = \begin{bmatrix} 2 \\ 1 \end{bmatrix}.$$

Then (3.1.3) in this case is

$$AB = \left[\begin{bmatrix} 1 & 2 \\ 3 & 4 \end{bmatrix}\begin{bmatrix} 4 \\ 6 \end{bmatrix} \quad \begin{bmatrix} 1 & 2 \\ 3 & 4 \end{bmatrix}\begin{bmatrix} 5 \\ 7 \end{bmatrix} \quad \begin{bmatrix} 1 & 2 \\ 3 & 4 \end{bmatrix}\begin{bmatrix} 2 \\ 1 \end{bmatrix} \right] = \begin{bmatrix} 16 & 19 & 4 \\ 36 & 43 & 10 \end{bmatrix}.$$

Example 3.1.6 The identity (3.1.2) permits us to construct a matrix that maps a given basis $\mathbf{x}_1, \mathbf{x}_2, \ldots, \mathbf{x}_n$ of $\mathbb{F}^n$ to another given basis $\mathbf{y}_1, \mathbf{y}_2, \ldots, \mathbf{y}_n$ of $\mathbb{F}^n$. The matrices $X = [\mathbf{x}_1 \ \mathbf{x}_2 \ \ldots \ \mathbf{x}_n]$ and $Y = [\mathbf{y}_1 \ \mathbf{y}_2 \ \ldots \ \mathbf{y}_n]$ are invertible and satisfy $X\mathbf{e}_i = \mathbf{x}_i$, $X^{-1}\mathbf{x}_i = \mathbf{e}_i$, and $Y\mathbf{e}_i = \mathbf{y}_i$ for $i = 1, 2, \ldots, n$. Thus, $YX^{-1}\mathbf{x}_i = Y\mathbf{e}_i = \mathbf{y}_i$ for $i = 1, 2, \ldots, n$.

Example 3.1.7 The identity (3.1.3) provides a short proof of the fact that $A \in \mathsf{M}_n$ is invertible if $A\mathbf{x} = \mathbf{y}$ is consistent for each $\mathbf{y} \in \mathbb{F}^n$. By hypothesis, there exist $\mathbf{b}_1, \mathbf{b}_2, \ldots, \mathbf{b}_n \in \mathbb{F}^n$ such that $A\mathbf{b}_i = \mathbf{e}_i$ for $i = 1, 2, \ldots, n$. Let $B = [\mathbf{b}_1 \ \mathbf{b}_2 \ \ldots \ \mathbf{b}_n]$. Then $AB = A[\mathbf{b}_1 \ \mathbf{b}_2 \ \ldots \ \mathbf{b}_n] = [A\mathbf{b}_1 \ A\mathbf{b}_2 \ \ldots \ A\mathbf{b}_n] = [\mathbf{e}_1 \ \mathbf{e}_2 \ \ldots \ \mathbf{e}_n] = I$, so Theorem 2.2.19 ensures that $B = A^{-1}$.

Example 3.1.8 We can combine (3.1.3) with the dimension theorem to give another proof of Theorem 2.2.19. If $A, B \in \mathsf{M}_n(\mathbb{F})$ and $AB = I$, then $A(B\mathbf{x}) = (AB)\mathbf{x} = \mathbf{x}$ for every $\mathbf{x} \in \mathbb{F}^n$. Therefore, $\operatorname{col} A = \mathbb{F}^n$ and $\dim \operatorname{col} A = n$. Corollary 2.5.4 ensures that $\dim \operatorname{null} A = 0$. Compute $A(I - BA) = A - (AB)A = A - IA = A - A = 0$. Since $\operatorname{null} A = \{\mathbf{0}\}$, (3.1.3) ensures that every column of $I - BA$ is the zero vector. Therefore, $BA = I$.

Column partitions permit us to give a short proof of Cramer's rule. Although conceptually elegant, Cramer's rule is not recommended for use in numerical algorithms.

Theorem 3.1.9 (Cramer's Rule) *Let* $A = [\mathbf{a}_1 \ \mathbf{a}_2 \ \ldots \ \mathbf{a}_n] \in \mathsf{M}_n$ *be invertible, let* $\mathbf{y} \in \mathbb{C}^n$, *and let*

$$A_i(\mathbf{y}) = [\mathbf{a}_1 \ \ldots \ \mathbf{a}_{i-1} \ \mathbf{y} \ \mathbf{a}_{i+1} \ \ldots \ \mathbf{a}_n] \in \mathsf{M}_n$$

denote the matrix obtained by replacing the ith column of A with $\mathbf{y}$. *Let* $\mathbf{x} = [x_1 \ x_2 \ \ldots \ x_n]^\mathsf{T}$, *in which*

$$x_1 = \frac{\det A_1(\mathbf{y})}{\det A}, \qquad x_2 = \frac{\det A_2(\mathbf{y})}{\det A}, \ldots \qquad x_n = \frac{\det A_n(\mathbf{y})}{\det A}. \tag{3.1.10}$$

Then $\mathbf{x}$ *is the unique solution to* $A\mathbf{x} = \mathbf{y}$.

Proof Let $\mathbf{x} = [x_i] \in \mathbb{C}^n$ be the unique solution to $A\mathbf{x} = \mathbf{y}$. For $i = 1, 2, \ldots, n$, let

$$X_i = [\mathbf{e}_1 \ \ldots \ \mathbf{e}_{i-1} \ \mathbf{x} \ \mathbf{e}_{i+1} \ \ldots \ \mathbf{e}_n] \in \mathsf{M}_n$$

denote the matrix obtained by replacing the ith column of I_n with $\mathbf{x}$. Perform a Laplace expansion along the ith row of X_i and obtain

$$\det X_i = x_i \det I_{n-1} = x_i.$$

Since $A\mathbf{e}_j = \mathbf{a}_j$ for $j \neq i$ and $A\mathbf{x} = \mathbf{y}$,

$$\begin{aligned}
X_i &= [\mathbf{e}_1 \ \ldots \ \mathbf{e}_{i-1} \ \mathbf{x} \ \mathbf{e}_{i+1} \ \ldots \ \mathbf{e}_n] \\
&= [A^{-1}\mathbf{a}_1 \ \ldots \ A^{-1}\mathbf{a}_{i-1} \ A^{-1}\mathbf{y} \ A^{-1}\mathbf{a}_{i+1} \ \ldots \ A^{-1}\mathbf{a}_n] \\
&= A^{-1}[\mathbf{a}_1 \ \ldots \ \mathbf{a}_{i-1} \ \mathbf{y} \ \mathbf{a}_{i+1} \ \ldots \ \mathbf{a}_n] \\
&= A^{-1}A_i(\mathbf{y}).
\end{aligned}$$

Since $\det A \neq 0$, for $i = 1, 2, \ldots, n$,

$$x_i = \det X_i = \det \left(A^{-1}A_i(\mathbf{y}) \right) = \det(A^{-1}) \det A_i(\mathbf{y}) = \frac{\det A_i(\mathbf{y})}{\det A}. \qquad \square$$

Example 3.1.11 Let

$$A = \begin{bmatrix} 1 & 2 & 3 \\ 8 & 9 & 4 \\ 7 & 6 & 5 \end{bmatrix}, \qquad \mathbf{x} = \begin{bmatrix} x_1 \\ x_2 \\ x_3 \end{bmatrix}, \qquad \text{and} \qquad \mathbf{y} = \begin{bmatrix} 2 \\ 2 \\ 2 \end{bmatrix}.$$

Since $\det A = -48$ is nonzero, $A\mathbf{x} = \mathbf{y}$ has a unique solution. To form A_1, A_2, and A_3, replace the first, second, and third columns of A by $\mathbf{y}$, respectively:

$$A_1(\mathbf{y}) = \begin{bmatrix} 2 & 2 & 3 \\ 2 & 9 & 4 \\ 2 & 6 & 5 \end{bmatrix}, \qquad A_2(\mathbf{y}) = \begin{bmatrix} 1 & 2 & 3 \\ 8 & 2 & 4 \\ 7 & 2 & 5 \end{bmatrix}, \qquad \text{and} \qquad A_3(\mathbf{y}) = \begin{bmatrix} 1 & 2 & 2 \\ 8 & 9 & 2 \\ 7 & 6 & 2 \end{bmatrix}.$$

Then

$$x_1 = \frac{\det A_1(\mathbf{y})}{\det A} = \frac{20}{-48} = -\frac{5}{12},$$

$$x_2 = \frac{\det A_2(\mathbf{y})}{\det A} = \frac{-16}{-48} = \frac{1}{3}, \qquad \text{and}$$

$$x_3 = \frac{\det A_3(\mathbf{y})}{\det A} = \frac{-28}{-48} = \frac{7}{12}.$$

What we have done for columns, we can also do for rows. Let $\mathbf{a}_1, \mathbf{a}_2, \ldots, \mathbf{a}_m \in \mathbb{F}^n$ and let

$$A = \begin{bmatrix} \mathbf{a}_1^{\mathsf{T}} \\ \mathbf{a}_2^{\mathsf{T}} \\ \vdots \\ \mathbf{a}_m^{\mathsf{T}} \end{bmatrix} \in \mathsf{M}_{m \times n}(\mathbb{F}). \tag{3.1.12}$$

For any $\mathbf{x} \in \mathbb{F}^m$,

$$\mathbf{x}^{\mathsf{T}} A = [x_1 \ x_2 \ \cdots \ x_m] \begin{bmatrix} \mathbf{a}_1^{\mathsf{T}} \\ \vdots \\ \mathbf{a}_m^{\mathsf{T}} \end{bmatrix} = x_1 \mathbf{a}_1^{\mathsf{T}} + x_2 \mathbf{a}_2^{\mathsf{T}} + \cdots + x_m \mathbf{a}_m^{\mathsf{T}} \tag{3.1.13}$$

and

$$AB = \begin{bmatrix} \mathbf{a}_1^{\mathsf{T}} B \\ \vdots \\ \mathbf{a}_m^{\mathsf{T}} B \end{bmatrix}. \tag{3.1.14}$$

These presentations partition A and AB *according to their rows* and make it clear that each row of AB is a linear combination of the rows of B. The coefficients in that linear combination are the entries of the corresponding row of A.

Example 3.1.15 With the matrices A and B in (3.1.5), $\mathbf{a}_1^{\mathsf{T}} = [1 \ 2]$ and $\mathbf{a}_2^{\mathsf{T}} = [3 \ 4]$. In this case (3.1.14) is

$$AB = \begin{bmatrix} [1 \ 2] \begin{bmatrix} 4 & 5 & 2 \\ 6 & 7 & 1 \end{bmatrix} \\ [3 \ 4] \begin{bmatrix} 4 & 5 & 2 \\ 6 & 7 & 1 \end{bmatrix} \end{bmatrix} = \begin{bmatrix} 16 & 19 & 4 \\ 36 & 43 & 10 \end{bmatrix}.$$

Using the partitions of A and B in (3.1.12) and (3.1.1),

$$AB = \begin{bmatrix} \mathbf{a}_1^{\mathsf{T}} \\ \vdots \\ \mathbf{a}_m^{\mathsf{T}} \end{bmatrix} [\mathbf{b}_1 \ \mathbf{b}_2 \ \cdots \ \mathbf{b}_n] = \begin{bmatrix} \mathbf{a}_1^{\mathsf{T}} \mathbf{b}_1 & \mathbf{a}_1^{\mathsf{T}} \mathbf{b}_2 & \cdots & \mathbf{a}_1^{\mathsf{T}} \mathbf{b}_n \\ \mathbf{a}_2^{\mathsf{T}} \mathbf{b}_1 & \mathbf{a}_2^{\mathsf{T}} \mathbf{b}_2 & \cdots & \mathbf{a}_2^{\mathsf{T}} \mathbf{b}_n \\ \vdots & \vdots & \ddots & \vdots \\ \mathbf{a}_m^{\mathsf{T}} \mathbf{b}_1 & \mathbf{a}_m^{\mathsf{T}} \mathbf{b}_2 & \cdots & \mathbf{a}_m^{\mathsf{T}} \mathbf{b}_n \end{bmatrix} \in \mathsf{M}_{m \times n}. \tag{3.1.16}$$

The scalars $\mathbf{a}_i^{\mathsf{T}} \mathbf{b}_j$ in (3.1.16) are often referred to as "inner products," though that term is strictly correct only for real matrices (the inner product of $\mathbf{b}_j$ and $\mathbf{a}_i$ in $\mathbb{C}^n$ is $\mathbf{a}_i^* \mathbf{b}_j$; see Example 4.4.3).

Example 3.1.17 With the matrices A and B in (3.1.5), the identity (3.1.16) is

$$AB = \begin{bmatrix} [1 \ 2] \begin{bmatrix} 4 \\ 6 \end{bmatrix} & [1 \ 2] \begin{bmatrix} 5 \\ 7 \end{bmatrix} & [1 \ 2] \begin{bmatrix} 2 \\ 1 \end{bmatrix} \\ [3 \ 4] \begin{bmatrix} 4 \\ 6 \end{bmatrix} & [3 \ 4] \begin{bmatrix} 5 \\ 7 \end{bmatrix} & [3 \ 4] \begin{bmatrix} 2 \\ 1 \end{bmatrix} \end{bmatrix} = \begin{bmatrix} 16 & 19 & 4 \\ 36 & 43 & 10 \end{bmatrix}.$$

Example 3.1.18 If $A = [\mathbf{a}_1 \ \mathbf{a}_2 \ \cdots \ \mathbf{a}_n] \in \mathsf{M}_n(\mathbb{R})$ and $\mathbf{a}_j^{\mathsf{T}} \mathbf{a}_i = \delta_{ij}$ for all $i, j = 1, 2, \ldots, n$, then

$$A^{\mathsf{T}} A = \begin{bmatrix} \mathbf{a}_1^{\mathsf{T}} \\ \mathbf{a}_2^{\mathsf{T}} \\ \vdots \\ \mathbf{a}_n^{\mathsf{T}} \end{bmatrix} [\mathbf{a}_1 \ \mathbf{a}_2 \ \cdots \ \mathbf{a}_n] = \begin{bmatrix} \mathbf{a}_1^{\mathsf{T}} \mathbf{a}_1 & \mathbf{a}_1^{\mathsf{T}} \mathbf{a}_2 & \cdots & \mathbf{a}_1^{\mathsf{T}} \mathbf{a}_n \\ \mathbf{a}_2^{\mathsf{T}} \mathbf{a}_1 & \mathbf{a}_2^{\mathsf{T}} \mathbf{a}_2 & \cdots & \mathbf{a}_2^{\mathsf{T}} \mathbf{a}_n \\ \vdots & \vdots & \ddots & \vdots \\ \mathbf{a}_n^{\mathsf{T}} \mathbf{a}_1 & \mathbf{a}_n^{\mathsf{T}} \mathbf{a}_2 & \cdots & \mathbf{a}_n^{\mathsf{T}} \mathbf{a}_n \end{bmatrix} = I.$$

Thus, A is invertible and $A^{-1} = A^{\mathsf{T}}$; matrices of this type are studied in Section 6.2.

There is another way to present a matrix product. Let $A \in M_{m \times r}$ and $B \in M_{r \times n}$. Partition $A = [\mathbf{a}_1 \ \mathbf{a}_2 \ \ldots \ \mathbf{a}_r]$ according to its columns, and partition B according to its rows, so

$$B = \begin{bmatrix} \mathbf{b}_1^T \\ \mathbf{b}_2^T \\ \vdots \\ \mathbf{b}_r^T \end{bmatrix}.$$

Then

$$AB = [\mathbf{a}_1 \ \mathbf{a}_2 \ \ldots \ \mathbf{a}_r] \begin{bmatrix} \mathbf{b}_1^T \\ \vdots \\ \mathbf{b}_r^T \end{bmatrix} = \mathbf{a}_1 \mathbf{b}_1^T + \mathbf{a}_2 \mathbf{b}_2^T + \cdots + \mathbf{a}_r \mathbf{b}_r^T \qquad (3.1.19)$$

is presented as a sum of $m \times n$ matrices, each of which has rank at most one. The summands in (3.1.19) are called *outer products*. For an application of this identity, see Example 9.7.9.

Example 3.1.20 With the matrices A and B in (3.1.5), the identity (3.1.19) is

$$AB = \begin{bmatrix} 1 \\ 3 \end{bmatrix} [4 \ 5 \ 2] + \begin{bmatrix} 2 \\ 4 \end{bmatrix} [6 \ 7 \ 1] = \begin{bmatrix} 4 & 5 & 2 \\ 12 & 15 & 6 \end{bmatrix} + \begin{bmatrix} 12 & 14 & 2 \\ 24 & 28 & 4 \end{bmatrix} = \begin{bmatrix} 16 & 19 & 4 \\ 36 & 43 & 10 \end{bmatrix}.$$

The identity (3.1.2) leads to a useful block identity. Suppose that each $\mathbf{a}_1, \mathbf{a}_2, \ldots, \mathbf{a}_r \in \mathbb{C}^n$ is a linear combination of $\mathbf{b}_1, \mathbf{b}_2, \ldots, \mathbf{b}_m \in \mathbb{C}^n$. Let $A = [\mathbf{a}_1 \ \mathbf{a}_2 \ \ldots \ \mathbf{a}_r] \in M_{n \times r}$ and $B = [\mathbf{b}_1 \ \mathbf{b}_2 \ \ldots \ \mathbf{b}_m] \in M_{n \times m}$. Then (3.1.2) says that there are $\mathbf{x}_1, \mathbf{x}_2, \ldots, \mathbf{x}_r \in \mathbb{C}^m$ such that $\mathbf{a}_i = B\mathbf{x}_i$, a linear combination of the columns of B, for $i = 1, 2, \ldots, r$. If $X = [\mathbf{x}_1 \ \mathbf{x}_2 \ \ldots \ \mathbf{x}_r] \in M_{m \times r}$, then

$$A = BX \qquad (3.1.21)$$

says that each column of A is a linear combination of the columns of B.

It can be useful to group together some adjacent columns and present $B \in M_{n \times r}$ in the partitioned form

$$B = [B_1 \ B_2 \ \ldots \ B_k]$$

in which

$$B_j \in M_{n \times r_j}, \quad j = 1, 2, \ldots, k, \quad \text{and} \quad r_1 + r_2 + \cdots + r_k = r.$$

If $A \in M_{m \times n}$, then $AB_j \in M_{m \times r_j}$ and

$$AB = [AB_1 \ AB_2 \ \ldots \ AB_k]. \qquad (3.1.22)$$

An analogous partition according to groups of rows is also possible.

Example 3.1.23 Partition the matrix B in (3.1.5) as

$$B = [B_1 \ B_2], \qquad B_1 = \begin{bmatrix} 4 & 5 \\ 6 & 7 \end{bmatrix}, \quad \text{and} \quad B_2 = \begin{bmatrix} 2 \\ 1 \end{bmatrix}.$$

Then

$$AB = [AB_1 \ AB_2] = \begin{bmatrix} \begin{bmatrix} 1 & 2 \\ 3 & 4 \end{bmatrix} \begin{bmatrix} 4 & 5 \\ 6 & 7 \end{bmatrix} & \begin{bmatrix} 1 & 2 \\ 3 & 4 \end{bmatrix} \begin{bmatrix} 2 \\ 1 \end{bmatrix} \end{bmatrix} = \begin{bmatrix} 16 & 19 & 4 \\ 36 & 43 & 10 \end{bmatrix}.$$

Example 3.1.24 The "side-by-side" method for matrix inversion can be justified with a block matrix calculation. If $A \in \mathsf{M}_n$ is invertible, then its reduced row echelon form is I. Let $R = E_k E_{k-1} \cdots E_1$ be the product of elementary matrices $E_1, E_2, \ldots, E_k$ that encode row operations that row reduce A to I. Then $RA = I$ and $R = A^{-1}$ (Theorem 2.2.19). Thus, $R[A \; I] = [RA \; R] = [I \; A^{-1}]$, so reducing the block matrix $[A \; I]$ to its reduced row echelon form reveals A^{-1}.

Our final observation about row and column partitions is that they provide a convenient way to express a property of determinants.

Example 3.1.25 Let $A = [\mathbf{a}_1 \; \mathbf{a}_2 \; \ldots \; \mathbf{a}_n] \in \mathsf{M}_n$, let $j \in \{1, 2, \ldots, n\}$, and partition $A = [A_1 \; \mathbf{a}_j \; A_2]$, in which $A_1 = [\mathbf{a}_1 \; \ldots \; \mathbf{a}_{j-1}]$ (not present if $j = 1$) and $A_2 = [\mathbf{a}_{j+1} \; \ldots \; \mathbf{a}_n]$ (not present if $j = n$). For $\mathbf{x}_1, \mathbf{x}_2, \ldots, \mathbf{x}_m \in \mathbb{C}^n$ and $c_1, c_2, \ldots, c_m \in \mathbb{C}$, we have

$$\det \left[A_1 \; \sum_{j=1}^{m} c_j \mathbf{x}_j \; A_2 \right] = \sum_{j=1}^{m} c_j \det[A_1 \; \mathbf{x}_j \; A_2], \tag{3.1.26}$$

which can be verified with a Laplace expansion of the determinant by minors along column j. The identity (3.1.26) implies that $\mathbf{x} \mapsto \det[A_1 \; \mathbf{x} \; A_2]$ is a linear function of $\mathbf{x}$. Since $\det A = \det A^\mathsf{T}$, there is an analogous result for rows.

3.2 Rank

The preceding partitions and presentations are not mere notational devices. Flexible use of block matrix operations can help shed light on important concepts. For example, it leads to a fundamental fact about two subspaces associated with $A \in \mathsf{M}_{m \times n}(\mathbb{F})$. One is its column space (see (1.3.8)); the other is its *row space*

$$\mathrm{row}\, A = \{A^\mathsf{T} \mathbf{x} : \mathbf{x} \in \mathbb{F}^m\} \subseteq \mathbb{F}^n,$$

which is the column space of A^T.

Theorem 3.2.1 *If* $A = [\mathbf{a}_1 \; \mathbf{a}_2 \; \ldots \; \mathbf{a}_n] \in \mathsf{M}_{m \times n}(\mathbb{F})$, *then*

$$\dim \mathrm{col}\, A = \dim \mathrm{row}\, A. \tag{3.2.2}$$

Proof If $A = 0$, then both $\mathrm{col}\, A$ and $\mathrm{row}\, A$ are zero dimensional. Suppose that $A \neq 0$. The presentation (3.1.2) reminds us that

$$\mathrm{col}\, A = \mathrm{span}\{\mathbf{a}_1, \mathbf{a}_2, \ldots, \mathbf{a}_n\}.$$

Let $\dim \mathrm{col}\, A = r$. Among $\mathbf{a}_1, \mathbf{a}_2, \ldots, \mathbf{a}_n$ there are r vectors

$$\mathbf{a}_{j_1}, \mathbf{a}_{j_2}, \ldots, \mathbf{a}_{j_r}, \quad 1 \leq j_1 < j_2 < \cdots < j_r \leq n,$$

that comprise a basis for $\mathrm{col}\, A$; see Theorem 2.2.7.a. Let $B = [\mathbf{a}_{j_1} \; \mathbf{a}_{j_2} \; \ldots \; \mathbf{a}_{j_r}] \in \mathsf{M}_{m \times r}$.

Since each $\mathbf{a}_1, \mathbf{a}_2, \ldots, \mathbf{a}_n$ is a linear combination of the columns of B, (3.1.21) ensures that there is an $X \in \mathsf{M}_{r \times n}$ such that $A = BX$. Now partition $X^\mathsf{T} = [\mathbf{u}_1 \; \mathbf{u}_2 \; \ldots \; \mathbf{u}_r] \in \mathsf{M}_{n \times r}$

according to its columns. The factorization $A = BX$ and the presentation (3.1.19) permit us to write

$$A^\mathsf{T} = X^\mathsf{T} B^\mathsf{T} = \mathbf{u}_1 \mathbf{a}_{j_1}^\mathsf{T} + \mathbf{u}_2 \mathbf{a}_{j_2}^\mathsf{T} + \cdots + \mathbf{u}_r \mathbf{a}_{j_r}^\mathsf{T}.$$

For any $\mathbf{x} \in \mathbb{F}^m$,

$$A^\mathsf{T}\mathbf{x} = X^\mathsf{T} B^\mathsf{T} \mathbf{x} = (\mathbf{a}_{j_1}^\mathsf{T}\mathbf{x})\mathbf{u}_1 + (\mathbf{a}_{j_2}^\mathsf{T}\mathbf{x})\mathbf{u}_2 + \cdots + (\mathbf{a}_{j_r}^\mathsf{T}\mathbf{x})\mathbf{u}_r \in \operatorname{span}\{\mathbf{u}_1, \mathbf{u}_2, \ldots, \mathbf{u}_r\}$$

and hence

$$\dim \operatorname{row} A \le r = \dim \operatorname{col} A. \tag{3.2.3}$$

Apply (3.2.3) to A^T and obtain

$$\dim \operatorname{col} A = \dim \operatorname{row}(A^\mathsf{T}) \le \dim \operatorname{col}(A^\mathsf{T}) = \dim \operatorname{row} A. \tag{3.2.4}$$

Now combine (3.2.3) and (3.2.4) to obtain (3.2.2). □

If $A \in \mathsf{M}_{m \times n}(\mathbb{F})$, then $\operatorname{rank} A = \dim \operatorname{col} A$ (Definition 2.2.6). The preceding theorem says that $\operatorname{rank} A$ is the common value of $\dim \operatorname{row} A$ and $\dim \operatorname{col} A$, as subspaces of $\mathbb{F}^n$ and $\mathbb{F}^m$, respectively. These equal quantities are the *row rank* and *column rank* of A, respectively.

Theorem 3.2.1 says that $\dim \operatorname{col} A = \dim \operatorname{col} A^\mathsf{T}$, so

$$\operatorname{rank} A = \operatorname{rank} A^\mathsf{T}. \tag{3.2.5}$$

Corollary 3.2.6 *Let $A \in \mathsf{M}_n(\mathbb{F})$. Then A is invertible if and only if A^T is invertible.*

Proof $\operatorname{rank} A = \operatorname{rank} A^\mathsf{T}$, so $\operatorname{rank} A = n$ if and only if $\operatorname{rank} A^\mathsf{T} = n$. The assertion follows from Corollary 2.4.12. □

Since $\dim \operatorname{row} A \le m$ and $\dim \operatorname{col} A \le n$, we have the upper bound

$$\operatorname{rank} A \le \min\{m, n\}.$$

This inequality can be strict, or it can be an equality.

Example 3.2.7 Consider

$$A = \begin{bmatrix} 1 & 2 & -1 \\ 2 & 4 & -2 \end{bmatrix} \quad \text{and} \quad B = \begin{bmatrix} 1 & 2 & 3 \\ 4 & 5 & 6 \end{bmatrix}.$$

Then $\operatorname{rank} A = 1 < \min\{2, 3\}$ and $\operatorname{rank} B = 2 = \min\{2, 3\}$. The rows of A are linearly dependent, while the rows of B are linearly independent. The columns of A are linearly dependent, as are the columns of B.

Definition 3.2.8 If $A \in \mathsf{M}_{m \times n}(\mathbb{F})$ and $\operatorname{rank} A = \min\{m, n\}$, then A *has full rank*. If $\operatorname{rank} A = n$, then A *has full column rank*; if $\operatorname{rank} A = m$, then A *has full row rank*.

If A has full column rank, the dimension theorem (Corollary 2.5.4) ensures that $\operatorname{null} A = \{\mathbf{0}\}$. This observation is the foundation of the following theorem about matrix products that preserve rank. Products with this property play an important role in reductions of matrices to various standard forms.

Theorem 3.2.9 *Let $A \in M_{m \times n}(\mathbb{F})$. If $X \in M_{p \times m}(\mathbb{F})$ has full column rank and $Y \in M_{n \times q}(\mathbb{F})$ has full row rank, then*

$$\operatorname{rank} A = \operatorname{rank} XAY. \tag{3.2.10}$$

In particular, (3.2.10) is valid if $X \in M_m(\mathbb{F})$ and $Y \in M_n(\mathbb{F})$ are invertible.

Proof First consider the product $XA \in M_{p \times n}(\mathbb{F})$. Since X has full column rank, the dimension theorem ensures that $\operatorname{null} X = \{\mathbf{0}\}$. If $\mathbf{u} \in \mathbb{F}^n$, then $(XA)\mathbf{u} = X(A\mathbf{u}) = \mathbf{0}$ if and only if $A\mathbf{u} = \mathbf{0}$. Thus, $\operatorname{null} XA = \operatorname{null} A$ and the dimension theorem tells us that $\operatorname{rank} XA = \operatorname{rank} A$ since XA and A have the same number of columns. To analyze the product $AY \in M_{m \times q}(\mathbb{F})$, consider its transpose, use (3.2.5) and apply Theorem 3.2.1:

$$\operatorname{rank} AY = \operatorname{rank}(AY)^{\mathsf{T}} = \operatorname{rank} Y^{\mathsf{T}}A^{\mathsf{T}} = \operatorname{rank} A^{\mathsf{T}} = \operatorname{rank} A. \qquad \square$$

Example 3.2.11 If the columns of $A = [\mathbf{a}_1 \ \mathbf{a}_2 \ \ldots \ \mathbf{a}_n] \in M_{m \times n}$ are linearly independent and $X \in M_m$ is invertible, then $\operatorname{rank} XA = \operatorname{rank} A = n$. This means that the columns of $XA = [X\mathbf{a}_1 \ X\mathbf{a}_2 \ \ldots \ X\mathbf{a}_n]$ are linearly independent.

The following theorem provides an upper bound on the rank of a product and a lower bound on the rank of an augmented matrix.

Theorem 3.2.12 *Let $A \in M_{m \times k}$, $B \in M_{k \times n}$, and $C \in M_{m \times p}$. Then*

$$\operatorname{rank} AB \leq \min\{\operatorname{rank} A, \operatorname{rank} B\}, \tag{3.2.13}$$

and $\operatorname{rank} A = \operatorname{rank} AB$ if and only if $\operatorname{col} A = \operatorname{col} AB$. Also,

$$\max\{\operatorname{rank} A, \operatorname{rank} C\} \leq \operatorname{rank}[A \ C], \tag{3.2.14}$$

and $\operatorname{rank} A = \operatorname{rank}[A \ C]$ if and only if $\operatorname{col} A = \operatorname{col} A + \operatorname{col} C$.

Proof Since $\operatorname{col} AB \subseteq \operatorname{col} A$, Theorem 2.2.9 ensures that

$$\operatorname{rank} AB = \dim \operatorname{col} AB \leq \dim \operatorname{col} A = \operatorname{rank} A,$$

with equality if and only if $\operatorname{col} A = \operatorname{col} AB$. Apply this inequality to the transpose of AB and obtain

$$\operatorname{rank} AB = \operatorname{rank}(AB)^{\mathsf{T}} = \operatorname{rank} B^{\mathsf{T}}A^{\mathsf{T}} \leq \operatorname{rank} B^{\mathsf{T}} = \operatorname{rank} B.$$

Finally,

$$\operatorname{col} A \subseteq \operatorname{col} A + \operatorname{col} C = \operatorname{col}[A \ C],$$

so Theorem 2.2.9 ensures that

$$\operatorname{rank} A = \dim \operatorname{col} A \leq \dim \operatorname{col}[A \ C] = \operatorname{rank}[A \ C],$$

with equality if and only if $\operatorname{col} A = \operatorname{col} A + \operatorname{col} C$. The same argument shows that $\operatorname{rank} C \leq \operatorname{rank}[A \ C]$. $\qquad \square$

The factorization that we introduced in the proof of Theorem 3.2.1 has many applications, and it is useful to have a clear statement of its properties.

Theorem 3.2.15 *Let $A \in M_{m \times n}(\mathbb{F})$, suppose that* $\operatorname{rank} A = r \geq 1$, *and let the columns of* $X \in M_{m \times r}(\mathbb{F})$ *be a basis for* $\operatorname{col} A$. *Then:*

(a) *There is a unique $Y \in M_{r \times n}(\mathbb{F})$ such that $A = XY$.*

(b) $\operatorname{rank} Y = \operatorname{rank} X = r$.

Proof Partition $A = [\mathbf{a}_1 \ \mathbf{a}_2 \ \ldots \ \mathbf{a}_n]$ according to its columns. Since the r columns of X are a basis for $\operatorname{col} A$, we have $\operatorname{rank} X = r$. For each $i = 1, 2, \ldots, n$, there is a unique $\mathbf{y}_i \in \mathbb{F}^r$ such that $\mathbf{a}_i = X\mathbf{y}_i$. If $Y = [\mathbf{y}_1 \ \mathbf{y}_2 \ \ldots \ \mathbf{y}_n] \in M_{r \times n}(\mathbb{F})$, then

$$A = [\mathbf{a}_1 \ \mathbf{a}_2 \ \ldots \ \mathbf{a}_n] = [X\mathbf{y}_1 \ X\mathbf{y}_2 \ \ldots \ X\mathbf{y}_n] = X[\mathbf{y}_1 \ \mathbf{y}_2 \ \ldots \ \mathbf{y}_n] = XY$$

and $\operatorname{rank} Y \leq r$. The preceding theorem ensures that

$$r = \operatorname{rank}(XY) \leq \min\{\operatorname{rank} X, \operatorname{rank} Y\} = \min\{r, \operatorname{rank} Y\} = \operatorname{rank} Y \leq r,$$

and hence $\operatorname{rank} Y = r$. $\qquad\square$

Definition 3.2.16 The factorization $A = XY$ described in the preceding theorem is a *full-rank factorization*.

We can use Theorem 3.2.12 to establish some facts about the ranks of powers of a matrix.

Theorem 3.2.17 *Let $A \in M_n$. Then:*

(a) $\operatorname{rank} A^k \geq \operatorname{rank} A^{k+1}$ *for each $k = 1, 2, \ldots$.*

(b) *If $k \in \{0, 1, 2, \ldots\}$ and $\operatorname{rank} A^k = \operatorname{rank} A^{k+1}$, then $\operatorname{col} A^k = \operatorname{col} A^{k+p}$ and $\operatorname{rank} A^k = \operatorname{rank} A^{k+p}$ for every $p \in \{1, 2, \ldots\}$.*

(c) *If A is not invertible, then $\operatorname{rank} A \leq n - 1$ and there is a least positive integer $q \in \{1, 2, \ldots, n\}$ such that $\operatorname{rank} A^q = \operatorname{rank} A^{q+1}$.*

Proof (a) Theorem 3.2.12 ensures that $\operatorname{rank} A^{k+1} \leq \min\{\operatorname{rank} A, \operatorname{rank} A^k\} \leq \operatorname{rank} A^k$.

(b) Suppose that $\operatorname{rank} A^k = \operatorname{rank} A^{k+1}$. Since $A^{k+1} = A^k A$, Theorem 3.2.12 ensures that

$$\operatorname{col} A^k = \operatorname{col} A^{k+1}. \tag{3.2.18}$$

We claim that $\operatorname{col} A^k = \operatorname{col} A^{k+p}$ (and hence $\operatorname{rank} A^k = \operatorname{rank} A^{k+p}$) for every $p \in \{1, 2, \ldots\}$. To prove this claim, proceed by induction on p. The base case $p = 1$ is (3.2.18). Suppose that $p \geq 1$ and

$$\operatorname{col} A^k = \operatorname{col} A^{k+p}. \tag{3.2.19}$$

Let $\mathbf{x} \in \mathbb{C}^n$ and let $\mathbf{y}, \mathbf{z} \in \mathbb{C}^n$ be such that $A^k \mathbf{x} = A^{k+p}\mathbf{y}$ (the induction hypothesis (3.2.19)) and $A^k \mathbf{y} = A^{k+1}\mathbf{z}$ (the base case (3.2.18)). Then

$$A^k \mathbf{x} = A^{k+p}\mathbf{y} = A^p(A^k\mathbf{y}) = A^p(A^{k+1}\mathbf{z}) = A^{k+p+1}\mathbf{z},$$

which shows that $\operatorname{col} A^k \subseteq \operatorname{col} A^{k+p+1}$. Since $\operatorname{col} A^{k+p+1} \subseteq \operatorname{col} A^k$, we conclude that $\operatorname{col} A^k = \operatorname{col} A^{k+p+1}$. This concludes the induction.

(c) For each $k = 1, 2, \ldots, n$, either $\operatorname{rank} A^k = \operatorname{rank} A^{k+1}$ or $\operatorname{rank} A^k - \operatorname{rank} A^{k+1} \geq 1$. Thus, the strict inequalities in

$$\underbrace{\operatorname{rank} A}_{\leq n-1} > \underbrace{\operatorname{rank} A^2}_{\leq n-2} > \cdots > \underbrace{\operatorname{rank} A^n}_{\leq 0} > \operatorname{rank} A^{n+1}$$

cannot all be correct. We conclude that $\operatorname{rank} A^k = \operatorname{rank} A^{k+1}$ for some $k \in \{1, 2, \ldots, n\}$. Let q be the least value of k for which this occurs. $\qquad \square$

Definition 3.2.20 Let $A \in \mathsf{M}_n$. If A is invertible, define its *index* to be 0. If A is not invertible, its *index* is the least positive integer q such that $\operatorname{rank} A^q = \operatorname{rank} A^{q+1}$.

The preceding theorem ensures that the index of each $A \in \mathsf{M}_n$ is at most n. For an example in which the index equals n, see P.3.22. Our definition of the index of an invertible matrix is consistent with the convention that $A^0 = I$ and $\operatorname{rank} A^0 = n$.

3.3 Block Partitions and Direct Sums

Consider the 2×2 block partitions

$$A = \begin{bmatrix} A_{11} & A_{12} \\ A_{21} & A_{22} \end{bmatrix} \in \mathsf{M}_{m \times n}(\mathbb{F}) \quad \text{and} \quad B = \begin{bmatrix} B_{11} & B_{12} \\ B_{21} & B_{22} \end{bmatrix} \in \mathsf{M}_{p \times q}(\mathbb{F}), \tag{3.3.1}$$

in which each A_{ij} is $m_i \times n_j$, each B_{ij} is $p_i \times q_j$,

$$m_1 + m_2 = m, \quad n_1 + n_2 = n, \quad p_1 + p_2 = p, \quad \text{and} \quad q_1 + q_2 = q.$$

To form $A + B$ we must have $m = p$ and $n = q$. To use block matrix operations to compute

$$A + B = \begin{bmatrix} A_{11} + B_{11} & A_{12} + B_{12} \\ A_{21} + B_{21} & A_{22} + B_{22} \end{bmatrix}$$

we must have $m_i = p_i$ and $n_i = q_i$ for each i. If these conditions are satisfied, then the partitions (3.3.1) are *conformal for addition*.

To form AB we must have $n = p$, in which case AB is $m \times q$. To use block matrix operations to compute

$$AB = \begin{bmatrix} A_{11} & A_{12} \\ A_{21} & A_{22} \end{bmatrix} \begin{bmatrix} B_{11} & B_{12} \\ B_{21} & B_{22} \end{bmatrix} = \begin{bmatrix} A_{11}B_{11} + A_{12}B_{21} & A_{11}B_{12} + A_{12}B_{22} \\ A_{21}B_{11} + A_{22}B_{21} & A_{21}B_{12} + A_{22}B_{22} \end{bmatrix}$$

we must have $n_i = p_i$ for each i. If these conditions are satisfied, then the partitions (3.3.1) are *conformal for multiplication* (or just *conformal*).

Partitioned matrices (3.3.1) are examples of *block matrices*; the submatrices A_{ij} are often called *blocks*. For notational simplicity we have described 2×2 block matrices, but all those ideas carry over to other conformally structured block matrices. For example,

$$[A_1 \ A_2] \begin{bmatrix} B_1 \\ B_2 \end{bmatrix} = A_1 B_1 + A_2 B_2 \tag{3.3.2}$$

if the respective partitions are conformal for multiplication. The following theorems about rank exploit the identity (3.3.2).

Theorem 3.3.3 *Let $A \in M_{m \times k}$ and $B \in M_{k \times n}$. Then*

$$\operatorname{rank} A + \operatorname{rank} B - k \leq \operatorname{rank} AB. \tag{3.3.4}$$

Proof First suppose that $AB = 0$. If $A = 0$, then (3.3.4) is equivalent to the (correct) inequality $\operatorname{rank} B \leq k$. If $\operatorname{rank} A \geq 1$, then every column of B is in $\operatorname{null} A$, so the dimension theorem tells us that

$$\operatorname{rank} B = \dim \operatorname{col} B \leq \dim \operatorname{null} A = k - \operatorname{rank} A.$$

Therefore, $\operatorname{rank} A + \operatorname{rank} B \leq k$ and the assertion is proved in the case $AB = 0$.

Now suppose that $\operatorname{rank} AB = r \geq 1$ and let $AB = XY$ be a full-rank factorization. Let

$$C = [A \ X] \in M_{m \times (k+r)} \quad \text{and} \quad D = \begin{bmatrix} B \\ -Y \end{bmatrix} \in M_{(k+r) \times n}.$$

Then $CD = AB - XY = 0$, so (3.2.14) and the preceding case ensure that

$$\operatorname{rank} A + \operatorname{rank} B \leq \operatorname{rank} C + \operatorname{rank} D \leq k + r = k + \operatorname{rank} AB. \qquad \square$$

The rank of a sum can also be bounded from above and below.

Theorem 3.3.5 *Let $A, B \in M_{m \times n}$. Then*

$$|\operatorname{rank} A - \operatorname{rank} B| \leq \operatorname{rank}(A + B) \leq \operatorname{rank} A + \operatorname{rank} B. \tag{3.3.6}$$

Proof The asserted inequalities are valid if either $A = 0$ or $B = 0$, so we may assume that $\operatorname{rank} A = r \geq 1$ and $\operatorname{rank} B = s \geq 1$. Let $A = X_1 Y_1$ and $B = X_2 Y_2$ be full-rank factorizations, and let

$$C = [X_1 \ X_2] \in M_{m \times (r+s)} \quad \text{and} \quad D = \begin{bmatrix} Y_1 \\ Y_2 \end{bmatrix} \in M_{(r+s) \times n}.$$

Then $A + B = X_1 Y_1 + X_2 Y_2 = CD$, so (3.2.13) ensures that

$$\operatorname{rank}(A + B) = \operatorname{rank} CD \leq \min\{\operatorname{rank} C, \operatorname{rank} D\} \leq \min\{r + s, r + s\} = r + s.$$

The inequality (3.3.4) says that

$$\operatorname{rank}(A + B) = \operatorname{rank} CD \geq \operatorname{rank} C + \operatorname{rank} D - (r + s).$$

Since $\operatorname{rank} C \geq \max\{r, s\}$ and $\operatorname{rank} D \geq \max\{r, s\}$ (see (3.2.14)), the pair of inequalities

$$\operatorname{rank} C + \operatorname{rank} D - (r + s) \geq r + r - (r + s) = r - s$$

and

$$\operatorname{rank} C + \operatorname{rank} D - (r + s) \geq s + s - (r + s) = s - r$$

imply the left-hand inequality in (3.3.6). $\qquad \square$

Example 3.3.7 Let $X \in M_{m \times n}$ and consider the block matrix

$$\begin{bmatrix} I_m & X \\ 0 & I_n \end{bmatrix} \in M_{m+n}.$$

Then

$$\begin{bmatrix} I & -X \\ 0 & I \end{bmatrix}\begin{bmatrix} I & X \\ 0 & I \end{bmatrix} = \begin{bmatrix} I & X - X \\ 0 & I \end{bmatrix} = \begin{bmatrix} I & 0 \\ 0 & I \end{bmatrix} = I,$$

and hence

$$\begin{bmatrix} I & X \\ 0 & I \end{bmatrix}^{-1} = \begin{bmatrix} I & -X \\ 0 & I \end{bmatrix}. \tag{3.3.8}$$

The following example presents a generalization of (3.3.8).

Example 3.3.9 Consider a 2×2 block matrix

$$\begin{bmatrix} Y & X \\ 0 & Z \end{bmatrix}, \tag{3.3.10}$$

in which $Y \in M_n$ and $Z \in M_m$ are invertible. Then

$$\begin{bmatrix} Y & X \\ 0 & Z \end{bmatrix}^{-1} = \begin{bmatrix} Y^{-1} & -Y^{-1}XZ^{-1} \\ 0 & Z^{-1} \end{bmatrix}, \tag{3.3.11}$$

which can be verified by block matrix multiplication:

$$\begin{bmatrix} Y & X \\ 0 & Z \end{bmatrix}\begin{bmatrix} Y^{-1} & -Y^{-1}XZ^{-1} \\ 0 & Z^{-1} \end{bmatrix} = \begin{bmatrix} YY^{-1} & Y(-Y^{-1}XZ^{-1}) + XZ^{-1} \\ 0 & ZZ^{-1} \end{bmatrix} = I.$$

The identity (3.3.11) implies a useful fact about upper triangular matrices.

Theorem 3.3.12 *Let $n \geq 2$ and suppose that $A = [a_{ij}] \in M_n$ is upper triangular and has nonzero diagonal entries. Then A is invertible, its inverse is upper triangular, and the diagonal entries of A^{-1} are $a_{11}^{-1}, a_{22}^{-1}, \ldots, a_{nn}^{-1}$, in that order.*

Proof We proceed by induction on n. In the base case $n = 2$, the identity (3.3.11) ensures that

$$\begin{bmatrix} a_{11} & a_{12} \\ 0 & a_{22} \end{bmatrix}^{-1} = \begin{bmatrix} a_{11}^{-1} & \star \\ 0 & a_{22}^{-1} \end{bmatrix},$$

which is upper triangular and has the asserted diagonal entries. The symbol $\star$ indicates an entry or block whose value is not relevant to the argument. For the induction step, let $n \geq 3$ and suppose that every upper triangular matrix of size less than n with nonzero diagonal entries has an inverse that is upper triangular and has the asserted diagonal entries. Let $A \in M_n$ and partition it as

$$A = \begin{bmatrix} B & \star \\ 0 & a_{nn} \end{bmatrix},$$

in which $B \in M_{n-1}$. It follows from (3.3.11) that

$$A^{-1} = \begin{bmatrix} B^{-1} & \star \\ 0 & a_{nn}^{-1} \end{bmatrix}.$$

The induction hypothesis ensures that B^{-1} is upper triangular and has the asserted diagonal entries. Thus, A^{-1} is upper triangular and has the asserted diagonal entries. □

The identity (3.3.8) leads to an enormously useful similarity of a 2×2 block upper triangular matrix.

Theorem 3.3.13 *Let* $B \in M_m$, *let* $C, X \in M_{m \times n}$, *and let* $D \in M_n$. *Then*

$$\begin{bmatrix} B & C \\ 0 & D \end{bmatrix} \quad \text{is similar to} \quad \begin{bmatrix} B & C + XD - BX \\ 0 & D \end{bmatrix}. \tag{3.3.14}$$

Proof Use (3.3.8) and compute the similarity

$$\begin{bmatrix} I & X \\ 0 & I \end{bmatrix} \begin{bmatrix} B & C \\ 0 & D \end{bmatrix} \begin{bmatrix} I & -X \\ 0 & I \end{bmatrix} = \begin{bmatrix} B & C + XD - BX \\ 0 & D \end{bmatrix}. \quad \square$$

Example 3.3.15 If $A \in M_n(\mathbb{F})$, $\mathbf{x}, \mathbf{y} \in \mathbb{F}^n$, and $c \in \mathbb{F}$, then

$$\begin{bmatrix} c & \mathbf{x}^\mathsf{T} \\ \mathbf{y} & A \end{bmatrix}, \quad \begin{bmatrix} \mathbf{x}^\mathsf{T} & c \\ A & \mathbf{y} \end{bmatrix}, \quad \begin{bmatrix} A & \mathbf{x} \\ \mathbf{y}^\mathsf{T} & c \end{bmatrix}, \quad \text{and} \quad \begin{bmatrix} \mathbf{x} & A \\ c & \mathbf{y}^\mathsf{T} \end{bmatrix} \tag{3.3.16}$$

are *bordered matrices*. They are obtained from A by *bordering*.

Example 3.3.17 The transpose and conjugate transpose (adjoint) operate on block matrices as follows:

$$\begin{bmatrix} A_{11} & A_{12} \\ A_{21} & A_{22} \end{bmatrix}^\mathsf{T} = \begin{bmatrix} A_{11}^\mathsf{T} & A_{21}^\mathsf{T} \\ A_{12}^\mathsf{T} & A_{22}^\mathsf{T} \end{bmatrix} \quad \text{and} \quad \begin{bmatrix} A_{11} & A_{12} \\ A_{21} & A_{22} \end{bmatrix}^* = \begin{bmatrix} A_{11}^* & A_{21}^* \\ A_{12}^* & A_{22}^* \end{bmatrix}.$$

If $A = [A_{ij}]$ is an $m \times n$ block matrix, then

$$\begin{bmatrix} A_{11} & A_{12} & \cdots & A_{1n} \\ A_{21} & A_{22} & \cdots & A_{2n} \\ \vdots & \vdots & \ddots & \vdots \\ A_{m1} & A_{m2} & \cdots & A_{mn} \end{bmatrix}^\mathsf{T} = \begin{bmatrix} A_{11}^\mathsf{T} & A_{21}^\mathsf{T} & \cdots & A_{m1}^\mathsf{T} \\ A_{12}^\mathsf{T} & A_{22}^\mathsf{T} & \cdots & A_{m2}^\mathsf{T} \\ \vdots & \vdots & \ddots & \vdots \\ A_{1n}^\mathsf{T} & A_{2n}^\mathsf{T} & \cdots & A_{mn}^\mathsf{T} \end{bmatrix}$$

and

$$\begin{bmatrix} A_{11} & A_{12} & \cdots & A_{1n} \\ A_{21} & A_{22} & \cdots & A_{2n} \\ \vdots & \vdots & \ddots & \vdots \\ A_{m1} & A_{m2} & \cdots & A_{mn} \end{bmatrix}^* = \begin{bmatrix} A_{11}^* & A_{21}^* & \cdots & A_{m1}^* \\ A_{12}^* & A_{22}^* & \cdots & A_{m2}^* \\ \vdots & \vdots & \ddots & \vdots \\ A_{1n}^* & A_{2n}^* & \cdots & A_{mn}^* \end{bmatrix}.$$

A *direct sum* of square matrices is a block matrix that is *block diagonal*, that is, every off-diagonal block is zero:

$$A \oplus B = \begin{bmatrix} A & 0 \\ 0 & B \end{bmatrix}, \qquad A_{11} \oplus A_{22} \oplus \cdots \oplus A_{kk} = \begin{bmatrix} A_{11} & & \\ & \ddots & \\ & & A_{kk} \end{bmatrix}. \tag{3.3.18}$$

The matrices $A_{11}, A_{22}, \ldots, A_{kk}$ in (3.3.18) are *direct summands*. Our convention is that if a direct summand has size 0, then it is omitted from the direct sum. For any scalars $\lambda_1, \lambda_2, \ldots, \lambda_n$, the direct sum

$$[\lambda_1] \oplus [\lambda_2] \oplus \cdots \oplus [\lambda_n]$$

of 1×1 matrices is often written as

$$\mathrm{diag}(\lambda_1, \lambda_2, \ldots, \lambda_n) = \begin{bmatrix} \lambda_1 & & \\ & \ddots & \\ & & \lambda_n \end{bmatrix} \in \mathsf{M}_n.$$

Linear combinations and products of conformally partitioned direct sums all act blockwise; so do powers and polynomials. Let A, B, C, D be matrices of appropriate sizes, let a, b be scalars, k be an integer, and let p be a polynomial.

(a) $a(A \oplus B) + b(C \oplus D) = (aA + bC) \oplus (aB + bD)$.

(b) $(A \oplus B)(C \oplus D) = AC \oplus BD$.

(c) $(A \oplus B)^k = A^k \oplus B^k$.

(d) $p(A \oplus B) = p(A) \oplus p(B)$.

The following fact about left and right multiplication by diagonal matrices comes up frequently in matrix manipulations. Let $A \in \mathsf{M}_{m \times n}$ be presented according to its entries, columns, and rows as

$$A = [a_{ij}] = [\mathbf{c}_1 \ \mathbf{c}_2 \ \ldots \ \mathbf{c}_n] = \begin{bmatrix} \mathbf{r}_1^\mathsf{T} \\ \vdots \\ \mathbf{r}_m^\mathsf{T} \end{bmatrix} \in \mathsf{M}_{m \times n}.$$

Let $\Lambda = \mathrm{diag}(\lambda_1, \lambda_2, \ldots, \lambda_n)$ and $M = \mathrm{diag}(\mu_1, \mu_2, \ldots, \mu_m)$. Then

$$A\Lambda = [\lambda_j a_{ij}] = [\lambda_1 \mathbf{c}_1 \ \lambda_2 \mathbf{c}_2 \ \ldots \ \lambda_n \mathbf{c}_n] \tag{3.3.19}$$

and

$$MA = [\mu_i a_{ij}] = \begin{bmatrix} \mu_1 \mathbf{r}_1^\mathsf{T} \\ \vdots \\ \mu_m \mathbf{r}_m^\mathsf{T} \end{bmatrix}. \tag{3.3.20}$$

Thus, *right* multiplication of A by a diagonal matrix Λ multiplies the *columns* of A by the corresponding diagonal entries of Λ. *Left* multiplication of A by a diagonal matrix M multiplies the *rows* of A by the corresponding diagonal entries of M.

We say that k scalars $\lambda_1, \lambda_2, \ldots, \lambda_k$ are *distinct* if $\lambda_i \neq \lambda_j$ whenever $i \neq j$. The following lemma about commuting block matrices is at the heart of several important results.

Lemma 3.3.21 *Let $\lambda_1, \lambda_2, \ldots, \lambda_k$ be scalars and let $\Lambda = \lambda_1 I_{n_1} \oplus \lambda_2 I_{n_2} \oplus \cdots \oplus \lambda_k I_{n_k} \in \mathsf{M}_n$. Partition the rows and columns of the $k \times k$ block matrix $A = [A_{ij}] \in \mathsf{M}_n$ conformally with Λ, so each $A_{ij} \in \mathsf{M}_{n_i \times n_j}$.*

(a) *If A is block diagonal, then $A\Lambda = \Lambda A$.*

(b) *If $A\Lambda = \Lambda A$ and $\lambda_1, \lambda_2, \ldots, \lambda_k$ are distinct, then A is block diagonal.*

Proof Use (3.3.19) and (3.3.20) to compute

$$A\Lambda = \begin{bmatrix} A_{11} & \cdots & A_{1k} \\ \vdots & \ddots & \vdots \\ A_{k1} & \cdots & A_{kk} \end{bmatrix} \begin{bmatrix} \lambda_1 I_{n_1} & & \\ & \ddots & \\ & & \lambda_k I_{n_k} \end{bmatrix} = \begin{bmatrix} \lambda_1 A_{11} & \cdots & \lambda_k A_{1k} \\ \vdots & \ddots & \vdots \\ \lambda_1 A_{k1} & \cdots & \lambda_k A_{kk} \end{bmatrix}$$

and

$$\Lambda A = \begin{bmatrix} \lambda_1 I_{n_1} & & \\ & \ddots & \\ & & \lambda_k I_{n_k} \end{bmatrix} \begin{bmatrix} A_{11} & \cdots & A_{1k} \\ \vdots & \ddots & \vdots \\ A_{k1} & \cdots & A_{kk} \end{bmatrix} = \begin{bmatrix} \lambda_1 A_{11} & \cdots & \lambda_1 A_{1k} \\ \vdots & \ddots & \vdots \\ \lambda_k A_{k1} & \cdots & \lambda_k A_{kk} \end{bmatrix}.$$

If A is block diagonal, then $A_{ij} = 0$ for $i \neq j$ and the preceding computation confirms that $A\Lambda = \Lambda A$. If $A\Lambda = \Lambda A$, then $\lambda_j A_{ij} = \lambda_i A_{ij}$ and hence $(\lambda_i - \lambda_j)A_{ij} = 0$ for $1 \leq i,j \leq k$. If $\lambda_i - \lambda_j \neq 0$, then $A_{ij} = 0$. Thus, every off-diagonal block is a zero matrix if $\lambda_1, \lambda_2, \ldots, \lambda_k$ are distinct. □

Example 3.3.22 The hypothesis that $\lambda_1, \lambda_2, \ldots, \lambda_k$ are distinct in Lemma 3.3.21.b is critical. Indeed, if $\lambda_1 = \lambda_2 = \cdots = \lambda_k = 1$, then $\Lambda = I$ commutes with every $A \in \mathsf{M}_k$. In this case, $A\Lambda = \Lambda A$ does not imply that A is a block diagonal matrix partitioned conformally with Λ.

A fundamental principle in linear algebra is that a linearly independent list of vectors in a finite-dimensional vector space can be extended to a basis; see Theorem 2.2.7.b. The following block matrix version of this principle is a workhorse of matrix theory.

Theorem 3.3.23 *If $X \in \mathsf{M}_{m \times n}(\mathbb{F})$ and rank $X = n < m$, then there is an $X' \in \mathsf{M}_{m \times (m-n)}(\mathbb{F})$ such that $A = [X \ X'] \in \mathsf{M}_m(\mathbb{F})$ is invertible.*

Proof Partition $X = [\mathbf{x}_1 \ \mathbf{x}_2 \ \ldots \ \mathbf{x}_n]$ according to its columns. Since rank $X = n$, the vectors $\mathbf{x}_1, \mathbf{x}_2, \ldots, \mathbf{x}_n \in \mathbb{F}^m$ are linearly independent. Theorem 2.2.7.b says that there are vectors $\mathbf{x}_{n+1}, \mathbf{x}_{n+2}, \ldots, \mathbf{x}_m \in \mathbb{F}^m$ such that the list $\mathbf{x}_1, \mathbf{x}_2, \ldots, \mathbf{x}_n, \mathbf{x}_{n+1}, \mathbf{x}_{n+2}, \ldots, \mathbf{x}_m$ is a basis of $\mathbb{F}^m$. Corollary 2.4.11 ensures that

$$A = [\mathbf{x}_1 \ \mathbf{x}_2 \ \ldots \ \mathbf{x}_n \ \mathbf{x}_{n+1} \ \mathbf{x}_{n+2} \ \ldots \ \mathbf{x}_m] \in \mathsf{M}_n(\mathbb{F})$$

is invertible. Let $X' = [\mathbf{x}_{n+1} \ \mathbf{x}_{n+2} \ \ldots \ \mathbf{x}_m]$ so that $A = [X \ X']$. □

3.4 Determinants of Block Matrices

The determinant of a diagonal or triangular matrix is the product of its main diagonal entries. This observation can be generalized to block diagonal or block triangular matrices. The key idea is a determinant identity for the direct sum of a square matrix and an identity matrix.

Lemma 3.4.1 *Let $A \in M_m$ and let $n \in \{0, 1, 2, \ldots\}$. Then*

$$\det \begin{bmatrix} I_n & 0 \\ 0 & A \end{bmatrix} = \det A = \det \begin{bmatrix} A & 0 \\ 0 & I_n \end{bmatrix}.$$

Proof Let

$$B_n = \begin{bmatrix} I_n & 0 \\ 0 & A \end{bmatrix} \quad \text{and} \quad C_n = \begin{bmatrix} A & 0 \\ 0 & I_n \end{bmatrix},$$

let S_n be the statement that $\det B_n = \det A = \det C_n$, and proceed by induction. In the base case, $n = 0$, there is nothing to prove since $B_0 = A = C_0$. For the inductive step, we assume that S_n is true and observe that

$$\det B_{n+1} = \det \begin{bmatrix} I_{n+1} & 0 \\ 0 & A \end{bmatrix} = \det \begin{bmatrix} 1 & 0 & 0 \\ 0 & I_n & 0 \\ 0 & 0 & A \end{bmatrix} = \det \begin{bmatrix} 1 & 0 \\ 0 & B_n \end{bmatrix} \qquad (3.4.2)$$

and

$$\det C_{n+1} = \det \begin{bmatrix} A & 0 \\ 0 & I_{n+1} \end{bmatrix} = \det \begin{bmatrix} A & 0 & 0 \\ 0 & I_n & 0 \\ 0 & 0 & 1 \end{bmatrix} = \det \begin{bmatrix} C_n & 0 \\ 0 & 1 \end{bmatrix}. \qquad (3.4.3)$$

In a Laplace expansion by minors along the first row (see Section 0.5) of the final determinant in (3.4.2), only one summand has a nonzero coefficient:

$$\det \begin{bmatrix} 1 & 0 \\ 0 & B_n \end{bmatrix} = 1 \cdot (-1)^{1+1} \det B_n = \det B_n.$$

In a Laplace expansion by minors along the last row of the final determinant in (3.4.3), only one summand has a nonzero coefficient:

$$\det \begin{bmatrix} C_n & 0 \\ 0 & 1 \end{bmatrix} = 1 \cdot (-1)^{(n+1)+(n+1)} \det C_n = \det C_n.$$

The induction hypothesis ensures that $\det B_n = \det A = \det C_n$, so we conclude that $\det B_{n+1} = \det A = \det C_{n+1}$. This completes the induction. $\qquad \square$

Theorem 3.4.4 *Let $A \in M_r$, $B \in M_{r \times (n-r)}$, and $D \in M_{n-r}$. Then*

$$\det \begin{bmatrix} A & B \\ 0 & D \end{bmatrix} = (\det A)(\det D).$$

Proof Write

$$\begin{bmatrix} A & B \\ 0 & D \end{bmatrix} = \begin{bmatrix} I_r & 0 \\ 0 & D \end{bmatrix} \begin{bmatrix} I_r & B \\ 0 & I_{n-r} \end{bmatrix} \begin{bmatrix} A & 0 \\ 0 & I_{n-r} \end{bmatrix}$$

and use the product rule for determinants:

$$\det \begin{bmatrix} A & B \\ 0 & D \end{bmatrix} = \left(\det \begin{bmatrix} I_r & 0 \\ 0 & D \end{bmatrix} \right) \left(\det \begin{bmatrix} I_r & B \\ 0 & I_{n-r} \end{bmatrix} \right) \left(\det \begin{bmatrix} A & 0 \\ 0 & I_{n-r} \end{bmatrix} \right).$$

The preceding lemma ensures that the first and last factors in this product are $\det D$ and $\det A$, respectively. The matrix in the middle factor is an upper triangular matrix with every main diagonal entry equal to 1, so its determinant is 1.   □

Block Gaussian elimination illustrates the utility of block matrix manipulations. Suppose that

$$M = \begin{bmatrix} A & B \\ C & D \end{bmatrix} \in \mathsf{M}_n, \quad A \in \mathsf{M}_r, \quad D \in \mathsf{M}_{n-r}, \tag{3.4.5}$$

and suppose further that A is invertible. The computation

$$\begin{bmatrix} I_r & 0 \\ -CA^{-1} & I_{n-r} \end{bmatrix} \begin{bmatrix} A & B \\ C & D \end{bmatrix} = \begin{bmatrix} A & B \\ 0 & D - CA^{-1}B \end{bmatrix} \tag{3.4.6}$$

transforms the block matrix M to block upper triangular form by multiplying it on the left with a matrix that has determinant 1. Apply Theorem 3.4.4 and the product rule for determinants to (3.4.6) to see that

$$\det \begin{bmatrix} A & B \\ C & D \end{bmatrix} = \det \begin{bmatrix} A & B \\ 0 & D - CA^{-1}B \end{bmatrix} = (\det A) \det(D - CA^{-1}B). \tag{3.4.7}$$

The expression

$$M/A = D - CA^{-1}B \tag{3.4.8}$$

is the *Schur complement of A in M* and the identity

$$\det \begin{bmatrix} A & B \\ C & D \end{bmatrix} = (\det A) \det(M/A) \tag{3.4.9}$$

is the *Schur determinant formula*. It permits us to evaluate the determinant of a large block matrix by computing the determinants of two smaller matrices, provided that a certain submatrix is invertible. The Schur complement of D in M can be defined in a similar manner; see P.3.7.

Example 3.4.10 Let $A \in \mathsf{M}_n$, let $\mathbf{x}, \mathbf{y} \in \mathbb{C}^n$, and let c be a nonzero scalar. The *reduction formula*

$$\det \begin{bmatrix} c & \mathbf{x}^\mathsf{T} \\ \mathbf{y} & A \end{bmatrix} = c \det \left(A - \frac{1}{c} \mathbf{y} \mathbf{x}^\mathsf{T} \right) \tag{3.4.11}$$

for the determinant of a bordered matrix is a special case of (3.4.7).

Example 3.4.12 Let $A \in \mathsf{M}_n$ be invertible, let $\mathbf{x}, \mathbf{y} \in \mathbb{C}^n$, and let c be a scalar. The *Cauchy expansion*

$$\det \begin{bmatrix} A & \mathbf{x} \\ \mathbf{y}^\mathsf{T} & c \end{bmatrix} = (c - \mathbf{y}^\mathsf{T} A^{-1} \mathbf{x}) \det A \tag{3.4.13}$$

$$= c \det A - \mathbf{y}^\mathsf{T} (\operatorname{adj} A) \mathbf{x} \tag{3.4.14}$$

for the determinant of a bordered matrix is also a special case of (3.4.7). The formulation (3.4.14) is valid even if A is not invertible; see (0.5.2).

We care about algorithms that transform general matrices into block triangular matrices because many computational problems are easier to solve after such a transformation. For example, if we need to solve a system of linear equations $A\mathbf{x} = \mathbf{b}$ and if A has the block triangular form

$$A = \begin{bmatrix} A_{11} & A_{12} \\ 0 & A_{22} \end{bmatrix},$$

partition the vectors $\mathbf{x} = [\mathbf{x}_1^\mathsf{T}\ \mathbf{x}_2^\mathsf{T}]^\mathsf{T}$ and $\mathbf{b} = [\mathbf{b}_1^\mathsf{T}\ \mathbf{b}_2^\mathsf{T}]^\mathsf{T}$ conformally with A and write the system as a pair of smaller systems:

$$A_{11}\mathbf{x}_1 = \mathbf{b}_1 - A_{12}\mathbf{x}_2$$
$$A_{22}\mathbf{x}_2 = \mathbf{b}_2.$$

First solve $A_{22}\mathbf{x}_2 = \mathbf{b}_2$ for $\mathbf{x}_2$, then solve $A_{11}\mathbf{x}_1 = \mathbf{b}_1 - A_{12}\mathbf{x}_2$.

3.5 Commutators and Shoda's Theorem

If T and S are linear operators on a vector space $\mathcal{V}$, their *commutator* is the linear operator $[T, S] = TS - ST$. If $A, B \in \mathsf{M}_n$, their *commutator* is the matrix $[A, B] = AB - BA$. The commutator of operators is zero if and only if the operators commute, so if we want to understand how noncommuting matrices or operators are related, we might begin by studying their commutator.

Example 3.5.1 In one-dimensional quantum mechanics, $\mathcal{V}$ is a vector space of suitably differentiable complex-valued functions of x (position) and t (time). The *position operator* T is defined by $(Tf)(x, t) = xf(x, t)$, and the *momentum operator* S is defined by $(Sf)(x, t) = -i\hbar\frac{\partial}{\partial x}f(x, t)$ ($\hbar$ is a physical constant). Compute

$$TSf = x\left(-i\hbar\frac{\partial f}{\partial x}\right) = -i\hbar x\frac{\partial f}{\partial x},$$

and

$$STf = -i\hbar\frac{\partial}{\partial x}(xf) = -i\hbar x\frac{\partial f}{\partial x} - i\hbar f.$$

We have $(TS - ST)f = i\hbar f$ for all $f \in \mathcal{V}$, that is,

$$TS - ST = i\hbar I.$$

This commutator identity is known to imply the *Heisenberg Uncertainty Principle*, which says that precise simultaneous measurement of position and momentum is impossible in a one-dimensional quantum mechanical system.

The commutator of the position and momentum operators is a nonzero scalar multiple of the identity, but this cannot happen for matrices. If $A, B \in M_n$, then (0.3.5) ensures that

$$\text{tr}(AB - BA) = \text{tr}\, AB - \text{tr}\, BA = \text{tr}\, AB - \text{tr}\, AB = 0. \tag{3.5.2}$$

However, $\text{tr}(cI_n) = nc \neq 0$ if $c \neq 0$. Consequently, a nonzero scalar matrix cannot be a commutator.

How can we decide whether a given matrix is a commutator? It must have trace zero, but is this necessary condition also sufficient? The following lemma is the first step toward proving that it is.

Lemma 3.5.3 *Let $A \in M_n(\mathbb{F})$ and suppose that $n \geq 2$. The list $\mathbf{x}, A\mathbf{x}$ is linearly dependent for all $\mathbf{x} \in \mathbb{F}^n$ if and only if A is a scalar matrix.*

Proof If $A = cI_n$ for some $c \in \mathbb{F}$, then the list $\mathbf{x}, A\mathbf{x}$ equals $\mathbf{x}, c\mathbf{x}$. It is linearly dependent for all $\mathbf{x} \in \mathbb{F}^n$. Conversely, suppose that the list $\mathbf{x}, A\mathbf{x}$ is linearly dependent for all $\mathbf{x} \in \mathbb{F}^n$. Since the list $\mathbf{e}_i, A\mathbf{e}_i$ is linearly dependent for each $i = 1, 2, \ldots, n$, there are scalars $a_1, a_2, \ldots, a_n$ such that each $A\mathbf{e}_i = a_i\mathbf{e}_i$. Consequently, A is a diagonal matrix. We must show that its diagonal entries are all equal. For each $i = 2, 3, \ldots, n$, we know that the list $\mathbf{e}_1 + \mathbf{e}_i, A(\mathbf{e}_1 + \mathbf{e}_i)$ is linearly dependent, so there are scalars b_i such that each $A(\mathbf{e}_1 + \mathbf{e}_i) = b_i(\mathbf{e}_1 + \mathbf{e}_i)$. Then

$$b_i\mathbf{e}_1 + b_i\mathbf{e}_i = A(\mathbf{e}_1 + \mathbf{e}_i) = A\mathbf{e}_1 + A\mathbf{e}_i = a_1\mathbf{e}_1 + a_i\mathbf{e}_i, \quad i = 2, 3, \ldots, n,$$

that is,

$$(b_i - a_1)\mathbf{e}_1 + (b_i - a_i)\mathbf{e}_i = \mathbf{0}, \quad i = 2, 3, \ldots, n.$$

The linear independence of $\mathbf{e}_1$ and $\mathbf{e}_i$ ensures that $b_i - a_1 = 0$ and $b_i - a_i = 0$ for $i = 1, 2, \ldots, n$. We conclude that $a_i = b_i = a_1$ for $i = 2, 3, \ldots, n$, so A is a scalar matrix. $\square$

The next step is to show that any nonscalar matrix is similar to a matrix with at least one zero diagonal entry.

Lemma 3.5.4 *Let $n \geq 2$ and let $A \in M_n(\mathbb{F})$. If A is not a scalar matrix, then it is similar over $\mathbb{F}$ to a matrix that has a zero entry in its $(1, 1)$ position.*

Proof The preceding lemma ensures that there is an $\mathbf{x} \in \mathbb{F}^n$ such that the list $\mathbf{x}, A\mathbf{x}$ is linearly independent. If $n = 2$, let $S = [\mathbf{x} \; A\mathbf{x}]$. If $n > 2$, invoke Theorem 3.3.23 and choose $S_2 \in M_{n\times(n-2)}$ such that $S = [\mathbf{x} \; A\mathbf{x} \; S_2]$ is invertible. Let $S^{-*} = [\mathbf{y} \; Y]$, in which $Y \in M_{n\times(n-1)}$. The $(1, 2)$ entry of $I_n = S^{-1}S$ is zero; this entry is $\mathbf{y}^*A\mathbf{x}$. Furthermore, the $(1, 1)$ entry of

$$S^{-1}AS = \begin{bmatrix} \mathbf{y}^* \\ Y^* \end{bmatrix} A \begin{bmatrix} \mathbf{x} & A\mathbf{x} & \mathbf{y}^*AS_2 \end{bmatrix} = \begin{bmatrix} 0 & \star \\ \star & A_1 \end{bmatrix} \tag{3.5.5}$$

is zero. $\square$

Because $\text{tr}\, A = \text{tr}\, S^{-1}AS = 0 + \text{tr}\, A_1$, the matrix A_1 in (3.5.5) has the same trace as A, so our construction suggests the induction in the following argument.

Theorem 3.5.6 *Let $A \in M_n(\mathbb{F})$. Then A is similar over $\mathbb{F}$ to a matrix, each of whose diagonal entries is equal to $\frac{1}{n} \operatorname{tr} A$.*

Proof If $n = 1$, then there is nothing to prove, so assume that $n \geq 2$. The matrix $B = A - (\frac{1}{n} \operatorname{tr} A)I_n$ has trace zero. If we can show that B is similar over $\mathbb{F}$ to a matrix C with zero diagonal entries, then (0.8.4) ensures that $A = B + (\frac{1}{n} \operatorname{tr} A)I_n$ is similar over $\mathbb{F}$ to $C + (\frac{1}{n} \operatorname{tr} A)I_n$, which is a matrix of the asserted form. The preceding lemma ensures that B is similar over $\mathbb{F}$ to a matrix of the form

$$\begin{bmatrix} 0 & \star \\ \star & B_1 \end{bmatrix}, \quad B_1 \in M_{n-1}(\mathbb{F}),$$

in which $\operatorname{tr} B_1 = 0$. We proceed by induction. Let P_k be the statement that B is similar over $\mathbb{F}$ to a matrix of the form

$$\begin{bmatrix} C_k & \star \\ \star & B_k \end{bmatrix}, \quad C_k \in M_k(\mathbb{F}), \quad B_k \in M_{n-k}(\mathbb{F}), \quad \operatorname{tr} B_k = 0,$$

in which C_k has zero diagonal entries. We have established the base case P_1. Suppose that $k < n - 1$ and P_k is true. If B_k is a scalar matrix, then $B_k = 0$ and the assertion is proved. If B_k is not a scalar matrix, then there is an invertible $S_k \in M_{n-k}(\mathbb{F})$ such that $S_k^{-1} B_k S_k$ has a zero entry in position $(1, 1)$. Then

$$\begin{bmatrix} I_k & 0 \\ 0 & S_k^{-1} \end{bmatrix} \begin{bmatrix} C_k & \star \\ \star & B_k \end{bmatrix} \begin{bmatrix} I_k & 0 \\ 0 & S_k \end{bmatrix} = \begin{bmatrix} C_k & \star \\ \star & S_k^{-1} B_k S_k \end{bmatrix} = \begin{bmatrix} C_{k+1} & \star \\ \star & B_{k+1} \end{bmatrix},$$

in which $C_{k+1} \in M_{k+1}(\mathbb{F})$ has zero diagonal entries, $B_{k+1} \in M_{n-k-1}(\mathbb{F})$, and $\operatorname{tr} B_{k+1} = 0$. This shows that P_k implies P_{k+1} if $k < n-1$, and we conclude that P_{n-1} is true. This completes the induction, so B is similar over $\mathbb{F}$ to a matrix of the form

$$\begin{bmatrix} C_{n-1} & \star \\ \star & b \end{bmatrix},$$

in which $C_{n-1} \in M_{n-1}(\mathbb{F})$ has zero diagonal entries. Thus,

$$0 = \operatorname{tr} B = \operatorname{tr} C_{n-1} + b = b,$$

which shows that B is similar to a matrix with zero diagonal. $\square$

We can now prove the following characterization of commutators.

Theorem 3.5.7 (Shoda) *Let $A \in M_n(\mathbb{F})$. Then A is a commutator of matrices in $M_n(\mathbb{F})$ if and only if $\operatorname{tr} A = 0$.*

Proof The necessity of the trace condition was established in (3.5.2), so we consider only its sufficiency. Suppose that $\operatorname{tr} A = 0$. The preceding corollary ensures that there is an invertible $S \in M_n(\mathbb{F})$ such that $S^{-1}AS = B = [b_{ij}]$ has zero diagonal entries. If $B = XY - YX$ for some $X, Y \in M_n(\mathbb{F})$, then A is the commutator of SXS^{-1} and SYS^{-1}. Let $X = \operatorname{diag}(1, 2, \ldots, n)$ and let $Y = [y_{ij}]$. Then

$$XY - YX = [iy_{ij}] - [jy_{ij}] = [(i - j)y_{ij}].$$

If we let

$$y_{ij} = \begin{cases} (i-j)^{-1}b_{ij} & \text{if } i \neq j, \\ 1 & \text{if } i = j, \end{cases}$$

then $XY - YX = B$.   □

3.6 Kronecker Products

Block matrices are central to a matrix "product" that finds applications in physics, signal processing, digital imaging, the solution of linear matrix equations, and many areas of pure mathematics.

Definition 3.6.1 Let $A = [a_{ij}] \in \mathsf{M}_{m \times n}$ and $B \in \mathsf{M}_{p \times q}$. The *Kronecker product* of A and B is the block matrix

$$A \otimes B = \begin{bmatrix} a_{11}B & a_{12}B & \cdots & a_{1n}B \\ a_{21}B & a_{22}B & \cdots & a_{2n}B \\ \vdots & \vdots & \ddots & \vdots \\ a_{m1}B & a_{m2}B & \cdots & a_{mn}B \end{bmatrix} \in \mathsf{M}_{mp \times nq}. \tag{3.6.2}$$

The Kronecker product is also called the *tensor product*. In the special case $n = q = 1$, the definition (3.6.2) says that the Kronecker product of $\mathbf{x} = [x_i] \in \mathbb{C}^m$ and $\mathbf{y} \in \mathbb{C}^p$ is

$$\mathbf{x} \otimes \mathbf{y} = \begin{bmatrix} x_1\mathbf{y} \\ x_2\mathbf{y} \\ \vdots \\ x_m\mathbf{y} \end{bmatrix} \in \mathbb{C}^{mp}.$$

Example 3.6.3 Using the matrices A and B in (3.1.5),

$$A \otimes B = \begin{bmatrix} B & 2B \\ 3B & 4B \end{bmatrix} = \left[\begin{array}{ccc|ccc} 4 & 5 & 2 & 8 & 10 & 4 \\ 6 & 7 & 1 & 12 & 14 & 2 \\ \hline 12 & 15 & 6 & 16 & 20 & 8 \\ 18 & 21 & 3 & 24 & 28 & 4 \end{array} \right] \in \mathsf{M}_{4 \times 6}$$

and

$$B \otimes A = \begin{bmatrix} 4A & 5A & 2A \\ 6A & 7A & A \end{bmatrix} = \left[\begin{array}{cc|cc|cc} 4 & 8 & 5 & 10 & 2 & 4 \\ 12 & 16 & 15 & 20 & 6 & 8 \\ \hline 6 & 12 & 7 & 14 & 1 & 2 \\ 18 & 24 & 21 & 28 & 3 & 4 \end{array} \right] \in \mathsf{M}_{4 \times 6}.$$

Although $A \otimes B \neq B \otimes A$, these two matrices have the same size and contain the same entries.

The Kronecker product satisfies many identities that one would like a product to obey:

$$c(A \otimes B) = (cA) \otimes B = A \otimes (cB) \tag{3.6.4}$$

$$(A + B) \otimes C = A \otimes C + B \otimes C \tag{3.6.5}$$

$$A \otimes (B + C) = A \otimes B + A \otimes C \tag{3.6.6}$$

$$(A \otimes B) \otimes C = A \otimes (B \otimes C) \tag{3.6.7}$$

$$(A \otimes B)^{\mathsf{T}} = A^{\mathsf{T}} \otimes B^{\mathsf{T}} \tag{3.6.8}$$

$$\overline{A \otimes B} = \overline{A} \otimes \overline{B} \tag{3.6.9}$$

$$(A \otimes B)^* = A^* \otimes B^* \tag{3.6.10}$$

$$I_m \otimes I_n = I_{mn}. \tag{3.6.11}$$

There is a relationship between the ordinary product and the Kronecker product.

Theorem 3.6.12 (Mixed Product Property) *Let $A \in \mathsf{M}_{m \times n}$, $B \in \mathsf{M}_{p \times q}$, $C \in \mathsf{M}_{n \times r}$, and $D \in \mathsf{M}_{q \times s}$. Then*

$$(A \otimes B)(C \otimes D) = (AC) \otimes (BD) \in \mathsf{M}_{mp \times rs}. \tag{3.6.13}$$

Proof If $A = [a_{ij}]$ and $C = [c_{ij}]$, then $A \otimes B = [a_{ij}B]$ and $C \otimes D = [c_{ij}D]$. The (i,j) block of $(A \otimes B)(C \otimes D)$ is

$$\sum_{k=1}^{n} (a_{ik}B)(c_{kj}D) = \left(\sum_{k=1}^{n} a_{ik}c_{kj} \right) BD = (AC)_{ij}BD,$$

in which $(AC)_{ij}$ denotes the (i,j) entry of AC. This identity shows that the (i,j) block of $(A \otimes B)(C \otimes D)$ is the (i,j) block of $(AC) \otimes (BD)$. $\qquad \square$

Corollary 3.6.14 *If $A \in \mathsf{M}_n$ and $B \in \mathsf{M}_m$ are invertible, then $A \otimes B$ is invertible and $(A \otimes B)^{-1} = A^{-1} \otimes B^{-1}$.*

Proof $(A \otimes B)(A^{-1} \otimes B^{-1}) = (AA^{-1}) \otimes (BB^{-1}) = I_n \otimes I_m = I_{nm}.$ $\qquad \square$

The following definition introduces a way to convert a matrix to a vector that is compatible with Kronecker product operations.

Definition 3.6.15 Let $X = [\mathbf{x}_1 \ \mathbf{x}_2 \ \dots \ \mathbf{x}_n] \in \mathsf{M}_{m \times n}$. The operator $\mathrm{vec} : \mathsf{M}_{m \times n} \to \mathbb{C}^{mn}$ is defined by

$$\mathrm{vec}\, X = \begin{bmatrix} \mathbf{x}_1 \\ \mathbf{x}_2 \\ \vdots \\ \mathbf{x}_n \end{bmatrix} \in \mathbb{C}^{mn},$$

that is, vec stacks the columns of X vertically.

Theorem 3.6.16 *Let $A \in \mathsf{M}_{m \times n}$, $X \in \mathsf{M}_{n \times p}$, and $B \in \mathsf{M}_{p \times q}$. Then $\mathrm{vec}\, AXB = (B^{\mathsf{T}} \otimes A)\,\mathrm{vec}\, X$.*

Proof Let $B = [b_{ij}] = [\mathbf{b}_1 \ \mathbf{b}_2 \ \dots \ \mathbf{b}_q]$ and $X = [\mathbf{x}_1 \ \mathbf{x}_2 \ \dots \ \mathbf{x}_p]$. The kth column of AXB is

$$AX\mathbf{b}_k = A \sum_{i=1}^{p} b_{ik}\mathbf{x}_i = [b_{1k}A \ b_{2k}A \ \dots \ b_{pk}A] \operatorname{vec} X = (\mathbf{b}_k^{\mathsf{T}} \otimes A) \operatorname{vec} X.$$

Stack these vectors vertically and obtain

$$\operatorname{vec} AXB = \begin{bmatrix} \mathbf{b}_1^{\mathsf{T}} \otimes A \\ \mathbf{b}_2^{\mathsf{T}} \otimes A \\ \vdots \\ \mathbf{b}_q^{\mathsf{T}} \otimes A \end{bmatrix} \operatorname{vec} X = (B^{\mathsf{T}} \otimes A) \operatorname{vec} X. \qquad \square$$

3.7 Problems

P.3.1 Verify that the outer product identity (3.1.19) is correct.

P.3.2 Let $A = [a_{ij}] \in \mathsf{M}_n$. Show that the entries of A are determined by the action of A on the standard basis via the identity $A = [\mathbf{e}_i^* A \mathbf{e}_j]$. *Hint*: $A = I^* A I$, in which $I = [\mathbf{e}_1 \ \mathbf{e}_2 \ \dots \ \mathbf{e}_n]$.

P.3.3 Let $X = [X_1 \ X_2] \in \mathsf{M}_{m \times n}$, in which $X_1 \in \mathsf{M}_{m \times n_1}$, $X_2 \in \mathsf{M}_{m \times n_2}$, and $n_1 + n_2 = n$. Compute $X^{\mathsf{T}} X$ and $X X^{\mathsf{T}}$.

P.3.4 Let $X \in \mathsf{M}_{m \times n}$. Show that the inverse of $\begin{bmatrix} I & 0 \\ X & I \end{bmatrix} \in \mathsf{M}_{m+n}$ is $\begin{bmatrix} I & 0 \\ -X & I \end{bmatrix}$.

P.3.5 (a) Partition

$$A = \begin{bmatrix} 2 & 2 & 3 \\ 2 & 9 & 7 \\ 4 & -3 & 8 \end{bmatrix} \qquad (3.7.1)$$

as a 2×2 block matrix $A = [A_{ij}] \in \mathsf{M}_3$ in which $A_{11} = [2]$ is 1×1. Verify that the reduced form (3.4.6) obtained by block Gaussian elimination is

$$\begin{bmatrix} 2 & 2 & 3 \\ 0 & 7 & 4 \\ 0 & -7 & 2 \end{bmatrix}. \qquad (3.7.2)$$

(b) Now perform ordinary Gaussian elimination on (3.7.1) to zero out all the entries in the first column below the $(1, 1)$ entry. Verify that you obtain the same reduced form (3.7.2). (c) Was this a coincidence, or do these two algorithms always lead to the same reduced form? Can you prove it?

P.3.6 Let $A = \begin{bmatrix} a & b \\ c & d \end{bmatrix} \in \mathsf{M}_2$ and suppose that $\dim \operatorname{row} A = 1$. If $[a \ b] = \lambda[c \ d]$, how are the columns $[a \ c]^{\mathsf{T}}$ and $[b \ d]^{\mathsf{T}}$ related?

P.3.7 Partition $M \in \mathsf{M}_n$ as a 2×2 block matrix as in (3.4.5). If D is invertible, then the *Schur complement of D in M* is $M/D = A - BD^{-1}C$. (a) Show that $\det M = (\det D)(\det M/D)$. (b) If A is invertible, show that the determinant of the bordered matrix (3.4.11) can be evaluated as

$$\det \begin{bmatrix} c & \mathbf{x}^{\mathsf{T}} \\ \mathbf{y} & A \end{bmatrix} = (c - \mathbf{y}^{\mathsf{T}} A^{-1} \mathbf{x}) \det A$$

$$= c \det A - \mathbf{y}^{\mathsf{T}} (\operatorname{adj} A) \mathbf{x}. \tag{3.7.3}$$

This is the *Cauchy expansion of the determinant* of a bordered matrix. The formulation (3.7.3) is valid even if A is not invertible.

P.3.8 Suppose that $A, B, C, D \in \mathsf{M}_n$. If A is invertible and commutes with B, show that

$$\det \begin{bmatrix} A & B \\ C & D \end{bmatrix} = \det(DA - CB). \tag{3.7.4}$$

If D is invertible and commutes with C, show that

$$\det \begin{bmatrix} A & B \\ C & D \end{bmatrix} = \det(AD - BC). \tag{3.7.5}$$

What can you say if $n = 1$?

P.3.9 Let $M = \begin{bmatrix} A & B \\ 0 & D \end{bmatrix} \in \mathsf{M}_n$ be block upper triangular and let p be a polynomial. Prove that $p(M) = \begin{bmatrix} p(A) & \star \\ 0 & p(D) \end{bmatrix}$.

P.3.10 Let $\mathcal{U}$ and $\mathcal{V}$ be finite-dimensional subspaces of a vector space. (a) Show that $\mathcal{U} + \mathcal{V} = \mathcal{U}$ if and only if $\mathcal{V} \subseteq \mathcal{U}$. (b) What does (a) say about the case of equality in (3.2.14)? Discuss.

P.3.11 A 2×2 block matrix $M = \begin{bmatrix} A & B \\ C & D \end{bmatrix} \in \mathsf{M}_{2n}(\mathbb{R})$, in which each block is $n \times n$, $A = D$, and $C = -B$ is a *matrix of complex type*. Let $J_{2n} = \begin{bmatrix} 0 & I_n \\ -I_n & 0 \end{bmatrix}$. Show that $M \in \mathsf{M}_{2n}(\mathbb{R})$ is a matrix of complex type if and only if J_{2n} commutes with M.

P.3.12 Suppose that $M, N \in \mathsf{M}_{2n}(\mathbb{R})$ are matrices of complex type. Show that $M + N$ and MN are matrices of complex type. If M is invertible, show that M^{-1} is also a matrix of complex type. *Hint*: Use the criterion in the preceding problem.

P.3.13 A 2×2 block matrix $M = \begin{bmatrix} A & B \\ C & D \end{bmatrix} \in \mathsf{M}_{2n}$, in which each block is $n \times n$, $A = D$, and $B = C$ is *block centrosymmetric*. Let $L_{2n} = \begin{bmatrix} 0 & I_n \\ I_n & 0 \end{bmatrix}$. Show that M is block centrosymmetric if and only if L_{2n} commutes with M.

P.3.14 Suppose that $M, N \in \mathsf{M}_{2n}$ are block centrosymmetric. Show that $M + N$ and MN are block centrosymmetric. If M is invertible, show that M^{-1} is also block centrosymmetric. *Hint*: Use the criterion in the preceding problem.

P.3.15 If $A \in \mathsf{M}_n$ is to be represented as $A = XY^{\mathsf{T}}$ for some $X, Y \in \mathsf{M}_{n \times r}$, explain why r cannot be smaller than $\operatorname{rank} A$.

P.3.16 Suppose that $1 \le r \le \min\{m, n\}$. Let $X \in \mathsf{M}_{m \times r}$ and $Y \in \mathsf{M}_{r \times n}$, and suppose that $\operatorname{rank} X = \operatorname{rank} Y = r$. Explain why there are $X_2 \in \mathsf{M}_{m \times (m-r)}$ and $Y_2 \in \mathsf{M}_{(n-r) \times n}$ such that

$$B = [X \ X_2] \in \mathsf{M}_m \quad \text{and} \quad C = \begin{bmatrix} Y \\ Y_2 \end{bmatrix} \in \mathsf{M}_n$$

are invertible. Verify that

$$XY = B \begin{bmatrix} I_r & 0 \\ 0 & 0 \end{bmatrix} C.$$

What sizes are the zero submatrices?

P.3.17 Let $A \in M_{m \times n}$. Use the preceding problems to show that rank $A = r$ if and only if there are invertible $B \in M_m$ and $C \in M_n$ such that

$$A = B \begin{bmatrix} I_r & 0 \\ 0 & 0 \end{bmatrix} C.$$

P.3.18 Let $x_1, x_2, \ldots, x_k \in \mathbb{R}^n$. Then $x_1, x_2, \ldots, x_k$ are real vectors in $\mathbb{C}^n$. Show that $x_1, x_2, \ldots, x_k$ are linearly independent in $\mathbb{R}^n$ if and only if they are linearly independent in $\mathbb{C}^n$.

P.3.19 Let $A \in M_{m \times n}(\mathbb{R})$. Definition 2.2.6 suggests that the rank of A (considered as an element of $M_{m \times n}(\mathbb{F})$) might depend on whether $\mathbb{F} = \mathbb{R}$ or $\mathbb{F} = \mathbb{C}$. Show that rank $A = r$ (considered as an element of $M_{m \times n}(\mathbb{R})$) if and only if rank $A = r$ (considered as an element of $M_{m \times n}(\mathbb{C})$).

P.3.20 Let $A \in M_{m \times n}(\mathbb{F})$ and let $x_1, x_2, \ldots, x_k \in \mathbb{F}^n$. (a) If $x_1, x_2, \ldots, x_k$ are linearly independent and rank $A = n$, show that $Ax_1, Ax_2, \ldots, Ax_k \in \mathbb{F}^m$ are linearly independent. (b) If $Ax_1, Ax_2, \ldots, Ax_k \in \mathbb{F}^m$ are linearly independent, show that $x_1, x_2, \ldots, x_k$ are linearly independent. (c) Although the hypothesis that A has full column rank is not needed in (b), show by example that it cannot be omitted in (a).

P.3.21 For $X \in M_{m \times n}$, let $\nu(X)$ denote the nullity of X, that is, $\nu(X) = \dim \text{null} X$. (a) Let $A \in M_{m \times k}$ and $B \in M_{k \times n}$. Show that the rank inequality (3.3.4) is equivalent to the inequality

$$\nu(AB) \leq \nu(A) + \nu(B). \tag{3.7.6}$$

(b) If $A, B \in M_n$, show that the rank inequality (3.2.13) is equivalent to the inequality

$$\max\{\nu(A), \nu(B)\} \leq \nu(AB). \tag{3.7.7}$$

The inequalities (3.7.6) and (3.7.7) are known as *Sylvester's Law of Nullity*. (c) Consider

$$A = [0 \; 1] \quad \text{and} \quad B = \begin{bmatrix} 1 \\ 0 \end{bmatrix}.$$

Show that the inequality (3.7.7) need not be valid for matrices that are not square.

P.3.22 Let $A \in M_n$ be the bidiagonal matrix in which the main diagonal entries are all 0 and the superdiagonal entries are all 1. Show that the index of A is n.

P.3.23 Let $A \in M_{m \times n}$ and suppose that $B \in M_{m \times n}$ is obtained by changing the value of exactly one entry of A. Show that rank B has one of the three values rank $A - 1$, rank A, or rank $A + 1$. Give examples to illustrate all three possibilities.

P.3.24 Use Theorem 2.2.10 to prove the upper bound in (3.3.6).

P.3.25 Let $A \in \mathsf{M}_{m \times k}$ and $B \in \mathsf{M}_{k \times n}$. Use (2.6.1) to show that

$$\operatorname{rank} AB = \operatorname{rank} A + \operatorname{rank} B - \dim(\operatorname{null} A \cap \operatorname{col} B). \tag{3.7.8}$$

Use this identity to prove the two inequalities in (3.3.6).

P.3.26 Let $A \in \mathsf{M}_{m \times k}$, $B \in \mathsf{M}_{k \times p}$, and $C \in \mathsf{M}_{p \times n}$. Show that

$$\operatorname{null} A \cap \operatorname{col} BC \subseteq \operatorname{null} A \cap \operatorname{col} B.$$

Use (3.7.8) to prove the *Frobenius rank inequality*

$$\operatorname{rank} AB + \operatorname{rank} BC \leq \operatorname{rank} B + \operatorname{rank} ABC.$$

Show that (3.2.13) and (3.3.4) are special cases of the Frobenius rank inequality.

P.3.27 If $A, B \in \mathsf{M}_n$, is $\operatorname{rank} AB = \operatorname{rank} BA$? Why?

P.3.28 If an invertible matrix M is partitioned as a 2×2 block matrix as in (3.4.5), there is a conformally partitioned presentation of its inverse:

$$M^{-1} = \begin{bmatrix} (A - BD^{-1}C)^{-1} & -A^{-1}B(D - CA^{-1}B)^{-1} \\ -D^{-1}C(A - BD^{-1}C)^{-1} & (D - CA^{-1}B)^{-1} \end{bmatrix}, \tag{3.7.9}$$

provided that all the indicated inverses exist. (a) Verify that (3.7.9) can be written as

$$M^{-1} = \begin{bmatrix} A^{-1} & 0 \\ 0 & D^{-1} \end{bmatrix} \begin{bmatrix} A & -B \\ -C & D \end{bmatrix} \begin{bmatrix} (M/D)^{-1} & 0 \\ 0 & (M/A)^{-1} \end{bmatrix}. \tag{3.7.10}$$

(b) Derive the identity

$$\begin{bmatrix} I_k & 0 \\ X & I_{n-k} \end{bmatrix}^{-1} = \begin{bmatrix} I_k & 0 \\ -X & I_{n-k} \end{bmatrix}$$

from (3.7.9). (c) If all the blocks in (3.7.9) are 1×1 matrices, show that it reduces to

$$M^{-1} = \begin{bmatrix} a & b \\ c & d \end{bmatrix}^{-1} = \frac{1}{\det M} \begin{bmatrix} d & -b \\ -c & a \end{bmatrix}.$$

P.3.29 Suppose that $A \in \mathsf{M}_{n \times m}$ and $B \in \mathsf{M}_{m \times n}$. Use the block matrix

$$\begin{bmatrix} I & -A \\ B & I \end{bmatrix}$$

to derive the *Sylvester determinant identity*

$$\det(I + AB) = \det(I + BA), \tag{3.7.11}$$

which relates the determinant of an $n \times n$ matrix to the determinant of an $m \times m$ matrix.

P.3.30 Let $\mathbf{u}, \mathbf{v} \in \mathbb{C}^n$ and let $z \in \mathbb{C}$. (a) Show that $\det(I + z\mathbf{u}\mathbf{v}^\mathsf{T}) = 1 + z\mathbf{v}^\mathsf{T}\mathbf{u}$. (b) If $A \in \mathsf{M}_n$ is invertible, show that

$$\det(A + z\mathbf{u}\mathbf{v}^\mathsf{T}) = \det A + z(\det A)(\mathbf{v}^\mathsf{T}A^{-1}\mathbf{u}). \tag{3.7.12}$$

P.3.31 Let $\mathcal{V}$ be a complex vector space. Let $T : \mathsf{M}_n \to \mathcal{V}$ be a linear transformation such that $T(XY) = T(YX)$ for all $X, Y \in \mathsf{M}_n$. Show that $T(A) = (\frac{1}{n} \operatorname{tr} A) T(I_n)$ for all $A \in \mathsf{M}_n$ and $\dim \ker T = n^2 - 1$. *Hint*: $A = (A - (\frac{1}{n} \operatorname{tr} A) I_n) + (\frac{1}{n} \operatorname{tr} A) I_n$.

P.3.32 Let $\Phi : \mathsf{M}_n \to \mathbb{C}$ be a linear transformation. Show that $\Phi = \operatorname{tr}$ if and only if $\Phi(I_n) = n$ and $\Phi(XY) = \Phi(YX)$ for all $X, Y \in \mathsf{M}_n$.

P.3.33 Use (3.1.2) and (3.1.3) (and their notation) to verify the associative identity

$$(AB)\mathbf{x} = A(B\mathbf{x}).$$

Hint: $[A\mathbf{b}_1 \ A\mathbf{b}_2 \ \ldots \ A\mathbf{b}_n]\mathbf{x} = \sum_i x_i A\mathbf{b}_i = A(\sum_i x_i \mathbf{b}_i)$.

P.3.34 Suppose that $A \in \mathsf{M}_{m \times r}$, $B \in \mathsf{M}_{r \times n}$, and $C \in \mathsf{M}_{n \times p}$. Use the preceding problem to verify the associative identity $(AB)C = A(BC)$. *Hint*: Let $\mathbf{x}$ be a column of C.

P.3.35 Let $A, B \in \mathsf{M}_{m \times n}$. Suppose that $r = \operatorname{rank} A \geq 1$ and let $s = \operatorname{rank} B$. Show that $\operatorname{col} A \subseteq \operatorname{col} B$ if and only if there is a full-rank $X \in \mathsf{M}_{m \times r}$ such that $A = XY$ and $B = [X \ X_2]Z$, in which $Y \in \mathsf{M}_{r \times n}$ and $Z \in \mathsf{M}_{s \times n}$ have full rank, $X_2 \in \mathsf{M}_{m \times (s-r)}$, and $\operatorname{rank}[X \ X_2] = s$. *Hint*: Theorem 3.2.15.

P.3.36 Let $A, B \in \mathsf{M}_{m \times n}$ and suppose that $r = \operatorname{rank} A \geq 1$. Show that the following are equivalent:

(a) $\operatorname{col} A = \operatorname{col} B$.

(b) There are full-rank matrices $X \in \mathsf{M}_{m \times r}$ and $Y, Z \in \mathsf{M}_{r \times n}$ such that $A = XY$ and $B = XZ$.

(c) There is an invertible $S \in \mathsf{M}_n$ such that $B = AS$.

Hint: If $A = XY$ is a full-rank factorization, then $A = [X \ 0_{n \times (n-r)}]W$, in which

$$W = \begin{bmatrix} Y \\ Y_2 \end{bmatrix} \in \mathsf{M}_n$$

is invertible.

P.3.37 Let $A, C \in \mathsf{M}_m$ and $B, D \in \mathsf{M}_n$. Show that there is an invertible $Z \in \mathsf{M}_{m+n}$ such that $A \oplus B = (C \oplus D)Z$ if and only if there are invertible matrices $X \in \mathsf{M}_m$ and $Y \in \mathsf{M}_n$ such that $A = CX$ and $B = DY$.

P.3.38 Let $A = [\mathbf{a}_1 \ \mathbf{a}_2 \ \ldots \ \mathbf{a}_p] \in \mathsf{M}_{n \times p}$ and $B = [\mathbf{b}_1 \ \mathbf{b}_2 \ \ldots \ \mathbf{b}_q] \in \mathsf{M}_{n \times q}$. If $\mathbf{a}_1, \mathbf{a}_2, \ldots, \mathbf{a}_p$ are linearly independent, $\mathbf{b}_1, \mathbf{b}_2, \ldots, \mathbf{b}_q$ are linearly independent, and $\operatorname{span}\{\mathbf{a}_1, \mathbf{a}_2, \ldots, \mathbf{a}_p\} = \operatorname{span}\{\mathbf{b}_1, \mathbf{b}_2, \ldots, \mathbf{b}_q\}$, use (3.1.21) and (0.3.5) to show that $p = q$. How is this related to Corollary 2.1.10? *Hint*: $A = BX$, $B = AY$, $A(I_p - YX) = 0$, and $\operatorname{tr} XY = \operatorname{tr} YX$.

P.3.39 If $A \otimes B = 0$, show that either $A = 0$ or $B = 0$. Is this true for the ordinary matrix product AB?

P.3.40 Choose two of the Kronecker product identities (3.6.4)–(3.6.10) and prove them.

P.3.41 Explain why $(A \otimes B \otimes C)(D \otimes E \otimes F) = (AD) \otimes (BE) \otimes (CF)$ if all the matrices have appropriate sizes.

P.3.42 If $A, B, C, D, R, S \in \mathsf{M}_n$, R and S are invertible, $A = RBR^{-1}$ and $C = SDS^{-1}$, show that $A \otimes C = (R \otimes S)(B \otimes D)(R \otimes S)^{-1}$.

P.3.43 Let $A = [a_{ij}] \in \mathsf{M}_m$ and $B = [b_{ij}] \in \mathsf{M}_n$ be upper triangular matrices. Prove the following:

(a) $A \otimes B \in \mathsf{M}_{mn}$ is an upper triangular matrix whose mn diagonal entries are (in some order) the mn scalars $a_{ii}b_{jj}$, for $i = 1, 2, \ldots, m$ and $j = 1, 2, \ldots, n$.

(b) $A \otimes I_n + I_m \otimes B$ is an upper triangular matrix whose mn diagonal entries are (in some order) the mn scalars $a_{ii} + b_{jj}$, for $i = 1, 2, \ldots, m$ and $j = 1, 2, \ldots, n$.

P.3.44 Let $A \in \mathsf{M}_m$ and $B \in \mathsf{M}_n$. Show that $A \otimes I_n$ commutes with $I_m \otimes B$.

3.8 Notes

Lemma 3.5.4 and Theorem 3.5.6 describe special structures that can be achieved for the diagonal of a matrix similar to a given $A \in \mathsf{M}_n$. Other results of this type are: If A is not a scalar matrix, then it is similar to a matrix whose diagonal entries are $0, 0, \ldots, 0, \operatorname{tr} A$. If $A \neq 0$, then it is similar to a matrix whose diagonal entries are all nonzero.

See Example 1.3.24 of [HJ13] for a version of P.3.30 that is valid for all $A \in \mathsf{M}_n$. For some historical comments about the Kronecker product and more of its properties, see [HJ94, Ch. 4].

3.9 Some Important Concepts

- Row and column partitions of a matrix.

- Conformal partitions for addition and multiplication of matrices.

- Row rank equals column rank.

- Index of a matrix.

- When does a matrix commute with a direct sum of scalar matrices? (Lemma 3.3.21)

- Schur complement and the determinant of a 2×2 block matrix.

- Determinant of a bordered matrix (Cauchy expansion).

- Kronecker product and its properties.

4 Inner Product Spaces

Many abstract concepts that make linear algebra a powerful mathematical tool have their roots in plane geometry, so we begin our study of inner product spaces by reviewing basic properties of lengths and angles in the real two-dimensional plane $\mathbb{R}^2$. Guided by these geometrical properties, we formulate axioms for inner products and norms, which provide generalized notions of length (norm) and perpendicularity (orthogonality) in abstract vector spaces.

4.1 The Pythagorean Theorem

Given the lengths of two orthogonal line segments that form the sides of a right triangle in the real Euclidean plane, the classical Pythagorean theorem describes how to find the length of its hypotenuse.

Theorem 4.1.1 (Classical Pythagorean Theorem) *If a and b are the lengths of two sides of a right triangle T, and if c is the length of its hypotenuse, then $a^2 + b^2 = c^2$.*

Proof Construct a square with side c and place four copies of T around it to make a larger square with side $a + b$; see Figure 4.1. The area of the larger square is equal to the area of the smaller square plus four times the area of T:

$$(a + b)^2 = c^2 + 4\left(\tfrac{1}{2}ab\right),$$

and hence

$$a^2 + 2ab + b^2 = c^2 + 2ab.$$

We conclude that $a^2 + b^2 = c^2$. $\qquad\square$

4.2 The Law of Cosines

The length of any one side of a plane triangle (it need not be a right triangle) is determined by the lengths of its other two sides and the cosine of the angle between them. This is a consequence of the Pythagorean theorem.

Theorem 4.2.1 (Law of Cosines) *Let a and b be the lengths of two sides of a plane triangle, let θ be the angle between the two sides, and let c be the length of the third side. Then $a^2 + b^2 - 2ab\cos\theta = c^2$. If $\theta = \pi/2$ (a right angle), then $a^2 + b^2 = c^2$.*

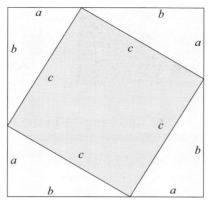

Figure 4.1 Proof of the classical Pythagorean theorem.

Proof See Figure 4.2. The Pythagorean theorem ensures that

$$c^2 = (a - b\cos\theta)^2 + (b\sin\theta)^2$$
$$= a^2 - 2ab\cos\theta + b^2\cos^2\theta + b^2\sin^2\theta$$
$$= a^2 - 2ab\cos\theta + b^2(\cos^2\theta + \sin^2\theta)$$
$$= a^2 - 2ab\cos\theta + b^2.$$

If $\theta = \pi/2$, then $\cos\theta = 0$; the triangle is a right triangle and the law of cosines reduces to the classical Pythagorean theorem. ☐

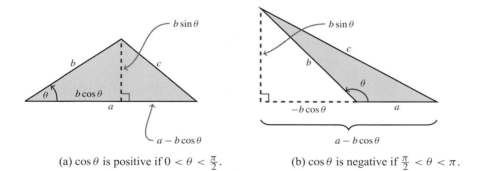

(a) $\cos\theta$ is positive if $0 < \theta < \frac{\pi}{2}$.

(b) $\cos\theta$ is negative if $\frac{\pi}{2} < \theta < \pi$.

Figure 4.2 Proof of the law of cosines.

The law of cosines implies a familiar fact about plane triangles: the length of one side is not greater than the sum of the lengths of the other two sides.

Corollary 4.2.2 (Triangle Inequality) *Let a, b, and c be the lengths of the sides of a plane triangle. Then*

$$c \le a + b. \tag{4.2.3}$$

Proof Let θ be the angle between the sides whose lengths are a and b. Since $-\cos\theta \le 1$, Theorem 4.2.1 tells us that

$$c^2 = a^2 - 2ab\cos\theta + b^2 \le a^2 + 2ab + b^2 = (a+b)^2.$$

Thus, $c \le a + b$. $\qquad\qquad\qquad\qquad\qquad\qquad\qquad\qquad\qquad\qquad\qquad\qquad\qquad\qquad\qquad$ $\square$

4.3 Angles and Lengths in the Plane

Consider the triangle in Figure 4.3 whose vertices are given by $\mathbf{0}$ and the real Cartesian coordinate vectors

$$\mathbf{a} = \begin{bmatrix} a_1 \\ a_2 \end{bmatrix} \quad \text{and} \quad \mathbf{b} = \begin{bmatrix} b_1 \\ b_2 \end{bmatrix}.$$

If we place

$$\mathbf{c} = \mathbf{a} - \mathbf{b} = \begin{bmatrix} a_1 - b_1 \\ a_2 - b_2 \end{bmatrix}$$

so that its initial point is at $\mathbf{b}$, it forms the third side of a triangle. Inspired by the Pythagorean theorem, we introduce the notation

$$\|\mathbf{a}\| = \sqrt{a_1^2 + a_2^2} \qquad\qquad\qquad\qquad (4.3.1)$$

to indicate the (Euclidean) length of the vector $\mathbf{a}$. Then

$$\|\mathbf{a}\|^2 = a_1^2 + a_2^2,$$
$$\|\mathbf{b}\|^2 = b_1^2 + b_2^2, \quad \text{and}$$
$$\|\mathbf{c}\|^2 = (a_1 - b_1)^2 + (a_2 - b_2)^2.$$

The law of cosines tells us that

$$\|\mathbf{a}\|^2 + \|\mathbf{b}\|^2 - 2\|\mathbf{a}\|\|\mathbf{b}\|\cos\theta = \|\mathbf{c}\|^2.$$

Thus,

$$a_1^2 + a_2^2 + b_1^2 + b_2^2 - 2\|\mathbf{a}\|\|\mathbf{b}\|\cos\theta = (a_1 - b_1)^2 + (a_2 - b_2)^2$$
$$= a_1^2 - 2a_1b_1 + b_1^2 + a_2^2 - 2a_2b_2 + b_2^2,$$

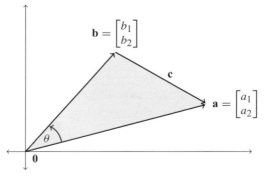

Figure 4.3 Angles and vectors.

and therefore,

$$a_1b_1 + a_2b_2 = \|\mathbf{a}\|\|\mathbf{b}\| \cos \theta. \tag{4.3.2}$$

The left-hand side of (4.3.2) is the *dot product* of **a** and **b**; we denote it by

$$\mathbf{a} \cdot \mathbf{b} = a_1b_1 + a_2b_2. \tag{4.3.3}$$

The Euclidean length and the dot product are related by (4.3.1), which we can write as

$$\|\mathbf{a}\| = \sqrt{\mathbf{a} \cdot \mathbf{a}}. \tag{4.3.4}$$

The identity (4.3.2) is

$$\mathbf{a} \cdot \mathbf{b} = \|\mathbf{a}\|\|\mathbf{b}\| \cos \theta, \tag{4.3.5}$$

in which θ is the angle (see Figure 4.4) between **a** and **b**; $\theta = \pi/2$ (a right angle) if and only if **a** and **b** are orthogonal, in which case $\mathbf{a} \cdot \mathbf{b} = 0$.

Because $|\cos \theta| \le 1$, the identity (4.3.5) implies that

$$|\mathbf{a} \cdot \mathbf{b}| \le \|\mathbf{a}\|\|\mathbf{b}\|. \tag{4.3.6}$$

After a few computations, one verifies that the dot product has the following properties for $\mathbf{a}, \mathbf{b}, \mathbf{c} \in \mathbb{R}^2$ and $c \in \mathbb{R}$:

(a) $\mathbf{a} \cdot \mathbf{a}$ is real and nonnegative.   *Nonnegativity*

(b) $\mathbf{a} \cdot \mathbf{a} = 0$ if and only if $\mathbf{a} = \mathbf{0}$.   *Positivity*

(c) $(\mathbf{a} + \mathbf{b}) \cdot \mathbf{c} = \mathbf{a} \cdot \mathbf{c} + \mathbf{b} \cdot \mathbf{c}$.   *Additivity*

(d) $(c\mathbf{a}) \cdot \mathbf{b} = c(\mathbf{a} \cdot \mathbf{b})$.   *Homogeneity*

(e) $\mathbf{a} \cdot \mathbf{b} = \mathbf{b} \cdot \mathbf{a}$.   *Symmetry*

This list of properties suggests that the first position in the dot product enjoys a favored status, but it does not. The symmetry property ensures that the dot product is additive and homogeneous in both positions:

$$\mathbf{a} \cdot (\mathbf{b} + \mathbf{c}) = (\mathbf{b} + \mathbf{c}) \cdot \mathbf{a} = \mathbf{b} \cdot \mathbf{a} + \mathbf{c} \cdot \mathbf{a} = \mathbf{a} \cdot \mathbf{b} + \mathbf{a} \cdot \mathbf{c}$$

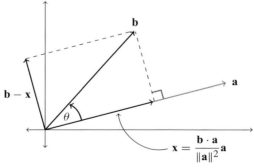

Figure 4.4 The orthogonal projection of one vector onto another.

and

$$\mathbf{a} \cdot (c\mathbf{b}) = (c\mathbf{b}) \cdot \mathbf{a} = c(\mathbf{b} \cdot \mathbf{a}) = c(\mathbf{a} \cdot \mathbf{b}).$$

In Figure 4.4,

$$\mathbf{x} = \|\mathbf{b}\| \cos\theta \frac{\mathbf{a}}{\|\mathbf{a}\|} = \frac{\mathbf{b} \cdot \mathbf{a}}{\|\mathbf{a}\|^2} \mathbf{a} \tag{4.3.7}$$

is the *projection of* **b** *onto* **a**. Then

$$\mathbf{b} - \mathbf{x} = \mathbf{b} - \frac{\mathbf{b} \cdot \mathbf{a}}{\|\mathbf{a}\|^2} \mathbf{a}$$

is orthogonal to **a** (and hence also to **x**) because

$$(\mathbf{b} - \mathbf{x}) \cdot \mathbf{a} = \left(\mathbf{b} - \frac{\mathbf{b} \cdot \mathbf{a}}{\|\mathbf{a}\|^2} \mathbf{a} \right) \cdot \mathbf{a}$$

$$= \mathbf{b} \cdot \mathbf{a} - \frac{\mathbf{b} \cdot \mathbf{a}}{\|\mathbf{a}\|^2}(\mathbf{a} \cdot \mathbf{a})$$

$$= \mathbf{b} \cdot \mathbf{a} - \mathbf{b} \cdot \mathbf{a}$$

$$= 0.$$

Thus,

$$\mathbf{b} = \mathbf{x} + (\mathbf{b} - \mathbf{x})$$

decomposes **b** into the sum of two vectors, one parallel to **a** and one orthogonal to it.

From the properties of the dot product, we deduce that the Euclidean length function has the following properties for $\mathbf{a}, \mathbf{b} \in \mathbb{R}^2$ and $c \in \mathbb{R}$:

(a) $\|\mathbf{a}\|$ is real and nonnegative.   *Nonnegativity*

(b) $\|\mathbf{a}\| = 0$ if and only if $\mathbf{a} = \mathbf{0}$.   *Positivity*

(c) $\|c\mathbf{a}\| = |c|\|\mathbf{a}\|$.   *Homogeneity*

(d) $\|\mathbf{a} + \mathbf{b}\| \le \|\mathbf{a}\| + \|\mathbf{b}\|$.   *Triangle Inequality*

(e) $\|\mathbf{a} + \mathbf{b}\|^2 + \|\mathbf{a} - \mathbf{b}\|^2 = 2\|\mathbf{a}\|^2 + 2\|\mathbf{b}\|^2$.   *Parallelogram Identity*

Nonnegativity and positivity follow directly from the corresponding properties of the dot product. Homogeneity follows from the homogeneity and symmetry of the dot product:

$$\|c\mathbf{a}\|^2 = c\mathbf{a} \cdot c\mathbf{a} = c(\mathbf{a} \cdot c\mathbf{a})$$

$$= c(c\mathbf{a} \cdot \mathbf{a}) = c^2(\mathbf{a} \cdot \mathbf{a})$$

$$= |c|^2(\mathbf{a} \cdot \mathbf{a}) = |c|^2\|\mathbf{a}\|^2.$$

The triangle inequality is Corollary 4.2.2, which follows from the law of cosines. The parallelogram identity (see Figure 4.5) says that the sum of the squares of the lengths of the two diagonals of a plane parallelogram is equal to the sum of the squares of the lengths of its four sides. This follows from additivity of the dot product in both positions:

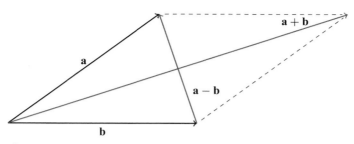

Figure 4.5 The parallelogram identity.

$$\|\mathbf{a} + \mathbf{b}\|^2 + \|\mathbf{a} - \mathbf{b}\|^2 = (\mathbf{a} + \mathbf{b}) \cdot (\mathbf{a} + \mathbf{b}) + (\mathbf{a} - \mathbf{b}) \cdot (\mathbf{a} - \mathbf{b})$$
$$= (\mathbf{a} \cdot \mathbf{a} + \mathbf{a} \cdot \mathbf{b} + \mathbf{b} \cdot \mathbf{a} + \mathbf{b} \cdot \mathbf{b})$$
$$+ (\mathbf{a} \cdot \mathbf{a} - \mathbf{a} \cdot \mathbf{b} - \mathbf{b} \cdot \mathbf{a} + \mathbf{b} \cdot \mathbf{b})$$
$$= 2(\mathbf{a} \cdot \mathbf{a} + \mathbf{b} \cdot \mathbf{b})$$
$$= 2(\|\mathbf{a}\|^2 + \|\mathbf{b}\|^2).$$

4.4 Inner Products

Guided by our experience with plane geometry and the dot product, we make the following definition.

Definition 4.4.1 Let $\mathbb{F} = \mathbb{R}$ or $\mathbb{C}$. An *inner product* on an $\mathbb{F}$-vector space $\mathcal{V}$ is a function

$$\langle \cdot, \cdot \rangle : \mathcal{V} \times \mathcal{V} \to \mathbb{F}$$

that satisfies the following axioms for any $\mathbf{u}, \mathbf{v}, \mathbf{w} \in \mathcal{V}$ and $c \in \mathbb{F}$:

(a) $\langle \mathbf{v}, \mathbf{v} \rangle$ is real and nonnegative. *Nonnegativity*

(b) $\langle \mathbf{v}, \mathbf{v} \rangle = 0$ if and only if $\mathbf{v} = \mathbf{0}$. *Positivity*

(c) $\langle \mathbf{u} + \mathbf{v}, \mathbf{w} \rangle = \langle \mathbf{u}, \mathbf{w} \rangle + \langle \mathbf{v}, \mathbf{w} \rangle$. *Additivity*

(d) $\langle c\mathbf{u}, \mathbf{v} \rangle = c \langle \mathbf{u}, \mathbf{v} \rangle$. *Homogeneity*

(e) $\langle \mathbf{u}, \mathbf{v} \rangle = \overline{\langle \mathbf{v}, \mathbf{u} \rangle}$. *Conjugate Symmetry*

The nonnegativity, positivity, additivity, and homogeneity axioms reflect familiar properties of the dot product on $\mathbb{R}^2$; the conjugate symmetry axiom (sometimes called the *Hermitian axiom*) looks like the symmetry of the dot product if $\mathbb{F} = \mathbb{R}$, but it is different if $\mathbb{F} = \mathbb{C}$. It ensures that

$$\langle a\mathbf{v}, a\mathbf{v} \rangle = a \langle \mathbf{v}, a\mathbf{v} \rangle = a \overline{\langle a\mathbf{v}, \mathbf{v} \rangle} = a(\overline{a \langle \mathbf{v}, \mathbf{v} \rangle}) = a\bar{a} \langle \mathbf{v}, \mathbf{v} \rangle = |a|^2 \langle \mathbf{v}, \mathbf{v} \rangle,$$

in agreement with (a).

The additivity, homogeneity, and conjugate symmetry axioms ensure that

$$\langle a\mathbf{u} + b\mathbf{v}, \mathbf{w} \rangle = \langle a\mathbf{u}, \mathbf{w} \rangle + \langle b\mathbf{v}, \mathbf{w} \rangle = a \langle \mathbf{u}, \mathbf{w} \rangle + b \langle \mathbf{v}, \mathbf{w} \rangle,$$

so an inner product is *linear* in its first position. However,

$$\langle \mathbf{u}, a\mathbf{v} + b\mathbf{w} \rangle = \overline{\langle a\mathbf{v} + b\mathbf{w}, \mathbf{u} \rangle} = \overline{\langle a\mathbf{v}, \mathbf{u} \rangle + \langle b\mathbf{w}, \mathbf{u} \rangle}$$

$$= \overline{\langle a\mathbf{v}, \mathbf{u} \rangle} + \overline{\langle b\mathbf{w}, \mathbf{u} \rangle} = \overline{a}\overline{\langle \mathbf{v}, \mathbf{u} \rangle} + \overline{b}\overline{\langle \mathbf{w}, \mathbf{u} \rangle}$$

$$= \overline{a}\langle \mathbf{u}, \mathbf{v} \rangle + \overline{b}\langle \mathbf{u}, \mathbf{w} \rangle.$$

If $\mathbb{F} = \mathbb{C}$, we can summarize the preceding computations as "the inner product is *conjugate linear* in its second position." If $\mathbb{F} = \mathbb{R}$, then $a = \overline{a}$ and $b = \overline{b}$, so the inner product is linear in its second position. Since an inner product on a complex vector space is linear in its first position and conjugate linear in its second position, one says that it is *sesquilinear* (one-and-a-half linear). An inner product on a real vector space is *bilinear* (twice linear).

Definition 4.4.2 Let $\mathbb{F} = \mathbb{R}$ or $\mathbb{C}$. An *inner product space* is an $\mathbb{F}$-vector space $\mathcal{V}$, endowed with an inner product $\langle \cdot, \cdot \rangle : \mathcal{V} \times \mathcal{V} \to \mathbb{F}$. We say that $\mathcal{V}$ is an $\mathbb{F}$-*inner product space* or that $\mathcal{V}$ is an *inner product space over* $\mathbb{F}$.

Some examples of inner product spaces are given below.

Example 4.4.3 Consider $\mathcal{V} = \mathbb{F}^n$ as a vector space over $\mathbb{F}$. For $\mathbf{u} = [u_i], \mathbf{v} = [v_i] \in \mathcal{V}$, let

$$\langle \mathbf{u}, \mathbf{v} \rangle = \mathbf{v}^*\mathbf{u} = \sum_{i=1}^{n} u_i\overline{v_i}.$$

This is the *standard inner product* on $\mathbb{F}^n$. If $\mathbb{F} = \mathbb{R}$, then $\langle \mathbf{u}, \mathbf{v} \rangle = \mathbf{v}^\mathsf{T}\mathbf{u}$. If $\mathbb{F} = \mathbb{R}$ and $n = 2$, then $\langle \mathbf{u}, \mathbf{v} \rangle = \mathbf{u} \cdot \mathbf{v}$ is the dot product on $\mathbb{R}^2$. If $n = 1$, the "vectors" in $\mathcal{V} = \mathbb{F}$ are scalars and $\langle c, d \rangle = c\overline{d}$.

Example 4.4.4 Let $\mathcal{V} = \mathcal{P}_n$ be the complex vector space of polynomials of degree at most n. Fix a finite nonempty real interval $[a, b]$, and define

$$\langle p, q \rangle = \int_a^b p(t)\overline{q(t)}\, dt.$$

This is known as the L^2 *inner product* on $\mathcal{P}_n$ *over the interval* $[a, b]$. Verification of the nonnegativity, additivity, homogeneity, and conjugate symmetry axioms is straightforward. Verification of the positivity axiom requires some analysis and algebra. If p is a polynomial such that

$$\langle p, p \rangle = \int_a^b p(t)\overline{p(t)}\, dt = \int_a^b |p(t)|^2\, dt = 0,$$

one can use properties of the integral and the continuity of the nonnegative function $|p|$ to show that $p(t) = 0$ for all $t \in [a, b]$. It follows that p is the zero polynomial, which is the zero element of the vector space $\mathcal{P}_n$.

Example 4.4.5 Let $\mathbb{F} = \mathbb{R}$ or $\mathbb{C}$. Let $\mathcal{V} = \mathsf{M}_{m \times n}(\mathbb{F})$ be the $\mathbb{F}$-vector space of $m \times n$ matrices over $\mathbb{F}$, let $A = [a_{ij}] \in \mathcal{V}$, let $B = [b_{ij}] \in \mathcal{V}$, and define the *Frobenius inner product*

$$\langle A, B \rangle_F = \operatorname{tr} B^*A. \tag{4.4.6}$$

Let $B^*A = [c_{ij}] \in M_n(\mathbb{F})$ and compute

$$\langle A, B \rangle_F = \operatorname{tr} B^*A = \sum_{j=1}^{n} c_{jj} = \sum_{j=1}^{n} \left(\sum_{i=1}^{m} \overline{b_{ij}} a_{ij} \right) = \sum_{i,j} a_{ij} \overline{b_{ij}}.$$

Since

$$\operatorname{tr} A^*A = \sum_{i,j} |a_{ij}|^2 \geq 0, \tag{4.4.7}$$

we see that $\operatorname{tr} A^*A = 0$ if and only if $A = 0$. Conjugate symmetry follows from the fact that $\operatorname{tr} X^* = \overline{\operatorname{tr} X}$ for any $X \in M_n$. Compute

$$\langle A, B \rangle_F = \operatorname{tr} B^*A = \operatorname{tr}(A^*B)^* = \overline{\operatorname{tr} A^*B} = \overline{\langle B, A \rangle_F}.$$

If $n = 1$, then $V = \mathbb{F}^m$ and the Frobenius inner product is the standard inner product on $\mathbb{F}^m$.

In the preceding examples, the vector spaces are finite dimensional. In the following examples, they are not. Infinite-dimensional inner product spaces play important roles in physics (quantum mechanics), aeronautics (model approximation), and engineering (signal analysis), as well as in mathematics itself.

Example 4.4.8 Let $V = C_{\mathbb{F}}[a, b]$ be the $\mathbb{F}$-vector space of continuous $\mathbb{F}$-valued functions on the finite, nonempty real interval $[a, b]$. For any $f, g \in V$, define

$$\langle f, g \rangle = \int_a^b f(t) \overline{g(t)} \, dt. \tag{4.4.9}$$

This is the L^2 *inner product* on $C_{\mathbb{F}}[a, b]$. Verification of the nonnegativity, additivity, homogeneity, and conjugate symmetry axioms is straightforward. Positivity follows in the same manner as in Example 4.4.4.

Example 4.4.10 If $\langle \cdot, \cdot \rangle$ is an inner product on V and if c is a positive real scalar, then $c \langle \cdot, \cdot \rangle$ is also an inner product on V. For example, the L^2 inner product (4.4.9) on $[-\pi, \pi]$ often appears in the modified form

$$\langle f, g \rangle = \frac{1}{\pi} \int_{-\pi}^{\pi} f(t) \overline{g(t)} \, dt \tag{4.4.11}$$

in the study of Fourier series; see Section 5.8.

Example 4.4.12 Let V be the complex vector space of finitely nonzero sequences $\mathbf{v} = (v_1, v_2, \ldots)$; see Example 1.2.7. For any $\mathbf{u}, \mathbf{v} \in V$, define

$$\langle \mathbf{u}, \mathbf{v} \rangle = \sum_{i=1}^{\infty} u_i \overline{v_i}.$$

The indicated sum involves only finitely many nonzero summands since each vector has only finitely many nonzero entries. Verification of the nonnegativity, additivity, homogeneity, and conjugate symmetry axioms is straightforward. To verify positivity, observe that if $\mathbf{u} \in \mathcal{V}$ and

$$0 = \langle \mathbf{u}, \mathbf{u} \rangle = \sum_{i=1}^{\infty} u_i \overline{u_i} = \sum_{i=1}^{\infty} |u_i|^2,$$

then each $u_i = 0$, so $\mathbf{u}$ is the zero vector in $\mathcal{V}$.

Based on our experience with orthogonal lines and dot products of real vectors in the plane, we define orthogonality in an inner product space.

Definition 4.4.13 Let $\mathcal{V}$ be an inner product space with inner product $\langle \cdot, \cdot \rangle$. Then $\mathbf{u}, \mathbf{v} \in \mathcal{V}$ are *orthogonal* if $\langle \mathbf{u}, \mathbf{v} \rangle = 0$. If $\mathbf{u}, \mathbf{v} \in \mathcal{V}$ are orthogonal we write $\mathbf{u} \perp \mathbf{v}$. Nonempty subsets $\mathscr{S}_1, \mathscr{S}_2 \subseteq \mathcal{V}$ are *orthogonal* if $\mathbf{u} \perp \mathbf{v}$ for every $\mathbf{u} \in \mathscr{S}_1$ and $\mathbf{v} \in \mathscr{S}_2$.

Three important properties of orthogonality follow from the definition and the axioms for an inner product.

Theorem 4.4.14 *Let $\mathcal{V}$ be an inner product space with inner product $\langle \cdot, \cdot \rangle$.*

(a) $\mathbf{u} \perp \mathbf{v}$ *if and only if* $\mathbf{v} \perp \mathbf{u}$.

(b) $\mathbf{0} \perp \mathbf{u}$ *for every* $\mathbf{u} \in \mathcal{V}$.

(c) *If* $\mathbf{v} \perp \mathbf{u}$ *for every* $\mathbf{u} \in \mathcal{V}$, *then* $\mathbf{v} = \mathbf{0}$.

Proof

(a) $\langle \mathbf{u}, \mathbf{v} \rangle = \overline{\langle \mathbf{v}, \mathbf{u} \rangle}$, so $\langle \mathbf{u}, \mathbf{v} \rangle = 0$ if and only if $\langle \mathbf{v}, \mathbf{u} \rangle = 0$.

(b) $\langle \mathbf{0}, \mathbf{u} \rangle = \langle 0\mathbf{0}, \mathbf{u} \rangle = 0 \langle \mathbf{0}, \mathbf{u} \rangle = 0$ for every $\mathbf{u} \in \mathcal{V}$.

(c) If $\mathbf{v} \perp \mathbf{u}$ for every $\mathbf{u} \in \mathcal{V}$, then $\langle \mathbf{v}, \mathbf{v} \rangle = 0$ and the positivity axiom implies that $\mathbf{v} = \mathbf{0}$. $\quad\square$

Corollary 4.4.15 *Let $\mathcal{V}$ be an inner product space and let $\mathbf{v}, \mathbf{w} \in \mathcal{V}$. If $\langle \mathbf{u}, \mathbf{v} \rangle = \langle \mathbf{u}, \mathbf{w} \rangle$ for all $\mathbf{u} \in \mathcal{V}$, then $\mathbf{v} = \mathbf{w}$.*

Proof If $\langle \mathbf{u}, \mathbf{v} \rangle = \langle \mathbf{u}, \mathbf{w} \rangle$ for all $\mathbf{u} \in \mathcal{V}$, then $\langle \mathbf{u}, \mathbf{v} - \mathbf{w} \rangle = 0$ for all $\mathbf{u}$. Part (c) of the preceding theorem ensures that $\mathbf{v} - \mathbf{w} = \mathbf{0}$. $\quad\square$

4.5 The Norm Derived from an Inner Product

By analogy with the dot product and Euclidean length in the plane, a generalized length can be defined in any inner product space.

Definition 4.5.1 Let $\mathcal{V}$ be an inner product space with inner product $\langle \cdot, \cdot \rangle$. The function $\| \cdot \| : \mathcal{V} \to [0, \infty)$ defined by

$$\|\mathbf{v}\| = \sqrt{\langle \mathbf{v}, \mathbf{v} \rangle} \tag{4.5.2}$$

is *the norm derived from the inner product* $\langle \cdot, \cdot \rangle$. For brevity, we refer to (4.5.2) as *the norm on* $\mathcal{V}$.

The definition ensures that $\|\mathbf{v}\| \geq 0$ for all $\mathbf{v} \in \mathcal{V}$.

Example 4.5.3 The norm on $\mathcal{V} = \mathbb{F}^n$ derived from the standard inner product is the *Euclidean norm*

$$\|\mathbf{u}\|_2 = \left(\mathbf{u}^*\mathbf{u}\right)^{1/2} = \left(\sum_{i=1}^{n} |u_i|^2\right)^{1/2}, \qquad \mathbf{u} = [u_i] \in \mathbb{F}^n. \tag{4.5.4}$$

The Euclidean norm is also called the ℓ_2 *norm*.

Example 4.5.5 The norm on $\mathcal{V} = \mathsf{M}_{m \times n}(\mathbb{F})$ derived from the Frobenius inner product is the *Frobenius norm*

$$\|A\|_F^2 = \langle A, A \rangle_F = \operatorname{tr} A^* A = \sum_{i,j} |a_{ij}|^2, \qquad A = [a_{ij}] \in \mathsf{M}_{m \times n}.$$

The Frobenius norm is sometimes called the *Schur norm* or the *Hilbert–Schmidt norm*. If $n = 1$, the Frobenius norm is the Euclidean norm on $\mathbb{F}^m$.

Example 4.5.6 The norm derived from the L^2 inner product on $C[a, b]$ (see Example 4.4.8) is the L^2 *norm*

$$\|f\| = \left(\int_a^b |f(t)|^2 \, dt\right)^{1/2}. \tag{4.5.7}$$

If we think of the integral as a limit of Riemann sums, there is a natural analogy between (4.5.7) and (4.5.4).

Example 4.5.8 Consider the complex inner product space $\mathcal{V} = C[-\pi, \pi]$ with the L^2 inner product and norm. The functions $\cos t$ and $\sin t$ are in $\mathcal{V}$, so

$$\|\sin t\|^2 = \int_{-\pi}^{\pi} \sin^2 t \, dt = \frac{1}{2} \int_{-\pi}^{\pi} (1 - \cos 2t) \, dt$$

$$= \frac{1}{2} \left(t - \frac{1}{2} \sin 2t\right)\Big|_{-\pi}^{\pi} = \pi$$

and

$$\langle \sin t, \cos t \rangle = \int_{-\pi}^{\pi} \sin t \, \cos t \, dt = \frac{1}{2} \int_{-\pi}^{\pi} \sin 2t \, dt$$

$$= -\frac{1}{4} \cos 2t\Big|_{-\pi}^{\pi} = 0.$$

Thus, with respect to the L^2 norm and inner product on $C[-\pi, \pi]$, $\sin t$ has norm $\sqrt{\pi}$ and is orthogonal to $\cos t$.

The derived norm (4.5.2) satisfies many of the properties of Euclidean length in the plane.

Theorem 4.5.9 *Let* V *be an* $\mathbb{F}$*-inner product space with inner product* $\langle \cdot, \cdot \rangle$ *and derived norm* $\| \cdot \|$. *Let* $\mathbf{u}, \mathbf{v} \in V$ *and* $c \in \mathbb{F}$.

(a) $\|\mathbf{u}\|$ *is real and nonnegative.*   *Nonnegativity*

(b) $\|\mathbf{u}\| = 0$ *if and only if* $\mathbf{u} = \mathbf{0}$.   *Positivity*

(c) $\|c\mathbf{u}\| = |c| \|\mathbf{u}\|$.   *Homogeneity*

(d) *If* $\langle \mathbf{u}, \mathbf{v} \rangle = 0$, *then* $\|\mathbf{u} + \mathbf{v}\|^2 = \|\mathbf{u}\|^2 + \|\mathbf{v}\|^2$.   *Pythagorean Theorem*

(e) $\|\mathbf{u} + \mathbf{v}\|^2 + \|\mathbf{u} - \mathbf{v}\|^2 = 2\|\mathbf{u}\|^2 + 2\|\mathbf{v}\|^2$.   *Parallelogram Identity*

Proof

(a) Nonnegativity is built in to Definition 4.5.1.

(b) Positivity follows from the positivity of the inner product. If $\|\mathbf{u}\| = 0$, then $\langle \mathbf{u}, \mathbf{u} \rangle = 0$, which implies that $\mathbf{u} = \mathbf{0}$.

(c) Homogeneity follows from homogeneity and conjugate symmetry of the inner product:

$$\|c\mathbf{u}\| = (\langle c\mathbf{u}, c\mathbf{u} \rangle)^{1/2} = (c\bar{c}\langle \mathbf{u}, \mathbf{u} \rangle)^{1/2}$$
$$= (|c|^2 \langle \mathbf{u}, \mathbf{u} \rangle)^{1/2} = |c| \langle \mathbf{u}, \mathbf{u} \rangle^{1/2}$$
$$= |c| \|\mathbf{u}\|.$$

(d) Compute

$$\|\mathbf{u} + \mathbf{v}\|^2 = \langle \mathbf{u} + \mathbf{v}, \mathbf{u} + \mathbf{v} \rangle$$
$$= \langle \mathbf{u}, \mathbf{u} \rangle + \langle \mathbf{u}, \mathbf{v} \rangle + \langle \mathbf{v}, \mathbf{u} \rangle + \langle \mathbf{v}, \mathbf{v} \rangle$$
$$= \langle \mathbf{u}, \mathbf{u} \rangle + 0 + 0 + \langle \mathbf{v}, \mathbf{v} \rangle$$
$$= \|\mathbf{u}\|^2 + \|\mathbf{v}\|^2.$$

(e) Use additivity in both positions of the inner product to compute

$$\|\mathbf{u} + \mathbf{v}\|^2 + \|\mathbf{u} - \mathbf{v}\|^2 = \langle \mathbf{u} + \mathbf{v}, \mathbf{u} + \mathbf{v} \rangle + \langle \mathbf{u} - \mathbf{v}, \mathbf{u} - \mathbf{v} \rangle$$
$$= \langle \mathbf{u}, \mathbf{u} \rangle + \langle \mathbf{u}, \mathbf{v} \rangle + \langle \mathbf{v}, \mathbf{u} \rangle + \langle \mathbf{v}, \mathbf{v} \rangle$$
$$+ \langle \mathbf{u}, \mathbf{u} \rangle - \langle \mathbf{u}, \mathbf{v} \rangle - \langle \mathbf{v}, \mathbf{u} \rangle + \langle \mathbf{v}, \mathbf{v} \rangle$$
$$= 2(\langle \mathbf{u}, \mathbf{u} \rangle + \langle \mathbf{v}, \mathbf{v} \rangle)$$
$$= 2(\|\mathbf{u}\|^2 + \|\mathbf{v}\|^2). \qquad \square$$

Definition 4.5.10 Let V be an inner product space with derived norm $\| \cdot \|$. Then $\mathbf{u} \in V$ is a *unit vector* if $\|\mathbf{u}\| = 1$.

Any nonzero $\mathbf{u} \in V$ can be *normalized* to create a unit vector $\mathbf{u}/\|\mathbf{u}\|$ that is proportional to $\mathbf{u}$:

$$\left\| \frac{\mathbf{u}}{\|\mathbf{u}\|} \right\| = \frac{\|\mathbf{u}\|}{\|\mathbf{u}\|} = 1 \quad \text{if} \quad \mathbf{u} \neq \mathbf{0}. \tag{4.5.11}$$

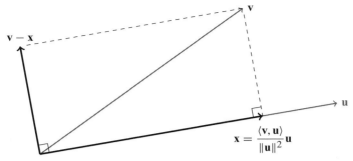

Figure 4.6 The orthogonal projection of one vector onto another.

In an inner product space $\mathcal{V}$, a generalization of (4.3.7) defines the projection of one vector onto another. Let $\mathbf{u} \in \mathcal{V}$ be nonzero and form

$$\mathbf{x} = \frac{\langle \mathbf{v}, \mathbf{u} \rangle}{\|\mathbf{u}\|^2} \mathbf{u} = \left\langle \mathbf{v}, \frac{\mathbf{u}}{\|\mathbf{u}\|} \right\rangle \frac{\mathbf{u}}{\|\mathbf{u}\|}, \tag{4.5.12}$$

which is the inner product of $\mathbf{v}$ with the unit vector in the direction of $\mathbf{u}$, times the unit vector in the direction of $\mathbf{u}$. This is *the projection of* $\mathbf{v}$ *onto* $\mathbf{u}$; see Figure 4.6. Since

$$\langle \mathbf{v} - \mathbf{x}, \mathbf{u} \rangle = \langle \mathbf{v}, \mathbf{u} \rangle - \langle \mathbf{x}, \mathbf{u} \rangle = \langle \mathbf{v}, \mathbf{u} \rangle - \left\langle \frac{\langle \mathbf{v}, \mathbf{u} \rangle}{\|\mathbf{u}\|^2} \mathbf{u}, \mathbf{u} \right\rangle$$

$$= \langle \mathbf{v}, \mathbf{u} \rangle - \langle \mathbf{v}, \mathbf{u} \rangle \frac{\langle \mathbf{u}, \mathbf{u} \rangle}{\|\mathbf{u}\|^2} = \langle \mathbf{v}, \mathbf{u} \rangle - \langle \mathbf{v}, \mathbf{u} \rangle = 0,$$

$\mathbf{v} - \mathbf{x}$ is orthogonal to $\mathbf{u}$ (and hence also to $\mathbf{x}$). Consequently,

$$\mathbf{v} = \mathbf{x} + (\mathbf{v} - \mathbf{x}) \tag{4.5.13}$$

is a decomposition of $\mathbf{v}$ as a sum of two orthogonal vectors, one proportional to $\mathbf{u}$ and the other orthogonal to it.

An important inequality that generalizes (4.3.6) is valid in any inner product space.

Theorem 4.5.14 (Cauchy–Schwarz Inequality) *Let* $\mathcal{V}$ *be an inner product space with inner product* $\langle \cdot, \cdot \rangle$ *and derived norm* $\| \cdot \|$. *Then*

$$|\langle \mathbf{u}, \mathbf{v} \rangle| \leq \|\mathbf{u}\| \|\mathbf{v}\| \tag{4.5.15}$$

for all $\mathbf{u}, \mathbf{v} \in \mathcal{V}$ *with equality if and only if* $\mathbf{u}$ *and* $\mathbf{v}$ *are linearly dependent, that is, if and only if one of them is a scalar multiple of the other.*

Proof If either $\mathbf{u} = \mathbf{0}$ or $\mathbf{v} = \mathbf{0}$ there is nothing to prove; both sides of the inequality (4.5.15) are zero and $\mathbf{u}, \mathbf{v}$ are linearly dependent. Thus, we may assume that $\mathbf{u}$ and $\mathbf{v}$ are both nonzero.

Define $\mathbf{x}$ as in (4.5.12) and write $\mathbf{v} = \mathbf{x} + (\mathbf{v} - \mathbf{x})$ as in (4.5.13). Since $\mathbf{x}$ and $\mathbf{v} - \mathbf{x}$ are orthogonal, Theorem 4.5.9.d ensures that

$$
\begin{aligned}
\|\mathbf{v}\|^2 &= \|\mathbf{x}\|^2 + \|\mathbf{v} - \mathbf{x}\|^2 \\
&\geq \|\mathbf{x}\|^2 = \left\| \frac{\langle \mathbf{v}, \mathbf{u} \rangle}{\|\mathbf{u}\|^2} \mathbf{u} \right\|^2 \\
&= \frac{|\langle \mathbf{v}, \mathbf{u} \rangle|^2}{\|\mathbf{u}\|^4} \|\mathbf{u}\|^2 = \frac{|\langle \mathbf{v}, \mathbf{u} \rangle|^2}{\|\mathbf{u}\|^2},
\end{aligned}
\tag{4.5.16}
$$

so $\|\mathbf{u}\|^2 \|\mathbf{v}\|^2 \geq |\langle \mathbf{v}, \mathbf{u} \rangle|^2 = |\overline{\langle \mathbf{u}, \mathbf{v} \rangle}|^2 = |\langle \mathbf{u}, \mathbf{v} \rangle|^2$. Consequently, $\|\mathbf{u}\| \|\mathbf{v}\| \geq |\langle \mathbf{u}, \mathbf{v} \rangle|$.

If (4.5.16) is an equality, then $\mathbf{v} - \mathbf{x} = \mathbf{0}$ and $\mathbf{v} = \mathbf{x} = \langle \mathbf{v}, \mathbf{u} \rangle \|\mathbf{u}\|^{-2} \mathbf{u}$, so $\mathbf{v}$ and $\mathbf{u}$ are linearly dependent. Conversely, if $\mathbf{v}$ and $\mathbf{u}$ are linearly dependent, then $\mathbf{v} = c\mathbf{u}$ for some nonzero scalar c, in which case

$$
\mathbf{x} = \frac{\langle \mathbf{v}, \mathbf{u} \rangle}{\|\mathbf{u}\|^2} \mathbf{u} = \frac{\langle c\mathbf{u}, \mathbf{u} \rangle}{\|\mathbf{u}\|^2} \mathbf{u} = c\mathbf{u} = \mathbf{v},
$$

and hence (4.5.16) is an equality. $\square$

Example 4.5.17 Let $\lambda_1, \lambda_2, \ldots, \lambda_n \in \mathbb{C}$. Consider the all-ones vector $\mathbf{e} \in \mathbb{C}^n$, the vector $\mathbf{u} = [\lambda_i] \in \mathbb{C}^n$, and the standard inner product on $\mathbb{C}^n$. The Cauchy–Schwarz inequality ensures that

$$
\left| \sum_{i=1}^{n} \lambda_i \right|^2 = |\langle \mathbf{u}, \mathbf{e} \rangle|^2 \leq \|\mathbf{u}\|_2^2 \|\mathbf{e}\|_2^2 = \left(\sum_{i=1}^{n} |\lambda_i|^2 \right) \left(\sum_{i=1}^{n} 1^2 \right) = n \sum_{i=1}^{n} |\lambda_i|^2
$$

with equality if and only if $\mathbf{e}$ and $\mathbf{u}$ are linearly dependent, that is, if and only if $\mathbf{u} = c\mathbf{e}$ for some $c \in \mathbb{C}$. Thus,

$$
\left| \sum_{i=1}^{n} \lambda_i \right| \leq \sqrt{n} \left(\sum_{i=1}^{n} |\lambda_i|^2 \right)^{1/2},
\tag{4.5.18}
$$

with equality if and only if $\lambda_1 = \lambda_2 = \cdots = \lambda_n$.

Example 4.5.19 Now consider a vector $\mathbf{p} = [p_i] \in \mathbb{R}^n$ with nonnegative entries that sum to 1, that is, $\mathbf{e}^\mathsf{T} \mathbf{p} = 1$. The Cauchy–Schwarz inequality ensures that

$$
\left| \sum_{i=1}^{n} p_i \lambda_i \right|^2 = \left| \sum_{i=1}^{n} \sqrt{p_i} \left(\sqrt{p_i} \lambda_i \right) \right|^2 \leq \left(\sum_{i=1}^{n} p_i \right) \left(\sum_{i=1}^{n} p_i |\lambda_i|^2 \right) = \sum_{i=1}^{n} p_i |\lambda_i|^2,
$$

with equality if and only if $\mathbf{p}$ and $\mathbf{u} = [\lambda_i] \in \mathbb{C}^n$ are linearly dependent.

In addition to the basic properties listed in Theorem 4.5.9, a derived norm also satisfies an inequality that generalizes (4.2.3); see Figure 4.7.

Corollary 4.5.20 (Triangle Inequality for a Derived Norm) *Let $\mathcal{V}$ be an inner product space and let $\mathbf{u}, \mathbf{v} \in \mathcal{V}$. Then*

$$
\|\mathbf{u} + \mathbf{v}\| \leq \|\mathbf{u}\| + \|\mathbf{v}\|,
\tag{4.5.21}
$$

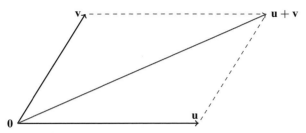

Figure 4.7 The triangle inequality.

with equality if and only if one of the vectors is a real nonnegative scalar multiple of the other.

Proof We invoke the additivity and conjugate symmetry of the inner product, together with the Cauchy–Schwarz inequality (4.5.15):

$$\begin{aligned}
\|\mathbf{u} + \mathbf{v}\|^2 &= \langle \mathbf{u} + \mathbf{v}, \mathbf{u} + \mathbf{v} \rangle \\
&= \langle \mathbf{u}, \mathbf{u} \rangle + \langle \mathbf{u}, \mathbf{v} \rangle + \langle \mathbf{v}, \mathbf{u} \rangle + \langle \mathbf{v}, \mathbf{v} \rangle \\
&= \langle \mathbf{u}, \mathbf{u} \rangle + \langle \mathbf{u}, \mathbf{v} \rangle + \overline{\langle \mathbf{u}, \mathbf{v} \rangle} + \langle \mathbf{v}, \mathbf{v} \rangle \\
&= \|\mathbf{u}\|^2 + 2\operatorname{Re}\langle \mathbf{u}, \mathbf{v} \rangle + \|\mathbf{v}\|^2 \\
&\leq \|\mathbf{u}\|^2 + 2|\langle \mathbf{u}, \mathbf{v} \rangle| + \|\mathbf{v}\|^2 \qquad (4.5.22) \\
&\leq \|\mathbf{u}\|^2 + 2\|\mathbf{u}\|\|\mathbf{v}\| + \|\mathbf{v}\|^2 \qquad (4.5.23) \\
&= \big(\|\mathbf{u}\| + \|\mathbf{v}\|\big)^2.
\end{aligned}$$

We conclude that $\|\mathbf{u} + \mathbf{v}\| \leq \|\mathbf{u}\| + \|\mathbf{v}\|$, with equality if and only if the inequalities (4.5.23) and (4.5.22) are equalities. Equality in the Cauchy–Schwarz inequality (4.5.23) occurs if and only if there is a scalar c such that either $\mathbf{u} = c\mathbf{v}$ or $\mathbf{v} = c\mathbf{u}$. Equality in (4.5.22) occurs if and only if c is real and nonnegative. $\qquad\square$

In an inner product space, the norm is determined by the inner product since $\|\mathbf{v}\|^2 = \langle \mathbf{v}, \mathbf{v} \rangle$. The following result shows that the inner product is determined by the norm.

Theorem 4.5.24 (Polarization Identities) *Let $\mathcal{V}$ be an $\mathbb{F}$-inner product space and let $\mathbf{u}, \mathbf{v} \in \mathcal{V}$.*

(a) *If $\mathbb{F} = \mathbb{R}$, then*

$$\langle \mathbf{u}, \mathbf{v} \rangle = \frac{1}{4}\big(\|\mathbf{u} + \mathbf{v}\|^2 - \|\mathbf{u} - \mathbf{v}\|^2\big). \qquad (4.5.25)$$

(b) *If $\mathbb{F} = \mathbb{C}$, then*

$$\langle \mathbf{u}, \mathbf{v} \rangle = \frac{1}{4}\big(\|\mathbf{u} + \mathbf{v}\|^2 - \|\mathbf{u} - \mathbf{v}\|^2 + i\|\mathbf{u} + i\mathbf{v}\|^2 - i\|\mathbf{u} - i\mathbf{v}\|^2\big). \qquad (4.5.26)$$

Proof

(a) If $\mathbb{F} = \mathbb{R}$, then

$$
\begin{aligned}
\|\mathbf{u} + \mathbf{v}\|^2 - \|\mathbf{u} - \mathbf{v}\|^2 &= (\|\mathbf{u}\|^2 + 2\langle \mathbf{u}, \mathbf{v}\rangle + \|\mathbf{v}\|^2) \\
&\quad - (\|\mathbf{u}\|^2 + 2\langle \mathbf{u}, -\mathbf{v}\rangle + \|-\mathbf{v}\|^2) \\
&= 2\langle \mathbf{u}, \mathbf{v}\rangle + \|\mathbf{v}\|^2 + 2\langle \mathbf{u}, \mathbf{v}\rangle - \|\mathbf{v}\|^2 \\
&= 4\langle \mathbf{u}, \mathbf{v}\rangle.
\end{aligned}
$$

(b) If $\mathbb{F} = \mathbb{C}$, then

$$
\begin{aligned}
\|\mathbf{u} + \mathbf{v}\|^2 - \|\mathbf{u} - \mathbf{v}\|^2 &= (\|\mathbf{u}\|^2 + 2\operatorname{Re}\langle \mathbf{u}, \mathbf{v}\rangle + \|\mathbf{v}\|^2) \\
&\quad - (\|\mathbf{u}\|^2 + 2\operatorname{Re}\langle \mathbf{u}, -\mathbf{v}\rangle + \|-\mathbf{v}\|^2) \\
&= 2\operatorname{Re}\langle \mathbf{u}, \mathbf{v}\rangle + \|\mathbf{v}\|^2 + 2\operatorname{Re}\langle \mathbf{u}, \mathbf{v}\rangle - \|\mathbf{v}\|^2 \\
&= 4\operatorname{Re}\langle \mathbf{u}, \mathbf{v}\rangle
\end{aligned}
$$

and

$$
\begin{aligned}
\|\mathbf{u} + i\mathbf{v}\|^2 - \|\mathbf{u} - i\mathbf{v}\|^2 &= (\|\mathbf{u}\|^2 + 2\operatorname{Re}\langle \mathbf{u}, i\mathbf{v}\rangle + \|i\mathbf{v}\|^2) \\
&\quad - (\|\mathbf{u}\|^2 + 2\operatorname{Re}\langle \mathbf{u}, -i\mathbf{v}\rangle + \|-i\mathbf{v}\|^2) \\
&= -2\operatorname{Re} i\langle \mathbf{u}, \mathbf{v}\rangle + \|\mathbf{v}\|^2 - 2\operatorname{Re} i\langle \mathbf{u}, \mathbf{v}\rangle - \|\mathbf{v}\|^2 \\
&= -4\operatorname{Re} i\langle \mathbf{u}, \mathbf{v}\rangle \\
&= 4\operatorname{Im}\langle \mathbf{u}, \mathbf{v}\rangle.
\end{aligned}
$$

Therefore, the right-hand side of (4.5.26) is

$$
\frac{1}{4}(4\operatorname{Re}\langle \mathbf{u}, \mathbf{v}\rangle + 4i\operatorname{Im}\langle \mathbf{u}, \mathbf{v}\rangle) = \operatorname{Re}\langle \mathbf{u}, \mathbf{v}\rangle + i\operatorname{Im}\langle \mathbf{u}, \mathbf{v}\rangle = \langle \mathbf{u}, \mathbf{v}\rangle. \qquad \square
$$

4.6 Normed Vector Spaces

In the preceding section, we showed how a generalized length function on a vector space can be derived from an inner product. We now introduce other kinds of generalized length functions that have proved to be useful in applications, but might not be derived from an inner product.

Definition 4.6.1 A *norm* on an $\mathbb{F}$-vector space $\mathcal{V}$ is a function $\|\cdot\| : \mathcal{V} \to [0, \infty)$ that has the following properties for any $\mathbf{u}, \mathbf{v} \in \mathcal{V}$ and $c \in \mathbb{F}$:

(a) $\|\mathbf{u}\|$ is real and nonnegative.   *Nonnegativity*

(b) $\|\mathbf{u}\| = 0$ if and only if $\mathbf{u} = 0$.   *Positivity*

(c) $\|c\mathbf{u}\| = |c|\|\mathbf{u}\|$.   *Homogeneity*

(d) $\|\mathbf{u} + \mathbf{v}\| \leq \|\mathbf{u}\| + \|\mathbf{v}\|$.   *Triangle Inequality*

In the following three examples, $\mathcal{V}$ is the $\mathbb{F}$-vector space $\mathbb{F}^n$ and $\mathbf{u} = [u_i], \mathbf{v} = [v_i] \in \mathcal{V}$; $\mathbf{e}_1$ and $\mathbf{e}_2$ are the first two standard unit basis vectors in $\mathbb{F}^n$.

Example 4.6.2 The function

$$\|\mathbf{u}\|_1 = |u_1| + |u_2| + \cdots + |u_n| \tag{4.6.3}$$

is the ℓ_1 *norm* (or *absolute sum norm*) on $\mathbb{F}^n$. It satisfies the nonnegativity, positivity, and homogeneity axioms in the preceding definition of a norm. To verify the triangle inequality for (4.6.3) we invoke the triangle inequality for the modulus function on $\mathbb{F}$ and compute

$$\begin{aligned}
\|\mathbf{u} + \mathbf{v}\|_1 &= |u_1 + v_1| + |u_2 + v_2| + \cdots + |u_n + v_n| \\
&\leq |u_1| + |v_1| + |u_2| + |v_2| + \cdots + |u_n| + |v_n| \\
&= \|\mathbf{u}\|_1 + \|\mathbf{v}\|_1.
\end{aligned}$$

Since

$$\|\mathbf{e}_1 + \mathbf{e}_2\|_1^2 + \|\mathbf{e}_1 - \mathbf{e}_2\|_1^2 = 8 > 4 = 2\|\mathbf{e}_1\|_1^2 + 2\|\mathbf{e}_2\|_1^2,$$

the ℓ_1 norm does not satisfy the parallelogram identity; see Theorem 4.5.9.e. Consequently, it is not derived from an inner product.

Example 4.6.4 The function

$$\|\mathbf{u}\|_\infty = \max\{|u_i| : 1 \leq i \leq n\} \tag{4.6.5}$$

is the ℓ_∞ *norm* (or *max norm*) on $\mathbb{F}^n$. Verification of the nonnegativity, positivity, and homogeneity axioms is straightforward. To verify the triangle inequality, let k be any index such that $\|\mathbf{u} + \mathbf{v}\|_\infty = |u_k + v_k|$. Use the triangle inequality for the modulus function on $\mathbb{F}$ to compute

$$\|\mathbf{u} + \mathbf{v}\|_\infty = |u_k + v_k| \leq |u_k| + |v_k| \leq \|\mathbf{u}\|_\infty + \|\mathbf{v}\|_\infty.$$

The computation

$$\|\mathbf{e}_1 + \mathbf{e}_2\|_\infty^2 + \|\mathbf{e}_1 - \mathbf{e}_2\|_\infty^2 = 2 < 4 = 2\|\mathbf{e}_1\|_\infty^2 + 2\|\mathbf{e}_2\|_\infty^2$$

shows that the ℓ_∞ norm does not satisfy the parallelogram identity. Hence it is not derived from an inner product.

Example 4.6.6 The Euclidean norm

$$\|\mathbf{u}\|_2 = (|u_1|^2 + |u_2|^2 + \cdots + |u_n|^2)^{1/2} \tag{4.6.7}$$

on $\mathbb{F}^n$ is derived from the standard inner product; see (4.5.4).

Definition 4.6.8 A *normed vector space* is a real or complex vector space $\mathcal{V}$, together with a norm $\|\cdot\| : \mathcal{V} \to [0, \infty)$. The *unit ball* of a normed space is $\{\mathbf{v} \in \mathcal{V} : \|\mathbf{v}\| \leq 1\}$.

The unit balls for the ℓ_1, ℓ_2 (Euclidean), and ℓ_∞ norms on $\mathbb{R}^2$ are illustrated in Figure 4.8.

Example 4.6.9 New norms on $\mathbb{F}^n$ can be constructed from old ones with the help of an invertible matrix. Let $\|\cdot\|$ be a norm on $\mathbb{F}^n$, let $A \in \mathsf{M}_n(\mathbb{F})$, and define the function $\|\cdot\|_A : \mathcal{V} \to [0, \infty)$ by $\|\mathbf{u}\|_A = \|A\mathbf{u}\|$. The nonnegativity, homogeneity, and triangle inequality axioms for a norm are always satisfied, but the positivity axiom is satisfied if and only if A is invertible; see P.4.26.

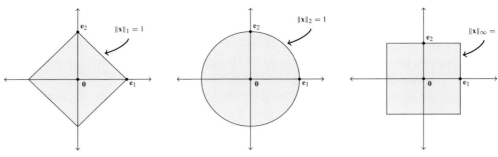

(a) Unit ball for the norm $\| \cdot \|_1$ on $\mathbb{R}^2$.

(b) Unit ball for the norm $\| \cdot \|_2$ on $\mathbb{R}^2$.

(c) Unit ball for the norm $\| \cdot \|_\infty$ on $\mathbb{R}^2$.

Figure 4.8 The ℓ_1, ℓ_2, and ℓ_∞ norms.

Definition 4.6.10 Let $\mathcal{V}$ be a normed vector space with norm $\| \cdot \|$. Then $\mathbf{v} \in \mathcal{V}$ is a *unit vector* if $\|\mathbf{v}\| = 1$.

Any nonzero vector $\mathbf{v}$ in a normed vector space can be scaled to create a unit vector $\mathbf{v}/\|\mathbf{v}\|$:

$$\text{If } \mathbf{v} \neq \mathbf{0}, \text{ then } \left\| \frac{\mathbf{v}}{\|\mathbf{v}\|} \right\| = \frac{\|\mathbf{v}\|}{\|\mathbf{v}\|} = 1.$$

The process of scaling a nonzero vector to a unit vector is called *normalization*.

4.7 Problems

P.4.1 Let $\mathcal{V}$ be a real inner product space and let $\mathbf{x}, \mathbf{y} \in \mathcal{V}$. Show that $\langle \mathbf{x}, \mathbf{y} \rangle = 0$ if and only if $\|\mathbf{x} + \mathbf{y}\| = \|\mathbf{x} - \mathbf{y}\|$.

P.4.2 Let $\mathcal{V}$ be a complex inner product space and let $\mathbf{x}, \mathbf{y} \in \mathcal{V}$. (a) If $\langle \mathbf{x}, \mathbf{y} \rangle = 0$, show that $\|\mathbf{x} + \mathbf{y}\| = \|\mathbf{x} - \mathbf{y}\|$. (b) What can you say about the converse? Consider the case $\mathbf{y} = i\mathbf{x}$.

P.4.3 Provide details for the following alternative proof of the Cauchy–Schwarz inequality in a real inner product space $\mathcal{V}$. For nonzero vectors $\mathbf{u}, \mathbf{v} \in \mathcal{V}$, consider the function $p(t) = \|t\mathbf{u} + \mathbf{v}\|^2$ of a real variable t. (a) Why is $p(t) \geq 0$ for all real t? (b) If $p(t) = 0$ has a real root, why are $\mathbf{u}$ and $\mathbf{v}$ linearly dependent? (c) Show that p is a polynomial

$$p(t) = \|\mathbf{u}\|^2 t^2 + 2\langle \mathbf{u}, \mathbf{v} \rangle t + \|\mathbf{v}\|^2$$

of degree 2 with real coefficients. (d) If $\mathbf{u}$ and $\mathbf{v}$ are linearly independent, why does $p(t) = 0$ have no real roots? (e) Use the quadratic formula to deduce the Cauchy–Schwarz inequality for a real inner product space.

P.4.4 Modify the argument in the preceding problem to prove the Cauchy–Schwarz inequality in a complex inner product space $\mathcal{V}$. Redefine $p(t) = \|t\mathbf{u} + e^{i\theta}\mathbf{v}\|^2$, in which θ is a real parameter such that $e^{-i\theta}\langle \mathbf{u}, \mathbf{v} \rangle = |\langle \mathbf{u}, \mathbf{v} \rangle|$. (a) Explain why such a choice of θ is possible and why

$$p(t) = \|\mathbf{u}\|^2 t^2 + 2|\langle \mathbf{u}, \mathbf{v} \rangle| t + \|\mathbf{v}\|^2,$$

a polynomial of degree 2 with real coefficients. (b) If **u** and **v** are linearly independent, why does $p(t) = 0$ have no real roots? (c) Use the quadratic formula to deduce the Cauchy–Schwarz inequality for a complex inner product space.

P.4.5 Let x and y be nonnegative real numbers. Use the fact that $(a - b)^2 \geq 0$ for real a, b to show that

$$\sqrt{xy} \leq \frac{x + y}{2}, \qquad (4.7.1)$$

with equality if and only if $x = y$. This inequality is known as the *arithmetic–geometric mean inequality*. The left-hand side of (4.7.1) is the *geometric mean* of x and y; the right-hand side is their *arithmetic mean*.

P.4.6 Let $\mathcal{V}$ be an inner product space and let $\mathbf{u}, \mathbf{v} \in \mathcal{V}$. (a) Expand the inequality $0 \leq \|\mathbf{x} - \mathbf{y}\|^2$ and choose suitable $\mathbf{x}, \mathbf{y} \in \mathcal{V}$ to obtain

$$|\langle \mathbf{u}, \mathbf{v} \rangle| \leq \frac{\lambda^2}{2} \|\mathbf{u}\|^2 + \frac{1}{2\lambda^2} \|\mathbf{v}\|^2 \qquad (4.7.2)$$

for all $\lambda > 0$. Be sure your proof covers both cases $\mathbb{F} = \mathbb{R}$ and $\mathbb{F} = \mathbb{C}$. (b) Use (4.7.2) to prove the arithmetic–geometric mean inequality (4.7.1). (c) Use (4.7.2) to prove the Cauchy–Schwarz inequality.

P.4.7 Let $x, y \geq 0$, and consider the vectors $\mathbf{u} = [\sqrt{x} \ \sqrt{y}]^\mathsf{T}$ and $\mathbf{v} = [\sqrt{y} \ \sqrt{x}]^\mathsf{T}$ in $\mathbb{R}^2$. Use the Cauchy–Schwarz inequality to prove the arithmetic–geometric mean inequality (4.7.1).

P.4.8 Let $\mathcal{V}$ be an inner product space and let $\mathbf{u}, \mathbf{v} \in \mathcal{V}$. Show that

$$2|\langle \mathbf{u}, \mathbf{v} \rangle| \leq \|\mathbf{u}\|^2 + \|\mathbf{v}\|^2.$$

P.4.9 Let $\mathcal{V}$ be an inner product space and let $\mathbf{u}, \mathbf{v} \in \mathcal{V}$. Show that

$$\|\mathbf{u} + \mathbf{v}\| \, \|\mathbf{u} - \mathbf{v}\| \leq \|\mathbf{u}\|^2 + \|\mathbf{v}\|^2.$$

P.4.10 Let $\mathcal{V}$ be an $\mathbb{F}$-inner product space and let $\mathbf{u}, \mathbf{v} \in \mathcal{V}$. Prove that $\mathbf{u}$ and $\mathbf{v}$ are orthogonal if and only if $\|\mathbf{v}\| \leq \|c\mathbf{u} + \mathbf{v}\|$ for all $c \in \mathbb{F}$. Be sure your proof covers both cases $\mathbb{F} = \mathbb{R}$ and $\mathbb{F} = \mathbb{C}$. Draw a diagram illustrating what this means if $\mathcal{V} = \mathbb{R}^2$.

P.4.11 (a) Let $f, g \in C_\mathbb{R}[0, 1]$ (see Example 4.4.8) be the functions depicted in Figure 4.9. Does there exist a $c \in \mathbb{R}$ such that $\|f + cg\| < \|f\|$? (b) Consider the functions $f(x) = x(1 - x)$ and $g(x) = \sin(2\pi x)$ in $C_\mathbb{R}[0, 1]$. Does there exist a $c \in \mathbb{R}$ such that $\|f + cg\| < \|f\|$?

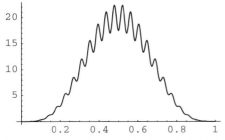
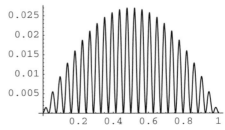

Figure 4.9 Graphs of $f(z)$ (left) and $g(x)$ (right).

P.4.12 Let $\mathcal{V}$ be a complex inner product space and let $\mathbf{u}, \mathbf{v} \in \mathcal{V}$. Show that

$$\langle \mathbf{u}, \mathbf{v} \rangle = \frac{1}{2\pi} \int_{-\pi}^{\pi} e^{i\theta} \|\mathbf{u} + e^{i\theta}\mathbf{v}\|^2 \, d\theta.$$

P.4.13 Let $\mathcal{V}$ be the real inner product space $C_{\mathbb{R}}[0, 1]$ and let $f \in \mathcal{V}$ be real-valued and strictly positive. Use the Cauchy–Schwarz inequality to deduce that

$$\frac{1}{\int_0^1 f(t)\,dt} \leq \int_0^1 \frac{1}{f(t)}\,dt.$$

P.4.14 Let $\mathcal{V}$ be the real inner product space $C_{\mathbb{R}}[0, 1]$ and let $a \in [0, 1]$. Show that there is no nonnegative function $f \in \mathcal{V}$ such that

$$\int_0^1 f(x)\,dx = 1, \qquad \int_0^1 xf(x)\,dx = a, \quad \text{and} \quad \int_0^1 x^2 f(x)\,dx = a^2.$$

P.4.15 Let $\langle \cdot, \cdot \rangle$ be an inner product on $\mathbb{F}^n$ and suppose that $A \in \mathsf{M}_n(\mathbb{F})$ is invertible. Define $\langle \mathbf{u}, \mathbf{v} \rangle_A = \langle A\mathbf{u}, A\mathbf{v} \rangle$. Show that $\langle \cdot, \cdot \rangle_A$ is an inner product on $\mathbb{F}^n$.

P.4.16 Let $\mathbf{x} \in \mathbb{C}^n$ and $A \in \mathsf{M}_n$. Let $\|\cdot\|_F$ denote the Frobenius norm. (a) Show that $\|A\mathbf{x}\|_2 \leq \|A\|_F\|\mathbf{x}\|_2$ and $\|AB\|_F \leq \|A\|_F\|B\|_F$; in particular, $\|A^2\|_F \leq \|A\|_F^2$. (b) If $A^2 = A \neq 0$, show that $\|A\|_F \geq 1$ and give an example to show that equality is possible.

P.4.17 Let $\mathbf{u}$, $\mathbf{v}$, and $\mathbf{w}$ be vectors in an inner product space. (a) If $\mathbf{u} \perp \mathbf{v}$ and $\mathbf{v} \perp \mathbf{w}$, is $\mathbf{u} \perp \mathbf{w}$? (b) If $\mathbf{u} \perp \mathbf{v}$ and $\mathbf{u} \perp \mathbf{w}$, is $\mathbf{u} \perp (\mathbf{v} + \mathbf{w})$?

P.4.18 Let $\mathbf{u}$ and $\mathbf{v}$ be vectors in a real inner product space $\mathcal{V}$. If $\|\mathbf{u} + \mathbf{v}\|^2 = \|\mathbf{u}\|^2 + \|\mathbf{v}\|^2$, show that $\mathbf{u} \perp \mathbf{v}$. What can you say if $\mathcal{V}$ is a complex inner product space?

P.4.19 Let $\mathbf{u}_1, \mathbf{u}_2, \ldots, \mathbf{u}_k$ be vectors in an inner product space, and let $c_1, c_2, \ldots, c_k$ be scalars. Suppose that $\mathbf{u}_i \perp \mathbf{u}_j$ for all $i \neq j$. Show that

$$\|c_1\mathbf{u}_1 + c_2\mathbf{u}_2 + \cdots + c_k\mathbf{u}_k\|^2 = |c_1|^2\|\mathbf{u}_1\|^2 + |c_2|^2\|\mathbf{u}_2\|^2 + \cdots + |c_k|^2\|\mathbf{u}_k\|^2.$$

P.4.20 Show that the triangle inequality $a + b \geq c$ in Corollary 4.2.2 is a strict inequality if the angle between the sides a and b is less than π. Sketch what the case of equality looks like.

P.4.21 Let $\mathcal{V}$ be an inner product space and let $\mathbf{u}, \mathbf{v}, \mathbf{w} \in \mathcal{V}$. Show that

$$\|\mathbf{u} + \mathbf{v} + \mathbf{w}\|^2 + \|\mathbf{u}\|^2 + \|\mathbf{v}\|^2 + \|\mathbf{w}\|^2 = \|\mathbf{u} + \mathbf{v}\|^2 + \|\mathbf{u} + \mathbf{w}\|^2 + \|\mathbf{v} + \mathbf{w}\|^2.$$

P.4.22 Let $\mathcal{V}$ be an inner product space and let $\mathbf{u}, \mathbf{v} \in \mathcal{V}$. Show that

$$\left| \|\mathbf{u}\| - \|\mathbf{v}\| \right| \leq \|\mathbf{u} - \mathbf{v}\|.$$

P.4.23 Let $\mathcal{V} = \mathcal{P}_n$. If $p(z) = \sum_{k=0}^n p_k z^k$ and $q(z) = \sum_{k=0}^n q_k z^k$, define

$$\langle p, q \rangle = \sum_{i,j=0}^n \frac{p_i \overline{q_j}}{i + j + 1}.$$

(a) Show that $\langle \cdot, \cdot \rangle : \mathcal{V} \times \mathcal{V} \to \mathbb{C}$ is an inner product. (b) Deduce that the matrix $A = [(i + j - 1)^{-1}] \in \mathsf{M}_n$ has the property that $\langle A\mathbf{x}, \mathbf{x} \rangle \geq 0$ for all $\mathbf{x} \in \mathbb{C}^n$.

P.4.24 Let $\mathcal{V}$ be a normed vector space with norm $\| \cdot \|$ that is derived from an inner product $\langle \cdot, \cdot \rangle_1$ on $\mathcal{V} \times \mathcal{V}$. Could there be a *different* inner product $\langle \cdot, \cdot \rangle_2$ on $\mathcal{V} \times \mathcal{V}$ such that $\|\mathbf{u}\| = \langle \mathbf{u}, \mathbf{u} \rangle_2^{1/2}$ for all $\mathbf{u} \in \mathcal{V}$?

P.4.25 Consider the function $\| \cdot \| : \mathbb{R}^2 \to [0, \infty)$ defined by $\|\mathbf{u}\| = |u_1|$. Show that it satisfies three of the axioms for a norm in Definition 4.6.1. Show that it does not satisfy the remaining axiom.

P.4.26 Verify the assertions in Example 4.6.9 about the four axioms for a norm and the function $\| \cdot \|_A$ on $\mathbb{F}^n$.

P.4.27 Let $\mathcal{V}$ be the real vector space $\mathbb{R}^2$ and let $\mathbf{u} = [u_1 \; u_2]^\mathsf{T} \in \mathcal{V}$. Show that the function $\|\mathbf{u}\| = 2|u_1| + 5|u_2|$ is a norm on $\mathcal{V}$. Is it derived from an inner product? If so, what is that inner product? Sketch the unit ball for this norm.

P.4.28 Let $\mathcal{V}$ be the real vector space $\mathbb{R}^2$ and let $\mathbf{u} = [u_1 \; u_2]^\mathsf{T} \in \mathcal{V}$. Show that the function $\|\mathbf{u}\| = (2|u_1|^2 + 5|u_2|^2)^{1/2}$ is a norm on $\mathcal{V}$. Is it derived from an inner product? If so, what is that inner product? Sketch the unit ball for this norm.

P.4.29 Let $\mathscr{B}$ be the unit ball of a normed vector space. If $\mathbf{u}, \mathbf{v} \in \mathscr{B}$ and $0 \le t \le 1$, show that $t\mathbf{u} + (1 - t)\mathbf{v} \in \mathscr{B}$. This shows that $\mathscr{B}$ is a *convex set*.

4.8 Notes

It is known that a norm is derived from an inner product if and only if it satisfies the parallelogram identity (Theorem 4.5.9.e); see [HJ13, 5.1.P12].

4.9 Some Important Concepts

- Axioms for an inner product space.

- Orthogonality.

- Parallelogram identity and derived norms.

- Cauchy–Schwarz inequality.

- Triangle inequality.

- Polarization identity and derived norms.

- Axioms for a normed vector space.

- Normalization and the unit ball.

The standard basis vectors in $\mathbb{R}^3$ are mutually orthogonal and of unit length; these two properties simplify many computations. In this chapter, we explore the role of orthonormal (orthogonal and normalized) vectors in the general setting of an $\mathbb{F}$-inner product space $\mathcal{V}$ ($\mathbb{F} = \mathbb{R}$ or $\mathbb{C}$) with inner product $\langle \cdot, \cdot \rangle$ and derived norm $\| \cdot \|$.

Basis representations of linear transformations with respect to orthonormal bases are of particular importance, and are intimately associated with the important notion of an adjoint transformation. In the final section of this chapter we give a brief introduction to Fourier series, which exploit the orthogonality properties of sine and cosine functions.

5.1 Orthonormal Systems

Definition 5.1.1 A sequence of vectors $\mathbf{u}_1, \mathbf{u}_2, \ldots$ (finite or infinite) in an inner product space is *orthonormal* if

$$\langle \mathbf{u}_i, \mathbf{u}_j \rangle = \delta_{ij} \text{ for all } i, j. \tag{5.1.2}$$

An orthonormal sequence of vectors is an *orthonormal system*.

If $\mathbf{u}_1, \mathbf{u}_2, \ldots$ is an orthonormal system and $1 \leq i_1 < i_2 < \cdots$, then $\mathbf{u}_{i_1}, \mathbf{u}_{i_2}, \ldots$ satisfy the conditions (5.1.2), so it is also an orthonormal system.

The vectors in an orthonormal system are mutually orthogonal and have unit norm. Any sequence $\mathbf{v}_1, \mathbf{v}_2, \ldots$ of nonzero mutually orthogonal vectors can be turned into an orthonormal system by normalizing each $\mathbf{v}_i$; see (4.5.11).

Example 5.1.3 The vectors

$$\mathbf{u}_1 = \begin{bmatrix} \frac{1}{\sqrt{2}} \\ \frac{1}{\sqrt{2}} \end{bmatrix} \quad \text{and} \quad \mathbf{u}_2 = \begin{bmatrix} -\frac{1}{\sqrt{2}} \\ \frac{1}{\sqrt{2}} \end{bmatrix}$$

are orthonormal in $\mathbb{F}^2$, endowed with the standard inner product.

Example 5.1.4 The vectors

$$\mathbf{u}_1 = (1, 0, 0, 0, \ldots), \quad \mathbf{u}_2 = (0, 1, 0, 0, \ldots), \quad \mathbf{u}_3 = (0, 0, 1, 0, \ldots), \ldots$$

comprise an orthonormal system in the inner product space of finitely nonzero sequences; see Example 4.4.12.

Example 5.1.5 We claim that the polynomials

$$f_1(x) = 1, \qquad f_2(x) = 2x - 1, \qquad \text{and} \qquad f_3(x) = 6x^2 - 6x + 1$$

are mutually orthogonal vectors in $C[0, 1]$, endowed with the L^2 inner product

$$\langle f, g \rangle = \int_0^1 f(t)\overline{g(t)}\, dt. \tag{5.1.6}$$

To verify this, compute

$$\langle f_1, f_2 \rangle = \int_0^1 (1)\overline{(2x-1)}\, dx = \int_0^1 (2x-1)\, dx = x^2 - x \Big|_0^1 = 0,$$

$$\langle f_1, f_3 \rangle = \int_0^1 (1)\overline{(6x^2 - 6x + 1)}\, dx = \int_0^1 (6x^2 - 6x + 1)\, dx$$

$$= 2x^3 - 3x^2 + x \Big|_0^1 = 2 - 3 + 1 = 0, \quad \text{and}$$

$$\langle f_2, f_3 \rangle = \int_0^1 (2x-1)\overline{(6x^2 - 6x + 1)}\, dx = \int_0^1 (12x^3 - 18x^2 + 8x - 1)\, dx$$

$$= 3x^4 - 6x^3 + 4x^2 - x \Big|_0^1 = 3 - 6 + 4 - 1 = 0.$$

A further computation shows that

$$\|f_1\| = 1, \quad \|f_2\| = \frac{1}{\sqrt{3}}, \quad \text{and} \quad \|f_3\| = \frac{1}{\sqrt{5}}.$$

Normalize the vectors f_i to obtain an orthonormal system in $C[0, 1]$:

$$u_1 = 1, \qquad u_2 = \sqrt{3}(2x - 1), \qquad u_3 = \sqrt{5}(6x^2 - 6x + 1). \tag{5.1.7}$$

The following theorem generalizes Theorem 4.5.9.d to an identity that involves n orthonormal vectors.

Theorem 5.1.8 *If* $\mathbf{u}_1, \mathbf{u}_2, \ldots, \mathbf{u}_n$ *is an orthonormal system, then*

$$\left\| \sum_{i=1}^n a_i \mathbf{u}_i \right\|^2 = \sum_{i=1}^n |a_i|^2 \tag{5.1.9}$$

for all $a_1, a_2, \ldots, a_n \in \mathbb{F}$.

Proof Compute

$$\left\| \sum_{i=1}^n a_i \mathbf{u}_i \right\|^2 = \left\langle \sum_{i=1}^n a_i \mathbf{u}_i, \sum_{j=1}^n a_j \mathbf{u}_j \right\rangle = \sum_{i=1}^n a_i \left\langle \mathbf{u}_i, \sum_{j=1}^n a_j \mathbf{u}_j \right\rangle$$

$$= \sum_{i=1}^n a_i \sum_{j=1}^n \overline{a_j} \langle \mathbf{u}_i, \mathbf{u}_j \rangle = \sum_{i=1}^n a_i \sum_{j=1}^n \overline{a_j} \delta_{ij}$$

$$= \sum_{i=1}^n a_i \overline{a_i} = \sum_{i=1}^n |a_i|^2. \qquad \square$$

A linearly independent list of vectors need not be an orthonormal system; its vectors need not be either normalized or orthogonal. However, a finite orthonormal system is linearly independent.

Theorem 5.1.10 *If* $\mathbf{u}_1, \mathbf{u}_2, \ldots, \mathbf{u}_n$ *is an orthonormal system, then* $\mathbf{u}_1, \mathbf{u}_2, \ldots, \mathbf{u}_n$ *is a linearly independent list.*

Proof If $a_1, a_2, \ldots, a_n \in \mathbb{F}$ and $a_1\mathbf{u}_1 + a_2\mathbf{u}_2 + \cdots + a_n\mathbf{u}_n = \mathbf{0}$, then (5.1.9) ensures that

$$|a_1|^2 + |a_2|^2 + \cdots + |a_n|^2 = \|\mathbf{0}\|^2 = 0,$$

so $a_1 = a_2 = \cdots = a_n = 0$. $\qquad\square$

5.2 Orthonormal Bases

Definition 5.2.1 An *orthonormal basis* for a finite-dimensional inner product space is a basis that is an orthonormal system.

Since a finite orthonormal system in an inner product space $\mathcal{V}$ is linearly independent (Theorem 5.1.10), it is a basis for its span. If that span is all of $\mathcal{V}$, then the orthonormal system is a basis for $\mathcal{V}$.

Example 5.2.2 In the $\mathbb{F}$-inner product space $\mathcal{V} = \mathsf{M}_{m \times n}(\mathbb{F})$ with the Frobenius inner product, the basis comprising the matrices E_{pq} defined in Example 2.2.3 is an orthonormal basis for $\mathcal{V}$.

Example 5.2.3 We claim that the vectors

$$\mathbf{u}_1 = \frac{1}{2}\begin{bmatrix} 1 \\ 1 \\ 1 \\ 1 \end{bmatrix}, \quad \mathbf{u}_2 = \frac{1}{2}\begin{bmatrix} 1 \\ i \\ -1 \\ -i \end{bmatrix}, \quad \mathbf{u}_3 = \frac{1}{2}\begin{bmatrix} 1 \\ -1 \\ 1 \\ -1 \end{bmatrix}, \quad \mathbf{u}_4 = \frac{1}{2}\begin{bmatrix} 1 \\ -i \\ -1 \\ i \end{bmatrix} \tag{5.2.4}$$

comprise an orthonormal basis for $\mathbb{C}^4$. Since $\langle \mathbf{u}_i, \mathbf{u}_j \rangle = \delta_{ij}$ for $1 \leq i, j \leq 4$, we see that $\mathbf{u}_1, \mathbf{u}_2, \mathbf{u}_3, \mathbf{u}_4$ is an orthonormal system; Theorem 5.1.10 ensures that it is linearly independent. Since it is a maximal linearly independent list in $\mathbb{C}^4$, Corollary 2.2.8 ensures that it is a basis. The vectors (5.2.4) are the columns of the 4×4 Fourier matrix; see (6.2.15).

The principle invoked in the preceding example is important. If $\mathcal{V}$ is an n-dimensional inner product space, then any n orthonormal vectors comprise an orthonormal basis for $\mathcal{V}$. An orthonormal basis is desirable for many reasons. One is that determining the basis representation of a given vector with respect to it is a straightforward task.

Theorem 5.2.5 *Let* $\mathcal{V}$ *be a finite-dimensional inner product space, let* $\beta = \mathbf{u}_1, \mathbf{u}_2, \ldots, \mathbf{u}_n$ *be an orthonormal basis, and let* $\mathbf{v} \in \mathcal{V}$. *Then*

$$\mathbf{v} = \sum_{i=1}^{n} \langle \mathbf{v}, \mathbf{u}_i \rangle \mathbf{u}_i, \tag{5.2.6}$$

$$\|\mathbf{v}\|^2 = \sum_{i=1}^{n} |\langle \mathbf{v}, \mathbf{u}_i \rangle|^2, \tag{5.2.7}$$

and the basis representation of $\mathbf{v}$ with respect to β is

$$[\mathbf{v}]_\beta = \begin{bmatrix} \langle \mathbf{v}, \mathbf{u}_1 \rangle \\ \langle \mathbf{v}, \mathbf{u}_2 \rangle \\ \vdots \\ \langle \mathbf{v}, \mathbf{u}_n \rangle \end{bmatrix}. \tag{5.2.8}$$

Proof Since β is a basis for $\mathcal{V}$, there exist scalars $a_i \in \mathbb{F}$ such that

$$\mathbf{v} = \sum_{i=1}^{n} a_i \mathbf{u}_i. \tag{5.2.9}$$

The inner product of both sides of (5.2.9) with $\mathbf{u}_j$ is

$$\langle \mathbf{v}, \mathbf{u}_j \rangle = \left\langle \sum_{i=1}^{n} a_i \mathbf{u}_i, \mathbf{u}_j \right\rangle = \sum_{i=1}^{n} a_i \langle \mathbf{u}_i, \mathbf{u}_j \rangle = \sum_{i=1}^{n} a_i \delta_{ij} = a_j,$$

which implies (5.2.6). The identity (5.2.7) follows from (5.2.6) and Theorem 5.1.8. The basis representation (5.2.8) is a restatement of (5.2.6). □

Example 5.2.10 The vectors

$$\mathbf{u}_1 = \frac{1}{3}[1 \ 2 \ 2]^\mathsf{T}, \quad \mathbf{u}_2 = \frac{1}{3}[-2 \ 2 \ -1]^\mathsf{T}, \quad \text{and} \quad \mathbf{u}_3 = \frac{1}{3}[-2 \ -1 \ 2]^\mathsf{T}$$

comprise an orthonormal basis for $\mathbb{R}^3$. How can we express $\mathbf{v} = [1 \ 2 \ 3]^\mathsf{T}$ as a linear combination of these basis vectors? The preceding theorem tells us that $\mathbf{v} = \sum_{i=1}^{3} \langle \mathbf{v}, \mathbf{u}_i \rangle \mathbf{u}_i$, so we compute

$$\langle \mathbf{v}, \mathbf{u}_1 \rangle = \tfrac{1}{3}(1 + 4 + 6) = \frac{11}{3},$$

$$\langle \mathbf{v}, \mathbf{u}_2 \rangle = \tfrac{1}{3}(-2 + 4 - 3) = -\frac{1}{3},$$

$$\langle \mathbf{v}, \mathbf{u}_3 \rangle = \tfrac{1}{3}(-2 - 2 + 6) = \frac{2}{3},$$

and find that $\mathbf{v} = \frac{11}{3}\mathbf{u}_1 - \frac{1}{3}\mathbf{u}_2 + \frac{2}{3}\mathbf{u}_3$.

5.3 The Gram–Schmidt Process

Example 5.3.1 Consider the set $\mathcal{U}$ in $\mathbb{R}^3$ defined by the equation

$$x + 2y + 3z = 0. \tag{5.3.2}$$

If $A = [1\ 2\ 3] \in \mathsf{M}_{1\times 3}$ and $\mathbf{x} = [x\ y\ z]^\mathsf{T}$, then (5.3.2) is equivalent to $A\mathbf{x} = \mathbf{0}$, which tells us that $\mathcal{U} = \text{null}\,A$. Consequently, $\mathcal{U}$ is a subspace of $\mathbb{R}^3$. Since $\text{rank}\,A = 1$, the dimension theorem (Corollary 2.5.4) ensures that $\dim \mathcal{U} = 2$.

The linearly independent vectors $\mathbf{v}_1 = [3\ 0\ -1]^\mathsf{T}$ and $\mathbf{v}_2 = [-2\ 1\ 0]^\mathsf{T}$ are in $\text{null}\,A$, so they comprise a basis for $\mathcal{U}$. Since $\|\mathbf{v}_1\|_2 = \sqrt{10}$, $\|\mathbf{v}_2\|_2 = \sqrt{5}$, and $\langle \mathbf{v}_1, \mathbf{v}_2 \rangle = -6$, our basis vectors are not normalized and they are not orthogonal, but we can use them to construct an orthonormal basis of $\mathcal{U}$.

First normalize $\mathbf{v}_1$ to obtain a unit vector $\mathbf{u}_1$:

$$\mathbf{u}_1 = \tfrac{1}{\sqrt{10}}\mathbf{v}_1 = \frac{1}{\sqrt{10}}\begin{bmatrix} 3 \\ 0 \\ -1 \end{bmatrix}. \tag{5.3.3}$$

In our derivation of (4.5.13), we discovered how to use $\mathbf{v}_2$ to construct a vector that is orthogonal to $\mathbf{u}_1$: find the projection of $\mathbf{v}_2$ onto $\mathbf{u}_1$ and subtract it from $\mathbf{v}_2$. This vector,

$$\mathbf{x}_2 = \mathbf{v}_2 - \underbrace{\langle \mathbf{v}_2, \mathbf{u}_1 \rangle \mathbf{u}_1}_{\text{Projection of } \mathbf{v}_2 \text{ onto } \mathbf{u}_1}, \tag{5.3.4}$$

belongs to $\mathcal{U}$ because it is a linear combination of vectors in $\mathcal{U}$. Using (5.3.3), we compute

$$\mathbf{x}_2 = \mathbf{v}_2 - \tfrac{1}{10}\langle \mathbf{v}_2, \mathbf{v}_1 \rangle \mathbf{v}_1 = \mathbf{v}_2 + \tfrac{3}{5}\mathbf{v}_1 = \begin{bmatrix} -2 \\ 1 \\ 0 \end{bmatrix} + \tfrac{3}{5}\begin{bmatrix} 3 \\ 0 \\ -1 \end{bmatrix} = \tfrac{1}{5}\begin{bmatrix} -1 \\ 5 \\ -3 \end{bmatrix}.$$

We have $\|\mathbf{x}_2\|_2^2 = \tfrac{35}{25}$, so the unit vector $\mathbf{u}_2 = \mathbf{x}_2/\|\mathbf{x}_2\|_2$ is

$$\mathbf{u}_2 = \frac{1}{\sqrt{35}}\begin{bmatrix} -1 \\ 5 \\ -3 \end{bmatrix}.$$

By construction, $\|\mathbf{u}_2\|_2 = 1$ and $\mathbf{u}_2$ is orthogonal to $\mathbf{u}_1$. As a check, we compute $\langle \mathbf{u}_1, \mathbf{u}_2 \rangle = \frac{1}{\sqrt{350}}(-3 + 0 + 3) = 0$.

The Gram–Schmidt process is a systematic implementation of the ideas employed in the preceding example. It starts with a linearly independent list of vectors and produces an orthonormal system of vectors with the same span. Details of the process are explained in the proof of the following theorem.

Theorem 5.3.5 (Gram–Schmidt) *Let V be an inner product space, and suppose that $\mathbf{v}_1, \mathbf{v}_2, \ldots, \mathbf{v}_n \in V$ are linearly independent. There is an orthonormal system $\mathbf{u}_1, \mathbf{u}_2, \ldots, \mathbf{u}_n$ such that*

$$\text{span}\{\mathbf{v}_1, \mathbf{v}_2, \ldots, \mathbf{v}_k\} = \text{span}\{\mathbf{u}_1, \mathbf{u}_2, \ldots, \mathbf{u}_k\}, \quad k = 1, 2, \ldots, n. \tag{5.3.6}$$

Proof We proceed by induction on n. In the base case $n = 1$, $\mathbf{u}_1 = \mathbf{v}_1/\|\mathbf{v}_1\|$ is a unit vector and $\text{span}\{\mathbf{v}_1\} = \text{span}\{\mathbf{u}_1\}$.

For the induction step, let $2 \leq m \leq n$. Suppose that, given $m - 1$ linearly independent vectors $\mathbf{v}_1, \mathbf{v}_2, \ldots, \mathbf{v}_{m-1}$, there are orthonormal vectors $\mathbf{u}_1, \mathbf{u}_2, \ldots, \mathbf{u}_{m-1}$ such that

$$\text{span}\{\mathbf{v}_1, \mathbf{v}_2, \ldots, \mathbf{v}_k\} = \text{span}\{\mathbf{u}_1, \mathbf{u}_2, \ldots, \mathbf{u}_k\} \quad \text{for } k = 1, 2, \ldots, m - 1. \tag{5.3.7}$$

Since $\mathbf{v}_1, \mathbf{v}_2, \ldots, \mathbf{v}_m$ are linearly independent,

$$\mathbf{v}_m \notin \text{span}\{\mathbf{v}_1, \mathbf{v}_2, \ldots, \mathbf{v}_{m-1}\} = \text{span}\{\mathbf{u}_1, \mathbf{u}_2, \ldots, \mathbf{u}_{m-1}\},$$

so

$$\mathbf{x}_m = \mathbf{v}_m - \sum_{i=1}^{m-1} \langle \mathbf{v}_m, \mathbf{u}_i \rangle \mathbf{u}_i \neq \mathbf{0}$$

and we may define $\mathbf{u}_m = \mathbf{x}_m / \|\mathbf{x}_m\|$.

We claim that $\mathbf{x}_m$ (and hence also $\mathbf{u}_m$) is orthogonal to $\mathbf{u}_1, \mathbf{u}_2, \ldots, \mathbf{u}_{m-1}$. Indeed, if $1 \leq j \leq m - 1$, then

$$\langle \mathbf{u}_j, \mathbf{x}_m \rangle = \left\langle \mathbf{u}_j, \mathbf{v}_m - \sum_{i=1}^{m-1} \langle \mathbf{v}_m, \mathbf{u}_i \rangle \mathbf{u}_i \right\rangle = \langle \mathbf{u}_j, \mathbf{v}_m \rangle - \left\langle \mathbf{u}_j, \sum_{i=1}^{m-1} \langle \mathbf{v}_m, \mathbf{u}_i \rangle \mathbf{u}_i \right\rangle$$

$$= \langle \mathbf{u}_j, \mathbf{v}_m \rangle - \sum_{i=1}^{m-1} \overline{\langle \mathbf{v}_m, \mathbf{u}_i \rangle} \langle \mathbf{u}_j, \mathbf{u}_i \rangle = \langle \mathbf{u}_j, \mathbf{v}_m \rangle - \sum_{i=1}^{m-1} \langle \mathbf{u}_i, \mathbf{v}_m \rangle \delta_{ij}$$

$$= \langle \mathbf{u}_j, \mathbf{v}_m \rangle - \langle \mathbf{u}_j, \mathbf{v}_m \rangle = 0.$$

Since $\mathbf{u}_m$ is a linear combination of $\mathbf{v}_1, \mathbf{v}_2, \ldots, \mathbf{v}_m$, the induction hypothesis (5.3.6) ensures that

$$\text{span}\{\mathbf{u}_1, \mathbf{u}_2, \ldots, \mathbf{u}_m\} \subseteq \text{span}\{\mathbf{v}_1, \mathbf{v}_2, \ldots, \mathbf{v}_m\}. \tag{5.3.8}$$

The vectors $\mathbf{u}_1, \mathbf{u}_2, \ldots, \mathbf{u}_m$ are orthonormal and hence linearly independent. Consequently, $\dim \text{span}\{\mathbf{u}_1, \mathbf{u}_2, \ldots, \mathbf{u}_m\} = m$. The containment (5.3.8) is an equality since both spans are m-dimensional vector spaces; see Theorem 2.2.9. $\square$

The Gram–Schmidt process takes a linearly independent list $\mathbf{v}_1, \mathbf{v}_2, \ldots, \mathbf{v}_n$ and constructs orthonormal vectors $\mathbf{u}_1, \mathbf{u}_2, \ldots, \mathbf{u}_n$ using the following algorithm. Start by setting $\mathbf{x}_1 = \mathbf{v}_1$ and then normalize it to obtain

$$\mathbf{u}_1 = \frac{\mathbf{x}_1}{\|\mathbf{x}_1\|}.$$

Then, for each $k = 2, 3, \ldots, n$, compute

$$\mathbf{x}_k = \mathbf{v}_k - \langle \mathbf{v}_k, \mathbf{u}_1 \rangle \mathbf{u}_1 - \cdots - \langle \mathbf{v}_k, \mathbf{u}_{k-1} \rangle \mathbf{u}_{k-1} \tag{5.3.9}$$

and normalize it:

$$\mathbf{u}_k = \frac{\mathbf{x}_k}{\|\mathbf{x}_k\|}.$$

What does the Gram–Schmidt process do to an orthogonal list of vectors?

Lemma 5.3.10 *Let $\mathcal{V}$ be an inner product space and suppose that $\mathbf{v}_1, \mathbf{v}_2, \ldots, \mathbf{v}_n \in \mathcal{V}$ are mutually orthogonal. The Gram–Schmidt process constructs the orthonormal system*

$$\mathbf{u}_1 = \frac{\mathbf{v}_1}{\|\mathbf{v}_1\|}, \qquad \mathbf{u}_2 = \frac{\mathbf{v}_2}{\|\mathbf{v}_2\|}, \ldots, \qquad \mathbf{u}_n = \frac{\mathbf{v}_n}{\|\mathbf{v}_n\|}.$$

If $\mathbf{v}_1, \mathbf{v}_2, \ldots, \mathbf{v}_n \in \mathcal{V}$ is an orthonormal system, then each $\mathbf{u}_i = \mathbf{v}_i$.

Proof It suffices to show that the vectors (5.3.9) constructed by the Gram–Schmidt process are $x_i = v_i$. We proceed by induction. In the base case $n = 1$, we have $x_1 = v_1$. For the induction step, suppose that $x_i = v_i$ whenever $1 \leq i \leq m < n$. Then

$$x_{m+1} = v_{m+1} - \sum_{i=1}^{m} \langle v_{m+1}, u_i \rangle u_i = v_{m+1} - \sum_{i=1}^{m} \left\langle v_{m+1}, \frac{v_i}{\|v_i\|} \right\rangle \frac{v_i}{\|v_i\|}$$

$$= v_{m+1} - \sum_{i=1}^{m} \langle v_{m+1}, v_i \rangle \frac{v_i}{\|v_i\|^2} = v_{m+1}.$$ $\square$

Example 5.3.11 The polynomials $1, x, x^2, x^3, \ldots, x^n$ are linearly independent in $C[0, 1]$ for each $n = 1, 2, \ldots$ (see Example 1.6.7), but they are not orthonormal with respect to the L^2 inner product (5.1.6). For example, $\|x\| = \frac{1}{\sqrt{3}}$ and $\langle x, x^2 \rangle = \frac{1}{4}$. To construct an orthonormal system with the same span as $1, x, x^2, x^3, \ldots, x^n$, we label $v_1 = 1$, $v_2 = x$, $v_3 = x^2$, $\ldots$, apply the Gram–Schmidt process, and obtain the orthonormal system of polynomials

$$u_1 = 1, \qquad u_2 = \sqrt{3}(2x - 1), \qquad u_3 = \sqrt{5}(6x^2 - 6x + 1), \ldots.$$

This is how the orthogonal polynomials in Example 5.1.5 were constructed.

An important consequence of the Gram–Schmidt process is the following.

Corollary 5.3.12 *Every finite-dimensional inner product space has an orthonormal basis.*

Proof Start with any basis $v_1, v_2, \ldots, v_n$ and apply the Gram–Schmidt process to obtain an orthonormal system $u_1, u_2, \ldots, u_n$, which is linearly independent and has the same span as $v_1, v_2, \ldots, v_n$. $\square$

Corollary 5.3.13 *Every orthonormal system in a finite-dimensional inner product space can be extended to an orthonormal basis.*

Proof Given an orthonormal system $v_1, v_2, \ldots, v_r$, extend it to a basis $v_1, v_2, \ldots, v_n$. Apply the Gram–Schmidt process to this basis and obtain an orthonormal basis $u_1, u_2, \ldots, u_n$. Since the vectors $v_1, v_2, \ldots, v_r$ are already orthonormal, Lemma 5.3.10 ensures that the Gram–Schmidt process leaves them unchanged, that is, $u_i = v_i$ for $i = 1, 2, \ldots, r$. $\square$

5.4 The Riesz Representation Theorem

Now that we know a finite-dimensional $\mathbb{F}$-inner product space $\mathcal{V}$ has an orthonormal basis, we can use (5.2.6) to obtain a remarkable representation for any linear transformation from $\mathcal{V}$ to $\mathbb{F}$.

Definition 5.4.1 Let $\mathcal{V}$ be an $\mathbb{F}$-vector space. A *linear functional* is a linear transformation $\phi : \mathcal{V} \to \mathbb{F}$.

Example 5.4.2 Let $\mathcal{V}$ be an $\mathbb{F}$-inner product space. If $w \in \mathcal{V}$, then $\phi(v) = \langle v, w \rangle$ defines a linear functional on $\mathcal{V}$.

Example 5.4.3 Let $V = C[0, 1]$. Then $\phi(f) = f(\frac{1}{2})$ defines a linear functional on V.

Theorem 5.4.4 (Riesz Representation Theorem) *Let V be a finite-dimensional $\mathbb{F}$-inner product space and let $\phi : V \to \mathbb{F}$ be a linear functional.*

(a) *There is a unique $\mathbf{w} \in V$ such that*

$$\phi(\mathbf{v}) = \langle \mathbf{v}, \mathbf{w} \rangle \quad \text{for all } \mathbf{v} \in V. \tag{5.4.5}$$

(b) *Let $\mathbf{u}_1, \mathbf{u}_2, \ldots, \mathbf{u}_n$ be an orthonormal basis of V. The vector $\mathbf{w}$ in (a) is*

$$\mathbf{w} = \overline{\phi(\mathbf{u}_1)}\mathbf{u}_1 + \overline{\phi(\mathbf{u}_2)}\mathbf{u}_2 + \cdots + \overline{\phi(\mathbf{u}_n)}\mathbf{u}_n. \tag{5.4.6}$$

Proof For any $\mathbf{v} \in V$ and for any orthonormal basis $\mathbf{u}_1, \mathbf{u}_2, \ldots, \mathbf{u}_n$ of V, use (5.2.6) to compute

$$\phi(\mathbf{v}) = \phi\left(\sum_{i=1}^{n} \langle \mathbf{v}, \mathbf{u}_i \rangle \mathbf{u}_i\right) = \sum_{i=1}^{n} \phi\left(\langle \mathbf{v}, \mathbf{u}_i \rangle \mathbf{u}_i\right) = \sum_{i=1}^{n} \langle \mathbf{v}, \mathbf{u}_i \rangle \phi(\mathbf{u}_i)$$

$$= \sum_{i=1}^{n} \langle \mathbf{v}, \overline{\phi(\mathbf{u}_i)}\mathbf{u}_i \rangle = \left\langle \mathbf{v}, \sum_{i=1}^{n} \overline{\phi(\mathbf{u}_i)}\mathbf{u}_i \right\rangle.$$

Therefore, $\mathbf{w}$ in (5.4.6) satisfies (5.4.5). If $\mathbf{y} \in V$ and $\phi(\mathbf{v}) = \langle \mathbf{v}, \mathbf{y} \rangle$ for all $\mathbf{v} \in V$, then $\langle \mathbf{v}, \mathbf{y} \rangle = \langle \mathbf{v}, \mathbf{w} \rangle$ for all $\mathbf{v} \in V$. Corollary 4.4.15 ensures that $\mathbf{y} = \mathbf{w}$. $\square$

Definition 5.4.7 The vector $\mathbf{w}$ in (5.4.6) is the *Riesz vector* for the linear functional ϕ.

The formula (5.4.6) may give the impression that the Riesz vector for ϕ depends on the choice of orthonormal basis $\mathbf{u}_1, \mathbf{u}_2, \ldots, \mathbf{u}_n$, but it does not.

Example 5.4.8 The Riesz vector for the linear functional $A \mapsto \operatorname{tr} A$ on M_n (with the Frobenius inner product) is the identity matrix since $\operatorname{tr} A = \operatorname{tr} I^* A = \langle A, I \rangle_F$.

5.5 Basis Representations

Let V and W be finite-dimensional vector spaces over the same field $\mathbb{F}$ and let $T \in \mathcal{L}(V, W)$. The following theorem provides a convenient way to compute the basis representation of T with respect to a basis $\beta = \mathbf{v}_1, \mathbf{v}_2, \ldots, \mathbf{v}_n$ of V and an orthonormal basis $\gamma = \mathbf{w}_1, \mathbf{w}_2, \ldots, \mathbf{w}_m$ of W.

If

$$\mathbf{v} = a_1 \mathbf{v}_1 + a_2 \mathbf{v}_2 + \cdots + a_n \mathbf{v}_n$$

is the unique representation of $\mathbf{v} \in V$ as a linear combination of the vectors in the basis β, then the β-coordinate vector of $\mathbf{v}$ is $[\mathbf{v}]_\beta = [a_1 \ a_2 \ \ldots \ a_n]^{\mathsf{T}}$. There is a coordinate vector for $\mathbf{v}$ associated with any basis. The coordinate vector for $\mathbf{v}$ associated with an orthonormal basis has the special form (5.2.8). What does this special form tell us about the basis representation of a linear transformation?

The $m \times n$ matrix that represents $T \in \mathcal{L}(\mathcal{V}, \mathcal{W})$ with respect to the bases β and γ is

$$_\gamma[T]_\beta = \begin{bmatrix} [T\mathbf{v}_1]_\gamma & [T\mathbf{v}_2]_\gamma & \cdots & [T\mathbf{v}_n]_\gamma \end{bmatrix}. \tag{5.5.1}$$

It is a consequence of linearity that

$$[T\mathbf{v}]_\gamma = {}_\gamma[T]_\beta[\mathbf{v}]_\beta \tag{5.5.2}$$

for each $\mathbf{v} \in \mathcal{V}$; see (2.3.18).

Theorem 5.5.3 *Let $\mathcal{V}$ and $\mathcal{W}$ be finite-dimensional vector spaces over the same field $\mathbb{F}$. Let $\beta = \mathbf{v}_1, \mathbf{v}_2, \ldots, \mathbf{v}_n$ be a basis for $\mathcal{V}$ and let $\gamma = \mathbf{w}_1, \mathbf{w}_2, \ldots, \mathbf{w}_m$ be an orthonormal basis for $\mathcal{W}$.*

(a) *If $T \in \mathcal{L}(\mathcal{V}, \mathcal{W})$, then*

$$_\gamma[T]_\beta = [\langle T\mathbf{v}_j, \mathbf{w}_i \rangle] = \begin{bmatrix} \langle T\mathbf{v}_1, \mathbf{w}_1 \rangle & \langle T\mathbf{v}_2, \mathbf{w}_1 \rangle & \cdots & \langle T\mathbf{v}_n, \mathbf{w}_1 \rangle \\ \langle T\mathbf{v}_1, \mathbf{w}_2 \rangle & \langle T\mathbf{v}_2, \mathbf{w}_2 \rangle & \cdots & \langle T\mathbf{v}_n, \mathbf{w}_2 \rangle \\ \vdots & \vdots & \ddots & \vdots \\ \langle T\mathbf{v}_1, \mathbf{w}_m \rangle & \langle T\mathbf{v}_2, \mathbf{w}_m \rangle & \cdots & \langle T\mathbf{v}_n, \mathbf{w}_m \rangle \end{bmatrix}. \tag{5.5.4}$$

(b) *Let $A \in \mathsf{M}_{m \times n}(\mathbb{F})$. There is a unique $T \in \mathcal{L}(\mathcal{V}, \mathcal{W})$ such that $_\gamma[T]_\beta = A$.*

Proof (a) It suffices to show that each column of $_\gamma[T]_\beta$ has the asserted form. Since γ is an orthonormal basis for $\mathcal{W}$, Theorem 5.2.5 ensures that

$$T\mathbf{v}_j = \langle T\mathbf{v}_j, \mathbf{w}_1 \rangle \mathbf{w}_1 + \langle T\mathbf{v}_j, \mathbf{w}_2 \rangle \mathbf{w}_2 + \cdots + \langle T\mathbf{v}_j, \mathbf{w}_m \rangle \mathbf{w}_m,$$

and hence

$$[T\mathbf{v}_j]_\gamma = \begin{bmatrix} \langle T\mathbf{v}_j, \mathbf{w}_1 \rangle \\ \langle T\mathbf{v}_j, \mathbf{w}_2 \rangle \\ \vdots \\ \langle T\mathbf{v}_j, \mathbf{w}_m \rangle \end{bmatrix}.$$

(b) Let $A = [\mathbf{a}_1 \ \mathbf{a}_2 \ \cdots \ \mathbf{a}_n]$. Because a vector in $\mathcal{W}$ is uniquely determined by its γ-coordinate vector, we may define a function $T : \mathcal{V} \to \mathcal{W}$ by $[T\mathbf{v}]_\gamma = A[\mathbf{v}]_\beta$. Then

$$[T(a\mathbf{v} + \mathbf{u})]_\gamma = A[a\mathbf{v} + \mathbf{u}]_\beta = A(a[\mathbf{v}]_\beta + [\mathbf{u}]_\beta)$$
$$= aA[\mathbf{v}]_\beta + A[\mathbf{u}]_\beta = a[T\mathbf{v}]_\gamma + [T\mathbf{u}]_\gamma,$$

so T is a linear transformation. Since

$$[T\mathbf{v}_i]_\gamma = A[\mathbf{v}_i]_\beta = A\mathbf{e}_i = \mathbf{a}_i, \qquad i = 1, 2, \ldots, n,$$

we have $_\gamma[T]_\beta = A$. If $S \in \mathcal{L}(\mathcal{V}, \mathcal{W})$ and $_\gamma[S]_\beta = A$, then

$$[S\mathbf{v}_i]_\gamma = {}_\gamma[S]_\beta[\mathbf{v}_i]_\beta = A[\mathbf{v}_i]_\beta = [T\mathbf{v}_i]_\gamma$$

and hence $S\mathbf{v}_i = T\mathbf{v}_i$ for each $i = 1, 2, \ldots, n$. Thus, S and T are identical linear transformations because they agree on a basis. $\qquad \square$

5.6 Adjoints of Linear Transformations and Matrices

In this section, $\mathcal{V}$ and $\mathcal{W}$ are inner product spaces over the same field $\mathbb{F}$. We denote the respective inner products on $\mathcal{V}$ and $\mathcal{W}$ by $\langle \cdot, \cdot \rangle_{\mathcal{V}}$ and $\langle \cdot, \cdot \rangle_{\mathcal{W}}$.

Definition 5.6.1 A function $f : \mathcal{W} \to \mathcal{V}$ is an *adjoint* of a linear transformation $T : \mathcal{V} \to \mathcal{W}$ if $\langle T\mathbf{v}, \mathbf{w} \rangle_{\mathcal{W}} = \langle \mathbf{v}, f(\mathbf{w}) \rangle_{\mathcal{V}}$ for all $\mathbf{v} \in \mathcal{V}$ and all $\mathbf{w} \in \mathcal{W}$.

Adjoints arise in differential equations, integral equations, and functional analysis. Our first observation is that if an adjoint exists, then it is unique.

Lemma 5.6.2 *Let f and g be functions from $\mathcal{W}$ to $\mathcal{V}$. If $\langle \mathbf{v}, f(\mathbf{w}) \rangle_{\mathcal{V}} = \langle \mathbf{v}, g(\mathbf{w}) \rangle_{\mathcal{V}}$ for all $\mathbf{v} \in \mathcal{V}$ and all $\mathbf{w} \in \mathcal{W}$, then $f = g$. In particular, if $T \in \mathcal{L}(\mathcal{V}, \mathcal{W})$, then it has at most one adjoint.*

Proof Corollary 4.4.15 ensures that $f(\mathbf{w}) = g(\mathbf{w})$ for all $\mathbf{w} \in \mathcal{W}$. □

Since a linear transformation T has at most one adjoint, we refer to it as *the* adjoint of T and denote it by T^*. Our second observation is that if a linear transformation has an adjoint, then that adjoint is a linear transformation.

Theorem 5.6.3 *If $T \in \mathcal{L}(\mathcal{V}, \mathcal{W})$ has an adjoint $T^* : \mathcal{W} \to \mathcal{V}$, then T^* is a linear transformation.*

Proof Let $\mathbf{u}, \mathbf{w} \in \mathcal{W}$ and $a \in \mathbb{F}$. Then for all $\mathbf{v} \in \mathcal{V}$,

$$\langle \mathbf{v}, T^*(a\mathbf{u} + \mathbf{w}) \rangle = \langle T\mathbf{v}, a\mathbf{u} + \mathbf{w} \rangle = \overline{a} \langle T\mathbf{v}, \mathbf{u} \rangle + \langle T\mathbf{v}, \mathbf{w} \rangle$$
$$= \overline{a} \langle \mathbf{v}, T^*(\mathbf{u}) \rangle + \langle \mathbf{v}, T^*(\mathbf{w}) \rangle = \langle \mathbf{v}, aT^*(\mathbf{u}) \rangle + \langle \mathbf{v}, T^*(\mathbf{w}) \rangle$$
$$= \langle \mathbf{v}, aT^*(\mathbf{u}) + T^*(\mathbf{w}) \rangle.$$

Corollary 4.4.15 ensures that $T^*(a\mathbf{u} + \mathbf{w}) = aT^*(\mathbf{u}) + T^*(\mathbf{w})$. □

Example 5.6.4 Let $\mathcal{V} = \mathbb{F}^n$ and $\mathcal{W} = \mathbb{F}^m$, both endowed with the standard inner product. Let $A \in \mathsf{M}_{m \times n}(\mathbb{F})$ and consider the linear transformation $T_A : \mathcal{V} \to \mathcal{W}$ induced by A; see Definition 2.3.9. For all $\mathbf{x} \in \mathcal{V}$ and $\mathbf{y} \in \mathcal{W}$,

$$\langle A\mathbf{x}, \mathbf{y} \rangle = \mathbf{y}^*(A\mathbf{x}) = (A^*\mathbf{y})^*\mathbf{x} = \langle \mathbf{x}, A^*\mathbf{y} \rangle; \tag{5.6.5}$$

see Example 4.4.3. We conclude that T_{A^*} is the adjoint of T_A. The terms *adjoint* and *conjugate transpose* of a matrix are synonyms.

Example 5.6.6 If $\mathbb{F} = \mathbb{C}$, $m = n = 1$, and $A = [a]$, the linear transformation $T_A : \mathbb{C} \to \mathbb{C}$ is $T_A(z) = az$. Its adjoint is $T_{A^*}(z) = \overline{a}z$.

Important matrix classes whose definitions involve adjoints include normal matrices (they commute with their adjoints; see Chapter 12), Hermitian matrices (they are equal to their adjoints; see Definition 5.6.9), and unitary matrices (their adjoints are their inverses; see Chapter 6).

Example 5.6.7 Let $V = W = \mathcal{P}$ be the inner product space of all polynomials, endowed with the L^2 inner product (5.1.6) on $[0, 1]$, and let $p \in \mathcal{P}$ be given. Consider $T \in \mathcal{L}(\mathcal{P})$ defined by $Tf = pf$ for each $f \in \mathcal{P}$. Then for all $g \in \mathcal{P}$,

$$\langle Tf, g \rangle = \int_0^1 p(t)f(t)\overline{g(t)}\,dt = \int_0^1 f(t)\overline{\overline{p(t)}g(t)}\,dt = \langle f, \bar{p}g \rangle. \tag{5.6.8}$$

Define the function $\Phi : \mathcal{P} \to \mathcal{P}$ by $\Phi g = \bar{p}g$. Then (5.6.8) says that $\langle Tf, g \rangle = \langle f, \Phi g \rangle$ for all $g \in \mathcal{P}$. We conclude that T has an adjoint, and $T^* = \Phi$. If p is a real polynomial, then $\bar{p} = p$ and $T^* = T$, and vice-versa.

Definition 5.6.9 Suppose that T^* is the adjoint of $T \in \mathcal{L}(V)$. If $T^* = T$, then T is *self-adjoint*. A square matrix A is *Hermitian* if $A^* = A$. The terms Hermitian and self-adjoint are used interchangeably for linear transformations and matrices.

The linear transformation $f \mapsto pf$ in Example 5.6.7 is self-adjoint if and only if p is a real polynomial. In Example 5.6.4, the linear transformation $\mathbf{x} \mapsto A\mathbf{x}$ is self-adjoint if and only if $A = A^*$, that is, if and only if A is square and Hermitian. It makes no sense to speak of a self-adjoint operator $T \in \mathcal{L}(V, W)$ unless $V = W$; nor does it make sense to speak of a Hermitian matrix $A \in \mathsf{M}_{m \times n}$ unless $m = n$.

Adjoints are unique when they exist, but existence cannot be taken for granted. The following example shows that some linear transformations do not have an adjoint.

Example 5.6.10 Let $\mathcal{P}$ be as in Example 5.6.7. Define $T \in \mathcal{L}(\mathcal{P})$ by $Tf = f'$, that is, the action of T on polynomials is differentiation. Suppose that T has an adjoint T^*, let g be the constant polynomial $g(t) = 1$, and let $T^*g = h$. Then for each $f \in \mathcal{P}$, the fundamental theorem of calculus ensures that

$$\langle Tf, g \rangle = \int_0^1 f'(t)\,dt = f(t)\Big|_0^1 = f(1) - f(0)$$

and therefore

$$f(1) - f(0) = \langle Tf, g \rangle = \langle f, T^*g \rangle = \langle f, h \rangle = \int_0^1 f(t)\overline{h(t)}\,dt.$$

Now let $f(t) = t^2(t - 1)^2 h(t)$, so $f(0) = f(1) = 0$ and

$$0 = \int_0^1 f(t)\overline{h(t)}\,dt = \int_0^1 t^2(t-1)^2|h(t)|^2\,dt = \|t(t-1)h(t)\|^2.$$

We conclude that $t(t - 1)h(t)$ is the zero polynomial (see P.0.14), which implies that h is the zero polynomial. It follows that

$$p(1) - p(0) = \langle Tp, g \rangle = \langle p, T^*g \rangle = \langle p, h \rangle = 0$$

for every $p \in \mathcal{P}$, which is false for $p(t) = t$. We conclude that T does not have an adjoint.

For an example of a linear functional that does not have an adjoint, see P.5.13. It is not an accident that our examples of linear transformations without adjoints involve infinite-dimensional inner product spaces: every linear transformation between finite-dimensional inner product spaces has an adjoint.

Theorem 5.6.11 *Let V and W be finite-dimensional inner product spaces over the same field $\mathbb{F}$ and let $T \in \mathcal{L}(V, W)$. Let $\beta = \mathbf{v}_1, \mathbf{v}_2, \ldots, \mathbf{v}_n$ be an orthonormal basis of V, let $\gamma = \mathbf{w}_1, \mathbf{w}_2, \ldots, \mathbf{w}_m$ be an orthonormal basis of W, and let $_\gamma[T]_\beta = [\langle T\mathbf{v}_j, \mathbf{w}_i \rangle]$ be the $\beta - \gamma$ basis representation of T, as in (5.5.4). Then:*

(a) *T has an adjoint.*

(b) *The basis representation of T^* is $_\gamma[T]_\beta^*$, that is,*

$$_\beta[T^*]_\gamma = {}_\gamma[T]_\beta^*. \tag{5.6.12}$$

Proof (a) For each $\mathbf{w} \in W$, $\phi_\mathbf{w}(\mathbf{v}) = \langle T\mathbf{v}, \mathbf{w} \rangle$ defines a linear functional on V, so Theorem 5.4.4 ensures that there is a unique vector $S(\mathbf{w}) \in V$ such that $\langle T\mathbf{v}, \mathbf{w} \rangle = \langle \mathbf{v}, S(\mathbf{w}) \rangle$ for all $\mathbf{v} \in V$. This construction defines a function $S : W \to V$ that is an adjoint of T, according to Definition 5.6.1. Theorem 5.6.3 ensures that $S \in \mathcal{L}(W, V)$ and $S = T^*$.

(b) Theorem 5.5.3 and the definition of S tell us that

$$_\beta[T^*]_\gamma = {}_\beta[S]_\gamma = [\langle S\mathbf{w}_j, \mathbf{v}_i \rangle] = [\overline{\langle \mathbf{v}_i, S\mathbf{w}_j \rangle}] = [\overline{\langle T\mathbf{v}_i, \mathbf{w}_j \rangle}] = {}_\gamma[T]_\beta^*. \qquad \square$$

The identity (5.6.12) ensures that the basis representation of the adjoint of a linear transformation T is the conjugate transpose of the basis representation of T, if both bases involved are orthonormal. The restriction to orthonormal bases is essential; see P.5.19.

The notation "*" is used in two different ways in (5.6.12). In the expression $_\beta[T^*]_\gamma$ it indicates the adjoint of a linear transformation; in the expression $_\gamma[T]_\beta^*$ it indicates the conjugate transpose of a matrix. This abuse of notation is forgivable, as it reminds us that the conjugate transpose of an $m \times n$ matrix A represents the adjoint of the linear transformation $T_A : \mathbb{F}^m \to \mathbb{F}^n$.

The conjugate transpose operation on matrices shares with the inverse operation the property that it is *product reversing*: $(AB)^* = B^*A^*$ and $(AB)^{-1} = B^{-1}A^{-1}$. Do not let this common property lead you to confuse these two operations, which coincide only for unitary matrices; see Chapter 6. Other basic properties of the conjugate transpose are listed in Section 0.3.

Example 5.6.13 Let V be a finite-dimensional $\mathbb{F}$-inner product space and let $\phi : V \to \mathbb{F}$ be a linear functional. The preceding theorem ensures that ϕ has an adjoint $\phi^* \in \mathcal{L}(\mathbb{F}, V)$. What is it? Theorem 5.4.4 says that there is a $\mathbf{w} \in V$ such that $\phi(\mathbf{v}) = \langle \mathbf{v}, \mathbf{w} \rangle_V$ for all $\mathbf{v} \in V$. Use the definition of the adjoint to compute

$$\langle \mathbf{v}, \phi^*(c) \rangle_V = \langle \phi(\mathbf{v}), c \rangle_\mathbb{F} = \overline{c}\phi(\mathbf{v}) = \overline{c}\langle \mathbf{v}, \mathbf{w} \rangle_V = \langle \mathbf{v}, c\mathbf{w} \rangle_V,$$

which is valid for all $\mathbf{v} \in V$. We conclude that $\phi^*(c) = c\mathbf{w}$ for all $c \in \mathbb{F}$, in which $\mathbf{w}$ is the Riesz vector for ϕ.

Example 5.6.14 Let $V = \mathsf{M}_n$, with the Frobenius inner product, and let $A \in \mathsf{M}_n$. What is the adjoint of the linear operator defined by $T(X) = AX$? Compute

$$\langle T(X), Y \rangle_F = \langle AX, Y \rangle_F = \text{tr}(Y^*AX) = \text{tr}((A^*Y)^*X) = \langle X, A^*Y \rangle_F.$$

We conclude that $T^*(Y) = A^*Y$. For the linear transformation $S(X) = XA$, we have

$$\langle S(X), Y \rangle_F = \langle XA, Y \rangle_F = \text{tr}(Y^*XA) = \text{tr}(AY^*X) = \text{tr}((YA^*)^*X) = \langle X, YA^* \rangle_F$$

and hence $S^*(Y) = YA^*$.

5.7 Parseval's Identity and Bessel's Inequality

Parseval's identity is a generalization of (5.2.7), and it plays a role in the theory of Fourier series. It says that an inner product of two abstract vectors in a finite-dimensional inner product space can be calculated as the standard inner product of their coordinate vectors, if those coordinate vectors are computed with respect to an orthonormal basis.

Theorem 5.7.1 (Parseval's Identity) *Let* $\beta = \mathbf{u}_1, \mathbf{u}_2, \ldots, \mathbf{u}_n$ *be an orthonormal basis for an inner product space* $\mathcal{V}$. *Then*

$$\langle \mathbf{v}, \mathbf{w} \rangle = \sum_{i=1}^{n} \langle \mathbf{v}, \mathbf{u}_i \rangle \overline{\langle \mathbf{w}, \mathbf{u}_i \rangle} = \big\langle [\mathbf{v}]_\beta, [\mathbf{w}]_\beta \big\rangle_{\mathbb{F}^n} \tag{5.7.2}$$

for all $\mathbf{v}, \mathbf{w} \in \mathcal{V}$.

Proof Theorem 5.2.5 ensures that

$$\mathbf{v} = \sum_{i=1}^{n} \langle \mathbf{v}, \mathbf{u}_i \rangle \mathbf{u}_i \quad \text{and} \quad \mathbf{w} = \sum_{j=1}^{n} \langle \mathbf{w}, \mathbf{u}_j \rangle \mathbf{u}_j.$$

Compute

$$\langle \mathbf{v}, \mathbf{w} \rangle = \left\langle \sum_{i=1}^{n} \langle \mathbf{v}, \mathbf{u}_i \rangle \mathbf{u}_i, \sum_{j=1}^{n} \langle \mathbf{w}, \mathbf{u}_j \rangle \mathbf{u}_j \right\rangle = \sum_{i,j=1}^{n} \langle \mathbf{v}, \mathbf{u}_i \rangle \overline{\langle \mathbf{w}, \mathbf{u}_j \rangle} \langle \mathbf{u}_i, \mathbf{u}_j \rangle$$

$$= \sum_{i,j=1}^{n} \langle \mathbf{v}, \mathbf{u}_i \rangle \overline{\langle \mathbf{w}, \mathbf{u}_j \rangle} \delta_{ij} = \sum_{i=1}^{n} \langle \mathbf{v}, \mathbf{u}_i \rangle \overline{\langle \mathbf{w}, \mathbf{u}_i \rangle} = \big\langle [\mathbf{v}]_\beta, [\mathbf{w}]_\beta \big\rangle_{\mathbb{F}^n}. \qquad \square$$

Corollary 5.7.3 *Let* $\beta = \mathbf{u}_1, \mathbf{u}_2, \ldots, \mathbf{u}_n$ *be an orthonormal basis for an inner product space* $\mathcal{V}$ *and let* $T \in \mathcal{L}(\mathcal{V})$. *Then*

$$\langle T\mathbf{v}, \mathbf{w} \rangle_{\mathcal{V}} = [\mathbf{w}]_{\beta\beta}^* [T]_\beta [\mathbf{v}]_\beta = \langle {}_\beta[T]_\beta [\mathbf{v}]_\beta, [\mathbf{w}]_\beta \rangle_{\mathbb{F}^n}$$

for all $\mathbf{v}, \mathbf{w} \in \mathcal{V}$.

Proof The assertion follows from (5.7.2) and (5.5.2). $\qquad \square$

Bessel's inequality is another generalization of (5.2.7). It is valid for all inner product spaces, and it has special significance for infinite-dimensional spaces.

Theorem 5.7.4 (Bessel's Inequality) *Let* $\mathbf{u}_1, \mathbf{u}_2, \ldots, \mathbf{u}_n$ *be an orthonormal system in an inner product space* $\mathcal{V}$. *Then for every* $\mathbf{v} \in \mathcal{V}$,

$$\sum_{i=1}^{n} |\langle \mathbf{v}, \mathbf{u}_i \rangle|^2 \le \|\mathbf{v}\|^2. \tag{5.7.5}$$

Proof Use Theorem 5.1.8 and compute

$$0 \le \left\| \mathbf{v} - \sum_{i=1}^{n} \langle \mathbf{v}, \mathbf{u}_i \rangle \mathbf{u}_i \right\|^2$$

$$= \|\mathbf{v}\|^2 - \left\langle \mathbf{v}, \sum_{i=1}^{n} \langle \mathbf{v}, \mathbf{u}_i \rangle \mathbf{u}_i \right\rangle - \left\langle \sum_{i=1}^{n} \langle \mathbf{v}, \mathbf{u}_i \rangle \mathbf{u}_i, \mathbf{v} \right\rangle + \left\| \sum_{i=1}^{n} \langle \mathbf{v}, \mathbf{u}_i \rangle \mathbf{u}_i \right\|^2$$

$$= \|\mathbf{v}\|^2 - 2 \operatorname{Re} \left\langle \mathbf{v}, \sum_{i=1}^{n} \langle \mathbf{v}, \mathbf{u}_i \rangle \mathbf{u}_i \right\rangle + \sum_{i=1}^{n} |\langle \mathbf{v}, \mathbf{u}_i \rangle|^2$$

$$= \|\mathbf{v}\|^2 - 2 \operatorname{Re} \sum_{i=1}^{n} \langle \mathbf{v}, \langle \mathbf{v}, \mathbf{u}_i \rangle \mathbf{u}_i \rangle + \sum_{i=1}^{n} |\langle \mathbf{v}, \mathbf{u}_i \rangle|^2$$

$$= \|\mathbf{v}\|^2 - 2 \sum_{i=1}^{n} |\langle \mathbf{v}, \mathbf{u}_i \rangle|^2 + \sum_{i=1}^{n} |\langle \mathbf{v}, \mathbf{u}_i \rangle|^2$$

$$= \|\mathbf{v}\|^2 - \sum_{i=1}^{n} |\langle \mathbf{v}, \mathbf{u}_i \rangle|^2.$$

This is (5.7.5). □

If $\mathcal{V}$ is infinite dimensional and $\mathbf{u}_1, \mathbf{u}_2, \ldots$ is an orthonormal system in $\mathcal{V}$, then (5.7.5) (and the fact that a bounded monotone sequence of real numbers converges) implies that the infinite series $\sum_{i=1}^{\infty} |\langle \mathbf{v}, \mathbf{u}_i \rangle|^2$ is convergent for each $\mathbf{v} \in \mathcal{V}$. A weaker (but easier to prove) consequence of (5.7.5) is the following corollary.

Corollary 5.7.6 *Let $\mathbf{u}_1, \mathbf{u}_2, \ldots$ be an orthonormal system in an inner product space $\mathcal{V}$. Then for each $\mathbf{v} \in \mathcal{V}$,*

$$\lim_{n \to \infty} \langle \mathbf{v}, \mathbf{u}_n \rangle = 0.$$

Proof Let $\mathbf{v} \in \mathcal{V}$ and $\varepsilon > 0$ be given. It suffices to show that there are only finitely many indices $i = 1, 2, \ldots$ such that $|\langle \mathbf{v}, \mathbf{u}_i \rangle| \ge \varepsilon$. Let $N(n, \varepsilon, \mathbf{v})$ denote the number of indices $i \in \{1, 2, \ldots, n\}$ such that $|\langle \mathbf{v}, \mathbf{u}_i \rangle| \ge \varepsilon$. Then

$$\|\mathbf{v}\|^2 \ge \sum_{i=1}^{n} |\langle \mathbf{v}, \mathbf{u}_i \rangle|^2 \ge N(n, \varepsilon, \mathbf{v}) \varepsilon^2,$$

so

$$N(n, \varepsilon, \mathbf{v}) \le \frac{\|\mathbf{v}\|^2}{\varepsilon^2}$$

for all $n = 1, 2, \ldots$. □

5.8 Fourier Series

In this section, we use inner product spaces to discuss Fourier series, which attempt to represent (or approximate) periodic functions as linear combinations of sines and cosines.

Definition 5.8.1 A function $f : \mathbb{R} \to \mathbb{C}$ is *periodic* if there is a nonzero $\tau \in \mathbb{R}$ such that $f(x) = f(x + \tau)$ for all $x \in \mathbb{R}$; τ is a *period* for f.

If f is periodic with period τ and $n \in \mathbb{N}$, then

$$f(x) = f(x + \tau) = f(x + \tau + \tau) = f(x + 2\tau) = \cdots = f(x + n\tau)$$

and

$$f(x - n\tau) = f(x - n\tau + \tau) = f(x - (n - 1)\tau) = \cdots = f(x - 2\tau) = f(x - \tau) = f(x).$$

Thus, $n\tau$ is a period for f for each nonzero $n \in \mathbb{Z}$.

Example 5.8.2 Since $\sin(x + 2\pi n) = \sin x$ for all $x \in \mathbb{R}$ and $n \in \mathbb{Z}$, the function $\sin x$ is periodic and $\tau = 2\pi n$ is a period for each nonzero $n \in \mathbb{Z}$.

Example 5.8.3 If $n \in \mathbb{Z}$ and $n \neq 0$, then $\sin nx = \sin(nx + 2\pi) = \sin(n(x + 2\pi/n))$. Therefore, $\sin nx$ is periodic with period $\tau = 2\pi/n$. It is also periodic with period $n\tau = 2\pi$.

Periodic functions are ubiquitous in nature: anything that vibrates (a musical instrument), trembles (an earthquake), oscillates (a pendulum), or is in any way connected to electromagnetism involves periodic functions.

The trigonometric functions $\sin nx$ and $\cos nx$ are familiar examples of periodic functions. They have period $\tau = 2\pi$, so their behavior everywhere is determined by their behavior on any real interval of length 2π. It is convenient to study them on the interval $[-\pi, \pi]$ and rescale the usual L^2 inner product on $[-\pi, \pi]$ so that they have unit norm; see Example 4.4.10. In the proof of the following lemma, we make repeated use of the facts $\sin n\pi = \sin(-n\pi)$ and $\cos n\pi = \cos(-n\pi)$ for integer n.

Lemma 5.8.4 *With respect to the inner product*

$$\langle f, g \rangle = \frac{1}{\pi} \int_{-\pi}^{\pi} f(x)g(x)\, dx \tag{5.8.5}$$

on $C_{\mathbb{R}}[-\pi, \pi]$,

$$\frac{1}{\sqrt{2}}, \cos x, \cos 2x, \cos 3x, \ldots, \sin x, \sin 2x, \sin 3x, \ldots \tag{5.8.6}$$

is an orthonormal system.

Proof We begin by verifying that

$$\left\langle \tfrac{1}{\sqrt{2}}, \tfrac{1}{\sqrt{2}} \right\rangle = \frac{1}{\pi} \int_{-\pi}^{\pi} \left(\tfrac{1}{\sqrt{2}}\right)^2 dx = 1.$$

If $n \neq 0$, then

$$\left\langle \tfrac{1}{\sqrt{2}}, \sin nx \right\rangle = \frac{1}{\pi\sqrt{2}} \int_{-\pi}^{\pi} \sin nx \, dx = -\left. \frac{1}{\pi n\sqrt{2}} \cos nx \right|_{-\pi}^{\pi} = 0.$$

A similar calculation shows that $\left\langle \tfrac{1}{\sqrt{2}}, \cos nx \right\rangle = 0$ for $n = 1, 2, \ldots$. The remaining assertions are that, if m and n are positive integers, then

$$\langle \sin mx, \cos nx \rangle = \frac{1}{\pi} \int_{-\pi}^{\pi} \sin mx \cos nx \, dx = 0, \qquad (5.8.7)$$

$$\langle \sin mx, \sin nx \rangle = \frac{1}{\pi} \int_{-\pi}^{\pi} \sin mx \sin nx \, dx = \delta_{mn}, \quad \text{and} \qquad (5.8.8)$$

$$\langle \cos mx, \cos nx \rangle = \frac{1}{\pi} \int_{-\pi}^{\pi} \cos mx \cos nx \, dx = \delta_{mn}. \qquad (5.8.9)$$

(a) Equation (5.8.7) follows from the fact that $\sin mx \cos nx$ is an odd function and the interval $[-\pi, \pi]$ is symmetric with respect to 0.

(b) To verify (5.8.8), suppose that $m \neq n$ and compute

$$\begin{aligned}
\langle \sin mx, \sin nx \rangle &= \frac{1}{\pi} \int_{-\pi}^{\pi} \sin mx \sin nx \, dx \\
&= \frac{1}{\pi} \int_{-\pi}^{\pi} \frac{1}{2}[\cos(m-n)x - \cos(m+n)x] \, dx \\
&= \left. \frac{\sin(m-n)x}{2\pi(m-n)} - \frac{\sin(m+n)x}{2\pi(m+n)} \right|_{-\pi}^{\pi} = 0.
\end{aligned}$$

If $m = n$, then

$$\begin{aligned}
\langle \sin mx, \sin nx \rangle &= \frac{1}{\pi} \int_{-\pi}^{\pi} \sin^2 nx \, dx = \frac{1}{\pi} \int_{-\pi}^{\pi} \left(\frac{1 - \cos 2nx}{2} \right) dx \\
&= \frac{1}{2\pi} \int_{-\pi}^{\pi} dx - \frac{1}{2\pi} \int_{-\pi}^{\pi} \cos 2nx \, dx \\
&= 1 - \left. \frac{1}{4\pi n} \sin 2nx \right|_{-\pi}^{\pi} = 1.
\end{aligned}$$

(c) Equation (5.8.9) can be verified in a similar fashion.   □

The equations (5.8.7), (5.8.8), and (5.8.9) are the *orthonormality relations* for the sine and cosine functions. The fact that n in the expressions $\cos nx$ and $\sin nx$ is an integer is important; it ensures that these functions have period 2π.

Because the vectors (5.8.6) comprise an orthonormal system, we label them for consistency with the notation in Section 5.1. Let

$$\mathbf{u}_0 = \tfrac{1}{\sqrt{2}}, \qquad \mathbf{u}_n = \cos nx, \qquad \mathbf{u}_{-n} = \sin nx, \qquad n = 1, 2, \ldots.$$

The preceding lemma says that $\beta = \ldots, \mathbf{u}_{-2}, \mathbf{u}_{-1}, \mathbf{u}_0, \mathbf{u}_1, \mathbf{u}_2, \ldots$ is an orthonormal system in the inner product space $C_{\mathbb{R}}[-\pi, \pi]$, endowed with the inner product (5.8.5).

For $N = 1, 2, \ldots$, the $2N + 1$ vectors $\mathbf{u}_{-N}, \mathbf{u}_{-N+1}, \ldots, \mathbf{u}_N$ form an orthonormal basis for the subspace $\mathcal{V}_N = \text{span}\{\mathbf{u}_{-N}, \mathbf{u}_{-N+1}, \ldots, \mathbf{u}_N\}$. Consider

$$f = \frac{a_0}{\sqrt{2}} + \sum_{n=1}^{N}(a_n \cos nx + a_{-n} \sin nx) = \sum_{n=-N}^{N} a_n \mathbf{u}_n.$$

Then $f \in C_\mathbb{R}[-\pi, \pi]$ and (5.2.6) ensures that the coefficients can be computed as

$$a_{\pm n} = \langle f, \mathbf{u}_{\pm n} \rangle, \qquad n = 0, 1, 2, \ldots. \tag{5.8.10}$$

These inner products are integrals, namely,

$$a_{-n} = \langle f, \mathbf{u}_{-n} \rangle = \frac{1}{\pi} \int_{-\pi}^{\pi} f(x) \sin nx \, dx, \tag{5.8.11}$$

$$a_0 = \langle f, \mathbf{u}_0 \rangle = \frac{1}{\pi} \int_{-\pi}^{\pi} \frac{f(x)}{\sqrt{2}} \, dx, \tag{5.8.12}$$

and

$$a_n = \langle f, \mathbf{u}_n \rangle = \frac{1}{\pi} \int_{-\pi}^{\pi} f(x) \cos nx \, dx \tag{5.8.13}$$

if $1 \leq n \leq N$. These integrals are defined not only for $f \in \mathcal{V}_N$, but also for any $f \in C_\mathbb{R}[-\pi, \pi]$. We might ask (as Fourier did) how the function

$$f_N = \sum_{n=-N}^{N} \langle f, \mathbf{u}_n \rangle \mathbf{u}_n, \qquad f \in C_\mathbb{R}[-\pi, \pi] \tag{5.8.14}$$

is related to f. We know that $f_N = f$ if $f \in \mathcal{V}_N$, so is f_N some sort of approximation to f if $f \notin \mathcal{V}$? For an answer, see Example 7.4.5. What happens as $N \to \infty$?

Definition 5.8.15 Let $f : [-\pi, \pi] \to \mathbb{R}$ and suppose that the integrals (5.8.11), (5.8.12), (5.8.13) exist and are finite. The *Fourier series* associated with f is the infinite series

$$\frac{a_0}{\sqrt{2}} + \sum_{n=1}^{\infty}(a_n \cos nx + a_{-n} \sin nx). \tag{5.8.16}$$

There are many questions to be asked about such a series. Does it converge? If so, does it converge to f? What does it mean for a series of functions to "converge"? What happens if f is not continuous but the integrals (5.8.10) are defined? Fourier was not able to answer these questions in his lifetime, and mathematicians have been working on them ever since. Their attempts to find answers have led to fundamental discoveries about sets, functions, measurement, integration, and convergence.

Definition 5.8.17 A function $f : [a, b] \to \mathbb{R}$ has a *jump discontinuity* at $c \in (a, b)$ if the one-sided limits $f(c^+) = \lim_{x \to c^+} f(x)$ and $f(c^-) = \lim_{x \to c^-} f(x)$ both exist and at least one of them is different from $f(c)$. A function $f : [a, b] \to \mathbb{R}$ is *piecewise continuous* if f is continuous on $[a, b]$ except possibly at finitely many points, each of which is a jump discontinuity.

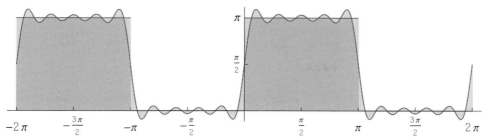

Figure 5.1 The graph of the function f from Example 5.8.19 and its Fourier approximation $\frac{\pi}{2} + 2\sin x + \frac{2}{3}\sin 3x + \frac{2}{5}\sin 5x + \frac{2}{7}\sin 7x + \frac{2}{9}\sin 9x$.

The following important theorem provides a partial answer to some of the preceding questions; for a proof see [Bha05, 2.3.10].

Theorem 5.8.18 *Let $f : \mathbb{R} \to \mathbb{R}$ be a periodic function with period 2π. Suppose that f and f' are both piecewise continuous on $[-\pi, \pi]$. If f is continuous at x, then the Fourier series (5.8.16) converges to $f(x)$. If f has a jump discontinuity at x, then the Fourier series (5.8.16) converges to $\frac{1}{2}(f(x^+) + f(x^-))$.*

Example 5.8.19 Consider the function $f : [-\pi, \pi] \to \mathbb{R}$ defined by

$$f(x) = \begin{cases} 0 & \text{if } -\pi < x < 0, \\ \pi & \text{if } 0 < x < \pi, \\ \frac{\pi}{2} & \text{if } x = 0 \text{ or } x = \pm\pi; \end{cases}$$

see Figure 5.1. Use the formulas (5.8.11), (5.8.12), and (5.8.13) to compute

$$a_0 = \left\langle f, \frac{1}{\sqrt{2}} \right\rangle = \frac{1}{\pi} \int_{-\pi}^{\pi} \frac{f(x)\,dx}{\sqrt{2}} = \frac{1}{\pi} \int_0^{\pi} \frac{\pi}{\sqrt{2}}\,dx = \frac{\pi}{\sqrt{2}},$$

$$a_n = \langle f, \cos nx \rangle = \frac{1}{\pi} \int_{-\pi}^{\pi} f(x) \cos nx\,dx = \frac{1}{\pi} \int_0^{\pi} \pi \cos nx\,dx = \int_0^{\pi} \cos nx = 0,$$

and

$$a_{-n} = \langle f, \sin nx \rangle = \frac{1}{\pi} \int_0^{\pi} \pi \sin nx\,dx = \frac{1}{n}(1 - \cos n\pi) = \begin{cases} 0 & \text{if } n \text{ is even}, \\ \frac{2}{n} & \text{if } n \text{ is odd}, \end{cases}$$

for $n = 1, 2, 3, \ldots$.

Since f and f' are piecewise continuous, Theorem 5.8.18 ensures that

$$f(x) = \frac{a_0}{\sqrt{2}} + \sum_{n=1}^{\infty}(a_n \cos nx + a_{-n} \sin nx) = \frac{\pi}{2} + \sum_{n=1}^{\infty} \frac{2\sin[(2n-1)x]}{2n-1}$$

for all x. Since $f(\frac{\pi}{2}) = \pi$, we find that

$$\pi = \frac{\pi}{2} + 2\sum_{n=1}^{\infty} \frac{(-1)^{n+1}}{(2n-1)},$$

which implies Leibniz's 1674 discovery that

$$1 - \frac{1}{3} + \frac{1}{5} - \frac{1}{7} + \cdots = \frac{\pi}{4}.$$

5.9 Problems

P.5.1 Show that

$$\mathbf{u}_1 = \begin{bmatrix} \frac{1}{\sqrt{2}} \\ \frac{1}{\sqrt{2}} \\ 0 \end{bmatrix}, \qquad \mathbf{u}_2 = \begin{bmatrix} \frac{1}{\sqrt{3}} \\ -\frac{1}{\sqrt{3}} \\ \frac{1}{\sqrt{3}} \end{bmatrix}, \qquad \mathbf{u}_3 = \begin{bmatrix} -\frac{1}{\sqrt{6}} \\ \frac{1}{\sqrt{6}} \\ \frac{2}{\sqrt{6}} \end{bmatrix}$$

form an orthonormal basis for $\mathbb{R}^3$ and find scalars a_1, a_2, a_3 such that $a_1\mathbf{u}_1 + a_2\mathbf{u}_2 + a_3\mathbf{u}_3 = [2\ 1\ 3]^\mathsf{T}$. Do not solve a 3×3 system of equations!

P.5.2 Use Example 5.2.2 to show that $\operatorname{tr} AB = \operatorname{tr} BA$ for all $A, B \in \mathsf{M}_{m \times n}(\mathbb{F})$.

P.5.3 Let $\mathcal{V} = \mathbb{R}^3$ and $\mathcal{W} = \mathbb{R}^2$, each with the standard inner product. Let $\beta = \mathbf{v}_1, \mathbf{v}_2, \mathbf{v}_3$, in which $\mathbf{v}_1 = [2\ 3\ 5]^\mathsf{T}$, $\mathbf{v}_2 = [7\ 11\ 13]^\mathsf{T}$, and $\mathbf{v}_3 = [17\ 19\ 23]^\mathsf{T}$, and let $\gamma = \mathbf{u}_1, \mathbf{u}_2$, in which $\mathbf{u}_1 = [\frac{1}{\sqrt{2}}\ \frac{1}{\sqrt{2}}]^\mathsf{T}$ and $\mathbf{u}_2 = [-\frac{1}{\sqrt{2}}\ \frac{1}{\sqrt{2}}]^\mathsf{T}$. Let $T \in \mathcal{L}(\mathcal{V}, \mathcal{W})$ be such that

$$T\mathbf{v}_1 = \begin{bmatrix} 2 \\ 3 \end{bmatrix}, \qquad T\mathbf{v}_2 = \begin{bmatrix} 7 \\ 11 \end{bmatrix}, \qquad T\mathbf{v}_3 = \begin{bmatrix} 17 \\ 19 \end{bmatrix}.$$

Show that

$$_\gamma[T]_\beta = \begin{bmatrix} \frac{5}{\sqrt{2}} & 9\sqrt{2} & 18\sqrt{2} \\ \frac{1}{\sqrt{2}} & 2\sqrt{2} & \sqrt{2} \end{bmatrix}.$$

P.5.4 What does the Gram–Schmidt process do to a linearly dependent list of nonzero vectors $\mathbf{v}_1, \mathbf{v}_2, \ldots, \mathbf{v}_n$ in an inner product space $\mathcal{V}$? Suppose that $q \in \{2, 3, \ldots, n\}$, $\mathbf{v}_1, \mathbf{v}_2, \ldots, \mathbf{v}_{q-1}$ are linearly independent, and $\mathbf{v}_1, \mathbf{v}_2, \ldots, \mathbf{v}_q$ are linearly dependent. (a) Show that the orthogonal vectors $\mathbf{x}_k$ in (5.3.9) can be computed for $k = 2, 3, \ldots, q - 1$, but $\mathbf{x}_q = \mathbf{0}$, so the Gram–Schmidt algorithm cannot proceed. Why not? (b) Describe how to use the equation $\mathbf{x}_q = \mathbf{0}$ to compute the coefficients in a linear combination $c_1\mathbf{v}_1 + c_2\mathbf{v}_2 + \cdots + c_{q-1}\mathbf{v}_{q-1} + \mathbf{v}_q = \mathbf{0}$. Why are these coefficients unique?

P.5.5 Let $A = [\mathbf{a}_1\ \mathbf{a}_2\ \ldots\ \mathbf{a}_n] \in \mathsf{M}_n(\mathbb{F})$ be invertible. Let $\mathcal{V} = \mathbb{F}^n$ with the inner product $\langle \mathbf{u}, \mathbf{v} \rangle_{A^{-1}} = \langle A^{-1}\mathbf{u}, A^{-1}\mathbf{v} \rangle$; see P.4.15. Let $A_i(\mathbf{u}) = [\mathbf{a}_1\ \ldots\ \mathbf{a}_{i-1}\ \mathbf{u}\ \mathbf{a}_{i+1}\ \ldots\ \mathbf{a}_n]$ denote the matrix obtained by replacing the ith column of A with $\mathbf{u} \in \mathbb{F}^n$. Let $\phi_i : \mathcal{V} \to \mathbb{F}$ be the linear functional

$$\phi_i(\mathbf{u}) = \frac{\det A_i(\mathbf{u})}{\det A} \qquad i = 1, 2, \ldots, n;$$

see Example 3.1.25.

(a) Show that $\mathbf{a}_1, \mathbf{a}_2, \ldots, \mathbf{a}_n$ is an orthonormal basis of $\mathcal{V}$.

(b) Show that $\phi_i(\mathbf{a}_j) = \delta_{ij}$ for all $i, j = 1, 2, \ldots, n$.

(c) For each $i = 1, 2, \ldots, n$, show that $\mathbf{a}_i$ is the Riesz vector for ϕ_i.

(d) Let $\mathbf{x}, \mathbf{y} \in V$ and $\mathbf{x} = [x_1\ x_2\ \ldots\ x_n]^\mathsf{T}$. Suppose that $A\mathbf{x} = \mathbf{y}$. Show that $\phi_i(\mathbf{y}) = \langle \mathbf{y}, \mathbf{a}_i \rangle_{A^{-1}} = x_i$ for each $i = 1, 2, \ldots, n$. Hint: (5.2.6).

(e) Conclude that $\mathbf{x} = [\phi_1(\mathbf{y})\ \phi_2(\mathbf{y})\ \ldots\ \phi_n(\mathbf{y})]^\mathsf{T}$. This is Cramer's rule.

P.5.6 Let $\mathbf{u}_1, \mathbf{u}_2, \ldots, \mathbf{u}_n$ be an orthonormal system in an $\mathbb{F}$-inner product space V.

(a) Prove that

$$\left\| \mathbf{v} - \sum_{i=1}^n a_i \mathbf{u}_i \right\|^2 = \|\mathbf{v}\|^2 - \sum_{i=1}^n |\langle \mathbf{v}, \mathbf{u}_i \rangle|^2 + \sum_{i=1}^n |\langle \mathbf{v}, \mathbf{u}_i \rangle - a_i|^2 \qquad (5.9.1)$$

for all $a_1, a_2, \ldots, a_n \in \mathbb{F}$.

(b) Let $\mathbf{v} \in V$ be given and let $W = \mathrm{span}\{\mathbf{u}_1, \mathbf{u}_2, \ldots, \mathbf{u}_n\}$. Prove that there is a unique $\mathbf{x} \in W$ such that $\|\mathbf{v} - \mathbf{x}\| \leq \|\mathbf{v} - \mathbf{w}\|$ for all $\mathbf{w} \in W$. Why is $\mathbf{x} = \sum_{i=1}^n \langle \mathbf{v}, \mathbf{u}_i \rangle \mathbf{u}_i$? This vector $\mathbf{x}$ is the orthogonal projection of $\mathbf{v}$ onto the subspace W; see Section 7.3.

(c) Deduce Bessel's inequality (5.7.5) from (5.9.1).

P.5.7 Let $\mathbf{u}_1, \mathbf{u}_2, \ldots, \mathbf{u}_n$ be an orthonormal system in an $\mathbb{F}$-inner product space V, let $\mathcal{U} = \mathrm{span}\{\mathbf{u}_1, \mathbf{u}_2, \ldots, \mathbf{u}_n\}$, and let $\mathbf{v} \in V$. Provide details for the following approach to Bessel's inequality (5.7.5):

(a) If $\mathbf{v} \in \mathcal{U}$, why is $\|\mathbf{v}\|^2 = \sum_{i=1}^n |\langle \mathbf{v}, \mathbf{u}_i \rangle|^2$?

(b) Suppose that $\mathbf{v} \notin \mathcal{U}$, let $W = \mathrm{span}\{\mathbf{u}_1, \mathbf{u}_2, \ldots, \mathbf{u}_n, \mathbf{v}\}$, and apply the Gram–Schmidt process to the linearly independent list $\mathbf{u}_1, \mathbf{u}_2, \ldots, \mathbf{u}_n, \mathbf{v}$. Why do you obtain an orthonormal basis of the form $\mathbf{u}_1, \mathbf{u}_2, \ldots, \mathbf{u}_n, \mathbf{u}_{n+1}$ for W?

(c) Why is $\|\mathbf{v}\|^2 = \sum_{i=1}^{n+1} |\langle \mathbf{v}, \mathbf{u}_i \rangle|^2 \geq \sum_{i=1}^n |\langle \mathbf{v}, \mathbf{u}_i \rangle|^2$?

P.5.8 Deduce the Cauchy–Schwarz inequality from Bessel's inequality.

P.5.9 Show that the functions $1, e^{\pm ix}, e^{\pm 2ix}, e^{\pm 3ix}, \ldots$ are an orthonormal system in the inner product space $C[-\pi, \pi]$ endowed with the inner product

$$\langle f, g \rangle = \frac{1}{2\pi} \int_{-\pi}^{\pi} f(x) \overline{g(x)}\, dx.$$

P.5.10 Let $A \in \mathsf{M}_n$ be invertible. Show that $(A^{-1})^* = (A^*)^{-1}$. Hint: Compute $(A^{-1}A)^*$.

P.5.11 Let $V = C[0, 1]$, endowed with the L^2 inner product (5.1.6), and let T be the *Volterra operator* on V defined by

$$(Tf)(t) = \int_0^t f(s)\, ds. \qquad (5.9.2)$$

Show that the adjoint of T is the linear operator on V defined by

$$(T^*g)(s) = \int_s^1 g(t)\, dt. \qquad (5.9.3)$$

P.5.12 Let $K(s, t)$ be a continuous complex-valued function on $[0, 1] \times [0, 1]$ and let $V = C[0, 1]$ with the L^2 inner product (5.1.6). Define $Tf = \int_0^1 K(s, t) f(t)\, dt$. Why does

T have an adjoint and what is it? What would you have to assume to make *T* self-adjoint?

P.5.13 Let $\mathcal{P}$ be the complex inner product space of polynomials of all degrees, endowed with the L^2 inner product (5.1.6). Let $\phi : \mathcal{P} \to \mathbb{C}$ be the linear functional defined by $\phi(p) = p(0)$. Suppose that ϕ has an adjoint, let $p \in \mathcal{P}$, and let $g = \phi^*(1)$.

(a) Explain why $p(0) = \langle p, \phi^*(1) \rangle = \int_0^1 p(t)\overline{g(t)}\, dt$.

(b) Let $p(t) = t^2 g(t)$ and explain why $0 = \int_0^1 t^2 |g(t)|^2\, dt = \|tg(t)\|^2$. Why must g be the zero polynomial?

(c) Explain why ϕ does not have an adjoint. *Hint*: Let $p = 1$.

P.5.14 Let V be a finite-dimensional inner product space over $\mathbb{F}$, let $\phi \in \mathcal{L}(V, \mathbb{F})$ be a linear functional, and let **w** be the Riesz vector (5.4.6) for ϕ. (a) Evaluate $\phi(\mathbf{w})$ and $\|\mathbf{w}\|$. (b) Show that $\max\{|\phi(\mathbf{v})| : \|\mathbf{v}\| = 1\} = \|\mathbf{w}\|$.

P.5.15 Let $V = \mathcal{P}_2$ be the real inner product space of real polynomials of degree at most two, endowed with the L^2 inner product (5.1.6). Let $\phi \in \mathcal{L}(V, \mathbb{R})$ be the linear functional defined by $\phi(p) = p(0)$. (a) Use the orthonormal basis $\beta = u_1, u_2, u_3$ in (5.1.7) to show that the Riesz vector for ϕ is $w(x) = 30x^2 - 36x + 9$. (b) If $p(x) = ax^2 + bx + c \in V$, verify that $\phi(p) = \langle p, w \rangle$.

P.5.16 Let $V = \mathcal{P}_2$ as in the preceding problem. Let $\psi \in \mathcal{L}(V, \mathbb{R})$ be the linear functional defined by $\psi(p) = \int_0^1 \sqrt{x}p(x)\, dx$. (a) Show that the Riesz vector for ψ is $w(x) = -\frac{4}{7}x^2 + \frac{48}{35}x + \frac{6}{35}$. (b) If $p(x) = ax^2 + bx + c \in V$, verify that $\psi(p) = \langle p, w \rangle$.

P.5.17 Does there exist a function $g : \mathbb{R} \to \mathbb{R}$ such that

$$f(-47) - 3f'(0) + 5f''(\pi) = \int_{-2}^{2} f(x)g(x)\, dx$$

for all functions of the form $f(x) = a_1 e^x + a_2 x e^x \cos x + a_3 x^2 e^x \sin x$?

P.5.18 Let ϕ be a linear functional on the inner product space M_n (with the Frobenius inner product) and suppose that $\phi(C) = 0$ for every commutator $C \in \mathsf{M}_n$. (a) Show that $\phi(AB) = \phi(BA)$ for all $A, B \in \mathsf{M}_n$. (b) If Y is the Riesz vector for ϕ, show that Y commutes with every matrix in M_n. (c) Deduce that Y is a scalar matrix and ϕ is a scalar multiple of the trace functional. (d) If $\phi(I) = 1$, show that $\phi(A) = \frac{1}{n}\operatorname{tr} A$ for all $A \in \mathsf{M}_n$.

P.5.19 Let V be a two-dimensional inner product space with orthonormal basis $\beta = \mathbf{u}_1, \mathbf{u}_2$. Define $T \in \mathcal{L}(V)$ by $T\mathbf{u}_1 = 2\mathbf{u}_1 + \mathbf{u}_2$ and $T\mathbf{u}_2 = \mathbf{u}_1 - \mathbf{u}_2$.

(a) Use Definition 5.6.1 to show that T is self-adjoint.

(b) Compute the basis representation $_\beta[T]_\beta$ of T with respect to the orthonormal basis β. Is it Hermitian?

(c) Define $\mathbf{v}_1 = \mathbf{u}_1$ and $\mathbf{v}_2 = \frac{1}{2}\mathbf{u}_1 + \frac{1}{2\sqrt{3}}\mathbf{u}_2$. Show that $\gamma = \mathbf{v}_1, \mathbf{v}_2$ is a basis of V.

(d) Compute the basis representation $_\gamma[T]_\gamma$ of T with respect to the basis γ. Is it Hermitian?

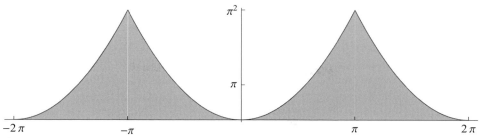

Figure 5.2 The graph of the function f in P.5.22.

(e) Let $V = \mathcal{P}_2$ (see P.5.15 and P.5.16) with the L^2 inner product (5.1.6) and let $\mathbf{u}_1 = 1$ and $\mathbf{u}_2 = \sqrt{3}(2x - 1)$ as in (5.1.7). What are $\mathbf{v}_1$ and $\mathbf{v}_2$? What are the actions of T and T^* on a polynomial $ax + b \in V$?

P.5.20 Let $f \in C_{\mathbb{R}}[-\pi, \pi]$. Show that

$$\int_{-\pi}^{\pi} f(x) \sin nx \, dx \to 0 \text{ as } n \to \infty, \qquad n \in \mathbb{N}.$$

This remarkable fact is an instance of the *Riemann–Lebesgue Lemma*. What happens if $\sin nx$ is replaced by $\cos nx$? *Hint*: Corollary 5.7.6.

P.5.21 Let f be a differentiable real-valued function on $[-\pi, \pi]$. Suppose that $f(-\pi) = f(\pi)$ and f' is continuous. Show that

$$n \int_{-\pi}^{\pi} f(x) \sin nx \, dx \to 0 \text{ as } n \to \infty.$$

What can you say if f is twice differentiable on $[-\pi, \pi]$, $f(-\pi) = f(\pi)$, $f'(-\pi) = f'(\pi)$, and f'' is continuous? What happens if $\sin nx$ is replaced by $\cos nx$? *Hint*: Integrate by parts.

P.5.22 Let $f : \mathbb{R} \to \mathbb{R}$ be periodic with period 2π and suppose that $f(x) = x^2$ for $x \in [-\pi, \pi]$; see Figure 5.2. (a) Use Theorem 5.8.18 to show that

$$f(x) = \frac{\pi^2}{3} + 4 \sum_{n=1}^{\infty} \frac{(-1)^n}{n^2} \cos nx, \qquad x \in \mathbb{R}. \tag{5.9.4}$$

(b) Use (5.9.4) to deduce Euler's 1735 discovery that $\sum_{n=1}^{\infty} 1/n^2 = \pi^2/6$.

P.5.23 Let V be an n-dimensional $\mathbb{F}$-inner product space. If $\mathbf{v}_1, \mathbf{v}_2, \ldots, \mathbf{v}_n \in V$ are linearly independent, show that there exists an inner product $\langle \cdot, \cdot \rangle$ on V with respect to which $\mathbf{v}_1, \mathbf{v}_2, \ldots, \mathbf{v}_n$ is an orthonormal basis (see P.4.15).

P.5.24 Let $\mathbf{v}_1, \mathbf{v}_2, \ldots, \mathbf{v}_n$ be linearly independent vectors in an inner product space V. The *modified Gram–Schmidt process* produces an orthonormal basis for $\text{span}\{\mathbf{v}_1, \mathbf{v}_2, \ldots, \mathbf{v}_n\}$ via the following algorithm. For $k = 1, 2, \ldots, n$, first replace $\mathbf{v}_k$ by $\mathbf{v}_k / \|\mathbf{v}_k\|$ and then for $j = k+1, k+2, \ldots, n$ replace $\mathbf{v}_j$ by $\mathbf{v}_j - \langle \mathbf{v}_j, \mathbf{v}_k \rangle \mathbf{v}_k$. At the conclusion of the algorithm, the vectors $\mathbf{v}_1, \mathbf{v}_2, \ldots, \mathbf{v}_n$ are orthonormal. (a) Let $n = 3$ and follow the steps in the modified Gram–Schmidt process. (b) Do the same for the classical Gram–Schmidt process described in the proof of Theorem 5.3.5. (c) Explain why the former involves the same computations as the latter, but in a different order.

5.10 Notes

Fourier series are named for Jean Baptiste Joseph Fourier (1768–1830), who trained for the priesthood, was imprisoned during the French Revolution, and was a scientific advisor to Napoleon. For an exposition of the fundamentals of Fourier series, see [Bha05].

The Gram–Schmidt process is a powerful theoretical tool, but is not the algorithm of choice for orthogonalizing large collections of vectors. Algorithms based on the QR and singular value decompositions (see Chapters 6 and 14) have proved to be more reliable in practice.

The two algorithms in P.5.24 produce the same lists of orthonormal vectors in exact arithmetic, but in floating point computations the modified Gram–Schmidt algorithm can sometimes produce better results than the classical algorithm. However, there are some problems for which the modified Gram–Schmidt algorithm and floating point arithmetic fail to produce a final set of vectors that are nearly orthonormal.

5.11 Some Important Concepts

- Orthonormal vectors and linear independence.

- Orthogonalization of a linearly independent list of vectors (Gram–Schmidt process).

- Linear functionals and the Riesz representation theorem.

- Orthonormal basis representation of a linear transformation (Theorem 5.5.3).

- Adjoint of a linear transformation or matrix.

- Parseval's identity, Bessel's inequality, and the Cauchy–Schwarz inequality.

- Fourier series and Theorem 5.8.18.

Unitary Matrices

Matrices whose adjoint and inverse are equal play an important role in both theory and computations. In this chapter, we explore their properties and study in detail an important special case: Householder matrices. We use Householder matrices to give constructive proofs of several matrix factorizations.

6.1 Isometries on an Inner Product Space

In this section, $\mathcal{V}$ is an $\mathbb{F}$-inner product space ($\mathbb{F} = \mathbb{R}$ or $\mathbb{C}$) with inner product $\langle \cdot, \cdot \rangle$ and derived norm $\| \cdot \|$.

Definition 6.1.1 Let $T \in \mathcal{L}(\mathcal{V})$. Then T is an *isometry* if $\|T\mathbf{u}\| = \|\mathbf{u}\|$ for all $\mathbf{u} \in \mathcal{V}$.

Example 6.1.2 If $\mathbb{F} = \mathbb{C}$, then the *complex rotation operators* $e^{i\theta}I$ are isometries for all $\theta \in \mathbb{R}$ since $\|e^{i\theta}\mathbf{u}\| = |e^{i\theta}| \|\mathbf{u}\| = \|\mathbf{u}\|$. If $\mathbb{F} = \mathbb{R}$, the operators $\pm I$ are isometries.

Theorem 6.1.3 *Let $\mathcal{V}$ be an inner product space and let $S, T \in \mathcal{L}(\mathcal{V})$ be isometries.*

(a) *ST is an isometry.*

(b) *T is one to one.*

(c) *If $\mathcal{V}$ is finite dimensional, then T is invertible and T^{-1} is an isometry.*

Proof

(a) $\|ST\mathbf{u}\| = \|T\mathbf{u}\| = \|\mathbf{u}\|$ for all $\mathbf{u} \in \mathcal{V}$.

(b) If $T\mathbf{u} = \mathbf{0}$, then $0 = \|T\mathbf{u}\| = \|\mathbf{u}\|$, so the positivity property of a norm ensures that $\mathbf{u} = \mathbf{0}$. Thus, $\ker T = \{\mathbf{0}\}$.

(c) If $\mathcal{V}$ is finite dimensional and T is one to one, then T is onto and invertible; see Corollary 2.5.3. Thus, for each $\mathbf{u} \in \mathcal{V}$, there is a unique $\mathbf{v} \in \mathcal{V}$ such that $\mathbf{u} = T\mathbf{v}$. Then $T^{-1}\mathbf{u} = \mathbf{v}$ and
$$\|T^{-1}\mathbf{u}\| = \|\mathbf{v}\| = \|T\mathbf{v}\| = \|\mathbf{u}\|,$$
so T^{-1} is an isometry. $\square$

Example 6.1.4 Consider the complex inner product space $\mathcal{V}$ of finitely nonzero sequences (see Example 4.4.12) and the right-shift operator $T(v_1, v_2, \ldots) = (0, v_1, v_2, \ldots)$. Then
$$\|T\mathbf{v}\|^2 = 0^2 + |v_1|^2 + |v_2|^2 + \cdots = |v_1|^2 + |v_2|^2 + \cdots = \|\mathbf{v}\|^2,$$

so T is an isometry. Theorem 6.1.3.b ensures that T is one to one. However, there is no $\mathbf{v} \in \mathcal{V}$ such that $T\mathbf{v} = (1, 0, 0, \ldots)$, so T is not onto. Consequently it is not invertible. The hypothesis of finite dimensionality in Theorem 6.1.3.c is necessary.

The polarization identities permit us to prove that isometries have a remarkable interaction with inner products.

Theorem 6.1.5 *Let $\mathcal{V}$ be an inner product space and let $T \in \mathcal{L}(\mathcal{V})$. Then $\langle T\mathbf{u}, T\mathbf{v} \rangle = \langle \mathbf{u}, \mathbf{v} \rangle$ for all $\mathbf{u}, \mathbf{v} \in \mathcal{V}$ if and only if T is an isometry.*

Proof If $\langle T\mathbf{u}, T\mathbf{v} \rangle = \langle \mathbf{u}, \mathbf{v} \rangle$ for all $\mathbf{u}, \mathbf{v} \in \mathcal{V}$, then

$$\| T\mathbf{u} \|^2 = \langle T\mathbf{u}, T\mathbf{u} \rangle = \langle \mathbf{u}, \mathbf{u} \rangle = \| \mathbf{u} \|^2,$$

for all $\mathbf{u} \in \mathcal{V}$, so T is an isometry.

Conversely, suppose that T is an isometry and let $\mathbf{u}, \mathbf{v} \in \mathcal{V}$. If $\mathbb{F} = \mathbb{R}$, the real polarization identity (4.5.25) says that

$$\langle T\mathbf{u}, T\mathbf{v} \rangle = \frac{1}{4} \left(\| T\mathbf{u} + T\mathbf{v} \|^2 - \| T\mathbf{u} - T\mathbf{v} \|^2 \right) = \frac{1}{4} \left(\| T(\mathbf{u} + \mathbf{v}) \|^2 - \| T(\mathbf{u} - \mathbf{v}) \|^2 \right)$$

$$= \frac{1}{4} \left(\| \mathbf{u} + \mathbf{v} \|^2 - \| \mathbf{u} - \mathbf{v} \|^2 \right) = \langle \mathbf{u}, \mathbf{v} \rangle.$$

If $\mathbb{F} = \mathbb{C}$, the complex polarization identity (4.5.26) says that

$$\langle T\mathbf{u}, T\mathbf{v} \rangle = \frac{1}{4} \sum_{k=1}^{4} i^k \| T\mathbf{u} + i^k T\mathbf{v} \|^2 = \frac{1}{4} \sum_{k=1}^{4} i^k \| T(\mathbf{u} + i^k \mathbf{v}) \|^2$$

$$= \frac{1}{4} \sum_{k=1}^{4} i^k \| \mathbf{u} + i^k \mathbf{v} \|^2 = \langle \mathbf{u}, \mathbf{v} \rangle. \qquad \square$$

An operator T on an inner product space $\mathcal{V}$ has an adjoint T^* if $\mathcal{V}$ is finite dimensional (Theorem 5.6.11). In this case, the following theorem says that $T \in \mathcal{L}(\mathcal{V})$ is an isometry if and only if T^*T is the identity operator.

Theorem 6.1.6 *Let $\mathcal{V}$ be a finite-dimensional inner product space and let $T \in \mathcal{L}(\mathcal{V})$. Then T is an isometry if and only if $T^*T = I$.*

Proof If $T^*T = I$, then

$$\| T\mathbf{u} \|^2 = \langle T\mathbf{u}, T\mathbf{u} \rangle = \langle T^*T\mathbf{u}, \mathbf{u} \rangle = \langle \mathbf{u}, \mathbf{u} \rangle = \| \mathbf{u} \|^2$$

for every $\mathbf{u} \in \mathcal{V}$, so T is an isometry. Conversely, suppose that T is an isometry and use the preceding theorem to compute

$$\langle T^*T\mathbf{u}, \mathbf{v} \rangle = \langle T\mathbf{u}, T\mathbf{v} \rangle = \langle \mathbf{u}, \mathbf{v} \rangle \text{ for all } \mathbf{u}, \mathbf{v} \in \mathcal{V}.$$

Corollary 4.4.15 ensures that $T^*T = I$. $\qquad \square$

6.2 Unitary Matrices

We are interested in matrices $U \in M_n$ such that the induced operators $T_U : \mathbb{F}^n \to \mathbb{F}^n$ are isometries with respect to the Euclidean norm. Theorems 6.1.5 and 6.1.6 reveal some characterizations of these matrices.

Definition 6.2.1 A square matrix U is *unitary* if $U^*U = I$. A real unitary matrix is *real orthogonal*; it satisfies $U^TU = I$.

A 1×1 unitary matrix is a complex number u such that $|u|^2 = \bar{u}u = 1$. Unitary matrices may be thought of as matrix analogs of complex numbers with modulus 1.

Since $U^*U = I$ if and only if U is invertible and U^* is its inverse, it is easy to invert a unitary matrix.

Example 6.2.2 Three 2×2 unitary matrices occur in the Pauli equation in quantum mechanics. The Pauli equation is a nonrelativistic version of the Schrödinger equation for spin-$\frac{1}{2}$ particles in an external electromagnetic field. It involves the three *Pauli spin matrices*

$$\sigma_x = \begin{bmatrix} 0 & 1 \\ 1 & 0 \end{bmatrix}, \quad \sigma_y = \begin{bmatrix} 0 & -i \\ i & 0 \end{bmatrix}, \quad \text{and} \quad \sigma_z = \begin{bmatrix} 1 & 0 \\ 0 & -1 \end{bmatrix}.$$

These matrices also arise in quantum information theory, where they are known, respectively, as the *Pauli-X gate*, *Pauli-Y gate*, and *Pauli-Z gate*.

Example 6.2.3 The *Hadamard gate*

$$H = \begin{bmatrix} \frac{1}{\sqrt{2}} & \frac{1}{\sqrt{2}} \\ \frac{1}{\sqrt{2}} & -\frac{1}{\sqrt{2}} \end{bmatrix}$$

is a real orthogonal matrix that arises in quantum information theory.

Example 6.2.4 The matrices $\mathrm{diag}(e^{i\theta_1}, e^{i\theta_2}, \ldots, e^{i\theta_n})$ are unitary for any $\theta_1, \theta_2, \ldots, \theta_n \in \mathbb{R}$. In particular, the complex rotation matrix $e^{i\theta}I \in M_n$ is unitary for any $\theta \in \mathbb{R}$.

Example 6.2.5 The *plane rotation* matrices (see Figure 6.1)

$$U(\theta) = \begin{bmatrix} \cos\theta & -\sin\theta \\ \sin\theta & \cos\theta \end{bmatrix}, \quad \theta \in \mathbb{R}, \tag{6.2.6}$$

are real orthogonal:

$$\begin{aligned} U(\theta)^T U(\theta) &= \begin{bmatrix} \cos\theta & \sin\theta \\ -\sin\theta & \cos\theta \end{bmatrix} \begin{bmatrix} \cos\theta & -\sin\theta \\ \sin\theta & \cos\theta \end{bmatrix} \\ &= \begin{bmatrix} \cos^2\theta + \sin^2\theta & -\cos\theta\sin\theta + \sin\theta\cos\theta \\ -\sin\theta\cos\theta + \cos\theta\sin\theta & \cos^2\theta + \sin^2\theta \end{bmatrix} \\ &= \begin{bmatrix} 1 & 0 \\ 0 & 1 \end{bmatrix}. \end{aligned}$$

Computations with plane rotations can be used to prove many trigonometric identities; see P.6.10.

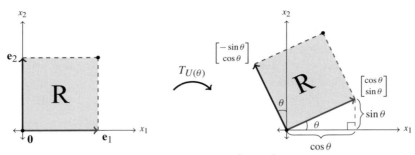

Figure 6.1 The linear transformation $T_{U(\theta)} : \mathbb{R}^2 \to \mathbb{R}^2$ induced by the matrix $U(\theta)$ defined by (6.2.6) is a rotation around the origin through an angle of θ.

An important fact about unitary matrices is that products and direct sums of unitary matrices are unitary.

Theorem 6.2.7 *Let $U, V \in \mathsf{M}_n$ and $W \in \mathsf{M}_m$.*

(a) *If U and V are unitary, then UV is unitary.*

(b) *$V \oplus W$ is unitary if and only if V and W are unitary.*

(c) *If U is unitary, then $|\det U| = 1$.*

Proof (a) Use Definition 6.2.1 to compute
$$(UV)^*(UV) = V^*U^*UV = V^*IV = V^*V = I.$$

(b) Compute
$$(V \oplus W)^*(V \oplus W) = (V^* \oplus W^*)(V \oplus W) = V^*V \oplus W^*W.$$
If V and W are unitary, then
$$V^*V \oplus W^*W = I_n \oplus I_m = I_{n+m},$$
so $V \oplus W$ is unitary. Conversely, if $V \oplus W$ is unitary, then
$$(V \oplus W)^*(V \oplus W) = I_{n+m} = V^*V \oplus W^*W,$$
so $V^*V = I_n$ and $W^*W = I_m$.

(c) Use Definition 6.2.1 and the product rule for the determinant to compute
$$1 = \det I = \det(U^*U) = (\det U^*)(\det U) = \overline{(\det U)}(\det U) = |\det U|^2. \qquad \square$$

Example 6.2.8 The $n \times n$ *reversal matrix*

$$K_n = \begin{bmatrix} & & & 1 \\ & & \reflectbox{$\ddots$} & \\ & 1 & & \\ 1 & & & \end{bmatrix} \tag{6.2.9}$$

is a real, symmetric, unitary involution: $K_n^\mathsf{T} K_n = K_n^2 = I$. Its action on the standard basis of $\mathbb{R}^n$ is $K_n \mathbf{e}_j = \mathbf{e}_{n-j+1}$, for $j = 1, 2, \ldots, n$.

Example 6.2.10 Combine the preceding example, Theorem 6.2.7.a, and Example 6.2.4 to conclude that matrices of the form

$$
DK_n = \begin{bmatrix} & & & e^{i\theta_n} \\ & & \ddots & \\ & e^{i\theta_2} & & \\ e^{i\theta_1} & & & \end{bmatrix}
$$

are unitary for any $\theta_1, \theta_2, \ldots, \theta_n \in \mathbb{R}$.

Theorem 6.2.11 *Let $U \in \mathsf{M}_n(\mathbb{F})$. The following are equivalent:*

(a) *U is unitary, that is, $U^*U = I$.*

(b) *The columns of U are orthonormal.*

(c) *$UU^* = I$.*

(d) *U^* is unitary.*

(e) *The rows of U are orthonormal.*

(f) *U is invertible and $U^{-1} = U^*$.*

(g) *For each $\mathbf{x} \in \mathbb{F}^n$, $\|\mathbf{x}\| = \|U\mathbf{x}\|$.*

(h) *For each $\mathbf{x}, \mathbf{y} \in \mathbb{F}^n$, $(U\mathbf{x})^*(U\mathbf{y}) = \mathbf{x}^*\mathbf{y}$.*

Proof (a) $\Leftrightarrow$ (b) Let $U = [\mathbf{u}_1 \ \mathbf{u}_2 \ \ldots \ \mathbf{u}_n]$. Then $U^*U = [\mathbf{u}_i^*\mathbf{u}_j]$, so $U^*U = I = [\delta_{ij}]$ if and only if $\mathbf{u}_i^*\mathbf{u}_j = \delta_{ij}$.

(a) $\Leftrightarrow$ (c) U^* is a left inverse of U if and only if it is a right inverse; see Theorem 2.2.19.

(c) $\Leftrightarrow$ (d) Write (c) as $(U^*)^*(U^*) = I$. The asserted equivalence follows from Definition 6.2.1.

(d) $\Leftrightarrow$ (e) The rows of U are the conjugates of the columns of U^*, so the equivalence of (d) and (e) follows from the equivalence of (a) and (b).

(f) $\Leftrightarrow$ (a) If $U^*U = I$, then U^* is a left inverse, and hence an inverse, of U. If $U^* = U^{-1}$, then $I = U^{-1}U = U^*U$.

(g) $\Leftrightarrow$ (a) This is Theorem 6.1.6.

(g) $\Leftrightarrow$ (h) The inner product in $\mathbb{F}^n$ is $\langle \mathbf{y}, \mathbf{x} \rangle = \mathbf{x}^*\mathbf{y}$, so this equivalence follows from Theorem 6.1.5. $\qquad\square$

Example 6.2.12 The 3×3 real matrix

$$
\frac{1}{3} \begin{bmatrix} 1 & -2 & -2 \\ 2 & 2 & -1 \\ 2 & -1 & 2 \end{bmatrix}
$$

has orthonormal columns, so it is real orthogonal.

Example 6.2.13 The $n \times n$ *Fourier matrix* is defined by

$$
F_n = \frac{1}{\sqrt{n}} \big[\omega^{(j-1)(k-1)} \big]_{j,k=1}^n, \qquad \omega = e^{2\pi i/n}. \tag{6.2.14}
$$

For example,

$$F_2 = \frac{1}{\sqrt{2}} \begin{bmatrix} \omega^0 & \omega^0 \\ \omega^0 & \omega^1 \end{bmatrix} = \frac{1}{\sqrt{2}} \begin{bmatrix} 1 & 1 \\ 1 & -1 \end{bmatrix}, \qquad \omega = e^{\pi i} = -1$$

and

$$F_4 = \frac{1}{2} \begin{bmatrix} \omega^0 & \omega^0 & \omega^0 & \omega^0 \\ \omega^0 & \omega^1 & \omega^2 & \omega^3 \\ \omega^0 & \omega^2 & \omega^4 & \omega^6 \\ \omega^0 & \omega^3 & \omega^6 & \omega^9 \end{bmatrix} = \frac{1}{2} \begin{bmatrix} 1 & 1 & 1 & 1 \\ 1 & i & -1 & -i \\ 1 & -1 & 1 & -1 \\ 1 & -i & -1 & i \end{bmatrix}, \qquad \omega = e^{\pi i/2} = i. \qquad (6.2.15)$$

These examples illustrate something that is apparent in the definition (6.2.14): Each F_n is symmetric.

The identity

$$\sum_{k=1}^{n} z^{k-1} = \begin{cases} \dfrac{1-z^n}{1-z} & \text{if } z \neq 1, \\ n & \text{if } z = 1, \end{cases}$$

for finite geometric series is P.0.8. Since $\omega^\ell = (e^{2\pi i/n})^\ell = 1$ if and only if ℓ is an integer multiple of n,

$$\sum_{k=1}^{n} \omega^{(k-1)\ell} = \sum_{k=1}^{n} (\omega^\ell)^{k-1} = \begin{cases} 0 & \text{if } \ell \neq pn, \\ n & \text{if } \ell = pn, \end{cases} \qquad \omega = e^{2\pi i/n}, \quad p = 0, \pm 1, \pm 2, \ldots .$$

Write $F_n = [\mathbf{f}_1 \; \mathbf{f}_2 \; \ldots \; \mathbf{f}_n]$, in which the jth column of F_n is

$$\mathbf{f}_j = \frac{1}{\sqrt{n}} \begin{bmatrix} 1 & \omega^{j-1} & \omega^{2(j-1)} & \omega^{3(j-1)} & \ldots & \omega^{(n-1)(j-1)} \end{bmatrix}^{\mathsf{T}}, \quad j = 1, 2, \ldots, n.$$

The (i, j) entry of $F_n^* F_n$ is

$$\mathbf{f}_i^* \mathbf{f}_j = \frac{1}{n} \sum_{k=1}^{n} \overline{\omega^{(i-1)(k-1)}} \omega^{(j-1)(k-1)}$$

$$= \frac{1}{n} \sum_{k=1}^{n} \omega^{-(i-1)(k-1)} \omega^{(j-1)(k-1)}$$

$$= \frac{1}{n} \sum_{k=1}^{n} (\omega^{1-i})^{k-1} (\omega^{j-1})^{k-1}$$

$$= \frac{1}{n} \sum_{k=1}^{n} (\omega^{j-i})^{k-1} = \begin{cases} 0 & \text{if } i \neq j, \\ 1 & \text{if } i = j, \end{cases} \qquad 1 \leq i, j \leq n.$$

This calculation shows that $F_n^* F_n = I$, so F_n is unitary. Since F_n is symmetric,

$$F_n^{-1} = F_n^* = \overline{F_n}.$$

In particular,

$$F_2^{-1} = \frac{1}{\sqrt{2}} \begin{bmatrix} 1 & 1 \\ 1 & -1 \end{bmatrix} \quad \text{and} \quad F_4^{-1} = \frac{1}{2} \begin{bmatrix} 1 & 1 & 1 & 1 \\ 1 & -i & -1 & i \\ 1 & -1 & 1 & -1 \\ 1 & i & -1 & -i \end{bmatrix}.$$

Because algorithms that employ unitary matrices are well behaved with respect to propagation and analysis of errors associated with floating point arithmetic, they are widely used in modern numerical linear algebra. It costs almost nothing (in time or precision) to invert a unitary matrix, so algorithms that involve matrix inverses (such as similarity transformations) are likely to be more efficient and stable if unitary matrices are employed.

Definition 6.2.16 $A, B \in \mathsf{M}_n$ are *unitarily similar* if there is a unitary $U \in \mathsf{M}_n$ such that $A = UBU^*$; they are *real orthogonally similar* if there is a real orthogonal $Q \in \mathsf{M}_n$ such that $A = QBQ^\mathsf{T}$.

An orthonormal list of vectors in a finite-dimensional inner product space can be extended to an orthonormal basis; see Corollary 5.3.13. The following theorem is a block matrix version of this principle.

Theorem 6.2.17 *If $X \in \mathsf{M}_{m \times n}(\mathbb{F})$ and $X^*X = I_n$, then there is an $X' \in \mathsf{M}_{m \times (m-n)}(\mathbb{F})$ such that $A = [X\ X'] \in \mathsf{M}_m(\mathbb{F})$ is unitary.*

Proof Partition $X = [\mathbf{x}_1\ \mathbf{x}_2\ \ldots\ \mathbf{x}_n]$ according to its columns. Since $X^*X = [\mathbf{x}_i^*\mathbf{x}_j] = I_n$, the vectors $\mathbf{x}_1, \mathbf{x}_2, \ldots, \mathbf{x}_n \in \mathbb{F}^m$ are orthonormal. Corollary 5.3.13 says that there are $\mathbf{x}_{n+1}, \mathbf{x}_{n+2}, \ldots, \mathbf{x}_m \in \mathbb{F}^m$ such that the list $\mathbf{x}_1, \mathbf{x}_2, \ldots, \mathbf{x}_n, \mathbf{x}_{n+1}, \ldots, \mathbf{x}_m$ is an orthonormal basis for $\mathbb{F}^m$. Let $X' = [\mathbf{x}_{n+1}\ \mathbf{x}_{n+2}\ \ldots\ \mathbf{x}_m]$. Theorem 6.2.11 ensures that $A = [X\ X'] \in \mathsf{M}_m(\mathbb{F})$ is unitary. $\qquad\square$

6.3 Permutation Matrices

Definition 6.3.1 A square matrix A is a *permutation matrix* if exactly one entry in each row and in each column is 1; all other entries are 0.

Multiplying a matrix on the left by a permutation matrix permutes rows:

$$\begin{bmatrix} 0 & 1 & 0 \\ 0 & 0 & 1 \\ 1 & 0 & 0 \end{bmatrix} \begin{bmatrix} 1 & 2 & 3 \\ 4 & 5 & 6 \\ 7 & 8 & 9 \end{bmatrix} = \begin{bmatrix} 4 & 5 & 6 \\ 7 & 8 & 9 \\ 1 & 2 & 3 \end{bmatrix}.$$

Multiplying a matrix on the right by a permutation matrix permutes columns:

$$\begin{bmatrix} 1 & 2 & 3 \\ 4 & 5 & 6 \\ 7 & 8 & 9 \end{bmatrix} \begin{bmatrix} 0 & 1 & 0 \\ 0 & 0 & 1 \\ 1 & 0 & 0 \end{bmatrix} = \begin{bmatrix} 3 & 1 & 2 \\ 6 & 4 & 5 \\ 9 & 7 & 8 \end{bmatrix}.$$

The columns of an $n \times n$ permutation matrix are a permutation of the standard basis vectors in $\mathbb{R}^n$. If $\sigma : \{1, 2, \ldots, n\} \to \{1, 2, \ldots, n\}$ is a function, then

$$P = [\mathbf{e}_{\sigma(1)}\ \mathbf{e}_{\sigma(2)}\ \ldots\ \mathbf{e}_{\sigma(n)}] \in \mathsf{M}_n \qquad (6.3.2)$$

is a permutation matrix if and only if σ is a permutation of the list $1, 2, \ldots, n$.

The transpose of a permutation matrix is a permutation matrix. If P is the permutation matrix (6.3.2), then

$$P^\mathsf{T} P = [e_{\sigma(i)}^\mathsf{T} e_{\sigma(j)}] = [\delta_{ij}] = I_n.$$

This says that P^T is the inverse of P, so permutation matrices are real orthogonal.

Definition 6.3.3 Square matrices A, B are *permutation similar* if there is a permutation matrix P such that $A = PBP^\mathsf{T}$.

A permutation similarity rearranges columns and rows:

$$\begin{bmatrix} 0 & 1 & 0 \\ 0 & 0 & 1 \\ 1 & 0 & 0 \end{bmatrix} \begin{bmatrix} 1 & 2 & 3 \\ 4 & 5 & 6 \\ 7 & 8 & 9 \end{bmatrix} \begin{bmatrix} 0 & 0 & 1 \\ 1 & 0 & 0 \\ 0 & 1 & 0 \end{bmatrix} = \begin{bmatrix} 5 & 6 & 4 \\ 8 & 9 & 7 \\ 2 & 3 & 1 \end{bmatrix}.$$

Entries that start on the diagonal remain on the diagonal, but their positions within the diagonal may be rearranged. Entries that start in a row together remain together in a row, but the row may be moved and the entries in the row may be rearranged. Entries that start in a column together remain together in a column, but the column may be moved and the entries in the column may be rearranged.

Permutation similarities can be used to permute the diagonal entries of a matrix to achieve a particular pattern. For example, we might want equal diagonal entries grouped together. If the matrix has real diagonal entries, we might want to rearrange them in increasing order. If $\Lambda = \mathrm{diag}(\lambda_1, \lambda_2, \ldots, \lambda_n)$, inspection of the permutation similarity

$$\begin{aligned} P^\mathsf{T} \Lambda P &= [e_{\sigma(1)} \ e_{\sigma(2)} \ \cdots \ e_{\sigma(n)}]^\mathsf{T} \, \mathrm{diag}(\lambda_1, \lambda_2, \ldots, \lambda_n)[e_{\sigma(1)} \ e_{\sigma(2)} \ \cdots \ e_{\sigma(n)}] \\ &= [\lambda_{\sigma(1)} e_{\sigma(1)} \ \lambda_{\sigma(2)} e_{\sigma(2)} \ \cdots \ \lambda_{\sigma(n)} e_{\sigma(n)}]^\mathsf{T} [e_{\sigma(1)} \ e_{\sigma(2)} \ \cdots \ e_{\sigma(n)}] \\ &= [\lambda_{\sigma(i)} e_{\sigma(i)}^\mathsf{T} e_{\sigma(j)}] = [\lambda_{\sigma(i)} \delta_{ij}] \\ &= \mathrm{diag}(\lambda_{\sigma(1)}, \lambda_{\sigma(2)}, \ldots, \lambda_{\sigma(n)}) \end{aligned} \tag{6.3.4}$$

reveals how to arrange the standard basis vectors in the columns of P to achieve a particular pattern for the rearranged diagonal entries of a square matrix whose diagonal is Λ.

Example 6.3.5 A permutation similarity via the permutation matrix $K_n = [e_n \ e_{n-1} \ \cdots \ e_1]$ in (6.2.9) reverses the order of diagonal entries:

$$K_n^\mathsf{T} \Lambda K_n = K_n \Lambda K_n = \mathrm{diag}(\lambda_n, \lambda_{n-1}, \ldots, \lambda_1).$$

Permutation matrices can be used to rearrange diagonal blocks of a matrix.

Example 6.3.6 Let $A = B \oplus C$, in which $B \in \mathsf{M}_p$ and $C \in \mathsf{M}_q$. Let

$$P = \begin{bmatrix} 0_{p \times q} & I_p \\ I_q & 0_{q \times p} \end{bmatrix} \in \mathsf{M}_{p+q}.$$

Then P is a permutation matrix and

$$P^\mathsf{T}AP = \begin{bmatrix} 0_{q\times p} & I_q \\ I_p & 0_{p\times q} \end{bmatrix} \begin{bmatrix} B & 0 \\ 0 & C \end{bmatrix} \begin{bmatrix} 0_{p\times q} & I_p \\ I_q & 0_{q\times p} \end{bmatrix}$$

$$= \begin{bmatrix} 0_{q\times p} & C \\ B & 0_{p\times q} \end{bmatrix} \begin{bmatrix} 0_{p\times q} & I_p \\ I_q & 0_{q\times p} \end{bmatrix}$$

$$= \begin{bmatrix} C & 0 \\ 0 & B \end{bmatrix}.$$

Another useful permutation similarity involves a 2×2 block matrix of diagonal matrices.

Example 6.3.7 The matrices

$$\begin{bmatrix} a & 0 & c & 0 \\ 0 & b & 0 & d \\ e & 0 & g & 0 \\ 0 & f & 0 & h \end{bmatrix} \quad \text{and} \quad \begin{bmatrix} a & c & 0 & 0 \\ e & g & 0 & 0 \\ 0 & 0 & b & d \\ 0 & 0 & f & h \end{bmatrix}$$

are permutation similar via $P = [\mathbf{e}_1\ \mathbf{e}_3\ \mathbf{e}_2\ \mathbf{e}_4]$. In general, if $\Lambda = \operatorname{diag}(\lambda_1, \lambda_2, \ldots, \lambda_n)$, $M = \operatorname{diag}(\mu_1, \mu_2, \ldots, \mu_n)$, $N = \operatorname{diag}(\nu_1, \nu_2, \ldots, \nu_n)$, and $T = \operatorname{diag}(\tau_1, \tau_2, \ldots, \tau_n)$, then

$$\begin{bmatrix} \Lambda & M \\ N & T \end{bmatrix} \quad \text{and} \quad \begin{bmatrix} \lambda_1 & \mu_1 \\ \nu_1 & \tau_1 \end{bmatrix} \oplus \cdots \oplus \begin{bmatrix} \lambda_n & \mu_n \\ \nu_n & \tau_n \end{bmatrix} \tag{6.3.8}$$

are permutation similar.

6.4 Householder Matrices and Rank-1 Projections

Householder matrices are an important family of unitary matrices that are used in many numerical linear algebra algorithms.

Definition 6.4.1 Let $\mathbf{u} \in \mathbb{F}^n$ be a unit vector. Then $P_\mathbf{u} = \mathbf{u}\mathbf{u}^* \in \mathsf{M}_n(\mathbb{F})$ is a *rank-1 projection matrix*.

For any $\mathbf{x} \in \mathbb{F}^n$,

$$P_\mathbf{u}\mathbf{x} = \mathbf{u}\mathbf{u}^*\mathbf{x} = (\mathbf{u}^*\mathbf{x})\mathbf{u} = \langle \mathbf{x}, \mathbf{u} \rangle \mathbf{u}$$

is the projection of $\mathbf{x}$ onto the unit vector $\mathbf{u}$; see (4.5.12) and Figure 6.2. A rank-1 projection matrix has the following properties:

(a) $\operatorname{col} P_\mathbf{u} = \operatorname{span}\{\mathbf{u}\}$ is one dimensional, so $\operatorname{rank} P_\mathbf{u} = 1$.

(b) If $\mathbf{n}$ is orthogonal to $\mathbf{u}$, then $P_\mathbf{u}\mathbf{n} = \langle \mathbf{n}, \mathbf{u} \rangle \mathbf{u} = \mathbf{0}$.

(c) If $\mathbf{v} = c\mathbf{u}$, then $P_\mathbf{u}\mathbf{v} = \langle \mathbf{v}, \mathbf{u} \rangle \mathbf{u} = c\langle \mathbf{u}, \mathbf{u} \rangle \mathbf{u} = c\mathbf{u} = \mathbf{v}$.

(d) $P_\mathbf{u}^* = (\mathbf{u}^*)^*\mathbf{u}^* = \mathbf{u}\mathbf{u}^* = P_\mathbf{u}$, so $P_\mathbf{u}$ is Hermitian. It is real symmetric if $\mathbf{u} \in \mathbb{R}^n$.

(e) $P_\mathbf{u}^2 = \mathbf{u}\mathbf{u}^*\mathbf{u}\mathbf{u}^* = \mathbf{u}(\mathbf{u}^*\mathbf{u})\mathbf{u}^* = \mathbf{u}(1)\mathbf{u}^* = \mathbf{u}\mathbf{u}^* = P_\mathbf{u}$, so $P_\mathbf{u}$ is idempotent.

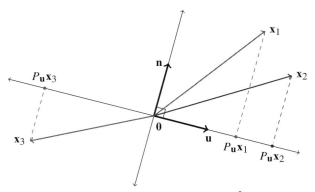

Figure 6.2 Projections of $\mathbf{x}_1, \mathbf{x}_2, \mathbf{x}_3$ onto $\mathbf{u}$ in $\mathbb{R}^2$.

Example 6.4.2 Let $\mathbf{u} = \frac{1}{\sqrt{5}}[1\ 2]^\mathsf{T} \in \mathbb{R}^2$. Then

$$P_\mathbf{u} = \frac{1}{5}\begin{bmatrix} 1 \\ 2 \end{bmatrix}[1\ 2] = \begin{bmatrix} \frac{1}{5} & \frac{2}{5} \\ \frac{2}{5} & \frac{4}{5} \end{bmatrix}$$

is a rank-1 projection matrix.

Definition 6.4.3 Let $\mathbf{w} \in \mathbb{F}^n$ be nonzero, let $\mathbf{u} = \mathbf{w}/\|\mathbf{w}\|_2$, and let $P_\mathbf{u} = \mathbf{u}\mathbf{u}^* = \mathbf{w}\mathbf{w}^*/\|\mathbf{w}\|_2^2$. The corresponding *Householder matrix* is

$$U_\mathbf{w} = I - 2P_\mathbf{u} \in \mathsf{M}_n(\mathbb{F}) \tag{6.4.4}$$

and the corresponding *Householder transformation* is $T_{U_\mathbf{w}}$, that is,

$$\mathbf{x} \mapsto U_\mathbf{w}\mathbf{x} = \mathbf{x} - 2\langle \mathbf{x}, \mathbf{u}\rangle\mathbf{u}.$$

If $\mathbf{w} \in \mathbb{R}^n$, then the rank-1 projection $P_\mathbf{u}$ and Householder matrix $U_\mathbf{w}$ are both real. The action of a Householder transformation $U_\mathbf{w}$ on $\mathbf{x} \in \mathbb{R}^3$ is illustrated in Figure 6.3. The transformation $U_\mathbf{w}$ reflects $\mathbf{x}$ across the plane that is orthogonal to $\mathbf{w}$ and contains $\mathbf{0}$.

In $\mathbb{C}^n$, the Householder transformation $U_\mathbf{w}$ acts on $\mathbf{x}$ by reflecting it across an $(n-1)$-dimensional subspace (namely, null $\mathbf{w}^*$) that is orthogonal to $\mathbf{w}$. The computation

$$U_\mathbf{w}\mathbf{x} = (I - 2P_\mathbf{u})\mathbf{x} = \mathbf{x} - 2P_\mathbf{u}\mathbf{x}$$
$$= -P_\mathbf{u}\mathbf{x} + (\mathbf{x} - P_\mathbf{u}\mathbf{x})$$

identifies $U_\mathbf{w}\mathbf{x}$ as the sum of two orthogonal vectors: $P_\mathbf{u}\mathbf{x}$ is a scalar multiple of $\mathbf{w}$ and $\mathbf{x} - P_\mathbf{u}\mathbf{x}$ is orthogonal to $\mathbf{w}$.

Theorem 6.4.5 *Householder matrices are unitary, Hermitian, and involutive. Real Householder matrices are real orthogonal, symmetric, and involutive.*

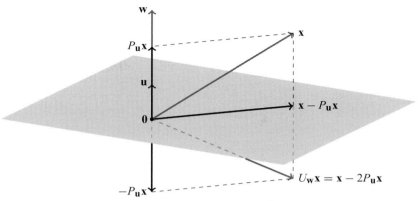

Figure 6.3 Householder transformation $\mathbf{x} \mapsto U_{\mathbf{w}}\mathbf{x} \in \mathbb{R}^3$; $\mathbf{u} = \mathbf{w}/\|\mathbf{w}\|_2$.

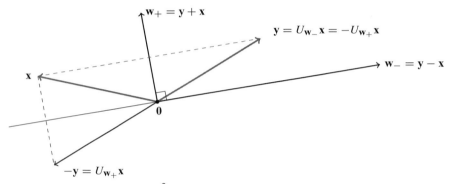

Figure 6.4 Theorem 6.4.7 in $\mathbb{R}^2$. $\mathbf{w}_+$ and $\mathbf{w}_-$ are orthogonal. $U_{\mathbf{w}_-}$ reflects $\mathbf{x}$ across the line spanned by $\mathbf{y} + \mathbf{x}$. $U_{\mathbf{w}_+}$ reflects $\mathbf{x}$ across the line spanned by $\mathbf{y} - \mathbf{x}$. $U_{\mathbf{w}_-}\mathbf{x} = \mathbf{y}$ and $-U_{\mathbf{w}_+}\mathbf{x} = \mathbf{y}$.

Proof Let $\mathbf{w} \in \mathbb{F}^n$ be nonzero and let $\mathbf{u} = \mathbf{w}/\|\mathbf{w}\|_2$. Then $U_{\mathbf{w}}^* = (I - 2P_{\mathbf{u}})^* = I - 2P_{\mathbf{u}}^* = I - 2P_{\mathbf{u}} = U_{\mathbf{w}}$ and $U_{\mathbf{w}}^* U_{\mathbf{w}} = U_{\mathbf{w}}^2 = I$. Since $P_{\mathbf{u}}$ is idempotent,

$$U_{\mathbf{w}}^2 = (I - 2P_{\mathbf{u}})(I - 2P_{\mathbf{u}}) = I - 2P_{\mathbf{u}} - 2P_{\mathbf{u}} + 4P_{\mathbf{u}}^2$$
$$= I - 4P_{\mathbf{u}} + 4P_{\mathbf{u}} = I.$$

Thus, $U_{\mathbf{w}}$ is an involution. $\qquad\square$

If $\mathbf{x}, \mathbf{y} \in \mathbb{F}^n$ and $\|\mathbf{x}\|_2 = \|\mathbf{y}\|_2 \neq 0$, a Householder matrix can be used to construct a unitary matrix in $\mathsf{M}_n(\mathbb{F})$ that maps $\mathbf{x}$ to $\mathbf{y}$; see Figure 6.4. The following example illustrates how to do this in a particular case.

Example 6.4.6 Let $\mathbf{x} = [4\ 0\ -3]^\mathsf{T} \in \mathbb{R}^3$, so $\|\mathbf{x}\|_2 = 5$. How can we construct a real orthogonal matrix that maps $\mathbf{x}$ to $\mathbf{y} = 5\mathbf{e}_1$? Let $\mathbf{w}_\pm = \mathbf{y} \pm \mathbf{x}$. Then

$$\mathbf{w}_+ = \mathbf{y} + \mathbf{x} = [9\ 0\ -3]^\mathsf{T} \quad \text{and} \quad \mathbf{w}_- = \mathbf{y} - \mathbf{x} = [1\ 0\ 3]^\mathsf{T},$$

and

$$\|\mathbf{w}_\pm\|_2^2 = \|\mathbf{y}\|_2^2 \pm 2\langle \mathbf{y}, \mathbf{x}\rangle + \|\mathbf{x}\|_2^2 = 25 \pm 40 + 25 = 50 \pm 40.$$

In this case, $\langle \mathbf{x}, \mathbf{y}\rangle > 0$ and $\|\mathbf{w}_+\|_2 > \|\mathbf{w}_-\|_2$. Compute

$$\mathbf{u}_+ = \frac{1}{\|\mathbf{w}_+\|_2}\mathbf{w}_+ = \frac{1}{\sqrt{90}}[9 \ 0 \ -3]^\mathsf{T} \quad \text{and} \quad \mathbf{u}_- = \frac{1}{\|\mathbf{w}_-\|_2}\mathbf{w}_- = \frac{1}{\sqrt{10}}[1 \ 0 \ 3]^\mathsf{T}.$$

Then

$$U_{\mathbf{w}_+} = I - \frac{2}{90}\begin{bmatrix} 9 \\ 0 \\ -3 \end{bmatrix}[9 \ \ 0 \ \ -3] = \begin{bmatrix} -\frac{4}{5} & 0 & \frac{3}{5} \\ 0 & 1 & 0 \\ \frac{3}{5} & 0 & \frac{4}{5} \end{bmatrix}$$

and

$$U_{\mathbf{w}_-} = I - \frac{2}{10}\begin{bmatrix} 1 \\ 0 \\ 3 \end{bmatrix}[1 \ \ 0 \ \ 3] = \begin{bmatrix} \frac{4}{5} & 0 & -\frac{3}{5} \\ 0 & 1 & 0 \\ -\frac{3}{5} & 0 & -\frac{4}{5} \end{bmatrix}.$$

Therefore

$$U_{\mathbf{w}_-}\mathbf{x} = \begin{bmatrix} \frac{4}{5} & 0 & -\frac{3}{5} \\ 0 & 1 & 0 \\ -\frac{3}{5} & 0 & -\frac{4}{5} \end{bmatrix}\begin{bmatrix} 4 \\ 0 \\ -3 \end{bmatrix} = \begin{bmatrix} 5 \\ 0 \\ 0 \end{bmatrix} = \mathbf{y}$$

and

$$-U_{\mathbf{w}_+}\mathbf{x} = \begin{bmatrix} \frac{4}{5} & 0 & -\frac{3}{5} \\ 0 & -1 & 0 \\ -\frac{3}{5} & 0 & -\frac{4}{5} \end{bmatrix}\begin{bmatrix} 4 \\ 0 \\ -3 \end{bmatrix} = \begin{bmatrix} 5 \\ 0 \\ 0 \end{bmatrix} = \mathbf{y}.$$

There is no theoretical reason to prefer $-U_{\mathbf{w}_+}$ to $U_{\mathbf{w}_-}$, but there is a computational reason to prefer $-U_{\mathbf{w}_+}$ in this case. To construct $U_{\mathbf{w}_\pm}$, we must perform the normalization $\mathbf{w}_\pm/\|\mathbf{w}_\pm\|_2$. If the denominator is small (because one of the vectors $\mathbf{y} + \mathbf{x}$ or $\mathbf{y} - \mathbf{x}$ is close to zero), we must divide one small number by another. To mitigate numerical instability, it is best to use $-U_{\mathbf{w}_+}$ if $\|\mathbf{w}_+\|_2 > \|\mathbf{w}_-\|_2$, and to use $U_{\mathbf{w}_-}$ if $\|\mathbf{w}_-\|_2 > \|\mathbf{w}_+\|_2$. Since

$$\|\mathbf{w}_\pm\|_2^2 = \|\mathbf{x} \pm \mathbf{y}\|_2^2 = \|\mathbf{x}\|_2^2 \pm 2\langle \mathbf{x}, \mathbf{y}\rangle + \|\mathbf{y}\|_2^2,$$

the former situation occurs if $\langle \mathbf{x}, \mathbf{y}\rangle > 0$, the latter if $\langle \mathbf{x}, \mathbf{y}\rangle < 0$.

Example 6.4.6 illustrates an algorithm that uses a scalar multiple of a Householder matrix to map a given vector in $\mathbb{F}^n$ to any other vector in $\mathbb{F}^n$ that has the same Euclidean norm. The following theorem states the algorithm in the real case; it employs real Householder matrices and incorporates a choice of signs that enhances numerical stability.

Theorem 6.4.7 *Let* $\mathbf{x}, \mathbf{y} \in \mathbb{R}^n$ *and suppose that* $\|\mathbf{x}\|_2 = \|\mathbf{y}\|_2 \neq 0$. *Let*

$$\sigma = \begin{cases} 1 & \text{if } \langle \mathbf{x}, \mathbf{y}\rangle \leq 0, \\ -1 & \text{if } \langle \mathbf{x}, \mathbf{y}\rangle > 0, \end{cases}$$

and let $\mathbf{w} = \mathbf{y} - \sigma\mathbf{x}$. *Then* $\sigma U_{\mathbf{w}}$ *is real orthogonal and* $\sigma U_{\mathbf{w}}\mathbf{x} = \mathbf{y}$.

Proof Observe that $\sigma \langle \mathbf{x}, \mathbf{y} \rangle = -|\langle \mathbf{x}, \mathbf{y} \rangle|$ and compute

$$\|\mathbf{w}\|_2^2 = \|\mathbf{y} - \sigma \mathbf{x}\|_2^2 = \|\mathbf{y}\|_2^2 - 2\sigma \langle \mathbf{x}, \mathbf{y} \rangle + \|\mathbf{x}\|_2^2$$
$$= 2(\|\mathbf{x}\|_2^2 - \sigma \langle \mathbf{x}, \mathbf{y} \rangle)$$
$$= 2(\|\mathbf{x}\|_2^2 + |\langle \mathbf{x}, \mathbf{y} \rangle|) > 0.$$

Therefore,

$$\sigma U_\mathbf{w} \mathbf{x} = \sigma(I - 2P_\mathbf{u})\mathbf{x} = \sigma \mathbf{x} - 2\sigma \frac{\mathbf{w}\mathbf{w}^\mathsf{T}}{\|\mathbf{w}\|_2^2}\mathbf{x}$$

$$= (\mathbf{y} - \mathbf{w}) - 2\sigma \frac{\langle \mathbf{x}, \mathbf{w} \rangle}{\|\mathbf{w}\|_2^2}\mathbf{w}$$

$$= \mathbf{y} - \left(\frac{\|\mathbf{w}\|_2^2 + 2\sigma \langle \mathbf{x}, \mathbf{w} \rangle}{\|\mathbf{w}\|_2^2}\right)\mathbf{w}. \qquad (6.4.8)$$

However,

$$2\sigma \langle \mathbf{x}, \mathbf{w} \rangle = 2\sigma \langle \mathbf{x}, \mathbf{y} - \sigma \mathbf{x} \rangle = -2(\|\mathbf{x}\|_2^2 + |\langle \mathbf{x}, \mathbf{y} \rangle|) = -\|\mathbf{w}\|_2^2,$$

so the second term in (6.4.8) vanishes and $\sigma U_\mathbf{w} \mathbf{x} = \mathbf{y}$. $\qquad\square$

Complex Householder matrices can be used to construct a unitary matrix that maps a vector in $\mathbb{C}^n$ to any other vector in $\mathbb{C}^n$ that has the same Euclidean norm. The algorithm in the following theorem incorporates a complex rotation that is an analog of the choice of sign in the real case. Once the rotation is defined, verification of the algorithm proceeds as in the real case.

Theorem 6.4.9 *Let* $\mathbf{x}, \mathbf{y} \in \mathbb{C}^n$ *and suppose that* $\|\mathbf{x}\|_2 = \|\mathbf{y}\|_2 \neq 0$. *Let*

$$\sigma = \begin{cases} 1 & \text{if } \langle \mathbf{x}, \mathbf{y} \rangle = 0, \\ -\overline{\langle \mathbf{x}, \mathbf{y} \rangle}/|\langle \mathbf{x}, \mathbf{y} \rangle| & \text{if } \langle \mathbf{x}, \mathbf{y} \rangle \neq 0, \end{cases}$$

and let $\mathbf{w} = \mathbf{y} - \sigma \mathbf{x}$. *Then* $\sigma U_\mathbf{w}$ *is unitary and* $\sigma U_\mathbf{w} \mathbf{x} = \mathbf{y}$.

Proof Observe that $\sigma \langle \mathbf{x}, \mathbf{y} \rangle = -|\langle \mathbf{x}, \mathbf{y} \rangle|$ and compute

$$\|\mathbf{w}\|_2^2 = \|\mathbf{y} - \sigma \mathbf{x}\|_2^2 = \|\mathbf{y}\|_2^2 - 2\operatorname{Re}(\sigma \langle \mathbf{x}, \mathbf{y} \rangle) + \|\mathbf{x}\|_2^2$$
$$= 2(\|\mathbf{x}\|_2^2 + |\langle \mathbf{x}, \mathbf{y} \rangle|) > 0.$$

Therefore,

$$\sigma U_\mathbf{w} \mathbf{x} = \sigma(I - 2P_\mathbf{u})\mathbf{x} = \sigma \mathbf{x} - 2\sigma \frac{\mathbf{w}\mathbf{w}^*}{\|\mathbf{w}\|_2^2}\mathbf{x}$$

$$= (\mathbf{y} - \mathbf{w}) - 2\sigma \frac{\langle \mathbf{x}, \mathbf{w} \rangle}{\|\mathbf{w}\|_2^2}\mathbf{w}$$

$$= \mathbf{y} - \left(\frac{\|\mathbf{w}\|_2^2 + 2\sigma \langle \mathbf{x}, \mathbf{w} \rangle}{\|\mathbf{w}\|_2^2}\right)\mathbf{w}.$$

However,

$$2\sigma \langle \mathbf{x}, \mathbf{w} \rangle = 2\sigma \langle \mathbf{x}, \mathbf{y} - \sigma \mathbf{x} \rangle = -2(\|\mathbf{x}\|_2^2 + |\langle \mathbf{x}, \mathbf{y} \rangle|) = -\|\mathbf{w}\|_2^2,$$

so $\sigma U_\mathbf{w} \mathbf{x} = \mathbf{y}$. $\qquad\square$

Theorems 6.4.7 and 6.4.9 are powerful tools. The following corollary provides easily computed unitary matrices (scalar multiples of Householder matrices) that either (a) maps a given nonzero vector to a positive scalar multiple of the standard basis vector $\mathbf{e}_1$, or (b) has a given unit vector as its first column. These matrices are used in many numerical linear algebra algorithms.

Corollary 6.4.10 *Let* $\mathbf{x} = [x_i] \in \mathbb{F}^n$ *be nonzero, let*

$$\sigma = \begin{cases} 1 & \text{if } x_1 = 0, \\ -\overline{x_1}/|x_1| & \text{if } x_1 \neq 0, \end{cases}$$

and let $\mathbf{w} = \|\mathbf{x}\|_2 \mathbf{e}_1 - \sigma \mathbf{x}$.

(a) $\sigma U_{\mathbf{w}} \mathbf{x} = \|\mathbf{x}\|_2 \mathbf{e}_1$.

(b) *If* $\mathbf{x}$ *is a unit vector, then it is the first column of the unitary matrix* $\overline{\sigma} U_{\mathbf{w}}$.

(c) *If* $\mathbf{x} \in \mathbb{R}^n$, *then* σ, $\mathbf{w}$, *and* $U_{\mathbf{w}}$ *are real.*

Proof (a) This is the special case $\mathbf{y} = \|\mathbf{x}\|_2 \mathbf{e}_1$ of the preceding two theorems.

(b) If $\mathbf{x}$ is a unit vector, then (a) ensures that the unitary matrix $\sigma U_{\mathbf{w}}$ maps $\mathbf{x}$ to $\mathbf{e}_1$. Therefore, $(\sigma U_{\mathbf{w}})^{-1}$ maps $\mathbf{e}_1$ to $\mathbf{x}$, that is, $\mathbf{x}$ is the first column of $(\sigma U_{\mathbf{w}})^{-1} = (\sigma U_{\mathbf{w}})^* = \overline{\sigma} U_{\mathbf{w}}^* = \overline{\sigma} U_{\mathbf{w}}$. In this computation we use the facts that $|\sigma| = 1$, $U_{\mathbf{w}}$ is unitary, and $U_{\mathbf{w}}$ is Hermitian.

(c) Inspect each term. $\qquad\qquad\qquad\qquad\qquad\qquad\qquad\qquad\qquad\qquad\qquad\qquad\square$

The matrix $\sigma U_{\mathbf{w}}$ described in (a) of the preceding corollary is often referred to as a *unitary annihilator of the lower entries of* $\mathbf{x}$; see Example 6.4.6.

6.5 The *QR* Factorization

The *QR factorization* of an $m \times n$ matrix A (also called the *QR decomposition*) presents it as a product of an $m \times n$ matrix Q with orthonormal columns and a square upper triangular matrix R; this presentation requires that $m \geq n$. If A is square, the factor Q is unitary. If A is real, both factors Q and R may be chosen to be real.

The *QR* factorization is an important tool for numerical solution of least squares problems, for transforming a basis into an orthonormal basis, and for computing eigenvalues.

Example 6.5.1 A computation verifies that

$$A = \begin{bmatrix} 3 & 1 \\ 4 & 2 \end{bmatrix} = \begin{bmatrix} \frac{3}{5} & -\frac{4}{5} \\ \frac{4}{5} & \frac{3}{5} \end{bmatrix} \begin{bmatrix} 5 & \frac{11}{5} \\ 0 & \frac{2}{5} \end{bmatrix}$$

is a *QR* factorization of A.

Householder transformations provide a stable method to compute *QR* factorizations. The algorithm described in the following theorem uses Householder matrices to annihilate the lower entries of a sequence of vectors, thereby transforming a given matrix into an upper triangular matrix.

Theorem 6.5.2 (*QR* Factorization) *Let $A \in M_{m \times n}(\mathbb{F})$ and suppose that $m \geq n$.*

(a) *There is a unitary $V \in M_m(\mathbb{F})$ and an upper triangular $R \in M_n(\mathbb{F})$ with real nonnegative diagonal entries such that*

$$A = V \begin{bmatrix} R \\ 0 \end{bmatrix}. \tag{6.5.3}$$

If $V = [Q \; Q']$, in which $Q \in M_{m \times n}(\mathbb{F})$ contains the first n columns of V, then Q has orthonormal columns and

$$A = QR. \tag{6.5.4}$$

(b) *If $\operatorname{rank} A = n$, then the factors Q and R in (6.5.4) are unique and R has positive diagonal entries.*

Proof (a) Let $\mathbf{a}_1$ be the first column of A. If $\mathbf{a}_1 = \mathbf{0}$, let $U_1 = I$. If $\mathbf{a}_1 \neq \mathbf{0}$, use Corollary 6.4.10.a to construct a unitary $U_1 \in M_m(\mathbb{F})$ such that $U_1 \mathbf{a}_1 = \|\mathbf{a}_1\|_2 \mathbf{e}_1 \in \mathbb{F}^m$, and let $r_{11} = \|\mathbf{a}_1\|_2$. Then $r_{11} \geq 0$ and

$$U_1 A = \begin{bmatrix} r_{11} & \star \\ 0 & A' \end{bmatrix}, \qquad A' \in M_{(m-1) \times (n-1)}(\mathbb{F}). \tag{6.5.5}$$

Let $\mathbf{a}'_1$ be the first column of A', let $U' \in M_{m-1}(\mathbb{F})$ be a unitary matrix such that $U' \mathbf{a}'_1 = \|\mathbf{a}'_1\|_2 \mathbf{e}_1 \in \mathbb{F}^{m-1}$, let $r_{22} = \|\mathbf{a}'_1\|_2$, and let $U_2 = I_1 \oplus U'$. Then $r_{22} \geq 0$ and

$$U_2 U_1 A = \begin{bmatrix} r_{11} & \star & \star \\ 0 & r_{22} & \star \\ 0 & 0 & A'' \end{bmatrix}, \qquad A'' \in M_{(m-2) \times (n-2)}(\mathbb{F}). \tag{6.5.6}$$

The direct sum structure of U_2 ensures that the reduction achieved in (6.5.6) affects only the lower-right block A' in (6.5.5). After n reduction steps we obtain

$$U_n \cdots U_2 U_1 A = \begin{bmatrix} R \\ 0 \end{bmatrix},$$

in which

$$R = \begin{bmatrix} r_{11} & \star & \star & \star \\ & r_{22} & \star & \star \\ & & \ddots & \star \\ 0 & & & r_{nn} \end{bmatrix} \in M_n(\mathbb{F})$$

is upper triangular and has diagonal entries that are nonnegative (each is the Euclidean length of some vector). Let $U = U_n \cdots U_2 U_1$, which is unitary, let $V = U^*$, and partition $V = [Q \; Q']$, in which $Q \in M_{m \times n}$. The block Q has orthonormal columns since it comprises the first n columns of the unitary matrix V. Then

$$A = V \begin{bmatrix} R \\ 0 \end{bmatrix} = [Q \; Q'] \begin{bmatrix} R \\ 0 \end{bmatrix} = QR.$$

(b) If $\operatorname{rank} A = n$, then the upper triangular matrix R is invertible and hence its diagonal entries are not just nonnegative, they are positive. Suppose that $A = Q_1 R_1 = Q_2 R_2$, in which Q_1 and

Q_2 have orthonormal columns, and R_1 and R_2 are upper triangular and have positive diagonal entries. Then $A^* = R_1^* Q_1^* = R_2^* Q_2^*$, so

$$A^* A = R_1^* Q_1^* Q_1 R_1 = R_1^* R_1$$

and

$$A^* A = R_2^* Q_2^* Q_2 R_2 = R_2^* R_2.$$

This means that $R_1^* R_1 = R_2^* R_2$, and hence

$$R_1 R_2^{-1} = R_1^{-*} R_2^* = (R_2 R_1^{-1})^*. \tag{6.5.7}$$

We know that R_1^{-1} and R_2^{-1} are upper triangular and have positive diagonal entries (Theorem 3.3.12). Thus, $R_1 R_2^{-1}$ and $R_2 R_1^{-1}$ (products of upper triangular matrices with positive diagonal entries) are upper triangular. The identity (6.5.7) says that an upper triangular matrix equals a lower triangular matrix. Therefore,

$$R_1 R_2^{-1} = R_1^{-*} R_2^* = D$$

is a diagonal matrix with positive diagonal entries. Then

$$D = R_1^{-*} R_2^* = (DR_2)^{-*} R_2^* = D^{-1} R_2^{-*} R_2^* = D^{-1},$$

so $D^2 = I$. Since D is diagonal and has positive diagonal entries, we conclude that $D = I$. Therefore, $R_1 = R_2$ and $Q_1 = Q_2$.   □

An important, if subtle, part of the statement of the preceding theorem is that if $A \in M_{m \times n}(\mathbb{R})$, then the matrices V, R, and Q in (6.5.3) and (6.5.4) are real. The last line of Corollary 6.4.10 is the key to understanding why. If A is real, then every step in the constructive proof of Theorem 6.5.2 involves only real vectors and matrices.

Definition 6.5.8 The factorization (6.5.3) is the *wide QR factorization* of A; (6.5.4) is the *(narrow) QR factorization* of A.

Example 6.5.9 The factorizations

$$A = \begin{bmatrix} 1 & -5 \\ -2 & 4 \\ 2 & 2 \end{bmatrix} = \begin{bmatrix} \frac{1}{3} & -\frac{2}{3} & \frac{2}{3} \\ -\frac{2}{3} & \frac{1}{3} & \frac{2}{3} \\ \frac{2}{3} & \frac{2}{3} & \frac{1}{3} \end{bmatrix} \begin{bmatrix} 3 & -3 \\ 0 & 6 \\ 0 & 0 \end{bmatrix} = V \begin{bmatrix} R \\ 0 \end{bmatrix} \tag{6.5.10}$$

and

$$A = \begin{bmatrix} 1 & -5 \\ -2 & 4 \\ 2 & 2 \end{bmatrix} = \begin{bmatrix} \frac{1}{3} & -\frac{2}{3} \\ -\frac{2}{3} & \frac{1}{3} \\ \frac{2}{3} & \frac{2}{3} \end{bmatrix} \begin{bmatrix} 3 & -3 \\ 0 & 6 \end{bmatrix} = QR \tag{6.5.11}$$

are examples of wide and narrow QR factorizations, respectively.

Example 6.5.12 The display (6.5.13) illustrates the algorithm in the preceding theorem with a 4×3 matrix A. The symbol $\star$ indicates an entry that is not necessarily zero; $\bigstar$ indicates an entry that has just been changed and is not necessarily zero; $\tilde{r}_{ii}$ indicates a nonnegative diagonal entry that has just been created; $\widetilde{0}$ indicates a zero entry that has just been created.

$$
\begin{bmatrix} \star & \star & \star \\ \star & \star & \star \\ \star & \star & \star \\ \star & \star & \star \end{bmatrix} \xrightarrow{U_1} \begin{bmatrix} \tilde{r}_{11} & \star & \star \\ \widetilde{0} & \star & \star \\ \widetilde{0} & \star & \star \\ \widetilde{0} & \star & \star \end{bmatrix} \xrightarrow{U_2} \begin{bmatrix} r_{11} & \star & \star \\ 0 & \tilde{r}_{22} & \star \\ 0 & \widetilde{0} & \star \\ 0 & \widetilde{0} & \star \end{bmatrix} \xrightarrow{U_3} \begin{bmatrix} r_{11} & \star & \star \\ 0 & r_{22} & \star \\ 0 & 0 & \tilde{r}_{33} \\ 0 & 0 & \widetilde{0} \end{bmatrix}
$$
$$
A \qquad\qquad U_1 A \qquad\qquad U_2 U_1 A \qquad\qquad U_3 U_2 U_1 A
$$

(6.5.13)

The 4×4 unitary matrices U_1, $U_2 = I_1 \oplus U'$, and $U_3 = I_2 \oplus U''$ are constructed from Householder matrices and complex rotations, as in Theorem 6.4.10.c. They annihilate the lower entries of successive columns of A without disturbing previously created lower column zeros. The transformation $A \mapsto U_3 U_2 U_1 A$ creates an upper triangular matrix with nonnegative diagonal entries that are positive if A has full column rank. In the QR decomposition of A,

- Q is the first three columns of the 4×4 unitary matrix $(U_3 U_2 U_1)^*$, and

- R is the upper 3×3 block of the 4×3 matrix $U_3 U_2 U_1 A$.

Example 6.5.14 Let us apply the algorithm in Theorem 6.5.2 to find the QR factorization of

$$
A = \begin{bmatrix} 3 & 1 \\ 4 & 2 \end{bmatrix}.
$$

The first step is to map $\mathbf{x} = [3\ 4]^\mathsf{T} = [x_i]$ to $\mathbf{y} = [5\ 0]^\mathsf{T}$ with the algorithm in Corollary 6.4.10.a. Since x_1 is real and positive, $\sigma = -1$, $\mathbf{w} = 5\mathbf{e}_1 + \mathbf{x} = [8\ 4]^\mathsf{T}$, $\|\mathbf{w}\|_2^2 = 80$, and $\mathbf{u} = [8\ 4]^\mathsf{T}/\sqrt{80}$. Then

$$
U_1 = -(I - 2\mathbf{u}\mathbf{u}^\mathsf{T}) = \begin{bmatrix} \frac{3}{5} & \frac{4}{5} \\ \frac{4}{5} & -\frac{3}{5} \end{bmatrix}
$$

and

$$
U_1 A = \begin{bmatrix} 5 & \frac{11}{5} \\ 0 & -\frac{2}{5} \end{bmatrix}.
$$

To make the (2,2) entry positive, multiply on the left by $U_2 = [1] \oplus [-1]$ and obtain

$$
U_2 U_1 A = \begin{bmatrix} \frac{3}{5} & \frac{4}{5} \\ -\frac{4}{5} & \frac{3}{5} \end{bmatrix} A = \begin{bmatrix} 5 & \frac{11}{5} \\ 0 & \frac{2}{5} \end{bmatrix} = R.
$$

The QR factorization is

$$
A = (U_2 U_1)^\mathsf{T} R = \begin{bmatrix} \frac{3}{5} & -\frac{4}{5} \\ \frac{4}{5} & \frac{3}{5} \end{bmatrix} \begin{bmatrix} 5 & \frac{11}{5} \\ 0 & \frac{2}{5} \end{bmatrix}.
$$

It is unique since $\operatorname{rank} A = 2$.

For a discussion of the role of QR factorizations in numerical computation of a least squares solution to a linear system, see Section 7.5. Their role in orthogonalizing a list of linearly independent vectors is revealed in the following corollary.

Corollary 6.5.15 *Let $A \in \mathsf{M}_{m \times n}(\mathbb{F})$. Suppose that $m \geq n$ and $\operatorname{rank} A = n$. Let $A = QR$, in which $Q \in \mathsf{M}_{m \times n}(\mathbb{F})$ has orthonormal columns and $R \in \mathsf{M}_n(\mathbb{F})$ is upper triangular and has positive diagonal entries. For each $k = 1, 2, \ldots, n$, let A_k and Q_k denote the submatrices of A and Q, respectively, comprising their first k columns. Then*

$$\operatorname{col} A_k = \operatorname{col} Q_k, \quad k = 1, 2, \ldots, n,$$

that is, the columns of each submatrix Q_k are an orthonormal basis for $\operatorname{col} A_k$.

Proof The preceding theorem ensures that a factorization of the stated form exists. Let R_k denote the leading $k \times k$ principal submatrix of R; it has positive diagonal entries, so it is invertible. Then

$$[A_k \; \star] = A = [Q_k \; \star] \begin{bmatrix} R_k & \star \\ 0 & \star \end{bmatrix} = [Q_k R_k \; \star],$$

so $A_k = Q_k R_k$ and $\operatorname{col} A_k \subseteq \operatorname{col} Q_k$ for each $k = 1, 2, \ldots, n$. Since $Q_k = A_k R_k^{-1}$, it follows that $\operatorname{col} Q_k \subseteq \operatorname{col} A_k$ for each $k = 1, 2, \ldots, n$. $\qquad\square$

Example 6.5.16 The (narrow) QR factorization (6.5.11) provides an orthonormal basis for $\operatorname{col} A$. In some applications, that might be all one needs. The wide QR factorization (6.5.10) provides an orthonormal basis for $\mathbb{R}^3$ that contains the orthonormal basis for $\operatorname{col} A$.

6.6 Upper Hessenberg Matrices

Reduction of a matrix to some standard form that contains many zero entries is a common ingredient in numerical linear algebra algorithms. One of those standard forms is described in the following definition.

Definition 6.6.1 Let $A = [a_{ij}]$. Then A is *upper Hessenberg* if $a_{ij} = 0$ whenever $i > j + 1$.

Example 6.6.2 A 5×5 upper Hessenberg matrix has the following pattern of zero entries:

$$\begin{bmatrix} \star & \star & \star & \star & \star \\ \star & \star & \star & \star & \star \\ 0 & \star & \star & \star & \star \\ 0 & 0 & \star & \star & \star \\ 0 & 0 & 0 & \star & \star \end{bmatrix}.$$

All entries below the first subdiagonal are zero.

Every square matrix is unitarily similar to an upper Hessenberg matrix, and the unitary similarity can be constructed from a sequence of Householder matrices and complex rotations. If the matrix is real, all the unitary similarities and rotations can be chosen to be real orthogonal similarities and real rotations. Just as in our construction of the QR factorization, the basic idea is unitary annihilation of the lower entries of a sequence of vectors. However, those vectors must now be chosen in a slightly different way because we need to be sure that previous annihilations are not disturbed by subsequent unitary similarities.

Theorem 6.6.3 *Let $A \in M_n(\mathbb{F})$. Then A is unitarily similar to an upper Hessenberg matrix $B = [b_{ij}] \in M_n$, in which each subdiagonal entry $b_{2,1}, b_{3,2}, \ldots, b_{n,n-1}$ is real and nonnegative.*

Proof We are not concerned with the entries in the first row of A. Let

$$A = \begin{bmatrix} \star \\ A' \end{bmatrix} \quad \text{and} \quad A' \in M_{(n-1) \times n}.$$

Let $\mathbf{a}_1' \in \mathbb{F}^{n-1}$ be the first column of A'. Let $V_1 \in M_{n-1}(\mathbb{F})$ be a unitary matrix that annihilates the lower entries in the first column of A', that is,

$$V_1 \mathbf{a}_1' = \begin{bmatrix} \|\mathbf{a}_1'\|_2 \\ \mathbf{0} \end{bmatrix}.$$

If $\mathbf{a}_1' = \mathbf{0}$, take $V_1 = I$; if $\mathbf{a}_1' \neq \mathbf{0}$, use Corollary 6.4.10.a. Let $U_1 = I_1 \oplus V_1$ and compute

$$U_1 A = \begin{bmatrix} \star & \star \\ \|\mathbf{a}_1'\|_2 & \star \\ \mathbf{0} & \star \end{bmatrix} \quad \text{and} \quad U_1 A U_1^* = \begin{bmatrix} \star & \star \\ \|\mathbf{a}_1'\|_2 & \star \\ \mathbf{0} & A'' \end{bmatrix}, \quad A'' \in M_{(n-2) \times (n-1)}.$$

Our construction ensures that right multiplication by $U_1^* = I_1 \oplus V_1^*$ does not disturb the first column of $U_1 A$.

Now construct a unitary $V_2 \in M_{n-2}(\mathbb{F})$ that annihilates the lower entries in the first column of A''. Let $U_2 = I_2 \oplus V_2$ and form $U_2 (U_1 A U_1^*) U_2^*$. This does not disturb any entries in the first column, puts a nonnegative entry in the $(3, 2)$ position, and puts zeros in the entries below it. After $n - 2$ steps, this algorithm produces an upper Hessenberg matrix with nonnegative entries in the first subdiagonal except possibly for the entry in position $(n, n - 1)$, which may not be real and nonnegative. If necessary, a unitary similarity by $I_{n-1} \oplus [e^{i\theta}]$ (for a suitable real θ) makes the entry in position $(n, n - 1)$ real and nonnegative. Each step is a unitary similarity, so the overall reduction is achieved with a unitary similarity. □

We illustrate the algorithm described in the preceding theorem with a 4×4 matrix A and the same symbol conventions that we used in the illustration (6.5.13).

$$\begin{bmatrix} \star & \star & \star & \star \\ \star & \star & \star & \star \\ \star & \star & \star & \star \\ \star & \star & \star & \star \end{bmatrix} \rightarrow \begin{bmatrix} \star & \star & \star & \star \\ \star & \star & \star & \star \\ \widetilde{0} & \star & \star & \star \\ \widetilde{0} & \star & \star & \star \end{bmatrix} \rightarrow \begin{bmatrix} \star & \star & \star & \star \\ \star & \star & \star & \star \\ 0 & \star & \star & \star \\ 0 & \widetilde{0} & \star & \star \end{bmatrix} \rightarrow \begin{bmatrix} \star & \star & \star & \star \\ \star & \star & \star & \star \\ 0 & \star & \star & \star \\ 0 & 0 & \star & \star \end{bmatrix}$$

$$\quad A \qquad\qquad U_1 A U_1^* \qquad\qquad U_2 U_1 A U_1^* U_2^* \qquad U_3 U_2 U_1 A U_1^* U_2^* U_3^*$$

The third and final unitary similarity uses a matrix of the form $U_3 = I_3 \oplus [e^{i\theta}]$; it rotates the $(4, 3)$ entry to the nonnegative real axis if necessary.

6.7 Problems

P.6.1 Let $U, V \in M_n$ be unitary. Is $U + V$ unitary?

P.6.2 Suppose that $U \in M_n(\mathbb{F})$ is unitary. If there is a nonzero $\mathbf{x} \in \mathbb{F}^n$ and a $\lambda \in \mathbb{F}$ such that $U\mathbf{x} = \lambda \mathbf{x}$, show that $|\lambda| = 1$.

P.6.3 Let V be an inner product space, let T be an isometry on V, and let $\mathbf{u}_1, \mathbf{u}_2, \ldots, \mathbf{u}_n$ be orthonormal vectors in V. Show that the vectors $T\mathbf{u}_1, T\mathbf{u}_2, \ldots, T\mathbf{u}_n$ are orthonormal.

P.6.4 Let $\mathbf{u}_1, \mathbf{u}_2, \ldots, \mathbf{u}_n$ be an orthonormal basis of an inner product space V and let $T \in \mathcal{L}(V)$. If $\|T\mathbf{u}_i\| = 1$ for each $i = 1, 2, \ldots, n$, must T be an isometry?

P.6.5 Show that $U \in M_2$ is unitary if and only if there are $a, b \in \mathbb{C}$ and $\phi \in \mathbb{R}$ such that $|a|^2 + |b|^2 = 1$ and
$$U = \begin{bmatrix} a & b \\ -e^{i\phi}\overline{b} & e^{i\phi}\overline{a} \end{bmatrix}.$$

P.6.6 Show that $Q \in M_2(\mathbb{R})$ is real orthogonal if and only if there is a $\theta \in \mathbb{R}$ such that either
$$Q = \begin{bmatrix} \cos\theta & \sin\theta \\ -\sin\theta & \cos\theta \end{bmatrix} \quad \text{or} \quad Q = \begin{bmatrix} \cos\theta & \sin\theta \\ \sin\theta & -\cos\theta \end{bmatrix}.$$

P.6.7 Show that $U \in M_2$ is unitary if and only if there are $\alpha, \beta, \theta, \phi \in \mathbb{R}$ such that
$$U = \begin{bmatrix} e^{i\alpha} & 0 \\ 0 & e^{i(\phi-\alpha)} \end{bmatrix} \begin{bmatrix} \cos\theta & \sin\theta \\ -\sin\theta & \cos\theta \end{bmatrix} \begin{bmatrix} e^{i\beta} & 0 \\ 0 & e^{-i\beta} \end{bmatrix}. \tag{6.7.1}$$

P.6.8 Complex numbers z with modulus 1 (that is, $|z| = 1$) are characterized by the identity $\overline{z}z = 1$. With this in mind, explain why it is reasonable to think of the unitary matrices as matrix analogs of complex numbers with unit modulus.

P.6.9 Let V be a finite-dimensional $\mathbb{F}$-inner product space and let $f : V \to V$ be a function. If $\langle f(\mathbf{u}), f(\mathbf{v}) \rangle = \langle \mathbf{u}, \mathbf{v} \rangle$ for all $\mathbf{u}, \mathbf{v} \in V$, show that f is (a) one to one, (b) linear, (c) onto, and (d) an isometry. *Hint:* Show that $\langle f(\mathbf{u}+c\mathbf{v}), f(\mathbf{w}) \rangle = \langle f(\mathbf{u})+cf(\mathbf{v}), f(\mathbf{w}) \rangle$ for all $\mathbf{u}, \mathbf{v}, \mathbf{w} \in V$ and all $c \in \mathbb{F}$. Compute $\|f(\mathbf{u}+c\mathbf{v}) - f(\mathbf{u}) - cf(\mathbf{v})\|^2$.

P.6.10 How are the plane rotation matrices $U(\theta)$ and $U(\phi)$ (see (6.2.6)) related to $U(\theta+\phi)$? Use these three matrices to prove the addition formula $\cos(\theta+\phi) = \cos\theta\cos\phi - \sin\theta\sin\phi$ and its analog for the sine function.

P.6.11 Let $\mathbf{u}_1, \mathbf{u}_2, \ldots, \mathbf{u}_n$ be an orthonormal basis for $\mathbb{C}^n$ with respect to the standard inner product, and let $c_1, c_2, \ldots, c_n$ be complex scalars with modulus 1. Show that $P_{\mathbf{u}_1} + P_{\mathbf{u}_2} + \cdots + P_{\mathbf{u}_n} = I$ and that $c_1 P_{\mathbf{u}_1} + c_2 P_{\mathbf{u}_2} + \cdots + c_n P_{\mathbf{u}_n}$ is unitary; see (6.4.1).

P.6.12 Let $\mathbf{a}_1, \mathbf{a}_2, \ldots, \mathbf{a}_n$ and $\mathbf{b}_1, \mathbf{b}_2, \ldots, \mathbf{b}_n$ be orthonormal bases of $\mathbb{F}^n$. Describe how to construct a unitary matrix U such that $U\mathbf{b}_k = \mathbf{a}_k$ for each $k = 1, 2, \ldots, n$.

P.6.13 What are all possible entries of a real diagonal $n \times n$ unitary matrix? How many such matrices are there?

P.6.14 If $U \in M_n$ is unitary, show that the matrices U^*, U^T, and $\overline{U}$ are unitary. What is the inverse of $\overline{U}$?

P.6.15 Review the discussion of the identity (3.1.21). Suppose that each of the matrices $A, B \in M_{m \times n}$ has orthonormal columns. Show that $\operatorname{col} A = \operatorname{col} B$ if and only if there is a unitary $U \in M_n$ such that $A = BU$.

P.6.16 If $U \in M_n$ is unitary, compute the Frobenius norm $\|U\|_F$ of U; see Example 4.5.5.

P.6.17 Show that $\beta = I_2, \sigma_x, \sigma_y, \sigma_z$ (the identity matrix and the three Pauli spin matrices) is an orthogonal basis for the real vector space of 2×2 complex Hermitian matrices, endowed with the Frobenius inner product.

P.6.18 Show that unitary similarity is an equivalence relation on M_n and real orthogonal similarity is an equivalence relation on $M_n(\mathbb{R})$. Is real orthogonal similarity an equivalence relation on $M_n(\mathbb{C})$?

P.6.19 Suppose that $U = [u_{ij}] \in M_2$ is unitary and $u_{21} = 0$. What can you say about the remaining three entries?

P.6.20 If $U \in M_n$ is upper triangular and unitary, what can you say about its entries?

P.6.21 What happens if the vectors $\mathbf{x}$ and $\mathbf{y}$ in Theorems 6.4.7 and 6.4.9 are linearly dependent? (a) Let $\mathbf{y} = \pm\mathbf{x} \in \mathbb{R}^n$ be nonzero. Show that the real orthogonal matrix produced by the algorithm in Theorem 6.4.7 is $\sigma U_{\mathbf{w}} = \mp U_{\mathbf{x}}$. (b) Let $\theta \in \mathbb{R}$ and let $\mathbf{y} = e^{i\theta}\mathbf{x} \in \mathbb{C}^n$ be nonzero. Show that the unitary matrix produced by the algorithm in Theorem 6.4.9 is $\sigma U_{\mathbf{w}} = -e^{i\theta}U_{\mathbf{x}}$. (c) In each case, compute $\sigma U_{\mathbf{w}}\mathbf{x}$ to verify that it achieves the asserted values.

P.6.22 Let $\mathbf{x}, \mathbf{y} \in \mathbb{F}^n$ and suppose that $\|\mathbf{x}\|_2 = \|\mathbf{y}\|_2$ but $\mathbf{x} \neq \pm\mathbf{y}$. (a) If $n = 2$, show that $U_{\mathbf{y}-\mathbf{x}} = -U_{\mathbf{y}+\mathbf{x}}$. (b) If $n \geq 3$, show that $-U_{\mathbf{y}+\mathbf{x}} \neq U_{\mathbf{y}-\mathbf{x}}$.

P.6.23 Use the algorithm in Theorem 6.4.9 to construct a unitary matrix that maps $\mathbf{x} = [1 \ i]^\mathsf{T}$ to $\mathbf{y} = [1 \ 1]^\mathsf{T}$.

P.6.24 Let $\mathbf{w} \in \mathbb{F}^n$ be nonzero, let $U_{\mathbf{w}}$ be the Householder matrix (6.4.4), and let $X = [\mathbf{x}_1 \ \mathbf{x}_2 \ \ldots \ \mathbf{x}_n] \in M_n(\mathbb{F})$. Show that $U_{\mathbf{w}}X = X - [\langle\mathbf{x}_1, \mathbf{u}\rangle\mathbf{u} \ \langle\mathbf{x}_2, \mathbf{u}\rangle\mathbf{u} \ \ldots \ \langle\mathbf{x}_n, \mathbf{u}\rangle\mathbf{u}]$.

P.6.25 Let $A \in M_n$ and let $A = QR$ be a QR factorization. Let $A = [\mathbf{a}_1 \ \mathbf{a}_2 \ \ldots \ \mathbf{a}_n]$, $Q = [\mathbf{q}_1 \ \mathbf{q}_2 \ \ldots \ \mathbf{q}_n]$, and $R = [\mathbf{r}_1 \ \mathbf{r}_2 \ \ldots \ \mathbf{r}_n] = [r_{ij}]$. (a) Explain why $|\det A| = \det R = r_{11}r_{22}\cdots r_{nn}$. (b) Show that $\|\mathbf{a}_i\|_2 = \|\mathbf{r}_i\|_2 \geq r_{ii}$ for each $i = 1, 2, \ldots, n$, with equality for some i if and only if $\mathbf{a}_i = r_{ii}\mathbf{q}_i$. (c) Conclude that

$$|\det A| \leq \|\mathbf{a}_1\|_2\|\mathbf{a}_2\|_2 \cdots \|\mathbf{a}_n\|_2 \qquad (6.7.2)$$

with equality if and only if either A has a zero column or A has orthogonal columns (that is, $A^*A = \mathrm{diag}(\|\mathbf{a}_1\|_2^2, \|\mathbf{a}_2\|_2^2, \ldots, \|\mathbf{a}_n\|_2^2)$). This is *Hadamard's Inequality*.

P.6.26 Suppose that $A \in M_{m \times n}(\mathbb{F})$ and $\mathrm{rank}\,A = n$. If $A = QR$ is a QR factorization, then the columns of Q are an orthonormal basis for $\mathrm{col}\,A$. Explain why this orthonormal basis is identical to the orthonormal basis produced by applying the Gram–Schmidt process to the list of columns of A.

P.6.27 If $A \in M_n(\mathbb{F})$ has orthonormal rows and the upper triangular factor in the wide QR factorization of $B \in M_n(\mathbb{F})$ is I, show that $|\mathrm{tr}\,ABAB| \leq n$.

P.6.28 Let $A \in M_n$ be Hermitian. Deduce from Theorem 6.6.3 that A is unitarily similar to a real symmetric tridiagonal matrix with nonnegative superdiagonal and subdiagonal entries.

P.6.29 Let $A \in M_n$ and let $A_0 = A = Q_0R_0$ be a QR factorization. Define $A_1 = R_0Q_0$ and let $A_1 = Q_1R_1$ be a QR factorization. Define $A_2 = R_1Q_1$. Continue this construction, so at the kth step $A_k = Q_kR_k$ is a QR factorization and $A_{k+1} = R_kQ_k$. Show that each A_k is unitarily similar to A, for $k = 1, 2, \ldots$. This construction is at the heart of the *QR algorithm* for computing eigenvalues; see Chapter 8. In practice, the algorithm in Theorem 6.6.3 is often employed to reduce A to upper Hessenberg form before the QR algorithm is started.

P.6.30 Let $\mathbf{x}, \mathbf{y} \in \mathbb{C}^n$. If there is a nonzero $\mathbf{w} \in \mathbb{C}^n$ such that $U_{\mathbf{w}}\mathbf{x} = \mathbf{y}$, show that $\|\mathbf{x}\|_2 = \|\mathbf{y}\|_2$ and $\mathbf{y}^*\mathbf{x}$ is real. What about the converse?

P.6.31 Let $\mathbf{u} \in \mathbb{C}^n$ be a unit vector and let c be a scalar with modulus 1. Show that $P_{\mathbf{u}} = P_{c\mathbf{u}}$ and $P_{\bar{\mathbf{u}}} = P_{\mathbf{u}}^\mathsf{T}$.

P.6.32 Let $\mathbf{a} \in \mathbb{F}^m$ be a unit vector and let $A = [\mathbf{a}] \in \mathsf{M}_{m \times 1}(\mathbb{F})$. Let

$$A = V \begin{bmatrix} R \\ 0 \end{bmatrix} \tag{6.7.3}$$

be a wide QR factorization of A. Explain why $R = [1]$ and $V \in \mathsf{M}_m(\mathbb{F})$ is a unitary matrix whose first column is $\mathbf{a}$.

P.6.33 Let $\mathbf{a}_1, \mathbf{a}_2, \ldots, \mathbf{a}_n \in \mathbb{F}^m$ be an orthonormal list of vectors and let $A = [\mathbf{a}_1\ \mathbf{a}_2\ \ldots\ \mathbf{a}_n] \in \mathsf{M}_{m \times n}(\mathbb{F})$. Let (6.7.3) be a wide QR factorization of A. Show that $R = I_n$ and $V \in \mathsf{M}_m(\mathbb{F})$ is a unitary matrix whose first n columns are the columns of A. Describe how one might use this result to create an algorithm that extends a given orthonormal list to an orthonormal basis.

P.6.34 If $A \in \mathsf{M}_{m \times n}(\mathbb{R})$ with $m \geq n$, why does the algorithm used to prove Theorem 6.5.2 produce *real* matrices V, Q, and R as factors in the wide and narrow QR factorizations (6.5.3) and (6.5.4)?

P.6.35 Let $A \in \mathsf{M}_{m \times n}(\mathbb{R})$ with $m \geq n$ and let $A = QR$, as in (6.5.4). Let $R = [r_{ij}] \in \mathsf{M}_n$, $Q = [\mathbf{q}_1\ \mathbf{q}_2\ \ldots\ \mathbf{q}_n]$, and $A = [\mathbf{a}_1\ \mathbf{a}_2\ \ldots\ \mathbf{a}_n]$. Show that $r_{ij} = \langle \mathbf{a}_i, \mathbf{q}_j \rangle$ for all $i,j = 1, 2, \ldots, n$. Why is $r_{ij} = 0$ if $i > j$?

P.6.36 Prove that a square matrix is similar to an upper triangular matrix if and only if it is unitarily similar to some upper triangular matrix. *Hint*: If $A = SBS^{-1}$, consider the QR factorization of S.

P.6.37 Consider the 4×4 Fourier matrix defined in (6.2.15). (a) Verify that

$$F_4^2 = \begin{bmatrix} 1 & 0 & 0 & 0 \\ 0 & 0 & 0 & 1 \\ 0 & 0 & 1 & 0 \\ 0 & 1 & 0 & 0 \end{bmatrix} = \begin{bmatrix} 1 & 0 \\ 0 & K_3 \end{bmatrix};$$

see (6.2.9). (b) Deduce that $F_4^4 = I$.

P.6.38 Consider the Fourier matrix F_n defined in (6.2.14). (a) Show that

$$F_n^2 = \frac{1}{n} \left[\sum_{k=1}^{n-1} \left(\omega^{i+j-2} \right)^k \right]_{i,j=1}^{n-1} = \begin{bmatrix} 1 & 0 \\ 0 & K_{n-1} \end{bmatrix};$$

see (6.2.9). (b) Deduce that $F_n^4 = I$.

P.6.39 Verify the following identity that connects the Fourier matrices F_4 and F_2:

$$F_4 = \frac{1}{\sqrt{2}} \begin{bmatrix} I & D_2 \\ I & -D_2 \end{bmatrix} \begin{bmatrix} F_2 & 0 \\ 0 & F_2 \end{bmatrix} P_4, \tag{6.7.4}$$

in which $D_2 = \mathrm{diag}(1, \omega)$, $\omega = e^{\pi i/2} = i$ and $P_4 = [\mathbf{e}_1\ \mathbf{e}_3\ \mathbf{e}_2\ \mathbf{e}_4]$ is the permutation matrix such that $P_4 [x_1\ x_2\ x_3\ x_4]^\mathsf{T} = [x_1\ x_3\ x_2\ x_4]^\mathsf{T}$.

P.6.40 Let $A \in \mathsf{M}_n$. If B is permutation similar to A, show that each entry of A is an entry of B and it occurs in the same number of distinct positions in both matrices.

P.6.41 Let $A \in \mathsf{M}_n$. Each entry of A is an entry of A^T and it occurs in the same number of distinct positions in both matrices. Are A and A^T always permutation similar? Why?

P.6.42 Let $A \in \mathsf{M}_n$ and suppose that $A \neq 0$.

(a) If $\operatorname{rank} A = r$, show that A has an $r \times r$ submatrix that is invertible. *Hint*: Why is there a permutation matrix P such that $PA = \begin{bmatrix} A_1 \\ A_2 \end{bmatrix}$ and $A_1 \in \mathsf{M}_{r \times n}$ has linearly independent rows? Why is there a permutation matrix Q such that $A_1 Q = [A_{11} \ A_{22}]$ and $A_{11} \in \mathsf{M}_r$ has linearly independent columns?

(b) If $k \geq 1$ and A has a $k \times k$ invertible submatrix, show that $\operatorname{rank} A \geq k$. *Hint*: Why are at least k rows of A linearly independent?

(c) Show that $\operatorname{rank} A$ is equal to the size of the largest square submatrix of A that has nonzero determinant, that is,

$$\operatorname{rank} A = \max\{k : B \text{ is a } k \times k \text{ submatrix of } A \text{ and } \det B \neq 0\}. \tag{6.7.5}$$

(d) If $E \in \mathsf{M}_n$ is the all-ones matrix, use (6.7.5) to show that $\operatorname{rank} E = 1$.

(e) If $n \geq 3$ and $\operatorname{rank} A \leq n - 2$, use (6.7.5) to show that $\operatorname{adj} A = 0$; see (0.5.2).

(f) If $n \geq 2$ and $\operatorname{rank} A = n - 1$, use (6.7.5) to show that $\operatorname{rank} \operatorname{adj} A \geq 1$. Use (0.5.2) to show that the nullity of $\operatorname{adj} A$ is at least $n - 1$ and conclude that $\operatorname{rank} \operatorname{adj} A = 1$.

P.6.43 If $U \in \mathsf{M}_m$ and $V \in \mathsf{M}_n$ are unitary, show that $U \otimes V \in \mathsf{M}_{mn}$ is unitary.

6.8 Notes

Use of the Gram–Schmidt process to orthogonalize a linearly independent list of vectors with floating point arithmetic on a computer can result in a list of vectors that is far from orthogonal. Despite the result in P.6.26, numerical orthogonalization is typically much more satisfactory via the *QR* factorization obtained with Householder transformations; see [GVL13].

The identity (6.7.4) is a special case of the identity

$$F_{2n} = \frac{1}{\sqrt{2}} \begin{bmatrix} I & D_n \\ I & -D_n \end{bmatrix} \begin{bmatrix} F_n & 0 \\ 0 & F_n \end{bmatrix} P_{2n}, \tag{6.8.1}$$

in which $D_n = \operatorname{diag}(1, \omega, \omega^2, \ldots, \omega^{n-1})$, $\omega = e^{i\pi/n}$, and P_{2n} is the permutation matrix such that

$$P_{2n}[x_1 \ x_2 \ \ldots \ x_{2n-1} \ x_{2n}]^\mathsf{T} = [x_1 \ x_3 \ x_5 \ \ldots \ x_{2n-1} \ x_2 \ x_4 \ x_6 \ \ldots \ x_{2n}]^\mathsf{T},$$

which places the odd-indexed entries first, followed by the even-indexed entries. The identity (6.8.1) suggests a recursive scheme to calculate the matrix-vector product $F_{2^m}\mathbf{x}$, in which $\mathbf{x}$ is a vector whose size is a power of 2. This scheme is at the heart of the celebrated fast Fourier transform (FFT) algorithm.

The factorization (6.7.1) of a 2×2 unitary matrix is a special case of the *CS* decomposition of an $n \times n$ unitary matrix that is presented as a 2×2 block matrix; see [HJ13, Sect. 2.7].

6.9 Some Important Concepts

- The adjoint of a unitary matrix is its inverse.

- Characterizations of unitary matrices (Theorem 6.2.11).

- Unitary similarity and permutation similarity.

- Fourier matrices, Householder matrices, and permutation matrices are unitary.

- Householder matrices can be used to construct a unitary matrix that maps a given vector to any other vector that has the same Euclidean norm.

- QR factorizations and orthogonalization of a linearly independent list of vectors.

- Each square matrix is unitarily similar to an upper Hessenberg matrix.

7 Orthogonal Complements and Orthogonal Projections

Many problems in applied mathematics involve finding a minimum norm solution or a best approximation, subject to certain constraints. Orthogonal subspaces are frequently encountered in solving such problems, so we study them carefully in this chapter. Among the topics we discuss are the minimum norm solution to a consistent linear system, a least squares solution to an inconsistent linear system, and orthogonal projections.

7.1 Orthogonal Complements

Orthogonal projections are simple operators that are used as the building blocks for a variety of other operators (see Section 12.9). They are also a fundamental tool in real-world optimization problems and in many applications. To define orthogonal projections, we require the following notion.

Definition 7.1.1 Let $\mathcal{U}$ be a subset of an inner product space $\mathcal{V}$. If $\mathcal{U}$ is nonempty, then

$$\mathcal{U}^\perp = \{\mathbf{v} \in \mathcal{V} : \langle \mathbf{u}, \mathbf{v} \rangle = 0 \text{ for all } \mathbf{u} \in \mathcal{U}\}.$$

If $\mathcal{U} = \varnothing$, then $\mathcal{U}^\perp = \mathcal{V}$. The set $\mathcal{U}^\perp$ (read as "$\mathcal{U}$-perp") is the *orthogonal complement* of $\mathcal{U}$ in $\mathcal{V}$.

Figure 7.1 and the following examples illustrate several basic properties of orthogonal complements.

Example 7.1.2 Let $\mathcal{V} = \mathbb{R}^3$ and $\mathcal{U} = \{[1 \; 2 \; 3]^\mathsf{T}\}$. Then $\mathbf{x} = [x_1 \; x_2 \; x_3]^\mathsf{T} \in \mathcal{U}^\perp$ if and only if $x_1 + 2x_2 + 3x_3 = 0$. Thus,

$$\mathcal{U}^\perp = \{[x_1 \; x_2 \; x_3]^\mathsf{T} : x_1 + 2x_2 + 3x_3 = 0\} \tag{7.1.3}$$

is the plane in $\mathbb{R}^3$ that passes through $\mathbf{0} = [0 \; 0 \; 0]^\mathsf{T}$ and has $[1 \; 2 \; 3]^\mathsf{T}$ as a normal vector. Observe that $\mathcal{U}^\perp$ is a subspace of $\mathbb{R}^n$ even though $\mathcal{U}$ is merely a subset of $\mathbb{R}^n$; see Theorem 7.1.5.

Example 7.1.4 Consider $\mathcal{V} = \mathsf{M}_n$, endowed with the Frobenius inner product. If $\mathcal{U}$ is the subspace of all upper triangular matrices, then $\mathcal{U}^\perp$ is the subspace of all strictly lower triangular matrices; see P.7.11.

The following theorem lists some important properties of orthogonal complements.

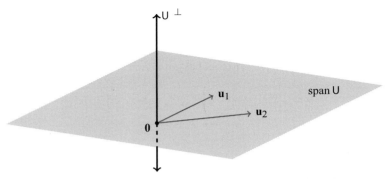

Figure 7.1 The orthogonal complement of the set $\mathscr{U} = \{\mathbf{u}_1, \mathbf{u}_2\}$ is the subspace $\mathscr{U}^\perp$.

Theorem 7.1.5 *Let $\mathscr{U}, \mathscr{W}$ be nonempty subsets of an inner product space $\mathcal{V}$.*

(a) $\mathscr{U}^\perp$ *is a subspace of $\mathcal{V}$. In particular, $\mathbf{0} \in \mathscr{U}^\perp$.*

(b) *If $\mathscr{U} \subseteq \mathscr{W}$, then $\mathscr{W}^\perp \subseteq \mathscr{U}^\perp$.*

(c) $\mathscr{U}^\perp = (\mathrm{span}\,\mathscr{U})^\perp$.

(d) *If $\mathscr{U} \cap \mathscr{U}^\perp \neq \varnothing$, then $\mathscr{U} \cap \mathscr{U}^\perp = \{\mathbf{0}\}$.*

(e) *If $\mathbf{0} \in \mathscr{U}$, then $\mathscr{U} \cap \mathscr{U}^\perp = \{\mathbf{0}\}$.*

(f) $\{\mathbf{0}\}^\perp = \mathcal{V}$.

(g) $\mathcal{V}^\perp = \{\mathbf{0}\}$.

(h) $\mathscr{U} \subseteq \mathrm{span}\,\mathscr{U} \subseteq (\mathscr{U}^\perp)^\perp$.

Proof (a) If $\mathbf{u}, \mathbf{v} \in \mathscr{U}^\perp$ and a is a scalar, then $\langle \mathbf{w}, \mathbf{u} + a\mathbf{v} \rangle = \langle \mathbf{w}, \mathbf{u} \rangle + \bar{a}\langle \mathbf{w}, \mathbf{v} \rangle = 0$ for all $\mathbf{w} \in \mathscr{U}$. Thus $\mathbf{u} + a\mathbf{v} \in \mathscr{U}^\perp$, which proves that $\mathscr{U}^\perp$ is a subspace.

(b) If $\mathbf{v} \in \mathcal{W}^\perp$, then $\langle \mathbf{u}, \mathbf{v} \rangle = 0$ for all $\mathbf{u} \in \mathcal{U}$ since $\mathcal{U} \subseteq \mathcal{W}$. Thus, $\mathbf{v} \in \mathcal{U}^\perp$ and we conclude that $\mathcal{W}^\perp \subseteq \mathcal{U}^\perp$.

(c) Since $\mathscr{U} \subseteq \mathrm{span}\,\mathscr{U}$, (b) ensures that $(\mathrm{span}\,\mathscr{U})^\perp \subseteq \mathscr{U}^\perp$. If $\mathbf{v} \in \mathscr{U}^\perp$, then $\langle \sum_{i=1}^r c_i \mathbf{u}_i, \mathbf{v} \rangle = \sum_{i=1}^r c_i \langle \mathbf{u}_i, \mathbf{v} \rangle = 0$ for any scalars $c_1, c_2, \ldots, c_r$ and any $\mathbf{u}_1, \mathbf{u}_2, \ldots, \mathbf{u}_r \in \mathcal{U}$. Thus, $\mathscr{U}^\perp \subseteq (\mathrm{span}\,\mathscr{U})^\perp$. We conclude that $\mathscr{U}^\perp = (\mathrm{span}\,\mathscr{U})^\perp$.

(d) If $\mathbf{u} \in \mathscr{U} \cap \mathscr{U}^\perp$, then $\mathbf{u}$ is orthogonal to itself. Therefore, $\|\mathbf{u}\|^2 = \langle \mathbf{u}, \mathbf{u} \rangle = 0$ and $\mathbf{u} = \mathbf{0}$.

(e) Since $\mathbf{0} \in \mathscr{U}^\perp$ by (a), $\mathscr{U} \cap \mathscr{U}^\perp \neq \varnothing$ and hence $\mathscr{U} \cap \mathscr{U}^\perp = \{\mathbf{0}\}$ by (d).

(f) Since $\langle \mathbf{v}, \mathbf{0} \rangle = 0$ for all $\mathbf{v} \in \mathcal{V}$, we see that $\mathcal{V} \subseteq \{\mathbf{0}\}^\perp \subseteq \mathcal{V}$, so $\mathcal{V} = \{\mathbf{0}\}^\perp$.

(g) Since $\mathbf{0} \in \mathcal{V}^\perp$ by (a), $\mathcal{V}^\perp = \mathcal{V} \cap \mathcal{V}^\perp = \{\mathbf{0}\}$ by (d).

(h) Since every vector in $\mathrm{span}\,\mathscr{U}$ is orthogonal to every vector in $(\mathrm{span}\,\mathscr{U})^\perp$, (c) ensures that $\mathscr{U} \subseteq \mathrm{span}\,\mathscr{U} \subseteq ((\mathrm{span}\,\mathscr{U})^\perp)^\perp = (\mathscr{U}^\perp)^\perp$. □

The following theorem asserts that an inner product space can always be decomposed as the direct sum of any finite-dimensional subspace and its orthogonal complement.

Theorem 7.1.6 *If $\mathcal{U}$ is a finite-dimensional subspace of an inner product space $\mathcal{V}$, then $\mathcal{V} = \mathcal{U} \oplus \mathcal{U}^\perp$. Consequently, for each $\mathbf{v} \in \mathcal{V}$ there is a unique $\mathbf{u} \in \mathcal{U}$ such that $\mathbf{v} - \mathbf{u} \in \mathcal{U}^\perp$.*

Proof Let $\mathbf{u}_1, \mathbf{u}_2, \ldots, \mathbf{u}_r$ be an orthonormal basis for $\mathcal{U}$. For any $\mathbf{v}$ in $\mathcal{V}$, write

$$\mathbf{v} = \underbrace{\left(\sum_{i=1}^{r} \langle \mathbf{v}, \mathbf{u}_i \rangle \mathbf{u}_i \right)}_{\mathbf{u}} + \underbrace{\left(\mathbf{v} - \sum_{i=1}^{r} \langle \mathbf{v}, \mathbf{u}_i \rangle \mathbf{u}_i \right)}_{\mathbf{v}-\mathbf{u}}, \tag{7.1.7}$$

in which $\mathbf{u} \in \mathcal{U}$. We claim that $\mathbf{v} - \mathbf{u} \in \mathcal{U}^\perp$. Indeed, since

$$\langle \mathbf{v} - \mathbf{u}, \mathbf{u}_j \rangle = \langle \mathbf{v}, \mathbf{u}_j \rangle - \langle \mathbf{u}, \mathbf{u}_j \rangle$$
$$= \langle \mathbf{v}, \mathbf{u}_j \rangle - \left\langle \sum_{i=1}^{r} \langle \mathbf{v}, \mathbf{u}_i \rangle \mathbf{u}_i, \mathbf{u}_j \right\rangle$$
$$= \langle \mathbf{v}, \mathbf{u}_j \rangle - \sum_{i=1}^{r} \langle \mathbf{v}, \mathbf{u}_i \rangle \langle \mathbf{u}_i, \mathbf{u}_j \rangle$$
$$= \langle \mathbf{v}, \mathbf{u}_j \rangle - \langle \mathbf{v}, \mathbf{u}_j \rangle$$
$$= 0,$$

Theorem 7.1.5.c ensures that

$$\mathbf{v} - \mathbf{u} \in \{\mathbf{u}_1, \mathbf{u}_2, \ldots, \mathbf{u}_r\}^\perp = \left(\mathrm{span}\{\mathbf{u}_1, \mathbf{u}_2, \ldots, \mathbf{u}_r\} \right)^\perp = \mathcal{U}^\perp.$$

Thus, $\mathcal{V} = \mathcal{U} + \mathcal{U}^\perp$. Since $\mathcal{U}$ is a subspace, $\mathbf{0} \in \mathcal{U}$ and Theorem 7.1.5.e ensures that $\mathcal{U} \cap \mathcal{U}^\perp = \{\mathbf{0}\}$. We conclude that $\mathcal{V} = \mathcal{U} \oplus \mathcal{U}^\perp$. The uniqueness of $\mathbf{u}$ follows from Theorem 1.5.9. $\square$

Theorem 7.1.8 *If $\mathcal{U}$ is a finite-dimensional subspace of an inner product space, then $(\mathcal{U}^\perp)^\perp = \mathcal{U}$.*

Proof Theorem 7.1.5.g asserts that $\mathcal{U} \subseteq (\mathcal{U}^\perp)^\perp$. Let $\mathbf{v} \in (\mathcal{U}^\perp)^\perp$ and use Theorem 7.1.6 to write $\mathbf{v} = \mathbf{u} + (\mathbf{v} - \mathbf{u})$, in which $\mathbf{u} \in \mathcal{U}$ and $(\mathbf{v} - \mathbf{u}) \in \mathcal{U}^\perp$. Since

$$0 = \langle \mathbf{v}, \mathbf{v} - \mathbf{u} \rangle = \langle \mathbf{u} + (\mathbf{v} - \mathbf{u}), \mathbf{v} - \mathbf{u} \rangle = \langle \mathbf{u}, \mathbf{v} - \mathbf{u} \rangle + \|\mathbf{v} - \mathbf{u}\|^2 = \|\mathbf{v} - \mathbf{u}\|^2,$$

it follows that $\mathbf{v} = \mathbf{u}$. Thus, $\mathbf{v} \in \mathcal{U}$ and $(\mathcal{U}^\perp)^\perp \subseteq \mathcal{U}$. $\square$

Corollary 7.1.9 *If $\mathscr{U}$ is a nonempty subset of an inner product space and if span $\mathscr{U}$ is finite dimensional, then $(\mathscr{U}^\perp)^\perp = \mathrm{span}\, \mathscr{U}$.*

Proof Since span $\mathscr{U}$ is a finite-dimensional subspace, Theorems 7.1.8 and 7.1.5.c tell us that span $\mathscr{U} = ((\mathrm{span}\, \mathscr{U})^\perp)^\perp = (\mathscr{U}^\perp)^\perp$. $\square$

Example 7.1.10 Let $\mathcal{U}$ be the set defined in Example 7.1.2. Why is $(\mathscr{U}^\perp)^\perp = \mathrm{span}\, \mathscr{U}$? Since $\mathscr{U}^\perp$ is the null space of the rank-1 matrix $[1\ 2\ 3]$, the dimension theorem ensures that $\dim \mathscr{U}^\perp = 2$. Therefore, any two linearly independent vectors in $\mathscr{U}^\perp$, such as $\mathbf{v}_1 = [3\ 0\ -1]^\mathsf{T}$ and $\mathbf{v}_2 = [2\ -1\ 0]^\mathsf{T}$, form a basis for $\mathscr{U}^\perp$. Thus $(\mathscr{U}^\perp)^\perp = (\mathrm{span}\{\mathbf{v}_1, \mathbf{v}_2\})^\perp = \{\mathbf{v}_1, \mathbf{v}_2\}^\perp$.

However, $\{\mathbf{v}_1, \mathbf{v}_2\}^\perp$ is the null space of the 2×3 matrix $[\mathbf{v}_1 \ \mathbf{v}_2]^\mathsf{T}$, which row reduction confirms is span $\mathscr{U}$.

If $\mathcal{V}$ is infinite dimensional and $\mathscr{U} \subseteq \mathcal{V}$, then span $\mathscr{U}$ can be a proper subset of $(\mathscr{U}^\perp)^\perp$; see P.7.33.

7.2 The Minimum Norm Solution of a Consistent Linear System

Definition 7.2.1 Let $A \in \mathsf{M}_{m \times n}(\mathbb{F})$ and suppose that $\mathbf{y} \in \operatorname{col} A$. A solution $\mathbf{s}$ of the consistent linear system $A\mathbf{x} = \mathbf{y}$ is a *minimum norm solution* if $\|\mathbf{s}\|_2 \leq \|\mathbf{u}\|_2$ whenever $A\mathbf{u} = \mathbf{y}$.

There are many reasons why we might be interested in finding minimum norm solutions to a consistent system of equations that has infinitely many solutions. For example, if the entries of its solution vectors represent economic quantities, perhaps total expenditures are minimized with a minimum norm solution. Fortunately, every consistent linear system has a unique minimum norm solution. To prove this, we begin with the following lemma.

Lemma 7.2.2 *If $\mathcal{V}, \mathcal{W}$ are finite-dimensional inner product spaces and $T \in \mathcal{L}(\mathcal{V}, \mathcal{W})$, then*

$$\ker T = (\operatorname{ran} T^*)^\perp \quad \text{and} \quad \operatorname{ran} T = (\ker T^*)^\perp. \tag{7.2.3}$$

Proof The equivalences

$$
\begin{aligned}
\mathbf{v} \in \ker T \quad &\Longleftrightarrow \quad T\mathbf{v} = \mathbf{0} \\
&\Longleftrightarrow \quad \langle T\mathbf{v}, \mathbf{w} \rangle_{\mathcal{W}} = 0 \text{ for all } \mathbf{w} \in \mathcal{W} \\
&\Longleftrightarrow \quad \langle \mathbf{v}, T^*\mathbf{w} \rangle_{\mathcal{V}} = 0 \text{ for all } \mathbf{w} \in \mathcal{W} \\
&\Longleftrightarrow \quad \mathbf{v} \in (\operatorname{ran} T^*)^\perp
\end{aligned}
$$

tell us that $\ker T = (\operatorname{ran} T^*)^\perp$. To see that $\operatorname{ran} T = (\ker T^*)^\perp$, replace T with T^* and use Theorem 7.1.8. $\qquad\square$

If $\mathcal{V} = \mathbb{F}^n$ and $\mathcal{W} = \mathbb{F}^m$, and if $T_A \in \mathcal{L}(\mathcal{V}, \mathcal{W})$ is the linear operator induced by $A \in \mathsf{M}_{m \times n}(\mathbb{F})$ (see Definition 2.3.9), the preceding lemma ensures that

$$\operatorname{null} A = (\operatorname{col} A^*)^\perp. \tag{7.2.4}$$

We can now prove our assertion about minimum norm solutions; see Figure 7.2.

Theorem 7.2.5 *Suppose that $\mathbf{y} \in \mathbb{F}^m$, $A \in \mathsf{M}_{m \times n}(\mathbb{F})$, and*

$$A\mathbf{x} = \mathbf{y} \tag{7.2.6}$$

is consistent.

(a) *Equation (7.2.6) has a unique minimum norm solution $\mathbf{s} \in \mathbb{F}^n$. Moreover, $\mathbf{s} \in \operatorname{col} A^*$.*

(b) $\mathbf{s}$ *is the only solution to (7.2.6) that lies in $\operatorname{col} A^*$.*

(c) $AA^*\mathbf{u} = \mathbf{y}$ *is consistent. If $\mathbf{u}_0 \in \mathbb{F}^m$ and $AA^*\mathbf{u}_0 = \mathbf{y}$, then $\mathbf{s} = A^*\mathbf{u}_0$.*

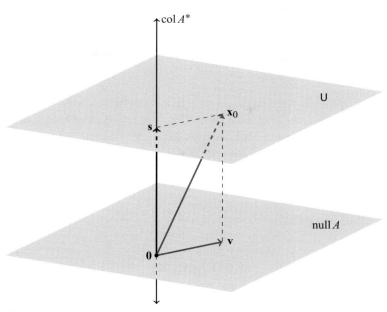

Figure 7.2 Graphical depiction of Theorem 7.2.5. The solution set to
$A\mathbf{x} = \mathbf{y}$ is $\mathcal{U}$, which is a copy of null A that has been translated by $\mathbf{x}_0$
(a solution to $A\mathbf{x} = \mathbf{y}$). The set $\mathcal{U}$ is *not* a subspace of $\mathbb{F}^n$ unless $\mathbf{y} = \mathbf{0}$, in
which case $\mathcal{U} = \text{null } A$. The minimum norm solution $\mathbf{s}$ to $A\mathbf{x} = \mathbf{y}$ is the
unique vector in $\mathcal{U}$ that belongs to $\text{col } A^* = (\text{null } A)^{\perp}$.

Proof (a) From (7.2.4) we obtain the orthogonal direct sum

$$\mathbb{F}^n = \text{col } A^* \oplus \underbrace{\text{null } A}_{=(\text{col } A^*)^{\perp}}.$$

Let $\mathbf{x}_0$ be a solution (any solution will do) to (7.2.6) and write $\mathbf{x}_0 = \mathbf{s} + \mathbf{v}$, in which $\mathbf{s} \in \text{col } A^*$
and $\mathbf{v} \in \text{null } A$. It follows that

$$\mathbf{y} = A\mathbf{x}_0 = A(\mathbf{s} + \mathbf{v}) = A\mathbf{s} + A\mathbf{v} = A\mathbf{s},$$

so $\mathbf{s}$ is a solution to (7.2.6). Every solution $\mathbf{x}$ to (7.2.6) has the form $\mathbf{x} = \mathbf{s} + \mathbf{v}$ for some
$\mathbf{v} \in \text{null } A$. The Pythagorean theorem ensures that

$$\|\mathbf{x}\|_2^2 = \|\mathbf{s} + \mathbf{v}\|_2^2 = \|\mathbf{s}\|_2^2 + \|\mathbf{v}\|_2^2 \geq \|\mathbf{s}\|_2^2,$$

so $\mathbf{s}$ is a minimum norm solution to (7.2.6).

If $\mathbf{s}'$ is any minimum norm solution to (7.2.6), then $\mathbf{s}' = \mathbf{s} + \mathbf{v}$, in which $\mathbf{v} \in \text{null } A$ and
$\|\mathbf{s}\|_2 = \|\mathbf{s}'\|_2$. The Pythagorean theorem ensures that

$$\|\mathbf{s}\|_2^2 = \|\mathbf{s}'\|_2^2 = \|\mathbf{s} + \mathbf{v}\|_2^2 = \|\mathbf{s}\|_2^2 + \|\mathbf{v}\|_2^2.$$

Consequently, $\mathbf{v} = \mathbf{0}$ and $\mathbf{s} = \mathbf{s}'$.

(b) Let $\mathbf{s}'$ be any vector in $\text{col } A^*$. Then $\mathbf{s} - \mathbf{s}' \in \text{col } A^*$ since $\text{col } A^*$ is a subspace of $\mathbb{F}^n$. If $\mathbf{s}'$ is
a solution to (7.2.6), then $A\mathbf{s} = A\mathbf{s}' = \mathbf{y}$ and $A(\mathbf{s} - \mathbf{s}') = \mathbf{0}$. Thus, $\mathbf{s} - \mathbf{s}' \in \text{col } A^* \cap \text{null } A = \{\mathbf{0}\}$.
We conclude that $\mathbf{s} = \mathbf{s}'$ and hence $\mathbf{s}$ is the only solution to (7.2.6) in $\text{col } A^*$.

(c) Because the minimum norm solution s to (7.2.6) is in $\operatorname{col} A^*$, there is a $\mathbf{w}$ such that $\mathbf{s} = A^*\mathbf{w}$. Since $AA^*\mathbf{w} = A\mathbf{s} = \mathbf{y}$, the linear system $AA^*\mathbf{u} = \mathbf{y}$ is consistent. If $\mathbf{u}_0$ is any solution to $AA^*\mathbf{u} = \mathbf{y}$, then $A^*\mathbf{u}_0$ is a solution to (7.2.6) that lies in $\operatorname{col} A^*$. It follows from (b) that $A^*\mathbf{u}_0 = \mathbf{s}$.   □

Theorem 7.2.5 provides a recipe to find the minimum norm solution of the consistent linear system (7.2.6). First find a $\mathbf{u}_0 \in \mathbb{F}^n$ such that $AA^*\mathbf{u}_0 = \mathbf{y}$; existence of a solution is guaranteed. Then $\mathbf{s} = A^*\mathbf{u}_0$ is the minimum norm solution to (7.2.6). If A and $\mathbf{y}$ are real, then $\mathbf{s}$ is real. If either A or $\mathbf{y}$ is not real, then $\mathbf{s}$ need not be real. An alternative approach to finding the minimum norm solution is in Theorem 15.5.9.

Example 7.2.7 Consider the real linear system

$$
\begin{array}{rcrcrcl}
x_1 & + & 2x_2 & + & 3x_3 & = & 3, \\
4x_1 & + & 5x_2 & + & 6x_3 & = & 3, \\
7x_1 & + & 8x_2 & + & 9x_3 & = & 3,
\end{array}
\qquad (7.2.8)
$$

which can be written in the form $A\mathbf{x} = \mathbf{y}$ with

$$
A = \begin{bmatrix} 1 & 2 & 3 \\ 4 & 5 & 6 \\ 7 & 8 & 9 \end{bmatrix}, \qquad \mathbf{x} = \begin{bmatrix} x_1 \\ x_2 \\ x_3 \end{bmatrix}, \quad \text{and} \quad \mathbf{y} = \begin{bmatrix} 3 \\ 3 \\ 3 \end{bmatrix}.
$$

The reduced row echelon form of the augmented matrix $[A \ \mathbf{y}]$ is

$$
\begin{bmatrix} 1 & 0 & -1 & -3 \\ 0 & 1 & 2 & 3 \\ 0 & 0 & 0 & 0 \end{bmatrix},
$$

from which we obtain the general solution

$$
\mathbf{x}(t) = \begin{bmatrix} x_1 \\ x_2 \\ x_3 \end{bmatrix} = \begin{bmatrix} -3 \\ 3 \\ 0 \end{bmatrix} + t \begin{bmatrix} 1 \\ -2 \\ 1 \end{bmatrix}, \qquad t \in \mathbb{R}.
$$

It is tempting to think that the solution with $t = 0$ might be the minimum norm solution. We have $\mathbf{x}(0) = [-3 \ 3 \ 0]^\mathsf{T}$ and $\|\mathbf{x}\|_2 = 3\sqrt{2} \approx 4.24264$. However, we can do better.

To find the minimum norm solution to (7.2.8), we begin by finding a solution to the (necessarily consistent) linear system $(AA^*)\mathbf{u} = \mathbf{y}$. We have

$$
AA^* = \begin{bmatrix} 14 & 32 & 50 \\ 32 & 77 & 122 \\ 50 & 122 & 194 \end{bmatrix}.
$$

The linear system

$$
\begin{array}{rcrcrcl}
14u_1 & + & 32u_2 & + & 50u_3 & = & 3, \\
32u_1 & + & 77u_2 & + & 122u_3 & = & 3, \\
50u_1 & + & 122u_2 & + & 194u_3 & = & 3.
\end{array}
\qquad (7.2.9)
$$

has infinitely many solutions, one of which is $\mathbf{u}_0 = [0 \ 4 \ -\frac{5}{2}]^\mathsf{T}$. The minimum norm solution to (7.2.8) is

$$\mathbf{s} = A^*\mathbf{u}_0 = \begin{bmatrix} 1 & 4 & 7 \\ 2 & 5 & 8 \\ 3 & 6 & 9 \end{bmatrix} \begin{bmatrix} 0 \\ 4 \\ -\frac{5}{2} \end{bmatrix} = \begin{bmatrix} -\frac{3}{2} \\ 0 \\ \frac{3}{2} \end{bmatrix}.$$

Note that $\|\mathbf{s}\|_2 = \frac{3\sqrt{2}}{2} \approx 2.12132$. The vector $\mathbf{u}_1 = [2 \ 0 \ -\frac{1}{2}]^\mathsf{T}$ is also a solution to (7.2.9), but $A^*\mathbf{u}_1 = \mathbf{s}$, as predicted by Theorem 7.2.5.

7.3 Orthogonal Projections

Throughout this section, $\mathcal{V}$ denotes an inner product space with inner product $\langle \cdot, \cdot \rangle$ and $\mathcal{U}$ is a subspace of $\mathcal{V}$. We showed in Theorem 7.1.6 that if $\mathcal{U}$ is a finite-dimensional subspace of $\mathcal{V}$, then $\mathcal{V} = \mathcal{U} \oplus \mathcal{U}^\perp$. Consequently, $\mathbf{v} \in \mathcal{V}$ can be written uniquely in the form

$$\mathbf{v} = \mathbf{u} + (\mathbf{v} - \mathbf{u}), \tag{7.3.1}$$

in which $\mathbf{u} \in \mathcal{U}$ and $(\mathbf{v} - \mathbf{u}) \in \mathcal{U}^\perp$ (Theorem 7.1.6). The vector $\mathbf{u} = P_\mathcal{U}\mathbf{v}$ is the *orthogonal projection* of $\mathbf{v}$ onto $\mathcal{U}$; see Figure 7.3.

Theorem 7.3.2 *Let $\mathcal{U}$ be a finite-dimensional subspace of $\mathcal{V}$ and suppose that $\mathbf{u}_1, \mathbf{u}_2, \ldots, \mathbf{u}_r$ is an orthonormal basis for $\mathcal{U}$. Then*

$$P_\mathcal{U}\mathbf{v} = \sum_{i=1}^r \langle \mathbf{v}, \mathbf{u}_i \rangle \mathbf{u}_i \tag{7.3.3}$$

for every $\mathbf{v} \in \mathcal{V}$.

Proof This follows from (7.1.7) and the uniqueness of the decomposition (7.3.1). □

The uniqueness of the representation (7.3.1) ensures that $P_\mathcal{U}\mathbf{v}$ is independent of the choice of orthonormal basis for $\mathcal{U}$, something that is not immediately obvious from (7.3.3). Furthermore, (7.3.3) tells us that the map $\mathbf{v} \mapsto P_\mathcal{U}\mathbf{v}$ is a linear transformation since each inner product $\langle \mathbf{v}, \mathbf{u}_i \rangle$ is linear in $\mathbf{v}$.

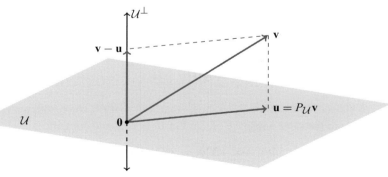

Figure 7.3 An illustration of the decomposition (7.3.1).

Definition 7.3.4 The linear operator $P_{\mathcal{U}} \in \mathcal{L}(V)$ defined by (7.3.3) is the *orthogonal projection* from V onto $\mathcal{U}$.

Example 7.3.5 Suppose that $\mathbf{u}_1, \mathbf{u}_2, \dots, \mathbf{u}_r$ is an orthonormal basis for a subspace $\mathcal{U}$ of $\mathbb{F}^n$ and let $U = [\mathbf{u}_1 \ \mathbf{u}_2 \ \dots \ \mathbf{u}_r] \in \mathsf{M}_{n \times r}$. For any $\mathbf{v} \in \mathbb{F}^n$,

$$P_{\mathcal{U}}\mathbf{v} = \sum_{i=1}^{r} \langle \mathbf{v}, \mathbf{u}_i \rangle \mathbf{u}_i = \sum_{i=1}^{r} \mathbf{u}_i(\mathbf{u}_i^*\mathbf{v}) = \sum_{i=1}^{r} (\mathbf{u}_i\mathbf{u}_i^*)\mathbf{v}$$

$$= \left(\sum_{i=1}^{r} \mathbf{u}_i\mathbf{u}_i^* \right)\mathbf{v} = UU^*\mathbf{v}, \tag{7.3.6}$$

in which we invoke (3.1.19) for the final equality. For example, if $A \in \mathsf{M}_{m \times n}(\mathbb{F})$ has full column rank and $A = QR$ is a QR factorization, then $\operatorname{col} A = \operatorname{col} Q$ (Corollary 6.5.15). Therefore,

$$P_{\operatorname{col} A} = QQ^* \tag{7.3.7}$$

is the orthogonal projection onto the column space of A.

Let $\beta = \mathbf{e}_1, \mathbf{e}_2, \dots, \mathbf{e}_n$ be the standard basis of $\mathbb{F}^n$. Theorem 5.5.3 tells us that $_\beta[P_{\mathcal{U}}]_\beta = [\langle P_{\mathcal{U}}\mathbf{e}_j, \mathbf{e}_i \rangle]$. Use (7.3.6) to compute

$$[\langle P_{\mathcal{U}}\mathbf{e}_j, \mathbf{e}_i \rangle] = [\langle UU^*\mathbf{e}_j, \mathbf{e}_i \rangle] = [\mathbf{e}_i^* UU^*\mathbf{e}_j] = UU^*.$$

Thus, UU^* is the standard basis representation of the orthogonal projection $P_{\mathcal{U}}$.

Example 7.3.8 If $\mathbf{u} \neq \mathbf{0}$, then $\mathbf{u}/\|\mathbf{u}\|_2$ is an orthonormal basis for $\mathcal{U} = \operatorname{span}\{\mathbf{u}\} \subseteq \mathbb{F}^n$. The preceding example includes as a special case the formula (4.5.12) for projecting one vector in $\mathbb{F}^n$ onto the span of the other. Indeed, for any $\mathbf{v} \in \mathbb{F}^n$ we have

$$P_{\mathcal{U}}\mathbf{v} = \frac{\mathbf{u}\mathbf{u}^*}{\|\mathbf{u}\|_2^2}\mathbf{v} = \frac{\mathbf{u}(\mathbf{u}^*\mathbf{v})}{\|\mathbf{u}\|_2^2} = \frac{\langle \mathbf{v}, \mathbf{u} \rangle \mathbf{u}}{\|\mathbf{u}\|_2^2}.$$

If an orthonormal basis for a subspace $\mathcal{U}$ of $\mathbb{F}^n$ is not readily available, one can apply the QR factorization or another orthogonalization algorithm to a basis for $\mathcal{U}$ to obtain the matrix representation for $P_{\mathcal{U}}$. Even though this orthogonalization can be avoided in some applications (see Sections 7.4 and 7.5), in actual numerical work it is often advisable to do it to enhance the stability of computations.

The following theorem returns to an issue that was addressed differently in P.6.15.

Theorem 7.3.9 *Let $X, Y \in \mathsf{M}_{n \times r}$ have orthonormal columns. Then $\operatorname{col} X = \operatorname{col} Y$ if and only if there is a unitary $U \in \mathsf{M}_r$ such that $X = YU$.*

Proof If $\operatorname{col} X = \operatorname{col} Y$, Example 7.3.5 shows that $XX^* = YY^*$, since both matrices represent the projection onto the same subspace of $\mathbb{C}^n$. Let $U = Y^*X$ and compute

$$X = XX^*X = YY^*X = YU.$$

Then

$$I = X^*X = (YU)^*(YU) = U^*Y^*YU = U^*U,$$

so U is unitary. Conversely, if $X = YU$ for some unitary $U \in M_r$, then $\operatorname{col} X \subseteq \operatorname{col} Y$; see (3.1.21). Since U is unitary, we also have $Y = XU^*$, which implies that $\operatorname{col} Y \subseteq \operatorname{col} X$. □

Orthogonal projections permit us to associate an operator (an algebraic object) with each finite-dimensional subspace (a geometric object) of an inner product space. Many statements about subspaces and their relationships can be translated into statements about the corresponding orthogonal projections. Moreover, orthogonal projections form the basic building blocks from which many other operators are built (see Section 12.9). Some of their properties are listed in the following two theorems.

Theorem 7.3.10 *Let $\mathcal{U}$ be a finite-dimensional subspace of $\mathcal{V}$.*

(a) $P_{\{0\}} = 0$ *and* $P_{\mathcal{V}} = I$.

(b) $\operatorname{ran} P_{\mathcal{U}} = \mathcal{U}$.

(c) $\ker P_{\mathcal{U}} = \mathcal{U}^{\perp}$.

(d) $\mathbf{v} - P_{\mathcal{U}}\mathbf{v} \in \mathcal{U}^{\perp}$ *for all* $\mathbf{v} \in \mathcal{V}$.

(e) $\|P_{\mathcal{U}}\mathbf{v}\| \le \|\mathbf{v}\|$ *for all* $\mathbf{v} \in \mathcal{V}$ *with equality if and only if* $\mathbf{v} \in \mathcal{U}$.

Proof (a), (b), (c), and (d) follow from the fact that if we represent $\mathbf{v} \in \mathcal{V}$ as in (7.3.1), then $P_{\mathcal{U}}\mathbf{v} = \mathbf{u}$. Using the fact that $\mathbf{u}$ and $\mathbf{v} - \mathbf{u}$ in (7.3.1) are orthogonal, (e) follows from the Pythagorean theorem and the calculation

$$\|P_{\mathcal{U}}\mathbf{v}\|^2 = \|\mathbf{u}\|^2 \le \|\mathbf{u}\|^2 + \|\mathbf{v} - \mathbf{u}\|^2 = \|\mathbf{u} + (\mathbf{v} - \mathbf{u})\|^2 = \|\mathbf{v}\|^2.$$

Finally, $\|P_{\mathcal{U}}\mathbf{v}\| = \|\mathbf{v}\|$ if and only if $\mathbf{v} - \mathbf{u} = \mathbf{0}$, which occurs if and only if $\mathbf{v} \in \mathcal{U}$. □

Theorem 7.3.11 *Let $\mathcal{V}$ be finite dimensional and let $\mathcal{U}$ be a subspace of $\mathcal{V}$.*

(a) $P_{\mathcal{U}^{\perp}} = I - P_{\mathcal{U}}$.

(b) $P_{\mathcal{U}}P_{\mathcal{U}^{\perp}} = P_{\mathcal{U}^{\perp}}P_{\mathcal{U}} = 0$.

(c) $P_{\mathcal{U}}^2 = P_{\mathcal{U}}$.

(d) $P_{\mathcal{U}} = P_{\mathcal{U}}^*$.

Proof Let $\mathbf{v} \in \mathcal{V}$ and write $\mathbf{v} = \mathbf{u} + \mathbf{w}$, in which $\mathbf{u} \in \mathcal{U}$ and $\mathbf{w} = \mathbf{v} - \mathbf{u} \in \mathcal{U}^{\perp}$.

(a) Compute

$$P_{\mathcal{U}^{\perp}}\mathbf{v} = \mathbf{v} - \mathbf{u} = \mathbf{v} - P_{\mathcal{U}}\mathbf{v} = (I - P_{\mathcal{U}})\mathbf{v}.$$

Consequently, $P_{\mathcal{U}^{\perp}} = I - P_{\mathcal{U}}$.

(b) Since $\mathbf{w} \in \mathcal{U}^{\perp}$, we have $P_{\mathcal{U}}P_{\mathcal{U}^{\perp}}\mathbf{v} = P_{\mathcal{U}}\mathbf{w} = \mathbf{0}$. Similarly, $P_{\mathcal{U}^{\perp}}P_{\mathcal{U}}\mathbf{v} = P_{\mathcal{U}^{\perp}}\mathbf{u} = \mathbf{0}$ since $\mathbf{u} \in \mathcal{U}$. Thus, $P_{\mathcal{U}}P_{\mathcal{U}^{\perp}} = P_{\mathcal{U}^{\perp}}P_{\mathcal{U}} = 0$.

(c) From (a) and (b) we see that $0 = P_{\mathcal{U}}P_{\mathcal{U}^{\perp}} = P_{\mathcal{U}}(I - P_{\mathcal{U}}) = P_{\mathcal{U}} - P_{\mathcal{U}}^2$, so $P_{\mathcal{U}}^2 = P_{\mathcal{U}}$.

(d) Let $\mathbf{v}_1, \mathbf{v}_2 \in \mathcal{V}$ and write $\mathbf{v}_1 = \mathbf{u}_1 + \mathbf{w}_1$ and $\mathbf{v}_2 = \mathbf{u}_2 + \mathbf{w}_2$, in which $\mathbf{u}_1, \mathbf{u}_2 \in \mathcal{U}$ and $\mathbf{w}_1, \mathbf{w}_2 \in \mathcal{U}^{\perp}$. Then

$$\langle P_{\mathcal{U}}\mathbf{v}_1, \mathbf{v}_2 \rangle = \langle \mathbf{u}_1, \mathbf{u}_2 + \mathbf{w}_2 \rangle = \langle \mathbf{u}_1, \mathbf{u}_2 \rangle = \langle \mathbf{u}_1 + \mathbf{w}_1, \mathbf{u}_2 \rangle = \langle \mathbf{v}_1, P_{\mathcal{U}}\mathbf{v}_2 \rangle,$$

so $P_{\mathcal{U}} = P_{\mathcal{U}}^*$. □

Example 7.3.12 Suppose that V is n-dimensional, let $\mathbf{u}_1, \mathbf{u}_2, \ldots, \mathbf{u}_r$ be an orthonormal basis for $\mathcal{U}$, and let $\mathbf{u}_{r+1}, \mathbf{u}_{r+2}, \ldots, \mathbf{u}_n$ be an orthonormal basis for $\mathcal{U}^{\perp}$. Let $\beta = \mathbf{u}_1, \mathbf{u}_2, \ldots, \mathbf{u}_r, \mathbf{u}_{r+1}, \mathbf{u}_{r+2}, \ldots, \mathbf{u}_n$. The β-β basis representation of $P_{\mathcal{U}}$ is

$$_{\beta}[P_{\mathcal{U}}]_{\beta} = [\langle P_{\mathcal{U}} \mathbf{u}_j, \mathbf{u}_i \rangle] = \begin{bmatrix} I_r & 0 \\ 0 & 0_{n-r} \end{bmatrix}$$

and the β-β basis representation of $P_{\mathcal{U}^{\perp}}$ is

$$_{\beta}[P_{\mathcal{U}^{\perp}}]_{\beta} = {}_{\beta}[I - P_{\mathcal{U}}]_{\beta} = I - {}_{\beta}[P_{\mathcal{U}}]_{\beta} = \begin{bmatrix} 0_r & 0 \\ 0 & I_{n-r} \end{bmatrix}.$$

Example 7.3.13 Let $\mathcal{U}$ be a nonzero subspace of $\mathbb{F}^n$ and let $\mathbf{u}_1, \mathbf{u}_2, \ldots, \mathbf{u}_r$ be an orthonormal basis for $\mathcal{U}$. Let $U = [\mathbf{u}_1 \ \mathbf{u}_2 \ \ldots \ \mathbf{u}_r] \in M_{n \times r}(\mathbb{F})$. Then $P_{\mathcal{U}} = UU^*$ (see Example 7.3.5), so $I - UU^*$ represents the projection $P_{\mathcal{U}^{\perp}}$ onto the orthogonal complement of $\mathcal{U}$.

Theorem 7.3.14 *Let V be finite dimensional and let $P \in \mathcal{L}(V)$. Then P is the orthogonal projection onto $\mathrm{ran}\,P$ if and only if it is self-adjoint and idempotent.*

Proof In light of Theorems 7.3.10.b, 7.3.11.c, and 7.3.11.d, it suffices to prove the reverse implication. Let $P \in \mathcal{L}(V)$ and let $\mathcal{U} = \mathrm{ran}\,P$. For any $\mathbf{v} \in V$,

$$\mathbf{v} = P\mathbf{v} + (\mathbf{v} - P\mathbf{v}), \tag{7.3.15}$$

in which $P\mathbf{v} \in \mathrm{ran}\,P$. Now assume that $P^2 = P$ and compute

$$P(\mathbf{v} - P\mathbf{v}) = P\mathbf{v} - P^2\mathbf{v} = P\mathbf{v} - P\mathbf{v} = \mathbf{0}.$$

This shows that $\mathbf{v} - P\mathbf{v} \in \ker P$, which is equal to $(\mathrm{ran}\,P^*)^{\perp}$ by Lemma 7.2.2. If we now assume that $P = P^*$, then $\mathbf{v} - P\mathbf{v} \in (\mathrm{ran}\,P)^{\perp}$. If P is both idempotent and self-adjoint, the decomposition (7.3.15) expresses $\mathbf{v}$ as the sum of something in $\mathcal{U}$, namely $P\mathbf{v}$, and something in $\mathcal{U}^{\perp}$, namely $\mathbf{v} - P\mathbf{v}$. It follows from the definition of an orthogonal projection that $P\mathbf{v} = P_{\mathcal{U}}\mathbf{v}$ for all $\mathbf{v} \in V$, so $P = P_{\mathcal{U}}$. $\qquad \square$

Our proof of the preceding theorem identifies the role of each of the two key assumptions. The idempotence of P ensures that (7.3.15) provides a way to express V as a direct sum, namely

$$V = \mathrm{ran}\,P \oplus \ker P = \mathrm{ran}\,P \oplus (\mathrm{ran}\,P^*)^{\perp}.$$

However, this direct sum need not have orthogonal direct summands. Orthogonality is ensured by the self-adjointness of P, for then we have

$$V = \mathrm{ran}\,P \oplus (\mathrm{ran}\,P)^{\perp}.$$

Example 7.3.16 Let $V = \mathbb{R}^2$, let

$$A = \begin{bmatrix} -1 & 1 \\ -2 & 2 \end{bmatrix},$$

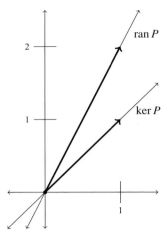

Figure 7.4 Range and kernel
of the idempotent operator in
Example 7.3.16.

and let $P = T_A : \mathcal{V} \to \mathcal{V}$. Since $A^2 = A$, the operator P is idempotent. Then $\ker P = \mathrm{span}\{[1 \ 1]^\mathsf{T}\}$ and $\mathrm{ran}\, P = \mathrm{span}\{[1 \ 2]^\mathsf{T}\}$. The vectors $[1 \ 1]^\mathsf{T}$ and $[1 \ 2]^\mathsf{T}$ are linearly independent, so $\mathrm{ran}\, P \oplus \ker P = \mathbb{R}^2$, but the direct summands are not orthogonal; see Figure 7.4 and P.7.4.

7.4 Best Approximation

Many practical applications of orthogonal projections stem from the following theorem. It states that the best approximation to a given vector $\mathbf{v}$ by a vector lying in a finite-dimensional subspace $\mathcal{U}$ is the orthogonal projection of $\mathbf{v}$ onto $\mathcal{U}$.

Theorem 7.4.1 (Best Approximation Theorem) *Let $\mathcal{U}$ be a finite-dimensional subspace of an inner product space $\mathcal{V}$ and let $P_{\mathcal{U}}$ be the orthogonal projection onto $\mathcal{U}$. Then*

$$\|\mathbf{v} - P_{\mathcal{U}}\mathbf{v}\| \le \|\mathbf{v} - \mathbf{u}\|$$

for all $\mathbf{v} \in \mathcal{V}$ and $\mathbf{u} \in \mathcal{U}$, with equality if and only if $\mathbf{u} = P_{\mathcal{U}}\mathbf{v}$; see Figure 7.5.

Proof For each $\mathbf{v} \in \mathcal{V}$, we have the orthogonal decomposition

$$\mathbf{v} = \underbrace{P_{\mathcal{U}}\mathbf{v}}_{\in\,\mathcal{U}} + \underbrace{(\mathbf{v} - P_{\mathcal{U}}\mathbf{v})}_{\in\,\mathcal{U}^{\perp}}.$$

The Pythagorean theorem asserts that

$$\|\mathbf{v} - P_{\mathcal{U}}\mathbf{v}\|^2 \le \|\mathbf{v} - P_{\mathcal{U}}\mathbf{v}\|^2 + \|P_{\mathcal{U}}\mathbf{v} - \mathbf{u}\|^2 \qquad (7.4.2)$$
$$= \|(\mathbf{v} - P_{\mathcal{U}}\mathbf{v}) + (P_{\mathcal{U}}\mathbf{v} - \mathbf{u})\|^2$$
$$= \|\mathbf{v} - \mathbf{u}\|^2$$

since $P_{\mathcal{U}}\mathbf{v} - \mathbf{u}$ belongs to $\mathcal{U}$, with equality in (7.4.2) if and only if $\mathbf{u} = P_{\mathcal{U}}\mathbf{v}$. $\square$

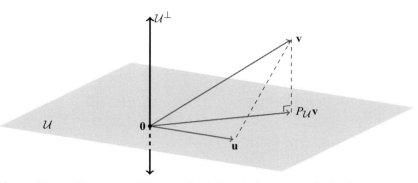

Figure 7.5 An illustration of Theorem 7.4.1. $P_{\mathcal{U}}\mathbf{v}$ is the vector in $\mathcal{U}$ that is closest to $\mathbf{v}$.

The preceding theorem can be interpreted as a statement about distances from a subspace. Let

$$d(\mathbf{v}, \mathcal{U}) = \min_{\mathbf{u} \in \mathcal{U}} \|\mathbf{v} - \mathbf{u}\| \tag{7.4.3}$$

denote the distance between $\mathbf{v}$ and the subspace $\mathcal{U}$. The best approximation theorem tells us that $d(\mathbf{v}, \mathcal{U}) = \|\mathbf{v} - P_{\mathcal{U}}\mathbf{v}\|$.

Example 7.4.4 Let $A \in \mathsf{M}_{m \times n}(\mathbb{F})$ have $\operatorname{rank} A = r \geq 1$. Suppose that $m < n$ and let $\mathcal{U} = \operatorname{null} A$. If $\mathbf{v} \in \mathbb{F}^n$, then

$$\mathbf{v} = P_{\mathcal{U}}\mathbf{v} + P_{\mathcal{U}^\perp}\mathbf{v}$$

and the distance from $\mathbf{v}$ to $\mathcal{U}$ is

$$\|\mathbf{v} - P_{\mathcal{U}}\mathbf{v}\|_2 = \|P_{\mathcal{U}^\perp}\mathbf{v}\|_2 .$$

The identity (7.2.4) tells us that $\mathcal{U}^\perp = \operatorname{col} A^*$. If $\beta = \mathbf{u}_1, \mathbf{u}_2, \ldots, \mathbf{u}_r$ is an orthonormal basis for $\operatorname{col} A^*$, then (7.3.3) ensures that

$$d(\mathbf{v}, \mathcal{U})^2 = \sum_{i=1}^{r} |\mathbf{u}_i^* \mathbf{v}|^2 .$$

We can obtain β by using the Gram–Schmidt process or another orthogonalization algorithm to orthogonalize the columns of A^*. For example, if A has full row rank and $A^* = QR$ is a QR factorization, then we may take β to be the columns of Q. In this case, $d(\mathbf{v}, \mathcal{U}) = \|Q^*\mathbf{v}\|_2$. See P.7.7 for a geometric application of this example.

Example 7.4.5 Let $\mathcal{V} = C_{\mathbb{R}}[-\pi, \pi]$ with the inner product (5.8.5), and let

$$\mathcal{U} = \operatorname{span}\left\{ \frac{1}{\sqrt{2}}, \cos x, \cos 2x, \ldots, \cos Nx, \sin x, \sin 2x, \ldots, \sin Nx \right\}.$$

Theorems 7.3.2 and 7.4.1 ensure that the function (5.8.14) (a *finite Fourier series*) is the function in $\mathcal{U}$ that is closest to f.

If $\mathbf{u}_1, \mathbf{u}_2, \ldots, \mathbf{u}_r$ is an orthonormal basis for a subspace $\mathcal{U}$ of $\mathcal{V}$, then (7.3.3) gives the orthogonal projection onto $\mathcal{U}$. If we have in hand only a basis of $\mathcal{U}$, or perhaps just a spanning list, we can use an orthogonalization algorithm to obtain an orthonormal basis, or we can proceed as follows:

Theorem 7.4.6 (Normal Equations) *Let $\mathcal{U}$ be a finite-dimensional subspace of an inner product space $\mathcal{V}$ and suppose that* $\mathrm{span}\{\mathbf{u}_1, \mathbf{u}_2, \ldots, \mathbf{u}_n\} = \mathcal{U}$. *The projection of* $\mathbf{v} \in \mathcal{V}$ *onto $\mathcal{U}$ is*

$$P_{\mathcal{U}}\mathbf{v} = \sum_{j=1}^{n} c_j \mathbf{u}_j, \tag{7.4.7}$$

in which $[c_1 \ c_2 \ \ldots \ c_n]^{\mathsf{T}}$ *is a solution of the* normal equations

$$\begin{bmatrix} \langle \mathbf{u}_1, \mathbf{u}_1 \rangle & \langle \mathbf{u}_2, \mathbf{u}_1 \rangle & \cdots & \langle \mathbf{u}_n, \mathbf{u}_1 \rangle \\ \langle \mathbf{u}_1, \mathbf{u}_2 \rangle & \langle \mathbf{u}_2, \mathbf{u}_2 \rangle & \cdots & \langle \mathbf{u}_n, \mathbf{u}_2 \rangle \\ \vdots & \vdots & \ddots & \vdots \\ \langle \mathbf{u}_1, \mathbf{u}_n \rangle & \langle \mathbf{u}_2, \mathbf{u}_n \rangle & \cdots & \langle \mathbf{u}_n, \mathbf{u}_n \rangle \end{bmatrix} \begin{bmatrix} c_1 \\ c_2 \\ \vdots \\ c_n \end{bmatrix} = \begin{bmatrix} \langle \mathbf{v}, \mathbf{u}_1 \rangle \\ \langle \mathbf{v}, \mathbf{u}_2 \rangle \\ \vdots \\ \langle \mathbf{v}, \mathbf{u}_n \rangle \end{bmatrix}. \tag{7.4.8}$$

The system (7.4.8) is consistent. If $\mathbf{u}_1, \mathbf{u}_2, \ldots, \mathbf{u}_n$ *are linearly independent, then (7.4.8) has a unique solution.*

Proof Since $P_{\mathcal{U}}\mathbf{v} \in \mathcal{U}$ and $\mathrm{span}\{\mathbf{u}_1, \mathbf{u}_2, \ldots, \mathbf{u}_n\} = \mathcal{U}$, there exist scalars $c_1, c_2, \ldots, c_n$ such that

$$P_{\mathcal{U}}\mathbf{v} = \sum_{j=1}^{n} c_j \mathbf{u}_j. \tag{7.4.9}$$

Then

$$\langle \mathbf{v}, \mathbf{u}_i \rangle = \langle \mathbf{v}, P_{\mathcal{U}}\mathbf{u}_i \rangle = \langle P_{\mathcal{U}}\mathbf{v}, \mathbf{u}_i \rangle = \left\langle \sum_{j=1}^{n} c_j \mathbf{u}_j, \mathbf{u}_i \right\rangle = \sum_{j=1}^{n} c_j \langle \mathbf{u}_j, \mathbf{u}_i \rangle$$

for $i = 1, 2, \ldots, n$. These equations are the system (7.4.8), which is consistent since $c_1, c_2, \ldots, c_n$ are already known to exist by (7.4.9). If the list $\mathbf{u}_1, \mathbf{u}_2, \ldots, \mathbf{u}_n$ is linearly independent, it is a basis for $\mathcal{U}$. In this case, (7.4.8) has a unique solution since each vector in $\mathcal{U}$ can be written as a linear combination of the basis vectors in exactly one way. $\square$

Example 7.4.10 Let $\mathbf{v} = [1 \ 1 \ 1]^{\mathsf{T}}$ and let

$$\mathcal{U} = \left\{ [x_1 \ x_2 \ x_3]^{\mathsf{T}} \in \mathbb{R}^3 : x_1 + 2x_2 + 3x_3 = 0 \right\},$$

which is spanned by the vectors $\mathbf{u}_1 = [3 \ 0 \ -1]^{\mathsf{T}}$ and $\mathbf{u}_2 = [2 \ -1 \ 0]^{\mathsf{T}}$ (see Examples 7.1.2 and 7.1.10). To find the projection of $\mathbf{v}$ onto $\mathcal{U}$, solve the normal equations (7.4.8), which are

$$\begin{bmatrix} 10 & 6 \\ 6 & 5 \end{bmatrix} \begin{bmatrix} c_1 \\ c_2 \end{bmatrix} = \begin{bmatrix} 2 \\ 1 \end{bmatrix}.$$

We obtain $c_1 = \frac{2}{7}$ and $c_2 = -\frac{1}{7}$, so (7.4.7) ensures that

$$P_{\mathcal{U}}\mathbf{v} = \frac{2}{7}\mathbf{u}_1 - \frac{1}{7}\mathbf{u}_2 = [\frac{4}{7} \ \frac{1}{7} \ -\frac{2}{7}]^{\mathsf{T}}.$$

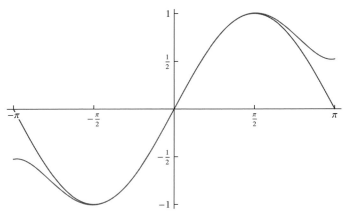

Figure 7.6 Graphs of $\sin x$ and its Taylor approximation $x - \frac{x^3}{6} + \frac{x^5}{120}$. The quality of the approximation deteriorates rapidly away from $x = 0$.

Example 7.4.11 Let $\mathcal{U}$, $\mathbf{v}$, $\mathbf{u}_1$, and $\mathbf{u}_2$ be as in the preceding example. Let $\mathbf{u}_3 = \mathbf{u}_1 + \mathbf{u}_2$, and consider the list $\mathbf{u}_1, \mathbf{u}_2, \mathbf{u}_3$, which spans $\mathcal{U}$ but is not a basis. The normal equations (7.4.8) are

$$
\begin{bmatrix} 10 & 6 & 16 \\ 6 & 5 & 11 \\ 16 & 11 & 27 \end{bmatrix} \begin{bmatrix} c_1 \\ c_2 \\ c_3 \end{bmatrix} = \begin{bmatrix} 2 \\ 1 \\ 3 \end{bmatrix}.
$$

A solution is $c_1 = \frac{5}{21}$, $c_2 = -\frac{4}{21}$, $c_3 = \frac{1}{21}$. Thus,

$$
P_{\mathcal{U}}\mathbf{v} = \tfrac{5}{21}\mathbf{u}_1 - \tfrac{4}{21}\mathbf{u}_2 + \tfrac{1}{21}\mathbf{u}_3 = [\tfrac{4}{7} \ \tfrac{1}{7} \ -\tfrac{2}{7}]^{\mathsf{T}}.
$$

Another solution is $c_1 = \frac{1}{3}$, $c_2 = -\frac{2}{21}$, $c_3 = -\frac{1}{21}$, which also gives

$$
P_{\mathcal{U}}\mathbf{v} = \tfrac{1}{3}\mathbf{u}_1 - \tfrac{2}{21}\mathbf{u}_2 - \tfrac{1}{21}\mathbf{u}_3 = [\tfrac{4}{7} \ \tfrac{1}{7} \ -\tfrac{2}{7}]^{\mathsf{T}}.
$$

The projection of $\mathbf{v}$ onto $\mathcal{U}$ is unique, even if the normal equations do not have a unique solution.

Example 7.4.12 Consider the function $f(x) = \sin x$ on the interval $[-\pi, \pi]$. How can it be approximated by real polynomials of degree 5 or less? The Taylor polynomial $x - \frac{x^3}{6} + \frac{x^5}{120}$ is one possibility. However, Taylor polynomials approximate well only near the center of the Taylor expansion; see Figure 7.6. We can do better by considering f to be an element of the inner product space $V = C_{\mathbb{R}}[-\pi, \pi]$ with the inner product (4.4.9) and derived norm (4.5.7).

Let $\mathcal{U} = \mathrm{span}\{1, x, x^2, x^3, x^4, x^5\}$ denote the subspace of V consisting of all polynomials of degree at most five. We want to find the unique polynomial $p(x) \in \mathcal{U}$ such that

$$
\| \sin x - p(x) \| = \min_{p(x) \in \mathcal{U}} \| \sin x - p(x) \|
$$

$$
= \min_{p(x) \in \mathcal{U}} \sqrt{\int_{-\pi}^{\pi} |\sin x - p(x)|^2 \, dx}.
$$

Theorem 7.4.1 tells us that

$$p(x) = P_{\mathcal{U}}(\sin x).$$

We can use the normal equations (Theorem 7.4.6) to compute $p(x)$. Let $\mathbf{u}_j = x^{j-1}$ for $j = 1$, $2, \ldots, 6$, so that

$$\langle \mathbf{u}_j, \mathbf{u}_i \rangle = \int_{-\pi}^{\pi} x^{i+j-2} \, dx = \left. \frac{x^{i+j-1}}{i+j-1} \right|_{-\pi}^{\pi} = \begin{cases} 0 & \text{if } i+j \text{ is odd,} \\ \dfrac{2\pi^{i+j-1}}{i+j-1} & \text{if } i+j \text{ is even.} \end{cases}$$

We have

$$\langle \sin x, 1 \rangle = \int_{-\pi}^{\pi} (\sin x)(1) \, dx = \left. - \cos x \right|_{-\pi}^{\pi} = 0,$$

which we could have deduced without a computation by observing that the function $1 \cdot \sin x$ is odd. This observation tells us that $\langle \sin x, x^2 \rangle = \langle \sin x, x^4 \rangle = 0$ as well, since $x^2 \sin x$ and $x^4 \sin x$ are odd functions. Some computations reveal that

$$\langle \sin x, x \rangle = 2\pi, \quad \langle \sin x, x^3 \rangle = 2\pi(\pi^2 - 6), \quad \langle \sin x, x^5 \rangle = 120 - 20\pi^2 + \pi^4.$$

We now have in hand all the entries of the 6×6 coefficient matrix and right-hand side of the linear system (7.4.8). Its solution is

$$P_{\mathcal{U}}(\sin x) = \frac{105(1485 - 153\pi^2 + \pi^4)}{8\pi^6} x - \frac{315(1155 - 125\pi^2 + \pi^4)}{4\pi^8} x^3$$

$$+ \frac{693(945 - 105\pi^2 + \pi^4)}{8\pi^{10}} x^5 \qquad\qquad (7.4.13)$$

$$\approx 0.987862x - 0.155271x^3 + 0.00564312x^5.$$

Figure 7.7 shows that (7.4.13) is an excellent approximation to $\sin x$ over the entire interval $[-\pi, \pi]$.

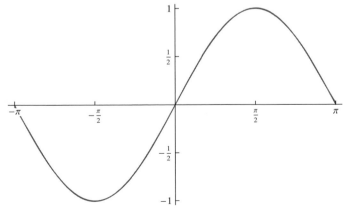

Figure 7.7 The graphs of $\sin x$ and $p(x) = P_{\mathcal{U}}(\sin x)$ are nearly indistinguishable on $[-\pi, \pi]$.

The matrix that appears in the normal equations (7.4.8) is an important one.

Definition 7.4.14 If $\mathbf{u}_1, \mathbf{u}_2, \ldots, \mathbf{u}_n$ are vectors in an inner product space, then

$$G(\mathbf{u}_1, \mathbf{u}_2, \ldots, \mathbf{u}_n) = \begin{bmatrix} \langle \mathbf{u}_1, \mathbf{u}_1 \rangle & \langle \mathbf{u}_2, \mathbf{u}_1 \rangle & \cdots & \langle \mathbf{u}_n, \mathbf{u}_1 \rangle \\ \langle \mathbf{u}_1, \mathbf{u}_2 \rangle & \langle \mathbf{u}_2, \mathbf{u}_2 \rangle & \cdots & \langle \mathbf{u}_n, \mathbf{u}_2 \rangle \\ \vdots & \vdots & \ddots & \vdots \\ \langle \mathbf{u}_1, \mathbf{u}_n \rangle & \langle \mathbf{u}_2, \mathbf{u}_n \rangle & \cdots & \langle \mathbf{u}_n, \mathbf{u}_n \rangle \end{bmatrix}$$

is the *Gram matrix* (or *Grammian*) of $\mathbf{u}_1, \mathbf{u}_2, \ldots, \mathbf{u}_n$. The *Gram determinant* of the list $\mathbf{u}_1, \mathbf{u}_2, \ldots, \mathbf{u}_n$ is

$$g(\mathbf{u}_1, \mathbf{u}_2, \ldots, \mathbf{u}_n) = \det G(\mathbf{u}_1, \mathbf{u}_2, \ldots, \mathbf{u}_n).$$

A Gram matrix is Hermitian ($G = G^*$) and positive semidefinite (see Chapter 13). Gram matrices are closely related to covariance matrices from statistics. If the vectors $\mathbf{u}_1, \mathbf{u}_2, \ldots, \mathbf{u}_n$ are centered random variables, then $G(\mathbf{u}_1, \mathbf{u}_2, \ldots, \mathbf{u}_n)$ is the corresponding covariance matrix.

7.5 A Least Squares Solution of an Inconsistent Linear System

Suppose that $A\mathbf{x} = \mathbf{y}$ is an inconsistent $m \times n$ system of linear equations. Since $\mathbf{y} - A\mathbf{x} \neq \mathbf{0}$ for all $\mathbf{x}$ in $\mathbb{F}^n$, we want to identify a vector $\mathbf{x}_0$ for which $\|\mathbf{y} - A\mathbf{x}_0\|_2$ is as small as possible. If we are able to find such a vector, we can regard it as a "best approximate solution" to the inconsistent system $A\mathbf{x} = \mathbf{y}$; see Figure 7.8.

Theorem 7.5.1 *If $A \in \mathsf{M}_{m \times n}(\mathbb{F})$ and $\mathbf{y} \in \mathbb{F}^m$, then $\mathbf{x}_0 \in \mathbb{F}^n$ satisfies*

$$\min_{\mathbf{x} \in \mathbb{F}^n} \|\mathbf{y} - A\mathbf{x}\|_2 = \|\mathbf{y} - A\mathbf{x}_0\|_2 \tag{7.5.2}$$

if and only if

$$A^*A\mathbf{x}_0 = A^*\mathbf{y}. \tag{7.5.3}$$

The system (7.5.3) is always consistent; it has a unique solution if $\operatorname{rank} A = n$.

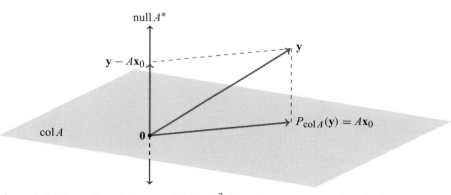

Figure 7.8 Illustration of Theorem 7.5.1 in $\mathbb{R}^3$. The closest vector to $\mathbf{y}$ in $\operatorname{col} A$ is $P_{\operatorname{col} A}(\mathbf{y})$, which is of the form $A\mathbf{x}_0$ for some $\mathbf{x}_0 \in \mathbb{R}^3$.

Proof Partition $A = [\mathbf{a}_1 \ \mathbf{a}_2 \ \ldots \ \mathbf{a}_n]$ according to its columns and let $\mathcal{U} = \operatorname{col} A = \operatorname{span}\{\mathbf{a}_1, \mathbf{a}_2, \ldots, \mathbf{a}_n\}$. We are asked to find an $A\mathbf{x}_0$ in $\mathcal{U}$ that is closest to $\mathbf{y}$ in the Euclidean norm. Theorem 7.4.1 ensures that $P_{\mathcal{U}}\mathbf{y}$ is the unique such vector, and Theorem 7.4.6 tells us how to compute it. First find a solution $\mathbf{x}_0 = [c_1 \ c_2 \ \ldots \ c_n]^{\mathsf{T}}$ of the normal equations

$$
\begin{bmatrix}
\mathbf{a}_1^*\mathbf{a}_1 & \mathbf{a}_1^*\mathbf{a}_2 & \cdots & \mathbf{a}_1^*\mathbf{a}_n \\
\mathbf{a}_2^*\mathbf{a}_1 & \mathbf{a}_2^*\mathbf{a}_2 & \cdots & \mathbf{a}_2^*\mathbf{a}_n \\
\vdots & \vdots & \ddots & \vdots \\
\mathbf{a}_n^*\mathbf{a}_1 & \mathbf{a}_n^*\mathbf{a}_2 & \cdots & \mathbf{a}_n^*\mathbf{a}_n
\end{bmatrix}
\begin{bmatrix}
c_1 \\ c_2 \\ \vdots \\ c_n
\end{bmatrix}
=
\begin{bmatrix}
\mathbf{a}_1^*\mathbf{y} \\ \mathbf{a}_2^*\mathbf{y} \\ \vdots \\ \mathbf{a}_n^*\mathbf{y}
\end{bmatrix},
$$

that is, $A^*A\mathbf{x}_0 = A^*\mathbf{y}$. Then

$$
P_{\mathcal{U}}\mathbf{y} = \sum_{i=1}^{n} c_i\mathbf{a}_i = A\mathbf{x}_0.
$$

Theorem 7.4.6 ensures that the normal equations (7.5.3) are consistent. They have a unique solution if the columns of A are linearly independent, that is, if $\operatorname{rank} A = n$. $\qquad\square$

If the null space of A is nontrivial, there can be many vectors $\mathbf{x}_0$ that satisfy (7.5.3). Fortunately $P_{\mathcal{U}}\mathbf{y} = A\mathbf{x}_0$ is the same for all of them.

If $\operatorname{rank} A = n$, then $A^*A \in \mathsf{M}_n$ is invertible (see P.7.25 or Theorem 13.1.10.a). We conclude that (7.5.3) has the unique solution $\mathbf{x}_0 = (A^*A)^{-1}A^*\mathbf{y}$. In this case, $P_{\mathcal{U}}\mathbf{y} = A\mathbf{x}_0 = A(A^*A)^{-1}A^*\mathbf{y}$, that is, the orthogonal projection onto the column space of A is

$$
P_{\operatorname{col} A} = A(A^*A)^{-1}A^*. \tag{7.5.4}
$$

Suppose that we are given some real data points $(x_1, y_1), (x_2, y_2), \ldots, (x_n, y_n)$ in the plane and we want to model them as a (perhaps approximate) linear relationship. If all the points lie on a vertical line $x = c$, there is nothing to do. If the points do not lie on a vertical line, consider the system of linear equations

$$
\begin{aligned}
y_1 &= ax_1 + b, \\
y_2 &= ax_2 + b, \\
&\vdots \quad \vdots \quad \vdots \\
y_n &= ax_n + b,
\end{aligned}
\tag{7.5.5}
$$

which we write as $A\mathbf{x} = \mathbf{y}$, with

$$
A = \begin{bmatrix} x_1 & 1 \\ x_2 & 1 \\ \vdots & \vdots \\ x_n & 1 \end{bmatrix}, \quad
\mathbf{x} = \begin{bmatrix} a \\ b \end{bmatrix}, \quad \text{and} \quad
\mathbf{y} = \begin{bmatrix} y_1 \\ y_2 \\ \vdots \\ y_n \end{bmatrix}.
$$

Because the data points do not lie on a vertical line, the columns of A are linearly independent and hence $\operatorname{rank} A = 2$. The system $A\mathbf{x} = \mathbf{y}$ involves n equations in two unknowns, so if $n > 2$

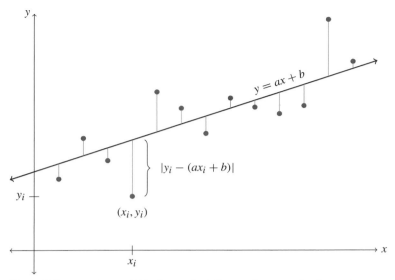

Figure 7.9 $|y_i - (ax_i + b)|$ is the vertical distance between the data point (x_i, y_i) and the graph of the line $y = ax + b$.

it might not have a solution. However, Theorem 7.5.1 ensures that any solution of the normal equations $A^*A\mathbf{x} = A^*\mathbf{y}$ minimizes

$$\|\mathbf{y} - A\mathbf{x}\|_2 = \left(\sum_{i=1}^{n} (y_i - (ax_i + b))^2 \right)^{1/2}; \tag{7.5.6}$$

see Figure 7.9. The normal equations have a unique solution because rank $A = 2$.

Example 7.5.7 Find a least squares line $y = ax + b$ to model the data

$$(0, 1), \ (1, 1), \ (2, 3), \ (3, 3), \ (4, 4).$$

According to the preceding recipe we have

$$A = \begin{bmatrix} 0 & 1 \\ 1 & 1 \\ 2 & 1 \\ 3 & 1 \\ 4 & 1 \end{bmatrix}, \qquad \mathbf{x} = \begin{bmatrix} a \\ b \end{bmatrix}, \qquad \mathbf{y} = \begin{bmatrix} 1 \\ 1 \\ 3 \\ 3 \\ 4 \end{bmatrix},$$

and we solve

$$\underbrace{\begin{bmatrix} 30 & 10 \\ 10 & 5 \end{bmatrix}}_{A^*A} \begin{bmatrix} a \\ b \end{bmatrix} = \underbrace{\begin{bmatrix} 32 \\ 12 \end{bmatrix}}_{A^*\mathbf{y}},$$

for $a = b = \frac{4}{5}$. Therefore, the least squares line is $y = \frac{4}{5}x + \frac{4}{5}$; see Figure 7.10.

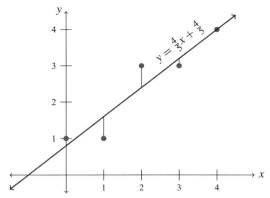

Figure 7.10 The least squares line corresponding to the data in Example 7.5.7.

In actual numerical work, there are some hazards associated with forming and then solving directly the normal equations (7.5.3) or (7.4.8) to obtain a solution to the least squares problem (7.5.2); see Section 15.6. A safer route is via the QR factorization of A; see Theorem 6.5.2. If $m \geq n$ and $A \in M_{m \times n}(\mathbb{F})$ has full column rank, then $A = QR$, in which $Q \in M_{m \times n}(\mathbb{F})$ has orthonormal columns and $R \in M_n(\mathbb{F})$ is upper triangular and has positive diagonal entries. Then $A^*A = R^*Q^*QR = R^*I_nR = R^*R$, so the normal equations (7.5.3) are

$$R^*R\mathbf{u} = R^*Q^*\mathbf{y}. \tag{7.5.8}$$

Since R is invertible, the system (7.5.8) has the same solutions as the upper triangular system

$$R\mathbf{u} = Q^*\mathbf{y}, \tag{7.5.9}$$

which can be solved by backward substitution.

The formula (7.5.4) should also be avoided in practical computations. Again, a safer route is via the QR factorization of A. If $A \in M_{m \times n}(\mathbb{F})$, rank $A = n$, and $A = QR$ is a QR factorization, then (7.3.7) ensures that $P_{\mathrm{col}\,A} = QQ^*$. With a QR factorization in hand, the orthogonal projection onto the column space of A can be computed without a matrix inversion.

7.6 Invariant Subspaces

If $V = \mathbb{F}^n$, then Theorem 7.3.14 says that $P \in M_n(\mathbb{F})$ is the matrix representation of an orthogonal projection with respect to an orthonormal basis of $\mathbb{F}^n$ if and only if P is Hermitian and idempotent. We therefore make the following definition.

Definition 7.6.1 A square matrix is an *orthogonal projection* if it is Hermitian and idempotent.

If an orthogonal projection $P \in M_n(\mathbb{F})$ is identified with the linear operator $T_P : \mathbb{F}^n \to \mathbb{F}^n$ induced by it, then P is the orthogonal projection onto col P.

Suppose that $\mathcal{U}$ is an r-dimensional subspace of $\mathbb{F}^n$ with $1 \leq r \leq n - 1$. Let $\mathbf{u}_1, \mathbf{u}_2, \ldots, \mathbf{u}_r$ be an orthonormal basis for $\mathcal{U}$ and let $U_1 = [\mathbf{u}_1 \ \mathbf{u}_2 \ \ldots \ \mathbf{u}_r] \in M_{n \times r}$. Then Example 7.3.5 tells us that $P_1 = U_1 U_1^*$ is the orthogonal projection onto $\mathcal{U}$. Similarly, if $\mathbf{u}_{r+1}, \mathbf{u}_{r+2}, \ldots, \mathbf{u}_n$ is an orthonormal basis for $\mathcal{U}^\perp$ and $U_2 = [\mathbf{u}_{r+1} \ \mathbf{u}_{r+2} \ \ldots \ \mathbf{u}_n] \in \mathsf{M}_{n \times (n-r)}$, then $P_2 = U_2 U_2^*$ is the orthogonal projection onto $\mathcal{U}^\perp$. The matrix $U = [U_1 \ U_2]$ is unitary since its columns form an orthonormal basis for $\mathbb{F}^n$.

It is instructive to use block matrices to rederive the algebraic properties of orthogonal projections from Theorem 7.3.11. For example,

$$P_1 + P_2 = U_1 U_1^* + U_2 U_2^* = [U_1 \ U_2] \begin{bmatrix} U_1^* \\ U_2^* \end{bmatrix} = UU^* = I_n,$$

which is Theorem 7.3.11.a. Now observe that the block matrix identity

$$\begin{bmatrix} I_r & 0 \\ 0 & I_{n-r} \end{bmatrix} = U^*U = \begin{bmatrix} U_1^* \\ U_2^* \end{bmatrix} [U_1 \ U_2] = \begin{bmatrix} U_1^*U_1 & U_1^*U_2 \\ U_2^*U_1 & U_2^*U_2 \end{bmatrix}$$

implies the matrix identities

$$U_1^*U_1 = I_r, \qquad U_2^*U_2 = I_{n-r}, \qquad U_1^*U_2 = 0, \qquad \text{and} \qquad U_2^*U_1 = 0. \tag{7.6.2}$$

Consequently,

$$P_1 P_2 = (U_1 U_1^*)(U_2 U_2^*) = U_1(U_1^*U_2)U_2^* = 0$$

and

$$P_2 P_1 = (U_2 U_2^*)(U_1 U_1^*) = U_2(U_2^*U_1)U_1^* = 0,$$

which implies Theorem 7.3.11.b. Moreover,

$$P_1^2 = (U_1 U_1^*)(U_1 U_1^*) = U_1(U_1^*U_1)U_1^* = U_1 I_r U_1^* = U_1 U_1^* = P_1,$$

which is Theorem 7.3.11.c. The matrices $P_1 = U_1 U_1^*$ and $P_2 = U_2 U_2^*$ are Hermitian, which is Theorem 7.3.11.d.

Another useful representation for an orthogonal projection follows from the observation

$$U \begin{bmatrix} I_r & 0 \\ 0 & 0 \end{bmatrix} U^* = [U_1 \ U_2] \begin{bmatrix} I_r & 0 \\ 0 & 0 \end{bmatrix} \begin{bmatrix} U_1^* \\ U_2^* \end{bmatrix} = U_1 U_1^* = P_1. \tag{7.6.3}$$

This is important enough to formalize as a theorem.

Theorem 7.6.4 *An orthogonal projection* $P \in \mathsf{M}_n$ *is unitarily similar to* $I_r \oplus 0_{n-r}$, *in which* $r = \dim \operatorname{col} P$.

Example 7.6.5 Consider the plane $\mathcal{U}$ in $\mathbb{R}^3$ determined by the equation $-2x_1 + 2x_2 - x_3 = 0$. An orthonormal basis for $\mathcal{U}$ is $\mathbf{u}_1 = \frac{1}{3}[1 \ 2 \ 2]^\mathsf{T}$, $\mathbf{u}_2 = \frac{1}{3}[-2 \ -1 \ 2]^\mathsf{T}$; an orthonormal basis for $\mathcal{U}^\perp$ is $\mathbf{u}_3 = \frac{1}{3}[-2 \ 2 \ -1]^\mathsf{T}$. We have

$$U = \frac{1}{3} \begin{bmatrix} 1 & -2 & -2 \\ 2 & -1 & 2 \\ 2 & 2 & -1 \end{bmatrix}, \qquad U_1 = \frac{1}{3} \begin{bmatrix} 1 & -2 \\ 2 & -1 \\ 2 & 2 \end{bmatrix}, \qquad \text{and} \qquad U_2 = \frac{1}{3} \begin{bmatrix} -2 \\ 2 \\ -1 \end{bmatrix}.$$

Consequently,

$$P_1 = U_1 U_1^* = \frac{1}{9} \begin{bmatrix} 5 & 4 & -2 \\ 4 & 5 & 2 \\ -2 & 2 & 8 \end{bmatrix} \quad \text{and} \quad P_2 = U_2 U_2^* = \frac{1}{9} \begin{bmatrix} 4 & -4 & 2 \\ -4 & 4 & -2 \\ 2 & -2 & 1 \end{bmatrix}.$$

Definition 7.6.6 Let $A \in M_n(\mathbb{F})$, let $\mathcal{U}$ be a subspace of $\mathbb{F}^n$, and let $A\mathcal{U} = \{A\mathbf{x} : \mathbf{x} \in \mathcal{U}\}$ (see Example 1.3.13). If $A\mathcal{U} \subseteq \mathcal{U}$, then $\mathcal{U}$ is *invariant under A*; alternatively, we say that $\mathcal{U}$ is *A-invariant*.

The subspaces $\{\mathbf{0}\}$ and $\mathbb{F}^n$ are A-invariant for any $A \in M_n(\mathbb{F})$.

Theorem 7.6.7 Let $A \in M_n(\mathbb{F})$, let $\mathcal{U}$ be an r-dimensional subspace of $\mathbb{F}^n$ such that $1 \leq r \leq n - 1$. Let $U = [U_1 \ U_2] \in M_n(\mathbb{F})$ be a unitary matrix such that the columns of $U_1 \in M_{n \times r}(\mathbb{F})$ and $U_2 \in M_{n \times (n-r)}(\mathbb{F})$ are orthonormal bases for $\mathcal{U}$ and $\mathcal{U}^\perp$, respectively. Let $P = U_1 U_1^*$ denote the orthogonal projection onto $\mathcal{U}$. Then the following are equivalent:

(a) $\mathcal{U}$ is A-invariant.

(b) $U^*AU = \begin{bmatrix} B & X \\ 0 & C \end{bmatrix}$, in which $B \in M_r(\mathbb{F})$ and $C \in M_{n-r}(\mathbb{F})$.

(c) $PAP = AP$.

Proof (a) $\Leftrightarrow$ (b) Since

$$U^*AU = \begin{bmatrix} U_1^* \\ U_2^* \end{bmatrix} A[U_1 \ U_2] = \begin{bmatrix} U_1^* \\ U_2^* \end{bmatrix} [AU_1 \ AU_2] = \begin{bmatrix} U_1^*AU_1 & U_1^*AU_2 \\ U_2^*AU_1 & U_2^*AU_2 \end{bmatrix},$$

it suffices to prove that $U_2^*AU_1 = 0$ if and only if $A\mathcal{U} \subseteq \mathcal{U}$. Let $U_1 = [\mathbf{u}_1 \ \mathbf{u}_2 \ \dots \ \mathbf{u}_r] \in M_{n \times r}(\mathbb{F})$. Using (7.2.4), we have

$$\text{null } U_2^* = (\text{col } U_2)^\perp = (\mathcal{U}^\perp)^\perp = \mathcal{U},$$

so

$$\begin{aligned} U_2^*AU_1 = 0 \quad &\Longleftrightarrow \quad U_2^*[A\mathbf{u}_1 \ A\mathbf{u}_2 \ \dots \ A\mathbf{u}_r] = 0 \\ &\Longleftrightarrow \quad [U_2^*(A\mathbf{u}_1) \ U_2^*(A\mathbf{u}_2) \ \dots \ U_2^*(A\mathbf{u}_r)] = 0 \\ &\Longleftrightarrow \quad \text{span}\{A\mathbf{u}_1, A\mathbf{u}_2, \dots, A\mathbf{u}_r\} \subseteq \text{null } U_2^* \\ &\Longleftrightarrow \quad A\mathcal{U} \subseteq \mathcal{U}. \end{aligned}$$

(b) $\Leftrightarrow$ (c) The representation (7.6.3) ensures that $U^*PU = I_r \oplus 0_{n-r}$. Partition

$$U^*AU = \begin{bmatrix} B & X \\ Y & C \end{bmatrix}$$

conformally with $I_r \oplus I_{n-r}$. It suffices to prove that $PAP = AP$ if and only if $Y = 0$, which follows from the block matrix computation

$$PAP = AP \iff (U^*PU)(U^*AU)(U^*PU) = (U^*AU)(U^*PU)$$

$$\iff \begin{bmatrix} I_r & 0 \\ 0 & 0_{n-r} \end{bmatrix} \begin{bmatrix} B & X \\ Y & C \end{bmatrix} \begin{bmatrix} I_r & 0 \\ 0 & 0_{n-r} \end{bmatrix} = \begin{bmatrix} B & X \\ Y & C \end{bmatrix} \begin{bmatrix} I_r & 0 \\ 0 & 0_{n-r} \end{bmatrix}$$

$$\iff \begin{bmatrix} B & 0 \\ 0 & 0_{n-r} \end{bmatrix} = \begin{bmatrix} B & 0 \\ Y & 0_{n-r} \end{bmatrix}$$

$$\iff Y = 0. \qquad \square$$

Corollary 7.6.8 *Maintain the notation of Theorem 7.6.7. The following are equivalent:*

(a) $\mathcal{U}$ *is invariant under A and A^*.*

(b) $U^*AU = B \oplus C$, *in which* $B \in M_r(\mathbb{F})$ *and* $C \in M_{n-r}(\mathbb{F})$.

(c) $PA = AP$.

Proof (a) $\Leftrightarrow$ (b) Theorem 7.6.7 ensures that $\mathcal{U}$ is invariant under A and A^* if and only if

$$U^*AU = \begin{bmatrix} B & X \\ 0 & C \end{bmatrix} \quad \text{and} \quad U^*A^*U = \begin{bmatrix} B' & X' \\ 0 & C' \end{bmatrix}, \qquad (7.6.9)$$

in which $B, B' \in M_r(\mathbb{F})$, $C, C' \in M_{n-r}(\mathbb{F})$, and $X, X' \in M_{r \times (n-r)}(\mathbb{F})$. Since the block matrices in (7.6.9) must be adjoints of each other, $B' = B^*$, $C' = C^*$, and $X = X' = 0$. Thus, $\mathcal{U}$ is invariant under A and A^* if and only if $U^*AU = B \oplus C$.

(a) $\Rightarrow$ (c) Since $\mathcal{U}$ is invariant under A and A^*, Theorem 7.6.7 ensures that $PAP = AP$ and $PA^*P = A^*P$. Therefore, $PA = (A^*P)^* = (PA^*P)^* = PAP = AP$.

(c) $\Rightarrow$ (a) Suppose that $PA = AP$. Then $PAP = AP^2 = AP$ and Theorem 7.6.7 ensures that $\mathcal{U}$ is A-invariant. Moreover, $A^*P = PA^*$ so that $PA^*P = A^*P^2 = A^*P$. Hence $\mathcal{U}$ is A^*-invariant (Theorem 7.6.7 again). $\qquad \square$

7.7 Problems

P.7.1 Consider

$$A_1 = \begin{bmatrix} 1 & 0 \\ 0 & 0 \end{bmatrix}, \quad A_2 = \begin{bmatrix} \frac{1}{2} & \frac{1}{2} \\ \frac{1}{2} & \frac{1}{2} \end{bmatrix}, \quad \text{and} \quad A_3 = \begin{bmatrix} 1 & 1 \\ 0 & 0 \end{bmatrix}.$$

(a) Verify that A_1, A_2, A_3 are idempotent. Draw a diagram illustrating null A_i for $i = 1, 2, 3$.

(b) Which, if any, of A_1, A_2, A_3 are orthogonal projections?

(c) Verify that the method of Example 7.3.5 produces the orthogonal projections found in (b).

P.7.2 If $A \in \mathsf{M}_n$, show that

$$M = \begin{bmatrix} A & A \\ I - A & I - A \end{bmatrix} \qquad (7.7.1)$$

is idempotent. When is M an orthogonal projection?

P.7.3 If $A \in \mathsf{M}_n$, show that

$$M = \begin{bmatrix} I & A \\ 0 & 0 \end{bmatrix} \qquad (7.7.2)$$

is idempotent. When is M an orthogonal projection?

P.7.4 Verify that the matrix A in Example 7.3.16 is idempotent, not of the form (7.7.1) or (7.7.2), and not an orthogonal projection.

P.7.5 Let V be a finite-dimensional inner product space, let $P \in \mathcal{L}(V)$, and suppose that $P^2 = P$. (a) Show that $V = \operatorname{ran} P \oplus \ker P$. (b) Show that $\operatorname{ran} P \perp \ker P$ if and only if $P = P^*$. *Hint*: See the proof of Theorem 7.3.11.d.

P.7.6 Let $\mathbf{u} = [a\ b\ c]^\mathsf{T} \in \mathbb{R}^3$ be a unit vector and let $\mathcal{U}$ denote the plane in $\mathbb{R}^3$ defined by the equation $ax_1 + bx_2 + cx_3 = 0$. Find the explicit 3×3 matrix P that represents, with respect to the standard basis of $\mathbb{R}^3$, the orthogonal projection from $\mathbb{R}^3$ onto $\mathcal{U}$. Verify that $P\mathbf{u} = \mathbf{0}$ and that P is Hermitian and idempotent.

P.7.7 Let $A = [a\ b\ c] \in \mathsf{M}_{1\times 3}(\mathbb{R})$ be nonzero, let $\mathcal{P}$ denote the plane in $\mathbb{R}^3$ determined by the equation $ax_1 + bx_2 + cx_3 + d = 0$, let $\mathbf{x}_0 \in \mathcal{P}$, and let $\mathbf{v} = [v_1\ v_2\ v_3]^\mathsf{T} \in \mathbb{R}^3$. Review Example 7.4.4 and show the following: (a) The distance from $\mathbf{v}$ to $\mathcal{P}$ is the projection of $\mathbf{v} - \mathbf{x}_0$ onto the subspace $\operatorname{col} A^\mathsf{T}$. (b) The vector $(a^2 + b^2 + c^2)^{-1/2}[a\ b\ c]^\mathsf{T}$ is an orthonormal basis for $\operatorname{col} A^\mathsf{T}$. (c) The distance from $\mathbf{v}$ to $\mathcal{P}$ is $(a^2 + b^2 + c^2)^{-1/2}|av_1 + bv_2 + cv_3 + d|$.

P.7.8 Find the minimum norm solution to the system (7.2.8) from Example 7.2.7 by minimizing $\|\mathbf{x}(t)\|_2$ using calculus. Does your answer agree with the answer obtained from Theorem 7.2.5?

P.7.9 Let

$$A = \begin{bmatrix} 1 & 2 & 1 \\ 2 & 4 & 2 \\ 0 & 1 & 0 \end{bmatrix} \quad \text{and} \quad \mathbf{y} = \begin{bmatrix} 1 \\ 2 \\ 1 \end{bmatrix}.$$

Show that $A\mathbf{x} = \mathbf{y}$ is consistent and show that $\mathbf{s} = [-\frac{1}{2}\ 1\ -\frac{1}{2}]^\mathsf{T}$ is its minimum norm solution.

P.7.10 Suppose that $A \in \mathsf{M}_{m\times n}$ has full row rank and let $A^* = QR$ be a narrow QR factorization. Show that the minimum norm solution of $A\mathbf{x} = \mathbf{y}$ is $\mathbf{s} = QR^{-*}\mathbf{y}$.

P.7.11 Let $V = \mathsf{M}_n(\mathbb{R})$ with the Frobenius inner product.

(a) Let $\mathcal{U}_+$ denote the subspace of all symmetric matrices in V and let $\mathcal{U}_-$ denote the subspace of all skew-symmetric matrices in V. Show that

$$A = \tfrac{1}{2}(A + A^\mathsf{T}) + \tfrac{1}{2}(A - A^\mathsf{T})$$

for any $A \in V$ and deduce that $V = \mathcal{U}_+ \oplus \mathcal{U}_-$.

(b) Show that $\mathcal{U}_- = \mathcal{U}_+^\perp$ and $\mathcal{U}_+ = \mathcal{U}_-^\perp$.

(c) Show that $P_{\mathcal{U}_+}A = \tfrac{1}{2}(A + A^\mathsf{T})$ and $P_{\mathcal{U}_-}A = \tfrac{1}{2}(A - A^\mathsf{T})$ for all $A \in \mathsf{M}_n(\mathbb{R})$.

P.7.12 Let $V = M_n(\mathbb{R})$ with the Frobenius inner product. Let $\mathcal{U}_1$ denote the subspace of upper triangular matrices in V and let $\mathcal{U}_2$ denote the subspace of all strictly lower triangular matrices in V. Show that $V = \mathcal{U}_1 \oplus \mathcal{U}_2$ and that $\mathcal{U}_2 = \mathcal{U}_1^\perp$.

P.7.13 Let $V = M_n(\mathbb{R})$ with the Frobenius inner product. Partition each $M \in M_n(\mathbb{R})$ as $M = \begin{bmatrix} A & B \\ C & D \end{bmatrix}$, in which $A \in M_{p \times q}$, and let $P(M) = \begin{bmatrix} A & 0 \\ 0 & 0 \end{bmatrix}$. Show that $P \in \mathcal{L}(M_n(\mathbb{R}))$ is an orthogonal projection.

P.7.14 Let $A \in M_n$ be an orthogonal projection. Show that the operator $T \in \mathcal{L}(\mathbb{F}^n)$ defined by $T(\mathbf{x}) = A\mathbf{x}$ is the orthogonal projection onto $\operatorname{ran} T$.

P.7.15 Let $P \in M_n$ be an orthogonal projection. Show that $\operatorname{tr} P = \dim \operatorname{col} P$.

P.7.16 Let V be a finite-dimensional $\mathbb{F}$-inner product space and let $P, Q \in \mathcal{L}(V)$ be orthogonal projections that commute. Show that PQ is the orthogonal projection onto $\operatorname{ran} P \cap \operatorname{ran} Q$. *Hint*: First show that PQ is Hermitian and idempotent.

P.7.17 Let V be a finite-dimensional $\mathbb{F}$-inner product space and let $P, Q \in \mathcal{L}(V)$ be orthogonal projections. Show that the following are equivalent:

(a) $\operatorname{ran} P \perp \operatorname{ran} Q$.

(b) $\operatorname{ran} Q \subseteq \ker P$.

(c) $PQ = 0$.

(d) $PQ + QP = 0$.

(e) $P + Q$ is an orthogonal projection.

Hint: (d) implies that $-PQP = QP^2 = QP$ is Hermitian.

P.7.18 Let V be a finite-dimensional $\mathbb{F}$-inner product space and let $P, Q \in \mathcal{L}(V)$ be orthogonal projections. Show that $(\operatorname{ran} P \cap \operatorname{ran} Q)^\perp = \ker P + \ker Q$.

P.7.19 Let V be a finite-dimensional $\mathbb{F}$-inner product space and let $P \in \mathcal{L}(V)$ be idempotent. Show that $V = \ker P \oplus \operatorname{ran} P$.

P.7.20 Let V be a finite-dimensional $\mathbb{F}$-inner product space and let $P \in \mathcal{L}(V)$ be idempotent. Show that if $\ker P \subseteq (\operatorname{ran} P)^\perp$, then P is the orthogonal projection onto $\operatorname{ran} P$.

P.7.21 Let V be a finite-dimensional $\mathbb{F}$-inner product space and let $P \in \mathcal{L}(V)$ be idempotent. Show that $\|P\mathbf{v}\| \leq \|\mathbf{v}\|$ for every $\mathbf{v} \in V$ if and only if P is an orthogonal projection. *Hint*: Use the preceding problem and P.4.10.

P.7.22 Let $P = [p_{ij}] \in M_n$ be idempotent and such that $p_{11} = p_{22} = \cdots = p_{nn} = 0$. Prove that $P = 0$.

P.7.23 Let $A \in M_{m \times n}$, suppose that $\operatorname{rank} A = n$, and let $A = QR$ be a QR factorization. If $\mathbf{y} \in \operatorname{col} A$, show that $\mathbf{x}_0 = R^{-1}Q^*\mathbf{y}$ is the unique solution of the linear system $A\mathbf{x} = \mathbf{y}$.

P.7.24 Let $A \in M_{m \times n}$, suppose that $\operatorname{rank} A = n$, and let $A = QR$ be a QR factorization. Show that $A(A^*A)^{-1}A^* = QQ^*$.

P.7.25 Let V be an inner product space and let $\mathbf{u}_1, \mathbf{u}_2, \ldots, \mathbf{u}_n \in V$ be linearly independent. Show that $G(\mathbf{u}_1, \mathbf{u}_2, \ldots, \mathbf{u}_n)$ is invertible. *Hint*: $\mathbf{x}^* G(\mathbf{u}_1, \mathbf{u}_2, \ldots, \mathbf{u}_n)\mathbf{x} = \|\sum_{i=1}^n x_i \mathbf{u}_i\|^2$.

P.7.26 Let $\mathcal{U}$ be a finite-dimensional subspace of an $\mathbb{F}$-inner product space V and define $d(\mathbf{v}, \mathcal{U})$ as in (7.4.3). Let $\mathbf{u}_1, \mathbf{u}_2, \ldots, \mathbf{u}_n$ be a basis for $\mathcal{U}$, let $\mathbf{v} \in V$, and suppose that $P_{\mathcal{U}}\mathbf{v} = \sum_{i=1}^n c_i \mathbf{u}_i$.

(a) Show that

$$d(\mathbf{v}, \mathcal{U})^2 = \|\mathbf{v}\|^2 - \sum_{i=1}^{n} c_i \langle \mathbf{u}_i, \mathbf{v} \rangle$$

for all $\mathbf{v} \in \mathcal{V}$. *Hint*: $\mathbf{v} - P_{\mathcal{U}}\mathbf{v} \in \mathcal{U}^{\perp}$.

(b) Combine (a) and the normal equations (7.4.8) to obtain an $(n+1) \times (n+1)$ linear system for the unknowns $c_1, c_2, \ldots, c_n, d(\mathbf{v}, \mathcal{U})^2$. Use Cramer's rule to obtain

$$d(\mathbf{v}, \mathcal{U})^2 = \frac{g(\mathbf{v}, \mathbf{u}_1, \mathbf{u}_2, \ldots, \mathbf{u}_n)}{g(\mathbf{u}_1, \mathbf{u}_2, \ldots, \mathbf{u}_n)},$$

which expresses $d(\mathbf{v}, \mathcal{U})^2$ as the quotient of two Gram determinants.

(c) Let $\mathbf{u}, \mathbf{v} \in \mathcal{V}$ and suppose that $\mathbf{u} \neq \mathbf{0}$. Show that

$$d(\mathbf{v}, \mathrm{span}\{\mathbf{u}\})^2 = \frac{g(\mathbf{v}, \mathbf{u})}{g(\mathbf{u})}$$

and conclude that $g(\mathbf{v}, \mathbf{u}) \geq 0$. Deduce that $|\langle \mathbf{u}, \mathbf{v} \rangle| \leq \|\mathbf{u}\| \|\mathbf{v}\|$, which is the Cauchy–Schwarz inequality.

P.7.27 In this problem, we approach the theory of least squares approximation from a different perspective. Let $\mathcal{V}, \mathcal{W}$ be finite-dimensional $\mathbb{F}$-inner product spaces with $\dim \mathcal{V} \leq \dim \mathcal{W}$ and let $T \in \mathcal{L}(\mathcal{V}, \mathcal{W})$.

(a) Show that $\ker T = \ker T^*T$.

(b) Prove that if $\dim \mathrm{ran}\, T = \dim \mathcal{V}$, then T^*T is invertible.

(c) Prove that if $\dim \mathrm{ran}\, T = \dim \mathcal{V}$, then $P = T(T^*T)^{-1}T^* \in \mathcal{L}(\mathcal{W})$ is the orthogonal projection onto $\mathrm{ran}\, T$.

(d) If $T \in \mathcal{L}(\mathcal{V}, \mathcal{W})$ and $\dim \mathrm{ran}\, T = \dim \mathcal{V}$, prove that there exists a unique vector $\mathbf{x} \in \mathcal{V}$ such that $\|T\mathbf{x} - \mathbf{y}\|$ is minimized. Show that this vector $\mathbf{x}$ satisfies $T^*T\mathbf{x} = T^*\mathbf{y}$.

P.7.28 Compute the least squares line for the data $(-2, -3)$, $(-1, -1)$, $(0, 1)$, $(1, 1)$, $(2, 3)$.

P.7.29 Find the quadratic $y = ax^2 + bx + c$ that minimizes $\sum_{i=1}^{5} (y_i - (ax_i^2 + bx_i + c))^2$ for the data $(-2, 3)$, $(-1, 1)$, $(0, 1)$, $(1, 2)$, $(2, 4)$.

P.7.30 In linear regression, one is given data points (x_1, y_1), (x_2, y_2), $\ldots$, (x_m, y_m) and must find parameters a and b such that $\sum_{i=1}^{m} (y_i - ax_i - b)^2$ is minimized. This is the same problem as in Example 7.5.7. Derive explicit formulas for a and b that involve the quantities

$$S_x = \frac{1}{m} \sum_{i=1}^{m} x_i, \qquad\qquad S_y = \frac{1}{m} \sum_{i=1}^{m} y_i,$$

$$S_{x^2} = \frac{1}{m} \sum_{i=1}^{m} x_i^2, \qquad\qquad S_{xy} = \frac{1}{m} \sum_{i=1}^{m} x_i y_i.$$

P.7.31 Let $A \in \mathsf{M}_{m \times n}$ and suppose that rank $A = n$. (a) Show that $P = A(A^*A)^{-1}A^*$ is well defined. (b) Show that P is Hermitian and idempotent. (c) Show that $\operatorname{col} P = \operatorname{col} A$, and conclude that P is the orthogonal projection onto $\operatorname{col} A$.

P.7.32 Let $\mathbf{u}_1, \mathbf{u}_2, \ldots, \mathbf{u}_n \in \mathbb{F}^m$ be linearly independent. Let $A = [\mathbf{u}_1 \ \mathbf{u}_2 \ \ldots \ \mathbf{u}_n] \in \mathsf{M}_{m \times n}(\mathbb{F})$ and let $A = QR$ be a (narrow) QR factorization in which $R = [r_{ij}] \in \mathsf{M}_n(\mathbb{F})$ is upper triangular. For each $k = 2, 3, \ldots, n$, show that r_{kk} is the distance from $\mathbf{u}_k$ to span$\{\mathbf{u}_1, \mathbf{u}_2, \ldots, \mathbf{u}_{k-1}\}$. Discuss how the QR factorization can be used to solve the minimization problem (7.4.3).

P.7.33 Let $\mathcal{V} = C_{\mathbb{R}}[0, 1]$ and let $\mathcal{P}$ be the subspace of $\mathcal{V}$ consisting of all real polynomials.

(a) Let $f \in \mathcal{V}$. The *Weierstrass Approximation Theorem* says that for any given $\varepsilon > 0$ there is a polynomial $p_\varepsilon \in \mathcal{P}$ such that $|f(t) - p_\varepsilon(t)| \le \varepsilon$ for all $t \in [0, 1]$. If $f \in \mathcal{P}^\perp$, show that the L^2 norm of f satisfies the inequality $\|f\| \le \varepsilon$ for every $\varepsilon > 0$; see (4.5.7). *Hint*: Consider $\|p_\varepsilon - f\|^2$.

(b) Show that $\mathcal{P}^\perp = \{0\}$ and conclude that $\mathcal{P} \ne (\mathcal{P}^\perp)^\perp$. This does not contradict Corollary 7.1.9 because span $\mathcal{P}$ is not finite dimensional.

P.7.34 Let $A \in \mathsf{M}_{m \times n}(\mathbb{F})$ and $\mathbf{y} \in \mathbb{F}^m$. Use (7.2.4) to prove the *Fredholm Alternative*: There is some $\mathbf{x}_0 \in \mathbb{F}^n$ such that $A\mathbf{x}_0 = \mathbf{y}$ if and only if $\mathbf{z}^*\mathbf{y} = 0$ for every $\mathbf{z} \in \mathbb{F}^m$ such that $A^*\mathbf{z} = \mathbf{0}$.

P.7.35 Let $A, B \in \mathsf{M}_n$. Show that null $A = $ null B if and only if there is an invertible $S \in \mathsf{M}_n$ such that $A = SB$. *Hint*: P.3.36 and (7.2.4).

7.8 Notes

For a proof of the Weierstrass approximation theorem, cited in P.7.33, see [Dav63, Thm. 6.1.1] or [Dur12, Ch. 6].

7.9 Some Important Concepts

- Orthogonal complements of sets and subspaces.

- Minimum norm solution of a consistent linear system.

- How to use an orthonormal basis to construct an orthogonal projection.

- Best approximation of a given vector by a vector in a given subspace.

- Least squares solution of an inconsistent linear system.

- Orthogonal projection matrices are Hermitian and idempotent.

- Invariant subspaces and block triangular matrices (Theorem 7.6.7).

Eigenvalues, Eigenvectors, and Geometric Multiplicity

In the next four chapters, we develop tools to show (in Chapter 11) that each square complex matrix is similar to an essentially unique direct sum of special bidiagonal matrices. The first step is to show that every square complex matrix has a one-dimensional invariant subspace and explore some consequences of that fact.

8.1 Eigenvalue–Eigenvector Pairs

Definition 8.1.1 Let $A \in \mathsf{M}_n$, let $\lambda \in \mathbb{C}$, and let $\mathbf{x}$ be a nonzero vector. Then $(\lambda, \mathbf{x})$ is an *eigenpair* of A if

$$\mathbf{x} \neq \mathbf{0} \quad \text{and} \quad A\mathbf{x} = \lambda\mathbf{x}. \tag{8.1.2}$$

If $(\lambda, \mathbf{x})$ is an eigenpair of A, then λ is an *eigenvalue* of A and $\mathbf{x}$ is an *eigenvector* of A.

We cannot emphasize strongly enough that an eigenvector must be a nonzero vector.

Although the eigenpair equation $A\mathbf{x} = \lambda\mathbf{x}$ looks a little like a linear system $A\mathbf{x} = \mathbf{b}$, it is fundamentally different. In the linear system, $\mathbf{b}$ on the right-hand side is known, but in the eigenpair identity, both $\mathbf{x}$ and λ are unknown.

Many different vectors can be eigenvectors of A associated with a given eigenvalue. For example, if $(\lambda, \mathbf{x})$ is an eigenpair of A and if $c \neq 0$, then $c\mathbf{x} \neq \mathbf{0}$ and $A(c\mathbf{x}) = cA\mathbf{x} = c\lambda\mathbf{x} = \lambda(c\mathbf{x})$, so $(\lambda, c\mathbf{x})$ is also an eigenpair of A. However, only one scalar can be an eigenvalue of A associated with a given eigenvector. If $(\lambda, \mathbf{x})$ and $(\mu, \mathbf{x})$ are eigenpairs of A, then $\lambda\mathbf{x} = A\mathbf{x} = \mu\mathbf{x}$, so $(\lambda - \mu)\mathbf{x} = \mathbf{0}$. Since $\mathbf{x} \neq \mathbf{0}$, it follows that $\lambda = \mu$.

Eigenvalues and eigenvectors are important tools that help us understand the behavior of matrices by resolving them into simple components that we can exploit for algorithms, data analysis, approximation, data compression, and other purposes.

Some questions to be answered in the rest of this chapter are: Does $A \in \mathsf{M}_n$ have any eigenvalues? How many different eigenvalues can A have? Is there something special about A if it has a maximal number of different eigenvalues? If A has some eigenvalues, where in the complex plane should we look (or not look) for them?

Before exploring the theory and applications of eigenvectors, we first consider several examples.

Example 8.1.3 Since $I\mathbf{x} = 1\mathbf{x}$ for all $\mathbf{x}$, it follows that $(1, \mathbf{x})$ is an eigenpair of I for any nonzero vector $\mathbf{x}$. Moreover, $I\mathbf{x} = \lambda\mathbf{x}$ for $\mathbf{x} \neq \mathbf{0}$ implies that $\mathbf{x} = \lambda\mathbf{x}$ and $\lambda = 1$.

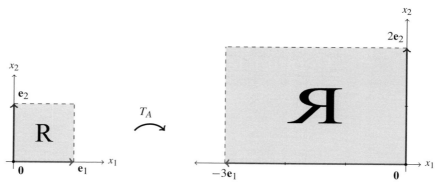

Figure 8.1 A graphical representation of the linear transformation $T_A : \mathbb{R}^2 \to \mathbb{R}^2$ induced by the matrix from Example 8.1.4; $(-3, \mathbf{e}_1)$ and $(2, \mathbf{e}_2)$ are eigenpairs.

Example 8.1.4 Since

$$A = \begin{bmatrix} -3 & 0 \\ 0 & 2 \end{bmatrix} = [-3\mathbf{e}_1 \ \ 2\mathbf{e}_2] \qquad (8.1.5)$$

satisfies $A\mathbf{e}_1 = -3\mathbf{e}_1$ and $A\mathbf{e}_2 = 2\mathbf{e}_2$, we see that $(-3, \mathbf{e}_1)$ and $(2, \mathbf{e}_2)$ are eigenpairs for A. For real matrices, it is often instructive to visualize eigenvectors and eigenvalues by studying the associated linear transformation (see Figure 8.1).

Example 8.1.6 Consider

$$A = \begin{bmatrix} 1 & 1 \\ 1 & 1 \end{bmatrix}$$

and observe that, unlike the case in Example 8.1.4, $\mathbf{e}_1$ and $\mathbf{e}_2$ are not eigenvectors of A. Let $\mathbf{x} = [x_1 \ x_2]^\mathsf{T}$ be nonzero and examine (8.1.2). We obtain

$$\begin{aligned} x_1 + x_2 &= \lambda x_1, \\ x_1 + x_2 &= \lambda x_2. \end{aligned} \qquad (8.1.7)$$

These two equations tell us that $\lambda x_1 = \lambda x_2$, that is, $\lambda(x_1 - x_2) = 0$. There are two cases. If $\lambda = 0$, then $x_1 + x_2 = 0$ is the only constraint imposed by (8.1.7); hence any nonzero $\mathbf{x}$ whose entries satisfy this equation, such as $\mathbf{v}_1 = [-1 \ 1]$, is an eigenvector corresponding to the eigenvalue 0. If $x_1 - x_2 = 0$, then x_1 and x_2 are equal and nonzero since $\mathbf{x} \neq \mathbf{0}$. Return to (8.1.7) and see that $2x_1 = \lambda x_1$, from which it follows that $\lambda = 2$. Any nonzero $\mathbf{x}$ whose entries satisfy $x_1 = x_2$, such as $\mathbf{v}_2 = [1 \ 1]$, is an eigenvector corresponding to the eigenvalue 2. These observations are illustrated in Figure 8.2. This example shows that, although an eigenvector may never be a zero vector, an eigenvalue may be a zero scalar.

Example 8.1.8 Consider

$$A = \begin{bmatrix} 1 & 1 \\ 0 & 2 \end{bmatrix}.$$

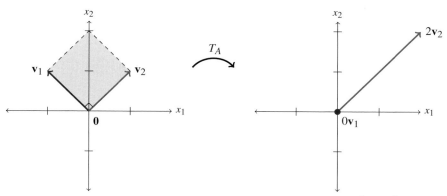

Figure 8.2 A graphical representation of the linear transformation $T_A : \mathbb{R}^2 \to \mathbb{R}^2$ induced by the matrix from Example 8.1.6; $(0, \mathbf{v}_1)$ and $(2, \mathbf{v}_2)$ are eigenpairs.

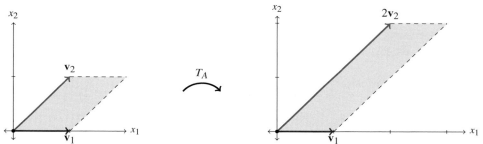

Figure 8.3 A graphical representation of the linear transformation $T_A : \mathbb{R}^2 \to \mathbb{R}^2$ induced by the matrix from Example 8.1.8; $(1, \mathbf{v}_1)$ and $(2, \mathbf{v}_2)$ are eigenpairs.

Let $\mathbf{x} = [x_1 \ x_2]^\mathsf{T}$ be nonzero and examine (8.1.2). We obtain

$$x_1 + x_2 = \lambda x_1,$$
$$2x_2 = \lambda x_2. \tag{8.1.9}$$

If $x_2 = 0$, then $x_1 \neq 0$ since $\mathbf{x} \neq \mathbf{0}$. Thus, the system reduces to $x_1 = \lambda x_1$, so 1 is an eigenvalue of A and every eigenvector corresponding to this eigenvalue is a nonzero multiple of $\mathbf{v}_1 = [1 \ 0]^\mathsf{T}$. If $x_2 \neq 0$, then the second equation in (8.1.9) tells us that $\lambda = 2$. The first equation in (8.1.9) reveals that $x_1 = x_2$, that is, every eigenvector corresponding to the eigenvalue 2 is a nonzero multiple of $\mathbf{v}_2 = [1 \ 1]^\mathsf{T}$; see Figure 8.3.

Example 8.1.10 Consider

$$A = \begin{bmatrix} 1 & 1 \\ 0 & 1 \end{bmatrix}. \tag{8.1.11}$$

Let $\mathbf{x} = [x_1 \ x_2]^\mathsf{T}$ be nonzero and examine (8.1.2). We obtain

$$x_1 + x_2 = \lambda x_1,$$
$$x_2 = \lambda x_2. \tag{8.1.12}$$

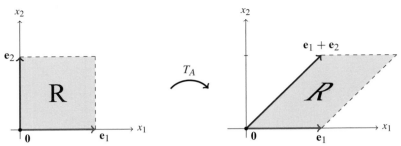

Figure 8.4 The linear transformation $T_A : \mathbb{R}^2 \to \mathbb{R}^2$ induced by the matrix A from Example 8.1.10 is a *shear* in the x_1-direction. The only eigenvalue of A is 1 and the corresponding eigenvectors are the nonzero multiples of $\mathbf{e}_1$.

If $x_2 \neq 0$, then the second equation in (8.1.12) tells us that $\lambda = 1$. Substitute this into the first equation and obtain $x_1 + x_2 = x_1$. We conclude that $x_2 = 0$, which is a contradiction. Therefore, $x_2 = 0$, from which we deduce that $x_1 \neq 0$ since $\mathbf{x} \neq \mathbf{0}$. The first equation in (8.1.12) ensures that $x_1 = \lambda x_1$, and hence $\lambda = 1$ is the only eigenvalue of A. The corresponding eigenvectors are the nonzero multiples $[x_1 \ 0]^T$ of $\mathbf{e}_1$ (see Figure 8.4).

Example 8.1.13 Consider

$$A = \begin{bmatrix} 1 & i \\ -i & 1 \end{bmatrix}.$$

Let $\mathbf{x} = [x_1 \ x_2]^T$ be nonzero and examine (8.1.2). We obtain the system

$$x_1 + ix_2 = \lambda x_1,$$
$$-ix_1 + x_2 = \lambda x_2.$$

Multiply the first equation by i and add the result to the second to get

$$0 = \lambda(x_2 + ix_1).$$

Theorem 1.1.2 ensures that there are two cases: either $\lambda = 0$, or $\lambda \neq 0$ and $x_2 + ix_1 = 0$. In the first case, the original system reduces to the single equation $x_2 = ix_1$, so $(0, [1 \ i]^T)$ is an eigenpair of A. In the second case, $x_2 = -ix_1$, to which $[1 \ -i]^T$ is a solution. Substitute this into the original system to find that $\lambda = 2$. Thus, $(2, [1 \ -i]^T)$ is an eigenpair of A.

Example 8.1.14 Consider

$$A = \begin{bmatrix} 0 & -1 \\ 1 & 0 \end{bmatrix}.$$

Let $\mathbf{x} = [x_1 \ x_2]^T$ be nonzero and examine (8.1.2). We obtain

$$-x_2 = \lambda x_1, \tag{8.1.15}$$
$$x_1 = \lambda x_2. \tag{8.1.16}$$

If $x_1 \neq 0$, substitute the first equation into the second and obtain $x_1 = -\lambda^2 x_1$, from which we conclude that any eigenvalue λ of A satisfies $\lambda^2 = -1$. If $x_2 \neq 0$, then substitute

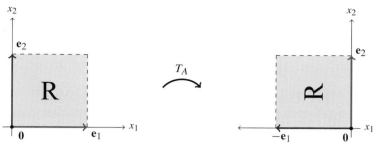

Figure 8.5 The linear transformation $T_A : \mathbb{R}^2 \to \mathbb{R}^2$ induced by the matrix from Example 8.1.14 is a rotation around the origin through an angle of $\frac{\pi}{2}$. A has no *real* eigenvalues, although it does have the non-real eigenpairs $(i, [1 \ -i]^\mathsf{T})$ and $(-i, [1 \ i]^\mathsf{T})$.

(8.1.16) into (8.1.15) to obtain the same conclusion. It follows that A has no real eigenvalues (see Figure 8.5). The equation $\lambda^2 = -1$ does have the two non-real solutions $\lambda_{\pm} = \pm i$. Substitute $\lambda = i$ into (8.1.16) and (8.1.15). Conclude that $-x_2 = ix_1$ and $x_1 = ix_2$. These two equations are multiples of each other; multiply the first equation by $-i$ to obtain the second. A nonzero solution to these equations is $[1 \ -i]^\mathsf{T}$, so $(i, [1 \ -i]^\mathsf{T})$ is an eigenpair of A. Similarly, $(-i, [1 \ i]^\mathsf{T})$ is an eigenpair of A. These two eigenpairs are complex conjugates; see P.8.4 for an explanation of this phenomenon.

The preceding example illustrates that eigenvalues and eigenvectors of a real matrix need not be real. That is one reason why complex numbers play a central role in linear algebra. Example 8.1.14 also suggests that the eigenvalues of a square matrix might be obtained by finding the zeros of an associated polynomial. We investigate this possibility in the following chapter.

The following variations on Definition 8.1.1 are useful in our exposition.

Theorem 8.1.17 *Let $A \in \mathsf{M}_n$ and let $\lambda \in \mathbb{C}$. The following statements are equivalent:*

(a) λ *is an eigenvalue of A.*

(b) $A\mathbf{x} = \lambda\mathbf{x}$ *for some nonzero $\mathbf{x} \in \mathbb{C}^n$.*

(c) $(A - \lambda I)\mathbf{x} = \mathbf{0}$ *has a nontrivial solution.*

(d) $A - \lambda I$ *is not invertible.*

(e) $A^\mathsf{T} - \lambda I$ *is not invertible.*

(f) λ *is an eigenvalue of A^T.*

Proof (a) $\Leftrightarrow$ (b) This is Definition 8.1.1.

(b) $\Leftrightarrow$ (c) These are restatements of each other.

(c) $\Leftrightarrow$ (d) See Corollary 2.5.4.

(d) $\Leftrightarrow$ (e) See Corollary 3.2.6.

(e) $\Leftrightarrow$ (f) Follows from (a) $\Leftrightarrow$ (d) applied to A^T. $\qquad\qquad\square$

Corollary 8.1.18 *Let $A \in M_n$. Then A is invertible if and only if 0 is not an eigenvalue of A.*

Proof This follows from the equivalence of (a) and (d) in the preceding theorem. $\square$

8.2 Every Square Matrix Has an Eigenvalue

The title of this section makes our aim clear. We want to prove that every square matrix has an eigenvalue. The relationship between matrices, eigenvalues, and polynomials is crucial to this endeavor. For a polynomial

$$p(z) = c_k z^k + c_{k-1} z^{k-1} + \cdots + c_1 z + c_0$$

and $A \in M_n$,

$$p(A) = c_k A^k + c_{k-1} A^{k-1} + \cdots + c_1 A + c_0 I; \tag{8.2.1}$$

see Section 0.8.

Example 8.2.2 The matrix

$$A = \begin{bmatrix} 0 & -1 \\ 1 & 0 \end{bmatrix}$$

from Example 8.1.14 satisfies $A^2 = -I$. The linear transformation induced by A is a rotation of $\mathbb{R}^2$ about the origin through an angle of $\frac{\pi}{2}$; see Figure 8.5. Thus, A^2 induces a rotation through an angle of π, which is represented by the matrix $-I$. If $p(z) = z^2 + 1$, then $p(A) = A^2 + I = 0$.

The preceding example motivates the following definition.

Definition 8.2.3 If $A \in M_n$ and p is a polynomial such that $p(A) = 0$, then p *annihilates* A (p is an *annihilating polynomial for A*).

The following lemma shows that each $A \in M_n$ is annihilated by some nonconstant polynomial; it relies on the fact that M_n is a vector space of dimension n^2.

Lemma 8.2.4 *Let $A \in M_n$. There is a nonconstant polynomial p of degree at most n^2 that annihilates A; if A is real, p may be chosen to have real coefficients.*

Proof The $\mathbb{F}$-vector space $M_n(\mathbb{F})$ has dimension n^2, so the $n^2 + 1$ matrices $I, A, A^2, \ldots, A^{n^2}$ are linearly dependent. Consequently, there are $c_0, c_1, c_2, \ldots, c_{n^2} \in \mathbb{F}$, not all zero, such that

$$c_0 I + c_1 A + c_2 A^2 + \cdots + c_{n^2} A^{n^2} = 0. \tag{8.2.5}$$

If $c_1 = c_2 = \cdots = c_{n^2} = 0$, then $c_0 I = 0$ and $c_0 = 0$, which is a contradiction. Let $r = \max\{k : 1 \leq k \leq n^2 \text{ and } c_k \neq 0\}$. Then $c_r \neq 0$ and

$$c_0 I + c_1 A + c_2 A^2 + \cdots + c_r A^r = 0,$$

so the nonconstant polynomial

$$p(z) = c_r z^r + c_{r-1} z^{r-1} + \cdots + c_1 z + c_0$$

has degree at most n^2 and annihilates A. Its coefficients belong to $\mathbb{F}$. $\square$

In the 2×2 case, there is an explicit polynomial p such that $p(A) = 0$; it is a special case of a general principle that is explained in Section 10.2.

Example 8.2.6 Consider the 2×2 matrix

$$A = \begin{bmatrix} a & b \\ c & d \end{bmatrix}.$$

The proof of Lemma 8.2.4 suggests a linear relationship among the matrices I, A, A^2, A^3, A^4. Compute

$$A^2 = \begin{bmatrix} a & b \\ c & d \end{bmatrix} \begin{bmatrix} a & b \\ c & d \end{bmatrix} = \begin{bmatrix} a^2 + bc & ab + bd \\ ac + cd & bc + d^2 \end{bmatrix} = \begin{bmatrix} a^2 + bc & b(a+d) \\ c(a+d) & bc + d^2 \end{bmatrix},$$

so

$$A^2 - (a+d)A = \begin{bmatrix} a^2 + bc & b(a+d) \\ c(a+d) & bc + d^2 \end{bmatrix} - \begin{bmatrix} a(a+d) & b(a+d) \\ c(a+d) & d(a+d) \end{bmatrix} = \begin{bmatrix} bc - ad & 0 \\ 0 & bc - ad \end{bmatrix},$$

from which it follows that

$$A^2 - (a+d)A + (ad - bc)I = 0. \tag{8.2.7}$$

Thus,

$$p(z) = z^2 - (\operatorname{tr} A)z + \det A = 0$$

annihilates A.

Our interest in polynomials that annihilate a given matrix stems from the following fact.

Lemma 8.2.8 *Let $A \in \mathsf{M}_n$ and let p be a nonconstant polynomial such that $p(A) = 0$. Then some root of $p(z) = 0$ is an eigenvalue of A.*

Proof Suppose that

$$p(z) = c_k z^k + c_{k-1} z^{k-1} + \cdots + c_1 z + c_0, \quad c_k \neq 0, \tag{8.2.9}$$

annihilates A and factor it as

$$p(z) = c_k (z - \lambda_1)(z - \lambda_2) \cdots (z - \lambda_k). \tag{8.2.10}$$

Equality of the two representations (8.2.9) and (8.2.10), together with the fact that powers of A commute, ensure that we can rewrite $p(A) = 0$ in the factored form

$$0 = p(A) = c_k (A - \lambda_1 I)(A - \lambda_2 I) \cdots (A - \lambda_k I). \tag{8.2.11}$$

Since $p(A) = 0$ is not invertible, at least one of the factors $A - \lambda_i I$ is not invertible. Thus, λ_i is an eigenvalue of A by Theorem 8.1.17. $\square$

Example 8.2.12 The identity (8.2.7) tells us that the matrix

$$A = \begin{bmatrix} 1 & 1 \\ 1 & 1 \end{bmatrix}$$

from Example 8.1.6, whose eigenvalues are 0 and 2, is annihilated by the polynomial $p(z) = z^2 - 2z = z(z - 2)$. In this case, the roots of $p(z) = 0$ are the eigenvalues of A. However, not every zero of a polynomial that annihilates A need be an eigenvalue of A. For example, $q(z) = z^3 - 3z^2 + 2z = z(z - 1)(z - 2) = (z - 1)p(z)$ satisfies

$$q(A) = A^3 - 3A^2 + 2A = (A - I)p(A) = (A - I)0 = 0,$$

but $q(1) = 0$ and 1 is not an eigenvalue of A.

Example 8.2.13 Example 8.2.6 ensures that $p(z) = z^2 - (a + d)z + (ad - bc)$ annihilates

$$A = \begin{bmatrix} a & b \\ c & d \end{bmatrix}.$$

The roots of $p(z) = 0$ are

$$\lambda_\pm = \frac{a + d \pm \sqrt{s}}{2}, \quad s = (a - d)^2 + 4bc. \tag{8.2.14}$$

Lemma 8.2.8 ensures that at least one of λ_+ and λ_- is an eigenvalue of A. If $s = 0$, then $\lambda_+ = \lambda_- = \frac{1}{2}(a + d)$ is an eigenvalue. If $s \neq 0$, λ_+ is an eigenvalue, but λ_- is not, then Theorem 8.1.17 ensures that $A - \lambda_- I$ is invertible. From the identity

$$0 = p(A) = (A - \lambda_- I)(A - \lambda_+ I),$$

we conclude that

$$0 = (A - \lambda_- I)^{-1}p(A) = A - \lambda_+ I.$$

Consequently, $A = \lambda_+ I$, so $b = c = 0$, $a = d$, and $s = 0$, which is a contradiction. If $s \neq 0$, λ_- is an eigenvalue of A, but λ_+ is not, then similar reasoning leads to another contradiction. Thus, both λ_+ and λ_- are eigenvalues of A.

The following theorem is the main result of this section.

Theorem 8.2.15 *Every square matrix has an eigenvalue.*

Proof Let $A \in M_n$. Lemma 8.2.4 ensures that there is a nonconstant polynomial p such that $p(A) = 0$. Lemma 8.2.8 says that among the roots of $p(z) = 0$, at least one is an eigenvalue of A. $\square$

Although every square matrix has at least one eigenvalue, there is much to learn. For example, (8.2.14) is a formula for eigenvalues of a 2×2 matrix, but are these the only eigenvalues? How many eigenvalues can an $n \times n$ matrix have?

8.3 How Many Eigenvalues are There?

Definition 8.3.1 The set of eigenvalues of $A \in M_n$ is the *spectrum* of A, denoted by spec A.

The spectrum of a matrix is a *set*, so it is characterized by its distinct elements. For example, the sets $\{1, 2\}$ and $\{1, 1, 2\}$ are identical.

Theorem 8.2.15 says that the spectrum of a square matrix is nonempty. How many elements are in spec A? The following lemma helps answer this question.

Lemma 8.3.2 *Let* $(\lambda, \mathbf{x})$ *be an eigenpair of* $A \in M_n$ *and let* p *be a polynomial. Then*

$$p(A)\mathbf{x} = p(\lambda)\mathbf{x},$$

that is, $(p(\lambda), \mathbf{x})$ *is an eigenpair of* $p(A)$.

Proof If $(\lambda, \mathbf{x})$ is an eigenpair of $A \in M_n$, proceed by induction to show that $A^j\mathbf{x} = \lambda^j\mathbf{x}$ for all $j \geq 0$. For the base case $j = 0$, observe that $A^0 = I$ and $\lambda^0 = 1$. For the inductive step, suppose that $A^j\mathbf{x} = \lambda^j\mathbf{x}$ for some $j \geq 0$. Then

$$A^{j+1}\mathbf{x} = A(A^j\mathbf{x}) = A(\lambda^j\mathbf{x}) = \lambda^j(A\mathbf{x}) = \lambda^j(\lambda\mathbf{x}) = \lambda^{j+1}\mathbf{x},$$

and the induction is complete. If p is given by (8.2.9), then

$$
\begin{aligned}
p(A)\mathbf{x} &= (c_k A^k + c_{k-1}A^{k-1} + \cdots + c_1 A + c_0 I)\mathbf{x} \\
&= c_k A^k\mathbf{x} + c_{k-1}A^{k-1}\mathbf{x} + \cdots + c_1 A\mathbf{x} + c_0 I\mathbf{x} \\
&= c_k \lambda^k\mathbf{x} + c_{k-1}\lambda^{k-1}\mathbf{x} + \cdots + c_1 \lambda\mathbf{x} + c_0\mathbf{x} \\
&= (c_k \lambda^k + c_{k-1}\lambda^{k-1} + \cdots + c_1 \lambda + c_0)\mathbf{x} \\
&= p(\lambda)\mathbf{x}.
\end{aligned}
$$
$\square$

One application of the preceding lemma is to provide a complement to Lemma 8.2.8. If $A \in M_n$ and p is a polynomial that annihilates A, then not only is some root of $p(z) = 0$ an eigenvalue, every eigenvalue of A is a root of $p(z) = 0$. Here is a formal statement of this observation.

Theorem 8.3.3 *Let* $A \in M_n$ *and let* p *be a polynomial that annihilates* A.

(a) *Every eigenvalue of* A *is a root of* $p(z) = 0$.

(b) *A has finitely many different eigenvalues.*

Proof (a) Let $(\lambda, \mathbf{x})$ be an eigenpair of A. Then

$$\mathbf{0} = 0\mathbf{x} = p(A)\mathbf{x} = p(\lambda)\mathbf{x},$$

so $p(\lambda) = 0$.

(b) Lemma 8.2.4 ensures that there is a polynomial of degree at most n^2 that annihilates A. Such a polynomial has at most n^2 zeros, so A has at most n^2 different eigenvalues. $\square$

The next application of Lemma 8.3.2 leads to a sharp bound for the number of different eigenvalues of a matrix.

Theorem 8.3.4 *Let* $(\lambda_1, \mathbf{x}_1), (\lambda_2, \mathbf{x}_2), \ldots, (\lambda_d, \mathbf{x}_d)$ *be eigenpairs of* $A \in \mathsf{M}_n$, *in which* $\lambda_1, \lambda_2, \ldots, \lambda_d$ *are distinct. Then* $\mathbf{x}_1, \mathbf{x}_2, \ldots, \mathbf{x}_d$ *are linearly independent.*

Proof The Lagrange Interpolation Theorem (Theorem 0.7.6) ensures that there are polynomials $p_1, p_2, \ldots, p_d$ such that

$$
p_i(\lambda_j) = \begin{cases} 1 & \text{if } i = j, \\ 0 & \text{if } i \neq j, \end{cases}
$$

for each $i = 1, 2, \ldots, d$. Suppose that $c_1, c_2, \ldots, c_d$ are scalars and

$$
c_1 \mathbf{x}_1 + c_2 \mathbf{x}_2 + \cdots + c_d \mathbf{x}_d = \mathbf{0}.
$$

We must show that each $c_i = 0$. Theorem 8.3.2 ensures that $p_i(A)\mathbf{x}_j = p_i(\lambda_j)\mathbf{x}_j$, so for each $i = 1, 2, \ldots, d$,

$$
\begin{aligned}
\mathbf{0} = p_i(A)\mathbf{0} &= p_i(A)(c_1 \mathbf{x}_1 + c_2 \mathbf{x}_2 + \cdots + c_d \mathbf{x}_d) \\
&= c_1 p_i(A)\mathbf{x}_1 + c_2 p_i(A)\mathbf{x}_2 + \cdots + c_d p_i(A)\mathbf{x}_d \\
&= c_1 p_i(\lambda_1)\mathbf{x}_1 + c_2 p_i(\lambda_2)\mathbf{x}_2 + \cdots + c_d p_i(\lambda_d)\mathbf{x}_d \\
&= c_i \mathbf{x}_i.
\end{aligned}
$$

Therefore, each $c_i = 0$. $\qquad\qquad\qquad\qquad\qquad\qquad\qquad\qquad\qquad\qquad\qquad\square$

Corollary 8.3.5 *Each* $A \in \mathsf{M}_n$ *has at most* n *distinct eigenvalues. If* A *has* n *distinct eigenvalues, then* $\mathbb{C}^n$ *has a basis consisting of eigenvectors of* A.

Proof The preceding theorem says that if $(\lambda_1, \mathbf{x}_1), (\lambda_2, \mathbf{x}_2), \ldots, (\lambda_d, \mathbf{x}_d)$ are eigenpairs of A and $\lambda_1, \lambda_2, \ldots, \lambda_d$ are distinct, then $\mathbf{x}_1, \mathbf{x}_2, \ldots, \mathbf{x}_d$ are linearly independent. Consequently, $d \leq n$. If $d = n$, then $\mathbf{x}_1, \mathbf{x}_2, \ldots, \mathbf{x}_n$ is a maximal linearly independent list in $\mathbb{C}^n$, so it is a basis; see Corollary 2.2.8. $\qquad\qquad\qquad\qquad\qquad\qquad\square$

Definition 8.3.6 If $A \in \mathsf{M}_n$ has n distinct eigenvalues, then A has *distinct eigenvalues*.

Example 8.3.7 The formula (8.2.14) ensures that

$$
A = \begin{bmatrix} 9 & 8 \\ 2 & -6 \end{bmatrix} \tag{8.3.8}
$$

has eigenvalues $\lambda_+ = 10$ and $\lambda_- = -7$. Corollary 8.3.5 says that these are the only eigenvalues of A, so $\operatorname{spec} A = \{10, -7\}$ and A has distinct eigenvalues. To find corresponding eigenvectors, solve the homogeneous systems

$$
(A - 10I)\mathbf{x} = \mathbf{0} \quad \text{and} \quad (A + 7I)\mathbf{x} = \mathbf{0}.
$$

This results in the eigenpairs $(10, [8\ 1]^T)$ and $(-7, [-1\ 2]^T)$ for A and the basis $\beta = [8\ 1]^T$, $[-1\ 2]^T$ for $\mathbb{C}^2$. Now use β to construct the (necessarily invertible) matrix

$$S = \begin{bmatrix} 8 & -1 \\ 1 & 2 \end{bmatrix}.$$

With respect to the standard basis of $\mathbb{C}^2$, the basis representation of $T_A : \mathbb{C}^2 \to \mathbb{C}^2$ is A; see Definition 2.3.9. Its representation with respect to β (see Corollary 2.4.17) is

$$
\begin{aligned}
S^{-1}AS &= \frac{1}{17} \begin{bmatrix} 2 & 1 \\ -1 & 8 \end{bmatrix} \begin{bmatrix} 9 & 8 \\ 2 & -6 \end{bmatrix} \begin{bmatrix} 8 & -1 \\ 1 & 2 \end{bmatrix} \\
&= \frac{1}{17} \begin{bmatrix} 2 & 1 \\ -1 & 8 \end{bmatrix} \begin{bmatrix} 80 & 7 \\ 10 & -14 \end{bmatrix} = \frac{1}{17} \begin{bmatrix} 170 & 0 \\ 0 & -119 \end{bmatrix} \\
&= \begin{bmatrix} 10 & 0 \\ 0 & -7 \end{bmatrix}.
\end{aligned}
$$

Thus, the existence of a basis of eigenvectors of A implies that A is similar to a diagonal matrix, in which the diagonal entries are the eigenvalues of A. An equivalent statement is that the β-basis representation of the linear operator T_A is a diagonal matrix. We have more to say about this in Section 9.4.

Example 8.3.9 The matrix

$$A_\theta = \begin{bmatrix} \cos\theta & -\sin\theta \\ \sin\theta & \cos\theta \end{bmatrix}, \quad 0 < \theta \le \frac{\pi}{2},$$

induces a counterclockwise rotation of $\mathbb{R}^2$ through θ radians around the origin (see Figure 8.6). The formula (8.2.14) ensures that

$$\lambda_\pm = \frac{2\cos\theta \pm \sqrt{-4\sin^2\theta}}{2} = \cos\theta \pm i\sin\theta = e^{\pm i\theta}$$

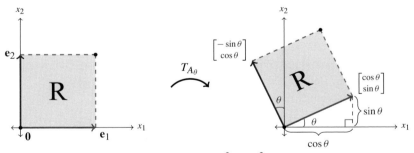

Figure 8.6 The linear transformation $T_{A_\theta} : \mathbb{R}^2 \to \mathbb{R}^2$ induced by the matrix A_θ from Example 8.3.9 is a counterclockwise rotation around the origin through θ radians, $0 < \theta \le \frac{\pi}{2}$. The matrix A_θ has the distinct non-real eigenvalues $\lambda_\pm = e^{\pm i\theta}$ and eigenpairs $(e^{i\theta}, [1\ -i]^T)$ and $(e^{-i\theta}, [1\ i]^T)$.

are distinct eigenvalues of A. Corollary 8.3.5 permits us to conclude that $\operatorname{spec} A = \{e^{i\theta}, e^{-i\theta}\}$. To find eigenvectors corresponding to $\lambda_+ = e^{i\theta}$, compute nonzero solutions of the homogeneous system $(A - \lambda_+ I)\mathbf{x} = \mathbf{0}$:

$$-i(\sin\theta)x_1 - (\sin\theta)x_2 = 0,$$
$$(\sin\theta)x_1 - i(\sin\theta)x_2 = 0,$$

in which $\mathbf{x} = [x_1 \ x_2]^{\mathsf{T}}$. Since i times the first equation equals the second equation and $\sin\theta \neq 0$, this system is equivalent to the single equation $x_1 = ix_2$. Thus $(e^{i\theta}, [1 \ -i]^{\mathsf{T}})$ is an eigenpair of A. A similar computation (or P.8.4) shows that $(e^{-i\theta}, [1 \ i]^{\mathsf{T}})$ is an eigenpair of A. Observe that $[1 \ i]^{\mathsf{T}}, [1 \ -i]^{\mathsf{T}}$ is a basis for $\mathbb{C}^2$. It is noteworthy that the eigenvalues of A_θ depend on $\theta \in (0, \frac{\pi}{2})$, but the associated eigenvectors do not. Each of the vectors $[1 \ \pm i]^{\mathsf{T}}$ is an eigenvector of all the matrices A_θ. Corollary 8.5.4 provides an explanation for this phenomenon.

Example 8.3.10 If $n \geq 2$, then I_n does not have distinct eigenvalues. Nevertheless, $\mathbb{C}^n$ has a basis consisting of eigenvectors of I_n. Any basis of $\mathbb{C}^n$ comprises eigenvectors of I_n since any nonzero vector in $\mathbb{C}^n$ is an eigenvector of I_n associated with the eigenvalue $\lambda = 1$.

Definition 8.3.11 Let $A \in \mathsf{M}_n$ and let $\lambda \in \mathbb{C}$. Then

$$\mathcal{E}_\lambda(A) = \operatorname{null}(A - \lambda I)$$

is the *eigenspace* of A *associated with* λ.

The term eigenspace suggests that $\mathcal{E}_\lambda(A)$ is a subspace of $\mathbb{C}^n$, which it is; the null space of any $n \times n$ matrix is a subspace of $\mathbb{C}^n$. If $\lambda \notin \operatorname{spec} A$, then $A - \lambda I$ is invertible (Theorem 8.1.17) and hence $\mathcal{E}_\lambda(A) = \{\mathbf{0}\}$ is the zero subspace. If $\lambda \in \operatorname{spec} A$, then $\mathcal{E}_\lambda(A)$ consists of the zero vector and all eigenvectors of A associated with λ (Theorem 8.1.17). Thus, $\mathcal{E}_\lambda(A) \neq \{\mathbf{0}\}$ if and only if $\lambda \in \operatorname{spec} A$.

If $\lambda \in \operatorname{spec} A$, every nonzero vector in $\mathcal{E}_\lambda(A)$ is an eigenvector of A associated with λ. Consequently, any basis of $\mathcal{E}_\lambda(A)$ is a linearly independent list of eigenvectors of A associated with λ.

Definition 8.3.12 Let $A \in \mathsf{M}_n$. The *geometric multiplicity* of λ as an eigenvalue of A is the dimension of the subspace $\mathcal{E}_\lambda(A)$.

If $\lambda \notin \operatorname{spec} A$, its geometric multiplicity is 0. If $A \in \mathsf{M}_n$, the geometric multiplicity of $\lambda \in \operatorname{spec} A$ is between 1 and n because $\mathcal{E}_\lambda(A)$ is a nonzero subspace of $\mathbb{C}^n$.

The dimension theorem (Corollary 2.5.4) says that

$$\dim \operatorname{null}(A - \lambda I) + \dim \operatorname{col}(A - \lambda I) = n,$$

that is,

$$\dim \mathcal{E}_\lambda(A) + \operatorname{rank}(A - \lambda I) = n.$$

Therefore, the geometric multiplicity of λ as an eigenvalue of $A \in \mathsf{M}_n$ is

$$\dim \mathcal{E}_\lambda(A) = n - \operatorname{rank}(A - \lambda I). \tag{8.3.13}$$

Example 8.3.14 Consider

$$A = \begin{bmatrix} 1 & 1 \\ 0 & 1 \end{bmatrix}.$$

The results of Example 8.1.10 show that $\operatorname{spec} A = \{1\}$ and $\mathcal{E}_1(A) = \operatorname{span}\{[1 \ 0]^\mathsf{T}\}$. Thus, the geometric multiplicity of 1 as an eigenvalue of A is one. As a check, observe that rank $(A - I) = 1$, so (8.3.13) says that the eigenvalue $\lambda = 1$ has geometric multiplicity 1.

Example 8.3.15 Consider

$$A = \begin{bmatrix} 1 & i \\ i & -1 \end{bmatrix}.$$

Let $\mathbf{x} = [x_1 \ x_2]^\mathsf{T}$ be nonzero and suppose that $A\mathbf{x} = \lambda\mathbf{x}$. Multiply the second equation in the system

$$x_1 + ix_2 = \lambda x_1, \tag{8.3.16}$$
$$ix_1 - x_2 = \lambda x_2, \tag{8.3.17}$$

by i and add it to the first. The result is that $0 = \lambda(x_1 + ix_2)$. Thus, either $\lambda = 0$ or $x_1 + ix_2 = 0$. If $\lambda = 0$, then (8.3.16) reduces to $x_1 + ix_2 = 0$. If $x_1 + ix_2 = 0$, then (8.3.16) and (8.3.17) reduce to $0 = \lambda x_1$ and $0 = -i\lambda x_2$. Then $\lambda = 0$ since $\mathbf{x} \neq \mathbf{0}$. Thus, $\operatorname{spec} A = \{0\}$ and $\mathcal{E}_0(A) = \operatorname{span}\{[1 \ i]^\mathsf{T}\}$. The geometric multiplicity of 0 as an eigenvalue of A is one. As a check, observe that rank $A = 1$ (its second column is i times its first column), so (8.3.13) says that the eigenvalue $\lambda = 0$ has geometric multiplicity 1.

Theorem 8.1.17 says that a square matrix and its transpose have the same eigenvalues. The following lemma refines this observation.

Lemma 8.3.18 *Let $A \in \mathsf{M}_n$ and let $\lambda \in \operatorname{spec} A$. Then $\dim \mathcal{E}_\lambda(A) = \dim \mathcal{E}_\lambda(A^\mathsf{T})$.*

Proof Compute

$$\dim \mathcal{E}_\lambda(A) = \dim \operatorname{null}(A - \lambda I) = n - \operatorname{rank}(A - \lambda I)$$
$$= n - \operatorname{rank}(A - \lambda I)^\mathsf{T} = n - \operatorname{rank}(A^\mathsf{T} - \lambda I)$$
$$= \dim \operatorname{null}(A^\mathsf{T} - \lambda I) = \dim \mathcal{E}_\lambda(A^\mathsf{T}). \qquad \square$$

8.4 Where are the Eigenvalues?

As illustrated in Examples 8.1.3 and 8.1.4, the diagonal entries of a diagonal matrix are eigenvalues; associated eigenvectors are the standard basis vectors. If we modify a diagonal matrix by inserting some nonzero off-diagonal entries, can we give quantitative bounds for how much each diagonal entry can differ from an eigenvalue? The following theorem provides an answer to this question.

Theorem 8.4.1 (Geršgorin) *If $n \geq 2$ and $A = [a_{ij}] \in \mathsf{M}_n$, then*

$$\operatorname{spec} A \subseteq G(A) = \bigcup_{k=1}^{n} G_k(A), \tag{8.4.2}$$

in which

$$G_k(A) = \left\{ z \in \mathbb{C} : |z - a_{kk}| \leq R'_k(A) \right\} \tag{8.4.3}$$

is a disk in the complex plane whose center is at the point a_{kk} and whose radius is the deleted absolute row sum

$$R'_k(A) = \sum_{j \neq k} |a_{kj}|. \tag{8.4.4}$$

Proof Let $(\lambda, \mathbf{x})$ be an eigenpair of $A = [a_{ij}] \in \mathsf{M}_n$, let $\mathbf{x} = [x_i]$, and assume that $n \geq 2$. The identity (8.1.2) is equivalent to the family of equations

$$\lambda x_i = \sum_{j=1}^{n} a_{ij} x_j = a_{ii} x_i + \sum_{j \neq i} a_{ij} x_j, \quad i = 1, 2, \ldots, n,$$

which we express as

$$(\lambda - a_{ii}) x_i = \sum_{j \neq i} a_{ij} x_j, \quad i = 1, 2, \ldots, n. \tag{8.4.5}$$

Let $k \in \{1, 2, \ldots, n\}$ be any index such that $|x_k| = \|\mathbf{x}\|_\infty$, which is nonzero since $\mathbf{x} \neq \mathbf{0}$; see (4.6.5). Set $i = k$ in (8.4.5) and divide by $\|\mathbf{x}\|_\infty$ to obtain

$$\lambda - a_{kk} = \sum_{j \neq k} a_{kj} \frac{x_j}{\|\mathbf{x}\|_\infty}. \tag{8.4.6}$$

The choice of k guarantees that each of the quotients in (8.4.6) has modulus at most 1. The triangle inequality ensures that

$$|\lambda - a_{kk}| = \left| \sum_{j \neq k} a_{kj} \frac{x_j}{\|\mathbf{x}\|_\infty} \right| \leq \sum_{j \neq k} |a_{kj}| \left| \frac{x_j}{\|\mathbf{x}\|_\infty} \right| \leq \sum_{j \neq k} |a_{kj}| = R'_k(A) \tag{8.4.7}$$

and hence $\lambda \in G_k(A) \subseteq G(A)$. $\qquad\qquad\square$

The disks $G_k(A)$ are *Geršgorin disks*; their boundaries are *Geršgorin circles*. The set $G(A)$ defined in (8.4.2) is the *Geršgorin region* of A. Each Geršgorin disk $G_k(A)$ is contained in the disk centered at the origin whose radius

$$R_k(A) = \sum_{j=1}^{n} |a_{kj}| = R'_k(A) + |a_{kk}|$$

is the kth *absolute row sum* of A. Therefore, the Geršgorin region $G(A)$ (and hence all of the eigenvalues of A) is contained in the single disk centered at the origin whose radius

$$R_{\max}(A) = \max_{1 \leq k \leq n} R_k(A) = \max \left\{ \sum_{j=1}^{n} |a_{kj}| : 1 \leq k \leq n \right\} \tag{8.4.8}$$

is the largest absolute row sum of A.

If we apply the preceding theorem to A^T, we obtain Geršgorin disks centered at the diagonal entries of A whose radii are the deleted absolute row sums of A^T.

Corollary 8.4.9 *If $A \in \mathsf{M}_n$, then*

$$\operatorname{spec} A \subseteq G(A) \cap G(A^\mathsf{T}) \subseteq \left\{ z \in \mathbb{C} : |z| \leq \min \left\{ R_{\max}(A), R_{\max}(A^\mathsf{T}) \right\} \right\}. \tag{8.4.10}$$

Proof Theorem 8.4.1 says that $\operatorname{spec} A \subseteq G(A)$, and Theorem 8.1.17 ensures that $\operatorname{spec} A = \operatorname{spec} A^\mathsf{T} \subseteq G(A^\mathsf{T})$. Now apply (8.4.8) to A and A^T. $\qquad\square$

Example 8.4.11 Consider the matrix

$$A = \begin{bmatrix} 1 & 1 \\ 0 & 2 \end{bmatrix} \tag{8.4.12}$$

from Example 8.1.8, for which $\operatorname{spec} A = \{1, 2\}$. Then

$$R_1'(A) = 1, \qquad R_2'(A) = 0, \qquad R_1'(A^\mathsf{T}) = 0, \quad \text{and} \quad R_2'(A^\mathsf{T}) = 1.$$

Consequently,

$$G_1(A) = \{z \in \mathbb{C} : |z - 1| \leq 1\}, \qquad G_1(A^\mathsf{T}) = \{1\},$$
$$G_2(A) = \{2\}, \qquad\qquad\qquad G_2(A^\mathsf{T}) = \{z \in \mathbb{C} : |z - 2| \leq 1\}.$$

The region $G(A) \cap G(A^\mathsf{T})$ contains $\operatorname{spec} A$ and is much smaller than either $G(A)$ or $G(A^\mathsf{T})$. Also,

$$R_1(A) = R_2(A) = R_1(A^\mathsf{T}) = 2 \quad \text{and} \quad R_2(A^\mathsf{T}) = 3,$$

so that $R_{\max}(A) = 2$ and $R_{\max}(A^\mathsf{T}) = 3$. Since $\min\{R_{\max}(A), R_{\max}(A^\mathsf{T})\} = 2$, $\operatorname{spec} A$ is contained in the disk $\{z \in \mathbb{C} : |z| \leq R_{\max}(A) = 2\}$; see Figure 8.7.

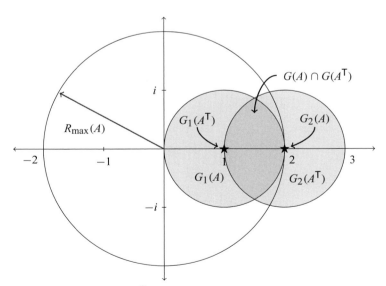

Figure 8.7 $G(A)$ and $G(A^\mathsf{T})$ for the matrix (8.4.12). The lens-shaped region is $G(A) \cap G(A^\mathsf{T})$. The eigenvalues $z = 1$ and $z = 2$ of A are denoted by the symbol $\star$.

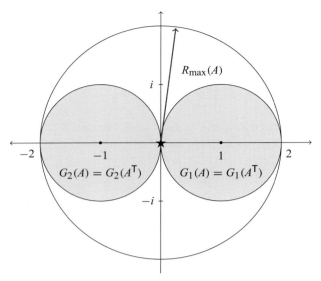

Figure 8.8 $G(A)$ for the matrix (8.4.14); $z = 0$ is an
eigenvalue; $R_{\max}(A) = 2$.

Example 8.4.13 The matrix

$$A = \begin{bmatrix} 1 & i \\ i & -1 \end{bmatrix} \tag{8.4.14}$$

in Example 8.3.15 has spec $A = \{0\}$. Here $R'_1(A) = R'_2(A) = 1$ and $R'_1(A^{\mathsf{T}}) = R'_2(A^{\mathsf{T}}) = 1$, so

$$G_1(A) = G_1(A^{\mathsf{T}}) = \{z \in \mathbb{C} : |z - 1| \le 1\}$$

and

$$G_2(A) = G_2(A^{\mathsf{T}}) = \{z \in \mathbb{C} : |z + 1| \le 1\}.$$

Also, $R_1(A) = R_2(A) = 2$ and $R_1(A^{\mathsf{T}}) = R_2(A^{\mathsf{T}}) = 2$, so $R_{\max}(A) = R_{\max}(A^{\mathsf{T}}) = 2$; see
Figure 8.8.

Example 8.4.15 Consider

$$A = \begin{bmatrix} 9 & 8 \\ 2 & -6 \end{bmatrix}, \tag{8.4.16}$$

for which spec $A = \{10, -7\}$; see Example 8.3.7. Then

$$R'_1(A) = 8, \qquad R'_2(A) = 2, \qquad R'_1(A^{\mathsf{T}}) = 2, \qquad R'_2(A^{\mathsf{T}}) = 8,$$

and hence

$$G_1(A) = \{z \in \mathbb{C} : |z - 9| \le 8\}, \qquad G_1(A^{\mathsf{T}}) = \{z \in \mathbb{C} : |z - 9| \le 2\},$$
$$G_2(A) = \{z \in \mathbb{C} : |z + 6| \le 2\}, \qquad G_2(A^{\mathsf{T}}) = \{z \in \mathbb{C} : |z + 6| \le 8\},$$

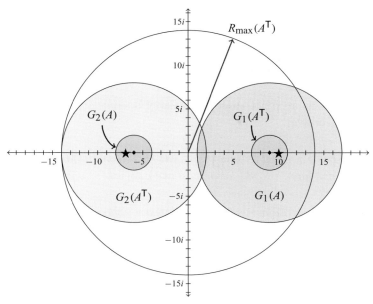

Figure 8.9 $G(A)$ and $G(A^\mathsf{T})$ for the matrix (8.4.16). The eigenvalues of A are $z = 10$ and $z = -7$.

$R_{\max}(A) = 17$, and $R_{\max}(A^\mathsf{T}) = 14$. Thus, spec $A \subseteq \{z \in \mathbb{C} : |z| \leq 14\}$ and, even better,

$$\text{spec } A \subseteq G(A) \cap G(A^\mathsf{T}) = G_2(A) \cup G_1(A^\mathsf{T}) \cup \big(G_1(A) \cap G_2(A^\mathsf{T})\big);$$

see Figure 8.9. The Geršgorin disks $G_1(A)$ and $G_2(A)$ do not contain 0, so 0 is not an eigenvalue of A. Corollary 8.1.18 ensures that A is invertible.

Definition 8.4.17 $A = [a_{ij}] \in \mathsf{M}_n$ is *diagonally dominant* if $|a_{kk}| \geq R'_k(A)$ for each $k = 1,$ $2, \ldots, n$; A is *strictly diagonally dominant* if $|a_{kk}| > R'_k(A)$ for each $k = 1, 2, \ldots, n$.

Corollary 8.4.18 *Let* $A \in \mathsf{M}_n$.

(a) *If A is strictly diagonally dominant, then it is invertible.*

(b) *Suppose that A has real eigenvalues and real nonnegative diagonal entries. If A is diagonally dominant, then all its eigenvalues are nonnegative. If A is strictly diagonally dominant, then all its eigenvalues are positive.*

Proof (a) The strict diagonal dominance of A ensures that $0 \notin G_k(A)$ for each $k = 1, 2, \ldots, n$. Theorem 8.4.1 tells us that 0 is not an eigenvalue of A and Corollary 8.1.18 says that A is invertible.

(b) The hypotheses ensure that each $G_k(A)$ is a disk in the right half plane. Theorem 8.4.1 ensures that the eigenvalues of A are in the right half plane. Since they are all real, they must be nonnegative. If A is strictly diagonally dominant, then it is invertible, so 0 is not an eigenvalue. $\qquad\square$

Example 8.4.19 The matrix in (8.4.16) is strictly diagonally dominant and invertible. The matrix (8.4.14) is diagonally dominant and not invertible. The matrix

$$\begin{bmatrix} 1 & i \\ 1 & 2 \end{bmatrix}$$

is diagonally dominant. It is not strictly diagonally dominant, but the following theorem ensures that it is invertible.

Theorem 8.4.20 *Let $n \geq 2$ and let $A = [a_{ij}] \in M_n$ be diagonally dominant. If $a_{ij} \neq 0$ for all $i, j \in \{1, 2, \ldots, n\}$ and $|a_{kk}| > R'_k(A)$ for at least one $k \in \{1, 2, \ldots, n\}$, then A is invertible.*

Proof Suppose that A is diagonally dominant and has no zero entries. If $0 \in \operatorname{spec} A$, then there is a nonzero $\mathbf{x} = [x_i] \in \mathbb{C}^n$ such that $A\mathbf{x} = \mathbf{0}$, that is,

$$-a_{ii}x_i = \sum_{j \neq i} a_{ij}x_j$$

for each $i = 1, 2, \ldots, n$. Let $k \in \{1, 2, \ldots, n\}$ be any index such that $|x_k| = \|\mathbf{x}\|_\infty$. Then

$$|a_{kk}| \|\mathbf{x}\|_\infty = |a_{kk}||x_k| = |-a_{kk}x_k| = \left| \sum_{j \neq i} a_{kj}x_j \right|$$

$$\leq \sum_{j \neq i} |a_{kj}||x_j| \leq \sum_{j \neq i} |a_{kj}| \|\mathbf{x}\|_\infty = R'_k(A)\|\mathbf{x}\|_\infty. \tag{8.4.21}$$

Since $\|\mathbf{x}\|_\infty \neq 0$, it follows that $|a_{kk}| \leq R'_k(A)$. However, $|a_{ii}| \geq R'_i(A)$ for every $i = 1, 2, \ldots, n$ (A is diagonally dominant), so $|a_{kk}| = R'_k(A)$ and the inequality in (8.4.21) is an equality. Therefore, each inequality

$$|a_{kj}| |x_j| \leq |a_{kj}| \|\mathbf{x}\|_\infty$$

is an equality, that is,

$$|a_{kj}||x_j| = |a_{kj}| \|\mathbf{x}\|_\infty, \qquad j = 1, 2, \ldots, n.$$

Because $a_{kj} \neq 0$ (A has no zero entries), it follows that $|x_j| = \|\mathbf{x}\|_\infty$ for each $j = 1, 2, \ldots, n$. The preceding argument shows that $|a_{jj}| = R'_j(A)$ for all $j = 1, 2, \ldots, n$.

If $|a_{kk}| > R'_k(A)$ for some $k \in \{1, 2, \ldots, n\}$, we conclude that $0 \notin \operatorname{spec} A$ and Theorem 8.1.17 ensures that A is invertible. $\qquad \square$

Example 8.4.22 The matrix

$$A = \begin{bmatrix} 3 & 1 & 1 \\ 0 & 1 & 1 \\ 0 & 1 & 1 \end{bmatrix} \tag{8.4.23}$$

is diagonally dominant and $|a_{11}| > R'_1(A)$. However, A has some zero entries and is not invertible. See Figure 8.10.

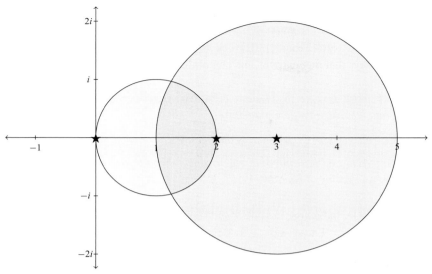

Figure 8.10 $G(A)$ for the matrix (8.4.23). A is diagonally dominant and $|a_{11}| > R'_1(A)$. Its eigenvalues are 0, 2, and 3.

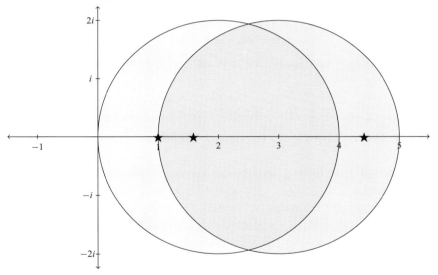

Figure 8.11 $G(A)$ for the matrix (8.4.25). A is diagonally dominant and $|a_{11}| > R'_1(A)$. To two decimal places, its eigenvalues are 1, $3 - \sqrt{2} = 1.59$, and $3 + \sqrt{2} = 4.41$.

Example 8.4.24 The matrix

$$A = \begin{bmatrix} 3 & 1 & 1 \\ 1 & 2 & 1 \\ 1 & 1 & 2 \end{bmatrix} \tag{8.4.25}$$

is diagonally dominant and $|a_{11}| > R'_1(A)$, so Theorem 8.4.20 ensures that it is invertible. See Figure 8.11.

8.5 Eigenvectors and Commuting Matrices

Commuting matrices often share properties of interest. The following generalization of Theorem 8.2.15 says that the matrices in a commuting family share a common eigenvector.

Theorem 8.5.1 *Let $k \geq 2$, let $A_1, A_2, \ldots, A_k \in M_n$, suppose that $A_i A_j = A_j A_i$ for all $i, j \in \{1, 2, \ldots, k\}$, and let λ be an eigenvalue of A_1.*

(a) *There is a nonzero $\mathbf{x}$ in $\mathcal{E}_\lambda(A_1)$ that is an eigenvector of each $A_2, A_3, \ldots, A_k$.*

(b) *If each A_i is real and has only real eigenvalues, then there is a real nonzero $\mathbf{x} \in \mathcal{E}_\lambda(A_1)$ that is an eigenvector of each $A_2, A_3, \ldots, A_k$.*

Proof (a) Proceed by induction. Let $k = 2$ and suppose that the geometric multiplicity of λ as an eigenvalue of A_1 is r. Let $\mathbf{x}_1, \mathbf{x}_2, \ldots, \mathbf{x}_r$ be a basis of $\mathcal{E}_\lambda(A_1)$ and let $X = [\mathbf{x}_1 \ \mathbf{x}_2 \ \ldots \ \mathbf{x}_r] \in M_{n \times r}$. For each $j = 1, 2, \ldots, r$,

$$A_1(A_2\mathbf{x}_j) = A_1 A_2 \mathbf{x}_j = A_2 A_1 \mathbf{x}_j = A_2 \lambda \mathbf{x}_j = \lambda(A_2 \mathbf{x}_j),$$

so $A_2 \mathbf{x}_j \in \mathcal{E}_\lambda(A_1)$. Consequently, $A_2 \mathbf{x}_j$ is a linear combination of the columns of X for each $j = 1, 2, \ldots, r$. Therefore, (3.1.21) ensures that there is a $C_2 \in M_r$ such that $A_2 X = XC_2$. Invoke Theorem 8.2.15 and let $(\mu, \mathbf{u})$ be an eigenpair of C_2. Observe that $X\mathbf{u} \neq \mathbf{0}$ because the columns of X are linearly independent and $\mathbf{u} \neq \mathbf{0}$. Thus,

$$A_2(X\mathbf{u}) = A_2 X \mathbf{u} = XC_2 \mathbf{u} = X\mu\mathbf{u} = \mu(X\mathbf{u}).$$

Finally,

$$A_1(X\mathbf{u}) = \lambda(X\mathbf{u})$$

because $X\mathbf{u} \in \mathcal{E}_\lambda(A_1)$. Then $\mathbf{x} = X\mathbf{u}$ is a common eigenvector of A_1 and A_2 with the asserted properties.

Suppose that for some $m \geq 2$, the theorem has been proved for commuting families of m or fewer matrices. Let $A_1, A_2, \ldots, A_{m+1} \in M_n$, suppose that $A_i A_j = A_j A_i$ for all $i, j \in \{1, 2, \ldots, m+1\}$, and let λ be an eigenvalue of A_1. Let $\mathbf{x}_1, \mathbf{x}_2, \ldots, \mathbf{x}_r$ be a basis for $\mathcal{E}_\lambda(A_1)$ and let $X = [\mathbf{x}_1 \ \mathbf{x}_2 \ \ldots \ \mathbf{x}_r] \in M_{n \times r}$. For each $j = 2, \ldots, m+1$, the same argument shows that there is a $C_j \in M_r$ such that $A_j X = XC_j$. Consequently,

$$A_i A_j X = A_i XC_j = XC_i C_j = A_j A_i X = A_j XC_i = XC_j C_i$$

and hence

$$X(C_i C_j - C_j C_i) = 0.$$

The linear independence of the columns of X ensures that $C_i C_j - C_j C_i = 0$, so the m matrices $C_2, C_3, \ldots, C_{m+1}$ are a commuting family. The induction hypothesis ensures the existence of a nonzero $\mathbf{u}$ that is an eigenvector of each $C_2, C_3, \ldots, C_{m+1}$. It follows that $\mathbf{x} = X\mathbf{u} \in \mathcal{E}_\lambda(A_1)$ is a common eigenvector of $A_1, A_2, \ldots, A_{m+1}$ with the required properties.

(b) Part (a) ensures that $A_1, A_2, \ldots, A_k$ have a common eigenvector $\mathbf{x} \in \mathbb{C}^n$. Write $\mathbf{x} = \mathbf{u} + i\mathbf{v}$ with $\mathbf{u}, \mathbf{v} \in \mathbb{R}^n$; at least one of $\mathbf{u}$ or $\mathbf{v}$ is nonzero. For each $j = 1, 2, \ldots, k$, there is a $\lambda_j \in \mathbb{R}$ such that

$$\lambda_j \mathbf{u} + i\lambda_j \mathbf{v} = \lambda_j \mathbf{x} = A_j \mathbf{x} = A_j \mathbf{u} + iA_j \mathbf{v}.$$

Therefore, $A_j \mathbf{u} = \lambda_j \mathbf{u}$ and $A_j \mathbf{v} = \lambda_j \mathbf{v}$ for all $j = 1, 2, \ldots, k$. If $\mathbf{u} \neq \mathbf{0}$, it is a common real eigenvector for $A_1, A_2, \ldots, A_k$; if $\mathbf{u} = \mathbf{0}$, then $\mathbf{v}$ is a common real eigenvector. $\qquad \square$

The preceding theorem concerns *finite* families of commuting matrices, but sometimes we encounter *infinite* families of commuting matrices.

Example 8.5.2 Consider the matrices A_θ in Example 8.3.9, one for each of the infinitely many values of θ in the interval $(0, \pi/2]$. A computation reveals that

$$A_\theta A_\phi = \begin{bmatrix} \cos\theta\cos\phi - \sin\theta\sin\phi & -(\sin\theta\cos\phi + \cos\theta\sin\phi) \\ \sin\theta\cos\phi + \cos\theta\sin\phi & \cos\theta\cos\phi - \sin\theta\sin\phi \end{bmatrix} = A_\phi A_\theta,$$

which reflects the geometric fact that any two rotations about the origin in $\mathbb{R}^2$ commute. Thus, $\mathcal{F} = \{A_\theta : 0 < \theta \leq \pi/2\}$ is an infinite commuting family. Moreover, each of the vectors $[1 \ \pm i]^{\mathsf{T}}$ is an eigenvector of every $A_\theta \in \mathcal{F}$ even though a key hypothesis of Theorem 8.5.1 is not satisfied. This is not an accident.

The key to extending Theorem 8.5.1 to infinite families is in the following lemma, which relies in an essential fashion on the finite dimensionality of M_n.

Lemma 8.5.3 *Let $\mathcal{F} \subseteq \mathsf{M}_n$ be a nonempty set of matrices.*

(a) *There are finitely many matrices in $\mathcal{F}$ whose span contains $\mathcal{F}$.*

(b) *Let $A_1, A_2, \ldots, A_k \in \mathsf{M}_n$. If a nonzero $\mathbf{x} \in \mathbb{C}^n$ is an eigenvector of each $A_1, A_2, \ldots, A_k$, then it is an eigenvector of each matrix in $\mathrm{span}\{A_1, A_2, \ldots, A_k\}$.*

Proof (a) If $\mathcal{F}$ is a finite set, there is nothing to prove. If $\mathcal{F}$ has infinitely many elements, then observe that $\mathcal{F} \subseteq \mathrm{span}\,\mathcal{F} \subseteq \mathsf{M}_n$. Theorem 2.2.9 ensures that there are at most n^2 elements of $\mathcal{F}$ that comprise a basis for $\mathrm{span}\,\mathcal{F}$.

(b) If $A_i \mathbf{x} = \lambda_i \mathbf{x}$ for each $i = 1, 2, \ldots, k$, then

$$\begin{aligned} (c_1 A_1 + c_2 A_2 + \cdots + c_k A_k)\mathbf{x} &= c_1 A_1 \mathbf{x} + c_2 A_2 \mathbf{x} + \cdots + c_k A_k \mathbf{x} \\ &= c_1 \lambda_1 \mathbf{x} + c_2 \lambda_2 \mathbf{x} + \cdots + c_k \lambda_k \mathbf{x} \\ &= (c_1 \lambda_1 + c_2 \lambda_2 + \cdots + c_k \lambda_k)\mathbf{x}, \end{aligned}$$

so $\mathbf{x}$ is an eigenvector of every matrix in $\mathrm{span}\{A_1, A_2, \ldots, A_k\}$. $\qquad \square$

Corollary 8.5.4 *Let $\mathcal{F} \subseteq \mathsf{M}_n$ be a nonempty set of commuting matrices, let $A \in \mathcal{F}$ be given, and let λ be an eigenvalue of A.*

(a) *Some nonzero vector in $\mathcal{E}_\lambda(A)$ is an eigenvector of every matrix in $\mathcal{F}$.*

(b) *If every matrix in $\mathcal{F}$ is real and has only real eigenvalues, then some real vector in $\mathcal{E}_\lambda(A)$ is an eigenvector of every matrix in $\mathcal{F}$.*

Proof Invoke Lemma 8.5.3.a to obtain finitely many matrices $A_2, A_3, \ldots, A_k \in \mathcal{F}$ whose span contains $\mathcal{F}$. Then

$$\mathcal{F} \subseteq \text{span}\{A_2, A_3, \ldots, A_k\} \subseteq \text{span}\{A, A_2, \ldots, A_k\}$$

and $A_i A_j = A_j A_i$ for all $i, j = 1, 2, \ldots, k$. Theorem 8.5.1 ensures that there is a nonzero vector $\mathbf{x}$ (real under the hypotheses of (b)) such that $A\mathbf{x} = \lambda \mathbf{x}$ and $\mathbf{x}$ is an eigenvector of each $A_2, A_3, \ldots, A_k$. Finally, Lemma 8.5.3.b ensures that $\mathbf{x}$ is an eigenvector of each matrix in $\text{span}\{A, A_2, \ldots, A_k\}$, which includes every matrix in $\mathcal{F}$. $\square$

In both Theorem 8.5.1 and Corollary 8.5.4, the initial choices of a matrix A and its eigenvalue λ are unrestricted. Once we have made these choices, all we know about the common eigenvector is that it is in $\mathcal{E}_\lambda(A)$. However, if $\dim \mathcal{E}_\lambda(A) = 1$, then it is unique up to a nonzero scalar.

Example 8.5.5 Among the matrices A_θ defined in Example 8.3.9,

$$A_{\frac{\pi}{2}} = \begin{bmatrix} 0 & -1 \\ 1 & 0 \end{bmatrix}$$

has eigenvalues $\lambda_\pm = \pm i$ and corresponding eigenvectors $\mathbf{x}_\pm = [1 \mp i]^\mathsf{T}$. Because the eigenspaces of $A_{\pi/2}$ are one dimensional, common eigenvectors for all the matrices A_θ can only be nonzero scalar multiples of $\mathbf{x}_\pm$.

Example 8.5.6 The matrices

$$A_1 = \begin{bmatrix} 1 & 0 & 0 \\ 0 & 1 & 0 \\ 0 & 0 & 2 \end{bmatrix} \quad \text{and} \quad A_2 = \begin{bmatrix} 3 & 1 & 0 \\ 0 & 3 & 0 \\ 0 & 0 & 1 \end{bmatrix}$$

commute and $\mathcal{E}_1(A_1) = \text{span}\{\mathbf{e}_1, \mathbf{e}_2\}$. However, not every nonzero vector in $\mathcal{E}_1(A_1)$ is an eigenvector of A_2. For example, $A_2\mathbf{e}_2 = [1 \ 3 \ 0]^\mathsf{T}$ is not a scalar multiple of $\mathbf{e}_2$. The only vectors in $\mathcal{E}_1(A_1)$ that are eigenvectors of A_2 are the nonzero vectors in $\text{span}\{\mathbf{e}_1\}$.

8.6 Real Similarity of Real Matrices

Let

$$A = \begin{bmatrix} 1 & 1 \\ 0 & 1 \end{bmatrix}, \qquad B = \begin{bmatrix} 1 & 2 \\ 0 & 1 \end{bmatrix}, \tag{8.6.1}$$

$$S = \begin{bmatrix} 1 & 1+i \\ 0 & 2 \end{bmatrix} \quad \text{and} \quad S^{-1} = \begin{bmatrix} 1 & -\frac{1}{2}(1+i) \\ 0 & \frac{1}{2} \end{bmatrix}. \tag{8.6.2}$$

Then $A = SBS^{-1}$, so the real matrices A and B are similar via the non-real similarity matrix S. Are they similar via a real similarity matrix?

Theorem 8.6.3 *If two real matrices are similar, then they are similar via a real matrix.*

Proof Let $A, B \in M_n(\mathbb{R})$. Let $S \in M_n$ be invertible and such that $A = SBS^{-1}$. For any $\theta \in (-\pi, \pi]$,

$$A = (e^{i\theta} S)B(e^{-i\theta} S^{-1}) = S_\theta B S_\theta^{-1},$$

in which $S_\theta = e^{i\theta} S$. Then $AS_\theta = S_\theta B$. The complex conjugate of this identity is $A\overline{S_\theta} = \overline{S_\theta} B$ since A and B are real. Add the preceding two identities to obtain

$$A(S_\theta + \overline{S_\theta}) = (S_\theta + \overline{S_\theta})B.$$

Since $R_\theta = S_\theta + \overline{S_\theta} = 2\operatorname{Re} S_\theta$ is real for all $\theta \in (-\pi, \pi]$, the computation

$$S_\theta^{-1} R_\theta = I + S_\theta^{-1}\overline{S_\theta} = I + e^{-2i\theta} S^{-1}\overline{S} = e^{-2i\theta}(e^{2i\theta} I + S^{-1}\overline{S})$$

shows that R_θ is invertible if $-e^{-2i\theta}$ is not an eigenvalue of $S^{-1}\overline{S}$. Since $S^{-1}\overline{S}$ has at most n distinct eigenvalues (Corollary 8.3.5), there is a $\phi \in (-\pi, \pi]$ such that $-e^{-2i\phi}$ is not an eigenvalue of $S^{-1}\overline{S}$. Then $AR_\phi = R_\phi B$ and $A = R_\phi B R_\phi^{-1}$. □

Example 8.6.4 With the matrices in (8.6.1) and (8.6.2), we have

$$S^{-1}\overline{S} = \begin{bmatrix} 1 & -2i \\ 0 & 1 \end{bmatrix} \quad \text{and} \quad \operatorname{spec} S^{-1}\overline{S} = \{1\},$$

so $R_\theta = S_\theta + \overline{S_\theta}$ is invertible for every $\theta \in (-\pi, \pi]$ except $\pm\pi/2$. With the choice $\phi = 0$ we have

$$R_0 = \begin{bmatrix} 2 & 2 \\ 0 & 4 \end{bmatrix} \quad \text{and} \quad R_0^{-1} = \frac{1}{4}\begin{bmatrix} 2 & -1 \\ 0 & 1 \end{bmatrix}.$$

A computation confirms that $A = R_0 B R_0^{-1}$.

8.7 Problems

P.8.1 Let $A \in M_8$ and suppose that $A^5 + 2A + I = 0$. Is A invertible?

P.8.2 The commutativity of powers of A is essential for the identity (8.2.11). Consider the two presentations $z^2 - w^2 = (z + w)(z - w)$ of a two-variable polynomial and the matrices

$$A = \begin{bmatrix} 0 & 1 \\ 0 & 0 \end{bmatrix} \quad \text{and} \quad B = \begin{bmatrix} 0 & 0 \\ 1 & 0 \end{bmatrix}.$$

Compute $A^2 - B^2$ and $(A + B)(A - B)$, and verify that they are not equal. Discuss.

P.8.3 Suppose that $A \in M_n$ is invertible and let $(\lambda, \mathbf{x})$ be an eigenpair of A. Explain why $\lambda \neq 0$ and show that $(\lambda^{-1}, \mathbf{x})$ is an eigenpair of A^{-1}.

P.8.4 Suppose that $A \in M_n(\mathbb{R})$ has a non-real eigenvalue λ. (a) Show that no eigenvector $\mathbf{x}$ associated with λ can be real. (b) If $(\lambda, \mathbf{x})$ is an eigenpair of A, then so is $(\overline{\lambda}, \overline{\mathbf{x}})$.

P.8.5 Suppose that $A \in M_n(\mathbb{R})$ has a real eigenvalue λ and an associated non-real eigenvector $\mathbf{x}$. Write $\mathbf{x} = \mathbf{u} + i\mathbf{v}$, in which $\mathbf{u}$ and $\mathbf{v}$ are real vectors. Show that at least one of $(\lambda, \mathbf{u})$ or $(\lambda, \mathbf{v})$ is a real eigenpair of A.

P.8.6 Suppose that $A \in M_n(\mathbb{R})$. Show that λ is an eigenvalue of A with geometric multiplicity k if and only if $\overline{\lambda}$ is an eigenvalue of A with geometric multiplicity k.

P.8.7 Let $A = [a_{ij}] \in M_n$, let $\mathbf{e} = [1 \ 1 \ \ldots \ 1]^T \in \mathbb{C}^n$, and let γ be a scalar. Show that $\sum_{j=1}^{n} a_{ij} = \gamma$ for each $i = 1, 2, \ldots, n$ (all row sums equal to γ) if and only if $(\gamma, \mathbf{e})$ is an eigenpair of A. What can you say about the column sums of A if $(\gamma, \mathbf{e})$ is an eigenpair of A^T?

P.8.8 Let $n \geq 2$, let $c \in \mathbb{C}$, and consider $A_c \in M_n$, whose off-diagonal entries are all 1, and whose diagonal entries are c. Sketch the Geršgorin region $G(A_c)$. Why is A_c invertible if $|c| > n - 1$? Show that A_{1-n} is not invertible.

P.8.9 Let $E = \mathbf{e}\mathbf{e}^T \in M_n$. (a) Why is $(n, \mathbf{e})$ an eigenpair of E? (b) If $\mathbf{v}$ is orthogonal to $\mathbf{e}$, why is $(0, \mathbf{v})$ an eigenpair of E? (c) Let $p(z) = z^2 - nz$ and show that $p(E) = 0$. (d) Use Theorem 8.3.3 to explain why 0 and n are the only eigenvalues of E. What are their respective geometric multiplicities?

P.8.10 Consider $A = [a_{ij}] \in M_n$, in which $a_{ii} = a$ for each $i = 1, 2, \ldots, n$ and $a_{ij} = b$ if $i \neq j$. Find the eigenvalues of A and their geometric multiplicities. *Hint*: Use the preceding problem.

P.8.11 Let A be the matrix in (8.4.14). If $(\lambda, \mathbf{x})$ is an eigenpair of A, simplify $A(A\mathbf{x})$ and show that $(\lambda^2, \mathbf{x})$ is an eigenpair of A^2. Compute A^2, deduce that $\lambda^2 = 0$, and explain why spec $A = \{0\}$.

P.8.12 Is the matrix

$$A = \begin{bmatrix} 5 & 1 & 2 & 1 & 0 \\ 1 & 7 & 1 & 0 & 1 \\ 1 & 1 & 12 & 1 & 1 \\ 1 & 0 & 1 & 13 & 1 \\ 0 & 1 & 2 & 1 & 14 \end{bmatrix}$$

invertible? Why? Do not row reduce A or compute $\det A$.

P.8.13 Suppose that each row and each column of $A \in M_{n+1}$ contains all of the entries $1, 2, 2^2, \ldots, 2^n$ in some order. Prove that A is invertible.

P.8.14 Let $A = [a_{ij}] \in M_n$ and let $\lambda \in \mathbb{C}$. Suppose that $a_{ij} \neq 0$ whenever $i \neq j$, $A - \lambda I$ is diagonally dominant, and $|a_{kk} - \lambda| > \sum_{j \neq k} |a_{kj}|$ for some $k \in \{1, 2, \ldots, n\}$. Prove that $\lambda \notin$ spec A.

P.8.15 Let $A \in M_n$, $B \in M_m$, and $C = A \oplus B \in M_{n+m}$. Use eigenpairs to show the following: (a) If λ is an eigenvalue of A, then it is an eigenvalue of C. (b) If λ is an eigenvalue of B, then it is an eigenvalue of C. (c) If λ is an eigenvalue of C, then it is an eigenvalue of either A or B (perhaps both). Deduce that spec C = spec $A \cup$ spec B.

P.8.16 Let $A \in M_n$, suppose that A is idempotent, and let $(\lambda, \mathbf{x})$ be an eigenpair of A. Show that $\lambda = 1$ or $\lambda = 0$.

P.8.17 Find the eigenvalues of the following matrices:

$$\begin{bmatrix} 1 & 2 \\ 3 & 4 \end{bmatrix}, \quad \begin{bmatrix} 1 & 3 \\ 2 & 4 \end{bmatrix}, \quad \begin{bmatrix} 3 & 4 \\ 1 & 2 \end{bmatrix}, \quad \begin{bmatrix} 2 & 1 \\ 4 & 3 \end{bmatrix}, \quad \begin{bmatrix} 4 & 3 \\ 2 & 1 \end{bmatrix}.$$

Discuss any patterns that you observe in your answers.

P.8.18 Consider the matrices

$$A = \begin{bmatrix} 1 & 1 \\ 0 & 1 \end{bmatrix} \quad \text{and} \quad B = \begin{bmatrix} 1 & 0 \\ 1 & 1 \end{bmatrix}$$

and let λ be an eigenvalue of A. (a) Show that there is no eigenvalue μ of B such that $\lambda + \mu$ is an eigenvalue of $A + B$. (b) Show that there is no eigenvalue μ of B such that $\lambda\mu$ is an eigenvalue of AB. (c) Do A and B commute?

P.8.19 Consider the matrices

$$A = \begin{bmatrix} 1 & 2 \\ 0 & 4 \end{bmatrix} \quad \text{and} \quad B = \begin{bmatrix} 2 & 3 \\ 0 & 1 \end{bmatrix}.$$

(a) Show that 1 is an eigenvalue of A and 2 is an eigenvalue of B. (b) Show that $1 + 2 = 3$ is an eigenvalue of $A + B$. (c) Show that $1 \cdot 2 = 2$ is an eigenvalue of AB. (d) Do A and B commute?

P.8.20 Let $A, B \in \mathsf{M}_n$, suppose that A and B commute, and let λ be an eigenvalue of A. Prove the following: (a) There is an eigenvalue μ of B such that $\lambda + \mu$ is an eigenvalue of $A + B$. (b) There is an eigenvalue μ of B such that $\lambda\mu$ is an eigenvalue of AB.

P.8.21 Using the notation of Example 8.3.9, show that $A_\theta A_\phi = A_{\theta+\phi}$. Deduce that A_θ commutes with A_ϕ.

P.8.22 Let $A_1, A_2, \ldots, A_k \in \mathsf{M}_n$ and suppose that $A_i A_j = A_j A_i$ for all $i, j \in \{1, 2, \ldots, k\}$. Show that span$\{A_1, A_2, \ldots, A_k\}$ is a commuting family of matrices.

P.8.23 Let A_1 be the matrix in Example 8.1.14 and let $A_2 = I \in \mathsf{M}_2$. Show that

$$\text{span}\{A_1, A_2\} = \{Z_{a,b} : a, b \in \mathbb{C}\},$$

in which $Z_{a,b} = \begin{bmatrix} a & -b \\ b & a \end{bmatrix}$, and explain why span$\{A_1, A_2\}$ is a commuting family. Show that $\text{spec}\, Z_{a,b} = \{a \pm ib\}$ with associated eigenvectors $\mathbf{x}_\pm = [1 \mp i]^\mathsf{T}$. What is $Z_{\cos\theta, \sin\theta}$?

P.8.24 Consider the Fourier matrix F_n defined in (6.2.14). Show that $\text{spec}\, F_n \subseteq \{\pm 1, \pm i\}$. *Hint*: Theorem 8.3.3 and P.6.38.

P.8.25 Let $n \geq 2$, let $n \geq m \geq 1$, and let $X = [\mathbf{x}_1 \; \mathbf{x}_2 \; \ldots \; \mathbf{x}_m] \in \mathsf{M}_{n \times m}$ have full column rank. Show that there is an invertible $B \in \mathsf{M}_m$ and distinct indices $i_1, i_2, \ldots, i_m \in \{1, 2, \ldots, n\}$ such that the columns of

$$XB = Y = [\mathbf{y}_1 \; \mathbf{y}_2 \; \ldots \; \mathbf{y}_m] \tag{8.7.1}$$

have the following properties: For each $j = 1, 2, \ldots, m$, (a) $\mathbf{y}_j$ has an entry with modulus 1 in position i_j; (b) If $m \geq 2$ then $\mathbf{y}_j$ has the entry 0 in position i_k for each $k \neq j$, and (c) $\|\mathbf{y}_j\|_\infty = 1$. *Hint*: Let j_1 be the index of any largest-modulus entry of $\mathbf{x}_1$. Let $\mathbf{y}_1 = \mathbf{x}_1/\|\mathbf{x}_1\|_\infty$ and add suitable scalar multiples of $\mathbf{y}_1$ to the other columns to give them a 0 entry in position j_1. Repeat with the second column.

P.8.26 Let $A \in \mathsf{M}_n$ and suppose that $\lambda \in \text{spec}\, A$ has geometric multiplicity at least $m \geq 1$. Adopt the notation of Theorem 8.4.1. (a) Show that there are distinct indices $k_1, k_2, \ldots, k_m \in \{1, 2, \ldots, n\}$ such that $\lambda \in G_{k_i}$ for each $i = 1, 2, \ldots, m$. (b) Show that λ is contained in the union of any $n - m + 1$ of the disks G_k. (c) What do (a) and (b) say if $m = 1$? if $m = 2$? if $m = n$? *Hint*: Let the columns of $X \in \mathsf{M}_{n \times m}$ be a basis for $\mathcal{E}_\lambda(A)$ and construct the matrix Y in (8.7.1). Use a column of Y as the eigenvector in the proof of Theorem 8.4.1.

P.8.27 The *Volterra operator* is the linear operator $T : C[0, 1] \to C[0, 1]$ defined by $(Tf)(t) = \int_0^t f(s)\, ds$. The pair (λ, f) is an *eigenpair* of T if $\lambda \in \mathbb{C}, f \in C[0, 1]$ is not the zero function, and $(Tf)(t) = \lambda f(t)$ for all $t \in [0, 1]$. (a) Show that 0 is not an eigenvalue of T. (b) Find the eigenpairs of the Volterra operator. Compare your results with the situation for eigenvalues of a matrix. *Hint*: Consider the equation $Tf = \lambda f$ and use the fundamental theorem of calculus.

P.8.28 Suppose that $(\lambda, \mathbf{x})$ and $(\mu, \mathbf{y})$ are eigenpairs of $A \in M_n$ and $B \in M_m$, respectively. Show that $(\lambda\mu, \mathbf{x} \otimes \mathbf{y})$ is an eigenpair of $A \otimes B \in M_{nm}$ and $(\lambda + \mu, \mathbf{x} \otimes \mathbf{y})$ is an eigenpair of $(A \otimes I_m) + (I_n \otimes B) \in M_{nm}$.

8.8 Notes

S. A. Geršgorin (1901–1933) was born in what is now Belarus. Theorem 8.4.1 is in a paper that he published (in German) in a Soviet journal in 1931. At the time, he was Professor at the St. Petersburg Machine-Construction Institute. For a photograph of Geršgorin, comments on his life and research, and a copy of his 1931 paper, see [Var04, Appendix A].

The term *spectrum* (in the context of eigenvalues) was introduced by D. Hilbert in his 1912 book on integral equations. In the mid 1920s, W. Heisenberg, M. Born, and P. Jordan discovered the matrix mechanics formulation of quantum mechanics, after which W. Pauli quickly identified the wavelengths of the spectral lines of hydrogen (Balmer series) with the eigenvalues of the energy matrix.

For refinements to and variations on Theorem 8.4.20, see [HJ13, Sect. 6.2 and 6.3].

8.9 Some Important Concepts

- Eigenpair of a matrix.

- Eigenvectors must be nonzero.

- Characterizations of eigenvalues (Theorem 8.1.17).

- Every square complex matrix has an eigenvalue.

- Eigenvectors corresponding to distinct eigenvalues are linearly independent.

- An $n \times n$ complex matrix has at most n distinct eigenvalues.

- Geometric multiplicity of an eigenvalue.

- The Geršgorin region of a matrix contains all its eigenvalues.

- A strictly diagonally dominant matrix is invertible.

- A diagonally dominant matrix with no zero entries is invertible if the dominance is strict in some row (Theorem 8.4.20).

- Commuting matrices share a common eigenvector.

9 The Characteristic Polynomial and Algebraic Multiplicity

In this chapter, we identify the eigenvalues of a square complex matrix as the zeros of its characteristic polynomial. We show that an $n \times n$ complex matrix is diagonalizable (similar to a diagonal matrix) if and only if it has n linearly independent eigenvectors. If A is a diagonalizable matrix and if f is a complex-valued function on the spectrum of A, we discuss a way to define $f(A)$ that has many desirable properties.

9.1 The Characteristic Polynomial

A systematic method to determine all the eigenvalues of $A = [a_{ij}] \in M_n$ is based on the observation that the conditions in Theorem 8.1.17 are equivalent to

$$\det(\lambda I - A) = 0. \tag{9.1.1}$$

This observation suggests that a careful study of the function

$$p_A(z) = \det(zI - A) \tag{9.1.2}$$

could be fruitful, since the roots of $p_A(z) = 0$ are the eigenvalues of A. What sort of function is $p_A(z)$?

For $n = 1$,

$$p_A(z) = \det(zI - A) = \det[z - a_{11}] = z - a_{11},$$

so $p_A(z)$ is a monic polynomial in z of degree 1.

If $n = 2$, then

$$
\begin{aligned}
p_A(z) = \det(zI - A) &= \det \begin{bmatrix} z - a_{11} & -a_{12} \\ -a_{21} & z - a_{22} \end{bmatrix} \\
&= (z - a_{11})(z - a_{22}) - a_{21}a_{12} \\
&= z^2 - (a_{11} + a_{22})z + (a_{11}a_{22} - a_{21}a_{12}) \\
&= z^2 - (\operatorname{tr} A)z + \det A,
\end{aligned} \tag{9.1.3}
$$

so $p_A(z)$ is a monic polynomial in z of degree 2 whose coefficients are polynomials in the entries of A.

The case $n = 3$ is a little more challenging, but worth exploring:

$$
p_A(z) = \det(zI - A) = \det \begin{bmatrix} z - a_{11} & -a_{12} & -a_{13} \\ -a_{21} & z - a_{22} & -a_{23} \\ -a_{31} & -a_{32} & z - a_{33} \end{bmatrix}
$$

$$= (z - a_{11})(z - a_{22})(z - a_{33}) - a_{12}a_{23}a_{31} - a_{21}a_{32}a_{13}$$
$$- (z - a_{22})a_{31}a_{13} - (z - a_{11})a_{32}a_{23} - (z - a_{33})a_{21}a_{12}$$
$$= z^3 + c_2 z^2 + c_1 z + c_0, \tag{9.1.4}$$

in which a calculation reveals that

$$c_2 = -(a_{11} + a_{22} + a_{33}) = -\operatorname{tr} A,$$
$$c_1 = a_{11}a_{22} - a_{12}a_{21} + a_{11}a_{33} - a_{13}a_{31} + a_{22}a_{33} - a_{23}a_{32}, \quad \text{and}$$
$$c_0 = -\det A.$$

The coefficients of the two highest-order terms in (9.1.4) are obtained solely from the summand

$$(z - a_{11})(z - a_{22})(z - a_{33}) = z^3 - (a_{11} + a_{22} + a_{33})z^2 + \cdots.$$

The other summands contribute only to the coefficients of lower-order terms.

Theorem 9.1.5 *Let $A \in \mathsf{M}_n$. The function $p_A(z) = \det(zI - A)$ is a monic polynomial*

$$p_A(z) = z^n + c_{n-1}z^{n-1} + \cdots + c_1 z + c_0 \tag{9.1.6}$$

in z of degree n, in which each coefficient is a polynomial function of the entries of A, $c_{n-1} = -\operatorname{tr} A$, and $c_0 = (-1)^n \det A$. Moreover, $p_A(\lambda) = 0$ if and only if λ is an eigenvalue of A.

Proof The function $p_A(z)$ is a sum of terms, each of which is ± 1 times a product of entries of the matrix $zI - A$. Each product contains exactly n factors, which are entries of $zI - A$ selected from distinct rows and columns. Each such entry is either of the form $-a_{ij}$ with $i \neq j$, or of the form $z - a_{ii}$, so each product of entries (and hence also their sum) is a polynomial in z of degree at most n whose coefficients are polynomial functions of the entries of A. The only product that contributes to the coefficient of the z^n term is

$$(z - a_{11})(z - a_{22}) \cdots (z - a_{nn}) = z^n + \cdots, \tag{9.1.7}$$

which tells us that the polynomial $p_A(z)$ is monic and has degree n. Any product that contributes to the coefficient c_{n-1} must contain at least $n - 1$ of the factors $(z - a_{ii})$ in (9.1.7). These factors come from $n - 1$ different main diagonal positions (distinct rows and columns), so the nth factor can come only from the remaining diagonal position. Thus, only the product (9.1.7) contributes to $c_{n-1} = -a_{11} - a_{22} - \cdots - a_{nn} = -\operatorname{tr} A$. The presentation (9.1.6) ensures that

$$c_0 = p_A(0) = \det(0I - A) = \det(-A) = (-1)^n \det A.$$

The final assertion is a restatement of the equivalence of (9.1.1) and the conditions in Theorem 8.1.17. $\square$

Definition 9.1.8 The polynomial (9.1.6) is the *characteristic polynomial* of A.

Example 9.1.9 In (9.1.3) we found that the characteristic polynomial of

$$A = \begin{bmatrix} a & b \\ c & d \end{bmatrix}$$

is $p_A(z) = z^2 - (a+d)z + (ad - bc) = z^2 - (\operatorname{tr} A)z + \det A$. This is the same polynomial that appears in Example 8.2.13. We showed that the two roots of $p_A(z) = 0$ (see (8.2.14)) are eigenvalues of A. Theorem 9.1.5 ensures that these two roots are the only eigenvalues of A.

Example 9.1.10 Suppose that $A = [a_{ij}] \in \mathsf{M}_n$ is upper triangular. Its determinant is the product of its main diagonal entries:

$$\det A = a_{11}a_{22}\cdots a_{nn}.$$

The matrix $zI - A$ is also upper triangular, so its determinant is

$$p_A(z) = \det(zI - A) = (z - a_{11})(z - a_{22})\cdots(z - a_{nn}).$$

Thus, λ is an eigenvalue of an upper triangular matrix if and only if it is a main diagonal entry. Lower triangular and diagonal matrices have the same property. For matrices without these special structures, a main diagonal entry need not be an eigenvalue; see Example 8.4.15.

9.2 Algebraic Multiplicity

The characteristic polynomial of $A \in \mathsf{M}_n$ has degree n and is monic, so

$$p_A(z) = (z - \lambda_1)(z - \lambda_2)\cdots(z - \lambda_n) \tag{9.2.1}$$
$$= (z - \mu_1)^{n_1}(z - \mu_2)^{n_2}\cdots(z - \mu_d)^{n_d}, \qquad \mu_i \neq \mu_j \text{ if } i \neq j, \tag{9.2.2}$$

in which $\lambda_1, \lambda_2, \ldots, \lambda_n$ is a list of the roots of $p_A(z) = 0$ and $\operatorname{spec} A = \{\mu_1, \mu_2, \ldots, \mu_d\}$.

Definition 9.2.3 The scalars $\mu_1, \mu_2, \ldots, \mu_d$ in (9.2.2) are the *distinct eigenvalues of A*. The exponents n_i in (9.2.2) are the number of times that each μ_i appears in the list $\lambda_1, \lambda_2, \ldots, \lambda_n$; n_i is the *algebraic multiplicity* of the eigenvalue μ_i. If $n_i = 1$, then μ_i is a *simple eigenvalue*. The scalars $\lambda_1, \lambda_2, \ldots, \lambda_n$ in (9.2.1) are the eigenvalues of A *including multiplicities*.

If we refer to the *multiplicity* of an eigenvalue without qualification (algebraic or geometric), we mean the algebraic multiplicity.

The highest-order term in (9.2.1) is z^n; the highest-order term in (9.2.2) is $z^{n_1+n_2+\cdots+n_d}$. These two terms must be equal, so

$$n = n_1 + n_2 + \cdots + n_d, \tag{9.2.4}$$

that is, the sum of the multiplicities of the eigenvalues of $A \in \mathsf{M}_n$ is n.

Example 9.2.5 The characteristic polynomial of the matrix in (8.1.11) is $p_A(z) = (z - 1)^2$, so $\lambda = 1$ is its only eigenvalue. It has algebraic multiplicity 2, but we saw in Example 8.1.10 that its geometric multiplicity is 1. Thus, the geometric multiplicity of an eigenvalue can be less than its algebraic multiplicity.

How are eigenvalues of $A^\mathsf{T}, \overline{A}$, and A^* related to the eigenvalues of A? The relationship is a consequence of the identities $\det A^\mathsf{T} = \det A$ and $\det \overline{A} = \overline{\det A}$.

Theorem 9.2.6 *Let $\lambda_1, \lambda_2, \ldots, \lambda_n$ be the eigenvalues of $A \in \mathsf{M}_n$.*

(a) *The eigenvalues of A^{T} are $\lambda_1, \lambda_2, \ldots, \lambda_n$.*

(b) *The eigenvalues of $\overline{A}$ and A^* are $\overline{\lambda_1}, \overline{\lambda_2}, \ldots, \overline{\lambda_n}$.*

(c) *If A has real entries and λ is a non-real eigenvalue of A with multiplicity k, then $\overline{\lambda}$ is an eigenvalue of A with multiplicity k.*

Proof (a) Compute

$$p_{A^{\mathsf{T}}}(z) = \det(zI - A^{\mathsf{T}}) = \det(zI - A)^{\mathsf{T}} = \det(zI - A) = p_A(z),$$

so A and A^{T} have the same characteristic polynomial. Therefore, they have the same eigenvalues with the same multiplicities.

(b) Observe that

$$p_{\overline{A}}(z) = \det(zI - \overline{A}) = \det(\overline{\overline{z}I - A}) = \overline{\det(\overline{z}I - A)} = \overline{p_A(\overline{z})}.$$

Use the presentation (9.2.1) to compute

$$
\begin{aligned}
p_{\overline{A}}(z) = \overline{p_A(\overline{z})} &= \overline{(\overline{z} - \lambda_1)(\overline{z} - \lambda_2) \cdots (\overline{z} - \lambda_n)} \\
&= (z - \overline{\lambda_1})(z - \overline{\lambda_2}) \cdots (z - \overline{\lambda_n}),
\end{aligned}
$$

which shows that the eigenvalues of $\overline{A}$ are $\overline{\lambda_1}, \overline{\lambda_2}, \ldots, \overline{\lambda_n}$. The eigenvalues of $A^* = \overline{A}^{\mathsf{T}}$ are the same as those of $\overline{A}$.

(c) If $A \in \mathsf{M}_n(\mathbb{R})$, then $A = \overline{A}$ and it follows from (b) that the elements of the two lists $\lambda_1, \lambda_2, \ldots, \lambda_n$ and $\overline{\lambda_1}, \overline{\lambda_2}, \ldots, \overline{\lambda_n}$ are the same, though one list might appear as a reordered presentation of the other. Thus, any non-real eigenvalues of A occur in complex conjugate pairs with equal multiplicities. $\qquad\square$

Two useful identities between matrix entries and eigenvalues can be deduced from a comparison of the presentations (9.1.6) and (9.2.1) of the characteristic polynomial.

Theorem 9.2.7 *Let $\lambda_1, \lambda_2, \ldots, \lambda_n$ be the eigenvalues of $A \in \mathsf{M}_n$. Then $\operatorname{tr} A = \lambda_1 + \lambda_2 + \cdots + \lambda_n$ and $\det A = \lambda_1 \lambda_2 \cdots \lambda_n$.*

Proof Expand (9.2.1) to obtain

$$p_A(z) = z^n - (\lambda_1 + \lambda_2 + \cdots + \lambda_n)z^{n-1} + \cdots + (-1)^n \lambda_1 \lambda_2 \cdots \lambda_n$$

and invoke Theorem 9.1.5 to write

$$p_A(z) = z^n - (\operatorname{tr} A)z^{n-1} + \cdots + (-1)^n \det A. \qquad\square$$

Block upper triangular matrices arise frequently, so it is important to understand how their eigenvalues are related to the eigenvalues of their diagonal blocks; a special case is discussed in Example 9.1.10.

Theorem 9.2.8 *Consider the block upper triangular matrix*

$$A = \begin{bmatrix} B & C \\ 0 & D \end{bmatrix}, \tag{9.2.9}$$

in which B and D are square. Then $\operatorname{spec} A = \operatorname{spec} B \cup \operatorname{spec} D$. *Moreover,* $p_A(z) = p_B(z)p_D(z)$, *so the eigenvalues of A are the eigenvalues of B together with the eigenvalues of D, including multiplicities in each case.*

Proof Use Theorem 3.4.4 and compute

$$p_A(z) = \det \begin{bmatrix} zI - B & -C \\ 0 & zI - D \end{bmatrix}$$

$$= \det(zI - B)\det(zI - D) = p_B(z)p_D(z).$$

The roots of $p_A(z) = 0$ are the roots of $p_B(z) = 0$ together with the roots of $p_D(z) = 0$, including their respective multiplicities. Each distinct root of $p_A(z) = 0$ is a root either of $p_B(z) = 0$ or of $p_D(z) = 0$. $\square$

Example 9.2.10 Consider the block matrix A in (9.2.9), in which

$$B = \begin{bmatrix} 1 & 1 \\ 1 & 1 \end{bmatrix} \quad \text{and} \quad D = \begin{bmatrix} 1 & i \\ i & -1 \end{bmatrix}$$

are the matrices in Examples 8.1.6 and 8.3.15, respectively. The eigenvalues of B are 2 and 0; the only eigenvalue of D is 0 (with multiplicity 2). Thus, the eigenvalues of A are 2 (with multiplicity 1) and 0 (with multiplicity 3), regardless of what C is.

9.3 Similarity and Eigenvalue Multiplicities

Similar matrices represent the same linear transformation with respect to different bases; see Corollary 2.4.17. Therefore, similar matrices can be expected to share many important properties. For example, they have the same characteristic polynomials, eigenvalues, and eigenvalue multiplicities.

Theorem 9.3.1 *Let* $A, S \in \mathsf{M}_n$, *suppose that S is invertible, and let* $B = SAS^{-1}$. *Then A and B have the same characteristic polynomial and eigenvalues. Moreover, their eigenvalues have the same algebraic and geometric multiplicities.*

Proof Compute

$$p_B(z) = \det(zI - B) = \det(zI - SAS^{-1})$$

$$= \det(zSS^{-1} - SAS^{-1}) = \det(S(zI - A)S^{-1})$$

$$= (\det S)\det(zI - A)(\det S^{-1}) = (\det S)\det(zI - A)(\det S)^{-1}$$

$$= \det(zI - A) = p_A(z).$$

Since A and B have the same characteristic polynomials, they have the same eigenvalues with the same algebraic multiplicities.

Theorems 2.4.18 and 3.2.9 ensure that $A - \lambda I$ and $B - \lambda I$ are similar and have the same rank. The identity (8.3.13) implies that

$$\dim \mathcal{E}_\lambda(A) = n - \text{rank}(A - \lambda I) = n - \text{rank}(B - \lambda I) = \dim \mathcal{E}_\lambda(B),$$

which is the asserted equality of geometric multiplicities.   □

The preceding result suggests a strategy that is often effective in proving theorems about eigenvalues: transform by similarity to a new matrix for which the theorem is easier to prove. We employ this strategy to prove an inequality between the algebraic and geometric multiplicities of an eigenvalue, of which Example 9.2.5 is a special case.

Theorem 9.3.2 *The geometric multiplicity of an eigenvalue is less than or equal to its algebraic multiplicity.*

Proof Let λ be an eigenvalue of $A \in \mathsf{M}_n$ and suppose that $\dim \mathcal{E}_\lambda(A) = k$. Let $\mathbf{x}_1, \mathbf{x}_2, \ldots, \mathbf{x}_k$ be a basis for $\mathcal{E}_\lambda(A)$, let $X = [\mathbf{x}_1 \ \mathbf{x}_2 \ \ldots \ \mathbf{x}_k] \in \mathsf{M}_{n \times k}$, and compute

$$AX = [A\mathbf{x}_1 \ A\mathbf{x}_2 \ \ldots \ A\mathbf{x}_k] = [\lambda\mathbf{x}_1 \ \lambda\mathbf{x}_2 \ \ldots \ \lambda\mathbf{x}_k] = \lambda X.$$

Let $S = [X \ X'] \in \mathsf{M}_n$ be invertible (see Theorem 3.3.23). Then

$$[S^{-1}X \ S^{-1}X'] = S^{-1}S = I_n = \begin{bmatrix} I_k & 0 \\ 0 & I_{n-k} \end{bmatrix},$$

so

$$S^{-1}X = \begin{bmatrix} I_k \\ 0 \end{bmatrix}.$$

Compute

$$S^{-1}AS = S^{-1}[AX \ AX'] = S^{-1}[\lambda X \ AX'] = [\lambda S^{-1}X \ S^{-1}AX']$$

$$= \begin{bmatrix} \lambda I_k & \star \\ 0 & C \end{bmatrix}, \tag{9.3.3}$$

in which $C \in \mathsf{M}_{n-k}$. Since similar matrices have the same characteristic polynomial, Theorem 9.2.8 ensures that

$$p_A(z) = p_{S^{-1}AS}(z) = p_{\lambda I_k}(z)p_C(z) = (z - \lambda)^k p_C(z).$$

Consequently, λ is a root of $p_A(z) = 0$ with multiplicity at least k.   □

9.4 Diagonalization and Eigenvalue Multiplicities

What is special about a matrix if every eigenvalue has equal geometric and algebraic multiplicities?

Lemma 9.4.1 *Let $A \in \mathsf{M}_n$ and suppose that for each $\lambda \in \text{spec } A$, the algebraic and geometric multiplicities of λ are equal. Then $\mathbb{C}^n$ has a basis consisting of eigenvectors of A.*

Proof Let $\mu_1, \mu_2, \ldots, \mu_d$ be the distinct eigenvalues of A and let $n_1, n_2, \ldots, n_d$ be their respective algebraic multiplicities. We assume that each eigenspace $\mathcal{E}_{\mu_i}(A)$ has dimension n_i, so that $n_1 + n_2 + \cdots + n_d = n$; see (9.2.4). For each $i = 1, 2, \ldots, d$, let the columns of $X_i \in \mathsf{M}_{n \times n_i}$ be a basis of $\mathcal{E}_{\mu_i}(A)$. Since $AX_i = \mu_i X_i$, we have $AX_i \mathbf{y}_i = \mu_i X_i \mathbf{y}_i$ for any $\mathbf{y}_i \in \mathbb{C}^{n_i}$. Moreover, X_i has full column rank, so $X_i \mathbf{y}_i = \mathbf{0}$ only if $\mathbf{y}_i = \mathbf{0}$; that is, $(\mu_i, X_i \mathbf{y}_i)$ is an eigenpair of A whenever $\mathbf{y}_i \neq \mathbf{0}$.

Let $X = [X_1 \ X_2 \ \ldots \ X_d] \in \mathsf{M}_n$. We claim that rank $X = n$, which ensures that the columns of X are a basis for $\mathbb{C}^n$ comprising eigenvectors of A. Let $\mathbf{y} \in \mathbb{C}^n$ and suppose that $X\mathbf{y} = \mathbf{0}$. We must show that $\mathbf{y} = \mathbf{0}$. Partition $\mathbf{y} = [\mathbf{y}_1^\mathsf{T} \ \mathbf{y}_2^\mathsf{T} \ \ldots \ \mathbf{y}_d^\mathsf{T}]^\mathsf{T}$ conformally with X. Then

$$\mathbf{0} = X\mathbf{y} = X_1 \mathbf{y}_1 + X_2 \mathbf{y}_2 + \cdots + X_d \mathbf{y}_d. \tag{9.4.2}$$

If not every summand is zero, then (9.4.2) is equal to a nontrivial linear combination of eigenvectors of A corresponding to distinct eigenvalues. This contradicts Theorem 8.3.4, so each $X_i \mathbf{y}_i = \mathbf{0}$, and hence each $\mathbf{y}_i = \mathbf{0}$. Thus, $\mathbf{y} = \mathbf{0}$. $\qquad \square$

We can now characterize the matrices $A \in \mathsf{M}_n$ such that $\mathbb{C}^n$ has a basis consisting of eigenvectors of A: they are similar to diagonal matrices.

Theorem 9.4.3 *Let $A \in \mathsf{M}_n$. Then $\mathbb{C}^n$ has a basis consisting of eigenvectors of A if and only if there are $S, \Lambda \in \mathsf{M}_n$ such that S is invertible, Λ is diagonal, and*

$$A = S\Lambda S^{-1}. \tag{9.4.4}$$

Proof Suppose that $\mathbf{s}_1, \mathbf{s}_2, \ldots, \mathbf{s}_n$ comprise a basis of $\mathbb{C}^n$ and $A\mathbf{s}_j = \lambda_j \mathbf{s}_j$ for each $j = 1, 2, \ldots, n$. Let $\Lambda = \mathrm{diag}(\lambda_1, \lambda_2, \ldots, \lambda_n)$ and $S = [\mathbf{s}_1 \ \mathbf{s}_2 \ \ldots \ \mathbf{s}_n]$. Then S is invertible and

$$AS = [A\mathbf{s}_1 \ A\mathbf{s}_2 \ \ldots \ A\mathbf{s}_n] = [\lambda_1 \mathbf{s}_1 \ \lambda_2 \mathbf{s}_2 \ \ldots \ \lambda_n \mathbf{s}_n]$$

$$= [\mathbf{s}_1 \ \mathbf{s}_2 \ \ldots \ \mathbf{s}_n] \begin{bmatrix} \lambda_1 & & 0 \\ & \ddots & \\ 0 & & \lambda_n \end{bmatrix}$$

$$= S\Lambda.$$

Thus, $A = S\Lambda S^{-1}$. Conversely, if $S = [\mathbf{s}_1 \ \mathbf{s}_2 \ \ldots \ \mathbf{s}_n]$ is invertible, $\Lambda = \mathrm{diag}(\lambda_1, \lambda_2, \ldots, \lambda_n)$, and $A = S\Lambda S^{-1}$, then the columns of S are a basis for $\mathbb{C}^n$ and

$$[A\mathbf{s}_1 \ A\mathbf{s}_2 \ \ldots \ A\mathbf{s}_n] = AS = S\Lambda = [\lambda_1 \mathbf{s}_1 \ \lambda_2 \mathbf{s}_2 \ \ldots \ \lambda_n \mathbf{s}_n].$$

Thus, $A\mathbf{s}_j = \lambda_j \mathbf{s}_j$ for each $j = 1, 2, \ldots, n$. Since S has no zero columns, $\mathbf{s}_j$ is an eigenvector of A for each $j = 1, 2, \ldots, n$. $\qquad \square$

Definition 9.4.5 A square matrix is *diagonalizable* if it is similar to a diagonal matrix, that is, if it can be factored as in (9.4.4).

Corollary 9.4.6 *If $A \in \mathsf{M}_n$ has distinct eigenvalues, then it is diagonalizable.*

Proof If A has distinct eigenvalues, Corollary 8.3.5 ensures that $\mathbb{C}^n$ has a basis of eigenvectors of A. The preceding theorem says that A is similar to a diagonal matrix. $\qquad \square$

Not every matrix is diagonalizable. One way to produce examples is via the following theorem, which provides a converse to Lemma 9.4.1.

Theorem 9.4.7 *Let $A \in M_n$. Then A is diagonalizable if and only if for each $\lambda \in \operatorname{spec} A$, the algebraic and geometric multiplicities of λ are equal.*

Proof If the algebraic and geometric multiplicities are equal, Lemma 9.4.1 and Theorem 9.4.3 ensure that A is diagonalizable. Conversely, suppose that $A = S\Lambda S^{-1}$, in which S is invertible and $\Lambda = \operatorname{diag}(\lambda_1, \lambda_2, \dots, \lambda_n)$. The geometric multiplicity of λ is

$$
\begin{aligned}
\dim \mathcal{E}_\lambda(A) &= \dim \operatorname{null}(A - \lambda I) \\
&= \dim \operatorname{null}(S\Lambda S^{-1} - \lambda S I S^{-1}) \\
&= \dim \operatorname{null}(S(\Lambda - \lambda I)S^{-1}) \\
&= \dim \operatorname{null}(\Lambda - \lambda I) \qquad \text{(Theorem 3.2.9)} \\
&= \dim \operatorname{null} \operatorname{diag}(\lambda_1 - \lambda, \lambda_2 - \lambda, \dots, \lambda_n - \lambda) \\
&= \text{the number of eigenvalues of } A \text{ that are equal to } \lambda \\
&= \text{algebraic multiplicity of } \lambda.
\end{aligned}
$$
$\square$

Example 9.4.8 The matrix A in Example 8.1.10 is not diagonalizable. Its eigenvalue $\lambda = 1$ has geometric multiplicity 1 and algebraic multiplicity 2. Alternatively, one could argue as follows: If A were diagonalizable and $A = S\Lambda S^{-1}$, then since the algebraic multiplicities of $\lambda = 1$ for the similar matrices A and Λ are the same, we would have $\Lambda = I$ and $A = S\Lambda S^{-1} = SIS^{-1} = SS^{-1} = I$, which is not the case.

Corollary 9.4.9 *Suppose that $A \in M_n$ is diagonalizable, let $\lambda_1, \lambda_2, \dots, \lambda_n$ be a list of its eigenvalues in any given order, and let $D = \operatorname{diag}(\lambda_1, \lambda_2, \dots, \lambda_n)$. Then there is an invertible $R \in M_n$ such that $A = RDR^{-1}$.*

Proof The hypothesis is that $A = S\Lambda S^{-1}$, in which S is invertible and Λ is a diagonal matrix. Moreover, there is a permutation matrix P such that $\Lambda = PDP^{-1}$; see (6.3.4). Then $A = S\Lambda S^{-1} = SPDP^{-1}S^{-1} = (SP)D(SP)^{-1}$. Let $R = SP$. $\square$

Example 9.4.10 If $A \in M_n$ is diagonalizable and has d distinct eigenvalues $\mu_1, \mu_2, \dots, \mu_d$ with respective multiplicities $n_1, n_2, \dots, n_d$, then there is an invertible $S \in M_n$ such that $A = S\Lambda S^{-1}$ and $\Lambda = \mu_1 I_{n_1} \oplus \mu_2 I_{n_2} \oplus \cdots \oplus \mu_d I_{n_d}$. Thus, we may group equal eigenvalues together as diagonal entries of Λ.

Theorem 9.4.11 *Let $A = A_1 \oplus A_2 \oplus \cdots \oplus A_k \in M_n$, in which each $A_i \in M_{n_i}$. Then A is diagonalizable if and only if each direct summand A_i is diagonalizable.*

Proof If each A_i is diagonalizable, then there are invertible matrices $R_i \in M_{n_i}$ and diagonal matrices $\Lambda_i \in M_{n_i}$ such that $A_i = R_i \Lambda_i R_i^{-1}$. Let $R = R_1 \oplus R_2 \oplus \cdots \oplus R_k$ and $\Lambda = \Lambda_1 \oplus \Lambda_2 \oplus \cdots \oplus \Lambda_k$. Then $A = R\Lambda R^{-1}$.

Conversely, if A is diagonalizable, let $A = S\Lambda S^{-1}$ as in (9.4.4) and partition

$$S = \begin{bmatrix} S_1 \\ S_2 \\ \vdots \\ S_k \end{bmatrix}, \qquad S_i \in \mathsf{M}_{n_i \times n} \text{ for each } i = 1, 2, \dots, k.$$

Then S has linearly independent rows (because it is invertible), so each S_i has linearly independent rows. Compute

$$\begin{bmatrix} A_1 S_1 \\ A_2 S_2 \\ \vdots \\ A_k S_k \end{bmatrix} = AS = S\Lambda = \begin{bmatrix} S_1 \Lambda \\ S_2 \Lambda \\ \vdots \\ S_k \Lambda \end{bmatrix}.$$

Each identity $A_i S_i = S_i \Lambda$ says that every nonzero column of S_i is an eigenvector of A_i. To show that A_i is diagonalizable, it suffices to show that S_i has n_i linearly independent columns. But $S_i \in \mathsf{M}_{n_i \times n}$ has linearly independent rows, so the identity (3.2.2) ensures that it has n_i linearly independent columns. $\qquad \square$

Similar matrices have the same rank, and the rank of a diagonal matrix is the number of nonzero entries on its diagonal. These observations are synthesized in the following theorem.

Theorem 9.4.12 *Suppose that $A \in \mathsf{M}_n$ is diagonalizable. Then the number of nonzero eigenvalues of A is equal to its rank.*

Proof According to the dimension theorem, it suffices to show that $\dim \operatorname{null} A$ (the geometric multiplicity of the eigenvalue $\lambda = 0$ of A) is equal to the number of zero eigenvalues of A (the algebraic multiplicity of $\lambda = 0$). But every eigenvalue (zero or not) of a diagonalizable matrix has equal geometric and algebraic multiplicities; see Theorem 9.4.7. $\qquad \square$

Definition 9.4.13 $A, B \in \mathsf{M}_n$ are *simultaneously diagonalizable* if there is an invertible $S \in \mathsf{M}_n$ such that both $S^{-1}AS$ and $S^{-1}BS$ are diagonal.

The simultaneous diagonalizability of A and B means that there is a basis of $\mathbb{C}^n$ in which each basis vector is an eigenvector of both A and B.

Lemma 9.4.14 *Let $A, B, S, X, Y \in \mathsf{M}_n$. Let S be invertible, and suppose that $A = SXS^{-1}$ and $B = SYS^{-1}$. Then $AB = BA$ if and only if $XY = YX$.*

Proof $AB = (SXS^{-1})(SYS^{-1}) = SXYS^{-1}$ and $BA = (SYS^{-1})(SXS^{-1}) = SYXS^{-1}$. Thus $AB = BA$ if and only if $SXYS^{-1} = SYXS^{-1}$, which is equivalent to $XY = YX$. $\qquad \square$

Theorem 9.4.15 *Let $A, B \in \mathsf{M}_n$ be diagonalizable. Then A commutes with B if and only if they are simultaneously diagonalizable.*

Proof Suppose that $A = SXS^{-1}$ and $B = SYS^{-1}$, in which $S \in \mathsf{M}_n$ is invertible and $X, Y \in \mathsf{M}_n$ are diagonal. Then $XY = YX$, so the preceding lemma ensures that $AB = BA$.

Conversely, suppose that $AB = BA$ and let $\mu_1, \mu_2, \ldots, \mu_d$ be the distinct eigenvalues of A, with respective multiplicities $n_1, n_2, \ldots, n_d$. Since A is diagonalizable, Corollary 9.4.9 and Example 9.4.10 ensure that there is an invertible matrix S such that $A = S \Lambda S^{-1}$ and

$$\Lambda = \mu_1 I_{n_1} \oplus \mu_2 I_{n_2} \oplus \cdots \oplus \mu_d I_{n_d}.$$

Since A commutes with B, we have $S \Lambda S^{-1} B = B S \Lambda S^{-1}$ and hence

$$\Lambda(S^{-1} B S) = (S^{-1} B S) \Lambda.$$

Now invoke Lemma 3.3.21.b and conclude that $S^{-1} B S$ is block diagonal and conformal with Λ; write $S^{-1} B S = B_1 \oplus B_2 \oplus \cdots \oplus B_d$, in which each $B_i \in \mathsf{M}_{n_i}$. Since B is diagonalizable, $S^{-1} B S$ is diagonalizable and Theorem 9.4.11 ensures that each B_i is diagonalizable. Write $B_i = R_i D_i R_i^{-1}$ for each $i = 1, 2, \ldots, d$, in which each $D_i \in \mathsf{M}_{n_i}$ is diagonal and each $R_i \in \mathsf{M}_{n_i}$ is invertible. Define $R = R_1 \oplus R_2 \oplus \cdots \oplus R_d$ and $D = D_1 \oplus D_2 \oplus \cdots \oplus D_d$. Then

$$\begin{aligned}
S^{-1} B S &= B_1 \oplus B_2 \oplus \cdots \oplus B_d \\
&= R_1 D_1 R_1^{-1} \oplus R_2 D_2 R_2^{-1} \oplus \cdots \oplus R_d D_d R_d^{-1} \\
&= R D R^{-1},
\end{aligned}$$

so

$$B = (SR) D (SR)^{-1}.$$

Because R is block diagonal and conformal with Λ, we have $R\Lambda = \Lambda R$ and hence

$$\begin{aligned}
A = S \Lambda S^{-1} &= S \Lambda R R^{-1} S^{-1} \\
&= S R \Lambda R^{-1} S^{-1} = (SR) \Lambda (SR)^{-1}.
\end{aligned}$$

Therefore, each of A and B is similar to a diagonal matrix via SR. $\square$

9.5 The Functional Calculus for Diagonalizable Matrices

The next theorem states the *polynomial functional calculus* for diagonalizable matrices.

Theorem 9.5.1 *Let $A \in \mathsf{M}_n$ be diagonalizable and write $A = S \Lambda S^{-1}$, in which $S \in \mathsf{M}_n$ is invertible and $\Lambda = \mathrm{diag}(\lambda_1, \lambda_2, \ldots, \lambda_n)$. If p is a polynomial, then*

$$p(A) = S \, \mathrm{diag}\left(p(\lambda_1), p(\lambda_2), \ldots, p(\lambda_n)\right) S^{-1}. \tag{9.5.2}$$

Proof Theorem 0.8.1 and (0.8.3) ensure that $p(A) = S p(\Lambda) S^{-1}$ and

$$p(\Lambda) = \mathrm{diag}(p(\lambda_1), p(\lambda_2), \ldots, p(\lambda_n)). \qquad \square$$

Example 9.5.3 Consider the following real symmetric matrix and a few of its powers:

$$A = \begin{bmatrix} 1 & 1 \\ 1 & 0 \end{bmatrix}, \quad A^2 = \begin{bmatrix} 2 & 1 \\ 1 & 1 \end{bmatrix}, \quad A^3 = \begin{bmatrix} 3 & 2 \\ 2 & 1 \end{bmatrix}, \quad \text{and} \quad A^4 = \begin{bmatrix} 5 & 3 \\ 3 & 2 \end{bmatrix}.$$

Let f_k denote the $(1, 1)$ entry of A^{k-1}. We have $f_1 = 1, f_2 = 1$, and

$$A^{k-1} = \begin{bmatrix} f_k & f_{k-1} \\ f_{k-1} & f_{k-2} \end{bmatrix} \tag{9.5.4}$$

for $k = 2, 3, 4, 5$. Since

$$A^k = AA^{k-1} = \begin{bmatrix} 1 & 1 \\ 1 & 0 \end{bmatrix} \begin{bmatrix} f_k & f_{k-1} \\ f_{k-1} & f_{k-2} \end{bmatrix} = \begin{bmatrix} f_k + f_{k-1} & f_{k-1} + f_{k-2} \\ f_k & f_{k-1} \end{bmatrix},$$

we see that

$$f_1 = 1, \quad f_2 = 1, \quad \text{and} \quad f_{k+1} = f_k + f_{k-1}, \quad \text{for } k = 2, 3, \ldots,$$

which is the recurrence relation that defines the *Fibonacci numbers*

$$1, 1, 2, 3, 5, 8, 13, 21, 34, 55, \ldots.$$

Theorem 9.5.1 provides a recipe to evaluate polynomials in A if A is diagonalizable.

The characteristic polynomial of A is $p_A(z) = z^2 - z - 1$, so the eigenvalues of A, are $\lambda_\pm = \frac{1}{2}(1 \pm \sqrt{5})$. Corollary 9.4.6 ensures that A is diagonalizable. The eigenpair equation $A\mathbf{x} = \lambda\mathbf{x}$ is

$$\begin{bmatrix} 1 & 1 \\ 1 & 0 \end{bmatrix} \begin{bmatrix} x_1 \\ x_2 \end{bmatrix} = \begin{bmatrix} x_1 + x_2 \\ x_1 \end{bmatrix} = \begin{bmatrix} \lambda_\pm x_1 \\ \lambda_\pm x_2 \end{bmatrix}.$$

It has solutions $\mathbf{x}_\pm = [\lambda_\pm \ 1]^\mathsf{T}$, so (9.4.4) tells us that $A = S\Lambda S^{-1}$, in which

$$S = \begin{bmatrix} \lambda_+ & \lambda_- \\ 1 & 1 \end{bmatrix}, \quad \Lambda = \begin{bmatrix} \lambda_+ & 0 \\ 0 & \lambda_- \end{bmatrix}, \quad \text{and} \quad S^{-1} = \frac{1}{\sqrt{5}} \begin{bmatrix} 1 & -\lambda_- \\ -1 & \lambda_+ \end{bmatrix}.$$

Theorem 9.5.1 shows how to express powers of A as a function of powers of Λ:

$$\begin{bmatrix} f_k & \star \\ \star & \star \end{bmatrix} = A^{k-1} = S\Lambda^{k-1}S^{-1}$$

$$= \frac{1}{\sqrt{5}} \begin{bmatrix} \lambda_+ & \lambda_- \\ 1 & 1 \end{bmatrix} \begin{bmatrix} \lambda_+^{k-1} & 0 \\ 0 & \lambda_-^{k-1} \end{bmatrix} \begin{bmatrix} 1 & -\lambda_- \\ -1 & \lambda_+ \end{bmatrix}$$

$$= \frac{1}{\sqrt{5}} \begin{bmatrix} \lambda_+ & \lambda_- \\ \star & \star \end{bmatrix} \begin{bmatrix} \lambda_+^{k-1} & \star \\ -\lambda_-^{k-1} & \star \end{bmatrix}$$

$$= \begin{bmatrix} \frac{1}{\sqrt{5}}(\lambda_+^k - \lambda_-^k) & \star \\ \star & \star \end{bmatrix}.$$

This identity reveals *Binet's formula*

$$f_k = \frac{(1 + \sqrt{5})^k - (1 - \sqrt{5})^k}{2^k \sqrt{5}}, \quad k = 1, 2, \ldots, \tag{9.5.5}$$

for the Fibonacci numbers. The number $\phi = \frac{1}{2}(1 + \sqrt{5}) = 1.618\ldots$ is known as the *golden ratio*.

For diagonalizable matrices it is possible to develop a broader functional calculus, in which expressions such as $\sin A$, $\cos A$, and e^A are unambiguously defined.

Let $A \in M_n$ be diagonalizable and write $A = S\Lambda S^{-1}$, in which $S \in M_n$ is invertible and $\Lambda = \text{diag}(\lambda_1, \lambda_2, \ldots, \lambda_n)$. Let f be any complex-valued function on spec A and define

$$f(\Lambda) = \text{diag}\left(f(\lambda_1), f(\lambda_2), \ldots, f(\lambda_n)\right).$$

The Lagrange Interpolation Theorem (Theorem 0.7.6) provides an algorithm to construct a polynomial p such that $f(\lambda_i) = p(\lambda_i)$ for each $i = 1, 2, \ldots, n$. Therefore, $f(\Lambda) = p(\Lambda)$. Moreover, if M is a diagonal matrix that is obtained by permuting the diagonal entries of Λ, then $f(M) = p(M)$.

Suppose that A is diagonalized in two ways:

$$A = S\Lambda S^{-1} = RMR^{-1},$$

in which $R, S \in M_n$ are invertible and $\Lambda, M \in M_n$ are diagonal. Then Theorem 0.8.1 ensures that

$$
\begin{aligned}
S\Lambda S^{-1} = RMR^{-1} &\implies (R^{-1}S)\Lambda = M(R^{-1}S) \\
&\implies (R^{-1}S)p(\Lambda) = p(M)(R^{-1}S) \\
&\implies (R^{-1}S)f(\Lambda) = f(M)(R^{-1}S) \\
&\implies Sf(\Lambda)S^{-1} = Rf(M)R^{-1}.
\end{aligned}
$$

Thus, we may define

$$f(A) = Sf(\Lambda)S^{-1}. \tag{9.5.6}$$

We obtain the same matrix no matter what diagonalization of A is chosen. Since $f(A)$ is a polynomial in A, Theorem 0.8.1 ensures that it commutes with any matrix that commutes with A. Theorem 9.5.1 ensures that if f is a polynomial, then the definitions of $f(A)$ in (9.5.6) and (8.2.1) are not in conflict.

Example 9.5.7 Suppose that $A \in M_n$ is diagonalizable and has real eigenvalues. Consider the functions $\sin t$ and $\cos t$ on spec A. Let $A = S\Lambda S^{-1}$, in which $\Lambda = \text{diag}(\lambda_1, \lambda_2, \ldots, \lambda_n)$ is real. Since $\cos^2 \lambda_i + \sin^2 \lambda_i = 1$ for each i, the definition (9.5.6) permits us to compute

$$
\begin{aligned}
\cos^2 A + \sin^2 A &= S(\cos^2 \Lambda)S^{-1} + S(\sin^2 \Lambda)S^{-1} = S(\cos^2 \Lambda + \sin^2 \Lambda)S^{-1} \\
&= S\,\text{diag}(\cos^2 \lambda_1 + \sin^2 \lambda_1, \ldots, \cos^2 \lambda_n + \sin^2 \lambda_n)S^{-1} \\
&= SIS^{-1} = SS^{-1} = I.
\end{aligned}
$$

Example 9.5.8 Let $A \in M_n$ be diagonalizable and suppose that $A = S\Lambda S^{-1}$, in which $\Lambda = \text{diag}(\lambda_1, \lambda_2, \ldots, \lambda_n)$. Consider the function $f(z) = e^z$ on spec A. Using the definition (9.5.6), we compute

$$e^A = Se^\Lambda S^{-1} = S\,\text{diag}(e^{\lambda_1}, e^{\lambda_2}, \ldots, e^{\lambda_n})S^{-1},$$

so Theorem 9.2.7 ensures that

$$\det e^A = (\det S)(e^{\lambda_1} e^{\lambda_2} \cdots e^{\lambda_n})(\det S^{-1}) = e^{\lambda_1 + \lambda_2 + \cdots + \lambda_n} = e^{\text{tr} A}.$$

For an important application of the functional calculus, in which the function is $f(t) = \sqrt{t}$ on $[0, \infty)$, see Section 13.2.

9.6 Commutants

Since $p(A)$ is a polynomial in A, it commutes with A. This observation and Theorem 9.4.15 motivate the following definition.

Definition 9.6.1 Let $\mathcal{F}$ be a nonempty subset of M_n. The *commutant* of $\mathcal{F}$ is the set $\mathcal{F}'$ of all matrices that commute with every element of $\mathcal{F}$, that is,

$$\mathcal{F}' = \{X \in \mathsf{M}_n : AX = XA \text{ for all } A \in \mathcal{F}\}.$$

The commutant of any nonempty subset $\mathcal{F}$ of M_n is a *subspace* of M_n. Indeed, if $X, Y \in \mathcal{F}'$ and $c \in \mathbb{C}$, then $(X + cY)A = XA + cYA = AX + cAY = A(X + cY)$ for all $A \in \mathcal{F}$ (see also P.9.30).

Theorem 9.6.2 *Let $A \in \mathsf{M}_n$ be diagonalizable and suppose that the distinct eigenvalues of A are $\mu_1, \mu_2, \ldots, \mu_d$ with multiplicities $n_1, n_2, \ldots, n_d$. Write $A = S\Lambda S^{-1}$, in which $S \in \mathsf{M}_n$ is invertible and $\Lambda = \mu_1 I_{n_1} \oplus \cdots \oplus \mu_d I_{n_d}$.*

(a) $B \in \{A\}'$ *if and only if $B = SXS^{-1}$, in which $X = X_{11} \oplus \cdots \oplus X_{dd}$ and each $X_{ii} \in \mathsf{M}_{n_i}$.*

(b) $\dim\{A\}' = n_1^2 + n_2^2 + \cdots + n_d^2.$

Proof Lemma 9.4.14 ensures that A and B commute if and only if Λ and $S^{-1}BS$ commute. It follows from Lemma 3.3.21 that $S^{-1}BS = X_{11} \oplus \cdots \oplus X_{kk}$, in which $X_{ii} \in \mathsf{M}_{n_i}$ for $i = 1, 2, \ldots, k$. Since $\dim \mathsf{M}_{n_i} = n_i^2$, the claim in (b) follows. $\qquad\square$

Corollary 9.6.3 *Suppose that $A \in \mathsf{M}_n$ has distinct eigenvalues.*

(a) $\{A\}' = \{p(A) : p \text{ is a polynomial}\}.$

(b) $\dim\{A\}' = n.$

Proof Since A has distinct eigenvalues, it is diagonalizable (Corollary 9.4.6). Write $A = S\Lambda S^{-1}$, in which $S \in \mathsf{M}_n$ is invertible and $\Lambda = \mathrm{diag}(\lambda_1, \lambda_2, \ldots, \lambda_n)$ has distinct diagonal entries.

(a) Theorem 9.6.2 ensures that B commutes with A if and only if $B = SXS^{-1}$, in which $X = \mathrm{diag}(\xi_1, \xi_2, \ldots, \xi_n)$ for some $\xi_1, \xi_2, \ldots, \xi_n \in \mathbb{C}$. The Lagrange Interpolation Theorem (Theorem 0.7.6) provides a polynomial p such that $p(\lambda_i) = \xi_i$ for $i = 1, 2, \ldots, n$. Thus, $p(A) = Sp(\Lambda)S^{-1} = SXS^{-1} = B$. Conversely, $p(A) \in \{A\}'$ for any polynomial p.

(b) Let E_{ii} denote the $n \times n$ matrix with entry 1 in position (i, i) and zero entries elsewhere. The argument in (a) shows that every $B \in \{A\}'$ has the form $B = \xi_1 SE_{11}S^{-1} + \cdots + \xi_n SE_{nn}S^{-1}$, which is a linear combination of linearly independent elements of M_n. $\qquad\square$

Example 9.6.4 In Corollary 9.6.3, the hypothesis that A has distinct eigenvalues cannot be omitted. For example, $\{I_n\}' = \mathsf{M}_n$, which has dimension n^2. Any matrix that is not a scalar multiple of I_n belongs to $\{I_n\}'$ yet is not a polynomial in I_n.

9.7 The Eigenvalues of *AB* and *BA*

The identity

$$\begin{bmatrix} I_k & 0 \\ X & I_{n-k} \end{bmatrix}^{-1} = \begin{bmatrix} I_k & 0 \\ -X & I_{n-k} \end{bmatrix}, \quad X \in M_{(n-k) \times k}, \tag{9.7.1}$$

can be verified by a computation:

$$\begin{bmatrix} I_k & 0 \\ -X & I_{n-k} \end{bmatrix} \begin{bmatrix} I_k & 0 \\ X & I_{n-k} \end{bmatrix} = \begin{bmatrix} I_k & 0 \\ 0 & I_{n-k} \end{bmatrix} = I_n.$$

We can use (9.7.1) to clarify a perhaps unexpected relationship between the eigenvalues of AB and those of BA. These two products need not be matrices of the same size, and even if they are, they need not be equal. Nevertheless, their nonzero eigenvalues are the same.

Theorem 9.7.2 *Suppose that $A \in M_{m \times n}$, $B \in M_{n \times m}$, and $n \geq m$.*

(a) *The nonzero eigenvalues of $AB \in M_m$ and $BA \in M_n$ are the same, with the same algebraic multiplicities.*

(b) *If 0 is an eigenvalue of AB with algebraic multiplicity $k \geq 0$, then 0 is an eigenvalue of BA with algebraic multiplicity $k + n - m$.*

(c) *If $m = n$, then the eigenvalues of AB and BA are the same, with the same algebraic multiplicities.*

Proof Let

$$X = \begin{bmatrix} AB & A \\ 0 & 0_n \end{bmatrix} \quad \text{and} \quad Y = \begin{bmatrix} 0_m & A \\ 0 & BA \end{bmatrix}$$

and consider the following similarity transformation applied to X:

$$\begin{bmatrix} I_m & 0 \\ B & I_n \end{bmatrix} \begin{bmatrix} AB & A \\ 0 & 0_n \end{bmatrix} \begin{bmatrix} I_m & 0 \\ B & I_n \end{bmatrix}^{-1} = \begin{bmatrix} I_m & 0 \\ B & I_n \end{bmatrix} \begin{bmatrix} AB & A \\ 0 & 0_n \end{bmatrix} \begin{bmatrix} I_m & 0 \\ -B & I_n \end{bmatrix}$$

$$= \begin{bmatrix} AB & A \\ BAB & BA \end{bmatrix} \begin{bmatrix} I_m & 0 \\ -B & I_n \end{bmatrix}$$

$$= \begin{bmatrix} AB - AB & A \\ BAB - BAB & BA \end{bmatrix}$$

$$= \begin{bmatrix} 0_m & A \\ 0 & BA \end{bmatrix}$$

$$= Y.$$

Because X and Y are similar, $p_X(z) = p_Y(z)$. Since

$$p_X(z) = p_{AB}(z) p_{0_n}(z) = z^n p_{AB}(z)$$

and

$$p_Y(z) = p_{0_m}(z)p_{BA}(z) = z^m p_{BA}(z),$$

it follows that $p_{BA}(z) = z^{n-m}p_{AB}(z)$. Thus, if

$$\lambda_1, \lambda_2, \ldots, \lambda_m \tag{9.7.3}$$

are the roots of $p_{AB}(z) = 0$ (the eigenvalues of AB), then the n roots of $p_{BA}(z) = 0$ (the eigenvalues of BA) are

$$\lambda_1, \lambda_2, \ldots, \lambda_m, \underbrace{0, \ldots, 0}_{n-m}. \tag{9.7.4}$$

The nonzero eigenvalues in the two lists (9.7.3) and (9.7.4) are identical, with the same multiplicities. However, if k is the multiplicity of 0 as an eigenvalue of AB, its multiplicity as an eigenvalue of BA is $k + n - m$.   □

Example 9.7.5 Consider

$$A = \begin{bmatrix} 1 & 0 \\ 0 & 0 \end{bmatrix}, \quad B = \begin{bmatrix} 0 & 1 \\ 0 & 0 \end{bmatrix}, \quad AB = \begin{bmatrix} 0 & 1 \\ 0 & 0 \end{bmatrix}, \quad \text{and} \quad BA = \begin{bmatrix} 0 & 0 \\ 0 & 0 \end{bmatrix}.$$

Although zero is an eigenvalue of both AB and BA with algebraic multiplicity 2, it has different geometric multiplicities (one and two, respectively). This does not happen for nonzero eigenvalues; see Theorem 11.9.1.

Example 9.7.6 Consider $\mathbf{e} = [1 \ 1 \ \ldots \ 1]^T \in \mathbb{R}^n$. Every entry of the $n \times n$ matrix $\mathbf{e}\mathbf{e}^T$ is 1. What are its eigenvalues? Theorem 9.7.2 tells us that they are the eigenvalue of the 1×1 matrix $\mathbf{e}^T\mathbf{e} = [n]$ (namely, n), together with $n - 1$ zeros.

Example 9.7.7 Let $\mathbf{r} = [1 \ 2 \ \ldots \ n]^T \in \mathbb{R}^n$. Every column of

$$A = \mathbf{r}\mathbf{e}^T = \begin{bmatrix} 1 & 1 & \cdots & 1 \\ 2 & 2 & \cdots & 2 \\ \vdots & \vdots & & \vdots \\ n & n & \cdots & n \end{bmatrix} \in \mathsf{M}_n \tag{9.7.8}$$

is equal to $\mathbf{r}$. Theorem 9.7.2 tells us that the eigenvalues of A are $n - 1$ zeros together with $\mathbf{e}^T\mathbf{r} = 1 + 2 + \cdots + n = n(n + 1)/2$.

Example 9.7.9 Let A be the matrix (9.7.8) and observe that

$$A + A^T = [i + j] = \begin{bmatrix} 2 & 3 & 4 & \cdots & n+1 \\ 3 & 4 & 5 & \cdots & n+2 \\ 4 & 5 & 6 & \cdots & n+3 \\ \vdots & \vdots & \vdots & \ddots & \vdots \\ n+1 & n+2 & n+3 & \cdots & 2n \end{bmatrix}. \tag{9.7.10}$$

Now use the presentation of a matrix product in (3.1.19) to write

$$A + A^T = \mathbf{r}\mathbf{e}^T + \mathbf{e}\mathbf{r}^T = XY^T,$$

in which $X = [\mathbf{r} \ \mathbf{e}] \in \mathsf{M}_{n \times 2}$ and $Y = [\mathbf{e} \ \mathbf{r}] \in \mathsf{M}_{n \times 2}$. Theorem 9.7.2 tells us that the n eigenvalues of XY^T are the two eigenvalues of

$$Y^\mathsf{T}X = \begin{bmatrix} \mathbf{e}^\mathsf{T}\mathbf{r} & \mathbf{e}^\mathsf{T}\mathbf{e} \\ \mathbf{r}^\mathsf{T}\mathbf{r} & \mathbf{r}^\mathsf{T}\mathbf{e} \end{bmatrix} \tag{9.7.11}$$

together with $n - 2$ zeros; see P.9.17.

The three preceding examples suggest a strategy for finding eigenvalues. If $A \in \mathsf{M}_n$, $\operatorname{rank} A \leq r$, and $A = XY^\mathsf{T}$ for some $X, Y \in \mathsf{M}_{n \times r}$, then the n eigenvalues of A are the r eigenvalues of $Y^\mathsf{T}X \in \mathsf{M}_r$, together with $n - r$ zeros. If r is much smaller than n, one might prefer to compute the eigenvalues of $Y^\mathsf{T}X$ instead of the eigenvalues of A. For example, $A = XY^\mathsf{T}$ could be a full-rank factorization.

9.8 Problems

Remember: The "multiplicity" of an eigenvalue, with no modifier, means *algebraic multiplicity*. "Geometric multiplicity" always has a modifier.

P.9.1 Suppose that $A \in \mathsf{M}_5$ has eigenvalues $-4, -1, 0, 1, 4$. Is there an $A \in \mathsf{M}_5$ such that $B^2 = A$? Justify your answer.

P.9.2 Let $\lambda_1, \lambda_2, \ldots, \lambda_n$ be the scalars in (9.2.1) and let $\mu_1, \mu_2, \ldots, \mu_d$ be the scalars in (9.2.2). Explain why $\{\lambda_1, \lambda_2, \ldots, \lambda_n\} = \{\mu_1, \mu_2, \ldots, \mu_d\}$.

P.9.3 Let $A, B \in \mathsf{M}_n$. (a) If $\operatorname{spec} A = \operatorname{spec} B$, do A and B have the same characteristic polynomials? Why? (b) If A and B have the same characteristic polynomials, is $\operatorname{spec} A = \operatorname{spec} B$? Why?

P.9.4 Suppose that $A \in \mathsf{M}_n$ is diagonalizable. (a) Show that $\operatorname{rank} A$ is equal to the number of its nonzero eigenvalues (including multiplicities). (b) Consider the matrix

$$B = \begin{bmatrix} 0 & 1 \\ 0 & 0 \end{bmatrix}.$$

What is its rank? How many nonzero eigenvalues does it have? Is it diagonalizable?

P.9.5 Show that the eigenvalues of an $n \times n$ nilpotent matrix are $0, 0, \ldots, 0$. What is the characteristic polynomial of a nilpotent matrix? If $n \geq 2$, give an example of a nonzero nilpotent matrix in M_n.

P.9.6 Suppose that $A \in \mathsf{M}_n$ is diagonalizable. Show that A is nilpotent if and only if $A = 0$.

P.9.7 Let $\lambda_1, \lambda_2, \ldots, \lambda_n$ be the eigenvalues of $A \in \mathsf{M}_n$ and let $c \in \mathbb{C}$. Show that $p_{A+cI}(z) = p_A(z - c)$ and deduce that the eigenvalues of $A + cI$ are $\lambda_1 + c, \lambda_2 + c, \ldots, \lambda_n + c$.

P.9.8 Show that the two eigenvalues of $A \in \mathsf{M}_2$ may be expressed as $\lambda_\pm = \frac{1}{2}(\operatorname{tr} A \pm \sqrt{r})$, in which $r = (\operatorname{tr} A)^2 - 4 \det A$ is the *discriminant* of A. If A has real entries, show that its eigenvalues are real if and only if its discriminant is nonnegative.

P.9.9 Verify that the asserted values of $\lambda_\pm, \mathbf{x}_\pm, S, \Lambda$, and S^{-1} in Example 9.5.3 are correct, and show that the kth Fibonacci number is

$$f_k = \frac{1}{\sqrt{5}}\left(\phi^k + (-1)^{k+1}\phi^{-k}\right), \quad k = 1, 2, 3, \ldots.$$

Since $\phi \approx 1.6180$, $\phi^{-k} \to 0$ as $k \to \infty$ and hence $\frac{1}{\sqrt{5}}\phi^k$ is a good approximation to f_k for large k. For example, $f_{10} = 55$, $\frac{1}{\sqrt{5}}\phi^{10} \approx 55.004$, $f_{11} = 89$, and $\frac{1}{\sqrt{5}}\phi^{11} \approx 88.998$.

P.9.10 Let $A, B \in \mathsf{M}_n$. (a) Use (3.7.4) and (3.7.5) to explain why the characteristic polynomial of the block matrix

$$C = \begin{bmatrix} 0 & A \\ B & 0 \end{bmatrix} \in \mathsf{M}_{2n} \tag{9.8.1}$$

is related to the characteristic polynomials of AB and BA by the identities

$$p_C(z) = \det(zI_{2n} - C) = p_{AB}(z^2) = p_{BA}(z^2).$$

(b) Deduce from this identity that AB and BA have the same eigenvalues. This is the square case of Theorem 9.7.2. (c) If λ is an eigenvalue of C with multiplicity k, why is $-\lambda$ also an eigenvalue of C with multiplicity k? If zero is an eigenvalue of C, why must it have even multiplicity? (d) Let $\pm\lambda_1, \pm\lambda_2, \ldots, \pm\lambda_n$ be the $2n$ eigenvalues of C. Explain why the n eigenvalues of AB are $\lambda_1^2, \lambda_2^2, \ldots, \lambda_n^2$. (e) Let $\mu_1, \mu_2, \ldots, \mu_n$ be the eigenvalues of AB. Show that the eigenvalues of C are $\pm\sqrt{\mu_1}, \pm\sqrt{\mu_2}, \ldots, \pm\sqrt{\mu_n}$. (f) Show that $\det C = (-1)^n (\det A)(\det B)$.

P.9.11 Let $A \in \mathsf{M}_n$. Let $A_1 = \mathrm{Re}\,A$ and $A_2 = \mathrm{Im}\,A$, so $A_1, A_2 \in \mathsf{M}_n(\mathbb{R})$ and $A = A_1 + iA_2$. Let

$$V = \frac{1}{\sqrt{2}}\begin{bmatrix} -iI_n & -iI_n \\ I_n & -I_n \end{bmatrix} \in \mathsf{M}_{2n} \quad \text{and} \quad C = \begin{bmatrix} A_1 & A_2 \\ A_2 & -A_1 \end{bmatrix} \in \mathsf{M}_{2n}(\mathbb{R}).$$

(a) Show that V is unitary and

$$V^*CV = \begin{bmatrix} 0 & A \\ \overline{A} & 0 \end{bmatrix}.$$

(b) Show that the characteristic polynomials of C, $A\overline{A}$, and $\overline{A}A$ are related by the identities

$$p_C(z) = p_{A\overline{A}}(z^2) = p_{\overline{A}A}(z^2).$$

(c) If λ is an eigenvalue of C with multiplicity k, why are $-\lambda$ and $\overline{\lambda}$ also eigenvalues of C, each with multiplicity k? If zero is an eigenvalue of C, why must it have even multiplicity? (d) If $\pm\lambda_1, \pm\lambda_2, \ldots, \pm\lambda_n$ are the eigenvalues of C, explain why the eigenvalues of $A\overline{A}$ are $\lambda_1^2, \lambda_2^2, \ldots, \lambda_n^2$. Why do the non-real eigenvalues of $A\overline{A}$ occur in conjugate pairs?

P.9.12 Let $A \in \mathsf{M}_n$. Let $A_1 = \mathrm{Re}\,A$ and $A_2 = \mathrm{Im}\,A$, so $A_1, A_2 \in \mathsf{M}_n(\mathbb{R})$ and $A = A_1 + iA_2$. Let $B = B_1 + iB_2 \in \mathsf{M}_n$ with $B_1, B_2 \in \mathsf{M}_n(\mathbb{R})$. Let $U = \frac{1}{\sqrt{2}}\begin{bmatrix} I_n & iI_n \\ iI_n & I_n \end{bmatrix}$ and consider

$$C(A) = \begin{bmatrix} A_1 & -A_2 \\ A_2 & A_1 \end{bmatrix}, \tag{9.8.2}$$

which is a matrix of complex type; see P.3.11 and P.3.12. (a) Show that U is unitary and $U^*C(A)U = A \oplus \overline{A}$. (b) If $\lambda_1, \lambda_2, \ldots, \lambda_n$ are the eigenvalues of A, show that $\lambda_1, \lambda_2, \ldots, \lambda_n, \overline{\lambda}_1, \overline{\lambda}_2, \ldots, \overline{\lambda}_n$ are the eigenvalues of $C(A)$. (c) Why is $\det C(A) \geq 0$? (d) Show that the characteristic polynomials of $C(A)$, A, and $\overline{A}$ satisfy the identity

$p_C(z) = p_A(z)p_{\bar{A}}(z)$. (e) What can you say if $n = 1$? (f) Show that $C(A + B) = C(A)+C(B)$ and $C(A)C(B) = C(AB)$. (g) Show that $C(I_n) = I_{2n}$. (h) If A is invertible, explain why the real and imaginary parts of A^{-1} are, respectively, the $(1,1)$ and $(1,2)$ blocks of the 2×2 block matrix $C(A)^{-1}$.

P.9.13 Let $A, B \in M_n$ and let $Q = \frac{1}{\sqrt{2}}\begin{bmatrix} I_n & I_n \\ I_n & -I_n \end{bmatrix} \in M_{2n}$. Consider

$$C = \begin{bmatrix} A & B \\ B & A \end{bmatrix} \in M_{2n}, \qquad (9.8.3)$$

which is a block centrosymmetric matrix; see P.3.13 and P.3.14. (a) Show that Q is real orthogonal and $Q^T C Q = (A + B) \oplus (A - B)$. (b) Show that every eigenvalue of $A+B$ and every eigenvalue of $A-B$ is an eigenvalue of C. What about the converse? (c) Show that $\det C = \det(A^2 - AB + BA - B^2)$. If A and B commute, compare this identity with (3.7.4) and (3.7.5).

P.9.14 Let $\lambda_1, \lambda_2, \ldots, \lambda_n$ be the eigenvalues of $A \in M_n$. Use the preceding problem to determine the eigenvalues of $C = \begin{bmatrix} A & A \\ A & A \end{bmatrix} \in M_{2n}$. What does P.8.28 say about the eigenvalues of C? Discuss.

P.9.15 Let $A, B \in M_n$ and suppose that either A or B is invertible. Show that AB is similar to BA, and conclude that these two products have the same eigenvalues.

P.9.16 (a) Consider the 2×2 matrices $A = \begin{bmatrix} 0 & 1 \\ 0 & 0 \end{bmatrix}$ and $B = \begin{bmatrix} 0 & 0 \\ 1 & 0 \end{bmatrix}$. Is AB similar to BA? Do these two products have the same eigenvalues? (b) Answer the same questions for $A = \begin{bmatrix} 0 & 1 \\ 0 & 0 \end{bmatrix}$ and $B = \begin{bmatrix} 1 & 0 \\ 0 & 0 \end{bmatrix}$.

P.9.17 Verify that the matrix (9.7.11) is equal to

$$\begin{bmatrix} \frac{1}{2}n(n+1) & n \\ \frac{1}{6}n(n+1)(2n+1) & \frac{1}{2}n(n+1) \end{bmatrix}$$

and that its eigenvalues are

$$n(n+1)\left(\frac{1}{2} \pm \sqrt{\frac{2n+1}{6(n+1)}}\right).$$

What are the eigenvalues of the matrix (9.7.10)?

P.9.18 Use the vectors **e** and **r** in Example 9.7.9. (a) Verify that

$$A = [i - j] = \begin{bmatrix} 0 & -1 & -2 & \cdots & -n+1 \\ 1 & 0 & -1 & \cdots & -n+2 \\ 2 & 1 & 0 & \cdots & -n+3 \\ \vdots & \vdots & \vdots & \ddots & \vdots \\ n-1 & n-2 & n-3 & \cdots & 0 \end{bmatrix}$$

$$= \mathbf{r}\mathbf{e}^T - \mathbf{e}\mathbf{r}^T = ZY^T,$$

in which $Y = [\mathbf{e}\ \mathbf{r}]$ and $Z = [\mathbf{r}\ -\mathbf{e}]$. (b) Show that the eigenvalues of A are the two eigenvalues of

$$Y^\mathsf{T}Z = \begin{bmatrix} \mathbf{e}^\mathsf{T}\mathbf{r} & -\mathbf{e}^\mathsf{T}\mathbf{e} \\ \mathbf{r}^\mathsf{T}\mathbf{r} & -\mathbf{r}^\mathsf{T}\mathbf{e} \end{bmatrix} = \begin{bmatrix} \frac{1}{2}n(n+1) & -n \\ \frac{1}{6}n(n+1)(2n+1) & -\frac{1}{2}n(n+1) \end{bmatrix},$$

together with $n-2$ zeros. (c) Show that the discriminant of $Y^\mathsf{T}Z$ is negative (see P.9.8) and explain what this implies about the eigenvalues. (d) Show that the eigenvalues of $Y^\mathsf{T}Z$ are

$$\pm i\frac{n}{2}\sqrt{\frac{n^2-1}{3}}.$$

P.9.19 Let $n \geq 3$ and consider

$$A = \begin{bmatrix} a & a & a & \cdots & a \\ a & b & b & \cdots & b \\ \vdots & \vdots & \vdots & \vdots & \vdots \\ a & b & b & \cdots & b \\ a & a & a & \cdots & a \end{bmatrix} \in \mathsf{M}_n, \quad B = \begin{bmatrix} 1 & 0 \\ 0 & 1 \\ \vdots & \vdots \\ 0 & 1 \\ 1 & 0 \end{bmatrix} \in \mathsf{M}_{n,2}, \quad C^\mathsf{T} = \begin{bmatrix} a & a \\ a & b \\ \vdots & \vdots \\ a & b \\ a & b \end{bmatrix} \in \mathsf{M}_{n,2}.$$

(a) Show that $A = BC$. (b) Show that the eigenvalues of A are $n-2$ zeros and the eigenvalues of

$$\begin{bmatrix} 2a & (n-2)a \\ a+b & (n-2)b \end{bmatrix}.$$

(c) If a and b are real, show that all the eigenvalues of A are real.

P.9.20 Let $n \geq 2$ and let $A = [(i-1)n+j] \in \mathsf{M}_n$. If $n = 3$, then

$$A = \begin{bmatrix} 1 & 2 & 3 \\ 4 & 5 & 6 \\ 7 & 8 & 9 \end{bmatrix}.$$

(a) What is A if $n = 4$? (b) Let $\mathbf{v} = [0\ 1\ 2\ \ldots\ n-1]^\mathsf{T}$ and $\mathbf{r} = [1\ 2\ 3\ \ldots\ n]^\mathsf{T}$. Let $X = [\mathbf{v}\ \mathbf{e}]^\mathsf{T}$ and $Y = [n\mathbf{e}\ \mathbf{r}]^\mathsf{T}$. Show that $A = XY^\mathsf{T}$ and rank $A = 2$. (c) Show that the eigenvalues of A are $n-2$ zeros and the eigenvalues of

$$\begin{bmatrix} n\mathbf{e}^\mathsf{T}\mathbf{v} & n^2 \\ \mathbf{r}^\mathsf{T}\mathbf{v} & \mathbf{e}^\mathsf{T}\mathbf{r} \end{bmatrix}.$$

(d) Why are all the eigenvalues of A real?

P.9.21 Let $n \geq 2$. Let λ and μ be eigenvalues of $A \in \mathsf{M}_n$. Let $(\lambda, \mathbf{x})$ be an eigenpair of A and let $(\overline{\mu}, \mathbf{y})$ be an eigenpair of A^*. (a) Show that $\mathbf{y}^*A = \mu\mathbf{y}^*$. (b) If $\lambda \neq \mu$, prove that $\mathbf{y}^*\mathbf{x} = 0$. This is known as the *principle of biorthogonality*.

P.9.22 Suppose that $A, B \in \mathsf{M}_n$ commute and A has distinct eigenvalues. Use Theorem 8.5.1 to show that B is diagonalizable. Moreover, show that there is an invertible $S \in \mathsf{M}_n$ such that $S^{-1}AS$ and $S^{-1}BS$ are both diagonal matrices.

P.9.23 Use Lemma 3.3.21 to give another proof of the assertion in the preceding problem. *Hint*: If $A = S\Lambda S^{-1}$ commutes with B, then Λ commutes with $S^{-1}BS$ and hence $S^{-1}BS$ is diagonal.

P.9.24 Find an $A \in M_3$ that has eigenvalues 0, 1, and -1 and associated eigenvectors $[0\ 1\ -1]^T$, $[1\ -1\ 1]^T$, and $[0\ 1\ 1]^T$, respectively.

P.9.25 If $A \in M_n(\mathbb{R})$ and n is odd, show that A has at least one real eigenvalue.

P.9.26 Let $f(z) = z^n + c_{n-1}z^{n-1} + \cdots + c_1 z + c_0$ and let

$$
C_f = \begin{bmatrix}
0 & 0 & \cdots & 0 & -c_0 \\
1 & 0 & \cdots & 0 & -c_1 \\
0 & 1 & \cdots & 0 & -c_2 \\
\vdots & \vdots & \ddots & \vdots & \vdots \\
0 & 0 & \cdots & 1 & -c_{n-1}
\end{bmatrix}.
$$

(a) Show that $p_{C_f} = f$. Begin as follows: Add z times row n of $zI - C_f$ to row $n - 1$; then add z times row $n - 1$ to row $n - 2$. (b) Show that $p_{C_f} = f$ by using induction and (3.4.11). (c) Theorems about location of eigenvalues of a matrix can be used to say something about zeros of a polynomial f. For example, every zero of f is in the disk with radius $\max\{|c_0|, 1 + |c_1|, \ldots, 1 + |c_{n-1}|\}$ centered at the origin. It is also in the disk with radius $\max\{1, |c_0| + |c_1| + \cdots + |c_{n-1}|\}$ centered at the origin. Why? Can you give better bounds?

P.9.27 Let $A \in M_3$. (a) Use the presentation (9.2.1) for $p_A(z)$ to show that its coefficient c_1 in the presentation (9.1.6) is equal to $\lambda_2\lambda_3 + \lambda_1\lambda_3 + \lambda_1\lambda_2$ (a sum of three terms, each obtained from $\lambda_1\lambda_2\lambda_3$ by omitting one factor). (b) Now examine the expression for c_1 derived in the computation for (9.1.4). Show that c_1 is equal to the sum of the determinants of three matrices obtained from A by omitting row i and column i, for $i = 1, 2, 3$. (c) Make a list of the alternative (but equal) sets of expressions for all the coefficients of $p_A(z)$ when $n = 3$. One set involves only eigenvalues of A; the other set involves only functions of entries of A that are determinants of submatrices of sizes 1, 2, and 3. Explain. What do you think happens for $n = 4$ and larger? See [HJ13, Section 1.2] for the rest of the story.

P.9.28 Someone hands you an $n \times n$ matrix and tells you that it is an orthogonal projection. Describe how to compute the rank of this matrix using at most n operations of addition, multiplication, and division of scalars.

P.9.29 Use Theorems 9.2.7 and 9.7.2 to prove the Sylvester determinant identity (3.7.11). *Hint*: What are the eigenvalues of $I + AB$ and $I + BA$?

P.9.30 Let $\mathcal{F}$ be a nonempty subset of M_n and let $\mathcal{F}'$ be its commutant. If $A, B \in \mathcal{F}'$, show that $AB \in \mathcal{F}'$.

P.9.31 Let $A \in M_n$ be diagonalizable. Show that e^A is invertible and e^{-A} is its inverse.

P.9.32 We say that $B \in M_n$ is a *square root* of $A \in M_n$ if $B^2 = A$.

(a) Show that $\begin{bmatrix} 1 & 1 \\ 0 & 1 \end{bmatrix}$ is a square root of $\begin{bmatrix} 1 & 2 \\ 0 & 1 \end{bmatrix}$.

(b) Show that $\begin{bmatrix} 0 & 1 \\ 0 & 0 \end{bmatrix}$ does not have a square root.

(c) Show that each of the three matrices in (a) and (b) is nondiagonalizable.

P.9.33 Show that every diagonalizable matrix has a square root.

P.9.34 Let $A \in M_n$ and assume that $\operatorname{tr} A = 0$. Incorporate the following ideas into a proof that A is a commutator of matrices in M_n (Shoda's theorem; Theorem 3.5.7). (a) Use induction and Lemma 3.5.4 to show that it suffices to consider a block matrix of the form

$$A = \begin{bmatrix} 0 & \mathbf{x}^* \\ \mathbf{y} & BC - CB \end{bmatrix} \quad \text{and} \quad B, C \in M_{n-1}.$$

(b) Consider $B + \lambda I$ and show that B may be assumed to be invertible. (c) Let

$$X = \begin{bmatrix} 0 & 0 \\ 0 & B \end{bmatrix} \quad \text{and} \quad Y = \begin{bmatrix} 0 & \mathbf{u}^* \\ \mathbf{v} & C \end{bmatrix}.$$

Show that there are vectors $\mathbf{u}$ and $\mathbf{v}$ such that $A = XY - YX$.

P.9.35 Let $A \in M_n$ be idempotent. Show that (a) $\operatorname{tr} A \in \{0, 1, 2, \ldots, n\}$, (b) $\operatorname{tr} A = n$ if and only if $A = I$, and (c) $\operatorname{tr} A = 0$ if and only if $A = 0$.

P.9.36 Let $A \in M_2$ and let f be a complex-valued function on $\operatorname{spec} A = \{\lambda, \mu\}$. (a) If $\lambda \neq \mu$ and $f(A)$ is defined by (9.5.6), show that

$$f(A) = \frac{f(\lambda) - f(\mu)}{\lambda - \mu} A + \frac{\lambda f(\mu) - \mu f(\lambda)}{\lambda - \mu} I. \tag{9.8.4}$$

(b) If $\lambda = \mu$ and A is diagonalizable, why is $f(A) = f(\lambda) I$?

9.9 Notes

None of the schemes discussed in this chapter should be programmed to compute eigenpairs of real-world data matrices. They are important conceptually, but different strategies have been implemented for numerical computations. Improving algorithms for numerical calculation of eigenvalues has been a top priority research goal for numerical analysts since the mid-twentieth century [GVL13, Chap. 7 and 8].

Example 9.6.4 reveals only part of the story about Corollary 9.6.3.a. The commutant of A is the set of all polynomials in A if and only if every eigenvalue of A has geometric multiplicity 1; see [HJ13, Thm. 3.2.4.2].

P.9.26 suggests two bounds on the zeros of a given polynomial. For many others, see [HJ13, 5.6.P27–35].

If the 2×2 matrix in P.9.36 is not diagonalizable and f is differentiable on $\operatorname{spec} A = \{\lambda\}$, it turns out that

$$f(A) = f'(\lambda) A + (f(\lambda) - \lambda f'(\lambda)) I. \tag{9.9.1}$$

For an explanation of this remarkable formula, see [HJ94, P.11, Sect. 6.1].

9.10 Some Important Concepts

- Characteristic polynomial of a matrix.

- Algebraic multiplicity of an eigenvalue.

- Trace, determinant, and eigenvalues (Theorem 9.2.7).

- Diagonalizable matrices and eigenvalue multiplicities (Theorem 9.4.7).
- Simultaneous diagonalization of commuting matrices (Theorem 9.4.15).
- Functional calculus for diagonalizable matrices (9.5.6).
- Eigenvalues of AB and BA (Theorem 9.7.2).

10 Unitary Triangularization and Block Diagonalization

Many facts about matrices can be revealed (or questions about them answered) by performing a suitable transformation that puts them into a special form. Such a form typically contains many zero entries in strategic locations. For example, Theorem 9.4.3 says that some matrices are similar to diagonal matrices, and Theorem 6.6.3 says that every square matrix is unitarily similar to an upper Hessenberg matrix. In this chapter, we show that every square complex matrix is unitarily similar to an upper triangular matrix. This is a powerful result with a host of important consequences.

10.1 Schur's Triangularization Theorem

Theorem 10.1.1 (Schur Triangularization) *Let the eigenvalues of $A \in \mathsf{M}_n$ be arranged in any given order $\lambda_1, \lambda_2, \ldots, \lambda_n$ (including multiplicities), and let $(\lambda_1, \mathbf{x})$ be an eigenpair of A, in which $\mathbf{x}$ is a unit vector.*

(a) *There is a unitary $U = [\mathbf{x}\ U_2] \in \mathsf{M}_n$ such that $A = UTU^*$, in which $T = [t_{ij}]$ is upper triangular and has diagonal entries $t_{ii} = \lambda_i$ for $i = 1, 2, \ldots, n$.*

(b) *If A is real, each eigenvalue $\lambda_1, \lambda_2, \ldots, \lambda_n$ is real, and $\mathbf{x}$ is real, then there is a real orthogonal $Q = [\mathbf{x}\ Q_2] \in \mathsf{M}_n(\mathbb{R})$ such that $A = QTQ^\mathsf{T}$, in which $T = [t_{ij}]$ is real upper triangular and has diagonal entries $t_{ii} = \lambda_i$ for $i = 1, 2, \ldots, n$.*

Proof (a) We proceed by induction on n. In the base case $n = 1$, there is nothing to prove. For our induction hypothesis, assume that $n \geq 2$ and every matrix in M_{n-1} can be factored as asserted. Suppose that $A \in \mathsf{M}_n$ has eigenvalues $\lambda_1, \lambda_2, \ldots, \lambda_n$ and let $\mathbf{x}$ be a unit eigenvector of A associated with the eigenvalue λ_1. Corollary 6.4.10 ensures that there is a unitary $V = [\mathbf{x}\ V_2] \in \mathsf{M}_n$ whose first column is $\mathbf{x}$. Since the columns of V are orthonormal, $V_2^* \mathbf{x} = \mathbf{0}$. Then

$$AV = [A\mathbf{x}\ AV_2] = [\lambda_1 \mathbf{x}\ AV_2]$$

and

$$V^*AV = \begin{bmatrix} \mathbf{x}^* \\ V_2^* \end{bmatrix} [\lambda_1 \mathbf{x}\ AV_2] = \begin{bmatrix} \lambda_1 \mathbf{x}^* \mathbf{x} & \mathbf{x}^* AV_2 \\ \lambda_1 V_2^* \mathbf{x} & V_2^* AV_2 \end{bmatrix} = \begin{bmatrix} \lambda_1 & \star \\ \mathbf{0} & A' \end{bmatrix}. \tag{10.1.2}$$

Because A and V^*AV are (unitarily) similar, they have the same eigenvalues. The 1×1 block $[\lambda_1]$ is one of the two diagonal blocks of (10.1.2); its other diagonal block $A' \in \mathsf{M}_{n-1}$ has eigenvalues $\lambda_2, \lambda_3, \ldots, \lambda_n$ (see Theorem 9.2.8).

The induction hypothesis ensures that there is a unitary $W \in \mathsf{M}_{n-1}$ such that $W^*A'W = T'$, in which $T' = [\tau_{ij}] \in \mathsf{M}_{n-1}$ is upper triangular and has diagonal entries $\tau_{ii} = \lambda_{i+1}$ for $i = 1, 2, \ldots, n-1$. Then

$$U = V \begin{bmatrix} 1 & 0 \\ 0 & W \end{bmatrix}$$

is unitary (it is a product of unitaries), has the same first column as V, and satisfies

$$
\begin{aligned}
U^*AU &= \begin{bmatrix} 1 & 0 \\ 0 & W \end{bmatrix}^* V^*AV \begin{bmatrix} 1 & 0 \\ 0 & W \end{bmatrix} \\
&= \begin{bmatrix} 1 & 0 \\ 0 & W^* \end{bmatrix} \begin{bmatrix} \lambda_1 & \star \\ 0 & A' \end{bmatrix} \begin{bmatrix} 1 & 0 \\ 0 & W \end{bmatrix} \\
&= \begin{bmatrix} \lambda_1 & \star \\ 0 & W^*A'W \end{bmatrix} = \begin{bmatrix} \lambda_1 & \star \\ 0 & T' \end{bmatrix},
\end{aligned}
$$

which has the asserted form. This completes the induction.

(b) If a real matrix has a real eigenvalue, then it has a real associated unit eigenvector (see P.8.5). Corollary 6.4.10 shows how to construct a real unitary matrix with a given real first column. Now proceed as in (a). □

As a first application of Theorem 10.1.1, we have a transparent demonstration (independent of the characteristic polynomial; see Theorem 9.2.7) that the trace and determinant of a matrix are the sum and product, respectively, of its eigenvalues.

Corollary 10.1.3 *Let $\lambda_1, \lambda_2, \ldots, \lambda_n$ be the eigenvalues of $A \in \mathsf{M}_n$. Then*

$$\operatorname{tr} A = \lambda_1 + \lambda_2 + \cdots + \lambda_n \quad and \quad \det A = \lambda_1\lambda_2 \cdots \lambda_n.$$

Proof Let $A = UTU^*$, as in the preceding theorem. Then

$$\operatorname{tr} A = \operatorname{tr} UTU^* = \operatorname{tr} U(TU^*) = \operatorname{tr}(TU^*)U = \operatorname{tr} T(U^*U)$$
$$= \operatorname{tr} T = \lambda_1 + \lambda_2 + \cdots + \lambda_n$$

and

$$\det A = \det(UTU^*) = (\det U)(\det T)(\det U^{-1})$$
$$= (\det U)(\det T)(\det U)^{-1} = \det T = \lambda_1\lambda_2 \cdots \lambda_n. \quad □$$

Lemma 8.3.2 says that if λ is an eigenvalue of $A \in \mathsf{M}_n$ and p is a polynomial, then $p(\lambda)$ is an eigenvalue of $p(A)$. What is its algebraic multiplicity? Schur's triangularization theorem permits us to answer that question. The key observation is that if $T = [t_{ij}] \in \mathsf{M}_n$ is upper triangular, then the diagonal entries of $p(T)$ are $p(t_{11}), p(t_{22}), \ldots, p(t_{nn})$. These are all the eigenvalues of $p(T)$, including multiplicities.

Corollary 10.1.4 *Let $A \in \mathsf{M}_n$ and let p be a polynomial. If $\lambda_1, \lambda_2, \ldots, \lambda_n$ are the eigenvalues of A, then $p(\lambda_1), p(\lambda_2), \ldots, p(\lambda_n)$ are the eigenvalues of $p(A)$ (including multiplicities in both cases).*

Proof Let $A = UTU^*$, as in Theorem 10.1.1. Since $U^* = U^{-1}$, (0.8.2) ensures that

$$p(A) = p(UTU^*) = Up(T)U^*,$$

so the eigenvalues of $p(A)$ are the same as the eigenvalues of $p(T)$, which are

$$p(\lambda_1), p(\lambda_2), \dots, p(\lambda_n). \qquad \square$$

Corollary 10.1.5 *Let $A \in M_n$ and let p be a polynomial. If $p(\lambda) \neq 0$ for every $\lambda \in \operatorname{spec} A$, then $p(A)$ is invertible.*

Proof The preceding corollary tells us that $0 \notin \operatorname{spec} p(A)$. Theorem 8.1.17.d ensures that $p(A)$ is invertible. $\qquad \square$

If $(\lambda, \mathbf{x})$ is an eigenpair of an invertible $A \in M_n$, then $A\mathbf{x} = \lambda\mathbf{x}$ implies that $A^{-1}\mathbf{x} = \lambda^{-1}\mathbf{x}$. We conclude that λ^{-1} is an eigenvalue of A^{-1}, but what is its algebraic multiplicity? Schur's triangularization theorem permits us to answer this question, too.

Corollary 10.1.6 *Let $A \in M_n$ be invertible. If $\lambda_1, \lambda_2, \dots, \lambda_n$ are the eigenvalues of A, then $\lambda_1^{-1}, \lambda_2^{-1}, \dots, \lambda_n^{-1}$ are the eigenvalues of A^{-1} (including multiplicities in both cases).*

Proof Let $A = UTU^*$, as in Theorem 10.1.1. Then $A^{-1} = UT^{-1}U^*$. Theorem 3.3.12 ensures that T^{-1} is upper triangular and its diagonal entries are $\lambda_1^{-1}, \lambda_2^{-1}, \dots, \lambda_n^{-1}$. These are the eigenvalues of A^{-1}. $\qquad \square$

10.2 The Cayley–Hamilton Theorem

Our development of the theory of eigenvalues and eigenvectors relies on the existence of annihilating polynomials. We now use Schur's theorem to construct an annihilating polynomial for $A \in M_n$ that has degree n.

Theorem 10.2.1 (Cayley–Hamilton) *Let*

$$p_A(z) = \det(zI - A) = z^n + c_{n-1}z^{n-1} + \cdots + c_1z + c_0 \qquad (10.2.2)$$

be the characteristic polynomial of $A \in M_n$. Then

$$p_A(A) = A^n + c_{n-1}A^{n-1} + c_{n-2}A^{n-2} + \cdots + c_1A + c_0I_n = 0.$$

Proof Let $\lambda_1, \lambda_2, \dots, \lambda_n$ be the eigenvalues of A and write

$$p_A(z) = (z - \lambda_1)(z - \lambda_2) \cdots (z - \lambda_n), \qquad (10.2.3)$$

as in (9.2.1). Schur's triangularization theorem says that there is a unitary $U \in M_n$ and an upper triangular $T = [t_{ij}] \in M_n$ such that $A = UTU^*$ and $t_{ii} = \lambda_i$ for $i = 1, 2, \dots, n$. Since (0.8.2) ensures that

$$p_A(A) = p_A(UTU^*) = Up_A(T)U^*,$$

it suffices to show that

$$p_A(T) = (T - \lambda_1 I)(T - \lambda_2 I) \cdots (T - \lambda_n I) = 0. \tag{10.2.4}$$

Our strategy is to show that for each $j = 1, 2, \ldots, n$,

$$P_j = (T - \lambda_1 I)(T - \lambda_2 I) \cdots (T - \lambda_j I)$$

is a block matrix of the form

$$P_j = [0_{n \times j} \; \star]. \tag{10.2.5}$$

If we can do this, then (10.2.4) follows as the case $j = n$ in (10.2.5).

We proceed by induction. In the base case, the upper triangularity of T and the presence of λ_1 in its $(1, 1)$ entry ensure that $P_1 = (T - \lambda_1 I) = [\mathbf{0} \; \star]$ has a zero first column. The induction hypothesis is that for some $j \in \{1, 2, \ldots, n - 1\}$,

$$P_j = \begin{bmatrix} 0_{j \times j} & \star \\ 0_{(n-j) \times j} & \star \end{bmatrix}.$$

The upper triangular matrix $T - \lambda_{j+1} I$ has zero entries in and below its $(j + 1, j + 1)$ diagonal position, so

$$P_{j+1} = P_j(T - \lambda_{j+1} I) = \begin{bmatrix} 0_{j \times j} & \star \\ 0_{(n-j) \times j} & \star \end{bmatrix} \begin{bmatrix} \star & \star & \star \\ 0_{(n-j) \times j} & \mathbf{0} & \star \end{bmatrix}$$

$$= \begin{bmatrix} 0_{j \times j} & \mathbf{0} & \star \\ 0_{(n-j) \times j} & \mathbf{0} & \star \end{bmatrix} = [0_{n \times (j+1)} \; \star].$$

The induction is complete. □

Example 10.2.6 If $A \in \mathsf{M}_n$, then p_A is a monic polynomial of degree n that annihilates A. But a monic polynomial of lesser degree might also annihilate A. For example, the characteristic polynomial of

$$A = \begin{bmatrix} 0 & 1 & 0 \\ 0 & 0 & 0 \\ 0 & 0 & 0 \end{bmatrix}$$

is $p_A(z) = z^3$, but z^2 also annihilates A.

Example 10.2.7 Consider

$$A = \begin{bmatrix} 1 & 2 \\ 3 & 4 \end{bmatrix} \tag{10.2.8}$$

and its characteristic polynomial $p_A(z) = z^2 - 5z - 2$. Then

$$A^2 - 5A - 2I = \begin{bmatrix} 7 & 10 \\ 15 & 22 \end{bmatrix} - \begin{bmatrix} 5 & 10 \\ 15 & 20 \end{bmatrix} - \begin{bmatrix} 2 & 0 \\ 0 & 2 \end{bmatrix} = \begin{bmatrix} 0 & 0 \\ 0 & 0 \end{bmatrix}.$$

Rewriting $p_A(A) = 0$ as $A^2 = 5A + 2I$ suggests other identities, for example

$$A^3 = A(A^2) = A(5A + 2I)$$
$$= 5A^2 + 2A = 5(5A + 2I) + 2A$$
$$= 27A + 10I$$

and

$$A^4 = (A^2)^2 = (5A + 2I)^2 = 25A^2 + 20A + 4I$$
$$= 25(5A + 2I) + 20A + 4I$$
$$= 145A + 54I.$$

Our first corollary of the Cayley–Hamilton theorem provides a systematic way to derive and understand identities like those in the preceding example. It relies on the division algorithm (see Section 0.7).

Corollary 10.2.9 *Let $A \in M_n$ and let f be a polynomial of degree n or more. Let q and r be polynomials such that $f = p_A q + r$, in which the degree of r is less than n. Then $f(A) = r(A)$.*

Proof The degree of p_A is n, so the division algorithm ensures that there are (unique) polynomials q and r such that $f = p_A q + r$ and the degree of r is less than n. Then

$$f(A) = p_A(A)q(A) + r(A) = 0q(A) + r(A) = r(A). \qquad \square$$

Example 10.2.10 Let A be the matrix (10.2.8). If $f(z) = z^3$, then $f = p_A q + r$ with $q(z) = z + 5$ and $r(z) = 27z + 10$. Consequently, $A^3 = 27A + 10I$. If $f(z) = z^4$, then $f = p_A q + r$ with $q(z) = z^2 + 5z + 27$ and $r(z) = 145z + 54$, so $A^4 = 145A + 54I$. Both of these computations agree with the identities in Example 10.2.7.

Corollary 10.2.11 *Let $A \in M_n$ and let $S = \text{span}\{I, A, A^2, ...\} \subseteq M_n$. Then $S = \text{span}\{I, A, A^2, \ldots, A^{n-1}\}$ and $\dim S \leq n$.*

Proof For each integer $k \geq 0$, the preceding corollary ensures that there is a polynomial r_k of degree at most $n - 1$ such that $A^{n+k} = r_k(A)$. $\qquad \square$

If A is invertible, then the Cayley–Hamilton theorem permits us to express A^{-1} as a polynomial in A that is closely related to its characteristic polynomial.

Corollary 10.2.12 *Let $A \in M_n$ have characteristic polynomial (10.2.2). If A is invertible, then*

$$A^{-1} = -c_0^{-1}(A^{n-1} + c_{n-1}A^{n-2} + \cdots + c_2 A + c_1 I). \qquad (10.2.13)$$

Proof Theorem 9.1.5 ensures that $c_0 = (-1)^n \det A$, which is nonzero since A is invertible. Thus, we may rewrite

$$A^n + c_{n-1}A^{n-1} + \cdots + c_2 A^2 + c_1 A + c_0 I = 0$$

as

$$I = A\left(-c_0^{-1}(A^{n-1} + c_{n-1}A^{n-2} + \cdots + c_2A + c_1I)\right)$$
$$= \left(-c_0^{-1}(A^{n-1} + c_{n-1}A^{n-2} + \cdots + c_2A + c_1I)\right)A. \qquad \square$$

Example 10.2.14 Let A be the matrix (10.2.8), for which $p_A(z) = z^2 - 5z - 2$. Then (10.2.13) ensures that

$$A^{-1} = \tfrac{1}{2}(A - 5I) = \begin{bmatrix} -2 & 1 \\ 3/2 & -1/2 \end{bmatrix}.$$

10.3 The Minimal Polynomial

Example 10.2.6 shows that a square matrix might be annihilated by a monic polynomial whose degree is less than the degree of the characteristic polynomial. The following theorem establishes basic properties of an annihilating polynomial of minimum degree.

Theorem 10.3.1 *Let $A \in M_n$ and let $\lambda_1, \lambda_2, \ldots, \lambda_d$ be its distinct eigenvalues.*

(a) *There is a unique monic polynomial m_A of minimum positive degree that annihilates A.*

(b) *The degree of m_A is at most n.*

(c) *If p is a nonconstant polynomial that annihilates A, then there is a polynomial f such that $p = m_A f$. In particular, m_A divides p_A.*

(d) *There are positive integers $q_1, q_2, \ldots, q_d$ such that*

$$m_A(z) = (z - \lambda_1)^{q_1}(z - \lambda_2)^{q_2} \cdots (z - \lambda_d)^{q_d}. \qquad (10.3.2)$$

Each q_i is at least 1 and is at most the algebraic multiplicity of λ_i. In particular, the degree of m_A is at least d.

Proof (a) Lemma 8.2.4 and Theorem 10.2.1 ensure that the set of monic nonconstant polynomials that annihilate A is nonempty. Let ℓ be the degree of a monic polynomial of minimum positive degree that annihilates A. Let g and h be monic polynomials of degree ℓ that annihilate A. The division algorithm ensures that there is a polynomial f and a polynomial r with degree less than ℓ such that $g = hf + r$. Then

$$0 = g(A) = h(A)f(A) + r(A) = 0 + r(A) = r(A),$$

so r is a polynomial of degree less than ℓ that annihilates A. The definition of ℓ ensures that r has degree 0, so $r(z) = a$ for some scalar a. Then $aI = r(A) = 0$, so $a = 0$ and $g = hf$. But g and h are both monic polynomials of the same degree, so $f(z) = 1$ and $g = h$. Let $m_A = h$.

(b) p_A annihilates A and has degree n, so $\ell \leq n$.

(c) If p is a nonconstant polynomial that annihilates A, then its degree must be greater than or equal to the degree of m_A. The division algorithm ensures that there are polynomials f and r such that $p = m_A f + r$ and the degree of r is less than the degree of m_A. The argument in (a) shows that r is the zero polynomial and hence $p = m_A f$.

(d) Because m_A divides p_A, the factorization (10.3.2) follows from (9.2.2). Each exponent q_i can be no larger than the corresponding algebraic multiplicity n_i. Theorem 8.3.3 tells us that $z - \lambda_i$ is a factor of m_A for each $i = 1, 2, \ldots, d$, so each q_i is at least 1. $\square$

Definition 10.3.3 Let $A \in M_n$. The *minimal polynomial* m_A is the unique monic polynomial of minimum positive degree that annihilates A.

The equation (10.3.2) reveals that every eigenvalue of A is a zero of m_A and that every zero of m_A is an eigenvalue of A.

Theorem 10.3.4 *If $A, B \in M_n$ are similar, then $p_A = p_B$ and $m_A = m_B$.*

Proof The first assertion is Theorem 9.3.1. If p is a polynomial, $S \in M_n$ is invertible, and $A = SBS^{-1}$, then (0.8.2) ensures that $p(A) = Sp(B)S^{-1}$. Thus, $p(A) = 0$ if and only if $p(B) = 0$. Theorem 10.3.1.c ensures that m_A divides m_B and that m_B divides m_A. Since both polynomials are monic, $m_A = m_B$. $\square$

Example 10.3.5 Similar matrices have the same minimal polynomials, but two matrices with the same minimal polynomial need not be similar. Let

$$A = \begin{bmatrix} 1 & 1 & 0 & 0 \\ 0 & 1 & 0 & 0 \\ 0 & 0 & 1 & 1 \\ 0 & 0 & 0 & 1 \end{bmatrix} \quad \text{and} \quad B = \begin{bmatrix} 1 & 1 & 0 & 0 \\ 0 & 1 & 0 & 0 \\ 0 & 0 & 1 & 0 \\ 0 & 0 & 0 & 1 \end{bmatrix}.$$

Then $p_A(z) = p_B(z) = (z - 1)^4$ and $m_A(z) = m_B(z) = (z - 1)^2$. However, $\text{rank}(A - I) = 2$ and $\text{rank}(B - I) = 1$, so A and B are not similar; see (0.8.4).

Example 10.3.6 Let A be the matrix in Example 10.2.6. Only three nonconstant monic polynomials divide p_A, namely z, z^2, and z^3. Since z does not annihilate A, but z^2 does, $m_A(z) = z^2$.

Example 10.3.7 Let $A \in M_5$ and suppose that $p_A(z) = (z-1)^3(z+1)^2$. Six distinct polynomials of the form $(z - 1)^{q_1}(z + 1)^{q_2}$ satisfy $1 \leq q_1 \leq 3$ and $1 \leq q_2 \leq 2$: one of degree 2, two of degree 3, two of degree 4, and one of degree 5. One of these six polynomials is m_A. If $(A - I)(A + I) = 0$, then $m_A(z) = (z - 1)(z + 1)$. If not, then check whether

$$\text{(a)} \ (A - I)^2(A + I) = 0 \quad \text{or} \quad \text{(b)} \ (A - I)(A + I)^2 = 0. \quad (10.3.8)$$

Theorem 10.3.1.a ensures that at most one of the identities (10.3.8) is true. If (a) is true, then $m_A(z) = (z - 1)^2(z + 1)$; if (b) is true, then $m_A(z) = (z - 1)(z + 1)^2$. If neither (a) nor (b) is true, then check whether

$$\text{(c)} \ (A - I)^3(A + I) = 0 \quad \text{or} \quad \text{(d)} \ (A - I)^2(A + I)^2 = 0. \quad (10.3.9)$$

At most one of the identities (10.3.9) is true. If (c) is true, then $m_A(z) = (z-1)^3(z+1)$; if (d) is true, then $m_A(z) = (z - 1)^2(z + 1)^2$. If neither (c) nor (d) is true, then $m_A(z) = (z-1)^3(z+1)^2$.

The trial and error method in the preceding example works, in principle, for matrices whose eigenvalues and algebraic multiplicities are known. However, the maximum number of trials required grows rapidly with the size of the matrix. See P.10.30 for another algorithm to compute the minimal polynomial.

Given a matrix, we can determine its minimal polynomial. Given a monic polynomial p, is there a matrix whose minimal polynomial is p?

Definition 10.3.10 If $n \geq 2$, the *companion matrix* of the monic polynomial $f(z) = z^n + c_{n-1}z^{n-1} + \cdots + c_1 z + c_0$ is

$$C_f = \begin{bmatrix} 0 & 0 & \cdots & 0 & -c_0 \\ 1 & 0 & \cdots & 0 & -c_1 \\ 0 & 1 & \cdots & 0 & -c_2 \\ \vdots & \vdots & \ddots & \vdots & \vdots \\ 0 & 0 & 0 & 1 & -c_{n-1} \end{bmatrix} = [\mathbf{e}_2 \ \mathbf{e}_3 \ \ldots \ \mathbf{e}_n \ -\mathbf{c}], \qquad (10.3.11)$$

in which $\mathbf{e}_1, \mathbf{e}_2, \ldots, \mathbf{e}_n$ is the standard basis of $\mathbb{C}^n$ and $\mathbf{c} = [c_0 \ c_1 \ \ldots \ c_{n-1}]^\mathsf{T}$. The companion matrix of $f(z) = z + c_0$ is $C_f = [-c_0]$.

Theorem 10.3.12 *The polynomial $f(z) = z^n + c_{n-1}z^{n-1} + \cdots + c_1 z + c_0$ is both the minimal polynomial and the characteristic polynomial of its companion matrix (10.3.11), that is, $f = p_{C_f} = m_{C_f}$.*

Proof We have $C_f \mathbf{e}_n = -\mathbf{c}$ and $C_f \mathbf{e}_{j-1} = \mathbf{e}_j$ for each $j = 2, 3, \ldots, n$, so $\mathbf{e}_j = C_f \mathbf{e}_{j-1} = C_f^2 \mathbf{e}_{j-2} = \cdots = C_f^{j-1} \mathbf{e}_1$ for each $j = 1, 2, \ldots, n$. Then

$$C_f^n \mathbf{e}_1 = C_f C_f^{n-1} \mathbf{e}_1 = C_f \mathbf{e}_n = -\mathbf{c}$$

$$= -c_0 \mathbf{e}_1 - c_1 \mathbf{e}_2 - c_2 \mathbf{e}_3 - \cdots - c_{n-1} \mathbf{e}_n$$

$$= -c_0 I \mathbf{e}_1 - c_1 C_f \mathbf{e}_1 - c_2 C_f^2 \mathbf{e}_2 + \cdots - c_{n-1} C_f^{n-1} \mathbf{e}_1,$$

which shows that

$$f(C_f)\mathbf{e}_1 = C_f^n \mathbf{e}_1 + c_{n-1} C_f^{n-1} \mathbf{e}_1 + \cdots + c_1 C_f \mathbf{e}_1 + c_0 I \mathbf{e}_1 = \mathbf{0}.$$

For each $j = 2, 3, \ldots, n$,

$$f(C_f)\mathbf{e}_j = f(C_f)C_f^{j-1}\mathbf{e}_1 = C_f^{j-1}f(C_f)\mathbf{e}_1 = \mathbf{0}.$$

Thus, $f(C_f) = 0$. If $g(z) = z^m + b_{m-1}z^{m-1} + \cdots + b_1 z + b_0$ and $m < n$, then

$$g(C_f)\mathbf{e}_1 = C_f^m \mathbf{e}_1 + b_{m-1}C_f^{m-1}\mathbf{e}_1 + \cdots + b_1 C_f \mathbf{e}_1 + b_0 \mathbf{e}_1$$

$$= \mathbf{e}_{m+1} + b_{m-1}\mathbf{e}_m + \cdots + b_1 \mathbf{e}_2 + b_0 \mathbf{e}_1.$$

The linear independence of $\mathbf{e}_1, \mathbf{e}_2, \ldots, \mathbf{e}_{m+1}$ ensures that $g(C_f)\mathbf{e}_1 \neq \mathbf{0}$, so g cannot annihilate C_f. We conclude that $f = m_{C_f}$. Since p_A is a monic polynomial that annihilates C_f (Theorem 10.2.1) and has the same degree as m_{C_f}, it must be m_{C_f}. $\qquad \square$

The final result in this section identifies the minimal polynomial of a diagonalizable matrix and provides a criterion for a matrix to be nondiagonalizable.

Theorem 10.3.13 *Let* $\lambda_1, \lambda_2, \ldots, \lambda_d$ *be the distinct eigenvalues of* $A \in M_n$ *and let*

$$p(z) = (z - \lambda_1)(z - \lambda_2) \cdots (z - \lambda_d). \tag{10.3.14}$$

If A *is diagonalizable, then* $p(A) = 0$ *and* p *is the minimal polynomial of* A.

Proof Let $n_1, n_2, \ldots, n_d$ be the respective multiplicities of $\lambda_1, \lambda_2, \ldots, \lambda_d$. There is an invertible $S \in M_n$ such that $A = S \Lambda S^{-1}$ and

$$\Lambda = \lambda_1 I_{n_1} \oplus \lambda_2 I_{n_2} \oplus \cdots \oplus \lambda_d I_{n_d}.$$

Theorem 9.5.1 tells us that

$$p(A) = Sp(\Lambda)S^{-1} = S\big(p(\lambda_1)I_{n_1} \oplus p(\lambda_2)I_{n_2} \oplus \cdots \oplus p(\lambda_d)I_{n_d}\big)S^{-1}. \tag{10.3.15}$$

Since $p(\lambda_j) = 0$ for each $j = 1, 2, \ldots, d$, each direct summand in (10.3.15) is a zero matrix. Therefore, $p(A) = 0$. Theorem 10.3.1.d says that no monic polynomial of positive degree less than d annihilates A. Since p is a monic polynomial of degree d that annihilates A, it is the minimal polynomial. $\qquad \square$

In the next section we learn that the converse of Theorem 10.3.13 is true. For now, we know that if p is the polynomial (10.3.14) and if $p(A) \neq 0$, then A is not diagonalizable.

Example 10.3.16 The characteristic polynomial of

$$A = \begin{bmatrix} 3 & i \\ i & 1 \end{bmatrix}$$

is $p_A(z) = z^2 - 4z + 4 = (z - 2)^2$. Since $p(z) = z - 2$ does not annihilate A, we conclude that A is not diagonalizable.

10.4 Linear Matrix Equations and Block Diagonalization

If a matrix has two or more distinct eigenvalues, it is similar to an upper triangular matrix that has some off-diagonal zero blocks. To prove this, we use a theorem that employs the Cayley–Hamilton theorem in a clever way.

Theorem 10.4.1 (Sylvester's Theorem on Linear Matrix Equations) *Let* $A \in M_m$ *and* $B \in M_n$, *and suppose that* $\operatorname{spec} A \cap \operatorname{spec} B = \varnothing$. *For each* $C \in M_{m \times n}$,

$$AX - XB = C \tag{10.4.2}$$

has a unique solution $X \in M_{m \times n}$. *In particular, the only solution to* $AX - XB = 0$ *is* $X = 0$.

Proof Define the linear operator $T : M_{m \times n} \rightarrow M_{m \times n}$ by $T(X) = AX - XB$. We claim that T is onto and one to one. Since $M_{m \times n}$ is a finite-dimensional vector space, Corollary 2.5.3 says that T is onto if and only if it is one to one. Thus, it suffices to show that $X = 0$ if $T(X) = 0$.

If $AX = XB$, then Theorems 0.8.1 and 10.2.1 ensure that

$$p_B(A)X = Xp_B(B) = X0 = 0. \tag{10.4.3}$$

The zeros of p_B are the eigenvalues of B, so the hypotheses imply that $p_B(\lambda) \neq 0$ for all $\lambda \in \operatorname{spec} A$. Corollary 10.1.5 tells us that $p_B(A)$ is invertible, so $X = 0$ follows from (10.4.3).   □

Our theorem on block diagonalization is the following.

Theorem 10.4.4 *Let* $\lambda_1, \lambda_2, \ldots, \lambda_d$ *be the distinct eigenvalues of* $A \in M_n$, *in any given order and with respective algebraic multiplicities* $n_1, n_2, \ldots, n_d$. *Then* A *is unitarily similar to a block upper triangular matrix*

$$T = \begin{bmatrix} T_{11} & T_{12} & \cdots & T_{1d} \\ 0 & T_{22} & \cdots & T_{2d} \\ \vdots & & \ddots & \vdots \\ 0 & 0 & \cdots & T_{dd} \end{bmatrix}, \quad T_{ii} \in M_{n_i}, \quad i = 1, 2, \ldots, d, \tag{10.4.5}$$

in which each diagonal block T_{ii} *in* (10.4.5) *is upper triangular and all of its diagonal entries are equal to* λ_i. *Moreover,* A *is similar to the block diagonal matrix*

$$T_{11} \oplus T_{22} \oplus \cdots \oplus T_{dd}, \tag{10.4.6}$$

which contains the diagonal blocks in (10.4.5).

Proof Schur's triangularization theorem ensures that A is unitarily similar to an upper triangular matrix T with the stated properties, so we must prove that T is similar to the block diagonal matrix (10.4.6). We proceed by induction on d. In the base case $d = 1$, there is nothing to prove. For the inductive step, assume that $d \geq 2$ and the asserted block diagonalization has been established for matrices with at most $d - 1$ distinct eigenvalues.

Partition (10.4.5) as

$$T = \begin{bmatrix} T_{11} & C \\ 0 & T' \end{bmatrix}, \quad C = [T_{12} \ T_{13} \ \cdots \ T_{1d}] \in M_{n_1 \times (n - n_1)}.$$

Then $\operatorname{spec} T_{11} = \{\lambda_1\}$ and $\operatorname{spec} T' = \{\lambda_2, \lambda_3, \ldots, \lambda_d\}$, so $\operatorname{spec} T_{11} \cap \operatorname{spec} T' = \varnothing$. Theorem 10.4.1 ensures that there is an $X \in M_{n_1 \times (n - n_1)}$ such that $T_{11} X - X T' = C$. Theorem 3.3.13 says that T is similar to

$$\begin{bmatrix} T_{11} & -T_{11} X + X T' + C \\ 0 & T' \end{bmatrix} = \begin{bmatrix} T_{11} & 0 \\ 0 & T' \end{bmatrix}.$$

The induction hypothesis is that there is an invertible matrix $S \in M_{n - n_1}$ such that $S^{-1} T' S = T_{22} \oplus T_{33} \oplus \cdots \oplus T_{dd}$. Thus, T is similar to

$$\begin{bmatrix} I & 0 \\ 0 & S \end{bmatrix}^{-1} \begin{bmatrix} T_{11} & 0 \\ 0 & T' \end{bmatrix} \begin{bmatrix} I & 0 \\ 0 & S \end{bmatrix} = T_{11} \oplus T_{22} \oplus \cdots \oplus T_{dd}. \quad □$$

Definition 10.4.7 $A \in M_n$ *is* unispectral *if* $\operatorname{spec} A = \{\lambda\}$ *for some scalar* λ.

Theorem 10.4.4 says something of great importance. A square matrix with d distinct eigenvalues is similar (but not necessarily unitarily similar) to a direct sum of d unispectral matrices whose spectra are pairwise disjoint. These direct summands need not be unique, but one choice of them can be computed using only unitary similarities; they are the diagonal blocks of the block upper triangular matrix (10.4.5).

Lemma 10.4.8 *A square matrix is unispectral and diagonalizable if and only if it is a scalar matrix.*

Proof If $A \in \mathsf{M}_n$ is diagonalizable, then there is an invertible $S \in \mathsf{M}_n$ and a diagonal $\Lambda \in \mathsf{M}_n$ such that $A = S\Lambda S^{-1}$. If A is also unispectral, then $\Lambda = \lambda I$ for some scalar λ, so $A = S\Lambda S^{-1} = S(\lambda I)S^{-1} = \lambda SS^{-1} = \lambda I$. Conversely, if $A = \lambda I$ for some scalar λ, then it is diagonal and $\operatorname{spec} A = \{\lambda\}$. $\square$

Theorem 10.4.9 *Suppose that $A \in \mathsf{M}_n$ is unitarily similar to a block upper triangular matrix (10.4.5), in which the diagonal blocks T_{ii} are unispectral and have pairwise disjoint spectra. Then A is diagonalizable if and only if each T_{ii} is a scalar matrix.*

Proof Theorem 10.4.4 says that A is similar to $T_{11} \oplus T_{22} \oplus \cdots \oplus T_{dd}$, in which each direct summand is unispectral. If each T_{ii} is a scalar matrix, then A is similar to a diagonal matrix. Conversely, if A is diagonalizable, then Theorem 9.4.11 tells us that each T_{ii} is diagonalizable. But T_{ii} is also unispectral, so the preceding lemma ensures that it is a scalar matrix. $\square$

The preceding theorem provides a criterion for diagonalizability of a given square complex matrix A. Via a sequence of unitary similarities (for example, use the algorithm in Schur's triangularization theorem), reduce A to upper triangular form, in which equal eigenvalues are grouped together. Examine the unispectral diagonal blocks with pairwise disjoint spectra. They are all diagonal if and only if A is diagonalizable.

We can use the same ideas in a different way to formulate a criterion for diagonalizability that involves the minimal polynomial rather than unitary similarities.

Theorem 10.4.10 *Let $\lambda_1, \lambda_2, \ldots, \lambda_d$ be the distinct eigenvalues of $A \in \mathsf{M}_n$ and let*

$$p(z) = (z - \lambda_1)(z - \lambda_2) \cdots (z - \lambda_d). \tag{10.4.11}$$

Then A is diagonalizable if and only if $p(A) = 0$.

Proof If A is diagonalizable, then Theorem 10.3.13 tells us that $p(A) = 0$. Conversely, suppose that $p(A) = 0$. If $d = 1$, then $p(z) = z - \lambda_1$ and $p(A) = A - \lambda_1 I = 0$, so $A = \lambda_1 I$. If $d > 1$, let

$$p_i(z) = \frac{p(z)}{z - \lambda_i}, \quad i = 1, 2, \ldots, d.$$

Thus, p_i is the polynomial of degree $d - 1$ obtained from p by omitting the factor $z - \lambda_i$. Then $p(z) = (z - \lambda_i)p_i(z)$ for each $i = 1, 2, \ldots, d$, and

$$p_i(\lambda_i) = \prod_{j \neq i}(\lambda_j - \lambda_i) \neq 0, \quad i = 1, 2, \ldots, d.$$

Theorem 10.4.4 ensures that A is similar to $T_{11} \oplus T_{22} \oplus \cdots \oplus T_{dd}$, in which spec $T_{ii} = \{\lambda_i\}$ for each $i = 1, 2, \ldots, d$. Therefore, $p(A)$ is similar to

$$p(T_{11}) \oplus p(T_{22}) \oplus \cdots \oplus p(T_{dd});$$

see (0.8.2). Since $p(A) = 0$, we must have $p(T_{ii}) = 0$ for each $i = 1, 2, \ldots, d$. Because $p_i(\lambda_i) \neq 0$, Corollary 10.1.5 ensures that each matrix $p_i(T_{ii})$ is invertible. But the invertibility of $p_i(T_{ii})$ and the identity

$$0 = p(T_{ii}) = (T_{ii} - \lambda_i I)p_i(T_{ii})$$

imply that $T_{ii} - \lambda_i I = 0$, that is, $T_{ii} = \lambda_i I$. Since A is similar to a direct sum of scalar matrices it is diagonalizable. $\qquad \square$

Corollary 10.4.12 *Let $A \in \mathsf{M}_n$ and let f be a polynomial, each of whose zeros has multiplicity 1. If $f(A) = 0$, then A is diagonalizable.*

Proof m_A divides f, so its zeros also have multiplicity 1. Therefore, the exponents in (10.3.2) are all equal to one and m_A is the polynomial in (10.4.11). $\qquad \square$

Example 10.4.13 If $A \in \mathsf{M}_n$ is idempotent, then $A^2 = A$. Hence $f(A) = 0$, in which $f(z) = z^2 - z = z(z - 1)$. The preceding corollary ensures that A is diagonalizable. Theorems 10.4.4 and 10.4.9 tell us that A is unitarily similar to a 2×2 block upper triangular matrix with unispectral diagonal blocks. That is, any idempotent matrix with rank $r \geq 1$ is unitarily similar to

$$\begin{bmatrix} I_r & X \\ 0 & 0_{n-r} \end{bmatrix}, \quad X \in \mathsf{M}_{r \times (n-r)}.$$

Since each square complex matrix is similar to a direct sum of unispectral matrices, it is natural to ask whether a single unispectral matrix is similar to a direct sum of simpler unispectral matrices in some useful way. We answer that question in the following chapter.

The direct sum in (10.4.6) motivates the following observation.

Theorem 10.4.14 *Let $A = A_1 \oplus A_2 \oplus \cdots \oplus A_d$, in which spec $A_i = \{\lambda_i\}$ for each $i = 1, 2, \ldots, d$ and $\lambda_i \neq \lambda_j$ for all $i \neq j$. Then $m_A = m_{A_1} m_{A_2} \cdots m_{A_d}$.*

Proof Let $f = m_{A_1} m_{A_2} \cdots m_{A_d}$. Then $f(A_i) = 0$ for each $i = 1, 2, \ldots, d$, and hence

$$f(A) = f(A_1) \oplus f(A_2) \oplus \cdots \oplus f(A_d) = 0.$$

Moreover, there are positive integers q_i so that $m_{A_i} = (z - \lambda_i)^{q_i}$ for each $i = 1, 2, \ldots, d$. Since m_A divides f,

$$m_A(z) = (z - \lambda_1)^{r_1} (z - \lambda_2)^{r_2} \cdots (z - \lambda_d)^{r_d},$$

in which $1 \leq r_i \leq q_i$ for each $i = 1, 2, \ldots, d$. Define

$$h_i(z) = \frac{m_A(z)}{(z - \lambda_i)^{r_i}}, \quad i = 1, 2, \ldots, d.$$

Then $h_i(\lambda_i) \neq 0$, so $h_i(A_i)$ is invertible; see Corollary 10.1.5. Since

$$0 = m_A(A_i) = (A_i - \lambda_i I)^{r_i} h_i(A_i), \quad i = 1, 2, \ldots, d,$$

it follows that each $(A_i - \lambda_i I)^{r_i} = 0$. The definition of the minimal polynomial ensures that each $r_i = q_i$, so $f = m_A$. ☐

A final consequence of Theorem 10.4.1 is a generalization of Lemma 3.3.21.

Corollary 10.4.15 *Let* $A = A_{11} \oplus A_{22} \oplus \cdots \oplus A_{dd}$, *in which each* $A_{ii} \in M_{n_i}$ *and* $n_1 + n_2 + \cdots + n_d = n$. *Partition* $B = [B_{ij}] \in M_n$ *conformally with* A. *Suppose that* $\mathrm{spec}\, A_{ii} \cap \mathrm{spec}\, A_{jj} = \varnothing$ *for all* $i \neq j$. *If* $AB = BA$, *then* $B_{ij} = 0$ *for all* $i \neq j$.

Proof Equate the (i, j) blocks of both sides of the identity $AB = BA$ and obtain

$$A_{ii}B_{ij} - B_{ij}A_{jj} = 0 \quad \text{for all } i \neq j.$$

This is a linear matrix equation of the form (10.4.2) with $C = 0$. Theorem 10.4.1 ensures that $B_{ij} = 0$ for all $i \neq j$. ☐

10.5 Commuting Matrices and Triangularization

The algorithm described in the proof of Schur's triangularization theorem can be performed simultaneously on two or more matrices, provided that they have a common eigenvector. Commuting matrices have common eigenvectors (Corollary 8.5.4). These two observations are the basis for the following theorem, which gives a sufficient condition for a family of matrices to be simultaneously unitarily upper triangularizable.

Theorem 10.5.1 *Let* $n \geq 2$ *and let* $\mathcal{F} \subseteq M_n$ *be a nonempty set of commuting matrices.*

(a) *There is a unitary* $U \in M_n$ *such that* U^*AU *is upper triangular for all* $A \in \mathcal{F}$.

(b) *If every matrix in* $\mathcal{F}$ *is real and has only real eigenvalues, then there is a real orthogonal* $Q \in M_n(\mathbb{R})$ *such that* $Q^\mathsf{T}AQ$ *is upper triangular for all* $A \in \mathcal{F}$.

Proof (a) Corollary 8.5.4 ensures the existence of a unit vector $\mathbf{x}$ that is an eigenvector of every matrix in $\mathcal{F}$. Let $V \in M_n$ be any unitary matrix whose first column is $\mathbf{x}$ (Corollary 6.4.10.b). If $A, B \in \mathcal{F}$, then $A\mathbf{x} = \lambda\mathbf{x}$ and $B\mathbf{x} = \mu\mathbf{x}$ for some scalars λ and μ. Just as in the proof of Schur's triangularization theorem,

$$V^*AV = \begin{bmatrix} \lambda & \star \\ 0 & A' \end{bmatrix} \quad \text{and} \quad V^*BV = \begin{bmatrix} \mu & \star \\ 0 & B' \end{bmatrix},$$

so V achieves a simultaneous reduction of every matrix in $\mathcal{F}$. Now compute

$$\begin{bmatrix} \lambda\mu & \star \\ 0 & A'B' \end{bmatrix} = V^*ABV = V^*BAV = \begin{bmatrix} \lambda\mu & \star \\ 0 & B'A' \end{bmatrix},$$

which tells us that $A'B' = B'A'$ for all $A, B \in \mathcal{F}$. An induction argument completes the proof.

(b) If every matrix in $\mathcal{F}$ is real and has only real eigenvalues, Corollary 8.5.4 ensures the existence of a real unit vector $\mathbf{x}$ that is an eigenvector of every matrix in $\mathcal{F}$. Corollary 6.4.10.b says that there is a real unitary matrix V whose first column is $\mathbf{x}$. The rest of the argument is the same as in (a); all the matrices involved are real. $\qquad\square$

Commuting matrices have the pleasant property that the eigenvalues of their sum and product are, respectively, the sum and product of their eigenvalues in some order.

Corollary 10.5.2 *Let $A, B \in \mathsf{M}_n$, and suppose that $AB = BA$. There is some ordering $\lambda_1, \lambda_2, \ldots, \lambda_n$ of the eigenvalues of A and some ordering $\mu_1, \mu_2, \ldots, \mu_n$ of the eigenvalues of B such that $\lambda_1 + \mu_1, \lambda_2 + \mu_2, \ldots, \lambda_n + \mu_n$ are the eigenvalues of $A + B$ and $\lambda_1\mu_1, \lambda_2\mu_2, \ldots, \lambda_n\mu_n$ are the eigenvalues of AB.*

Proof The preceding theorem says that there is a unitary U such that $U^*AU = T = [t_{ij}]$ and $U^*BU = R = [r_{ij}]$ are upper triangular. The entries $t_{11}, t_{22}, \ldots, t_{nn}$ are the eigenvalues of A in some order and the entries $r_{11}, r_{22}, \ldots, r_{nn}$ are the eigenvalues of B in some order. Then

$$A + B = U(T + R)U^* = U[t_{ij} + r_{ij}]U^*,$$

so $A + B$ is similar to the upper triangular matrix $T + R$. Its diagonal entries are $t_{ii} + r_{ii}$ for $i = 1, 2, \ldots, n$; they are the eigenvalues of $A + B$. Also,

$$AB = U(TR)U^*,$$

so AB is similar to the upper triangular matrix TR. Its diagonal entries are $t_{ii}r_{ii}$ for $i = 1, 2, \ldots, n$; they are the eigenvalues of AB. $\qquad\square$

Example 10.5.3 If A and B do not commute, then the eigenvalues of $A + B$ need not be sums of eigenvalues of A and B. For example, all the eigenvalues of the noncommuting matrices

$$A = \begin{bmatrix} 0 & 1 \\ 0 & 0 \end{bmatrix} \quad \text{and} \quad B = \begin{bmatrix} 0 & 0 \\ 1 & 0 \end{bmatrix}$$

are zero. The eigenvalues of $A + B$ are ± 1, neither of which is a sum of eigenvalues of A and B. The eigenvalue 1 of AB is not a product of eigenvalues of A and B.

Example 10.5.4 If A and B do not commute, the eigenvalues of $A + B$ might nevertheless be sums of eigenvalues of A and B. Consider the noncommuting matrices

$$A = \begin{bmatrix} 1 & 1 \\ 0 & 2 \end{bmatrix} \quad \text{and} \quad B = \begin{bmatrix} 3 & 2 \\ 0 & 6 \end{bmatrix}.$$

The eigenvalues of $A + B$ are $1 + 3 = 4$ and $2 + 6 = 8$. The eigenvalues of AB are $1 \cdot 3 = 3$ and $2 \cdot 6 = 12$.

10.6 Eigenvalue Adjustments and the Google Matrix

In some applications, we would like to adjust an eigenvalue of a matrix by adding a suitable rank-1 matrix. The following theorem describes one way to do this.

Theorem 10.6.1 (Brauer) *Let* $(\lambda, \mathbf{x})$ *be an eigenpair of* $A \in \mathsf{M}_n$ *and let* $\lambda, \lambda_2, \ldots, \lambda_n$ *be its eigenvalues. For any* $\mathbf{y} \in \mathbb{F}^n$, *the eigenvalues of* $A + \mathbf{xy}^*$ *are* $\lambda + \mathbf{y}^*\mathbf{x}, \lambda_2, \ldots, \lambda_n$ *and* $(\lambda + \mathbf{y}^*\mathbf{x}, \mathbf{x})$ *is an eigenpair of* $A + \mathbf{xy}^*$.

Proof Let $\mathbf{u} = \mathbf{x}/\|\mathbf{x}\|_2$. Then $(\lambda, \mathbf{u})$ is an eigenpair of A, in which $\mathbf{u}$ is a unit vector. Schur's triangularization theorem says that there is a unitary $U = [\mathbf{u} \ U_2] \in \mathsf{M}_n$ whose first column is $\mathbf{u}$ and is such that

$$U^*AU = T = \begin{bmatrix} \lambda & \star \\ \mathbf{0} & T' \end{bmatrix}$$

is upper triangular. The eigenvalues (diagonal entries) of T' are $\lambda_2, \lambda_3, \ldots, \lambda_n$. Compute

$$U^*(\mathbf{xy}^*)U = (U^*\mathbf{x})(\mathbf{y}^*U) = \begin{bmatrix} \mathbf{u}^*\mathbf{x} \\ U_2^*\mathbf{x} \end{bmatrix} [\mathbf{y}^*\mathbf{u} \ \ \mathbf{y}^*U_2]$$

$$= \begin{bmatrix} \|\mathbf{x}\|_2 \\ \mathbf{0} \end{bmatrix} [\mathbf{y}^*\mathbf{u} \ \ \mathbf{y}^*U_2] = \begin{bmatrix} \mathbf{y}^*\mathbf{x} & \star \\ \mathbf{0} & 0 \end{bmatrix}.$$

Therefore,

$$U^*(A + \mathbf{xy}^*)U = \begin{bmatrix} \lambda & \star \\ \mathbf{0} & T' \end{bmatrix} + \begin{bmatrix} \mathbf{y}^*\mathbf{x} & \star \\ \mathbf{0} & 0 \end{bmatrix} = \begin{bmatrix} \lambda + \mathbf{y}^*\mathbf{x} & \star \\ \mathbf{0} & T' \end{bmatrix},$$

which has eigenvalues $\lambda + \mathbf{y}^*\mathbf{x}, \lambda_2, \ldots, \lambda_n$. These are also the eigenvalues of $A + \mathbf{xy}^*$, and $(A + \mathbf{xy}^*)\mathbf{x} = A\mathbf{x} + \mathbf{xy}^*\mathbf{x} = (\lambda + \mathbf{y}^*\mathbf{x})\mathbf{x}$. □

A notable application of Brauer's theorem arises in understanding a famous matrix that played a key role in early website ranking methods.

Corollary 10.6.2 *Let* $(\lambda, \mathbf{x})$ *be an eigenpair of* $A \in \mathsf{M}_n$ *and let* $\lambda, \lambda_2, \ldots, \lambda_n$ *be the eigenvalues of* A. *Let* $\mathbf{y} \in \mathbb{C}^n$ *be such that* $\mathbf{y}^*\mathbf{x} = 1$ *and let* $\tau \in \mathbb{C}$. *Then the eigenvalues of* $A_\tau = \tau A + (1 - \tau)\lambda \mathbf{xy}^*$ *are* $\lambda, \tau\lambda_2, \ldots, \tau\lambda_n$.

Proof The eigenvalues of τA are $\tau\lambda, \tau\lambda_2, \ldots, \tau\lambda_n$. The preceding theorem says that the eigenvalues of

$$A_\tau = \tau A + \mathbf{x}\big((1 - \overline{\tau})\overline{\lambda}\mathbf{y}\big)^*$$

are

$$\tau\lambda + \big((1 - \overline{\tau})\overline{\lambda}\mathbf{y}\big)^*\mathbf{x} = \tau\lambda + (1 - \tau)\lambda = \lambda,$$

together with $\tau\lambda_2, \tau\lambda_3, \ldots, \tau\lambda_n$. □

Example 10.6.3 Let $A = [a_{ij}] \in \mathsf{M}_n(\mathbb{R})$ have nonnegative entries and suppose that all of its row sums are equal to 1. Then $(1, \mathbf{e})$ is an eigenpair of A. Let $1, \lambda_2, \ldots, \lambda_n$ be the eigenvalues of A. Corollary 8.4.9 ensures that $|\lambda_i| \leq 1$ for each $i = 2, 3, \ldots, n$. Some of the eigenvalues $\lambda_2, \lambda_3, \ldots, \lambda_n$ can have modulus 1. For example,

$$\begin{bmatrix} 0 & 0 & 1 \\ 1 & 0 & 0 \\ 0 & 1 & 0 \end{bmatrix} \tag{10.6.4}$$

has nonnegative entries and row sums equal to 1; its eigenvalues are 1 and $e^{\pm 2\pi i/3}$, all of which have modulus 1. For computational reasons, we might wish to adjust our matrix A so that $\lambda = 1$ is the only eigenvalue of maximum modulus, and we would like the adjusted matrix to continue to have nonnegative entries and row sums equal to 1. Let $\mathbf{x} = \mathbf{e}$, let $\mathbf{y} = \mathbf{e}/n$, let $E = \mathbf{e}\mathbf{e}^\mathsf{T}$, and suppose that $0 < \tau < 1$. Then $\mathbf{y}^\mathsf{T}\mathbf{x} = 1$, so the preceding corollary says that the eigenvalues of

$$A_\tau = \tau A + (1 - \tau)\mathbf{x}\mathbf{y}^\mathsf{T} = \tau A + \frac{1 - \tau}{n}E \tag{10.6.5}$$

are $1, \tau\lambda_2, \ldots, \tau\lambda_n$. Thus, $A_\tau \in M_n(\mathbb{R})$ has positive entries and row sums equal to 1. Its only maximum-modulus eigenvalue is $\lambda = 1$ (with algebraic multiplicity 1); its other $n - 1$ eigenvalues have modulus at most $\tau < 1$.

The matrix $A_\tau \in M_n(\mathbb{R})$ in (10.6.5) is often called the *Google matrix* because of its association with the problem of ranking web pages. Folklore has it that the founders of Google used the value $\tau = 0.85$.

10.7 Problems

P.10.1 Let $A \in M_3$ and suppose that $\operatorname{spec} A = \{1\}$. Show that A is invertible and express A^{-1} as a linear combination of I, A, and A^2.

P.10.2 Let $A \in M_n$. Prove in two ways that A is nilpotent if and only if $\operatorname{spec} A = \{0\}$.
(a) Use Theorem 10.1.1 and consider powers of a strictly upper triangular matrix.
(b) Use Theorems 10.2.1 and 8.3.3.

P.10.3 Let $A \in M_n$. Show that the following statements are equivalent:

(a) A is nilpotent.

(b) A is unitarily similar to a strictly upper triangular matrix.

(c) A is similar to a strictly upper triangular matrix.

P.10.4 Suppose that an upper triangular matrix $T \in M_n$ has ν nonzero diagonal entries. Show that $\operatorname{rank} T \geq \nu$ and give an example for which $\operatorname{rank} T > \nu$.

P.10.5 Suppose that $A \in M_n$ has ν nonzero eigenvalues. Explain why $\operatorname{rank} A \geq \nu$ and give an example for which $\operatorname{rank} A > \nu$.

P.10.6 Let $A = [a_{ij}] \in M_n$ and write $A = UTU^*$, in which U is unitary, $T = [t_{ij}]$ is upper triangular, and $|t_{11}| \geq |t_{22}| \geq \cdots \geq |t_{nn}|$. Suppose that A has exactly $k \geq 1$ nonzero eigenvalues $\lambda_1, \lambda_2, \ldots, \lambda_k$ (including multiplicities). Explain why

$$\left|\sum_{i=1}^{k} \lambda_i\right|^2 \leq k\sum_{i=1}^{k} |\lambda_i|^2 = k\sum_{i=1}^{k} |t_{ii}|^2 \leq k\sum_{i,j=1}^{n} |t_{ij}|^2 = k\sum_{i,j=1}^{n} |a_{ij}|^2$$

and deduce that $\operatorname{rank} A \geq k \geq |\operatorname{tr} A|^2/(\operatorname{tr} A^*A)$, with equality if and only if $T = cI_k \oplus 0_{n-k}$ for some nonzero scalar c.

P.10.7 Let $\lambda_1, \lambda_2, \ldots, \lambda_n$ be the eigenvalues of $A \in M_n$.

(a) Show that $\operatorname{tr} A^k = \lambda_1^k + \lambda_2^k + \cdots + \lambda_n^k$ for each $k = 1, 2, \ldots$.

(b) Use the presentation (10.2.3) of the characteristic polynomial (10.2.2) to show that the coefficient of its z^{n-2} term is $c_{n-2} = \sum_{i<j} \lambda_i \lambda_j$.

(c) Show that $c_{n-2} = \frac{1}{2}\big((\operatorname{tr} A)^2 - \operatorname{tr} A^2\big)$.

(d) If $A \in M_3$, show that $\det A = \frac{1}{6}\big(2 \operatorname{tr} A^3 - 3(\operatorname{tr} A)(\operatorname{tr} A^2) + (\operatorname{tr} A)^3\big)$. *Hint*: Compute $\operatorname{tr} p_A(A)$.

P.10.8 If $A \in M_n$ is invertible and has characteristic polynomial (10.2.2), show that the characteristic polynomial of A^{-1} is $z^n p_A(z^{-1})/c_0$.

P.10.9 Suppose that $A \in M_n$ is invertible and let $S = \operatorname{span}\{I, A, A^{-1}, A^2, A^{-2}, \ldots\} \subseteq M_n$. Show that $S = \operatorname{span}\{I, A, A^2, \ldots, A^{n-1}\}$ and $\dim S \leq n$.

P.10.10 Let $A \in M_m$ and $B \in M_n$. Show that if $\operatorname{spec} A \cap \operatorname{spec} B \neq \varnothing$, then there is a nonzero $X \in M_{m \times n}$ such that $AX - XB = 0$. *Hint*: You may find it helpful to investigate $X = \mathbf{x}\mathbf{y}^*$, in which $(\lambda, \mathbf{x})$ is an eigenpair of A and $(\bar{\lambda}, \mathbf{y})$ is an eigenpair of B^*.

P.10.11 Show that the Google matrix A_τ in (10.6.5) has nonnegative real entries and row sums equal to 1. What is A_τ for the matrix in (10.6.4)? What are its eigenvalues if $\tau = 0.85$? What are their moduli?

P.10.12 If you wanted to use the method in Theorem 10.6.1 to adjust an eigenvalue λ of A, you would have to choose the vector $\mathbf{y}$ somehow. Why should you avoid choosing it to be an eigenvector of A^* associated with an eigenvalue of A^* different from $\bar{\lambda}$? Why is it always safe to take $\mathbf{y}$ to be a nonzero scalar multiple of $\mathbf{x}$? *Hint*: P.9.21.

P.10.13 Suppose that $A, B \in M_n$ commute and $\operatorname{spec} A \cap \operatorname{spec}(-B) = \varnothing$. Explain why $A + B$ is invertible.

P.10.14 Consider

$$A = \begin{bmatrix} 1 & 0 \\ 2 & 1 \end{bmatrix} \quad \text{and} \quad B = \begin{bmatrix} 1 & 2 \\ 0 & 1 \end{bmatrix}.$$

Show that $\operatorname{spec} A \cap \operatorname{spec}(-B) = \varnothing$, but $A + B$ is not invertible. Does this contradict the preceding problem?

P.10.15 Let $A = [a_{ij}] \in M_n$ be upper triangular and invertible. Use Corollary 10.2.12 to show that A^{-1} is upper triangular. What are its diagonal entries?

P.10.16 Let $A, B \in M_n$ and suppose that there is an invertible $S \in M_n$ such that $S^{-1}AS$ and $S^{-1}BS$ are upper triangular. Show that there is a unitary $U \in M_n$ such that U^*AU and U^*BU are upper triangular. *Hint*: Consider the QR factorization of S.

P.10.17 Explain what is wrong with the following false "proof" of the Cayley–Hamilton theorem: $p_A(z) = \det(zI - A)$, so $p_A(A) = \det(AI - A) = \det(A - A) = \det 0 = 0$.

P.10.18 If $A \in M_n$, $\operatorname{spec} A = \{-1, 1\}$, and $p(z) = (z-1)^{n-1}(z+1)^{n-1}$, show that $p(A) = 0$.

P.10.19 Let $A, B^{\mathsf{T}} \in M_{m \times n}$. Use Corollary 10.1.3 and Theorem 9.7.2 to explain why $\operatorname{tr} AB = \operatorname{tr} BA$. Can you explain this identity in an elementary way that does not involve eigenvalues?

P.10.20 Let $A \in M_{10}$, let $f(z) = z^4 + 11z^3 - 7z^2 + 5z + 3$, and suppose that $f(A) = 0$. Prove that A is invertible and find a polynomial g of degree 3 or less such that $A^{-1} = g(A)$.

P.10.21 Let $A, B \in M_n$ and suppose that $AB = BA$. Define $\operatorname{spec} A + \operatorname{spec} B = \{\lambda + \mu : \lambda \in \operatorname{spec} A \text{ and } \mu \in \operatorname{spec} B\}$ and $(\operatorname{spec} A)(\operatorname{spec} B) = \{\lambda \mu : \lambda \in \operatorname{spec} A \text{ and } \mu \in \operatorname{spec} B\}$.

(a) Show that $\operatorname{spec}(A + B) \subseteq \operatorname{spec} A + \operatorname{spec} B$.

(b) Give an example of diagonal matrices A, B such that $\operatorname{spec}(A + B) \neq \operatorname{spec} A + \operatorname{spec} B$.

(c) What can you say about $(\operatorname{spec} A)(\operatorname{spec} B)$?

P.10.22 Let $A = A_1 \oplus A_2 \oplus \cdots \oplus A_d$ and $B = B_1 \oplus B_2 \oplus \cdots \oplus B_d$ be $n \times n$ matrices that are conformally partitioned and block diagonal. Suppose that $\operatorname{spec} A_i \cap \operatorname{spec} B_j = \varnothing$ for all $i \neq j$. If $C \in M_n$ and $AC = CB$, show that $C = C_1 \oplus C_2 \oplus \cdots \oplus C_d$ is block diagonal and conformal with A and B.

P.10.23 Suppose that $A, B \in M_n$ commute and let $\mu_1, \mu_2, \ldots, \mu_d$ be the distinct eigenvalues of A. Prove that there is an invertible $S \in M_n$ such that $S^{-1}AS = T_1 \oplus T_2 \oplus \cdots \oplus T_d$ and $S^{-1}BS = B_1 \oplus B_2 \oplus \cdots \oplus B_d$ are conformally partitioned and block diagonal, each T_j is upper triangular, and $\operatorname{spec} T_j = \{\mu_j\}$ for each $j = 1, 2, \ldots, d$.

P.10.24 Let $A \in M_n$ be diagonalizable. Prove that $\operatorname{rank} A = \operatorname{rank} A^k$ for $k = 1, 2, \ldots$.

P.10.25 Let $C = \begin{bmatrix} 0 & I_n \\ I_n & 0 \end{bmatrix}$. Use both P.10.39.f and P.9.10.e to evaluate $\det C$.

P.10.26 Use Theorem 10.4.14 to show that each exponent q_i in the factorization (10.3.2) of the minimal polynomial is the index of $A - \lambda_i I$.

P.10.27 If $A \in M_5$ is diagonalizable and $p_A(z) = (z-2)^3(z-3)^2$, show that $m_A(z) = (z-2)(z-3)$.

P.10.28 Let $\lambda_1, \lambda_2, \ldots, \lambda_d$ be the distinct eigenvalues of $A \in M_n$. Show that (10.4.11) is the minimal polynomial of A if and only if A is diagonalizable.

P.10.29 If $A \in M_n$ is an involution, show that it is unitarily similar to a block matrix of the form

$$\begin{bmatrix} I_k & X \\ 0 & -I_{n-k} \end{bmatrix}, \quad X \in M_{k \times (n-k)}.$$

P.10.30 It is possible to determine the minimal polynomial of a matrix without knowing its eigenvalues or characteristic polynomial. Let $A \in M_n$, let m_A be its minimal polynomial, and suppose that m_A has degree ℓ. Let $\mathbf{v}_1 = \operatorname{vec} I$, $\mathbf{v}_2 = \operatorname{vec} A$, $\mathbf{v}_3 = \operatorname{vec} A^2, \ldots, \mathbf{v}_{n+1} = \operatorname{vec} A^n$.

(a) Show that $\mathbf{v}_1, \mathbf{v}_2, \ldots, \mathbf{v}_{n+1}$ is a linearly dependent list of vectors in $\mathbb{C}^{n^2}$.

(b) Show that

$$\min\{k : \mathbf{v}_1, \mathbf{v}_2, \ldots, \mathbf{v}_{k+1} \text{ is linearly dependent and } 1 \leq k \leq n\}$$

is equal to ℓ and let $c_1, c_2, \ldots, c_{k+1}$ be scalars, not all zero, such that $c_1\mathbf{v}_1 + c_2\mathbf{v}_2 + \cdots + c_{\ell+1}\mathbf{v}_{\ell+1} = \mathbf{0}$.

(c) Why is $c_{\ell+1} \neq 0$ and what are the coefficients of m_A?

(d) Formulate an algorithm, based on (b) and an orthogonalization process, to compute m_A; see P.5.4.

P.10.31 Use the algorithm in the preceding problem to compute the minimal polynomials of

$$\begin{bmatrix} 0 & 1 \\ 0 & 1 \end{bmatrix}, \quad \begin{bmatrix} 1 & 1 \\ 0 & 1 \end{bmatrix}, \quad \text{and} \quad \begin{bmatrix} 1 & 0 \\ 0 & 1 \end{bmatrix}.$$

P.10.32 Let $A \in \mathsf{M}_n$ and $B \in \mathsf{M}_m$. (a) Show that the minimal polynomial of $A \oplus B$ is the least common multiple of m_A and m_B (the monic polynomial of minimum degree that is divisible by both m_A and m_B). (b) If $A \oplus B$ is diagonalizable, use (a) to prove that A and B are diagonalizable.

P.10.33 If $A \in \mathsf{M}_n$ and the degree of m_A is ℓ, show that dim span$\{I, A, A^2, \ldots\} = \ell$.

P.10.34 Let $p(z) = z^2 + 4$. Is there an $A \in \mathsf{M}_3(\mathbb{R})$ with $m_A(z) = p(z)$? Is there an $A \in \mathsf{M}_2(\mathbb{R})$ with $m_A(z) = p(z)$? Is there an $A \in \mathsf{M}_3(\mathbb{C})$ with $m_A(z) = p(z)$? In each case, provide a proof or an example.

P.10.35 Let $A \in \mathsf{M}_n$. Use Theorem 10.4.4 to show that there is a diagonalizable $B \in \mathsf{M}_n$ and a nilpotent $C \in \mathsf{M}_n$ such that $A = B + C$ and $BC = CB$. *Hint:* $T_{ii} = \lambda_i I + (T_{ii} - \lambda_i I)$.

P.10.36 Let f be a monic polynomial of degree n, let $\lambda \in \mathbb{C}$, and let $\mathbf{x}_\lambda = [1 \ \lambda \ \lambda^2 \ \cdots \ \lambda^{n-1}]^{\mathsf{T}}$. (a) Show that $\mathbf{x}_\lambda$ is an eigenvector of C_f^{T} if and only if $\lambda \in \operatorname{spec} C_f$. (b) Show that $\mathbf{y} \in \mathbb{C}^n$ is an eigenvector of C_f^{T} if and only if $\mathbf{y}$ is a nonzero scalar multiple of $\mathbf{x}_\lambda$ for some $\lambda \in \operatorname{spec} C_f$. (c) Deduce from (b) that every eigenvalue of C_f has geometric multiplicity 1.

P.10.37 Let f and g be monic polynomials of the same degree. Prove that C_f commutes with C_g if and only if $f = g$. *Hint:* What is column $n - 1$ of $C_f C_g$?

P.10.38 Use the companion matrix to show that the Fundamental Theorem of Algebra is equivalent to the statement "every square complex matrix has an eigenvalue."

P.10.39 This problem relies on results in P.3.42, P.3.43, P.3.44, and P.6.43. Let $A \in \mathsf{M}_m$ and $B \in \mathsf{M}_n$ have eigenvalues $\lambda_1, \lambda_2, \ldots, \lambda_m$ and $\mu_1, \mu_2, \ldots, \mu_n$, respectively. Factor $A = UTU^*$ and $B = VT'V^*$ as in Theorem 10.1.1, in which U, V are unitary and T, T' are upper triangular.

(a) Explain why $(U \otimes V)^*(A \otimes B)(U \otimes V) = T \otimes T'$ is upper triangular, and its diagonal entries are the eigenvalues of $A \otimes B$.

(b) Deduce that the mn eigenvalues of $A \otimes B$ are $\lambda_i \mu_j$ for all $i = 1, 2, \ldots, m$ and $j = 1, 2, \ldots, n$.

(c) Show that $(U \otimes V)^*(A \otimes I_n)(U \otimes V) = T \otimes I_n$ and $(U \otimes V)^*(I_m \otimes B)(U \otimes V) = I_m \otimes T'$, which are upper triangular.

(d) Deduce that the mn eigenvalues of $A \otimes I_n + I_m \otimes B$ (the *Kronecker sum* of A and B) are $\lambda_i + \mu_j$ for all $i = 1, 2, \ldots, m$ and $j = 1, 2, \ldots, n$.

(e) What can you deduce by applying Corollary 10.5.2 to the commuting matrices $A \otimes I_n$ and $I_m \otimes B$? Is this as good a result as the one in (d)? Explain.

(f) Deduce from (b) that $\det(A \otimes B) = (\det A)^n (\det B)^m = \det(B \otimes A)$.

P.10.40 Let A, B, X, C be as in Theorem 10.4.1.

(a) Use Theorem 3.6.16 to show that $\operatorname{vec}(AX - XB) = (A^{\mathsf{T}} \otimes I_n - I_m \otimes B) \operatorname{vec} X = \operatorname{vec} C$.

(b) Let $K = A^{\mathsf{T}} \otimes I_n - I_m \otimes B$ and explain why the mn eigenvalues of K are $\lambda_i - \mu_j$ for all $i = 1, 2, \ldots, m$ and $j = 1, 2, \ldots, n$. *Hint:* P.10.39.

(c) Deduce that K is invertible if and only if $\operatorname{spec} A \cap \operatorname{spec}(-B) = \varnothing$.

(d) Deduce Theorem 10.4.1 from (c). *Hint:* The linear matrix equation (10.4.2) is equivalent to the linear system $K \operatorname{vec} X = \operatorname{vec} C$.

10.8 Notes

The decomposition described in P.10.35 is unique. If $A = B + C = D + E$, in which B and D are diagonalizable, C and E are nilpotent, $BC = CB$, and $DE = ED$, then $B = D$ and $C = E$. For a proof, see [HJ13, 3.2.P18].

For more information about Kronecker sums and the vec operator, see [HJ94, Ch. 4].

10.9 Some Important Concepts

- Each square complex matrix is unitarily similar to an upper triangular matrix.
- Each square complex matrix is annihilated by its characteristic polynomial.
- Minimal polynomial.
- Similar matrices have the same characteristic and minimal polynomials.
- Companion matrix.
- Sylvester's theorem on linear matrix equations.
- Each square complex matrix is similar to a direct sum of unispectral matrices.
- A square complex matrix is diagonalizable if and only if every zero of its minimal polynomial has multiplicity 1.
- Commuting matrices can be simultaneously unitarily upper triangularized.

11   Jordan Canonical Form

In Chapter 10 we found that each square complex matrix A is similar to a direct sum of unispectral matrices that may be taken to be upper triangular. We now show that A is similar to a direct sum of Jordan blocks (unispectral upper bidiagonal matrices with 1s in the superdiagonal) that is unique up to permutation of its direct summands. This direct sum (the Jordan canonical form of A) reveals many interesting properties of A. For example, A is similar to A^{T}; $A^p \to 0$ as $p \to \infty$ if and only if every eigenvalue of A has modulus less than 1; and the invertible Jordan blocks of AB and BA are the same (a generalization of Theorem 9.7.2).

11.1 Jordan Blocks and Jordan Matrices

How can we tell if two square matrices are similar? Similar matrices share the same characteristic and minimal polynomials, but so do some matrices that are not similar (see Theorem 10.3.4 and Example 10.3.5). A definitive test for similarity must involve something more than the characteristic and minimal polynomials.

A promising approach to finding a test for similarity is suggested by an important result in the preceding chapter: Each square complex matrix is similar to a direct sum of unispectral matrices with pairwise disjoint spectra (Theorem 10.4.4). This result permits us to focus on a more specific question: When are two unispectral matrices similar?

If $A, B \in \mathsf{M}_n$ and $\operatorname{spec} A = \operatorname{spec} B = \{\lambda\}$, then A is similar to B if and only if $(A - \lambda I)$ and $(B - \lambda I)$ are similar (Theorem 2.4.18.b). Since $\operatorname{spec}(A - \lambda I) = \operatorname{spec}(B - \lambda I) = \{0\}$, we can focus on an even more specific question: When are two nilpotent matrices similar?

Theorem 11.1.1 *Let $A \in \mathsf{M}_n$. The following are equivalent:*

(a) *A is nilpotent, that is, $A^k = 0$ for some positive integer k.*

(b) $\operatorname{spec} A = \{0\}$.

(c) $p_A(z) = z^n$.

(d) $A^n = 0$.

(e) $m_A(z) = z^q$ *for some $q \in \{1, 2, \ldots, n\}$.*

Proof (a) $\Rightarrow$ (b) Let $(\lambda, \mathbf{x})$ be an eigenpair of A. Suppose that $k \geq 1$ and $A^k = 0$. Lemma 8.3.2 ensures that $\mathbf{0} = 0_n \mathbf{x} = A^k \mathbf{x} = \lambda^k \mathbf{x}$, so $\lambda^k = 0$ and hence $\lambda = 0$.

(b) $\Rightarrow$ (c) Since 0 is the only eigenvalue of A, its algebraic multiplicity is n and (9.2.2) says that $p_A(z) = z^n$.

(c) $\Rightarrow$ (d) The Cayley–Hamilton theorem ensures that $A^n = p_A(A) = 0$.

(d) $\Rightarrow$ (e) The polynomial $p(z) = z^n$ annihilates A and m_A divides p, so $m_A(z) = z^q$ for some $q \in \{1, 2, \ldots, n\}$.

(e) $\Rightarrow$ (a) Take $k = q$ in (a). $\qquad\qquad\qquad\qquad\qquad\qquad\qquad\qquad\qquad\qquad\square$

Suppose that $A \in M_n$ is nilpotent. Theorem 3.2.17 ensures that

$$n - 1 \geq \operatorname{rank} A \geq \operatorname{rank} A^2 \geq \cdots \geq \operatorname{rank} A^n = 0.$$

It also ensures that

$$\operatorname{rank} A^k = \operatorname{rank} A^{k+1} \quad \Longrightarrow \quad \operatorname{rank} A^k = \operatorname{rank} A^{k+1} = \cdots = \operatorname{rank} A^n = 0.$$

According to Definition 3.2.20, the index of A is the smallest k such that $\operatorname{rank} A^k = 0$. The exponent in $m_A(z) = z^q$ is the smallest positive integer $r \in \{1, 2, \ldots, n\}$ such that $A^r = 0$. Since $A^k = 0$ if and only if $\operatorname{rank} A^k = 0$, the exponent q in $m_A(z) = z^q$ is the index of A.

Example 11.1.2 A strictly upper triangular matrix is unispectral with eigenvalue 0, so it is nilpotent. Such a matrix has a lot of zeros, but some nilpotent matrices have no zero entries. For example,

$$A = \begin{bmatrix} 1 & i \\ i & -1 \end{bmatrix} \quad \text{and} \quad B = \begin{bmatrix} 1 & 1 \\ -1 & -1 \end{bmatrix} \tag{11.1.3}$$

are nilpotent and have index two.

Definition 11.1.4 A $k \times k$ *Jordan block with eigenvalue* λ is the upper bidiagonal matrix

$$J_k(\lambda) = \begin{bmatrix} \lambda & 1 & & & \\ & \lambda & 1 & & \\ & & \ddots & \ddots & \\ & & & \lambda & 1 \\ & & & & \lambda \end{bmatrix} \in M_k. \tag{11.1.5}$$

Every main diagonal entry of $J_k(\lambda)$ is λ, every entry in the first superdiagonal is 1, and all other entries are 0.

Definition 11.1.6 A *Jordan matrix* J is a direct sum of Jordan blocks. If J has r direct summands, then

$$J = J_{n_1}(\lambda_1) \oplus J_{n_2}(\lambda_2) \oplus \cdots \oplus J_{n_r}(\lambda_r), \tag{11.1.7}$$

in which $n_1, n_2, \ldots, n_r$ are positive integers. If $r > 1$, the scalars $\lambda_1, \lambda_2, \ldots, \lambda_r$ need not be distinct. The number of repetitions of a Jordan block $J_k(\lambda)$ in the direct sum (11.1.7) is its *multiplicity*.

Example 11.1.8 Jordan blocks of sizes 1, 2, and 3 are

$$J_1(\lambda) = [\lambda], \quad J_2(\lambda) = \begin{bmatrix} \lambda & 1 \\ 0 & \lambda \end{bmatrix}, \quad \text{and} \quad J_3(\lambda) = \begin{bmatrix} \lambda & 1 & 0 \\ 0 & \lambda & 1 \\ 0 & 0 & \lambda \end{bmatrix}.$$

Example 11.1.9 In the 9×9 Jordan matrix

$$J_3(1) \oplus J_2(0) \oplus J_3(1) \oplus J_1(4),$$

the Jordan block $J_3(1)$ has multiplicity 2; each of the blocks $J_2(0)$ and $J_1(4) = [4]$ has multiplicity 1.

Example 11.1.10 An $r \times r$ diagonal matrix is a Jordan matrix (11.1.7) in which $n_i = 1$ for $i = 1, 2, \ldots, r$.

Jordan blocks of the form $J_k(0)$ and Jordan matrices that are direct sums of such blocks are of special importance.

Definition 11.1.11 A *nilpotent Jordan block* is a Jordan block with eigenvalue zero. We often denote the $k \times k$ nilpotent Jordan block $J_k(0)$ by J_k. A *nilpotent Jordan matrix J* is a direct sum

$$J = J_{n_1} \oplus J_{n_2} \oplus \cdots \oplus J_{n_r} \tag{11.1.12}$$

of nilpotent Jordan blocks.

We claim that each nilpotent matrix is similar to a nilpotent Jordan matrix that is unique up to permutation of its direct summands. Before we can prove this claim, we have some work to do.

Each Jordan block

$$J_k(\lambda) = \lambda I_k + J_k$$

is the sum of a scalar matrix and a nilpotent Jordan block. If we partition J_k according to its columns, then

$$J_k = [\mathbf{0}\ \mathbf{e}_1\ \mathbf{e}_2\ \ldots\ \mathbf{e}_{k-1}] \in \mathsf{M}_k. \tag{11.1.13}$$

This helps us visualize the *left-shift identities*

$$J_k\mathbf{e}_1 = \mathbf{0} \quad \text{and} \quad J_k\mathbf{e}_j = \mathbf{e}_{j-1}, \quad j = 2, 3, \ldots, k. \tag{11.1.14}$$

Example 11.1.15 If $k = 3$, then

$$J_3\mathbf{e}_2 = \begin{bmatrix} 0 & 1 & 0 \\ 0 & 0 & 1 \\ 0 & 0 & 0 \end{bmatrix} \begin{bmatrix} 0 \\ 1 \\ 0 \end{bmatrix} = \begin{bmatrix} 1 \\ 0 \\ 0 \end{bmatrix} = \mathbf{e}_1.$$

Lemma 11.1.16 *For each $p = 1, 2, \ldots, k - 1$,*

$$J_k^p = [\underbrace{\mathbf{0}\ \ldots\ \mathbf{0}}_{p}\ \mathbf{e}_1\ \ldots\ \mathbf{e}_{k-p}] \tag{11.1.17}$$

and

$$\operatorname{rank} J_k^p = \begin{cases} k - p & \text{if } p = 1, 2, \ldots, k, \\ 0 & \text{if } p = k, k + 1, \ldots. \end{cases} \tag{11.1.18}$$

The index of J_k is k. The geometric multiplicity of 0 as an eigenvalue of J_k is 1; its algebraic multiplicity is k.

Proof We proceed by induction to prove (11.1.17). The representation (11.1.13) establishes the base case $p = 1$. For the induction step, suppose that (11.1.17) is valid and $p < k$. Compute

$$J_k^{p+1} = J_k J_k^p = J_k[\mathbf{0} \ \underbrace{\cdots \ \mathbf{0}}_{p} \ \mathbf{e}_1 \ \mathbf{e}_2 \ \cdots \ \mathbf{e}_{k-p}]$$

$$= [\underbrace{\mathbf{0} \ \cdots \ \mathbf{0}}_{p} \ J_k\mathbf{e}_1 \ J_k\mathbf{e}_2 \ \cdots \ J_k\mathbf{e}_{k-p}]$$

$$= [\underbrace{\mathbf{0} \ \cdots \ \mathbf{0} \ \mathbf{0}}_{p+1} \ \mathbf{e}_1 \ \cdots \ \mathbf{e}_{k-(p+1)}].$$

For the case $p = k$, compute $J_k^k = J_k J_k^{k-1} = [\mathbf{0} \ \cdots \ \mathbf{0} \ J_k\mathbf{e}_1] = 0$.

The assertion (11.1.18) follows because $\mathbf{e}_1, \mathbf{e}_2, \ldots, \mathbf{e}_{k-p}$ are linearly independent. The index of J_k is k because $J_k^{k-1} = [\mathbf{0} \ \cdots \ \mathbf{0} \ \mathbf{e}_1] \neq 0$ and $J_k^k = 0$. Since J_k is unispectral and has size k, the algebraic multiplicity of its eigenvalue 0 is k. The geometric multiplicity of 0 is $\dim \operatorname{null} J_k = k - \operatorname{rank} J_k = 1$. ∎

For each $i = 1, 2, \ldots, r$, the index of the direct summand J_{n_i} in (11.1.12) is n_i. Since

$$J^p = J_{n_1}^p \oplus J_{n_2}^p \oplus \cdots \oplus J_{n_r}^p, \quad p = 1, 2, \ldots,$$

$J^p = 0$ if and only if each $J_{n_i}^p = 0$.

(a) The index of J is $q = \max\{n_1, n_2, \ldots, n_r\}$, so $m_J(z) = z^q$.

(b) The geometric multiplicity of 0 as an eigenvalue of J is r (one for each block).

(c) The algebraic multiplicity of 0 as an eigenvalue of J is $n_1 + n_2 + \cdots + n_r$.

In computations that involve nilpotent Jordan matrices, it can be convenient to recognize them as bordered matrices.

Example 11.1.19 Here are two ways to partition J_3 as a bordered matrix:

$$\begin{bmatrix} 0 & 1 & 0 \\ 0 & 0 & 1 \\ 0 & 0 & 0 \end{bmatrix} = \begin{bmatrix} 0 & \mathbf{e}_1^\mathsf{T} \\ \mathbf{0} & J_2 \end{bmatrix} \quad \text{and} \quad \begin{bmatrix} 0 & 1 & 0 \\ 0 & 0 & 1 \\ 0 & 0 & 0 \end{bmatrix} = \begin{bmatrix} \mathbf{0} & I_2 \\ 0 & \mathbf{0}^\mathsf{T} \end{bmatrix}, \quad \mathbf{0}, \mathbf{e}_1 \in \mathbb{R}^2.$$

Inspection of (11.1.5) reveals analogous partitions of any nilpotent Jordan block:

$$J_{k+1} = \begin{bmatrix} 0 & \mathbf{e}_1^\mathsf{T} \\ \mathbf{0} & J_k \end{bmatrix} = \begin{bmatrix} \mathbf{0} & I_k \\ 0 & \mathbf{0}^\mathsf{T} \end{bmatrix}, \quad \mathbf{0}, \mathbf{e}_1 \in \mathbb{R}^k, \quad k = 1, 2, \ldots. \qquad (11.1.20)$$

11.2 Existence of a Jordan Form

Our next goal is to show that every nilpotent matrix is similar to a nilpotent Jordan matrix. We begin with some technical lemmas.

Lemma 11.2.1 *For each* $k = 2, 3, \ldots,$

$$I_k - J_k^\mathsf{T} J_k = \mathrm{diag}(1, 0, 0, \ldots, 0). \qquad (11.2.2)$$

Proof Use (11.1.20) to compute

$$J_k^\mathsf{T} J_k = \begin{bmatrix} \mathbf{0}^\mathsf{T} & 0 \\ I_{k-1} & \mathbf{0} \end{bmatrix} \begin{bmatrix} \mathbf{0} & I_{k-1} \\ 0 & \mathbf{0}^\mathsf{T} \end{bmatrix} = \begin{bmatrix} 0 & \mathbf{0}^\mathsf{T} \\ \mathbf{0} & I_{k-1} \end{bmatrix} = \mathrm{diag}(0, 1, 1, \ldots, 1).$$

Then $I_k - J_k^\mathsf{T} J_k = I_k - \mathrm{diag}(0, 1, 1, \ldots, 1) = \mathrm{diag}(1, 0, 0, \ldots, 0).$ $\qquad \square$

Lemma 11.2.3 *Let* $S \in \mathsf{M}_k$ *be invertible and let* $\mathbf{z} \in \mathbb{C}^k$. *Then*

$$\begin{bmatrix} 1 & -\mathbf{z}^\mathsf{T} S \\ \mathbf{0} & S \end{bmatrix}^{-1} = \begin{bmatrix} 1 & \mathbf{z}^\mathsf{T} \\ \mathbf{0} & S^{-1} \end{bmatrix}.$$

Proof Compute

$$\begin{bmatrix} 1 & -\mathbf{z}^\mathsf{T} S \\ \mathbf{0} & S \end{bmatrix} \begin{bmatrix} 1 & \mathbf{z}^\mathsf{T} \\ \mathbf{0} & S^{-1} \end{bmatrix} = \begin{bmatrix} 1 & \mathbf{z}^\mathsf{T} - \mathbf{z}^\mathsf{T} S S^{-1} \\ \mathbf{0} & S S^{-1} \end{bmatrix} = I_{k+1}. \qquad \square$$

A variant of the similarity (3.3.14) is useful.

Lemma 11.2.4 *Let* $n \geq 2$ *and* $\mathbf{x} \in \mathbb{C}^{n-1}$. *Suppose that* $S, B \in \mathsf{M}_{n-1}$, *S is invertible, and* $SBS^{-1} = J_{n-1}$. *Let*

$$A = \begin{bmatrix} 0 & \mathbf{x}^\mathsf{T} \\ \mathbf{0} & B \end{bmatrix}.$$

Then there is a $\mathbf{z} \in \mathbb{C}^{n-1}$ *and a scalar* c *such that*

$$\begin{bmatrix} 1 & -\mathbf{z}^\mathsf{T} S \\ \mathbf{0} & S \end{bmatrix} \begin{bmatrix} 0 & \mathbf{x}^\mathsf{T} \\ \mathbf{0} & B \end{bmatrix} \begin{bmatrix} 1 & \mathbf{z}^\mathsf{T} \\ \mathbf{0} & S^{-1} \end{bmatrix} = \begin{bmatrix} 0 & c\mathbf{e}_1^\mathsf{T} \\ \mathbf{0} & J_{n-1} \end{bmatrix}.$$

If $c = 0$, *then* A *is similar to* $J_1 \oplus J_{n-1}$; *otherwise,* A *is similar to* J_n.

Proof Let $\mathbf{x}^\mathsf{T} S^{-1} = [x_1 \; x_2 \; \ldots \; x_{n-1}]$ and let $\mathbf{z}^\mathsf{T} = \mathbf{x}^\mathsf{T} S^{-1} J_{n-1}^\mathsf{T}$. Use the preceding lemma and (11.2.2) to show that A is similar to

$$\begin{bmatrix} 1 & -\mathbf{z}^\mathsf{T} S \\ \mathbf{0} & S \end{bmatrix} \begin{bmatrix} 0 & \mathbf{x}^\mathsf{T} \\ \mathbf{0} & B \end{bmatrix} \begin{bmatrix} 1 & \mathbf{z}^\mathsf{T} \\ \mathbf{0} & S^{-1} \end{bmatrix}$$

$$= \begin{bmatrix} 0 & \mathbf{x}^\mathsf{T} S^{-1} - \mathbf{z}^\mathsf{T} S B S^{-1} \\ \mathbf{0} & S B S^{-1} \end{bmatrix} = \begin{bmatrix} 0 & \mathbf{x}^\mathsf{T} S^{-1} - \mathbf{x}^\mathsf{T} S^{-1} J_{n-1}^\mathsf{T} J_{n-1} \\ \mathbf{0} & J_{n-1} \end{bmatrix}$$

$$= \begin{bmatrix} 0 & \mathbf{x}^\mathsf{T} S^{-1} (I - J_{n-1}^\mathsf{T} J_{n-1}) \\ \mathbf{0} & J_{n-1} \end{bmatrix} = \begin{bmatrix} 0 & \mathbf{x}^\mathsf{T} S^{-1} \mathrm{diag}(1, 0, 0, \ldots, 0) \\ \mathbf{0} & J_{n-1} \end{bmatrix}$$

$$= \begin{bmatrix} 0 & x_1 \mathbf{e}_1^\mathsf{T} \\ \mathbf{0} & J_{n-1} \end{bmatrix}.$$

If $x_1 = 0$, then A is similar to $J_1 \oplus J_{n-1}$. If $x_1 \neq 0$, then the bordering identity (11.1.20) and the similarity

$$\begin{bmatrix} x_1^{-1} & 0 \\ 0 & I \end{bmatrix} \begin{bmatrix} 0 & x_1 e_1^T \\ 0 & J_{n-1} \end{bmatrix} \begin{bmatrix} x_1 & 0 \\ 0 & I \end{bmatrix} = \begin{bmatrix} 0 & e_1^T \\ 0 & J_{n-1} \end{bmatrix} = J_n$$

show that A is similar to J_n.   □

A final lemma invokes the left-shift identities (11.1.14) for a nilpotent Jordan block. We use the standard basis $e_1, e_2, \ldots, e_{k+1}$ for $\mathbb{C}^{k+1}$ and a $y \in \mathbb{C}^m$ to construct rank-1 matrices of the form $e_j y^T \in M_{(k+1) \times m}$; they have y^T in the jth row and zeros elsewhere.

Lemma 11.2.5 *Let k and m be positive integers, let $y \in \mathbb{C}^m$, let $e_1 \in \mathbb{C}^{k+1}$, and let $B \in M_m$ be nilpotent and have index k or less. Then*

$$A = \begin{bmatrix} J_{k+1} & e_1 y^T \\ 0 & B \end{bmatrix} \quad \text{is similar to} \quad \begin{bmatrix} J_{k+1} & 0 \\ 0 & B \end{bmatrix}.$$

Proof For any $X \in M_{(k+1) \times m}$, (3.3.14) ensures that A is similar to

$$\begin{bmatrix} I & X \\ 0 & I \end{bmatrix} \begin{bmatrix} J_{k+1} & e_1 y^T \\ 0 & B \end{bmatrix} \begin{bmatrix} I & -X \\ 0 & I \end{bmatrix}$$

$$= \begin{bmatrix} J_{k+1} & e_1 y^T + XB - J_{k+1}X \\ 0 & B \end{bmatrix}.$$

Let $X = \sum_{j=1}^{k} e_{j+1} y^T B^{j-1}$ and use the fact that $B^k = 0$ to compute

$$XB = \sum_{j=1}^{k} e_{j+1} y^T B^j = \sum_{j=1}^{k-1} e_{j+1} y^T B^j = \sum_{j=2}^{k} e_j y^T B^{j-1}.$$

The left-shift identities (11.1.14) tell us that $J_{k+1} e_{j+1} = e_j$, so

$$J_{k+1} X = \sum_{j=1}^{k} J_{k+1} e_{j+1} y^T B^{j-1} = \sum_{j=1}^{k} e_j y^T B^{j-1} = e_1 y^T + XB.$$

Then $e_1 y^T + XB - J_{k+1} X = 0$, which shows that A is similar to $J_{k+1} \oplus B$.   □

Once we have proved the following theorem, it is only a short additional step to show that every square complex matrix is similar to a Jordan matrix (a direct sum of Jordan blocks with various eigenvalues).

Theorem 11.2.6 *Each nilpotent matrix is similar to a nilpotent Jordan matrix.*

Proof Let $A \in M_n$ be nilpotent. Since $\operatorname{spec} A = \{0\}$, Theorem 10.1.1 ensures that A is unitarily similar to an upper triangular matrix with all diagonal entries equal to 0. Thus, we may assume that A is strictly upper triangular. We proceed by induction.

If $n = 1$, then $A = J_1 = [0]$, so A is equal to a nilpotent Jordan block.

Assume that $n \geq 2$ and that every nilpotent matrix of size $n - 1$ or less is similar to a nilpotent Jordan matrix. Partition

$$A = \begin{bmatrix} 0 & \mathbf{a}^T \\ \mathbf{0} & B \end{bmatrix},$$

in which $\mathbf{a} \in \mathbb{C}^{n-1}$ and $B \in \mathsf{M}_{n-1}$ is strictly upper triangular. Let g be the geometric multiplicity of 0 as an eigenvalue of B and let q be the index of B. The induction hypothesis ensures that there is an invertible $S \in \mathsf{M}_{n-1}$ such that $S^{-1}BS = J_{k_1} \oplus J_{k_2} \oplus \cdots \oplus J_{k_g}$. After a block permutation similarity, if necessary, we may assume that $k_1 = q$, in which case $q \geq \max\{k_2, k_3, \ldots, k_g\}$.

If $g = 1$, then B is similar to J_{n-1} and Lemma 11.2.4 ensures that A is similar to one of the two nilpotent Jordan matrices $J_1 \oplus J_{n-1}$ or J_n.

If $g \geq 2$, let $J = J_{k_2} \oplus \cdots \oplus J_{k_g}$ and observe that $S^{-1}BS = J_q \oplus J$. Every direct summand in J has size q or less, so $J^q = 0$. Then A is similar to

$$\begin{bmatrix} 1 & \mathbf{0}^T \\ \mathbf{0} & S^{-1} \end{bmatrix} \begin{bmatrix} 0 & \mathbf{a}^T \\ \mathbf{0} & B \end{bmatrix} \begin{bmatrix} 1 & \mathbf{0}^T \\ \mathbf{0} & S \end{bmatrix}$$

$$= \begin{bmatrix} 0 & \mathbf{a}^T S \\ \mathbf{0} & S^{-1}BS \end{bmatrix} = \begin{bmatrix} 0 & \mathbf{a}^T S \\ \mathbf{0} & J_q \oplus J \end{bmatrix} \tag{11.2.7}$$

$$= \begin{bmatrix} 0 & \mathbf{x}^T & \mathbf{y}^T \\ \mathbf{0} & J_q & 0 \\ \mathbf{0} & \mathbf{0} & J \end{bmatrix}, \tag{11.2.8}$$

in which we have partitioned $\mathbf{a}^T S = [\mathbf{x}^T \ \mathbf{y}^T]$ with $\mathbf{x} = [x_1 \ x_2 \ \ldots \ x_q]^T \in \mathbb{C}^q$ and $\mathbf{y} \in \mathbb{C}^{n-q-1}$.

Lemma 11.2.4 tells us that there is a scalar c such that the upper-left 2×2 block matrix in (11.2.8) is similar to

$$\begin{bmatrix} 0 & c\mathbf{e}_1^T \\ \mathbf{0} & J_q \end{bmatrix}, \quad \mathbf{e}_1 \in \mathbb{C}^q$$

via a similarity matrix of the form

$$\begin{bmatrix} 1 & -\mathbf{z}^T \\ \mathbf{0} & I \end{bmatrix}, \quad \mathbf{z} \in \mathbb{C}^q.$$

This observation leads us to a similarity of the 3×3 block matrix (11.2.8):

$$\begin{bmatrix} 1 & -\mathbf{z}^T & \mathbf{0}^T \\ \mathbf{0} & I & 0 \\ \mathbf{0} & \mathbf{0} & I \end{bmatrix} \begin{bmatrix} 0 & \mathbf{x}^T & \mathbf{y}^T \\ \mathbf{0} & J_q & 0 \\ \mathbf{0} & \mathbf{0} & J \end{bmatrix} \begin{bmatrix} 1 & \mathbf{z}^T & \mathbf{0}^T \\ \mathbf{0} & I & 0 \\ \mathbf{0} & \mathbf{0} & I \end{bmatrix} = \begin{bmatrix} 0 & c\mathbf{e}_1^T & \mathbf{y}^T \\ \mathbf{0} & J_q & 0 \\ \mathbf{0} & \mathbf{0} & J \end{bmatrix}. \tag{11.2.9}$$

If $c = 0$, then (11.2.9) is block permutation similar to

$$\begin{bmatrix} J_q & \mathbf{0} & 0 \\ \mathbf{0}^T & 0 & \mathbf{y}^T \\ 0 & \mathbf{0} & J \end{bmatrix} = J_q \oplus \begin{bmatrix} 0 & \mathbf{y}^T \\ \mathbf{0} & J \end{bmatrix}. \tag{11.2.10}$$

The nilpotent 2×2 block matrix in (11.2.10) has size $n - q$, so the induction hypothesis ensures that it is similar to a nilpotent Jordan matrix. Thus, the matrix (11.2.9), and therefore A itself, is similar to a nilpotent Jordan matrix.

If $c \neq 0$, use the bordering identity (11.1.20) to demonstrate that (11.2.9) is similar to

$$\begin{bmatrix} c^{-1} & \mathbf{0}^\mathsf{T} & \mathbf{0}^\mathsf{T} \\ \mathbf{0} & I & 0 \\ \mathbf{0} & 0 & c^{-1}I \end{bmatrix} \begin{bmatrix} 0 & c\mathbf{e}_1^\mathsf{T} & \mathbf{y}^\mathsf{T} \\ \mathbf{0} & J_q & 0 \\ \mathbf{0} & 0 & J \end{bmatrix} \begin{bmatrix} c & \mathbf{0}^\mathsf{T} & \mathbf{0}^\mathsf{T} \\ \mathbf{0} & I & 0 \\ \mathbf{0} & 0 & cI \end{bmatrix}$$

$$= \begin{bmatrix} 0 & \mathbf{e}_1^\mathsf{T} & \mathbf{y}^\mathsf{T} \\ \mathbf{0} & J_q & 0 \\ \mathbf{0} & 0 & J \end{bmatrix}, \quad \mathbf{e}_1 \in \mathbb{C}^q$$

$$= \begin{bmatrix} J_{q+1} & \mathbf{e}_1\mathbf{y}^\mathsf{T} \\ 0 & J \end{bmatrix}, \quad \mathbf{e}_1 \in \mathbb{C}^{q+1}. \tag{11.2.11}$$

Lemma 11.2.5 ensures that (11.2.11) (and hence also A) is similar to the nilpotent Jordan matrix $J_{k_1+1} \oplus J$. The induction hypothesis ensures that J is similar to a nilpotent Jordan matrix. $\qquad\square$

Definition 11.2.12 A *Jordan form* for $A \in \mathsf{M}_n$ is a Jordan matrix to which A is similar.

We have just proved that any nilpotent matrix has a Jordan form; it is a direct sum of nilpotent Jordan blocks.

Example 11.2.13 The 2×2 nilpotent matrices A and B in (11.1.3) have Jordan forms. What are they? There are only two possibilities: $J_1 \oplus J_1 = 0_2$ or J_2. The first is excluded since neither A nor B is the zero matrix, so J_2 must be the Jordan form for A and for B. Consequently, A and B are similar because each is similar to the same Jordan matrix.

Finally, we can show that every square matrix has a Jordan form.

Theorem 11.2.14 *Each $A \in \mathsf{M}_n$ is similar to a Jordan matrix.*

Proof We have proved that A is similar to a direct sum (10.4.6) of unispectral matrices, so it suffices to show that every unispectral matrix has a Jordan form. If B is unispectral and $\operatorname{spec} B = \{\lambda\}$, then $B - \lambda I$ is nilpotent. Theorem 11.2.6 says that there is a nilpotent Jordan matrix

$$J = J_{k_1} \oplus J_{k_2} \oplus \cdots \oplus J_{k_g}$$

and an invertible matrix S such that $B - \lambda I = SJS^{-1}$. Therefore, $B = \lambda I + SJS^{-1} = S(\lambda I + J) S^{-1}$. The computation

$$\lambda I + J = (\lambda I_{k_1} + J_{k_1}) \oplus (\lambda I_{k_2} + J_{k_2}) \oplus \cdots \oplus (\lambda I_{k_g} + J_{k_g})$$

$$= J_{k_1}(\lambda) \oplus J_{k_2}(\lambda) \oplus \cdots \oplus J_{k_g}(\lambda) \tag{11.2.15}$$

shows that (11.2.15) is a Jordan form for B. $\qquad\square$

11.3 Uniqueness of a Jordan Form

In this section, we show that two Jordan matrices are similar if and only if one can be obtained from the other by permuting its direct summands.

In the preceding section, we constructed a Jordan form for a square complex matrix by combining (as a direct sum) Jordan forms of unispectral matrices with different eigenvalues. If we could show that a Jordan form for a unispectral matrix is unique up to a permutation of its direct summands, the same would be true for a Jordan form for a general square matrix. Once again, it suffices to consider nilpotent unispectral matrices. Ranks of powers of nilpotent Jordan blocks are at the heart of the matter.

Example 11.3.1 Consider

$$J = J_3 \oplus J_3 \oplus J_2 \oplus J_2 \oplus J_2 \oplus J_1 \in \mathsf{M}_{13}, \qquad (11.3.2)$$

which is a direct sum of six nilpotent Jordan blocks:

> one nilpotent Jordan block of size 1,
> three nilpotent Jordan blocks of size 2,
> two nilpotent Jordan blocks of size 3.

There are

> six blocks of size 1 or greater,
> five blocks of size 2 or greater,
> two blocks of size 3 or greater.

Use Lemma 11.1.16 to compute

$$\operatorname{rank} J^0 = 13,$$
$$\operatorname{rank} J^1 = 2 + 2 + 1 + 1 + 1 = 7,$$
$$\operatorname{rank} J^2 = 1 + 1 = 2,$$
$$\operatorname{rank} J^3 = \operatorname{rank} J^4 = 0.$$

Now let $w_i = \operatorname{rank} J^{i-1} - \operatorname{rank} J^i$ and compute

$$w_1 = 13 - 7 = 6,$$
$$w_2 = 7 - 2 = 5,$$
$$w_3 = 2 - 0 = 2,$$
$$w_4 = 0 - 0 = 0.$$

Notice that $w_1 = 6$ is the total number of blocks in (11.3.2), which is the geometric multiplicity of 0 as an eigenvalue of J. Moreover, each w_p is equal to the number of Jordan blocks in J that have size p or greater. Finally, $w_3 > 0$ and $w_4 = 0$, so the largest nilpotent block in (11.3.2) is 3×3. Now compute the differences

$$w_1 - w_2 = 1,$$
$$w_2 - w_3 = 3,$$
$$w_3 - w_4 = 2,$$

and observe that each difference $w_p - w_{p+1}$ is equal to the multiplicity of J_p. This is not an accident.

The preceding example illuminates our path forward, but we pause to establish a useful fact.

Lemma 11.3.3 *Let p and k be positive integers. Then for each $p = 1, 2, \ldots$*

$$\operatorname{rank} J_k^{p-1} - \operatorname{rank} J_k^p = \begin{cases} 1 & \text{if } p \leq k, \\ 0 & \text{if } p > k. \end{cases} \tag{11.3.4}$$

Proof If $p \leq k$, then Lemma 11.1.16 ensures that $\operatorname{rank} J_k^{p-1} - \operatorname{rank} J_k^p = (k - (p-1)) - (k - p) = 1$, but both ranks are 0 if $p > k$. $\qquad\square$

The identity (11.3.4) leads to an algorithm that determines the number of nilpotent blocks of each size in a Jordan matrix.

Theorem 11.3.5 *Let $A \in M_n$ and suppose that 0 is an eigenvalue of A with geometric multiplicity $g \geq 1$. Let*

$$J_{n_1} \oplus J_{n_2} \oplus \cdots \oplus J_{n_g} \oplus J \in M_n \tag{11.3.6}$$

be a Jordan form for A, in which J is a direct sum of Jordan blocks with nonzero eigenvalues. Let

$$w_p = \operatorname{rank} A^{p-1} - \operatorname{rank} A^p, \quad p = 1, 2, \ldots, n+1. \tag{11.3.7}$$

For each $p = 1, 2, \ldots, n$, the number of nilpotent blocks in (11.3.6) that have size p or larger is w_p. The multiplicity of J_p is $w_p - w_{p+1}$.

Proof Since J is invertible, $\operatorname{rank} J = \operatorname{rank} J^p$ for all $p = 0, 1, 2, \ldots$ and

$$\operatorname{rank} A^{p-1} - \operatorname{rank} A^p = \left(\sum_{i=1}^g \operatorname{rank} J_{n_i}^{p-1} + \operatorname{rank} J^{p-1} \right) - \left(\sum_{i=1}^g \operatorname{rank} J_{n_i}^p + \operatorname{rank} J^p \right)$$

$$= \left(\sum_{i=1}^g \operatorname{rank} J_{n_i}^{p-1} + \operatorname{rank} J \right) - \left(\sum_{i=1}^g \operatorname{rank} J_{n_i}^p + \operatorname{rank} J \right)$$

$$= \sum_{i=1}^g \left(\operatorname{rank} J_{n_i}^{p-1} - \operatorname{rank} J_{n_i}^p \right). \tag{11.3.8}$$

The identity (11.3.4) tells us that a summand in (11.3.8) is 1 if and only if its corresponding nilpotent block has size p or larger; otherwise it is 0. Therefore, (11.3.8) counts the number of nilpotent blocks that have size p or larger. The difference $w_p - w_{p+1}$ is the number of nilpotent blocks that have size p or larger, minus the number that have size $p + 1$ or larger; this is the number of nilpotent blocks whose size is exactly p. $\qquad\square$

If q is the size of the largest nilpotent block in the Jordan matrix (11.3.6), then $J_q^{q-1} \neq 0$ and $J_q^q = 0$. Therefore, $\operatorname{rank} A^q = \operatorname{rank} A^{q+1} = \cdots = \operatorname{rank} J$. That is, q is the index of A. The ranks of powers of A decrease monotonically to a stable value, which is the sum of the sizes of the invertible Jordan blocks in a Jordan form for A. Consequently, the rank differences in (11.3.7) are eventually all zero. In fact, $w_q > 0$ and $w_{q+1} = w_{q+2} = \cdots = 0$.

Definition 11.3.9 Let $A \in M_n$, suppose that $0 \in \operatorname{spec} A$, let q be the index of A, and let $w_p = \operatorname{rank} A^{p-1} - \operatorname{rank} A^p$, for each $p = 1, 2, \ldots$. The list of positive integers

$$w_1, w_2, \ldots, w_q,$$

$A =$	J_4	$J_3 \oplus J_1$	$J_2 \oplus J_2$	$J_2 \oplus J_1 \oplus J_1$	$J_1 \oplus J_1 \oplus J_1 \oplus J_1$
$\operatorname{rank} A^0 =$	4	4	4	4	4
$w_1 =$	1	2	2	3	4
$\operatorname{rank} A^1 =$	3 [0]	2 [1]	2 [0]	1 [2]	0 [4]
$w_2 =$	1	1	2	1	0
$\operatorname{rank} A^2 =$	2 [0]	1 [0]	0 [2]	0 [1]	0
$w_3 =$	1	1	0	0	
$\operatorname{rank} A^3 =$	1 [0]	0 [1]	0	0	
$w_4 =$	1	0			
$\operatorname{rank} A^4 =$	0 [1]	0			
$w_5 =$	0				
$\operatorname{rank} A^5 =$	0				

Figure 11.1 Weyr characteristics of 4×4 Jordan matrices. Block multiplicities are boxed.

is the *Weyr characteristic of* A. If $\lambda \in \operatorname{spec} A$, the Weyr characteristic of $A - \lambda I$ is the *Weyr characteristic of* A *associated with* λ.

Example 11.3.10 Figure 11.1 tabulates the Weyr characteristics of the five different 4×4 Jordan matrices. The multiplicities of the Jordan blocks (in boxes) are computed from the Weyr characteristics, as described in Theorem 11.3.5. In each column, the integers in the Weyr characteristic have the following properties:

(a) w_1 is the number of blocks in A.

(b) $w_q > 0$ and $w_{q+1} = 0$, in which q is the index of A.

(c) $w_1 + w_2 + \cdots + w_q = 4$.

(d) $w_i \geq w_{i+1}$ for $i = 1, 2, \ldots, q$.

If $A \in \mathsf{M}_n$ and $\lambda \in \operatorname{spec} A$, the first integer in the Weyr characteristic of $A - \lambda I$ is

$$\begin{aligned} w_1 &= \operatorname{rank}(A - \lambda I)^0 - \operatorname{rank}(A - \lambda I) = n - \operatorname{rank}(A - \lambda I) \\ &= \dim \operatorname{null}(A - \lambda I) = \dim \mathcal{E}_\lambda(A). \end{aligned}$$

Therefore, w_1 is the geometric multiplicity of λ (the number of blocks with eigenvalue λ in a Jordan form for A). For each $i = 1, 2, \ldots, q$, the integer w_i is positive because is it equal to the number of blocks of size i or greater. The differences $w_i - w_{i+1}$ are nonnegative (that is, the sequence w_i is decreasing) because it is equal to the number of blocks of size i (a nonnegative integer). The sum

$$\begin{aligned} w_1 + w_2 + \cdots + w_q &= \sum_{p=1}^{q} \left(\operatorname{rank}(A - \lambda I)^{p-1} - \operatorname{rank}(A - \lambda I)^p \right) \\ &= \operatorname{rank}(A - \lambda I)^0 - \operatorname{rank}(A - \lambda I)^q \\ &= n - \operatorname{rank}(A - \lambda I)^q \\ &= \dim \operatorname{null}(A - \lambda I)^q \end{aligned}$$

is the algebraic multiplicity of λ (the sum of the sizes of the blocks with eigenvalue λ in a Jordan form for A); see P.11.21.

Similar matrices A and B have the same spectrum, and corresponding powers A^p and B^p are similar (see (0.8.3)), so rank $A^p = $ rank B^p for all $p = 1, 2, \ldots$. Consequently, similar matrices have the same Weyr characteristic associated with corresponding eigenvalues. The following corollary provides a converse of this assertion.

Corollary 11.3.11 *Let* $J, J' \in M_n$ *be Jordan matrices. Then* J *is similar to* J' *if and only if* spec $J = $ spec J' *and* rank$(J - \lambda I)^p = $ rank$(J' - \lambda I)^p$ *for each* $\lambda \in$ spec J *and each* $p = 1$, $2, \ldots, n$.

Proof For each $\lambda \in$ spec $J = $ spec J', the hypotheses ensure that the Weyr characteristics of J and J' associated with λ are the same. It follows that the multiplicities of the Jordan blocks $J_k(\lambda)$ in J are the same as those in J' for each $k = 1, 2, \ldots, n$. $\qquad\qquad\square$

Theorem 11.3.12 *If* $A \in M_n$ *and if* J *and* J' *are Jordan matrices that are similar to* A, *then* J' *can be obtained from* J *by permuting its direct summands.*

Proof Each Jordan form of A is similar to A, so any two Jordan forms of A are similar. Therefore, they have the same Weyr characteristics associated with each of their eigenvalues. It follows that their block sizes and multiplicities are the same for each of their eigenvalues. Consequently, one can be obtained from the other by permuting its direct summands. $\qquad\square$

Example 11.3.13 Suppose that $A \in M_4$ has eigenvalues $3, 2, 2, 2$. A Jordan form for A must contain the direct summand $J_1(3) = [3]$ as well as a 3×3 Jordan matrix whose spectrum is $\{2\}$. There are only three possibilities for the latter: $J_3(2)$, $J_1(2) \oplus J_2(2)$, or $J_1(2) \oplus J_1(2) \oplus J_1(2)$. Therefore, A is similar to exactly one of

$$\begin{bmatrix} 3 & 0 & 0 & 0 \\ 0 & 2 & 0 & 0 \\ 0 & 0 & 2 & 0 \\ 0 & 0 & 0 & 2 \end{bmatrix} = J_1(3) \oplus J_1(2) \oplus J_1(2) \oplus J_1(2),$$

$$\begin{bmatrix} 3 & 0 & 0 & 0 \\ 0 & 2 & 1 & 0 \\ 0 & 0 & 2 & 0 \\ 0 & 0 & 0 & 2 \end{bmatrix} = J_1(3) \oplus J_2(2) \oplus J_1(2), \text{ or}$$

$$\begin{bmatrix} 3 & 0 & 0 & 0 \\ 0 & 2 & 1 & 0 \\ 0 & 0 & 2 & 1 \\ 0 & 0 & 0 & 2 \end{bmatrix} = J_1(3) \oplus J_3(2).$$

These matrices correspond to the three possibilities rank$(A - 2I) = 1, 2,$ or 3, respectively.

Corollary 11.3.14 *If* $A, B \in M_n$ *and* spec $A = $ spec B, *then* A *is similar to* B *if and only if* rank$(A - \lambda I)^p = $ rank$(B - \lambda I)^p$ *for each* $\lambda \in$ spec A *and each* $p = 1, 2, \ldots, n$.

Proof Theorem 11.2.14 ensures that there are Jordan matrices J_A and J_B such that A is similar to J_A and B is similar to J_B. It follows that $\mathrm{rank}(J_A - \lambda I)^p = \mathrm{rank}(J_B - \lambda I)^p$ for each $\lambda \in \mathrm{spec}\,A$ and each $p = 1, 2, \ldots, n$. Corollary 11.3.11 tells us that J_A is similar to J_B, so it follows from the transitivity of similarity that A is similar to B. $\qquad\square$

11.4 The Jordan Canonical Form

Theorem 11.2.14 says that each square complex matrix A is similar to a Jordan matrix J. Theorem 11.3.12 says that J is unique up to permutations of its direct summands. Therefore, if we agree that permuting the direct summands of a Jordan matrix is inessential, we may speak of *the* Jordan matrix that is similar to A. It is often called the *Jordan canonical form* of A.

When presenting the Jordan canonical form of A, it is a useful convention to group together the Jordan blocks with the same eigenvalue, and to arrange those blocks in decreasing order of their sizes. For example, if $\lambda_1, \lambda_2, \ldots, \lambda_d$ are the distinct eigenvalues of A, we can present its Jordan canonical form as a direct sum

$$J = J(\lambda_1) \oplus J(\lambda_2) \oplus \cdots \oplus J(\lambda_d) \qquad (11.4.1)$$

of Jordan matrices with different eigenvalues. The number of Jordan blocks in a direct summand $J(\lambda_i)$ in (11.4.1) is equal to the geometric multiplicity of λ_i, which we denote by g_i. If the index of $A - \lambda_i I$ is q_i, each direct summand in (11.4.1) is a direct sum of g_i Jordan blocks with eigenvalue λ_i, the largest of which has size q_i.

In principle, one can determine the Jordan canonical form of a given $A \in \mathsf{M}_n$ by doing the following for each $\lambda \in \mathrm{spec}\,A$:

(a) Compute $r_p = \mathrm{rank}(A - \lambda I)^p$ for $p = 0, 1, 2, \ldots$. Stop when the ranks stabilize; the index of $A - \lambda I$ (this is q) is the first value of p for which $r_p = r_{p+1}$. By definition, $r_0 = n$.

(b) Compute $w_p = r_{p-1} - r_p$ for $p = 1, 2, \ldots, q + 1$; the list of integers $w_1, w_2, \ldots, w_q$ is the Weyr characteristic of $A - \lambda I$.

(c) For each $k = 1, 2, \ldots, q$ there are $w_k - w_{k+1}$ blocks of the form $J_k(\lambda)$ in the Jordan canonical form of A.

This algorithm is an elegant conceptual tool, but it is not recommended for numerical computations. The following example illustrates why any attempt to compute the Jordan canonical form of a matrix in finite precision arithmetic is fraught with danger.

Example 11.4.2 For a Jordan block $J_k(\lambda)$ with $k \geq 2$ and for any $\varepsilon > 0$, the matrix $J_k(\lambda) + \mathrm{diag}(\varepsilon, 2\varepsilon, \ldots, k\varepsilon)$ has distinct eigenvalues $\lambda + \varepsilon, \lambda + 2\varepsilon, \ldots, \lambda + k\varepsilon$, so Corollary 9.4.6 ensures that it is diagonalizable. Thus, its Jordan canonical form is $[\lambda + \varepsilon] \oplus [\lambda + 2\varepsilon] \oplus \cdots \oplus [\lambda + k\varepsilon]$ and all the blocks are 1×1.

Small changes in the entries of a matrix can result in major changes in its Jordan canonical form. For another example of this phenomenon, see P.11.15. The Jordan canonical form is a powerful conceptual tool, but it is not a good numerical tool.

11.5 Differential Equations and the Jordan Canonical Form

Let $A = [a_{ij}] \in M_n$ and consider the problem of finding a vector-valued function $\mathbf{x}(t) = [x_i(t)] \in \mathbb{C}^n$ of a real variable t such that

$$\mathbf{x}'(t) = A\mathbf{x}(t), \quad \mathbf{x}(0) \text{ is given.} \tag{11.5.1}$$

A theorem from differential equations guarantees that for each choice of the initial condition $\mathbf{x}(0)$, the initial value problem (11.5.1) has a unique solution. The unknown functions $x_i(t)$ satisfy the coupled scalar equations

$$\frac{dx_i}{dt} = \sum_{j=1}^{n} a_{ij}x_j, \quad x_i(0) \text{ is given,} \quad i = 1, 2, \ldots, n.$$

If A is diagonalizable, a change of dependent variables decouples the equations and significantly simplifies the problem. Suppose that $A = S\Lambda S^{-1}$ with $\Lambda = \text{diag}(\lambda_1\lambda_2, \ldots, \lambda_n)$ and $S = [s_{ij}]$. Let $S^{-1}\mathbf{x}(t) = \mathbf{y}(t) = [y_i(t)]$. Then

$$S^{-1}\mathbf{x}'(t) = S^{-1}A\mathbf{x}(t) = S^{-1}AS\mathbf{y}(t) = \Lambda\mathbf{y}(t),$$

so in the new dependent variables our problem is

$$\mathbf{y}'(t) = \Lambda\mathbf{y}(t), \quad \mathbf{y}(0) = S^{-1}\mathbf{x}(0) \text{ is given.} \tag{11.5.2}$$

The initial value problem (11.5.2) is equivalent to n uncoupled scalar equations

$$\frac{dy_i}{dt} = \lambda_i y_i, \quad y_i(0) = \sum_{j=1}^{n} \sigma_{ij}x_j(0), \quad i = 1, 2, \ldots, n, \tag{11.5.3}$$

in which $S^{-1} = [\sigma_{ij}]$. Each of these equations can be solved separately. Their solutions are

$$y_i(t) = y_i(0)e^{\lambda_i t}, \quad i = 1, 2, \ldots, n.$$

If $\lambda_j = \alpha_j + i\beta_j$, in which α_j, β_j are real, then

$$e^{\lambda_j t} = e^{\alpha_j t + i\beta_j t} = e^{\alpha_j t}e^{i\beta_j t} = e^{\alpha_j t}(\cos \beta_j t + i \sin \beta_j t). \tag{11.5.4}$$

If λ_j is real, then $\beta_j = 0$ and $e^{\lambda_j t} = e^{\alpha_j t}$ is a real exponential function. If λ_j is not real, then $e^{\lambda_j t}$ is a linear combination of damped oscillating functions $e^{\alpha_j t} \cos \beta_j t$ and $e^{\alpha_j t} \sin \beta_j t$; see Figure 11.2. Therefore, the entries of the solution $\mathbf{x}(t) = [x_i(t)] = S\mathbf{y}(t)$ to (11.5.1) are linear combinations of functions of the form (11.5.4). The parameters λ_j are the eigenvalues of A.

If A is not diagonalizable, let $A = SJS^{-1}$, in which J is the Jordan canonical form of A. A change of variables partially decouples the equations and permits us to solve several smaller and simpler initial value problems, one for each block in the Jordan canonical form of A. Let $\mathbf{y}(t) = S^{-1}\mathbf{x}(t)$. Then $S^{-1}\mathbf{x}'(t) = S^{-1}A\mathbf{x}(t) = S^{-1}AS\mathbf{y}(t) = J\mathbf{y}(t)$, so in the new dependent variables our problem is

$$\mathbf{y}'(t) = J\mathbf{y}(t), \quad \mathbf{y}(0) = S^{-1}\mathbf{x}(0) \text{ is given.} \tag{11.5.5}$$

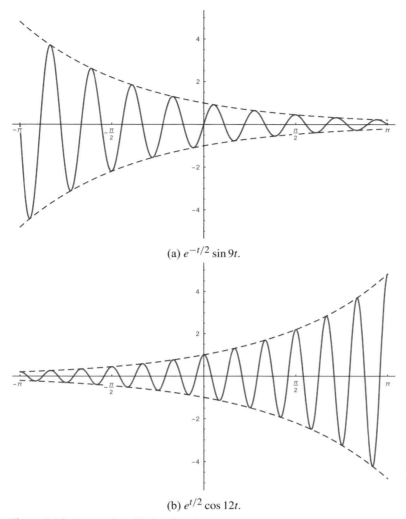

(a) $e^{-t/2} \sin 9t$.

(b) $e^{t/2} \cos 12t$.

Figure 11.2 Damped oscillating functions.

If J is a direct sum (11.1.7) of r Jordan blocks (their eigenvalues need not be distinct), the system (11.5.5) is equivalent to

$$\mathbf{y}_i'(t) = J_{n_i}(\lambda_i)\mathbf{y}_i(t), \quad \mathbf{y}_i(0) \in \mathbb{C}^{n_i} \text{ is given}, \quad i = 1, 2, \dots, r, \qquad (11.5.6)$$

in which $n_1 + n_2 + \cdots + n_r = n$. Each of these initial value problems can be solved separately. The following example illustrates how to solve a problem of the form (11.5.6).

Example 11.5.7 The system $\mathbf{y}'(t) = J_3(\lambda)\mathbf{y}(t)$ is

$$\begin{array}{rcl}
y_1' &=& \lambda y_1 + y_2, \\
y_2' &=& \lambda y_2 + y_3, \\
y_3' &=& \lambda y_3.
\end{array}$$

Suppose that the initial value $\mathbf{y}(0) = [y_1(0)\ y_2(0)\ y_3(0)]^T$ is given. Initial value problems of this kind can be solved sequentially, starting at the bottom. The solution to the last equation is $y_3(t) = e^{\lambda t}y_3(0)$. The next equation is $y_2' = \lambda y_2 + e^{\lambda t}y_3(0)$, which has the solution

$$y_2(t) = e^{\lambda t}(y_3(0)t + y_2(0)).$$

The remaining equation $y_1' = \lambda y_1 + e^{\lambda t}(y_3(0)t + y_2(0))$ has the solution

$$y_1(t) = e^{\lambda t}\left(y_3(0)\frac{t^2}{2} + y_2(0)t + y_1(0)\right).$$

The entries of the solution $\mathbf{y}(t)$ are functions of the form $e^{\lambda t}p(t)$, in which p is a polynomial whose degree is less than three (the size of the Jordan block).

The preceding example illustrates the principal features of the general case. The solution to each of the initial value problems (11.5.6) is a vector-valued function whose entries are functions of the form $e^{\lambda_i t}p_i(t)$, in which p_i is a polynomial of degree $n_i - 1$ or less. If $\lambda_1, \lambda_2, \ldots, \lambda_d$ are the distinct eigenvalues of A, then each entry $x_i(t)$ in the solution $\mathbf{x}(t) = [x_i(t)]$ of (11.5.1) is a linear combination of the entries of $\mathbf{y}(t)$, so it has the form

$$x_i(t) = e^{\lambda_1 t}p_1(t) + e^{\lambda_2 t}p_2(t) + \cdots + e^{\lambda_d t}p_d(t), \tag{11.5.8}$$

in which each p_i is a polynomial whose degree is less than the index of $A - \lambda_i I$.

If A has some non-real eigenvalues, the exponential functions $e^{\lambda_i t}$ and the polynomials $p_i(t)$ in the solution (11.5.8) need not be real-valued. If A is real, however, then $\operatorname{Re} \mathbf{x}(t)$ and $\operatorname{Im} \mathbf{x}(t)$ are both solutions of the differential equations in (11.5.1), with respective initial conditions $\operatorname{Re} \mathbf{x}(0)$ and $\operatorname{Im} \mathbf{x}(0)$. If the initial condition $\mathbf{x}(0)$ is real, then $\operatorname{Im} \mathbf{x}(0) = \mathbf{0}$ and the uniqueness of the solution of (11.5.1) implies that $\operatorname{Im} \mathbf{x}(t) = \mathbf{0}$ for all $t \in \mathbb{R}$. Consequently, the solution $\mathbf{x}(t) = S\mathbf{y}(t)$ must be real even if neither S nor $\mathbf{y}(t)$ is real.

Example 11.5.9 Consider the initial value problem (11.5.1), in which

$$A = \begin{bmatrix} 0 & 1 \\ -1 & 0 \end{bmatrix} \quad \text{and} \quad \mathbf{x}(0) = \begin{bmatrix} 1 \\ 2 \end{bmatrix}.$$

Then $\operatorname{spec} A = \{i, -i\}$ and $A = S\Lambda S^{-1}$, in which

$$S = \begin{bmatrix} 1 & 1 \\ i & -i \end{bmatrix}, \quad S^{-1} = \frac{1}{2}\begin{bmatrix} 1 & -i \\ 1 & i \end{bmatrix}, \quad \text{and} \quad \Lambda = \begin{bmatrix} i & 0 \\ 0 & -i \end{bmatrix}.$$

After the change of variables $\mathbf{y}(t) = S^{-1}\mathbf{x}(t)$, the decoupled equations (11.5.3) are

$$\frac{dy_1}{dt} = iy_1, \qquad\qquad y_1(0) = \tfrac{1}{2} - i,$$

$$\frac{dy_2}{dt} = -iy_2, \qquad\qquad y_2(0) = \tfrac{1}{2} + i.$$

The solution is

$$\mathbf{y}(t) = \frac{1}{2}\begin{bmatrix} (1 - 2i)\,e^{it} \\ (1 + 2i)\,e^{-it} \end{bmatrix}.$$

In the original variables, the solution to (11.5.1) is

$$\mathbf{x}(t) = S\mathbf{y}(t) = \begin{bmatrix} y_1 + y_2 \\ iy_1 - iy_2 \end{bmatrix} = \begin{bmatrix} \cos t + 2\sin t \\ 2\cos t - \sin t \end{bmatrix}.$$

11.6 Convergent Matrices

Let $A \in \mathsf{M}_n$ and let p be a positive integer. Then

$$(I + A + A^2 + \cdots + A^{p-1})(I - A)$$
$$= (I + A + A^2 + \cdots + A^{p-1}) - (A + A^2 + \cdots + A^{p-1} + A^p)$$
$$= I - A^p.$$

If $1 \notin \operatorname{spec} A$, then

$$\sum_{k=0}^{p-1} A^k = (I - A^p)(I - A)^{-1}. \tag{11.6.1}$$

If $\lim_{p \to \infty} A^p = 0$ (that is, each entry of A^p converges to 0 as $p \to \infty$), then

$$\sum_{k=0}^{\infty} A^k = \lim_{p \to \infty} \sum_{k=0}^{p-1} A^k = \lim_{p \to \infty} \left((I - A^p)(I - A)^{-1} \right)$$
$$= \left(I - \lim_{p \to \infty} A^p \right)(I - A)^{-1} = I(I - A)^{-1}$$
$$= (I - A)^{-1}.$$

The preceding calculations show that if $A \in \mathsf{M}_n$ and $1 \notin \operatorname{spec} A$, then

$$(I - A)^{-1} = \sum_{k=0}^{\infty} A^k$$

if $\lim_{k \to \infty} A^k = 0$. How can we decide if a matrix satisfies this condition?

Definition 11.6.2 Let $A = [a_{ij}] \in \mathsf{M}_n$. For each $p = 1, 2, \ldots$, let $A^p = [a_{ij}^{(p)}]$. Then A is *convergent* if $\lim_{p \to \infty} a_{ij}^{(p)} = 0$ for all $i, j = 1, 2, \ldots, n$. If A is convergent, we write $A^p \to 0$.

The property of being convergent (or not) is invariant under similarity.

Lemma 11.6.3 *If $A, B \in \mathsf{M}_n$ are similar, then A is convergent if and only if B is convergent.*

Proof Suppose that $S = [s_{ij}] \in \mathsf{M}_n$ is invertible, let $S^{-1} = [\sigma_{ij}]$, and suppose that $A = SBS^{-1}$. Let $A^p = [a_{ij}^{(p)}]$ and $B^p = [b_{ij}^{(p)}]$. Then $A^p = SB^pS^{-1}$, so each entry

$$a_{ij}^{(p)} = \sum_{k,\ell=1}^{n} s_{ik} b_{k\ell}^{(p)} \sigma_{\ell j}, \quad i, j = 1, 2, \ldots, n, \tag{11.6.4}$$

of A^p is a fixed linear combination of the entries of B^p, for each $p = 1, 2, \ldots$. If $\lim_{p \to \infty} b_{ij}^{(p)} = 0$ for each $i, j = 1, 2, \ldots, n$, then

$$
\lim_{p \to \infty} a_{ij}^{(p)} = \lim_{p \to \infty} \sum_{k,\ell=1}^{n} s_{ik} b_{k\ell}^{(p)} \sigma_{\ell j}
$$

$$
= \sum_{k,\ell=1}^{n} s_{ik} \left(\lim_{p \to \infty} b_{k\ell}^{(p)} \right) \sigma_{\ell j} = 0, \quad i, j = 1, 2, \ldots, n.
$$

The converse follows from interchanging A and B in this argument.   □

Our criterion for convergence involves an eigenvalue that has the largest modulus.

Definition 11.6.5 The *spectral radius* of $A \in M_n$ is

$$
\rho(A) = \max \left\{ |\lambda| : \lambda \in \operatorname{spec} A \right\}.
$$

If $n \geq 2$, it is possible to have $A \neq 0$ and $\rho(A) = 0$. For example, $\rho(A) = 0$ for any nilpotent $A \in M_n$.

Theorem 11.6.6 $A \in M_n$ *is convergent if and only if* $\rho(\lambda) < 1$.

Proof It follows from the preceding lemma that A is convergent if and only if its Jordan canonical form is convergent. The Jordan canonical form of A is a direct sum (11.1.7) of Jordan blocks, so it is convergent if and only if each of its direct summands $J_k(\lambda)$ is convergent. If $k = 1$, then $J_1(\lambda) = [\lambda]$ is convergent if and only if $|\lambda| < 1$. Suppose that $k \geq 2$. If $\lambda = 0$, then $J_k(0)^p = 0$ for all $p \geq k$, so $J_k(0)$ is convergent. Suppose $\lambda \neq 0$, let $p > k$, and use the binomial theorem to compute

$$
J_k(\lambda)^p = (\lambda I + J_k)^p = \sum_{j=0}^{p} \binom{p}{p-j} \lambda^{p-j} J_k^j
$$

$$
= \lambda^p I + \sum_{j=1}^{k-1} \binom{p}{p-j} \lambda^{p-j} J_k^j. \tag{11.6.7}
$$

Each matrix J_k^j in the sum (11.6.7) has ones on its jth superdiagonal and zero entries elsewhere. Consequently, every entry in the jth superdiagonal of $J_k(\lambda)^j$ is $\binom{p}{p-j} \lambda^{p-j}$. The diagonal entries of $J_k(\lambda)^p$ are all λ^p, so $|\lambda| < 1$ is a necessary condition for $J_k(\lambda)$ to be convergent.

To show that it is sufficient as well, we must show that if $|\lambda| < 1$, then

$$
\binom{p}{p-j} \lambda^{p-j} \to 0 \text{ as } p \to \infty
$$

for each $j = 1, 2, \ldots, k - 1$. We have

$$
\left| \binom{p}{p-j} \lambda^{p-j} \right| = \left| \frac{p(p-1)(p-2) \cdots (p-j+1) \lambda^p}{j! \, \lambda^j} \right| \leq \frac{1}{j! \, |\lambda|^j} p^j |\lambda|^p,
$$

so it is sufficient to show that $p^j|\lambda|^p \to 0$ as $p \to \infty$. Equivalently, we can show that $\log(p^j|\lambda|^p) \to -\infty$ as $p \to \infty$. Since $\log|\lambda| < 0$ and l'Hôpital's rule ensures that $(\log p)/p \to 0$ as $p \to \infty$,

$$\log(p^j|\lambda|^p) = j\log p + p\log|\lambda| = p\left(j\frac{\log p}{p} + \log|\lambda|\right) \to -\infty \text{ as } p \to \infty.$$

Finally, observe that $|\lambda| < 1$ for all $\lambda \in \operatorname{spec} A$ if and only if $\rho(A) < 1$. $\quad\square$

11.7 Power Bounded and Markov Matrices

Conceptually related to the notion of a convergent matrix is the notion of a power bounded matrix.

Definition 11.7.1 A square matrix $A = [a_{ij}]$ is *power bounded* if there is some $L > 0$ such that $|a_{ij}^{(p)}| \leq L$ for all $p = 1, 2, \ldots$.

Theorem 11.7.2 $A \in M_n$ *is power bounded if and only if* $\rho(A) \leq 1$ *and each* $\lambda \in \operatorname{spec} A$ *with* $|\lambda| = 1$ *has equal geometric and algebraic multiplicities.*

Proof It follows from the identity (11.6.4) that A is power bounded if and only if its Jordan canonical form is power bounded, so it suffices to consider power boundedness for a Jordan block $J_k(\lambda)$.

Suppose that $J_k(\lambda)$ is power bounded. The main diagonal entries of $J_k(\lambda)^p$ are λ^p, so $|\lambda| \leq 1$. Suppose that $|\lambda| = 1$. If $k \geq 2$, then the $(1, 2)$ entry of $J_k(\lambda)^p$ is $p\lambda^{p-1}$. Since $J_k(\lambda)$ is power bounded, we conclude that $k = 1$. Therefore, every Jordan block of A with eigenvalue λ is 1×1, which means that the algebraic and geometric multiplicities of λ are equal.

Conversely, suppose that λ has modulus 1, and that its geometric and algebraic multiplicities are equal to m. Then the Jordan matrix $J(\lambda)$ in the Jordan canonical form of A is $J(\lambda) = \lambda I_m$, which is power bounded. $\quad\square$

Definition 11.7.3 A real row or column vector is a *probability vector* if its entries are nonnegative and sum to 1.

Example 11.7.4 Some examples of probability vectors are

$$\begin{bmatrix} 0.25 \\ 0.75 \end{bmatrix}, \quad [0.25 \ 0.75], \quad \text{and} \quad \begin{bmatrix} 0.2 \\ 0.3 \\ 0.5 \end{bmatrix}.$$

A column vector $\mathbf{x} \in \mathbb{R}^n$ is a probability vector if and only if its entries are nonnegative and $\mathbf{x}^T\mathbf{e} = 1$, in which $\mathbf{e} = [1 \ 1 \ \ldots \ 1]^T \in \mathbb{R}^n$ is the all-ones vector.

Matrices whose rows or columns are probability vectors are an important class of power-bounded matrices.

Definition 11.7.5 $A = [a_{ij}] \in \mathsf{M}_n(\mathbb{R})$ is a *Markov matrix* if its entries are nonnegative and

$$\sum_{j=1}^{n} a_{ij} = 1 \qquad \text{for each } i = 1, 2, \ldots, n. \tag{11.7.6}$$

Example 11.7.7 Some examples of Markov matrices are

$$\begin{bmatrix} 0 & 1 \\ 1 & 0 \end{bmatrix}, \quad \begin{bmatrix} 0.25 & 0.75 \\ 0.4 & 0.6 \end{bmatrix}, \quad \text{and} \quad \begin{bmatrix} 0.4 & 0 & 0.6 \\ 0.2 & 0.7 & 0.1 \\ 0 & 0.5 & 0.5 \end{bmatrix}.$$

The definition ensures that the entries of a Markov matrix are nonnegative and that the sum of the entries in each row is 1.

Lemma 11.7.8 *Let* $A = [a_{ij}] \in \mathsf{M}_n$.

(a) *A satisfies* (11.7.6) *if and only if* $A\mathbf{e} = \mathbf{e}$, *that is,* $(1, \mathbf{e})$ *is an eigenpair of A.*

(b) *If A has real nonnegative entries and satisfies* (11.7.6), *then* $0 \leq a_{ij} \leq 1$ *for all* $i, j \in \{1, 2, \ldots, n\}$.

(c) *If A is a Markov matrix, then* A^p *is a Markov matrix for* $p = 0, 1, 2, \ldots$.

Proof (a) The ith entry of $A\mathbf{e}$ is the sum of the entries in the ith row of A, which is 1.

(b) For $i, j \in \{1, 2, \ldots, n\}$,

$$0 \leq a_{ij} \leq \sum_{k=1}^{n} a_{ik} = 1.$$

(c) For each $p \in \{1, 2, \ldots\}$, A^p has nonnegative entries and $A^p \mathbf{e} = \mathbf{e}$ (Lemma 8.3.2). □

Theorem 11.7.9 *Let* $A \in \mathsf{M}_n(\mathbb{R})$ *be a Markov matrix. Then* $\rho(A) = 1$. *Moreover, if* $\lambda \in \operatorname{spec} A$ *and* $|\lambda| = 1$, *then the geometric and algebraic multiplicities of* λ *are equal.*

Proof For each $p = 1, 2, \ldots$, the preceding lemma ensures that every entry of A^p is between 0 and 1. Therefore, A is power bounded and the assertion follows from Theorem 11.7.2. □

Example 11.7.10 The matrix

$$A = \begin{bmatrix} 0 & I_n \\ I_n & 0 \end{bmatrix}$$

is a Markov matrix. Since

$$A^p = \begin{cases} I_{2n} & \text{if } p \text{ is even,} \\ A & \text{if } p \text{ is odd,} \end{cases}$$

A is not convergent, although it is power bounded. Its eigenvalues are $\lambda = \pm 1$, each with geometric and algebraic multiplicity n. Its Jordan canonical form is $I_n \oplus (-I_n)$.

The preceding example shows that a Markov matrix A can have several eigenvalues with modulus 1. This cannot happen if every entry of A is positive.

Theorem 11.7.11 *Let* $n \geq 2$ *and let* $A \in \mathsf{M}_n(\mathbb{R})$ *be a Markov matrix with all entries positive.*

(a) *If* $\lambda \in \operatorname{spec} A$ *and* $|\lambda| = 1$, *then* $\lambda = 1$ *and* $\mathcal{E}_1(A) = \operatorname{span}\{\mathbf{e}\}$.

(b) $\lambda = 1$ *has algebraic multiplicity* 1.

(c) *The Jordan canonical form of* A *is* $[1] \oplus J$, *in which* $J \in \mathsf{M}_{n-1}$ *is a convergent Jordan matrix.*

Proof (a) Let $(\lambda, \mathbf{y})$ be an eigenpair of $A = [a_{ij}] \in \mathsf{M}_n$, in which $|\lambda| = 1$ and $\mathbf{y} = [y_i] \in \mathbb{C}^n$ is nonzero. Let $k \in \{1, 2, \dots, n\}$ be an index such that $|y_k| = \|\mathbf{y}\|_\infty$. Since $A\mathbf{y} = \lambda\mathbf{y}$,

$$\|\mathbf{y}\|_\infty = |\lambda||y_k| = |\lambda y_k| = \left| \sum_{j=1}^n a_{kj} y_j \right|$$

$$\leq \sum_{j=1}^n |a_{kj} y_j| = \sum_{j=1}^n a_{kj} |y_j| \tag{11.7.12}$$

$$\leq \sum_{j=1}^n a_{kj} \|\mathbf{y}\|_\infty = \|\mathbf{y}\|_\infty. \tag{11.7.13}$$

The inequalities (11.7.12) and (11.7.13) must be equalities. Since each $a_{kj} > 0$, equality at (11.7.13) means that $|y_j| = \|\mathbf{y}\|_\infty$ for each $j = 1, 2, \dots, n$. Equality at (11.7.12) is the equality case in the triangle inequality (A.2.9), so every entry of $\mathbf{y}$ is a nonnegative real multiple of y_1. Since every entry of $\mathbf{y}$ has the same positive modulus, $\mathbf{y} = y_1\mathbf{e}$. Therefore, $A\mathbf{y} = \mathbf{y}$ and $\lambda = 1$. Since $\mathbf{y} \in \operatorname{span}\{\mathbf{e}\}$, we conclude that $\mathcal{E}_1(A) = \operatorname{span}\{\mathbf{e}\}$ and $\dim \mathcal{E}_1(A) = 1$.

(b) The preceding theorem ensures that the eigenvalue $\lambda = 1$ has equal geometric and algebraic multiplicities, and (a) tells us that its geometric multiplicity is 1.

(c) Let $J_k(\lambda)$ be a Jordan block in the Jordan canonical form of A. Theorem 11.7.9.b says that $|\lambda| \leq 1$. If $|\lambda| = 1$, then (a) tells us that $\lambda = 1$ and there is only one such block; (b) tells us that $k = 1$. Therefore, $J_k(1) = [1]$ and the Jordan canonical form of A is $[1] \oplus J$, in which J is a direct sum of Jordan blocks whose eigenvalues have moduli strictly less than 1. Theorem 11.6.6 ensures that J is convergent. $\qquad\square$

Theorem 11.7.14 *Let* $n \geq 2$ *and let* $A \in \mathsf{M}_n(\mathbb{R})$ *be a Markov matrix with all entries positive. Then there is a unique* $\mathbf{x} \in \mathbb{R}^n$ *such that* $A^\mathsf{T}\mathbf{x} = \mathbf{x}$ *and* $\mathbf{x}^\mathsf{T}\mathbf{e} = 1$. *All the entries of* $\mathbf{x}$ *are positive and* $\lim_{p\to\infty} A^p = \mathbf{e}\mathbf{x}^\mathsf{T}$.

Proof The Jordan canonical form of A is $[1] \oplus J$, in which $\rho(J) < 1$. Therefore, there is an invertible $S \in \mathsf{M}_n$ such that

$$A = S([1] \oplus J)S^{-1}. \tag{11.7.15}$$

Partition $S = [\mathbf{s} \ Y]$, in which $Y \in \mathsf{M}_{n\times(n-1)}$. Then

$$[A\mathbf{s} \ AY] = AS = S([1] \oplus J) = [\mathbf{s} \ Y]([1] \oplus J) = [\mathbf{s} \ YJ],$$

so $A\mathbf{s} = \mathbf{s} \in \mathcal{E}_1(A)$. Therefore, $\mathbf{s} = c\mathbf{e}$ for some nonzero scalar c. Since (11.7.15) is equivalent to

$$A = (cS)\big([1] \oplus J\big)(cS)^{-1},$$

in which

$$cS = [c\mathbf{s} \ cY] = [\mathbf{e} \ cY],$$

we may assume that $S = [\mathbf{e} \ Y]$ from the start.

Partition $S^{-\mathsf{T}} = [\mathbf{x} \ X]$, in which $X \in \mathsf{M}_{n \times (n-1)}$ and $\mathbf{x} \neq \mathbf{0}$. Then

$$A^{\mathsf{T}} = S^{-\mathsf{T}}\big([1] \oplus J^{\mathsf{T}}\big)S^{\mathsf{T}}$$

and

$$[A^{\mathsf{T}}\mathbf{x} \ A^{\mathsf{T}}X] = A^{\mathsf{T}}S^{-\mathsf{T}} = S^{-\mathsf{T}}\big([1] \oplus J^{\mathsf{T}}\big) = [\mathbf{x} \ X]\big([1] \oplus J^{\mathsf{T}}\big) = [\mathbf{x} \ XJ^{\mathsf{T}}],$$

which shows that $A^{\mathsf{T}}\mathbf{x} = \mathbf{x} \in \mathcal{E}_1(A^{\mathsf{T}})$. However,

$$\begin{bmatrix} 1 & \mathbf{0}^{\mathsf{T}} \\ \mathbf{0} & I_{n-1} \end{bmatrix} = I_n = S^{-1}S = \begin{bmatrix} \mathbf{x}^{\mathsf{T}} \\ X^{\mathsf{T}} \end{bmatrix}[\mathbf{e} \ S_2] = \begin{bmatrix} \mathbf{x}^{\mathsf{T}}\mathbf{e} & \mathbf{x}^{\mathsf{T}}X \\ X^{\mathsf{T}}\mathbf{e} & X^{\mathsf{T}}X \end{bmatrix},$$

so $\mathbf{x}^{\mathsf{T}}\mathbf{e} = 1$. Theorem 11.7.11.a and Lemma 8.3.18 ensure that $\mathcal{E}_1(A^{\mathsf{T}})$ is one-dimensional, so $\mathbf{x}$ is unique. Since each A^p has only positive entries and

$$A^p = S\big([1] \oplus J^p\big)S^{-1} = [\mathbf{e} \ S_2]\begin{bmatrix} 1 & \mathbf{0}^{\mathsf{T}} \\ \mathbf{0} & J^p \end{bmatrix}\begin{bmatrix} \mathbf{x}^{\mathsf{T}} \\ Y^{\mathsf{T}} \end{bmatrix}$$

$$\rightarrow [\mathbf{e} \ S_2]\begin{bmatrix} 1 & \mathbf{0}^{\mathsf{T}} \\ \mathbf{0} & 0 \end{bmatrix}\begin{bmatrix} \mathbf{x}^{\mathsf{T}} \\ Y^{\mathsf{T}} \end{bmatrix} = \mathbf{e}\mathbf{x}^{\mathsf{T}}$$

as $p \rightarrow \infty$, it follows that $\mathbf{e}\mathbf{x}^{\mathsf{T}}$ (and therefore $\mathbf{x}$) has real nonnegative entries. Thus, $\mathbf{x}$ is a probability vector. Since A^{T} has positive entries and $A^{\mathsf{T}}\mathbf{x} = \mathbf{x}$, we conclude that $\mathbf{x}$ has positive entries; see P.11.7. $\qquad\square$

Definition 11.7.16 Let $A \in \mathsf{M}_n$ be a Markov matrix with positive entries. The *stationary distribution* of A is the unique probability vector $\mathbf{x}$ such that $A^{\mathsf{T}}\mathbf{x} = \mathbf{x}$.

The preceding theorem ensures that a Markov matrix with positive entries has a stationary distribution with positive entries.

Example 11.7.17 Suppose that there are cities C_1, C_2, and C_3 with initial respective populations n_1, n_2, and n_3, and let $\mathbf{y} = [n_1 \ n_2 \ n_3]^{\mathsf{T}}$. Then $N = n_1 + n_2 + n_3$ is the total population of the three cities. On the first day of each month, for each $i \neq j$, a fixed percentage p_{ij} of the population of C_i migrates to C_j. For each $i = 1, 2, 3$, a fixed percentage p_{ii} of the population of C_i remains in C_i. The entries of the matrix $P = [p_{ij}] \in \mathsf{M}_3(\mathbb{R})$ are real and nonnegative, and the row sums are all equal to 100. Suppose that $p_{ij} = 100a_{ij}$, in which

$$A = \begin{bmatrix} 0.1 & 0.2 & 0.7 \\ 0.3 & 0.6 & 0.1 \\ 0.5 & 0.3 & 0.2 \end{bmatrix}.$$

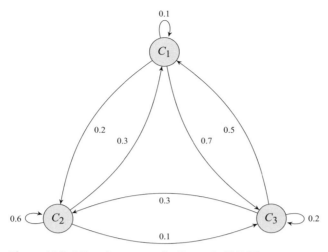

Figure 11.3 Migration pattern for Example 11.7.17.

After the first migration, the respective populations of the three cities are the entries of $\mathbf{y}^{\mathsf{T}}A$. After the second migration, the respective populations are the entries of $\mathbf{y}^{\mathsf{T}}A^2$. Since

$$A^{12} = \begin{bmatrix} 0.3021 & 0.3854 & 0.3125 \\ 0.3021 & 0.3854 & 0.3125 \\ 0.3021 & 0.3854 & 0.3125 \end{bmatrix},$$

after one year, the respective populations of the three cities are essentially stable. C_1 has a population of about $0.3N$, C_2 has a population of about $0.39N$, and C_3 has a population of about $0.31N$. These stable populations are independent of the initial distribution of population among the cities. See Figure 11.3 for an illustration of the migration pattern in this example.

11.8 Similarity of a Matrix and its Transpose

A square matrix A and its transpose have the same eigenvalues (Theorem 9.2.6), but this is only part of the story. With the help of the Jordan canonical form, we can show that A and A^{T} are similar.

Theorem 11.8.1 *Let $A \in \mathsf{M}_n$.*

(a) *A is similar to A^{T} via a symmetric similarity matrix.*

(b) *$A = BC = DE$, in which A, B, C, and D are symmetric, and B and E are invertible.*

Proof (a) The identity (11.1.5) tells us that

$$J_m = \begin{bmatrix} \mathbf{0} & I_{m-1} \\ 0 & \mathbf{0}^{\mathsf{T}} \end{bmatrix} \quad \text{and} \quad J_m^{\mathsf{T}} = \begin{bmatrix} \mathbf{0}^{\mathsf{T}} & 0 \\ I_{m-1} & \mathbf{0} \end{bmatrix}.$$

The reversal matrix K_m defined in (6.2.9) may be written in block form as

$$K_m = \begin{bmatrix} 0 & K_{m-1} \\ 1 & 0^\mathsf{T} \end{bmatrix} = \begin{bmatrix} 0^\mathsf{T} & 1 \\ K_{m-1} & 0 \end{bmatrix}.$$

Now compute

$$K_m J_m^\mathsf{T} = \begin{bmatrix} 0 & K_{m-1} \\ 1 & 0^\mathsf{T} \end{bmatrix} \begin{bmatrix} 0^\mathsf{T} & 0 \\ I_{m-1} & 0 \end{bmatrix} = \begin{bmatrix} K_{m-1} & 0 \\ 0^\mathsf{T} & 0 \end{bmatrix}$$

and

$$J_m K_m = \begin{bmatrix} 0 & I_{m-1} \\ 0 & 0^\mathsf{T} \end{bmatrix} \begin{bmatrix} 0^\mathsf{T} & 1 \\ K_{m-1} & 0 \end{bmatrix} = \begin{bmatrix} K_{m-1} & 0 \\ 0^\mathsf{T} & 0 \end{bmatrix}. \tag{11.8.2}$$

Therefore, $K_m J_m^\mathsf{T} = J_m K_m$, so $J_m^\mathsf{T} = K_m^{-1} J_m K_m$. The computation

$$J_m(\lambda)^\mathsf{T} = (\lambda I_m + J_m)^\mathsf{T} = \lambda I_m + J_m^\mathsf{T} = \lambda I_m + K_m^{-1} J_m K_m$$
$$= K_m^{-1}(\lambda I_m + J_m)K_m = K_m^{-1} J_m(\lambda) K_m$$

shows that every Jordan block is similar to its transpose via a reversal matrix. It follows that every Jordan matrix (a direct sum of Jordan blocks) is similar to its transpose via a direct sum of reversal matrices.

Let $A \in \mathsf{M}_n$ and let $A = SJS^{-1}$, in which $S \in \mathsf{M}_n$ is invertible and J is a Jordan matrix. Let K denote a direct sum of reversal matrices such that $J^\mathsf{T} = K^{-1}JK$. Then $J = S^{-1}AS$ and

$$A^\mathsf{T} = S^{-\mathsf{T}} J^\mathsf{T} S^\mathsf{T} = S^{-\mathsf{T}}(K^{-1}JK)S^\mathsf{T} = S^{-\mathsf{T}}K^{-1}(S^{-1}AS)KS^\mathsf{T}$$
$$= (SKS^\mathsf{T})^{-1} A (SKS^\mathsf{T}).$$

Therefore, A^T is similar to A via the similarity matrix SKS^T. The matrix K is a direct sum of reversal matrices, each of which is symmetric, so K and SKS^T are symmetric.

(b) The identity (11.8.2) and the computation

$$K_m J_m = \begin{bmatrix} 0^\mathsf{T} & 1 \\ K_{m-1} & 0 \end{bmatrix} \begin{bmatrix} 0 & I_{m-1} \\ 0 & 0^\mathsf{T} \end{bmatrix} = \begin{bmatrix} 0 & 0^\mathsf{T} \\ 0 & K_{m-1} \end{bmatrix}$$

imply that JK and KJ are both symmetric. A reversal matrix is an involution, so $K^2 = I$. The computations

$$A = SJS^{-1} = SK^2 JS^{-1} = (SKS^\mathsf{T})(S^{-\mathsf{T}}KJS^{-1})$$

and

$$A = SJS^{-1} = SJK^2 S^{-1} = (SJKS^\mathsf{T})(S^{-\mathsf{T}}KS^{-1})$$

confirm the asserted factorizations.   □

11.9 The Invertible Jordan Blocks of *AB* and *BA*

If A and B are square matrices of the same size, then AB and BA have the same eigenvalues with the same multiplicities (Theorem 9.7.2). However, something stronger is true. The Jordan canonical forms of AB and BA contain the same invertible Jordan blocks with the same multiplicities.

Theorem 11.9.1 *Let* $A, B \in \mathsf{M}_n$. *If* $\operatorname{spec} AB \neq \{0\}$, *then for each nonzero* $\lambda \in \operatorname{spec} AB$ *and each* $k = 1, 2, \ldots, n$, *the Jordan canonical forms of AB and BA contain the same Jordan blocks* $J_k(\lambda)$ *with the same multiplicities.*

Proof In the proof of Theorem 9.7.2 we showed that

$$X = \begin{bmatrix} AB & A \\ 0 & 0 \end{bmatrix} \quad \text{and} \quad Y = \begin{bmatrix} 0 & A \\ 0 & BA \end{bmatrix} \tag{11.9.2}$$

are similar. Let λ be a nonzero eigenvalue of AB (and hence of BA also). Apply $f(z) = (z - \lambda)^p$ to X and Y, then use (0.8.2) to find that

$$(X - \lambda I)^p = \begin{bmatrix} (AB - \lambda I)^p & \star \\ 0 & (-\lambda)^p I \end{bmatrix} \tag{11.9.3}$$

and

$$(Y - \lambda I)^p = \begin{bmatrix} (-\lambda)^p I & \star \\ 0 & (BA - \lambda I)^p \end{bmatrix}$$

are similar. Partition (11.9.3) as $(X - \lambda I)^p = [X_1 \ X_2]$, in which $X_1, X_2 \in \mathsf{M}_{2n \times n}$. Because $\lambda \neq 0$,

$$\operatorname{rank} X_2 = \operatorname{rank} \begin{bmatrix} \star \\ (-\lambda)^p I_n \end{bmatrix} = n.$$

Let $\mathbf{x}$ be a column of X_1. If $\mathbf{x} \in \operatorname{col} X_2$, then there is a $\mathbf{y} \in \mathbb{C}^n$ such that

$$\mathbf{x} = \begin{bmatrix} \star \\ \mathbf{0} \end{bmatrix} = X_2 \mathbf{y} = \begin{bmatrix} \star \\ (-\lambda)^p \mathbf{y} \end{bmatrix}.$$

It follows that $\mathbf{y} = \mathbf{0}$ and $\mathbf{x} = X_2 \mathbf{y} = \mathbf{0}$. We conclude that $\operatorname{col} X_1 \cap \operatorname{col} X_2 = \{\mathbf{0}\}$, so

$$\operatorname{rank}(X - \lambda I)^p = \operatorname{rank} X_1 + n = \operatorname{rank}(AB - \lambda I)^p + n.$$

A similar examination of the rows of $(Y - \lambda I)^p$ shows that

$$\operatorname{rank}(Y - \lambda I)^p = \operatorname{rank}(BA - \lambda I)^p + n.$$

It follows from the similarity of $(X - \lambda I)^p$ and $(Y - \lambda I)^p$ that $\operatorname{rank}(X - \lambda I)^p = \operatorname{rank}(Y - \lambda I)^p$, so

$$\operatorname{rank}(AB - \lambda I)^p = \operatorname{rank}(BA - \lambda I)^p, \quad p = 1, 2, \ldots.$$

Therefore, the Weyr characteristics of AB and BA associated with each nonzero $\lambda \in \operatorname{spec} AB$ are the same, so the multiplicities of $J_k(\lambda)$ in their respective Jordan canonical forms are the same for each $k = 1, 2, \ldots$. $\qquad \square$

The preceding theorem says nothing about the nilpotent Jordan blocks of AB and BA, which can be different.

Example 11.9.4 Let $A = J_2$ and $B = \operatorname{diag}(1, 0)$. The Jordan forms of $AB = J_1 \oplus J_1$ and $BA = J_2$ do not have the same nilpotent Jordan blocks.

There is a simple definitive test for the similarity of AB and BA; it focuses only on their nilpotent Jordan blocks.

Corollary 11.9.5 *Let $A, B \in \mathsf{M}_n$. Then AB is similar to BA if and only if $\operatorname{rank}(AB)^p = \operatorname{rank}(BA)^p$ for each $p = 1, 2, \ldots, n$.*

Proof If AB is similar to BA, then their corresponding powers are similar and hence have the same rank. Conversely, if the ranks of their corresponding powers are equal, then the Weyr characteristics of AB and BA are the same. Consequently, the multiplicities of each nilpotent block J_k in their respective Jordan canonical forms are the same. The preceding theorem ensures that the invertible Jordan blocks of AB and BA are the same, with the same multiplicities. Therefore, the Jordan canonical forms of AB and BA are the same.   □

Example 11.9.6 Consider

$$A = \begin{bmatrix} 0 & 0 & 0 & 0 \\ 0 & 0 & 1 & 0 \\ 0 & 0 & 0 & 1 \\ 0 & 1 & 0 & 0 \end{bmatrix} \quad \text{and} \quad B = \begin{bmatrix} 0 & 0 & 0 & 1 \\ 0 & 1 & 0 & 0 \\ 0 & 0 & 0 & 0 \\ 1 & 0 & 0 & 0 \end{bmatrix},$$

whose products are the nilpotent matrices

$$AB = \begin{bmatrix} 0 & 0 & 0 & 0 \\ 0 & 0 & 0 & 0 \\ 1 & 0 & 0 & 0 \\ 0 & 1 & 0 & 0 \end{bmatrix} \quad \text{and} \quad BA = \begin{bmatrix} 0 & 1 & 0 & 0 \\ 0 & 0 & 1 & 0 \\ 0 & 0 & 0 & 0 \\ 0 & 0 & 0 & 0 \end{bmatrix} = J_3 \oplus J_1.$$

A computation reveals that $\operatorname{rank} AB = \operatorname{rank} BA = 2$, but $\operatorname{rank}(AB)^2 = 0$ and $\operatorname{rank}(BA)^2 = 1$. The preceding corollary tells us that AB is not similar to BA. In fact, the Jordan canonical form of AB is $J_2 \oplus J_2$.

Corollary 11.9.7 *Let $A, H, K \in \mathsf{M}_n$, and suppose that H and K are Hermitian. Then HK is similar to KH and $A\bar{A}$ is similar to $\bar{A}A$.*

Proof For each $p = 1, 2, \ldots,$

$$\operatorname{rank}(HK)^p = \operatorname{rank}((HK)^p)^* = \operatorname{rank}((HK)^*)^p = \operatorname{rank}(K^*H^*)^p = \operatorname{rank}(KH)^p$$

and

$$\operatorname{rank}(A\bar{A})^p = \operatorname{rank} \overline{(A\bar{A})^p} = \operatorname{rank}(\overline{A\bar{A}})^p = \operatorname{rank}(\bar{A}A)^p.$$

The preceding corollary ensures that HK is similar to KH and $A\bar{A}$ is similar to $\bar{A}A$.   □

Example 11.9.8 Consider

$$A = \begin{bmatrix} -5 & i \\ 2i & 0 \end{bmatrix} \quad \text{and} \quad A\bar{A} = \begin{bmatrix} 27 & 5i \\ -10i & 2 \end{bmatrix}.$$

A calculation reveals that $A\overline{A} = SRS^{-1}$ and $\overline{A}A = \overline{S}R\overline{S}^{-1}$, in which

$$R = \begin{bmatrix} 7 & 10 \\ 15 & 22 \end{bmatrix} \quad \text{and} \quad S = \begin{bmatrix} -i & -i \\ 1 & -1 \end{bmatrix}.$$

This confirms that $A\overline{A}$ is similar to its complex conjugate, since it is similar to the real matrix R.

11.10 Similarity of a Matrix and its Complex Conjugate

Not every complex number is equal to its complex conjugate, so it is no surprise that not every square matrix A is similar to $\overline{A}$. If A is real, then A is similar to $\overline{A}$, but there are many non-real matrices with this property. For example, any matrix of the form $A\overline{A}$ is similar to its complex conjugate (Corollary 11.9.7), as is any product of two Hermitian matrices (Theorem 11.10.4).

If J is the Jordan canonical form of A, then $\overline{J}$ is the Jordan canonical form of $\overline{A}$. Therefore, A is similar to $\overline{A}$ if and only if J is similar to $\overline{J}$, which can be verified by using the test in the following theorem. The key observation is that for any Jordan block $J_k(\lambda)$, we have $\overline{J_k(\lambda)} = \overline{\lambda I_k + J_k} = \overline{\lambda}I + J_k = J_k(\overline{\lambda})$.

Theorem 11.10.1 *Let $A \in M_n$. The following statements are equivalent:*

(a) *A is similar to $\overline{A}$.*

(b) *If $J_k(\lambda)$ is a direct summand of the Jordan canonical form of A, then so is $J_k(\overline{\lambda})$, with the same multiplicity.*

Proof Let $\lambda_1, \lambda_2, \ldots, \lambda_d$ be the distinct eigenvalues of A and suppose that the Jordan canonical form of A is

$$J = J(\lambda_1) \oplus J(\lambda_2) \oplus \cdots \oplus J(\lambda_d), \tag{11.10.2}$$

in which each $J(\lambda_i)$ is a Jordan matrix.

If A is similar to $\overline{A}$, then J is similar to $\overline{J}$. Moreover,

$$\overline{J} = J(\overline{\lambda_1}) \oplus J(\overline{\lambda_2}) \oplus \cdots \oplus J(\overline{\lambda_d}). \tag{11.10.3}$$

The uniqueness of the Jordan canonical form implies that the direct sum in (11.10.3) can be obtained from the direct sum in (11.10.2) by permuting its direct summands. Thus, there is a permutation σ of the list $1, 2, \ldots, d$ such that Jordan matrices $J(\lambda_j)$ and $J(\overline{\lambda_{\sigma(j)}})$ have the same Jordan canonical forms (the direct summands of $J(\overline{\lambda_{\sigma(j)}})$ are obtained by permuting the direct summands of $J(\lambda_j)$). If λ_j is real, then $\sigma(j) = j$, but if λ_j is not real then $\sigma(j) \neq j$. Therefore, the non-real Jordan matrices that are direct summands in (11.10.2) must occur in conjugate pairs.

Conversely, if the non-real Jordan matrices in (11.10.2) occur in conjugate pairs, then so do the non-real direct summands in (11.10.3); they can be obtained by interchanging pairs of direct summands of (11.10.2). This means that the Jordan canonical forms for A and $\overline{A}$ are the same, so A is similar to $\overline{A}$. $\square$

The following theorem introduces some characterizations that do not involve inspection of the Jordan canonical form.

Theorem 11.10.4 *Let $A \in M_n$. The following statements are equivalent:*

(a) A *is similar to* $\overline{A}$.

(b) A *is similar to* A^*.

(c) A *is similar to* A^* *via a Hermitian similarity matrix.*

(d) $A = HK = LM$, *in which* H, K, L, *and* M *are Hermitian, and both* H *and* M *are invertible.*

(e) $A = HK$, *in which* H *and* K *are Hermitian.*

Proof (a) $\Rightarrow$ (b) Let $S, R \in M_n$ be invertible and such that $A = S\overline{A}S^{-1}$ and $A = RA^{\mathsf{T}}R^{-1}$ (Theorem 11.8.1). Then $\overline{A} = \overline{R}\,\overline{A}^{\mathsf{T}}\overline{R}^{-1}$ and

$$A = S\overline{A}S^{-1} = S\overline{R}\,\overline{A}^{\mathsf{T}}\overline{R}^{-1}S^{-1} = (S\overline{R})A^*(S\overline{R})^{-1}.$$

(b) $\Rightarrow$ (c) Let $S \in M_n$ be invertible and such that $A = SA^*S^{-1}$. For any $\theta \in \mathbb{R}$, we have $A = (e^{i\theta}S)A^*(e^{-i\theta}S^{-1}) = S_\theta A^* S_\theta^{-1}$, in which $S_\theta = e^{i\theta}S$. Then $AS_\theta = S_\theta A^*$. The adjoint of this identity is $AS_\theta^* = S_\theta^* A^*$. Add the preceding two identities to obtain

$$A(S_\theta + S_\theta^*) = (S_\theta + S_\theta^*)A^*.$$

The matrix $H_\theta = S_\theta + S_\theta^*$ is Hermitian, and the computation

$$S_\theta^{-1}H_\theta = I + S_\theta^{-1}S_\theta^* = I + e^{-2i\theta}S^{-1}S^* = e^{-2i\theta}(e^{2i\theta}I + S^{-1}S^*)$$

tells us that H_θ is invertible for any θ such that $-e^{2i\theta} \notin \mathrm{spec}(S^{-1}S^*)$. For such a choice of θ, we have $A = H_\theta A^* H_\theta^{-1}$.

(c) $\Rightarrow$ (d) Let S be invertible and Hermitian, and such that $A = SA^*S^{-1}$. Then $AS = SA^*$. It follows that $(SA^*)^* = (AS)^* = S^*A^* = SA^*$, that is, SA^* is Hermitian. Then $A = (SA^*)S^{-1}$ is a product of two Hermitian factors, the second of which is invertible. Since $S^{-1}A = A^*S^{-1}$, we also have $(A^*S^{-1})^* = (S^{-1}A)^* = A^*S^{-*} = A^*S^{-1}$, so A^*S^{-1} is Hermitian and $A = S(A^*S^{-1})$ is a product of two Hermitian factors, the first of which is invertible.

(d) $\Rightarrow$ (e) There is nothing to prove.

(e) $\Rightarrow$ (a) Corollary 11.9.7 ensures that HK is similar to KH. Theorem 11.8.1 tells us that $A = KH$ is similar to $(KH)^{\mathsf{T}} = H^{\mathsf{T}}K^{\mathsf{T}} = \overline{HK} = \overline{A}$. $\square$

11.11 Problems

P.11.1 Find the Jordan canonical forms of

$$\begin{bmatrix} 0 & 1 & 1 & 1 \\ 0 & 0 & 1 & 1 \\ 0 & 0 & 0 & 1 \\ 0 & 0 & 0 & 0 \end{bmatrix}, \quad \begin{bmatrix} 0 & 1 & 1 & 0 \\ 0 & 0 & 0 & 0 \\ 0 & 0 & 0 & 1 \\ 0 & 0 & 0 & 0 \end{bmatrix}, \quad \begin{bmatrix} 0 & 1 & 0 & 1 \\ 0 & 0 & 0 & 0 \\ 0 & 0 & 0 & 1 \\ 0 & 0 & 0 & 0 \end{bmatrix},$$

$$\begin{bmatrix} 0 & 0 & 1 & 1 \\ 0 & 0 & 0 & 1 \\ 0 & 0 & 0 & 0 \\ 0 & 0 & 0 & 0 \end{bmatrix}, \quad \begin{bmatrix} 0 & 0 & 1 & 1 \\ 0 & 0 & 1 & 1 \\ 0 & 0 & 0 & 0 \\ 0 & 0 & 0 & 0 \end{bmatrix}, \quad \begin{bmatrix} 0 & 0 & 1 & 1 \\ 0 & 0 & 0 & 1 \\ 0 & 1 & 0 & 0 \\ 0 & 0 & 0 & 0 \end{bmatrix}.$$

Which of these matrices are similar?

P.11.2   Show that

$$J_{k+1} = \begin{bmatrix} e_2^\mathsf{T} & 0 \\ J_k^2 & e_k \end{bmatrix} = \begin{bmatrix} J_k & e_k \\ 0^\mathsf{T} & 0 \end{bmatrix}.$$

P.11.3   Suppose that $A \in M_n$ and $B \in M_m$. (a) Show that $A \oplus B$ is nilpotent if and only if both A and B are nilpotent. (b) If A and B are nilpotent and $m = n$, are $A + B$ and AB nilpotent?

P.11.4   If $A \in M_5$, $(A - 2I)^3 = 0$, and $(A - 2I)^2 \neq 0$, what are the possible Jordan canonical forms for A?

P.11.5   Let $A, B \in M_n$. Prove, or give a counterexample to, the assertion that $ABAB = 0$ implies $BABA = 0$. Does the dimension matter?

P.11.6   Let $A \in M_n$ be nilpotent. (a) Show that

$$(I - A)^{-1} = I + A + A^2 + \cdots + A^{n-1}.$$

(b) Compute $(I - AB)^{-1}$, in which AB is the product in Example 11.9.6. (c) Compute $(I - J_n)^{-1}$ and $(I + J_n)^{-1}$.

P.11.7   Let $A \in M_n(\mathbb{R})$ and suppose that each of its entries is positive. If $\mathbf{x} \in \mathbb{R}^n$ is nonzero and has nonnegative entries, show that $A\mathbf{x}$ has positive entries.

P.11.8   Define the *partition function* $p : \mathbb{N} \to \mathbb{N}$ as follows: Let $p(n)$ be the number of different ways to write a positive integer n as a sum of positive integers (ignore the order of summands). For example, $p(1) = 1, p(2) = 2, p(3) = 3$, and $p(4) = 5$. It is known that $p(10) = 42$, $p(100) = 190,569,292$, and

$$p(1,000) = 24,061,467,864,032,622,473,692,149,727,991.$$

(a) Show that $p(5) = 7$, $p(6) = 11$, and $p(7) = 15$. (b) Show that there are $p(n)$ possible Jordan canonical forms for an $n \times n$ nilpotent matrix.

P.11.9   Let $A \in M_4$ and let $r_k = \operatorname{rank} A^k$ for $k = 0, 1, 2, 3, 4$. Show that $r_0 = 4, r_1 = 2$, $r_2 = 1, r_3 = 0$ is possible, but $r_0 = 4, r_1 = 3, r_2 = 0$ is not possible.

P.11.10   Suppose that $A \in M_n$ and $\lambda \in \operatorname{spec} A$. Show that the multiplicity of the block $J_k(\lambda)$ in the Jordan form of A is

$$\operatorname{rank}(A - \lambda I)^{k-1} - 2 \operatorname{rank}(A - \lambda I)^k + \operatorname{rank}(A - \lambda I)^{k+1}.$$

P.11.11   Let $A \in M_n$ be nilpotent. What are the possible Weyr characteristics and Jordan canonical forms for A if (a) $n = 2$? (b) $n = 3$? (c) $n = 4$? (d) $n = 5$?

P.11.12   Give examples of matrices $A, B \in M_n$ that are not similar, but that have the same characteristic polynomials and the same minimal polynomials. Why must your examples have size greater than 3?

P.11.13   Let p and q be positive integers and let $A \in M_{pq}$. If the Weyr characteristic of A is $w_1, w_2, \ldots, w_q$ and $w_1 = w_2 = \cdots = w_q = p$, show that the Jordan canonical form of A is $J_q \oplus \cdots \oplus J_q$ (p direct summands). For a special case of this phenomenon, see Example 11.9.6.

P.11.14   Suppose that $B \in M_m$ and $D \in M_n$ are nilpotent, with respective indices q_1 and q_2. Let $C \in M_{m \times n}$ and let

$$A = \begin{bmatrix} B & C \\ 0 & D \end{bmatrix}.$$

(a) Show that the $(1, 2)$ block of A^k is $\sum_{j=0}^{k} B^j C D^{k-j}$. (b) Prove that A is nilpotent, with index at most $\max\{q_1, q_2\}$.

P.11.15 Let $\varepsilon \geq 0$ and construct $A_\varepsilon \in M_n$ by setting the $(n, 1)$ entry of the nilpotent Jordan block $J_n(0)$ equal to ε. (a) Show that $p_{A_\varepsilon}(z) = z^n - \varepsilon$. (b) If $\varepsilon > 0$, show that A is diagonalizable and the modulus of every eigenvalue is $\varepsilon^{1/n}$. If $n = 64$ and $\varepsilon = 2^{-32}$, then every eigenvalue of A_ε has modulus $2^{-1/2} > 0.7$, while every eigenvalue of A_0 is equal to 0.

P.11.16 Let K_n denote the $n \times n$ reversal matrix and let $S_n = 2^{-1/2}(I_n + iK_n)$. Prove the following: (a) S_n is symmetric and unitary. (b) The matrix

$$S_n J_n(\lambda) S_n^* = \lambda I_n + \frac{1}{2}(J_n + K_n J_n K_n) + \frac{i}{2}(K_n J_n - J_n K_n)$$

is symmetric and unitarily similar to $J_n(\lambda)$. (c) Every Jordan matrix is unitarily similar to a complex symmetric matrix. (d) Every square complex matrix is similar to a complex symmetric matrix.

P.11.17 Suppose that $A, B \in M_6$, B has eigenvalues $1, 1, 2, 2, 3, 3$, $AB = BA$, and $A^3 = 0$. Show that there are exactly four possible Jordan canonical forms for A. What are they?

P.11.18 Consider the 4×4 matrices A and B in Example 11.9.6. Show that $p_A(z) = m_A(z) = z^4 - z$, $p_B(z) = z^4 - z^3 - z^2 + z$, and $m_B(z) = z^3 - z$.

P.11.19 Verify that $\mathbf{x}(t) = [\cos t + 2 \sin t \quad 2 \cos t - \sin t]^\mathsf{T}$ satisfies the differential equations $\mathbf{x}'(t) = A\mathbf{x}(t)$ and initial condition in Example 11.5.9.

P.11.20 Let $A \in M_n$ be diagonalizable, use the notation in our discussion of (11.5.1), and let $E(t) = \mathrm{diag}(e^{\lambda_1 t}, e^{\lambda_2 t}, \ldots, e^{\lambda_n t})$. (a) Show that $\mathbf{x}(t) = SE(t)S^{-1}\mathbf{x}(0)$ is the solution to (11.5.1). (b) Use (9.5.6) to explain why this solution can be written as $\mathbf{x}(t) = e^{At}\mathbf{x}(0)$. (c) What does this say if $n = 1$? (d) It is a fact that

$$e^{At} = \sum_{k=0}^{\infty} \frac{t^k}{k!} A^k,$$

in which the infinite series converges entrywise for every $A \in M_n$ and every $t \in \mathbb{R}$. Use this fact to show that if A and the initial values $\mathbf{x}(0)$ are real, then the solution $\mathbf{x}(t)$ is real for all $t \in \mathbb{R}$ even if some of the eigenvalues of A (and hence also $E(t)$ and S) are not real. (e) Let A and $\mathbf{x}(0)$ be the matrix and initial vector in Example 11.5.9. What are $(tA)^{4k}$, $(tA)^{4k+1}$, $(tA)^{4k+2}$, and $(tA)^{4k+3}$ for $k = 0, 1, 2, \ldots$? Use the infinite series in (c) to show that $e^{At} = (\sin t)A + (\cos t)I$. This method does not require knowledge of the eigenvalues of A. Compute $e^{At}\mathbf{x}(0)$ and discuss. (f) Let A be the matrix in Example 11.5.9. Use (9.8.4) to compute e^{At}. This method requires knowledge of the eigenvalues of A, but it does not involve infinite series.

P.11.21 Let $A \in M_n$ and let $\lambda \in \mathrm{spec}\, A$. Let $w_1, w_2, \ldots, w_q$ be the Weyr characteristic of $A - \lambda I$. Show that $w_1 + w_2 + \cdots + w_q$ is the algebraic multiplicity of λ. *Hint*: If $J_k(\mu)$ is a direct summand of the Jordan canonical form of A and $\mu \neq \lambda$, why is $\mathrm{rank}\, J_k(\mu - \lambda)^q = k$?

P.11.22 Let $A \in M_n$, suppose that $\mathrm{Re}\,\lambda < 0$ for every $\lambda \in \mathrm{spec}\, A$, and let $\mathbf{x}(t)$ be the solution to (11.5.1). Use the analysis in Section 11.5 to show that $\|\mathbf{x}(t)\|_2 \to 0$ as $t \to \infty$.

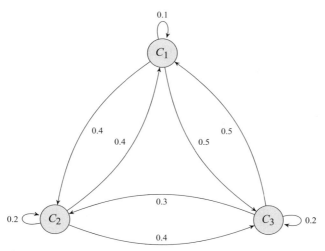

Figure 11.4 Migration pattern for P.11.25.

P.11.23 Cecil Sagehen is either happy or sad. If he is happy one day, then he is happy the next day three times out of four. If he is sad one day, then he is sad the next day one time out of three. During the coming year, how many days do you expect Cecil to be happy?

P.11.24 Gaul is divided into three parts, Gallia Aquitania (A), Gallia Belgica (B), and Gallia Lugdunensis (L). Each year, 5% of the residents of A move to B, and 5% move to L. Each year 15% of the residents of B of them move to A and 10% move to L. Finally, each year 10% of the residents of L move to A and 5% move to B. What percentage of the population should we expect to reside in each of the three regions after 50 years?

P.11.25 Each year, suppose that people migrate between cities C_1, C_2, and C_3 according to Figure 11.4. (a) What is the long-term prognosis if the initial populations are 10 million, 6 million, and 2 million, respectively? (b) What is the long-term prognosis if the initial populations are 1 million, 2 million, and 12 million, respectively? (c) How long must one wait before the population distribution among the three regions stabilizes?

P.11.26 Let $A \in M_n$ have all real positive entries. Suppose that all the row and column sums of A are 1. Show that $\lim_{p \to \infty} A^p = \frac{1}{n}E$, in which $E \in M_n$ is the all-ones matrix.

P.11.27 Let $A \in M_2$. (a) Show that A is unitarily similar to a unique matrix of the form

$$\begin{bmatrix} \lambda_1 & \alpha \\ 0 & \lambda_2 \end{bmatrix}, \tag{11.11.1}$$

in which:

(i) $\alpha = (\|A\|_F^2 - |\lambda_1|^2 - |\lambda_2|^2)^{1/2} \geq 0$,

(ii) $\operatorname{Re} \lambda_1 \geq \operatorname{Re} \lambda_2$, and

(iii) $\operatorname{Im} \lambda_1 \geq \operatorname{Im} \lambda_2$ if $\operatorname{Re} \lambda_1 = \operatorname{Re} \lambda_2$.

(b) Prove that every 2×2 complex matrix is unitarily similar to its transpose. However, see the following problem for the 3×3 case.

P.11.28 Let $A \in M_n$. (a) If $B \in M_n$ is unitarily similar to A, show that $\text{tr}(A^2(A^*)^2AA^*) = \text{tr}(B^2(B^*)^2BB^*)$. (b) If A is unitarily similar to A^T, show that $\text{tr}(A^2(A^*)^2AA^*) = \text{tr}((A^*)^2A^2A^*A)$. (c) Show that

$$A = \begin{bmatrix} 0 & 1 & 0 \\ 0 & 0 & 2 \\ 0 & 0 & 0 \end{bmatrix}$$

is not unitarily similar to A^T. Contrast this result with Theorem 11.8.1 and the preceding problem.

P.11.29 Let f be a monic polynomial of degree n and let $\lambda_1, \lambda_2, \ldots, \lambda_d$ be the distinct zeros of f, with respective multiplicities $n_1, n_2, \ldots, n_d$. (a) Show that the Jordan canonical form of the companion matrix C_f is

$$J_{n_1}(\lambda_1) \oplus J_{n_2}(\lambda_2) \oplus \cdots \oplus J_{n_d}(\lambda_d),$$

which has exactly one Jordan block corresponding to each distinct eigenvalue. (b) Show that each $\lambda \in \text{spec } C_f$ has geometric multiplicity 1.

P.11.30 Let $A \in M_n$. Show that the following statements are equivalent:

(a) The degree of m_A is equal to the degree of p_A.

(b) $p_A = m_A$.

(c) A is similar to the companion matrix of p_A.

(d) Each eigenvalue of A has geometric multiplicity 1.

(e) The Jordan canonical form of A contains exactly one block corresponding to each distinct eigenvalue.

(f) The Weyr characteristic associated with each eigenvalue of A has the form $1, 1, \ldots, 1$.

We say that A is *nonderogatory* if any of these statements is true.

P.11.31 If $A, B \in M_n$ are nonderogatory, show that A is similar to B if and only if $p_A = p_B$.

P.11.32 Let $D = \text{diag}(d_1, d_2, \ldots, d_n) \in M_n$ be an invertible diagonal matrix. (a) Use Corollary 11.3.11 to show that the Jordan canonical form of $J_n D$ is J_n. (b) Find a diagonal matrix E such that $EJ_n DE^{-1} = J_n$. (c) Suppose that $\varepsilon \neq 0$. What does $\lambda I + \varepsilon J_n$ look like, and why is it similar to $J_n(\lambda)$?

P.11.33 Show that $(I - J_n)$ is similar to $(I + J_n)$.

P.11.34 If $\lambda \neq 0$, show that $J_k(\lambda)^{-1}$ is similar to $J_k(\lambda^{-1})$.

P.11.35 If $\lambda \neq 0$ and p is a positive or negative integer, show that $J_k(\lambda)^p$ is similar to $J_k(\lambda^p)$. In particular, $J_k(\lambda)^2$ is similar to $J_k(\lambda^2)$.

P.11.36 Suppose that $\lambda \neq 0$ and let $\lambda^{1/2}$ be either choice of the square root. The preceding problem ensures that $J_k(\lambda^{1/2})^2$ is similar to $J_k(\lambda)$. Let $S \in M_n$ be invertible and such that $J_k(\lambda) = SJ_k(\lambda^{1/2})^2S^{-1}$. (a) Show that $(SJ_k(\lambda^{1/2})S^{-1})^2 = J_k(\lambda)$, so $SJ_k(\lambda^{1/2})S^{-1}$ is a square root of $J_k(\lambda)$. (b) Show that every invertible matrix has a square root.

P.11.37 Let $\lambda \neq 0$ and define

$$
R_4(\lambda) = \begin{bmatrix}
\lambda^{1/2} & \frac{1}{2}\lambda^{-1/2} & \frac{-1}{8}\lambda^{-3/2} & \frac{1}{16}\lambda^{-5/2} \\
0 & \lambda^{1/2} & \frac{1}{2}\lambda^{-1/2} & \frac{-1}{8}\lambda^{-3/2} \\
0 & 0 & \lambda^{1/2} & \frac{1}{2}\lambda^{-1/2} \\
0 & 0 & 0 & \lambda^{1/2}
\end{bmatrix}.
$$

(a) Verify that $R_4(\lambda)$ is a square root of $J_4(\lambda)$, that is, $R_4(\lambda)^2 = J_4(\lambda)$. The entries in the first row of $R_4(\lambda)$ are $f(\lambda), f'(\lambda), \frac{1}{2!}f''(\lambda)$, and $\frac{1}{3!}f'''(\lambda)$, in which $f(\lambda) = \lambda^{1/2}$.
(b) Show that $R_4(\lambda)$ is a polynomial in $J_k(\lambda)$. (c) Show that the 3×3 leading principal submatrix of $R_4(\lambda)$ is a square root of $J_3(\lambda)$. (d) Can you find a square root of $J_5(\lambda)$? (e) What do you get if you compute a square root of $J_2(\lambda)$ with (9.9.1) and $f(t) = t^{1/2}$?

P.11.38 Show that J_{2k}^2 is similar to $J_k \oplus J_k$ and J_{2k+1}^2 is similar to $J_{k+1} \oplus J_k$.

P.11.39 Of the three possible Jordan canonical forms for a 3×3 nilpotent matrix, show that $J_1 \oplus J_1 \oplus J_1$ and $J_2 \oplus J_1$ have square roots, but J_3 does not.

P.11.40 Of the five possible Jordan canonical forms for a 4×4 nilpotent matrix, show that $J_3 \oplus J_1$ and J_4 do not have square roots, but the other three do.

P.11.41 Let $T \in \mathcal{L}(\mathcal{P}_4)$ be the differentiation operator $T : p \to p'$ on the complex vector space of polynomials of degree 4 or less, and consider the basis $\beta = \{1, z, z^2, z^3, z^4\}$ of $\mathcal{P}_4$.
(a) Show that $_\beta[T]_\beta = [\mathbf{0} \ \mathbf{e}_1 \ 2\mathbf{e}_2 \ 3\mathbf{e}_3 \ 4\mathbf{e}_4] = J_5 D$, in which $D = \text{diag}(0, 1, 2, 3, 4)$.
(b) Find the Jordan canonical form of $_\beta[T]_\beta$.

P.11.42 Let $\beta = \{1, y, z, y^2, z^2, yz\}$ and let $\mathcal{V} = \text{span}\,\beta$ be the complex vector space of polynomials in two variables y and z that have degree at most 2. Let $T \in \mathcal{L}(\mathcal{V})$ be the partial differentiation operator $T : p \to \partial p/\partial y$. Find the Jordan canonical form of $_\beta[T]_\beta$.

P.11.43 Let $A, B \in M_m$ and $C, D \in M_n$. (a) If A is similar to B and C is similar to D, show that $(A \oplus C)$ and $(B \oplus D)$ are similar. (b) Give an example to show that similarity of $(A \oplus C)$ and $(B \oplus D)$ does not imply that A is similar to B or C is similar to D. (c) If $(A \oplus C)$ is similar to $(A \oplus D)$, show that C and D are similar. *Hint*: If $J(A)$ and $J(C)$ are the respective Jordan canonical forms of A and C, why is $J(A) \oplus J(C)$ the Jordan canonical form of both $A \oplus C$ and $A \oplus D$?

P.11.44 Let $A, B \in M_n$. Let $C = A \oplus \cdots \oplus A$ (k direct summands) and $D = B \oplus \cdots \oplus B$ (k direct summands). Show that A is similar to B if and only if C is similar to D.

P.11.45 If $A \in M_9$ and $A^7 = A$, what can you say about the Jordan canonical form of A?

P.11.46 Let $\lambda_1, \lambda_2, \ldots, \lambda_d$ be the distinct eigenvalues of $A \in M_n$. If $\text{rank}(A - \lambda_1 I)^{n-2} > \text{rank}(A - \lambda_1 I)^{n-1}$, prove that $d \leq 2$.

P.11.47 Let $A \in M_5$. If $p_A(z) = (z - 2)^3(z + 4)^2$ and $m_A(z) = (z - 2)^2(z + 4)$, what is the Jordan canonical form of A?

P.11.48 Let $\lambda_1, \lambda_2, \ldots, \lambda_d$ be the distinct eigenvalues of $A \in M_n$, let $p_A(z) = (z - \lambda_1)^{n_1}(z - \lambda_2)^{n_2} \cdots (z - \lambda_d)^{n_d}$ be its characteristic polynomial, and suppose that its minimal polynomial (10.3.2) has each exponent q_i equal either to n_i or to $n_i - 1$. What is the Jordan canonical form of A?

P.11.49 Let $A, B \in M_n$. If $p_A = p_B = m_A = m_B$, show that A is similar to B.

P.11.50 If $A \in M_n$ and $A = BC$, in which $B, C \in M_n$ are symmetric (Theorem 11.8.1), show that BC is similar to CB.

P.11.51 Let

$$B = \begin{bmatrix} 1 & 2 \\ 3 & 4 \end{bmatrix}.$$

Review Example 11.9.8 and adopt its notation. Verify that $A = SB\overline{S}^{-1}$ and $R = B^2$, so $A\overline{A}$ is similar to the square of a real matrix.

P.11.52 Let $A, S, R \in M_n$, suppose that S is invertible, R is real, and $A = SR\overline{S}^{-1}$. (a) Show that $A\overline{A}$ is similar to the square of a real matrix and deduce that $A\overline{A}$ is similar to $\overline{A}A$. (b) Show that any non-real eigenvalues of $A\overline{A}$ occur in conjugate pairs and that any negative eigenvalues of $A\overline{A}$ have even multiplicity.

P.11.53 Let $A = [a_{ij}] \in M_n$. Then A is a *Toeplitz matrix* if there are scalars $t_0, t_{\pm 1}, t_{\pm 2}, \dots,$ $t_{\pm(n-1)}$ such that $a_{ij} = t_{i-j}$. (a) Give an example of a 4×4 Toeplitz matrix. (b) Give an example of an upper triangular 4×4 Toeplitz matrix.

P.11.54 Let

$$J = \begin{bmatrix} J_2(\lambda) & 0 \\ 0 & J_2(\lambda) \end{bmatrix} \quad \text{and} \quad W = \begin{bmatrix} \lambda I_2 & I_2 \\ 0 & \lambda I_2 \end{bmatrix}, \tag{11.11.2}$$

and let $X \in M_4$. (a) Show that the Weyr characteristics of $J - \lambda I$ and $W - \lambda I$ are $w_1 = 2, w_2 = 2$. Deduce that J is similar to W. (b) Show that $JX = XJ$ if and only if

$$X = \begin{bmatrix} B & C \\ D & E \end{bmatrix},$$

in which $B, C, D, E \in M_2$ are upper triangular Toeplitz matrices. (c) Show that $WX = XW$ if and only if

$$X = \begin{bmatrix} F & G \\ 0 & F \end{bmatrix}, \quad F, G \in M_2,$$

which is block upper triangular. (d) Show that J and W are permutation similar.

P.11.55 What are the possible values of $\dim\{A\}'$ if $A \in M_3$? If $A \in M_4$? If $A \in M_5$ is unispectral?

P.11.56 Let $n \geq 2$, let $A \in M_n$, and let k be a positive integer. (a) Show that

$$\mathbb{C}^n = \text{null } A^k \oplus \text{col } A^k \tag{11.11.3}$$

need not be true if $k = 1$. (b) Show that (11.11.3) is true if $k = n$. (c) What is the smallest k for which (11.11.3) is true?

P.11.57 If $A \in M_n$ and $B \in M_m$ and if either A or B is nilpotent, show that $A \otimes B$ is nilpotent.

P.11.58 Let λ be a nonzero scalar. (a) Show that the Jordan canonical form of $J_p(0) \otimes J_q(\lambda)$ is $J_p(0) \oplus \cdots \oplus J_p(0)$ (q direct summands). (b) Show that the Jordan canonical form of $J_q(\lambda) \otimes J_p(0)$ is $J_q(0) \oplus \cdots \oplus J_q(0)$ (p direct summands).

11.12 Notes

For additional information about the Jordan canonical form, see [HJ13, Ch. 3].

If $A \in M_n$ and p is any polynomial, then $p(A)$ commutes with A. However, a matrix that commutes with A need not be a polynomial in A. For example, $A = I_2$ commutes with J_2, which is not a polynomial in I_2 (such a matrix must be diagonal). If A is nonderogatory, however, any matrix that commutes with A must be a polynomial in A; see P.11.30 and [HJ13, Sect. 3.2.4].

P.11.20 and P.11.37 introduce ways to define $f(A)$ for $f(t) = e^t$ and $f(t) = \sqrt{t}$. They are examples of *primary matrix functions*; see [HJ94, Chap. 6].

It is known that every square complex matrix A can be factored as $A = S R \bar{S}^{-1}$, in which R is real. Consequently, the pairing of negative and non-real eigenvalues of $A\bar{A}$ discussed in P.11.52 is true for all $A \in \mathsf{M}_n$; see [HJ13, Cor. 4.4.13, Cor. 4.6.15].

The nilpotent Jordan blocks of AB and BA need not be the same, but they cannot differ by much. Let $m_1 \geq m_2 \geq \cdots$ and $n_1 \geq n_2 \geq \cdots$ be the decreasingly ordered sizes of the nilpotent Jordan blocks of AB and BA, respectively; if necessary, append zero sizes to one list to get lists of equal lengths. Harley Flanders showed in 1951 that $|m_i - n_i| \leq 1$ for all i. Example 11.9.4 illustrates Flanders' theorem. For a discussion and proof see [JS96].

The Jordan canonical form was announced by Camille Jordan in an 1870 book. It is an excellent tool for problems involving powers and functions of matrices, but it is not the only way to classify matrices by similarity. In an 1885 paper, Eduard Weyr described the Weyr characteristic and used it to construct an alternative canonical form for similarity. Weyr's canonical form for $A \in \mathsf{M}_n$ is a direct sum of blocks $W(\lambda)$, one for each $\lambda \in \operatorname{spec} A$. Weyr's unispectral blocks are block upper bidiagonal matrices that directly display the Weyr characteristic of $A - \lambda I$:

$$
W(\lambda) =
\begin{bmatrix}
\lambda I_{w_1} & G_{w_1,w_2} & & & & \\
& \lambda I_{w_2} & G_{w_2,w_3} & & & \\
& & \ddots & & & \\
& & & \ddots & G_{w_{q-1},w_q} & \\
& & & & \lambda I_{w_q} &
\end{bmatrix},
\tag{11.12.1}
$$

in which

$$
G_{w_{j-1},w_j} =
\begin{bmatrix}
I_{w_j} \\
0
\end{bmatrix}
\in \mathsf{M}_{w_{j-1}j}, \quad j = 2, 3, \ldots, q.
$$

The Weyr and Jordan canonical forms for A are permutation similar. Weyr's canonical form is an excellent tool for problems involving commuting matrices, as illustrated in P.11.54. A matrix that commutes with a unispectral Weyr block (11.12.1) must be block upper triangular; matrices that are not block upper triangular can commute with a unispectral Jordan matrix. For more information about the Weyr canonical form and a historical summary, see [HJ13, Sect. 3.4].

For an approach to Theorem 11.7.14 that does not involve the Jordan canonical form and is valid for some matrices that are not Markov matrices, see [ĆJ15]. For example,

$$
A = \frac{1}{5}
\begin{bmatrix}
0 & -1 & 6 \\
2 & -1 & 4 \\
-4 & 0 & 9
\end{bmatrix}
$$

satisfies the condition $A\mathbf{e} = \mathbf{e}$, but does not have positive entries. Nevertheless, $\lim_{p \to \infty} A^p = \mathbf{e}\mathbf{x}^\mathsf{T}$, in which $\mathbf{x} = \frac{1}{3}[-6 \ 1 \ 8]^\mathsf{T}$ and $A^\mathsf{T}\mathbf{x} = \mathbf{x}$.

11.13 Some Important Concepts

- Index of a nilpotent matrix.

- Jordan block and Jordan matrix.

- Weyr characteristic.

- Each square complex matrix is similar to a Jordan matrix (its Jordan canonical form), which is unique up to permutation of its direct summands.

- The solution $\mathbf{x}(t)$ of $\mathbf{x}'(t) = A\mathbf{x}(t)$, in which $\mathbf{x}(0)$ is given, is a combination (11.5.8) of polynomials and exponentials determined by the Jordan canonical form of A.

- Convergent matrices (Theorem 11.6.6).

- Power bounded matrices (Theorem 11.7.2).

- Markov matrices.

- Stationary distribution of a Markov matrix with no zero entries (Theorem 11.7.14).

- Every square complex matrix is similar to its transpose.

- The Jordan canonical forms of AB and BA contain the same invertible Jordan blocks with the same multiplicities.

Normal Matrices and the Spectral Theorem

Every square complex matrix is unitarily similar to an upper triangular matrix, but which matrices are unitarily similar to a diagonal matrix? The answer is the main result of this chapter: the spectral theorem for normal matrices. Hermitian, skew-Hermitian, unitary, and circulant matrices are all unitarily diagonalizable. As a consequence, they have special properties that we investigate in this and following chapters.

12.1 Normal Matrices

Definition 12.1.1 A square matrix A is *normal* if $A^*A = AA^*$.

The term "normal," as used here, is not related to the "normal equations" or to the notion of a "normal vector" to a plane. Many mathematical objects are called "normal." Normal matrices are just one instance of this usage.

Example 12.1.2 Diagonal matrices are normal. If $A = \mathrm{diag}(\lambda_1, \lambda_2, \ldots, \lambda_n)$, then $A^* = \mathrm{diag}(\overline{\lambda_1}, \overline{\lambda_2}, \ldots, \overline{\lambda_n})$ and hence $A^*A = \mathrm{diag}(|\lambda_1|^2, |\lambda_2|^2, \ldots, |\lambda_n|^2) = AA^*$.

Example 12.1.3 Unitary matrices are normal. If U is unitary, then $U^*U = I = UU^*$.

Example 12.1.4 Hermitian matrices are normal. If $A = A^*$, then $A^*A = A^2 = AA^*$.

Example 12.1.5 Real symmetric, real skew-symmetric, and real orthogonal matrices are all normal.

Example 12.1.6 For $a, b \in \mathbb{C}$,

$$A = \begin{bmatrix} a & -b \\ b & a \end{bmatrix} \tag{12.1.7}$$

is normal since

$$A^*A = \begin{bmatrix} |a|^2 + |b|^2 & 2i \, \mathrm{Im}\, b\overline{a} \\ -2i \, \mathrm{Im}\, b\overline{a} & |a|^2 + |b|^2 \end{bmatrix} = AA^*.$$

If $a, b \in \mathbb{R}$, then A is the matrix (A.1.7). In this case, $A^*A = (a^2 + b^2)I = AA^*$. If $a, b \in \mathbb{R}$ and $a^2 + b^2 = 1$, then A is real orthogonal; see Example 8.3.9.

Example 12.1.8 A computation confirms that

$$\begin{bmatrix} 0 & 1 \\ 0 & 0 \end{bmatrix}, \quad \begin{bmatrix} 1 & 1 \\ 0 & 0 \end{bmatrix}, \quad \begin{bmatrix} 1 & 1 \\ 0 & 2 \end{bmatrix}, \quad \text{and} \quad \begin{bmatrix} 1 & 2 \\ 3 & 4 \end{bmatrix}$$

are not normal.

Theorem 12.1.9 *Let $A \in M_n$ be normal and let $\lambda \in \mathbb{C}$.*

(a) *If p is a polynomial, then $p(A)$ is normal and it commutes with A.*

(b) *$\|A\mathbf{x}\|_2 = \|A^*\mathbf{x}\|_2$ for all $\mathbf{x} \in \mathbb{C}^n$.*

(c) *$A\mathbf{x} = \lambda\mathbf{x}$ if and only if $A^*\mathbf{x} = \bar{\lambda}\mathbf{x}$.*

(d) *A^* is normal.*

(e) *$\operatorname{null} A = \operatorname{null} A^*$ and $\operatorname{col} A = \operatorname{col} A^*$.*

Proof (a) We first show that $A^*A^j = A^jA^*$ for $j = 0, 1, \ldots$, by induction on j. The base case $j = 0$ is $A^*I = IA^*$. Suppose that $A^*A^j = A^jA^*$ for some j. Then $A^*A^{j+1} = (A^*A^j)A = (A^jA^*)A = A^j(A^*A) = A^j(AA^*) = A^{j+1}A^*$, which completes the induction. A second induction confirms that $(A^*)^iA^j = A^j(A^*)^i$ for $i, j = 0, 1, \ldots$.

Let $p(z) = c_k z^k + \cdots + c_1 z + c_0$. Then

$$\left(p(A)\right)^* p(A) = \left(\sum_{i=0}^{k} \bar{c}_i(A^*)^i\right)\left(\sum_{j=0}^{k} c_j A^j\right) = \sum_{i,j=0}^{k} \bar{c}_i c_j (A^*)^i A^j$$

$$= \sum_{i,j=0}^{k} \bar{c}_i c_j A^j (A^*)^i = \left(\sum_{j=0}^{k} c_j A^j\right)\left(\sum_{i=0}^{k} \bar{c}_i(A^*)^i\right)$$

$$= p(A)\left(p(A)\right)^*$$

and hence $p(A)$ is normal. The fact that $p(A)$ commutes with A is Theorem 0.8.1.

(b) Let A be normal and let $\mathbf{x} \in \mathbb{C}^n$. Then

$$\|A\mathbf{x}\|_2^2 = \langle A\mathbf{x}, A\mathbf{x}\rangle = \langle A^*A\mathbf{x}, \mathbf{x}\rangle = \langle AA^*\mathbf{x}, \mathbf{x}\rangle = \langle A^*\mathbf{x}, A^*\mathbf{x}\rangle = \|A^*\mathbf{x}\|_2^2.$$

(c) It follows from (a) and (b) that $A - \lambda I$ is normal and

$$A\mathbf{x} = \lambda\mathbf{x} \iff \|(A - \lambda I)\mathbf{x}\|_2 = 0 \iff \|(A - \lambda I)^*\mathbf{x}\|_2 = 0$$
$$\iff \|(A^* - \bar{\lambda}I)\mathbf{x}\|_2 = 0 \iff A^*\mathbf{x} = \bar{\lambda}\mathbf{x}.$$

(d) $(A^*)^*(A^*) = AA^* = A^*A = A^*(A^*)^*$.

(e) The first assertion is the case $\lambda = 0$ in (c). The second assertion follows from the first and a computation: $\operatorname{col} A = \operatorname{col} A^{**} = (\operatorname{null} A^*)^\perp = (\operatorname{null} A)^\perp = \operatorname{col} A^*$. $\square$

Sums and products of normal matrices need not be normal. For example, the non-normal matrix

$$\begin{bmatrix} 0 & 1 \\ 0 & 0 \end{bmatrix} = \begin{bmatrix} 1 & 0 \\ 0 & 0 \end{bmatrix}\begin{bmatrix} 0 & 1 \\ 1 & 0 \end{bmatrix} = \begin{bmatrix} 0 & \frac{1}{2} \\ \frac{1}{2} & 0 \end{bmatrix} + \begin{bmatrix} 0 & \frac{1}{2} \\ -\frac{1}{2} & 0 \end{bmatrix}$$

is both a product of two normal matrices and a sum of two normal matrices. However, if $A, B \in M_n$ are normal and commute, then AB and $A + B$ are normal; see Corollary 12.2.10.

For any $A \in M_n$, the eigenvalues of A^* are the conjugates of the eigenvalues of A (Theorem 9.2.6), although A and A^* need not share the same eigenvectors. The content of Theorem 12.1.9.c is that the eigenvectors of A and A^* are the same if A is normal.

Conditions (b) and (c) in the preceding theorem are each equivalent to normality; see Theorem 12.8.1 and P.12.5.

Theorem 12.1.10 *Direct sums of normal matrices are normal.*

Proof If $A = A_{11} \oplus A_{22} \oplus \cdots \oplus A_{kk}$ and each A_{ii} is normal, then

$$A^*A = A_{11}^*A_{11} \oplus A_{22}^*A_{22} \oplus \cdots \oplus A_{kk}^*A_{kk}$$
$$= A_{11}A_{11}^* \oplus A_{22}A_{22}^* \oplus \cdots \oplus A_{kk}A_{kk}^* = AA^*,$$

so A is normal.   □

Lemma 12.1.11 *A matrix that is unitarily similar to a normal matrix is normal.*

Proof Let $A \in M_n$ be normal, let $U \in M_n$ be unitary, and let $B = UAU^*$. Then $B^* = UA^*U^*$ and

$$B^*B = (UA^*U^*)(UAU^*) = UA^*AU^* = UAA^*U^*$$
$$= (UAU^*)(UA^*U^*) = BB^*.$$ □

Certain patterns of zero entries are not allowed for a normal matrix.

Lemma 12.1.12 *Let B and C be square matrices and let*

$$A = \begin{bmatrix} B & X \\ 0 & C \end{bmatrix}.$$

Then A is normal if and only if B and C are normal and X = 0.

Proof If A is normal, then

$$A^*A = \begin{bmatrix} B^* & 0 \\ X^* & C^* \end{bmatrix} \begin{bmatrix} B & X \\ 0 & C \end{bmatrix} = \begin{bmatrix} B^*B & B^*X \\ X^*B & X^*X + C^*C \end{bmatrix} \qquad (12.1.13)$$

equals

$$AA^* = \begin{bmatrix} B & X \\ 0 & C \end{bmatrix} \begin{bmatrix} B^* & 0 \\ X^* & C^* \end{bmatrix} = \begin{bmatrix} BB^* + XX^* & XC^* \\ CX^* & CC^* \end{bmatrix}. \qquad (12.1.14)$$

Compare the $(1, 1)$ blocks in (12.1.13) and (12.1.14) and take traces to obtain

$$\operatorname{tr} B^*B = \operatorname{tr}(BB^* + XX^*) = \operatorname{tr} BB^* + \operatorname{tr} XX^* = \operatorname{tr} B^*B + \operatorname{tr} X^*X,$$

which implies that $\operatorname{tr} X^*X = 0$. It follows from (4.4.7) that $X = 0$ and hence $A = B \oplus C$. The $(1, 1)$ and $(2, 2)$ blocks of (12.1.13) and (12.1.14) tell us that B and C are normal.

If B and C are normal and $X = 0$, then Theorem 12.1.10 ensures that A is normal.   □

Lemma 12.1.12 can be generalized to normal block upper triangular matrices with any number of diagonal blocks.

Theorem 12.1.15 *A block upper triangular matrix is normal if and only if it is block diagonal and each of its diagonal blocks is normal. In particular, an upper triangular matrix is normal if and only if it is diagonal.*

Proof Let $A = [A_{ij}] \in M_n$ be normal and $k \times k$ block upper triangular, so $n_1 + n_2 + \cdots + n_k = n$, each $A_{ii} \in M_{n_i}$, and $A_{ij} = 0$ for $i > j$. Let S_k be the statement that $A_{11}, A_{22}, \ldots, A_{kk}$ are normal and $A_{ij} = 0$ for $i < j$. Proceed by induction on k. The preceding lemma shows that the base case S_2 is true. If $k > 2$ and S_{k-1} is true, write

$$A = \begin{bmatrix} A_{11} & X \\ 0 & A' \end{bmatrix},$$

in which $A' \in M_{n-n_1}$ is a $(k-1) \times (k-1)$ block upper triangular matrix and $X = [A_{12} \; A_{13} \; \ldots \; A_{1k}] \in M_{n_1 \times (n-n_1)}$. Lemma 12.1.12 ensures that A_{11} and A' are normal and that $X = 0$. The induction hypothesis ensures that A' is block diagonal and each of $A_{22}, A_{33}, \ldots, A_{kk}$ is normal. Therefore, $A = A_{11} \oplus A_{22} \oplus \cdots \oplus A_{kk}$, in which each direct summand is normal.

The converse is Theorem 12.1.10. $\qquad \square$

12.2 The Spectral Theorem

Definition 12.2.1 A square matrix is *unitarily diagonalizable* if it is unitarily similar to a diagonal matrix.

The criterion for normality is easy to check, but it has profound consequences. Normal matrices are not only diagonalizable, they are unitarily diagonalizable. The following theorem is the main result about normal matrices.

Theorem 12.2.2 (Spectral Theorem) *Let $A \in M_n$. The following are equivalent:*

(a) *A is normal, that is, $A^*A = AA^*$.*

(b) *A is unitarily diagonalizable.*

(c) *$\mathbb{C}^n$ has an orthonormal basis consisting of eigenvectors of A.*

If A is real, the following are equivalent:

(a′) *A is symmetric.*

(b′) *A is real orthogonally diagonalizable.*

(c′) *$\mathbb{R}^n$ has an orthonormal basis consisting of eigenvectors of A.*

Proof (a) $\Rightarrow$ (b) Suppose that A is normal. Use Theorem 10.1.1 to write $A = UTU^*$, in which U is unitary and T is upper triangular. Since T is normal (Lemma 12.1.11) and upper triangular, Theorem 12.1.15 ensures that T is diagonal. Thus, A is unitarily diagonalizable.

(b) $\Rightarrow$ (c) Suppose that A is unitarily diagonalizable and write $A = U\Lambda U^*$, in which $U = [\mathbf{u}_1 \ \mathbf{u}_2 \ \ldots \ \mathbf{u}_n]$ is unitary and $\Lambda = \mathrm{diag}(\lambda_1, \lambda_2, \ldots, \lambda_n)$. Because U is unitary, its columns are an orthonormal basis of $\mathbb{C}^n$. Moreover,

$$[A\mathbf{u}_1 \ A\mathbf{u}_2 \ \ldots \ A\mathbf{u}_n] = AU = U\Lambda = [\lambda_1\mathbf{u}_1 \ \lambda_2\mathbf{u}_2 \ \ldots \ \lambda_n\mathbf{u}_n],$$

so $A\mathbf{u}_i = \lambda_i\mathbf{u}_i$ for $i = 1, 2, \ldots, n$. Thus, the columns of U are eigenvectors of A that comprise an orthonormal basis of $\mathbb{C}^n$.

(c) $\Rightarrow$ (a) Suppose that $\mathbf{u}_1, \mathbf{u}_2. \ldots, \mathbf{u}_n$ are an orthonormal basis of $\mathbb{C}^n$ and $\lambda_1, \lambda_2, \ldots, \lambda_n$ are scalars such that $A\mathbf{u}_i = \lambda_i\mathbf{u}_i$ for each $i = 1, 2, \ldots, n$. Then $U = [\mathbf{u}_1 \ \mathbf{u}_2 \ \ldots \ \mathbf{u}_n]$ is unitary and satisfies $AU = U\Lambda$, in which $\Lambda = \mathrm{diag}(\lambda_1, \lambda_2, \ldots, \lambda_n)$. Lemma 12.1.11 ensures that $A = U\Lambda U^*$ is normal.

Now suppose that A is real.

(a') $\Rightarrow$ (b') Let $(\lambda, \mathbf{x})$ be an eigenpair of A, in which $\mathbf{x}$ is a real unit vector. Then $\langle A\mathbf{x}, \mathbf{x}\rangle = \langle \lambda\mathbf{x}, \mathbf{x}\rangle = \lambda\langle \mathbf{x}, \mathbf{x}\rangle = \lambda$, so the symmetry of A ensures that

$$\bar{\lambda} = \overline{\langle A\mathbf{x}, \mathbf{x}\rangle} = \langle \mathbf{x}, A\mathbf{x}\rangle = \langle A\mathbf{x}, \mathbf{x}\rangle = \lambda.$$

Therefore λ is real. The real case of Theorem 10.1.1 says that there is a real orthogonal matrix Q such that $Q^\mathsf{T}AQ = T$ is upper triangular and real. But $Q^\mathsf{T}AQ$ is symmetric, so T is symmetric and hence is real and diagonal.

(b') $\Rightarrow$ (c') The proof is the same as in the complex case, except that the orthonormal vectors $\mathbf{u}_i$ are real and $\Lambda = Q^\mathsf{T}AQ$ is real.

(c') $\Rightarrow$ (a') The proof is the same as in the complex case, except that the unitary matrix is real orthogonal, so $A = U^\mathsf{T}\Lambda U$ is symmetric. $\qquad\square$

An alternative formulation of the spectral theorem concerns the representation of a normal matrix as a linear combination of orthogonal projections; see Theorem 12.9.8.

Example 12.2.3 The matrix A in (12.1.7) has eigenvalues $\lambda_1 = a + bi$ and $\lambda_2 = a - bi$. Let $\Lambda = \mathrm{diag}(\lambda_1, \lambda_2)$. Corresponding unit eigenvectors are $\mathbf{u}_1 = [\frac{1}{\sqrt{2}} \ -\frac{i}{\sqrt{2}}]^\mathsf{T}$ and $\mathbf{u}_2 = [\frac{1}{\sqrt{2}} \ \frac{i}{\sqrt{2}}]^\mathsf{T}$. The matrix $U = [\mathbf{u}_1 \ \mathbf{u}_2]$ is unitary and

$$\underbrace{\begin{bmatrix} a & -b \\ b & a \end{bmatrix}}_{A} = \underbrace{\begin{bmatrix} \frac{1}{\sqrt{2}} & \frac{1}{\sqrt{2}} \\ -\frac{i}{\sqrt{2}} & \frac{i}{\sqrt{2}} \end{bmatrix}}_{U} \underbrace{\begin{bmatrix} a+bi & 0 \\ 0 & a-bi \end{bmatrix}}_{\Lambda} \underbrace{\begin{bmatrix} \frac{1}{\sqrt{2}} & \frac{i}{\sqrt{2}} \\ \frac{1}{\sqrt{2}} & -\frac{i}{\sqrt{2}} \end{bmatrix}}_{U^*}. \tag{12.2.4}$$

Example 12.2.5 The matrix

$$A = \begin{bmatrix} 1 & 8 & -4 \\ 8 & 1 & 4 \\ -4 & 4 & 7 \end{bmatrix} \tag{12.2.6}$$

is real symmetric and hence is real orthogonally diagonalizable. Its eigenvalues are $9, 9, -9$; corresponding orthonormal eigenvectors are $\mathbf{u}_1 = [\frac{1}{3} \ \frac{2}{3} \ \frac{2}{3}]^\mathsf{T}$, $\mathbf{u}_2 = [-\frac{2}{3} \ -\frac{1}{3} \ \frac{2}{3}]^\mathsf{T}$, and $\mathbf{u}_3 = [-\frac{2}{3} \ \frac{2}{3} \ -\frac{1}{3}]^\mathsf{T}$. The matrix $U = [\mathbf{u}_1 \ \mathbf{u}_2 \ \mathbf{u}_3]$ is real orthogonal and

$$
\underbrace{\begin{bmatrix} 1 & 8 & -4 \\ 8 & 1 & 4 \\ -4 & 4 & 7 \end{bmatrix}}_{A} = \underbrace{\begin{bmatrix} \frac{1}{3} & -\frac{2}{3} & -\frac{2}{3} \\ \frac{2}{3} & -\frac{1}{3} & \frac{2}{3} \\ \frac{2}{3} & \frac{2}{3} & -\frac{1}{3} \end{bmatrix}}_{U} \underbrace{\begin{bmatrix} 9 & 0 & 0 \\ 0 & 9 & 0 \\ 0 & 0 & -9 \end{bmatrix}}_{\Lambda} \underbrace{\begin{bmatrix} \frac{1}{3} & \frac{2}{3} & \frac{2}{3} \\ -\frac{2}{3} & -\frac{1}{3} & \frac{2}{3} \\ -\frac{2}{3} & \frac{2}{3} & -\frac{1}{3} \end{bmatrix}}_{U^{\mathsf{T}}}.
$$

Definition 12.2.7 Let $A \in \mathsf{M}_n$ be normal. A *spectral decomposition* of A is a factorization $A = U\Lambda U^*$, in which $U \in \mathsf{M}_n$ is unitary and $\Lambda \in \mathsf{M}_n$ is diagonal.

If $A \in \mathsf{M}_n$ is normal, here is a recipe for how one might find a spectral decomposition for it.

(a) Find the distinct eigenvalues $\mu_1, \mu_2, \ldots, \mu_d$ of A and their respective multiplicities $n_1, n_2, \ldots, n_d$.

(b) For each $j = 1, 2, \ldots, d$, find an orthonormal basis of $\mathcal{E}_{\mu_j}(A)$ as follows:

   (i) Find linearly independent solutions $\mathbf{x}_1, \mathbf{x}_2, \ldots, \mathbf{x}_{n_j}$ of the homogeneous system $(A - \mu_j I)\mathbf{x} = \mathbf{0}$. Construct $X_j = [\mathbf{x}_1 \ \mathbf{x}_2 \ \ldots \ \mathbf{x}_{n_j}] \in \mathsf{M}_{n \times n_j}$. Then rank $X_j = n_j$ and $AX_j = \mu_j X_j$.

   (ii) Find $U_j \in \mathsf{M}_{n \times n_j}$ with orthonormal columns, whose column space is the same as that of X_j. For example, one could use the QR algorithm (Theorem 6.5.2). Another option would be to apply the Gram–Schmidt process (Theorem 5.3.5) to the columns of X_j.

(c) Form $U = [U_1 \ U_2 \ \ldots \ U_d]$ and $\Lambda = \mu_1 I_{n_1} \oplus \mu_2 I_{n_2} \oplus \cdots \oplus \mu_d I_{n_d}$. Then U is unitary and $A = U\Lambda U^*$.

An important corollary of the spectral theorem says that if A is normal, then A and A^* are simultaneously unitarily diagonalizable.

Corollary 12.2.8 *Let $A \in \mathsf{M}_n$ be normal and write $A = U\Lambda U^*$, in which U is unitary and Λ is diagonal. Then $A^* = U\overline{\Lambda} U^*$.*

Proof $A^* = (U\Lambda U^*)^* = U\Lambda^* U^* = U\overline{\Lambda} U^*$ since Λ is diagonal. $\qquad\square$

The preceding corollary is a special case (in which $\mathcal{F} = \{A, A^*\}$) of a broad generalization of the spectral theorem to any set of commuting normal matrices.

Theorem 12.2.9 *Let $\mathcal{F} \subseteq \mathsf{M}_n$ be a nonempty set of matrices.*

(a) *Suppose that each matrix in $\mathcal{F}$ is normal. Then $AB = BA$ for all $A, B \in \mathcal{F}$ if and only if there is a unitary $U \in \mathsf{M}_n$ such that U^*AU is diagonal for every $A \in \mathcal{F}$.*

(b) *Suppose that each matrix in $\mathcal{F}$ is real and symmetric. Then $AB = BA$ for all $A, B \in \mathcal{F}$ if and only if there is a real orthogonal $Q \in \mathsf{M}_n(\mathbb{R})$ such that $Q^{\mathsf{T}}AQ$ is diagonal for every $A \in \mathcal{F}$.*

Proof (a) Theorem 10.5.1 ensures that there is a unitary $U \in \mathsf{M}_n$ such that U^*AU is upper triangular for every $A \in \mathcal{F}$. However, each of these upper triangular matrices is normal (Lemma 12.1.11), so Theorem 12.1.15 ensures that each is diagonal. Conversely, any pair of simultaneously diagonalizable matrices commutes; see Theorem 9.4.15.

(b) Since all the eigenvalues of a real symmetric matrix are real, Theorem 10.5.1 ensures that there is a real orthogonal $Q \in M_n(\mathbb{R})$ such that $Q^T A Q$ is real and upper triangular for every $A \in \mathcal{F}$. All of these upper triangular matrices are symmetric, so they are diagonal. The converse follows as in (a). □

Corollary 12.2.10 *Let $A, B \in M_n$ be normal and suppose that $AB = BA$. Then AB and $A + B$ are normal.*

Proof The preceding theorem provides a unitary U and diagonal matrices Λ and M such that $A = U\Lambda U^*$ and $B = UMU^*$. Thus, $AB = (U\Lambda U^*)(UMU^*) = U(\Lambda M)U^*$ and $A + B = U\Lambda U^* + UMU^* = U(\Lambda + M)U^*$ are unitarily diagonalizable, hence normal. □

Example 12.2.11 Let

$$ A = \begin{bmatrix} 1 & 0 \\ 0 & -1 \end{bmatrix} \quad \text{and} \quad B = \begin{bmatrix} 0 & 1 \\ 1 & 0 \end{bmatrix}. $$

Then A, B, AB, BA, and $A + B$ are normal, but $AB \neq BA$. Thus, commutativity is a sufficient, but not necessary, condition for the product or sum of normal matrices to be normal.

12.3 The Defect from Normality

Theorem 12.3.1 (Schur's Inequality) *Let $A \in M_n$ have eigenvalues $\lambda_1, \lambda_2, \ldots, \lambda_n$. Then*

$$ \sum_{i=1}^{n} |\lambda_i|^2 \leq \|A\|_F^2, \tag{12.3.2} $$

with equality if and only if A is normal.

Proof Use Theorem 10.1.1 to write $A = UTU^*$, in which $T = [t_{ij}] \in M_n$ is upper triangular and $t_{ii} = \lambda_i$ for $i = 1, 2, \ldots, n$. Then $A^*A = UT^*TU^*$ and

$$ \sum_{i=1}^{n} |\lambda_i|^2 \leq \sum_{i=1}^{n} |\lambda_i|^2 + \sum_{i<j} |t_{ij}|^2 = \operatorname{tr} T^*T = \operatorname{tr} UT^*TU^* = \operatorname{tr} A^*A = \|A\|_F^2, $$

with equality if and only if $\sum_{i<j} |t_{ij}|^2 = 0$, that is, if and only if T is diagonal. Thus, Schur's inequality is an equality if and only if A is unitarily similar to a diagonal matrix, that is, if and only if A is normal. □

Definition 12.3.3 Let $\lambda_1, \lambda_2, \ldots, \lambda_n$ be the eigenvalues of $A \in M_n$. The quantity

$$ \Delta(A) = \|A\|_F^2 - \sum_{i=1}^{n} |\lambda_i|^2 = \operatorname{tr} A^*A - \sum_{i=1}^{n} |\lambda_i|^2 $$

is the *defect from normality* of A.

Schur's inequality (12.3.2) says that $\Delta(A) \geq 0$ for every $A \in M_n$, with equality if and only if A is normal.

Theorem 12.3.4 *Let $A, B \in M_n$. If A, B, and AB are normal, then BA is normal.*

Proof Since AB is normal, $\Delta(AB) = 0$. However, BA has the same eigenvalues as AB with the same multiplicities (Theorem 9.7.2). Thus, $\Delta(BA) = 0$ if and only if $\operatorname{tr}(BA)^*(BA) = \operatorname{tr}(AB)^*(AB)$. Since A and B are normal, this is confirmed by the computation

$$\operatorname{tr}(BA)^*(BA) = \operatorname{tr} A^* B^* BA = \operatorname{tr} A^* BB^* A = \operatorname{tr} BB^* AA^*$$
$$= \operatorname{tr} BB^* A^* A = \operatorname{tr} B^* A^* AB = \operatorname{tr}(AB)^*(AB). \qquad \square$$

12.4 The Fuglede–Putnam Theorem

Lemma 12.4.1 *Let $A_i \in M_{n_i}$ be normal for $i = 1, 2, \ldots, k$. Then there exists a polynomial p such that $A_i^* = p(A_i)$ for $i = 1, 2, \ldots, k$.*

Proof Theorem 0.7.6 constructs a polynomial p such that $p(\lambda) = \bar{\lambda}$ for all $\lambda \in \bigcup_{i=1}^{k} \operatorname{spec} A_i$. For $i = 1, 2, \ldots, k$, write $A_i = U_i \Lambda_i U_i^*$, in which $U_i \in M_{n_i}$ is unitary and $\Lambda_i \in M_{n_i}$ is diagonal. Theorem 9.5.1 ensures that $p(A_i) = U_i p(\Lambda_i) U_i^* = U_i \bar{\Lambda}_i U_i^* = A_i^*$ for each $i = 1, 2, \ldots, k$. $\qquad \square$

Theorem 12.4.2 (Fuglede–Putnam) *Let $A \in M_m$ and $B \in M_n$ be normal and let $X \in M_{m \times n}$. Then $AX = XB$ if and only if $A^* X = XB^*$.*

Proof Suppose that $AX = XB$. Lemma 12.4.1 provides a polynomial p such that $p(A) = A^*$ and $p(B) = B^*$. Theorem 0.8.1 ensures that $A^* X = p(A)X = Xp(B) = XB^*$. Now apply the same argument with A^*, B^* in place of A, B. $\qquad \square$

For other proofs of the Fuglede–Putnam theorem, see P.12.16 and P.12.17. Corollary 12.2.10 can also be proved using the Fuglede–Putnam theorem; see P.12.20.

Corollary 12.4.3 *Let $A \in M_{m \times n}$ and $B \in M_{n \times m}$. Then AB and BA are normal if and only if $A^* AB = BAA^*$ and $ABB^* = B^* BA$.*

Proof If AB and BA are normal, then $(AB)^*$ and $(BA)^*$ are normal (Theorem 12.1.9.d). Since $(BA)^* A^* = A^* B^* A^* = A^*(AB)^*$, the preceding theorem ensures that $BAA^* = A^* AB$. Since the roles of A and B are interchangeable, $ABB^* = B^* BA$.

Conversely, suppose that $A^* AB = BAA^*$ and $ABB^* = B^* BA$. Then

$$(AB)^*(AB) = B^*(A^* AB) = B^*(BAA^*)$$
$$= (B^* BA)A^* = (ABB^*)A^*$$
$$= (AB)(AB)^*,$$

that is, AB is normal. Since the roles of A and B are interchangeable, BA is also normal. $\qquad \square$

12.5 Circulant Matrices

Circulant matrices are structured matrices that arise in signal processing, finite Fourier analysis, cyclic codes for error correction, and the asymptotic analysis of Toeplitz matrices. The set of $n \times n$ circulant matrices is an example of a commuting family of normal matrices.

Definition 12.5.1 An $n \times n$ *circulant matrix* is a complex matrix of the form

$$\begin{bmatrix} c_0 & c_1 & c_2 & \cdots & c_{n-1} \\ c_{n-1} & c_0 & c_1 & \cdots & c_{n-2} \\ c_{n-2} & c_{n-1} & c_0 & \cdots & c_{n-3} \\ \vdots & \vdots & \vdots & \ddots & \vdots \\ c_1 & c_2 & c_3 & \cdots & c_0 \end{bmatrix}. \tag{12.5.2}$$

The entries in each row are shifted one position to the right compared to the row above, with wraparound to the left.

Linear combinations of circulant matrices are circulant matrices, so the circulant matrices comprise a subspace of M_n.

The $n \times n$ *cyclic permutation matrix*

$$S_n = \begin{bmatrix} 0 & 1 & 0 & \cdots & 0 \\ 0 & 0 & 1 & \cdots & 0 \\ \vdots & \vdots & \vdots & \ddots & \vdots \\ 0 & 0 & 0 & \cdots & 1 \\ 1 & 0 & 0 & \cdots & 0 \end{bmatrix} \tag{12.5.3}$$

(sometimes called a *circular shift matrix*) is unitary and hence normal (see Example 12.1.3). Since

$$S_n[x_1 \ x_2 \ \ldots \ x_n]^\mathsf{T} = [x_2 \ x_3 \ \ldots \ x_1]^\mathsf{T},$$

it follows that $S_n^n = I$. Thus, S_n is annihilated by the polynomial $p(z) = z^n - 1$, whose zeros are the nth roots of unity $1, \omega, \omega^2, \ldots, \omega^{n-1}$, in which $\omega = e^{2\pi i/n}$. Theorem 8.3.3 ensures that $\operatorname{spec} S_n \subseteq \{1, \omega, \ldots, \omega^{n-1}\}$. We now show that this containment is an equality.

Let $\mathbf{x} = [x_1 \ x_2 \ \ldots \ x_n]^\mathsf{T}$ and consider the equation $S_n \mathbf{x} = \omega^j \mathbf{x}$, which is equivalent to

$$[x_2 \ x_3 \ \ldots \ x_1]^\mathsf{T} = \omega^j [x_1 \ x_2 \ \ldots \ x_n]^\mathsf{T}. \tag{12.5.4}$$

One checks that the identity (12.5.4) is satisfied if

$$\mathbf{x} = \mathbf{f}_j = \frac{1}{\sqrt{n}}[1 \ \omega^j \ \ldots \ \omega^{j(n-1)}]^\mathsf{T}.$$

Thus, $(\omega^j, \mathbf{f}_j)$ is an eigenpair of S_n for $j = 0, 1, \ldots, n-1$ and we write $S_n = F_n \Omega F_n^*$, in which $\Omega = \operatorname{diag}(1, \omega, \omega^2, \ldots, \omega^{n-1})$ and $F_n = [\mathbf{f}_1 \ \mathbf{f}_2 \ \ldots \ \mathbf{f}_n]$ is the $n \times n$ Fourier matrix (6.2.14). The Fourier matrix is unitary; see Example 6.2.13. The matrix S_n is unitary and has distinct eigenvalues.

The powers of S_n can be computed readily. For example, S_4^0, S_4^1, S_4^2, and S_4^3 are

$$
\begin{bmatrix} 1 & 0 & 0 & 0 \\ 0 & 1 & 0 & 0 \\ 0 & 0 & 1 & 0 \\ 0 & 0 & 0 & 1 \end{bmatrix}, \quad
\begin{bmatrix} 0 & 1 & 0 & 0 \\ 0 & 0 & 1 & 0 \\ 0 & 0 & 0 & 1 \\ 1 & 0 & 0 & 0 \end{bmatrix}, \quad
\begin{bmatrix} 0 & 0 & 1 & 0 \\ 0 & 0 & 0 & 1 \\ 1 & 0 & 0 & 0 \\ 0 & 1 & 0 & 0 \end{bmatrix}, \quad
\begin{bmatrix} 0 & 0 & 0 & 1 \\ 1 & 0 & 0 & 0 \\ 0 & 1 & 0 & 0 \\ 0 & 0 & 1 & 0 \end{bmatrix},
$$

respectively. If

$$p(z) = c_{n-1}z^{n-1} + \cdots + c_1 z + c_0, \tag{12.5.5}$$

then $p(S_n)$ is the circulant matrix (12.5.2). Theorem 9.5.1 tells us that

$$p(S_n) = F_n \operatorname{diag}\left(p(1), p(\omega), \ldots, p(\omega^{n-1})\right) F_n^*, \tag{12.5.6}$$

in which $\omega = e^{2\pi i/n}$.

Corollary 9.6.3 ensures that X commutes with S_n if and only if $X = p(S_n)$ for some polynomial p. Thus, a matrix commutes with S_n if and only if it is a circulant matrix. Since each $n \times n$ circulant matrix is a polynomial in S_n, Theorem 0.8.1 ensures that the product of circulant matrices is a circulant matrix and that any two circulant matrices commute.

We summarize the preceding discussion in the following theorem.

Theorem 12.5.7 *Let $C \in \mathsf{M}_n$ be the circulant matrix (12.5.2), let p be the polynomial (12.5.5), let $\omega = e^{2\pi i/n}$, and let $\Omega = \operatorname{diag}(1, \omega, \omega^2, \ldots, \omega^{n-1})$.*

(a) *The eigenvalues of C are $p(1), p(\omega), \ldots, p(\omega^{n-1})$. Corresponding unit eigenvectors are $\mathbf{f}_j = \frac{1}{\sqrt{n}}[1 \ \omega^j \ \ldots \ \omega^{j(n-1)}]^{\mathsf{T}}$ for $j = 1, 2, \ldots, n$, that is,*

$$C = F_n p(\Omega) F_n^*,$$

in which $F_n = [\mathbf{f}_1 \ \mathbf{f}_2 \ \ldots \ \mathbf{f}_n]$ is the $n \times n$ Fourier matrix.

(b) *$C = p(S_n)$, in which S_n denotes the cyclic permutation matrix (12.5.3).*

(c) *The set of all $n \times n$ circulant matrices is the subspace $\{S_n\}'$ of M_n. It is a commuting family of normal matrices that are simultaneously diagonalized by the unitary matrix F_n.*

Any n complex numbers can be the first row of an $n \times n$ circulant matrix, and the preceding theorem identifies its eigenvalues. What about the converse? Can any n complex numbers be the eigenvalues of an $n \times n$ circulant matrix? If so, what is its first row?

Theorem 12.5.8 *Let $\lambda_1, \lambda_2, \ldots, \lambda_n \in \mathbb{C}$, let $\omega = e^{2\pi i/n}$, and let $p(z) = c_{n-1}z^{n-1} + c_{n-2}z^{n-2} + \cdots + c_1 z + c_0$ be a polynomial of degree $n - 1$ or less such that $p(\omega^{j-1}) = \lambda_j$ for all $j = 1, 2, \ldots, n$. Then $\lambda_1, \lambda_2, \ldots, \lambda_n$ are the eigenvalues of the circulant matrix with first row $[c_0 \ c_1 \ \ldots \ c_{n-1}]$.*

Proof Maintain the notation of the preceding theorem. Since the roots of unity $1, \omega, \omega^2, \ldots, \omega^{n-1}$ are distinct, Theorem 0.7.6 ensures the existence of a unique polynomial p of degree at most $n - 1$ such that $p(\omega^{j-1}) = \lambda_j$ for all $j = 1, 2, \ldots, n$. The first row of the circulant matrix $p(S_n)$ is $[c_0 \ c_1 \ \ldots \ c_{n-1}]$ and since

$$p(S_n) = F_n p(\Omega) F_n^* = F_n \operatorname{diag}(\lambda_1, \lambda_2, \ldots, \lambda_n) F_n^*,$$

its eigenvalues are $\lambda_1, \lambda_2, \ldots, \lambda_n$. $\square$

12.6 Some Special Classes of Normal Matrices

We have encountered several classes of normal matrices already: Hermitian matrices (Chapter 5), unitary matrices (Chapter 6), and orthogonal projections (Chapter 7). Each of these classes has an elegant spectral characterization.

Theorem 12.6.1 *Let $A \in \mathsf{M}_n$ be normal.*

(a) *A is unitary if and only if $|\lambda| = 1$ for all $\lambda \in \operatorname{spec} A$.*

(b) *A is Hermitian if and only if $\lambda \in \mathbb{R}$ for all $\lambda \in \operatorname{spec} A$.*

(c) *A is an orthogonal projection if and only if $\lambda \in \{0, 1\}$ for all $\lambda \in \operatorname{spec} A$.*

(d) *A is skew Hermitian if and only if λ is purely imaginary for all $\lambda \in \operatorname{spec} A$.*

Proof Suppose that $(\lambda, \mathbf{x})$ is an eigenpair of A and $\|\mathbf{x}\| = 1$. Write $A = U\Lambda U^*$, in which U is unitary and Λ is diagonal. The diagonal entries of Λ are the eigenvalues of A.

(a) If A is unitary, then $|\lambda| = \|\lambda \mathbf{x}\| = \|A\mathbf{x}\| = \|\mathbf{x}\| = 1$. Conversely, if $|\lambda| = 1$ for all $\lambda \in \operatorname{spec} A$, then $\Lambda^{-1} = \overline{\Lambda} = \Lambda^*$. Thus, $A^{-1} = (U\Lambda U^*)^{-1} = U\Lambda^{-1}U^* = U\Lambda^*U^* = (U\Lambda U^*)^* = A^*$.

(b) If $A = A^*$, then

$$\lambda = \langle \lambda \mathbf{x}, \mathbf{x} \rangle = \langle A\mathbf{x}, \mathbf{x} \rangle = \langle \mathbf{x}, A^*\mathbf{x} \rangle = \langle \mathbf{x}, A\mathbf{x} \rangle = \langle \mathbf{x}, \lambda \mathbf{x} \rangle = \overline{\lambda}.$$

Thus, $\lambda = \overline{\lambda}$ and $\lambda \in \mathbb{R}$. Conversely, if $\lambda \in \mathbb{R}$ for all $\lambda \in \operatorname{spec} A$, then $\Lambda = \overline{\Lambda} = \Lambda^*$ and $A^* = U\Lambda^*U^* = U\Lambda U^* = A$.

(c) If A is an orthogonal projection, then $\lambda^2 \mathbf{x} = A^2 \mathbf{x} = A\mathbf{x} = \lambda \mathbf{x}$. Thus $\lambda(\lambda - 1)\mathbf{x} = (\lambda^2 - \lambda)\mathbf{x} = \mathbf{0}$, so $\lambda(\lambda - 1) = 0$ and hence $\lambda \in \{0, 1\}$. Conversely, if $\operatorname{spec} A \subseteq \{0, 1\}$, then $\Lambda^2 = \Lambda$. Thus, $A^2 = (U\Lambda U^*)(U\Lambda U^*) = U\Lambda^2 U^* = U\Lambda U^* = A$, so A is idempotent. Part (b) shows that A is Hermitian, and Theorem 7.3.14 ensures that A is an orthogonal projection.

(d) If A is skew Hermitian, then $B = iA$ is Hermitian. Thus, (b) ensures that $i\lambda \in \mathbb{R}$ for all $\lambda \in \operatorname{spec} A$ and hence λ is purely imaginary for all $\lambda \in \operatorname{spec} A$. $\qquad \square$

Normal matrices whose eigenvalues are real and nonnegative are particularly important. They are the positive semidefinite matrices; see Chapter 13.

Example 12.6.2 The matrix

$$A = \begin{bmatrix} 1 & 1 \\ 0 & 1 \end{bmatrix}$$

has $\operatorname{spec} A = \{1\}$, but it is neither unitary nor Hermitian. The hypothesis in Theorem 12.6.1 that A is normal cannot be omitted.

Example 12.6.3 The matrix

$$A = \begin{bmatrix} 1 & 1 \\ 0 & 0 \end{bmatrix}$$

has distinct eigenvalues 0 and 1, so it is diagonalizable but not Hermitian. The hypothesis in Theorem 12.6.1 that A is normal cannot be weakened to assume that A is diagonalizable.

Theorem 12.6.4 *If $A \in M_n$, then there are unique Hermitian matrices $H, K \in M_n$ such that $A = H + iK$.*

Proof We have

$$A = \frac{1}{2}(A + A^*) + i\frac{1}{2i}(A - A^*) = H + iK,$$

in which the matrices

$$H = \frac{1}{2}(A + A^*) \quad \text{and} \quad K = \frac{1}{2i}(A - A^*) \tag{12.6.5}$$

are Hermitian. Conversely, if $A = X + iY$, in which X and Y are Hermitian, then $A^* = X - iY$. Consequently, $2X = A + A^*$ and $2iY = A - A^*$, which confirm that $X = H$ and $Y = K$. $\qquad\square$

Definition 12.6.6 The *Cartesian decomposition* of $A \in M_n$ is $A = H + iK$, in which $H, K \in M_n$ are Hermitian. The matrix $H = H(A) = \frac{1}{2}(A + A^*)$ is the *Hermitian part* of A and $iK = \frac{1}{2}(A - A^*)$ is the *skew-Hermitian part* of A.

The Cartesian decomposition of A is analogous to the decomposition (A.2.4) of a complex number $z = a + bi$, in which $a = \operatorname{Re} z = \frac{1}{2}(z + \bar{z})$ and $b = \operatorname{Im} z = \frac{1}{2i}(z - \bar{z})$.

Example 12.6.7 An example of a Cartesian decomposition is:

$$\underbrace{\begin{bmatrix} 0 & 1 \\ 0 & 0 \end{bmatrix}}_{A} = \underbrace{\begin{bmatrix} 0 & \frac{1}{2} \\ \frac{1}{2} & 0 \end{bmatrix}}_{H} + i \underbrace{\begin{bmatrix} 0 & -\frac{i}{2} \\ \frac{i}{2} & 0 \end{bmatrix}}_{K}.$$

The matrix A is not normal. Observe that the Hermitian matrices H and K do not commute.

Theorem 12.6.8 *Let $A \in M_n$ with Cartesian decomposition $A = H + iK$. Then A is normal if and only if $HK = KH$.*

Proof Observe that

$$A^*A = (H - iK)(H + iK) = H^2 + K^2 + i(HK - KH)$$

and

$$AA^* = (H + iK)(H - iK) = H^2 + K^2 - i(HK - KH).$$

Thus, $A^*A = AA^*$ if and only if $HK = KH$. $\qquad\square$

Corollary 12.6.9 *Let $A \in M_n$ be symmetric.*

(a) *A is normal if and only if there is a real orthogonal $Q \in M_n(\mathbb{R})$ and a diagonal $D \in M_n$ such that $A = QDQ^T$.*

(b) *A is unitary if and only if there is a real orthogonal $Q \in M_n(\mathbb{R})$ and a unitary diagonal $D = \text{diag}(e^{i\theta_1}, e^{i\theta_2}, \ldots, e^{i\theta_n}) \in M_n$ such that each $\theta_j \in [0, 2\pi)$ and $A = QDQ^\mathsf{T}$. The real numbers θ_j are uniquely determined by A.*

Proof (a) Let $A = H + iK$ be the Cartesian decomposition of A. Since A is symmetric,

$$H^\mathsf{T} + iK^\mathsf{T} = A^\mathsf{T} = A = H + iK,$$

so the uniqueness assertion in Theorem 12.6.4 ensures that $H^\mathsf{T} = H$. But $H = H^* = \overline{H}^\mathsf{T}$, so H is real and Hermitian, that is, it is real symmetric. The same argument shows that K is real symmetric. Now suppose that A is normal. The preceding theorem says that $HK = KH$, so Theorem 12.2.9.b ensures that H and K are simultaneously real orthogonally diagonalizable. Thus, there is a real orthogonal matrix Q and real diagonal matrices Λ and M such that $H = Q\Lambda Q^\mathsf{T}$ and $K = QMQ^\mathsf{T}$. Then $A = H + iK = Q\Lambda Q^\mathsf{T} + iQMQ^\mathsf{T} = Q(\Lambda + iM)Q^\mathsf{T} = QDQ^\mathsf{T}$, in which $D = \Lambda + iM$. Conversely, if D is diagonal, then QDQ^T is symmetric and normal (Lemma 12.1.11).

(b) Since a unitary matrix is normal, it follows from (a) that $A = QDQ^\mathsf{T}$. Then $Q^\mathsf{T}AQ = D$ is unitary since it is a product of unitaries. Since $DD^* = D\overline{D} = I$, its diagonal entries have modulus 1 and hence each entry can be written as $e^{i\theta}$ for a unique $\theta \in [0, 2\pi)$. $\square$

The spectral decomposition of a symmetric unitary matrix in the preceding corollary leads to a basic result about certain matrix square roots.

Lemma 12.6.10 *Let $V \in M_n$ be unitary and symmetric. There is a polynomial p such that $S = p(V)$ is unitary and symmetric, and $S^2 = V$.*

Proof According to the preceding corollary, there is a real orthogonal matrix Q and a diagonal unitary matrix $D = \text{diag}(e^{i\theta_1}, e^{i\theta_2}, \ldots, e^{i\theta_n})$ such that each $\theta_j \in [0, 2\pi)$ and $V = QDQ^\mathsf{T}$. Let $E = \text{diag}(e^{i\theta_1/2}, e^{i\theta_2/2}, \ldots, e^{i\theta_n/2})$. Theorem 0.7.6 ensures that there is a polynomial p such that $p(e^{i\theta_j}) = e^{i\theta_j/2}$ for each $j = 1, 2, \ldots, n$, so $p(D) = E$. The matrix $S = QEQ^\mathsf{T}$ is symmetric and unitary (it is a product of unitary matrices), $S^2 = QE^2Q^\mathsf{T} = QDQ^\mathsf{T} = V$, and

$$S = QEQ^\mathsf{T} = Qp(D)Q^\mathsf{T} = p(QDQ^\mathsf{T}) = p(V). \qquad \square$$

We now apply this lemma to show that any unitary matrix is the product of a real orthogonal matrix and a symmetric unitary matrix.

Theorem 12.6.11 (QS Decomposition of a Unitary Matrix) *Let $U \in M_n$ be unitary. There are unitary $Q, S \in M_n$ and a polynomial p such that $U = QS$, Q is real orthogonal, S is symmetric, and $S = p(U^\mathsf{T}U)$.*

Proof The matrix $U^\mathsf{T}U$ is unitary and symmetric. The preceding lemma ensures that there is a polynomial p such that $S = p(U^\mathsf{T}U)$ is unitary and symmetric, and $S^2 = U^\mathsf{T}U$. Let $Q = US^* = U\overline{S}$. Then Q is unitary and

$$Q^\mathsf{T}Q = S^*U^\mathsf{T}US^* = S^*S^2S^* = (S^*S)(SS^*) = I.$$

Therefore, $Q^\mathsf{T} = Q^{-1} = Q^* = \overline{Q}^\mathsf{T}$, so Q^T (and hence also Q) is real. Consequently, Q is real orthogonal and $U = (US^*)S = QS$. □

The preceding theorem plays a key role in a result about unitary similarity of real matrices.

Corollary 12.6.12 *If two real matrices are unitarily similar, then they are real orthogonally similar.*

Proof Let $A, B \in \mathsf{M}_n(\mathbb{R})$. Let $U \in \mathsf{M}_n$ be unitary and such that $A = UBU^*$. Then

$$UBU^* = A = \overline{A} = \overline{UBU^*} = \overline{U}B\overline{U}^\mathsf{T},$$

so $U^\mathsf{T}UB = BU^\mathsf{T}U$. The preceding theorem ensures that there is a factorization $U = QS$ such that Q is real orthogonal, S is unitary and symmetric, and there is a polynomial p such that $S = p(U^\mathsf{T}U)$. Since B commutes with $U^\mathsf{T}U$, Theorem 0.8.1 ensures that B commutes with S. Compute

$$A = UBU^* = QSBS^*Q^\mathsf{T} = QBSS^*Q^\mathsf{T} = QBQ^\mathsf{T},$$

which is a real orthogonal similarity. □

12.7 Similarity of Normal and Other Diagonalizable Matrices

Theorem 12.7.1 *Suppose that $A, B \in \mathsf{M}_n$ have the same eigenvalues, including multiplicities.*

(a) *If A and B are diagonalizable, then they are similar.*

(b) *If A and B are normal, then they are unitarily similar.*

(c) *If A and B are real and normal, then they are real orthogonally similar.*

(d) *If A and B are diagonal, then they are permutation similar.*

Proof Let $\lambda_1, \lambda_2, \ldots, \lambda_n$ be the eigenvalues of A and B in any order, and let $\Lambda = \mathrm{diag}(\lambda_1, \lambda_2, \ldots, \lambda_n)$.

(a) Corollary 9.4.9 ensures that there are invertible $R, S \in \mathsf{M}_n$ such that $A = R\Lambda R^{-1}$ and $B = S\Lambda S^{-1}$. Then $\Lambda = S^{-1}BS$ and

$$A = R\Lambda R^{-1} = RS^{-1}BSR^{-1} = (RS^{-1})B(RS^{-1})^{-1},$$

so A is similar to B via RS^{-1}.

(b) Theorem 12.2.2 ensures that there is a unitary $U \in \mathsf{M}_n$ and a diagonal $D \in \mathsf{M}_n$ such that $A = UDU^*$. The diagonal entries of D are the eigenvalues of A, but they might not appear in the same order as in the diagonal of Λ. However, there is a permutation matrix P (a real orthogonal matrix) such that $D = P\Lambda P^\mathsf{T}$; see (6.3.4). Therefore,

$$A = UDU^* = UP\Lambda P^\mathsf{T}U^* = (UP)\Lambda(UP)^*,$$

that is, A is unitarily similar to Λ. The same argument shows that B is unitarily similar to Λ. Let V and W be unitary matrices such that $A = V\Lambda V^*$ and $B = W\Lambda W^*$. Then

$$A = V\Lambda V^* = VW^*BWV^* = (VW^*)B(VW^*)^*,$$

so A is unitarily similar to B via VW^*.

(c) Part (b) ensures that A and B are unitarily similar; Corollary 12.6.12 says that they are real orthogonally similar because they are real and unitarily similar.

(d) Since the main diagonal entries of A and B are their eigenvalues, the main diagonal entries of B can be obtained by permuting the main diagonal entries of A. The identity (6.3.4) shows how to construct a permutation similarity of A that effects this permutation of its diagonal entries. $\qquad\square$

12.8 Some Characterizations of Normality

There are many equivalent characterizations of normality. Here are several of them.

Theorem 12.8.1 *Let $A = [a_{ij}] \in \mathsf{M}_n$ have eigenvalues $\lambda_1, \lambda_2, \ldots, \lambda_n$. The following are equivalent:*

(a) *A is normal.*

(b) *$A = H + iK$ in which H, K are Hermitian and $HK = KH$.*

(c) *$\sum_{i,j=1}^{n} |a_{ij}|^2 = \sum_{i=1}^{n} |\lambda_i|^2$.*

(d) *There is a polynomial p such that $p(A) = A^*$.*

(e) *$\|A\mathbf{x}\|_2 = \|A^*\mathbf{x}\|_2$ for all $\mathbf{x} \in \mathbb{C}^n$.*

(f) *A commutes with A^*A.*

(g) *$A^* = AW$ for some unitary W.*

Proof (a) $\Leftrightarrow$ (b) This is Theorem 12.6.8.

(a) $\Leftrightarrow$ (c) This is the case of equality in Theorem 12.3.1.

(a) $\Rightarrow$ (d) Let $A \in \mathsf{M}_n$ be normal and write $A = U\Lambda U^*$, in which $\Lambda = \operatorname{diag}(\lambda_1, \lambda_2, \ldots, \lambda_n)$ and U is unitary. Theorem 0.7.6 provides a polynomial p such that $p(\lambda_i) = \overline{\lambda_i}$ for $i = 1, 2, \ldots, n$. Thus,

$$p(A) = Up(\Lambda)U^* = U \operatorname{diag}\big(p(\lambda_1), p(\lambda_2), \ldots, p(\lambda_n)\big)U^*$$
$$= U \operatorname{diag}(\overline{\lambda_1}, \overline{\lambda_2}, \ldots, \overline{\lambda_n})U^* = U\Lambda^*U^* = (U\Lambda U^*)^* = A^*.$$

(d) $\Rightarrow$ (a) If $A^* = p(A)$, then $A^*A = p(A)A = Ap(A) = AA^*$, so A is normal.

(a) $\Rightarrow$ (e) This is Theorem 12.1.9.b.

(e) $\Rightarrow$ (a) Let $S = A^*A - AA^*$ and let $\mathbf{x} \in \mathbb{C}^n$. Observe that A is normal if and only if $S = 0$. Since $\|A\mathbf{x}\|_2 = \|A^*\mathbf{x}\|_2$, we have

$$0 = \|A\mathbf{x}\|_2^2 - \|A^*\mathbf{x}\|_2^2 = \langle A\mathbf{x}, A\mathbf{x}\rangle - \langle A^*\mathbf{x}, A^*\mathbf{x}\rangle$$
$$= \langle A^*A\mathbf{x}, \mathbf{x}\rangle - \langle AA^*\mathbf{x}, \mathbf{x}\rangle = \langle S\mathbf{x}, \mathbf{x}\rangle.$$

Because $S = S^*$, it follows that $\langle S^2\mathbf{x}, \mathbf{x}\rangle = \langle S\mathbf{x}, S\mathbf{x}\rangle = \|S\mathbf{x}\|_2^2$ and

$$0 = \langle S(\mathbf{x} + S\mathbf{x}), \mathbf{x} + S\mathbf{x}\rangle = \langle S\mathbf{x}, \mathbf{x}\rangle + \langle S\mathbf{x}, S\mathbf{x}\rangle + \langle S^2\mathbf{x}, \mathbf{x}\rangle + \langle S(S\mathbf{x}), S\mathbf{x}\rangle$$
$$= 0 + \|S\mathbf{x}\|_2^2 + \|S\mathbf{x}\|_2^2 + 0 = 2\|S\mathbf{x}\|_2^2.$$

We conclude that $S\mathbf{x} = \mathbf{0}$ for all $\mathbf{x} \in \mathbb{C}^n$, so $S = 0$ and A is normal.

(a) $\Rightarrow$ (f) If A is normal, then $A(A^*A) = (AA^*)A = (A^*A)A$.

(f) $\Rightarrow$ (a) Suppose that A commutes with A^*A. Then $(A^*A)^2 = A^*AA^*A = (A^*)^2A^2$ and $(AA^*)^2 = AA^*AA^* = A^*AAA^*$. Let $S = A^*A - AA^*$. Since $S = S^*$,

$$\operatorname{tr} S^*S = \operatorname{tr} S^2 = \operatorname{tr}(A^*A - AA^*)^2$$
$$= \operatorname{tr}(A^*A)^2 - \operatorname{tr} A^*AAA^* - \operatorname{tr} AA^*A^*A + \operatorname{tr}(AA^*)^2$$
$$= \operatorname{tr}(A^*)^2A^2 - \operatorname{tr} A^*A^*AA - \operatorname{tr} A^*AAA^* + \operatorname{tr} A^*AAA^*$$
$$= 0.$$

(g) $\Rightarrow$ (a) If $A^* = AW$ for some unitary W, then $A^*A = A^*(A^*)^* = (AW)(AW)^* = AWW^*A^* = AA^*$.

(a) $\Rightarrow$ (g) Let A be normal and write $A = U\Lambda U^*$, in which $\Lambda = \operatorname{diag}(\lambda_1, \lambda_2, \ldots, \lambda_n)$ and U is unitary. Define $X = \operatorname{diag}(\xi_1, \xi_2, \ldots, \xi_n)$, in which

$$\xi_i = \begin{cases} \overline{\lambda_i}/\lambda_i & \text{if } \lambda_i \neq 0, \\ 1 & \text{if } \lambda_i = 0, \end{cases}$$

and let $W = U\operatorname{diag}(\xi_1, \xi_2, \ldots, \xi_n)U^*$, which is unitary (Theorem 12.6.1.a). Then $AW = (U\Lambda U^*)(UXU^*) = U\Lambda XU^* = U\operatorname{diag}(\overline{\lambda_1}, \overline{\lambda_2}, \ldots, \overline{\lambda_n})U^* = (U\Lambda U^*)^* = A^*$. $\square$

12.9 Spectral Resolutions

We present another version of the spectral theorem based upon orthogonal projections. This version of the spectral theorem is used in quantum computing and it generalizes readily to the infinite-dimensional setting.

If $P \in \mathsf{M}_n$ is an orthogonal projection ($P = P^*$ and $P^2 = P$), then $\mathbb{C}^n = \operatorname{col} P \oplus \operatorname{null} P$ is an orthogonal decomposition and the induced linear transformation $T_P : \mathbb{C}^n \to \mathbb{C}^n$ is the orthogonal projection onto $\operatorname{col} P$.

Definition 12.9.1 Nonzero orthogonal projections $P_1, P_2, \ldots, P_d \in \mathsf{M}_n$ are a *resolution of the identity* if $P_1 + P_2 + \cdots + P_d = I$.

If $P_1, P_2, \ldots, P_d \in \mathsf{M}_n$ are a resolution of the identity, then $\mathbf{x} = P_1\mathbf{x} + P_2\mathbf{x} + \cdots + P_d\mathbf{x}$ for all $\mathbf{x} \in \mathbb{C}^n$. Thus,

$$\mathbb{C}^n = \operatorname{col} P_1 + \cdots + \operatorname{col} P_d. \tag{12.9.2}$$

The following lemma tells us that (12.9.2) is actually a direct sum of pairwise orthogonal subspaces.

Lemma 12.9.3 *Suppose that* $P_1, P_2, \ldots, P_d \in \mathsf{M}_n$ *are a resolution of the identity. Then* $P_iP_j = 0$ *for* $i \neq j$. *In particular,* $\operatorname{col} P_i \perp \operatorname{col} P_j$ *for* $i \neq j$ *and*

$$\mathbb{C}^n = \operatorname{col} P_1 \oplus \cdots \oplus \operatorname{col} P_d$$

is an orthogonal direct sum.

Proof For any $j \in \{1, 2, \ldots, d\}$,

$$0 = P_j - P_j = P_j - P_j^2 = (I - P_j)P_j = \left(\sum_{i \neq j} P_i \right) P_j = \sum_{i \neq j} P_i P_j = \sum_{i \neq j} P_i^2 P_j^2. \quad (12.9.4)$$

Take the trace of both sides of (12.9.4) and obtain the identity

$$0 = \sum_{i \neq j} \operatorname{tr} P_i^2 P_j^2 = \sum_{i \neq j} \operatorname{tr} P_j P_i P_i P_j$$

$$= \sum_{i \neq j} \operatorname{tr} \left((P_i P_j)^* (P_i P_j) \right) = \sum_{i \neq j} \| P_i P_j \|_F^2. \quad (12.9.5)$$

Thus, $\| P_i P_j \|_F = 0$ for all $i \neq j$, and hence $P_i P_j = 0$. If $i \neq j$, then

$$\operatorname{col} P_j \subseteq \operatorname{null} P_i = (\operatorname{col} P_i)^\perp,$$

and hence $\operatorname{col} P_i \perp \operatorname{col} P_j$. $\qquad \square$

The following lemma provides a method to produce resolutions of the identity. In fact, all resolutions of the identity arise in this manner; see P.12.21.

Lemma 12.9.6 *Let $U = [U_1 \ U_2 \ \ldots \ U_d] \in \mathsf{M}_n$ be a unitary matrix, in which each $U_i \in \mathsf{M}_{n \times n_i}$ and $n_1 + n_2 + \cdots + n_d = n$. Then $P_i = U_i U_i^*$ for $i = 1, 2, \ldots, d$ are orthogonal projections that form a resolution of the identity.*

Proof We have

$$I = UU^* = [U_1 \ U_2 \ \ldots \ U_d] \begin{bmatrix} U_1^* \\ U_2^* \\ \vdots \\ U_d^* \end{bmatrix} = U_1 U_1^* + U_2 U_2^* + \cdots + U_d U_d^*,$$

in which each $P_i = U_i U_i^*$ is an orthogonal projection (see Example 7.3.5). $\qquad \square$

Example 12.9.7 Consider the unitary matrix $U = [U_1 \ U_2] \in \mathsf{M}_3$, in which

$$U = \frac{1}{3} \begin{bmatrix} 1 & -2 & -2 \\ 2 & -1 & 2 \\ 2 & 2 & -1 \end{bmatrix}, \qquad U_1 = \frac{1}{3} \begin{bmatrix} 1 & -2 \\ 2 & -1 \\ 2 & 2 \end{bmatrix}, \quad \text{and} \quad U_2 = \frac{1}{3} \begin{bmatrix} -2 \\ 2 \\ -1 \end{bmatrix}.$$

Then

$$P_1 = U_1 U_1^* = \frac{1}{9} \begin{bmatrix} 5 & 4 & -2 \\ 4 & 5 & 2 \\ -2 & 2 & 8 \end{bmatrix} \quad \text{and} \quad P_2 = U_2 U_2^* = \frac{1}{9} \begin{bmatrix} 4 & -4 & 2 \\ -4 & 4 & -2 \\ 2 & -2 & 1 \end{bmatrix}$$

are orthogonal projections that form a resolution of the identity; P_1 is the orthogonal projection onto $\operatorname{span}\{[1 \ 2 \ 2]^\mathsf{T}, [-2 \ -1 \ 2]^\mathsf{T}\}$ and P_2 is the orthogonal projection onto $\operatorname{span}\{[-2 \ 2 \ -1]^\mathsf{T}\}$.

Theorem 12.9.8 (Spectral Theorem, Version II) *Let the distinct eigenvalues of $A \in \mathsf{M}_n$ be $\mu_1, \mu_2, \ldots, \mu_d$ with multiplicities $n_1, n_2, \ldots, n_d$. The following are equivalent:*

(a) *A is normal.*

(b) $A = \sum_{i=1}^{d} \mu_i P_i$, *in which* $P_1, P_2, \ldots, P_d \in \mathsf{M}_n$ *are orthogonal projections that are a resolution of the identity.*

In (b), $\operatorname{col} P_j = \mathcal{E}_{\mu_j}(A)$.

Proof (a) $\Rightarrow$ (b) Suppose that A is normal. Theorem 12.2.2 ensures that $A = U\Lambda U^*$, in which $\Lambda = \mu_1 I_{n_1} \oplus \cdots \oplus \mu_d I_{n_d}$. Write $U = [U_1 \ \ldots \ U_d]$, in which each $U_i \in \mathsf{M}_{n \times n_i}$. Then

$$A = U\Lambda U^* = [U_1 \ \ldots \ U_d] \begin{bmatrix} \mu_1 I_{n_1} & & \\ & \ddots & 0 \\ & & \mu_d I_{n_d} \end{bmatrix} \begin{bmatrix} U_1^* \\ \vdots \\ U_d^* \end{bmatrix}$$

$$= [\mu_1 U_1 \ \ldots \ \mu_d U_d] \begin{bmatrix} U_1^* \\ \vdots \\ U_d^* \end{bmatrix} = \mu_1 U_1 U_1^* + \cdots + \mu_d U_d U_d^*$$

$$= \mu_1 P_1 + \cdots + \mu_d P_d.$$

Lemma 12.9.6 ensures that the matrices $P_i = U_i U_i^*$ are a resolution of the identity.

(b) $\Rightarrow$ (a) Suppose that $A = \sum_{i=1}^{d} \mu_i P_i$, in which $P_1, P_2, \ldots, P_d \in \mathsf{M}_n$ are a resolution of the identity. Since $P_i P_j = 0$ for $i \neq j$ (Lemma 12.9.3) and $A^* = \sum_{j=1}^{d} \overline{\mu_j} P_j$, we have

$$A^*A = \sum_{i,j=1}^{d} \overline{\mu_j} \mu_i P_j P_i = \sum_{i=1}^{d} |\mu_i|^2 P_i = \sum_{i,j=1}^{d} \mu_i \overline{\mu_j} P_i P_j = AA^*.$$

We claim that $\mathcal{E}_{\mu_j}(A) = \operatorname{col} P_j$. For each $j \in \{1, 2, \ldots, d\}$ and any $\mathbf{x} \in \mathbb{C}^n$,

$$(A - \mu_j I)\mathbf{x} = \sum_{i=1}^{n} \mu_i P_i \mathbf{x} - \sum_{i=1}^{n} \mu_j P_i \mathbf{x} = \sum_{i \neq j} (\mu_i - \mu_j) P_i \mathbf{x}.$$

Since $P_i \mathbf{x} \perp P_j \mathbf{x}$ if $i \neq j$, the Pythagorean theorem ensures that

$$\|(A - \mu_j I)\mathbf{x}\|_2^2 = \left\| \sum_{i \neq j} (\mu_i - \mu_j) P_i \mathbf{x} \right\|_2^2 = \sum_{i \neq j} |\mu_i - \mu_j|^2 \|P_i \mathbf{x}\|_2^2,$$

in which $\mu_i - \mu_j \neq 0$ for $i \neq j$. Therefore, $\mathbf{x} \in \mathcal{E}_{\mu_j}(A)$ if and only if $P_i \mathbf{x} = \mathbf{0}$ for all $i \neq j$. Since $\mathbf{x} = \sum_{i=1}^{d} P_i \mathbf{x}$, this occurs if and only if $\mathbf{x} = P_j \mathbf{x}$. Thus, $\mathcal{E}_{\mu_j}(A) = \operatorname{col} P_j$. $\qquad\square$

Definition 12.9.9 The orthogonal projections $P_1, P_2, \ldots, P_d$ in the preceding theorem are the *spectral projections of A*. The representation $A = \sum_{i=1}^{d} \mu_i P_i$ is the *spectral resolution of A*.

Theorem 12.9.8 says that a matrix is normal if and only if it is a linear combination of orthogonal projections that are a resolution of the identity. The latter condition implies that the orthogonal projections must be mutually orthogonal (Lemma 12.9.3).

Example 12.9.10 The spectral resolution of an orthogonal projection $P \in \mathsf{M}_n$ is

$$P = 1P + 0(I - P).$$

Example 12.9.11 Consider the matrix A in (12.1.7) and Example 12.2.3. The orthogonal projections P_1, P_2 onto the eigenspaces $\mathcal{E}_{\lambda_1}(A) = \text{span}\{[\frac{1}{\sqrt{2}} \ -\frac{i}{\sqrt{2}}]^\mathsf{T}\}$ and $\mathcal{E}_{\lambda_2}(A) = \text{span}\{[\frac{1}{\sqrt{2}} \ \frac{i}{\sqrt{2}}]^\mathsf{T}\}$ are

$$P_1 = \begin{bmatrix} \frac{1}{\sqrt{2}} \\ -\frac{i}{\sqrt{2}} \end{bmatrix} [\frac{1}{\sqrt{2}} \ \frac{i}{\sqrt{2}}] = \begin{bmatrix} \frac{1}{2} & \frac{i}{2} \\ -\frac{i}{2} & \frac{1}{2} \end{bmatrix}$$

and

$$P_2 = \begin{bmatrix} \frac{1}{\sqrt{2}} \\ \frac{i}{\sqrt{2}} \end{bmatrix} [\frac{1}{\sqrt{2}} \ -\frac{i}{\sqrt{2}}] = \begin{bmatrix} \frac{1}{2} & -\frac{i}{2} \\ \frac{i}{2} & \frac{1}{2} \end{bmatrix}.$$

That $\mathcal{E}_1(A) \perp \mathcal{E}_2(A)$ and $\mathbb{C}^2 = \mathcal{E}_1(A) \oplus \mathcal{E}_2(A)$ is reflected in the equation $P_1 + P_2 = I$:

$$\begin{bmatrix} \frac{1}{2} & \frac{i}{2} \\ -\frac{i}{2} & \frac{1}{2} \end{bmatrix} + \begin{bmatrix} \frac{1}{2} & -\frac{i}{2} \\ \frac{i}{2} & \frac{1}{2} \end{bmatrix} = \begin{bmatrix} 1 & 0 \\ 0 & 1 \end{bmatrix}.$$

The spectral theorem ensures that $A = \lambda_1 P_1 + \lambda_2 P_2$:

$$\begin{bmatrix} a & -b \\ b & a \end{bmatrix} = (a + bi) \begin{bmatrix} \frac{1}{2} & \frac{i}{2} \\ -\frac{i}{2} & \frac{1}{2} \end{bmatrix} + (a - bi) \begin{bmatrix} \frac{1}{2} & -\frac{i}{2} \\ \frac{i}{2} & \frac{1}{2} \end{bmatrix}.$$

Theorem 12.9.12 Let $A = \sum_{i=1}^{d} \mu_i P_i$ be the spectral resolution of a normal matrix $A \in \mathsf{M}_n$ and let f be a polynomial. Then $f(A) = \sum_{i=1}^{d} f(\mu_i) P_i$.

Proof It suffices to show that $A^\ell = \sum_{i=1}^{d} \mu_i^\ell P_i$ for $\ell = 0, 1, \ldots$. We proceed by induction on ℓ. The base case $\ell = 0$ is $I = P_1 + \cdots + P_d$, which is the definition of a resolution of the identity. If $A^\ell = \sum_{i=1}^{d} \mu_i^\ell P_i$ is valid for some ℓ, then

$$A^{\ell+1} = AA^\ell = \left(\sum_{i=1}^{d} \mu_i P_i \right) \left(\sum_{j=1}^{d} \mu_j^\ell P_j \right)$$

$$= \sum_{i,j=1}^{d} \mu_i \mu_j^\ell P_i P_j = \sum_{i=1}^{d} \mu_i^{\ell+1} P_i^2$$

$$= \sum_{i=1}^{d} \mu_i^{\ell+1} P_i,$$

since $P_i P_j = 0$ for $i \neq j$ (Lemma 12.9.3). This completes the induction. $\square$

Corollary 12.9.13 Let $A = \sum_{i=1}^{d} \mu_i P_i$ be the spectral resolution of a normal matrix $A \in \mathsf{M}_n$. There are polynomials $p_1, p_2, \ldots, p_d$ such that $p_j(A) = P_j$ for each $j \in \{1, 2, \ldots, d\}$.

Proof Theorem 0.7.6 ensures that there are unique polynomials p_j of degree at most $d - 1$ such that $p_j(\mu_i) = \delta_{ij}$ for $i, j \in \{1, 2, \ldots, d\}$. Therefore, $p_j(A) = \sum_{i=1}^{d} p_j(\mu_i) P_i = P_j$. $\square$

Example 12.9.14 The matrix (12.2.6) has eigenvalues $\mu_1 = 9$ and $\mu_2 = -9$, with multiplicities $n_1 = 2$ and $n_2 = 1$, respectively. The polynomials $p_1(z) = \frac{1}{18}(z+9)$ and $p_2(z) = -\frac{1}{18}(z-9)$ satisfy $p_j(\mu_i) = \delta_{ij}$ for $i, j \in \{1, 2\}$. Let $P_1 = p_1(A)$ and $P_2 = p_2(A)$. The spectral resolution of A is

$$
\begin{bmatrix} 1 & 8 & -4 \\ 8 & 1 & 4 \\ -4 & 4 & 7 \end{bmatrix} = 9 \underbrace{\begin{bmatrix} \frac{5}{9} & \frac{4}{9} & -\frac{2}{9} \\ \frac{4}{9} & \frac{5}{9} & \frac{2}{9} \\ -\frac{2}{9} & \frac{2}{9} & \frac{8}{9} \end{bmatrix}}_{P_1} - 9 \underbrace{\begin{bmatrix} \frac{4}{9} & -\frac{4}{9} & \frac{2}{9} \\ -\frac{4}{9} & \frac{4}{9} & -\frac{2}{9} \\ \frac{2}{9} & -\frac{2}{9} & \frac{1}{9} \end{bmatrix}}_{P_2}.
$$

If $A \in M_n$ is normal, the geometric content of Theorem 12.1.9.c is that each one-dimensional invariant subspace of A is also an invariant subspace of A^*. This property of normal matrices generalizes to invariant subspaces of any dimension.

Theorem 12.9.15 *Let $A \in M_n$ be normal.*

(a) *Every A-invariant subspace is A^*-invariant.*

(b) *If P is the orthogonal projection onto an A-invariant subspace, then $AP = PA$.*

Proof (a) There is nothing to prove if the A-invariant subspace is $\{0\}$ or $\mathbb{C}^n$. Suppose that $\mathcal{U} \subseteq \mathbb{C}^n$ is a k-dimensional A-invariant subspace with $1 \le k \le n-1$. Let $U = [U_1 \ U_2] \in M_n$ be a unitary matrix, in which the columns of $U_1 \in M_{n \times k}$ are a basis for $\mathcal{U}$ and the columns of $U_2 \in M_{n \times (n-k)}$ are a basis for $\mathcal{U}^\perp$. Since $\mathcal{U}$ is A-invariant, Theorem 7.6.7 ensures that

$$
U^*AU = \begin{bmatrix} B & X \\ 0 & C \end{bmatrix},
$$

in which $B \in M_k$ and $C \in M_{n-k}$. Lemma 12.1.11 ensures that U^*AU is normal. Lemma 12.1.12 tells us that B and C are normal and $X = 0$. Thus, $U^*AU = B \oplus C$ and $\mathcal{U}$ is A^*-invariant by Corollary 7.6.8.

(b) This follows from (a) and Corollary 7.6.8.   □

Example 12.9.16 Let

$$
A = \begin{bmatrix} 0 & 1 \\ 0 & 0 \end{bmatrix},
$$

which is not normal. The subspace span$\{e_1\}$ is an A-invariant subspace that is not A^*-invariant, and span$\{e_2\}$ is an A^*-invariant subspace that is not A-invariant.

For another proof of Theorem 12.9.15, see P.12.48. Complications arise in the infinite-dimensional setting; see P.12.47.

12.10 Problems

P.12.1 Can the product of two nonzero non-normal matrices be normal?

P.12.2 Let $B \in M_n$ be invertible and let $A = B^{-1}B^*$. Show that A is unitary if and only if B is normal.

P.12.3 Let $A \in M_n$. If there is a nonzero polynomial p such that $p(A)$ is normal, does it follow that A is normal?

P.12.4 Let $A, B \in M_n$ be normal. Prove that A and B are similar if and only if they are unitarily similar.

P.12.5 Prove that $A \in M_n$ is normal if and only if every eigenvector of A is an eigenvector of A^*. *Hint*: If $\mathbf{x}$ is a unit eigenvector of both A and A^*, then $\mathbf{x}^*AV_2 = \mathbf{0}^\mathsf{T}$ in (10.1.2).

P.12.6 (a) Give an example of a real orthogonal matrix with some non-real eigenvalues. (b) If two real orthogonal matrices are similar (perhaps via a non-real similarity) prove that they are real orthogonally similar.

P.12.7 Let $A, B \in M_n$. (a) If A and B are normal, show that $AB = 0$ if and only if $BA = 0$. (b) If A is normal and B is not, does $AB = 0$ imply that $BA = 0$?

P.12.8 If $A \in M_{m \times n}$ and A^*A is an orthogonal projection, is AA^* an orthogonal projection?

P.12.9 Let $U, V \in M_n$ be unitary. Show that $|\det(U+V)| \le 2^n$. Find conditions for equality.

P.12.10 (a) Let $A \in M_n$ be normal. Use Theorem 12.1.9.c to show that eigenvectors of A that correspond to distinct eigenvalues are orthogonal. (b) Suppose that $A \in M_n$ is diagonalizable. If eigenvectors of A that correspond to distinct eigenvalues are orthogonal, show that A is normal. (c) Show that the hypothesis of diagonalizability cannot be omitted in (b).

P.12.11 (a) If $B \in M_n$, show that

$$A = \begin{bmatrix} B & B^* \\ B^* & B \end{bmatrix}$$

is normal. (b) Let $B = H + iK$ be a Cartesian decomposition (12.6.5) and let

$$U = \frac{1}{\sqrt{2}} \begin{bmatrix} I_n & I_n \\ -I_n & I_n \end{bmatrix}.$$

Show that U is unitary and $UAU^* = 2H \oplus 2iK$. (c) Show that A has n real eigenvalues and n purely imaginary eigenvalues. (d) Show that $\det A = (4i)^n (\det H)(\det K)$. (e) If B is normal and has eigenvalues $\lambda_1, \lambda_2, \ldots, \lambda_n$, show that $\det A = (4i)^n \prod_{j=1}^n (\operatorname{Re} \lambda_j)(\operatorname{Im} \lambda_j)$.

P.12.12 If $A \in M_n$ is normal and nilpotent, prove that $A = 0$.

P.12.13 If $A \in M_n$ is normal and $\operatorname{spec} A = \{1\}$, prove that $A = I$.

P.12.14 If $A \in M_n$ is normal and $A^2 = A$, prove that A is an orthogonal projection. Can the hypothesis of normality be omitted?

P.12.15 (a) Show that $A \in M_n$ is Hermitian if and only if $\operatorname{tr} A^2 = \operatorname{tr} A^*A$. (b) Show that Hermitian matrices $A, B \in M_n$ commute if and only if $\operatorname{tr}(AB)^2 = \operatorname{tr} A^2B^2$.

P.12.16 Let $A \in M_m$ and $B \in M_n$ be normal and write $A = U \Lambda U^*$ and $B = VMV^*$ in which $U \in M_m$ and $V \in M_n$ are unitary, $\Lambda = \operatorname{diag}(\lambda_1, \lambda_2, \ldots, \lambda_m)$ and $M = \operatorname{diag}(\mu_1, \mu_2, \ldots, \mu_n)$. Suppose that $X \in M_{m \times n}$ and $AX = XB$. (a) Show that $\Lambda(U^*XV) = (U^*XV)M$. (b) Let $U^*XV = [\xi_{ij}]$ and show that $\overline{\lambda_i}\xi_{ij} = \xi_{ij}\overline{\mu_j} = 0$ for all i, j. (c) Deduce that $A^*X = XB^*$. This provides another proof of the Fuglede–Putnam theorem.

P.12.17 Let $N \in M_{m+n}$ be normal with spectral resolution $N = \sum_{i=1}^d \mu_i P_i$ and let $Y \in M_{m+n}$. (a) If $NY = YN$, show that $P_iY = YP_i$ for $i = 1, 2, \ldots, d$. (b) Show that $N^*Y = YN^*$. (c) Now suppose that $A \in M_m$ and $B \in M_n$ are normal, $X \in M_{m \times n}$, and $AX = XB$. Use the matrices

$$N = \begin{bmatrix} A & 0 \\ 0 & B \end{bmatrix} \quad \text{and} \quad Y = \begin{bmatrix} 0 & X \\ 0 & 0 \end{bmatrix}$$

to deduce the Fuglede–Putnam theorem.

P.12.18 In the statement of the Fuglede–Putnam theorem, do we have to assume that both A and B are normal? Consider

$$A = X = \begin{bmatrix} 0 & 1 \\ 0 & 0 \end{bmatrix} \quad \text{and} \quad B = \begin{bmatrix} 1 & 0 \\ 0 & 0 \end{bmatrix}.$$

P.12.19 Let $A \in \mathsf{M}_n$ be normal. Show that $A\bar{A} = \bar{A}A$ if and only if $AA^\mathsf{T} = A^\mathsf{T}A$.

P.12.20 Let $A, B \in \mathsf{M}_n$ be normal and suppose that $AB = BA$. Use the Fuglede–Putnam theorem to prove Corollary 12.2.10. *Hint*: First show that $A^*B = BA^*$ and $AB^* = B^*A$.

P.12.21 Suppose that $P_1, P_2, \ldots, P_d \in \mathsf{M}_n$ are a resolution of the identity. Prove that there exists a partitioned unitary matrix $U = [U_1 \ U_2 \ \cdots \ U_d] \in \mathsf{M}_n$ such that $P_i = U_i U_i^*$ for $i = 1, 2, \ldots, d$. Conclude that every resolution of the identity arises in the manner described in Lemma 12.9.6.

P.12.22 Find all possible values of $\dim\{A\}'$ (a) if $A \in \mathsf{M}_5$ is normal, or (b) if $A \in \mathsf{M}_6$ is normal.

P.12.23 Let $A = \begin{bmatrix} a & b \\ c & d \end{bmatrix} \in \mathsf{M}_2(\mathbb{R})$. Show that A is normal if and only if either A is symmetric or $c = -b$ and $a = d$.

P.12.24 Let $A = \begin{bmatrix} a & -b \\ b & a \end{bmatrix}$, in which $a, b \in \mathbb{C}$. Show that $\{A\}' = \mathsf{M}_2$ if $b = 0$ and

$$\{A\}' = \left\{ \begin{bmatrix} z & -w \\ w & z \end{bmatrix} : z, w \in \mathbb{C} \right\}$$

if $b \neq 0$.

P.12.25 Consider $A = \begin{bmatrix} a & -b \\ b & a \end{bmatrix}$, in which $a, b \in \mathbb{R}$. The eigenvalues of A are $\lambda_1 = a + bi$ and $\lambda_2 = a - bi$. If p is a real polynomial, show that $\overline{p(\lambda_1)} = p(\bar{\lambda_1}) = p(\lambda_2)$ and

$$p(A) = \begin{bmatrix} \mathrm{Re}\, p(\lambda_1) & -\mathrm{Im}\, p(\lambda_1) \\ \mathrm{Im}\, p(\lambda_1) & \mathrm{Re}\, p(\lambda_1) \end{bmatrix}.$$

P.12.26 Consider the *Volterra operator*

$$(Tf)(t) = \int_0^t f(s)\, ds$$

on $C[0, 1]$ (see P.5.11 and P.8.27). Write $T = H + iK$, in which $H = \frac{1}{2}(T + T^*)$ and $K = \frac{1}{2i}(T - T^*)$. (a) Show that $2H$ is the orthogonal projection onto the subspace of constant functions. (b) Compute the eigenvalues and eigenvectors of K. *Hint*: Use the fundamental theorem of calculus. Be sure that your prospective eigenvectors satisfy your original integral equation. (c) Verify that the eigenvalues of K are real and that the eigenvectors corresponding to distinct eigenvalues are orthogonal.

P.12.27 Let $A \in \mathsf{M}_n$ and suppose that $-1 \notin \mathrm{spec}\, A$. The *Cayley transform* of A is the matrix $(I - A)(I + A)^{-1}$. (a) If A is skew Hermitian, show that its Cayley transform is unitary. (b) If A is unitary, show that its Cayley transform is skew Hermitian.

P.12.28 Let $A = [a_{ij}] \in M_n$ be normal. (a) Examine the diagonal entries of the identity $AA^* = A^*A$ and show that, for each $i = 1, 2, \ldots, n$, the Euclidean norms of the ith row and ith column of A are equal. (b) If $n = 2$, show that $|a_{21}| = |a_{12}|$. What can you say about the four matrices in Example 12.1.8? (c) If A is tridiagonal, show that $|a_{i,i+1}| = |a_{i+1,i}|$ for each $i = 1, 2, \ldots, n$.

P.12.29 Let $A \in M_n$ be symmetric, let $B = \mathrm{Re}\, A$ and $C = \mathrm{Im}\, A$. Show that (a) B and C are real symmetric; (b) A is normal if and only if $BC = CB$.

P.12.30 Both of the matrices $\begin{bmatrix} 1 & i \\ i & 1 \end{bmatrix}$ and $\begin{bmatrix} i & i \\ i & -1 \end{bmatrix}$ are symmetric, but only one is normal. Which one?

P.12.31 Show that the following are equivalent: (a) A is normal; (b) A^* is normal; (c) $\overline{A}$ is normal; (d) A^T is normal.

P.12.32 If $A \in M_n$ is normal and has distinct eigenvalues, prove that $\{A\}'$ contains only normal matrices.

P.12.33 Let $A \in M_n$ be a Markov matrix. If A is normal, show that A^T is a Markov matrix.

P.12.34 Let $A = [a_{ij}] \in M_n$ be normal. (a) Show that $\mathrm{spec}\, A$ lies on a line in the complex plane if and only if $A = cI + e^{i\theta}H$, in which $c \in \mathbb{C}$, $\theta \in [0, 2\pi)$, and $H \in M_n$ is Hermitian. (b) Show that $\mathrm{spec}\, A$ lies on a circle in the complex plane if and only if $A = cI + rU$, in which $c \in \mathbb{C}$, $r \in [0, \infty)$, and $U \in M_n$ is unitary. (c) Let $n = 2$. Show that $\mathrm{spec}\, A$ lies on a line and $A = cI + e^{i\theta}H$ as in (a), and deduce that $|a_{21}| = |a_{12}|$. If $a_{12} \neq 0$, show that $(a_{11} - a_{22})^2/a_{12}a_{21}$ is real and nonnegative. (d) Let $n = 3$. Show that $A = cI + rU$ as in (b).

P.12.35 Show that the matrices A and B in Example 12.2.11 satisfy the criteria in Corollary 12.4.3 for AB and BA to be normal.

P.12.36 Suppose that $A = [a_{ij}] \in M_n$ is normal. (a) If $a_{11}, a_{22}, \ldots, a_{nn}$ are the eigenvalues of A (including multiplicities), show that A is diagonal. (b) If $n = 2$ and a_{11} is an eigenvalue of A, show that A is diagonal. (c) Give an example to show that the assertion in (b) need not be correct if A is not normal.

P.12.37 Let $A, B \in M_n$ be Hermitian. Prove that $\mathrm{rank}(AB)^k = \mathrm{rank}(BA)^k$ for all $k = 1, 2, \ldots$. Deduce from Corollary 11.9.5 that AB is similar to BA.

P.12.38 Let $A, B \in M_n$ be normal. Let $r_1 = \mathrm{rank}\, A$ and $r_2 = \mathrm{rank}\, B$. Let $A = U\Lambda U^*$ and $B = VMV^*$, in which U and V are unitary, $\Lambda = \Lambda_1 \oplus 0_{n-r_1}$ and $M = M_1 \oplus 0_{n-r_2}$ are diagonal, and Λ_1 and M_1 are invertible. Let $W = U^*V$ and let W_{11} denote the $r_1 \times r_2$ upper-left corner of W. (a) Show that $\mathrm{rank}\, AB = \mathrm{rank}\, W_{11}$ and $\mathrm{rank}\, BA = \mathrm{rank}\, W_{11}^*$. (b) Deduce that $\mathrm{rank}\, AB = \mathrm{rank}\, BA$. See P.13.37 for a different proof.

P.12.39 Show that the matrices A and B in Example 11.9.6 are normal. Although AB and BA are not similar, observe that $\mathrm{rank}\, AB = \mathrm{rank}\, BA$, as shown in the preceding problem.

P.12.40 Let $A \in M_n$. Show that the following three statements are equivalent:

(a) $\mathrm{col}\, A = \mathrm{col}\, A^*$.

(b) $\mathrm{null}\, A = \mathrm{null}\, A^*$.

(c) There is a unitary $U \in M_n$ such that $A = U(B \oplus 0_{n-r})U^*$, in which $r = \mathrm{rank}\, A$ and $B \in M_r$ is invertible.

A matrix that satisfies any one of these conditions is an *EP matrix*. Why is every normal matrix an EP matrix?

P.12.41 Let $A, B \in \mathsf{M}_n$ be EP matrices (see the preceding problem). Use the argument in P.12.38 to show that rank AB = rank BA.

P.12.42 Why do we assume that each $\theta_j \in [0, 2\pi)$ in the statement of Lemma 12.6.10? Is the lemma correct if this assumption is omitted? Discuss the example $V = I_2 = \mathrm{diag}(e^{i\theta_1}, e^{i\theta_2})$ with $\theta_1 = 0$ and $\theta_2 = 2\pi$.

P.12.43 The $n \times n$ circulant matrices are a subspace of M_n. What is its dimension?

P.12.44 Let $\lambda_1, \lambda_2, \ldots, \lambda_n \in \mathbb{R}$ and suppose that $\lambda_1 + \lambda_2 + \cdots + \lambda_n = 0$. Show that there is an $n \times n$ Hermitian matrix with zero main diagonal and eigenvalues $\lambda_1, \lambda_2, \ldots, \lambda_n$.

P.12.45 Let $A \in \mathsf{M}_n$. (a) Suppose that $A + I$ is unitary. Sketch the circle Γ in the complex plane that must contain every eigenvalue of A. (b) Suppose that $A + I$ and $A^2 + I$ are unitary. What portion of Γ cannot contain any eigenvalue of A? (c) If $A + I$, $A^2 + I$, and $A^3 + I$ are unitary, show that $A = 0$.

P.12.46 Let $A \in \mathsf{M}_n$ be normal and let $\mathcal{U}$ be a k-dimensional A-invariant subspace with $1 \leq k \leq n - 1$. Prove the following generalization of Theorem 12.1.9.c. There is a $V \in \mathsf{M}_{n \times k}$ and a diagonal $\Lambda \in \mathsf{M}_k$ such that the columns of V are an orthonormal basis for $\mathcal{U}$, $AV = V\Lambda$, and $A^*V = V\overline{\Lambda}$.

P.12.47 Let $\mathcal{V}$ denote the inner product space of finitely nonzero bilateral complex sequences $\mathbf{a} = (\ldots, a_{-1}, \underline{a_0}, a_1, \ldots)$, in which the underline denotes the zeroth position. The inner product on $\mathcal{V}$ is $\langle \mathbf{a}, \mathbf{b} \rangle = \sum_{n=-\infty}^{\infty} a_n \overline{b_n}$. Let $U \in \mathcal{L}(\mathcal{V})$ be the *right-shift operator* $U(\ldots, a_{-1}, \underline{a_0}, a_1, \ldots) = (\ldots, a_{-2}, \underline{a_{-1}}, a_0, \ldots)$. (a) Compute U^*, the adjoint of U, and show that $U^*U = UU^*$. (b) Show that there is a subspace $\mathcal{U}$ of $\mathcal{V}$ that is invariant under U, but not under U^*. Compare with Theorem 12.9.15.

P.12.48 Suppose that $A \in \mathsf{M}_n$ is normal and $\mathcal{U}$ is A-invariant. Let P denote the orthogonal projection onto $\mathcal{U}$. (a) Show that $\mathrm{tr}(AP - PA)^*(AP - PA) = 0$. (b) Conclude that $AP = PA$. (c) Deduce Theorem 12.9.15 from (b).

P.12.49 Let $A \in \mathsf{M}_m$ and $B \in \mathsf{M}_n$. If $A \oplus B$ is normal, show that A and B are normal.

P.12.50 Let $A \in \mathsf{M}_m$ and $B \in \mathsf{M}_n$ be normal. (a) Show that $A \otimes B \in \mathsf{M}_{mn}$ is normal. (b) If $A \otimes B$ is normal and neither A nor B is the zero matrix, show that A and B are normal.

12.11 Notes

A real normal matrix with some non-real eigenvalues can be diagonalized by a unitary similarity, but not by a real orthogonal similarity. However, it is real orthogonally similar to a direct sum of a real diagonal matrix and 2×2 real matrices of the form (12.1.7), in which $b \neq 0$. Examples of such matrices are nonnzero real skew-symmetric matrices and real orthogonal matrices that are not idempotent. For a discussion of real normal matrices, see [HJ13, Sect. 2.5].

More than 80 equivalent characterizations of normality are known; Theorem 12.8.1 lists several of them.

12.12 Some Important Concepts

- Definition and alternative characterizations of a normal matrix.

- Spectral theorem for complex normal and real symmetric matrices.

- Commuting normal matrices are simultaneously unitarily diagonalizable.
- Commuting real symmetric matrices are simultaneously real orthogonally diagonalizable.
- Defect from normality.
- Fuglede–Putnam theorem.
- Circulant matrices are a commuting family of normal matrices that are diagonalized simultaneously by the Fourier matrix.
- Characterization of Hermitian, skew-Hermitian, unitary, and orthogonal projection matrices as normal matrices with certain spectra.
- Cartesian decomposition.
- Spectral theorem for symmetric normal and symmetric unitary matrices (Corollary 12.6.9).
- Normal matrices are similar if and only if they are unitarily similar.
- Spectral projections and the spectral resolution of a normal matrix.

13 Positive Semidefinite Matrices

Many interesting mathematical ideas evolve from analogies. If we think of matrices as analogs of complex numbers, then the representation $z = a + bi$ suggests the Cartesian decomposition $A = H + iK$ of a square complex matrix, in which Hermitian matrices play the role of real numbers; see Definition 12.6.6.

Hermitian matrices with nonnegative eigenvalues are natural analogs of nonnegative real numbers. They arise in statistics (correlation matrices and the normal equations of least squares problems), Lagrangian mechanics (the kinetic energy functional), and quantum mechanics (density matrices). They are the subject of this chapter.

13.1 Positive Semidefinite Matrices

Definition 13.1.1 Let $A \in M_n(\mathbb{F})$.

(a) *A is positive semidefinite if it is Hermitian and $\langle A\mathbf{x}, \mathbf{x} \rangle \geq 0$ for all $\mathbf{x} \in \mathbb{F}^n$.*

(b) *A is positive definite if it is Hermitian and $\langle A\mathbf{x}, \mathbf{x} \rangle > 0$ for all nonzero $\mathbf{x} \in \mathbb{F}^n$.*

(c) *A is negative semidefinite if $-A$ is positive semidefinite.*

(d) *A is negative definite if $-A$ is positive definite.*

Whenever we speak of a positive semidefinite complex matrix, it is always with the understanding that it is Hermitian. A positive semidefinite real matrix is symmetric.

Theorem 13.1.2 *Let $A \in M_n(\mathbb{F})$. The following are equivalent:*

(a) *A is positive semidefinite.*

(b) *A is Hermitian and all its eigenvalues are nonnegative.*

(c) *There is a $B \in M_n(\mathbb{F})$ such that $A = B^*B$.*

(d) *For some positive integer m, there is a $B \in M_{m \times n}(\mathbb{F})$ such that $A = B^*B$.*

Proof (a) $\Rightarrow$ (b) Let $(\lambda, \mathbf{x})$ be an eigenpair of A, in which $\mathbf{x} \in \mathbb{F}^n$ is a unit vector. Then

$$\lambda = \lambda \langle \mathbf{x}, \mathbf{x} \rangle = \langle \lambda \mathbf{x}, \mathbf{x} \rangle = \langle A\mathbf{x}, \mathbf{x} \rangle \geq 0.$$

(b) $\Rightarrow$ (c) The spectral theorem (Theorem 12.2.2) says that there is a unitary $U \in M_n(\mathbb{F})$ and a real diagonal $\Lambda = \mathrm{diag}(\lambda_1, \lambda_2, \ldots, \lambda_n)$ such that $A = U\Lambda U^*$. Since each λ_i is nonnegative, $D = \mathrm{diag}(\lambda_1^{1/2}, \lambda_2^{1/2}, \ldots, \lambda_n^{1/2}) \in M_n(\mathbb{R})$. Let $B = UDU^* \in M_n(\mathbb{F})$. Then

$$B^*B = (UDU^*)^*(UDU^*) = UD^*U^*UDU^* = UDDU^*$$
$$= UD^2U^* = U\Lambda U^* = A.$$

(c) $\Rightarrow$ (d) Take $m = n$.

(d) $\Rightarrow$ (a) If $B \in M_{m \times n}(\mathbb{F})$ and $A = B^*B$, then A is Hermitian. If $\mathbf{x} \in \mathbb{F}^n$, then

$$\langle A\mathbf{x}, \mathbf{x} \rangle = \langle B^*B\mathbf{x}, \mathbf{x} \rangle = \langle B\mathbf{x}, B\mathbf{x} \rangle = \|B\mathbf{x}\|_2^2 \geq 0,$$

so A is positive semidefinite. $\square$

Example 13.1.3 The matrix

$$A = \begin{bmatrix} 1 & -1 \\ -1 & 1 \end{bmatrix} \in M_2(\mathbb{R}) \tag{13.1.4}$$

is real symmetric. Let us verify that it satisfies the criteria in the preceding theorem.

(a) For any $\mathbf{x} = [x_1\ x_2]^\mathsf{T} \in \mathbb{R}^2$,

$$\langle A\mathbf{x}, \mathbf{x} \rangle = (x_1 - x_2)x_1 + (-x_1 + x_2)x_2 = (x_1 - x_2)^2 \geq 0,$$

so A is positive semidefinite. Since $\langle A\mathbf{x}, \mathbf{x} \rangle = 0$ for $\mathbf{x} = [1\ 1]^\mathsf{T}$, A is not positive definite.

(b) The eigenvalues of A are the roots of $p_A(z) = z(z - 2) = 0$, which are 0 and 2.

(c) The matrix $B = \frac{1}{\sqrt{2}}A$ satisfies $A = B^*B$. Moreover, $C = QB$ satisfies $A = C^*C$ for any real orthogonal $Q \in M_2(\mathbb{R})$. There are many ways to represent A as in Theorem 13.1.2.c.

Example 13.1.5 Let A be the matrix (13.1.4) and let $\mathbf{x} = [x_1\ x_2]^\mathsf{T} \in \mathbb{C}^2$. Then

$$\langle A\mathbf{x}, \mathbf{x} \rangle = (x_1 - x_2)\overline{x_1} + (-x_1 + x_2)\overline{x_2} = |x_1 - x_2|^2 \geq 0.$$

This illustrates a general principle. If $A \in M_n(\mathbb{R})$ is symmetric and $\langle A\mathbf{x}, \mathbf{x} \rangle \geq 0$ for all $\mathbf{x} \in \mathbb{R}^n$, then A is Hermitian and Theorem 13.1.2.b ensures that it has nonnegative eigenvalues. Consequently, $\langle A\mathbf{x}, \mathbf{x} \rangle \geq 0$ for all $\mathbf{x} \in \mathbb{C}^n$.

Example 13.1.6 The matrix

$$A = \begin{bmatrix} 0 & 1 \\ -1 & 0 \end{bmatrix}$$

satisfies $\langle A\mathbf{x}, \mathbf{x} \rangle = x_1x_2 - x_1x_2 = 0$ for all $\mathbf{x} = [x_1\ x_2]^\mathsf{T} \in \mathbb{R}^2$. For $\mathbf{x} = [1\ i]^\mathsf{T}$, however, $\langle A\mathbf{x}, \mathbf{x} \rangle = 2i$. The principle stated in the preceding example requires that the real matrix A be symmetric.

Example 13.1.7 An orthogonal projection $P \in M_n$ is Hermitian, so Theorem 7.6.4 ensures that $\operatorname{spec} P \subseteq \{0, 1\}$. If $a \geq 0$, then aP is Hermitian and $\operatorname{spec}(aP) \subseteq \{0, a\}$. Thus, any real nonnegative scalar multiple of an orthogonal projection is positive semidefinite. In Example 13.1.3, $A = 2\mathbf{v}\mathbf{v}^*$, in which $\mathbf{v} = \frac{1}{\sqrt{2}}[1\ -1]^\mathsf{T}$.

There is a version of Theorem 13.1.2 for positive definite matrices.

Theorem 13.1.8 *Let $A \in M_n(\mathbb{F})$. The following are equivalent:*

(a) *A is positive definite.*

(b1) *A is Hermitian and all its eigenvalues are positive.*

(b2) *A is positive semidefinite and invertible.*

(b3) *A is invertible and A^{-1} is positive definite.*

(c) *There is an invertible $B \in M_n(\mathbb{F})$ such that $A = B^*B$.*

(d) *For some $m \geq n$, there is a $B \in M_{m \times n}(\mathbb{F})$ such that* rank $B = n$ *and $A = B^*B$.*

Proof We indicate how the proof of each implication in Theorem 13.1.2 can be modified to obtain a proof of the corresponding implication in the positive definite case.

(a) $\Rightarrow$ (b1) $\lambda = \langle Ax, x \rangle > 0$.

(b1) $\Leftrightarrow$ (b2) A matrix with nonnegative eigenvalues is invertible if and only if the eigenvalues are positive.

(b1) $\Leftrightarrow$ (b3) A is Hermitian if and only if A^{-1} is Hermitian. The eigenvalues of A are the reciprocals of the eigenvalues of A^{-1}.

(b1) $\Rightarrow$ (c) D has positive diagonal entries, so $B = UDU^* \in M_n(\mathbb{F})$ is invertible and $B^*B = A$.

(c) $\Rightarrow$ (d) Take $m = n$.

(d) $\Rightarrow$ (a) Since rank $B = n$, the dimension theorem ensures that null $B = \{0\}$. If $x \neq 0$, then $Bx \neq 0$ and hence $\langle Ax, x \rangle = \|Bx\|_2^2 > 0$. $\qquad\square$

Theorem 13.1.2.c says that $A \in M_n(\mathbb{F})$ is positive semidefinite if and only if it is the Gram matrix of n vectors in some $\mathbb{F}^m$; see Definition 7.4.14. Theorem 13.1.8.c says that A is positive definite if and only if those vectors are linearly independent.

Theorem 13.1.9 *Let $A \in M_n$ be positive semidefinite.*

(a) tr $A \geq 0$, *with strict inequality if and only if $A \neq 0$.*

(b) det $A \geq 0$, *with strict inequality if and only if A is positive definite.*

Proof Let $\lambda_1, \lambda_2, \ldots, \lambda_n$ be the eigenvalues of A, which are nonnegative. They are positive if A is positive definite.

(a) tr $A = \lambda_1 + \lambda_2 + \cdots + \lambda_n \geq 0$, with equality if and only if each $\lambda_i = 0$. A Hermitian matrix is zero if and only if all its eigenvalues are zero.

(b) det $A = \lambda_1 \lambda_2 \cdots \lambda_n$ is nonnegative. It is positive if and only if every eigenvalue is positive, which means that A is positive definite. $\qquad\square$

Definition 13.1.1 has an appealing geometric interpretation for real positive definite matrices. If θ is the angle between the real vectors x and Ax, then the condition

$$0 < \langle Ax, x \rangle = \|Ax\|_2 \|x\|_2 \cos \theta$$

means that $\cos \theta > 0$, that is, $|\theta| < \pi/2$. See Figure 13.1 for an illustration in $\mathbb{R}^2$.

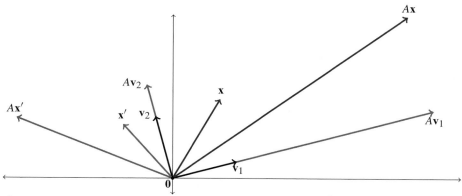

Figure 13.1 $A \in M_2(\mathbb{R})$ is positive definite. For any nonzero $\mathbf{x} \in \mathbb{R}^2$, the angle between $\mathbf{x}$ and $A\mathbf{x}$ is less than $\frac{\pi}{2}$.

If $A \in M_n(\mathbb{F})$ is positive semidefinite and nonzero, then Example 13.1.3.c ensures that there are many $B \in M_n(\mathbb{F})$ such that $A = B^*B$. The following theorem says that they all have the same null space and the same rank.

Theorem 13.1.10 *Suppose that $B \in M_{m \times n}(\mathbb{F})$ and let $A = B^*B$.*

(a) $\operatorname{null} A = \operatorname{null} B$.

(b) $\operatorname{rank} A = \operatorname{rank} B$.

(c) $A = 0$ *if and only if $B = 0$.*

(d) $\operatorname{col} A = \operatorname{col} B^*$.

(e) *If B is normal, then $\operatorname{col} A = \operatorname{col} B$.*

Proof (a) If $B\mathbf{x} = \mathbf{0}$, then $A\mathbf{x} = B^*B\mathbf{x} = \mathbf{0}$ and hence $\operatorname{null} B \subseteq \operatorname{null} A$. Conversely, if $A\mathbf{x} = \mathbf{0}$, then

$$0 = \langle A\mathbf{x}, \mathbf{x} \rangle = \langle B^*B\mathbf{x}, \mathbf{x} \rangle = \langle B\mathbf{x}, B\mathbf{x} \rangle = \|B\mathbf{x}\|_2^2, \qquad (13.1.11)$$

so $B\mathbf{x} = \mathbf{0}$. It follows that $\operatorname{null} A \subseteq \operatorname{null} B$ and hence $\operatorname{null} A = \operatorname{null} B$.

(b) Since A and B both have n columns, the dimension theorem ensures that

$$\operatorname{rank} A = n - \dim \operatorname{null} A = n - \dim \operatorname{null} B = \operatorname{rank} B.$$

(c) The assertion follows from (a).

(d) Since A is Hermitian,

$$\operatorname{col} A = (\operatorname{null} A^*)^\perp = (\operatorname{null} A)^\perp = (\operatorname{null} B)^\perp = \operatorname{col} B^*.$$

(e) This follows from (d) and Theorem 12.1.9.e. $\qquad \square$

One consequence of the factorization $A = B^*B$ of a positive semidefinite matrix is known as *column inclusion*.

Corollary 13.1.12 *Let $A \in \mathsf{M}_n$ be positive semidefinite and partitioned as*

$$A = \begin{bmatrix} A_{11} & A_{12} \\ A_{12}^* & A_{22} \end{bmatrix}, \quad A_{11} \in \mathsf{M}_k. \tag{13.1.13}$$

Then A_{11} is positive semidefinite and

$$\operatorname{col} A_{12} \subseteq \operatorname{col} A_{11}. \tag{13.1.14}$$

Consequently,

$$\operatorname{rank} A_{12} \leq \operatorname{rank} A_{11}. \tag{13.1.15}$$

If A is positive definite, then A_{11} is positive definite.

Proof Let $B \in \mathsf{M}_n$ be such that $A = B^*B$ (Theorem 13.1.2.c) and partition $B = [B_1 \ B_2]$, in which $B_1 \in \mathsf{M}_{n \times k}$. Then

$$A = B^*B = \begin{bmatrix} B_1^* \\ B_2^* \end{bmatrix} \begin{bmatrix} B_1 & B_2 \end{bmatrix} = \begin{bmatrix} B_1^*B_1 & B_1^*B_2 \\ B_2^*B_1 & B_2^*B_2 \end{bmatrix},$$

so $A_{11} = B_1^*B_1$ is positive semidefinite (Theorem 13.1.2.c) and

$$\operatorname{col} A_{12} = \operatorname{col} B_1^*B_2 \subseteq \operatorname{col} B_1^*.$$

The preceding theorem ensures that

$$\operatorname{col} B_1^* = \operatorname{col} B_1^*B_1 = \operatorname{col} A_{11},$$

which proves (13.1.14). It follows that

$$\operatorname{rank} A_{12} = \dim \operatorname{col} A_{12} \leq \dim \operatorname{col} A_{11} = \operatorname{rank} A_{11}.$$

If A is positive definite, then Theorem 13.1.8.c ensures that B is invertible. Consequently, $\operatorname{rank} B_1 = k$ and $B_1^*B_1$ is invertible (Theorem 13.1.8.d). $\qquad\square$

Example 13.1.16 Suppose that $A = [a_{ij}] \in \mathsf{M}_n$ is positive semidefinite and partitioned as in (13.1.13). If $A_{11} = [a_{11}] = 0$, then (13.1.14) ensures that $A_{12} = [a_{12} \ a_{13} \ \cdots \ a_{1n}] = 0$.

Example 13.1.17 Suppose that $A \in \mathsf{M}_n$ is positive semidefinite and partitioned as in (13.1.13). If

$$A_{11} = \begin{bmatrix} 1 & 1 \\ 1 & 1 \end{bmatrix},$$

then $\operatorname{col} A_{11} = \operatorname{span}\{[1 \ 1]^\mathsf{T}\}$. Therefore, each column of A_{12} is a scalar multiple of $[1 \ 1]^\mathsf{T}$, so $A_{12} \in \mathsf{M}_{2 \times (n-2)}$ has equal rows.

Corollary 13.1.18 *Let $A \in \mathsf{M}_n$ be positive semidefinite and let $\mathbf{x} \in \mathbb{C}^n$. Then $\langle A\mathbf{x}, \mathbf{x} \rangle = 0$ if and only if $A\mathbf{x} = \mathbf{0}$.*

Proof If $A\mathbf{x} = \mathbf{0}$, then $\langle A\mathbf{x}, \mathbf{x} \rangle = \langle \mathbf{0}, \mathbf{x} \rangle = 0$. Conversely, if $\langle A\mathbf{x}, \mathbf{x} \rangle = 0$ and $A = B^*B$, then (13.1.11) ensures that $B\mathbf{x} = \mathbf{0}$, which implies that $A\mathbf{x} = B^*B\mathbf{x} = \mathbf{0}$. $\qquad\square$

Example 13.1.19 The Hermitian matrix $D = \mathrm{diag}(1, -1)$ and the nonzero vector $\mathbf{x} = [1\ 1]^\mathsf{T}$ satisfy $\langle D\mathbf{x}, \mathbf{x} \rangle = 0$ and $D\mathbf{x} \neq \mathbf{0}$. The hypothesis "positive semidefinite" cannot be weakened to "Hermitian" in the preceding corollary.

A motivating analogy for the following theorem is that real linear combinations of real numbers are real, and nonnegative linear combinations of nonnegative real numbers are nonnegative.

Theorem 13.1.20 *Let $A, B \in \mathsf{M}_n(\mathbb{F})$ and let $a, b \in \mathbb{R}$.*

(a) *If A and B are Hermitian, then $aA + bB$ is Hermitian.*

(b) *If a and b are nonnegative and if A and B are positive semidefinite, then $aA + bB$ is positive semidefinite.*

(c) *If a and b are positive, if A and B are positive semidefinite, and if at least one of A and B is positive definite, then $aA + bB$ is positive definite.*

Proof (a) $(aA + bB)^* = \bar{a}A^* + \bar{b}B^* = aA + bB$.

(b) Part (a) ensures that $aA + bB$ is Hermitian, and

$$\langle (aA + bB)\mathbf{x}, \mathbf{x} \rangle = a\langle A\mathbf{x}, \mathbf{x} \rangle + b\langle B\mathbf{x}, \mathbf{x} \rangle \geq 0 \qquad (13.1.21)$$

for every $\mathbf{x} \in \mathbb{F}^n$ because both summands are nonnegative. Consequently, $aA + bB$ is positive semidefinite.

(c) The assertion follows from (13.1.21) because both summands are nonnegative and at least one of them is positive if $\mathbf{x} \neq \mathbf{0}$. $\qquad\square$

The preceding theorem asserts something that is not obvious: the sum of two Hermitian matrices with nonnegative eigenvalues has nonnegative eigenvalues.

Example 13.1.22 Consider

$$A = \begin{bmatrix} 1 & 3 \\ 0 & 1 \end{bmatrix}, \quad B = \begin{bmatrix} 1 & 0 \\ 3 & 1 \end{bmatrix}, \quad \text{and} \quad A + B = \begin{bmatrix} 2 & 3 \\ 3 & 2 \end{bmatrix}.$$

Then A and B have (positive) eigenvalues 1 and 1, but the eigenvalues of $A + B$ are -1 and 5. Since A and B are not Hermitian, this does not contradict the preceding theorem.

Here are two ways to make new positive semidefinite matrices from old ones.

Theorem 13.1.23 *Let $A, S \in \mathsf{M}_n$ and $B \in \mathsf{M}_m$, and suppose that A and B are positive semidefinite.*

(a) *$A \oplus B$ is positive semidefinite; it is positive definite if and only if both A and B are positive definite.*

(b) *S^*AS is positive semidefinite; it is positive definite if and only if A is positive definite and S is invertible.*

Proof Theorem 13.1.2.c says that there are $P \in \mathsf{M}_n$ and $Q \in \mathsf{M}_m$ such that $A = P^*P$ and $B = Q^*Q$.

(a) Let $C = A \oplus B$. Then $C^* = A^* \oplus B^* = A \oplus B = C$, so C is Hermitian. Compute

$$C = (P^*P) \oplus (Q^*Q) = (P \oplus Q)^*(P \oplus Q).$$

It follows from Theorem 13.1.2.c that C is positive semidefinite. A direct sum is invertible if and only if all of its summands are invertible, so C is positive definite if and only if both A and B are positive definite.

(b) S^*AS is Hermitian and $S^*AS = S^*P^*PS = (PS)^*(PS)$ is positive semidefinite. Theorem 13.1.8.c says that S^*AS is positive definite if and only if PS is invertible, which happens if and only if S and P are both invertible. Finally, P is invertible if and only if A is positive definite. □

A useful sufficient condition for positive semidefiniteness follows from Theorem 8.4.1 (Geršgorin's theorem).

Theorem 13.1.24 *Let $A = [a_{ij}] \in \mathsf{M}_n$ be Hermitian and suppose that it has nonnegative diagonal entries.*

(a) *If A is diagonally dominant, then it is positive semidefinite.*

(b) *If A is diagonally dominant and invertible, then it is positive definite.*

(c) *If A is strictly diagonally dominant, then it is positive definite.*

(d) *If A is diagonally dominant, has no zero entries, and $|a_{kk}| > R'_k(A)$ for at least one $k \in \{1, 2, \ldots, n\}$, then A is positive definite.*

Proof (a) Since A has real eigenvalues, Corollary 8.4.18.b ensures that $\operatorname{spec} A \subseteq [0, \infty)$. The assertion follows from Theorem 13.1.2.

(b) The assertion follows from (a) and Theorem 13.1.8.b2.

(c) The assertion follows from Corollary 8.4.18.a and (b).

(d) The assertion follows from Theorem 8.4.20. □

Example 13.1.25 Let

$$A = \begin{bmatrix} 2 & 1 & 0 \\ 1 & 3 & 1 \\ 0 & 1 & 2 \end{bmatrix}; \tag{13.1.26}$$

see Figure 13.2.

Example 13.1.27 Let

$$A = \begin{bmatrix} 5 & 2 & 0 \\ 2 & 3 & 2 \\ 0 & 2 & 5 \end{bmatrix}; \tag{13.1.28}$$

see Figure 13.3.

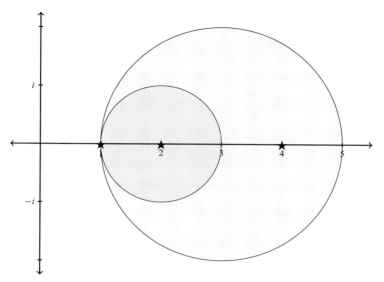

Figure 13.2 The Hermitian matrix (13.1.26) is strictly diagonally dominant, so it is positive definite. Its Geršgorin region is contained in the right half plane. Its eigenvalues are 4, 2, and 1.

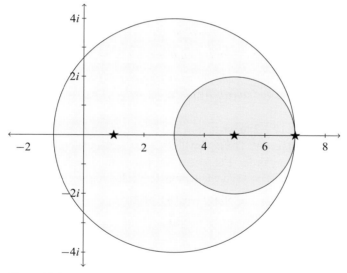

Figure 13.3 The Hermitian matrix (13.1.28) is positive definite even though it is not diagonally dominant. Its Geršgorin region extends into the left half plane. Its eigenvalues are 7, 5, and 1.

13.2 The Square Root of a Positive Semidefinite Matrix

Two nonnegative real numbers are equal if and only if their squares are equal. An analogous statement is valid for Hermitian matrices with nonnegative eigenvalues, but both conditions (Hermitian as well as nonnegative eigenvalues) are essential. For example, the unequal (but non-Hermitian) matrices

$$A = \begin{bmatrix} 0 & 1 \\ 0 & 0 \end{bmatrix} \quad \text{and} \quad B = \begin{bmatrix} 0 & 0 \\ 1 & 0 \end{bmatrix}$$

have nonnegative eigenvalues and $A^2 = B^2$. The unequal Hermitian (but not positive semidefinite) matrices

$$A = \begin{bmatrix} 1 & 0 \\ 0 & -1 \end{bmatrix} \quad \text{and} \quad B = \begin{bmatrix} 0 & 1 \\ 1 & 0 \end{bmatrix}$$

also have $A^2 = B^2$.

Lemma 13.2.1 *Let $B, C \in M_n$ be positive semidefinite diagonal matrices. If $B^2 = C^2$, then $B = C$.*

Proof Let $B = \text{diag}(b_1, b_2, \ldots, b_n)$ and $C = \text{diag}(c_1, c_2, \ldots, c_n)$ have real nonnegative diagonal entries. If $B^2 = C^2$, then $b_i^2 = c_i^2$ for each $i = 1, 2, \ldots, n$. Since $b_i \geq 0$ and $c_i \geq 0$, we have $b_i = c_i$ for each $i = 1, 2, \ldots, n$.   □

The square root function on positive semidefinite matrices is a special case of the functional calculus for diagonalizable matrices discussed in Section 9.5. For an alternative approach to uniqueness of the square root, see P.13.9.

Theorem 13.2.2 *Let $A \in M_n(\mathbb{F})$ be positive semidefinite. There is a unique positive semidefinite matrix $B \in M_n(\mathbb{F})$ such that $B^2 = A$. Moreover,*

(a) $B = p(A)$ *for some real polynomial p that is determined only by* spec A,

(b) null $A = $ null B *and* col $A = $ col B, *and*

(c) *A is positive definite if and only if B is positive definite.*

Proof (a) Use Theorem 12.2.2 to write $A = U\Lambda U^*$, in which $U \in M_n(\mathbb{F})$ is unitary and $\Lambda = \text{diag}(\lambda_1, \lambda_2, \ldots, \lambda_n)$ is diagonal. The eigenvalues λ_i are real and nonnegative; they may appear on the diagonal of Λ in any order. Let p be a real polynomial such that $p(\lambda_i) = \lambda_i^{1/2} \geq 0$ for $i = 1, 2, \ldots, n$, and let

$$B = Up(\Lambda)U^*, \tag{13.2.3}$$

which has eigenvalues $\lambda_1^{1/2}, \lambda_2^{1/2}, \ldots, \lambda_n^{1/2}$. Then

$$B^2 = Up(\Lambda)U^*Up(\Lambda)U^* = Up(\Lambda)^2U^* = U\Lambda U^* = A.$$

Moreover,

$$B = Up(\Lambda)U^* = p(U\Lambda U^*) = p(A) \in M_n(\mathbb{F}).$$

Suppose that $C \in M_n$ is positive semidefinite and $C^2 = A$. Then $B = p(A) = p(C^2)$ is a polynomial in C, so B commutes with C. Theorem 12.2.9 tells us that B and C are simultaneously unitarily diagonalizable. Let $V \in M_n$ be unitary and such that $B = VEV^*$ and $C = VFV^*$, in which E and F are diagonal and have nonnegative diagonal entries. Since

$$VE^2V^* = (VEV^*)^2 = B^2 = A = C^2 = (VFV^*)^2 = VF^2V^*,$$

it follows that $E^2 = V^*AV = F^2$. Lemma 13.2.1 ensures that $E = F$ and hence $C = B$.

(b) See Theorem 13.1.10.

(c) Since $B^2 = A$, it follows that A is invertible if and only if B is invertible. Since all the eigenvalues of an invertible positive semidefinite matrix are positive, the assertion follows. □

Definition 13.2.4 If $A \in M_n$ is positive semidefinite, its *positive semidefinite square root* is the unique positive semidefinite matrix $A^{1/2}$ such that $(A^{1/2})^2 = A$. If A is positive definite, we define $(A^{1/2})^{-1} = A^{-1/2}$.

A single polynomial can take several positive semidefinite matrices to their positive semidefinite square roots. The following corollary shows how this can be done for two matrices.

Corollary 13.2.5 *Let $A \in M_n$ and $B \in M_m$ be positive semidefinite. There is a real polynomial p such that $A^{1/2} = p(A)$ and $B^{1/2} = p(B)$.*

Proof Theorem 13.1.23 ensures that $A \oplus B$ and $A^{1/2} \oplus B^{1/2}$ are positive semidefinite. Since $(A^{1/2} \oplus B^{1/2})^2 = A \oplus B$, it follows from the uniqueness assertion in Theorem 13.2.2 that $(A \oplus B)^{1/2} = A^{1/2} \oplus B^{1/2}$. Moreover, there is a polynomial p such that $(A \oplus B)^{1/2} = p(A \oplus B)$, so

$$A^{1/2} \oplus B^{1/2} = (A \oplus B)^{1/2} = p(A \oplus B) = p(A) \oplus p(B),$$

which implies that $A^{1/2} = p(A)$ and $B^{1/2} = p(B)$. □

The proof of Theorem 13.2.2 contains a recipe to compute the positive semidefinite square root of a positive semidefinite matrix. Diagonalize $A = U\Lambda U^*$, form $\Lambda^{1/2}$, and obtain $A^{1/2} = U\Lambda^{1/2}U^*$. There is a way to bypass the diagonalization for 2×2 positive semidefinite matrices.

Proposition 13.2.6 *Let $A \in M_2$ be positive semidefinite and nonzero, and let*

$$\tau = (\operatorname{tr} A + 2(\det A)^{1/2})^{1/2}.$$

Then $\tau > 0$ and

$$A^{1/2} = \frac{1}{\tau}(A + (\det A)^{1/2}I). \tag{13.2.7}$$

Proof Theorem 13.1.9 ensures that $\det A \geq 0$ and $\operatorname{tr} A > 0$, so $\tau^2 \geq \operatorname{tr} A > 0$. Theorem 13.1.20 guarantees that (13.2.7) is positive semidefinite since it is a nonnegative linear combination of positive semidefinite matrices; it suffices to show that its square is A. Compute

$$\frac{1}{\tau^2}(A + (\det A)^{1/2}I)^2$$

$$= \frac{1}{\tau^2}(A^2 + 2(\det A)^{1/2}A + (\det A)I) \tag{13.2.8}$$

$$= \frac{1}{\tau^2}((\operatorname{tr} A)A - (\det A)I + 2(\det A)^{1/2}A + (\det A)I) \tag{13.2.9}$$

$$= \frac{1}{\tau^2}(\operatorname{tr} A + 2(\det A)^{1/2})A$$

$$= A.$$

To obtain (13.2.9), we used Theorem 10.2.1 to make the substitution $A^2 = (\operatorname{tr} A)A - (\det A)I$ in (13.2.8); see (9.1.3). □

Example 13.2.10 Consider the real diagonally dominant symmetric matrix

$$A = \begin{bmatrix} 2 & 1 \\ 1 & 1 \end{bmatrix}.$$

Theorem 13.1.24.d ensures that A is positive definite. To find $A^{1/2}$, Proposition 13.2.6 tells us to compute

$$\tau = \left(\operatorname{tr} A + 2(\det A)^{1/2}\right)^{1/2} = (3+2)^{1/2} = \sqrt{5}$$

and

$$A^{1/2} = \frac{1}{\sqrt{5}}(A + I) = \frac{1}{\sqrt{5}} \begin{bmatrix} 3 & 1 \\ 1 & 2 \end{bmatrix}. \tag{13.2.11}$$

Although (13.2.11) is the only positive semidefinite square root of A, there are other real symmetric square roots. For example,

$$B = \begin{bmatrix} 1 & 1 \\ 1 & 0 \end{bmatrix}$$

satisfies $B^2 = A$. The eigenvalues of B are $\frac{1}{2}(1 \pm \sqrt{5})$, one of which is negative; see Example 9.5.3.

The product of two positive real numbers is positive, but the product of two positive definite matrices need not be positive definite. For example, the product

$$\begin{bmatrix} 2 & 1 \\ 1 & 1 \end{bmatrix} \begin{bmatrix} 1 & 1 \\ 1 & 2 \end{bmatrix} = \begin{bmatrix} 3 & 4 \\ 2 & 3 \end{bmatrix} \tag{13.2.12}$$

of two positive definite matrices is not Hermitian, so it is not positive definite. However, it has positive eigenvalues $3 \pm 2\sqrt{2}$. The final theorem in this section provides an explanation for this observation.

Theorem 13.2.13 *Let $A, B \in \mathsf{M}_n$ and suppose that A is positive definite.*

(a) *If B is Hermitian, then AB is diagonalizable and has real eigenvalues.*

(b) *If B is positive semidefinite, then AB is diagonalizable and has nonnegative real eigenvalues. If $B \neq 0$, then AB has at least one positive eigenvalue.*

(c) *If B is positive definite, then AB is diagonalizable and has positive real eigenvalues.*

Proof Consider the similarity

$$A^{-1/2}(AB)A^{1/2} = A^{-1/2}A^{1/2}A^{1/2}BA^{1/2} = A^{1/2}BA^{1/2} = (A^{1/2})^*B(A^{1/2}),$$

in which $C = (A^{1/2})^*B(A^{1/2})$ is Hermitian and therefore diagonalizable. Any matrix that is similar to C (in particular, AB) is diagonalizable and has the same eigenvalues as C.

(a) The eigenvalues of the Hermitian matrix C are real.

(b) Theorem 13.1.23.b says that C is positive semidefinite, so its eigenvalues are real and nonnegative. If they are zero, then $C = 0$ since it is diagonalizable. If $C = 0$, then $B = A^{-1/2}CA^{-1/2} = 0$.

(c) AB is invertible and has nonnegative eigenvalues, so they are positive. $\qquad\square$

Corollary 13.2.14 *Let $A, B \in \mathsf{M}_n$. Suppose that A is positive definite and B is positive semidefinite. Then $\det(A + B) \geq \det A$ with equality if and only if $B = 0$.*

Proof The preceding theorem says that the eigenvalues $\lambda_1, \lambda_2, \ldots, \lambda_n$ of $A^{-1}B$ are nonnegative and $\lambda_1 = \lambda_2 = \cdots = \lambda_n = 0$ only if $B = 0$. Compute

$$\det(A + B) = \det\left(A(I + A^{-1}B)\right) = (\det A)\det(I + A^{-1}B)$$

$$= (\det A)\prod_{i=1}^{n}(1 + \lambda_i) \geq \det A,$$

with equality if and only if $\lambda_1 = \lambda_2 = \cdots = \lambda_n = 0$. $\qquad\square$

13.3 The Cholesky Factorization

If a square matrix A is invertible and triangular, then the solution to the linear system $A\mathbf{x} = \mathbf{y}$ can be computed by forward or backward substitution.

Example 13.3.1 A lower triangular linear system, such as

$$L = \begin{bmatrix} 1 & 0 \\ 2 & 3 \end{bmatrix}\begin{bmatrix} w_1 \\ w_2 \end{bmatrix} = \begin{bmatrix} 14 \\ 64 \end{bmatrix} = \mathbf{y},$$

can be solved by forward substitution. Substitute the solution $w_1 = 14$ of the first equation into the second equation to obtain $3w_2 = 64 - 28 = 36$, so $w_2 = 12$.

Example 13.3.2 An upper triangular linear system, such as

$$R\mathbf{x} = \begin{bmatrix} 4 & 5 \\ 0 & 6 \end{bmatrix}\begin{bmatrix} x_1 \\ x_2 \end{bmatrix} = \begin{bmatrix} 14 \\ 12 \end{bmatrix} = \mathbf{w},$$

can be solved by backward substitution. Substitute the solution $x_2 = 2$ of the second equation into the first equation to obtain $4x_1 = 14 - 10 = 4$, so $x_1 = 1$.

Example 13.3.3 Let L and R be the triangular matrices in the preceding two examples and let

$$A = LR = \begin{bmatrix} 4 & 5 \\ 8 & 28 \end{bmatrix}.$$

The linear system

$$L(R\mathbf{x}) = A\mathbf{x} = \begin{bmatrix} 4 & 5 \\ 8 & 28 \end{bmatrix}\begin{bmatrix} x_1 \\ x_2 \end{bmatrix} = \begin{bmatrix} 16 \\ 64 \end{bmatrix} = \mathbf{y}$$

can be solved in two steps. First do forward substitution to solve $L\mathbf{w} = \mathbf{y}$ and obtain $\mathbf{w} = [14\ 12]^\mathsf{T}$, then do backward substitution to solve $R\mathbf{x} = \mathbf{w}$ and obtain $\mathbf{x} = [1\ 2]^\mathsf{T}$.

The strategy employed in Example 13.3.3 to solve $A\mathbf{x} = \mathbf{y}$ relies on a factorization $A = LR$, in which L is lower triangular and R is upper triangular. This is not always possible.

Example 13.3.4 If

$$\begin{bmatrix} 0 & 1 \\ 1 & 0 \end{bmatrix} = \begin{bmatrix} \ell_{11} & 0 \\ \ell_{21} & \ell_{22} \end{bmatrix} \begin{bmatrix} r_{11} & r_{12} \\ 0 & r_{22} \end{bmatrix} = \begin{bmatrix} \ell_{11}r_{11} & \ell_{11}r_{12} \\ r_{11}\ell_{21} & \star \end{bmatrix},$$

then $\ell_{11}r_{11} = 0$. Consequently, either $\ell_{11} = 0$ (in which case $\ell_{11}r_{12} = 1$ is impossible), or $r_{11} = 0$ (in which case $r_{11}\ell_{21} = 1$ is impossible). Thus, no such factorization exists.

Every positive semidefinite matrix can be factored as a product of a lower triangular matrix and its adjoint, so linear systems $A\mathbf{x} = \mathbf{y}$ in which A is positive definite can be solved by forward and backward substitutions, as in Example 13.3.3.

Theorem 13.3.5 (Cholesky Factorization) *Let $A \in M_n(\mathbb{F})$. If A is positive semidefinite, then there is a lower triangular $L \in M_n(\mathbb{F})$ with nonnegative diagonal entries such that $A = LL^*$. If A is positive definite, then L is unique and has positive diagonal entries.*

Proof Since A is positive semidefinite, it has a unique positive semidefinite square root $A^{1/2}$. Theorem 6.5.2.a says that there is a unitary $Q \in M_n(\mathbb{F})$ and an upper triangular $R \in M_n(\mathbb{F})$ with nonnegative diagonal entries such that $A^{1/2} = QR$. Then $A = (A^{1/2})^2 = (A^{1/2})^*A^{1/2} = R^*Q^*QR = R^*R$.

Now assume that A is positive definite. Then R is invertible, so its diagonal entries are positive. Suppose that there is an upper triangular $S \in M_n(\mathbb{F})$ with positive diagonal entries such that $A = S^*S$. Then

$$(SA^{-1/2})^*(SA^{-1/2}) = A^{-1/2}S^*SA^{-1/2} = A^{-1/2}AA^{-1/2} = I,$$

which shows that $V = SA^{-1/2}$ is unitary. Then $A^{1/2} = V^*S$ and $A^{1/2} = QR$ are QR factorizations. Theorem 6.5.2.b ensures that $S = R$. Let $L = R^*$. $\qquad\square$

Example 13.3.6 Let L be the 4×4 lower triangular matrix with all entries 1 on and below the diagonal. Then

$$LL^* = \begin{bmatrix} 1 & 0 & 0 & 0 \\ 1 & 1 & 0 & 0 \\ 1 & 1 & 1 & 0 \\ 1 & 1 & 1 & 1 \end{bmatrix} \begin{bmatrix} 1 & 1 & 1 & 1 \\ 0 & 1 & 1 & 1 \\ 0 & 0 & 1 & 1 \\ 0 & 0 & 0 & 1 \end{bmatrix} = \begin{bmatrix} 1 & 1 & 1 & 1 \\ 1 & 2 & 2 & 2 \\ 1 & 2 & 3 & 3 \\ 1 & 2 & 3 & 4 \end{bmatrix} = [\min\{i,j\}]$$

presents the Cholesky factorization of the *4×4 min matrix* and shows that it is positive definite. Since

$$\begin{bmatrix} 1 & -1 & 0 & 0 \\ 0 & 1 & -1 & 0 \\ 0 & 0 & 1 & -1 \\ 0 & 0 & 0 & 1 \end{bmatrix} \begin{bmatrix} 1 & 1 & 1 & 1 \\ 0 & 1 & 1 & 1 \\ 0 & 0 & 1 & 1 \\ 0 & 0 & 0 & 1 \end{bmatrix} = \begin{bmatrix} 1 & 0 & 0 & 0 \\ 0 & 1 & 0 & 0 \\ 0 & 0 & 1 & 0 \\ 0 & 0 & 0 & 1 \end{bmatrix},$$

$$\underbrace{}_{L^{-*}} \quad \underbrace{}_{L^*}$$

the inverse of L^* is the upper bidiagonal matrix with main diagonal entries 1 and entries -1 on the first superdiagonal. Therefore, the inverse of the 4×4 min matrix is the tridiagonal matrix

$$(LL^*)^{-1} = L^{-*}L^{-1}$$

$$= \begin{bmatrix} 1 & -1 & 0 & 0 \\ 0 & 1 & -1 & 0 \\ 0 & 0 & 1 & -1 \\ 0 & 0 & 0 & 1 \end{bmatrix} \begin{bmatrix} 1 & 0 & 0 & 0 \\ -1 & 1 & 0 & 0 \\ 0 & -1 & 1 & 0 \\ 0 & 0 & -1 & 1 \end{bmatrix} = \begin{bmatrix} 2 & -1 & 0 & 0 \\ -1 & 2 & -1 & 0 \\ 0 & -1 & 2 & -1 \\ 0 & 0 & -1 & 1 \end{bmatrix}.$$

See P.13.20 for the $n \times n$ min matrix.

13.4 Simultaneous Diagonalization of Quadratic Forms

A real quadratic form in n variables is a real-valued function on $\mathbb{R}^n$ of the form

$$\mathbf{x} \mapsto \langle A\mathbf{x}, \mathbf{x} \rangle = \sum_{i,j=1}^{n} a_{ij} x_i x_j,$$

in which $A \in \mathsf{M}_n(\mathbb{R})$ is symmetric and $\mathbf{x} = [x_i] \in \mathbb{R}^n$. The spectral theorem says that there is a real orthogonal $Q = [q_{ij}] \in \mathsf{M}_n(\mathbb{R})$ and a real diagonal $\Lambda = \mathrm{diag}(\lambda_1, \lambda_2, \ldots, \lambda_n)$ such that $A = Q\Lambda Q^\mathsf{T}$. If $\mathbf{y} = Q^\mathsf{T}\mathbf{x}$, then

$$\sum_{i,j=1}^{n} a_{ij} x_i x_j = \langle A\mathbf{x}, \mathbf{x} \rangle = \langle Q\Lambda Q^\mathsf{T}\mathbf{x}, \mathbf{x} \rangle = \langle \Lambda Q^\mathsf{T}\mathbf{x}, Q^\mathsf{T}\mathbf{x} \rangle$$

$$= \langle \Lambda\mathbf{y}, \mathbf{y} \rangle = \sum_{i=1}^{n} \lambda_i y_i^2 \tag{13.4.1}$$

is a sum of squares of the new variables $\mathbf{y} = [y_i]$. The coefficients λ_i in (13.4.1) are the eigenvalues of A.

Example 13.4.2 In the study of central conic sections in plane geometry, there is a real symmetric invertible $A = [a_{ij}] \in \mathsf{M}_2(\mathbb{R})$ and a real quadratic form

$$a_{11}x_1^2 + 2a_{12}x_1x_2 + a_{22}x_2^2 = \langle A\mathbf{x}, \mathbf{x} \rangle = \langle Q\Lambda Q^\mathsf{T}\mathbf{x}, \mathbf{x} \rangle$$

$$= \langle \Lambda Q^\mathsf{T}\mathbf{x}, Q^\mathsf{T}\mathbf{x} \rangle = \lambda_1 y_1^2 + \lambda_2 y_2^2.$$

The plane curve $\langle A\mathbf{x}, \mathbf{x} \rangle = 1$ is an ellipse if the eigenvalues of A are both positive; it is a hyperbola if one eigenvalue is positive and the other is negative. The change of variables $\mathbf{x} \mapsto Q^\mathsf{T}\mathbf{x}$ is from one orthonormal basis (the standard basis of $\mathbb{R}^2$) to the orthonormal basis comprising the columns of Q^T.

What happens if we have *two* real quadratic forms $\langle A\mathbf{x}, \mathbf{x} \rangle$ and $\langle B\mathbf{x}, \mathbf{x} \rangle$? For example, the first might represent the kinetic energy of a system of masses and the second might represent the potential energy of the system. Or they might represent plane ellipses corresponding to the orbits of two different planets. Can we find a real orthogonal matrix Q such that $Q^\mathsf{T}AQ$ and $Q^\mathsf{T}BQ$ are both diagonal? Not unless A commutes with B, which is a strong requirement;

this is Theorem 12.2.9.b. However, if we are willing to accept a change of variables to a new basis that is not necessarily orthonormal, then we need not require that A and B commute. It is sufficient that one of them be positive definite, and that is often the case in applications. The following theorem explains how this works.

Theorem 13.4.3 *Let $A, B \in M_n(\mathbb{F})$ be Hermitian, and suppose that A is positive definite. There is an invertible $S \in M_n(\mathbb{F})$ such that $A = SIS^*$, $B = S\Lambda S^*$, and $\Lambda \in M_n(\mathbb{R})$ is a real diagonal matrix whose diagonal entries are the eigenvalues of $A^{-1}B$.*

Proof The matrix $A^{-1/2}BA^{-1/2} \in M_n(\mathbb{F})$ is Hermitian, so there is a unitary $U \in M_n(\mathbb{F})$ and a real diagonal matrix Λ such that $A^{-1/2}BA^{-1/2} = U\Lambda U^*$. Let $S = A^{1/2}U$. Then $SIS^* = A^{1/2}UU^*A^{1/2} = A^{1/2}A^{1/2} = A$ and $S\Lambda S^* = A^{1/2}U\Lambda U^*A^{1/2} = A^{1/2}A^{-1/2}BA^{-1/2}A^{1/2} = B$. The eigenvalues of $A^{-1/2}BA^{-1/2}$ are real; they are the same as the eigenvalues of Λ and $A^{-1/2}A^{-1/2}B = A^{-1}B$ (Theorems 9.7.2 and 13.2.13.a). $\qquad\qquad\square$

Example 13.4.4 Consider the real symmetric matrices

$$A = \begin{bmatrix} 20 & -20 \\ -20 & 40 \end{bmatrix} \quad \text{and} \quad B = \begin{bmatrix} -34 & 52 \\ 52 & -56 \end{bmatrix}. \tag{13.4.5}$$

Theorem 13.1.24.d ensures that A is positive definite. We would like to find an invertible matrix S and a diagonal matrix Λ such that $A = SIS^*$ and $B = S\Lambda S^*$. First, use (13.2.7) to compute

$$A^{1/2} = \begin{bmatrix} 4 & -2 \\ -2 & 6 \end{bmatrix} \quad \text{and} \quad A^{-1/2} = \frac{1}{10}\begin{bmatrix} 3 & 1 \\ 1 & 2 \end{bmatrix}.$$

Then

$$A^{-1/2}BA^{-1/2} = \frac{1}{2}\begin{bmatrix} -1 & 3 \\ 3 & -1 \end{bmatrix} = U\Lambda U^*,$$

in which

$$\Lambda = \begin{bmatrix} 1 & 0 \\ 0 & -2 \end{bmatrix} \quad \text{and} \quad U = \frac{1}{\sqrt{2}}\begin{bmatrix} 1 & 1 \\ 1 & -1 \end{bmatrix}.$$

The preceding theorem says that if

$$S = A^{1/2}U = \sqrt{2}\begin{bmatrix} 1 & 3 \\ 2 & -4 \end{bmatrix},$$

then

$$A = SIS^*, \quad \text{and} \quad B = S\Lambda S^*.$$

The quadratic forms

$$20x^2 - 40xy + 40y^2 \quad \text{and} \quad -34x^2 + 104xy - 56y^2$$

become

$$\xi^2 + \eta^2 \quad \text{and} \quad \xi^2 - 2\eta^2$$

in the new variables

$$\xi = \sqrt{2}x + 2\sqrt{2}y \quad \text{and} \quad \eta = 3\sqrt{2}x - 4\sqrt{2}y.$$

13.5 The Schur Product Theorem

Definition 13.5.1 Let $A = [a_{ij}]$ and $B = [b_{ij}] \in M_{m \times n}$. The *Hadamard product* of A and B is $A \circ B = [a_{ij}b_{ij}] \in M_{m \times n}$.

Example 13.5.2 The Hadamard product of

$$A = \begin{bmatrix} 1 & 2 & 3 \\ 4 & 5 & 6 \end{bmatrix} \quad \text{and} \quad B = \begin{bmatrix} 6 & 5 & 0 \\ 3 & 2 & 1 \end{bmatrix}$$

is

$$A \circ B = \begin{bmatrix} 6 & 10 & 0 \\ 12 & 10 & 6 \end{bmatrix}.$$

The Hadamard product is also called the *entrywise product* or the *Schur product*. The following theorems list some of its properties.

Theorem 13.5.3 *Let $A = [a_{ij}] \in M_{m \times n}$, let $B, C \in M_{m \times n}$, and let $\gamma \in \mathbb{C}$.*

(a) $A \circ B = B \circ A$.

(b) $A \circ (\gamma B + C) = (\gamma B + C) \circ A = \gamma(A \circ B) + A \circ C$.

(c) $A \circ E = A$, *in which $E \in M_{m \times n}$ is the all-ones matrix. If $a_{ij} \neq 0$ for all i, j, then $A \circ [a_{ij}^{-1}] = E$.*

(d) $(A \circ B)^* = A^* \circ B^*$.

(e) *If $A, B \in M_n$ are Hermitian, then $A \circ B$ is Hermitian.*

(f) $A \circ I_n = \mathrm{diag}(a_{11}, a_{22}, \ldots, a_{nn})$.

(g) *If $\Lambda = \mathrm{diag}(\lambda_1, \lambda_2, \ldots, \lambda_n)$, then*

$$\Lambda A + A\Lambda = [\lambda_i + \lambda_j] \circ A = A \circ [\lambda_i + \lambda_j].$$

Proof To prove (g), compute
$$\Lambda A + A\Lambda = [\lambda_i a_{ij} + a_{ij}\lambda_j] = [(\lambda_i + \lambda_j)a_{ij}] = [\lambda_i + \lambda_j] \circ A$$
and use (a). See P.13.40. □

Theorem 13.5.4 *Let $\mathbf{x} = [x_i], \mathbf{y} = [y_i] \in \mathbb{C}^n$ and let $A \in M_n$.*

(a) $\mathbf{x} \circ \mathbf{y} = [x_i y_i] \in \mathbb{C}^n$.

(b) $\mathbf{x}\mathbf{x}^* \circ \mathbf{y}\mathbf{y}^* = (\mathbf{x} \circ \mathbf{y})(\mathbf{x} \circ \mathbf{y})^*$.

(c) *Let $D = \mathrm{diag}(x_1, x_2, \ldots, x_n)$. Then $\mathbf{x}\mathbf{x}^* \circ A = DAD^* = A \circ \mathbf{x}\mathbf{x}^*$.*

Proof To prove (b), compute $\mathbf{x}\mathbf{x}^* = [x_i \overline{x_j}] \in M_n$ and

$$\mathbf{x}\mathbf{x}^* \circ \mathbf{y}\mathbf{y}^* = [x_i \overline{x_j} y_i \overline{y_j}] = [x_i y_i \overline{x_j y_j}] = (\mathbf{x} \circ \mathbf{y})(\mathbf{x} \circ \mathbf{y})^*.$$

See P.13.41. □

The ordinary matrix product of two positive semidefinite matrices need not be positive semidefinite; see (13.2.12). However, the following theorem shows that the Hadamard product of two positive semidefinite matrices is positive semidefinite; see also P.13.48.

Theorem 13.5.5 (Schur Product Theorem) *If $A, B \in M_n$ are positive semidefinite, then $A \circ B$ is positive semidefinite.*

Proof Theorem 13.1.2.c ensures that there are $X, Y \in M_n$ such that $A = XX^*$ and $B = YY^*$. Partition $X = [\mathbf{x}_1 \ \mathbf{x}_2 \ \dots \ \mathbf{x}_n]$ and $Y = [\mathbf{y}_1 \ \mathbf{y}_2 \ \dots \ \mathbf{y}_n]$ according to their columns. Then

$$A = XX^* = \sum_{i=1}^{n} \mathbf{x}_i \mathbf{x}_i^* \quad \text{and} \quad B = YY^* = \sum_{i=1}^{n} \mathbf{y}_i \mathbf{y}_i^*. \tag{13.5.6}$$

Therefore,

$$A \circ B = \left(\sum_{i=1}^{n} \mathbf{x}_i \mathbf{x}_i^* \right) \circ \left(\sum_{j=1}^{n} \mathbf{y}_j \mathbf{y}_j^* \right) = \sum_{i,j=1}^{n} \left((\mathbf{x}_i \mathbf{x}_i^*) \circ (\mathbf{y}_j \mathbf{y}_j^*) \right)$$

$$= \sum_{i,j=1}^{n} (\mathbf{x}_i \circ \mathbf{y}_j)(\mathbf{x}_i \circ \mathbf{y}_j)^*,$$

in which we have used properties of the Hadamard product listed in the preceding two theorems. Thus, $A \circ B$ equals a sum of rank-1 positive semidefinite matrices. Each summand is positive semidefinite, so Theorem 13.1.20 ensures that $A \circ B$ is positive semidefinite. □

A Hadamard product of positive semidefinite matrices can be positive definite even if some of its factors are not invertible.

Example 13.5.7 Let

$$A = \begin{bmatrix} 1 & 1 \\ 1 & 2 \end{bmatrix} \quad \text{and} \quad B = \begin{bmatrix} 1 & \sqrt{2} \\ \sqrt{2} & 2 \end{bmatrix}.$$

Then A is positive definite, B is positive semidefinite and not invertible, and

$$A \circ B = \begin{bmatrix} 1 & \sqrt{2} \\ \sqrt{2} & 4 \end{bmatrix}$$

is positive definite; see Theorem 13.1.8.b2.

Corollary 13.5.8 *Let $A, B \in M_n$ be positive semidefinite.*

(a) *If A is positive definite and every main diagonal entry of B is positive, then $A \circ B$ is positive definite.*

(b) *If A and B are positive definite, then $A \circ B$ is positive definite.*

Proof (a) Let $\lambda_1 \leq \lambda_2 \leq \dots \leq \lambda_n$ be the eigenvalues of A. Then $\lambda_1 > 0$ since A is positive definite. The matrix $A - \lambda_1 I$ is Hermitian and has eigenvalues

$$0, \ \lambda_2 - \lambda_1, \ \lambda_3 - \lambda_1, \ \dots, \ \lambda_n - \lambda_1,$$

all of which are nonnegative. Therefore, $A - \lambda_1 I$ is positive semidefinite. The preceding theorem ensures that $(A - \lambda_1 I) \circ B$ is positive semidefinite. Let $\mathbf{x} = [x_i] \in \mathbb{C}^n$ be nonzero, let $B = [b_{ij}]$, let $\beta = \min_{1 \leq i \leq n} b_{ii}$, and compute

$$
\begin{aligned}
0 &\leq \langle ((A - \lambda_1 I) \circ B)\mathbf{x}, \mathbf{x} \rangle \\
&= \langle ((A \circ B) - \lambda_1 I \circ B)\mathbf{x}, \mathbf{x} \rangle \\
&= \langle (A \circ B)\mathbf{x}, \mathbf{x} \rangle - \langle (\lambda_1 I \circ B)\mathbf{x}, \mathbf{x} \rangle \\
&= \langle (A \circ B)\mathbf{x}, \mathbf{x} \rangle - \lambda_1 \langle \operatorname{diag}(b_{11}, b_{22}, \ldots, b_{nn})\mathbf{x}, \mathbf{x} \rangle \\
&= \langle (A \circ B)\mathbf{x}, \mathbf{x} \rangle - \lambda_1 \sum_{i=1}^{n} b_{ii}|x_i|^2 \\
&\leq \langle (A \circ B)\mathbf{x}, \mathbf{x} \rangle - \lambda_1 \beta \sum_{i=1}^{n} |x_i|^2 \\
&= \langle (A \circ B)\mathbf{x}, \mathbf{x} \rangle - \lambda_1 \beta \|\mathbf{x}\|_2^2.
\end{aligned}
$$

Therefore,

$$
\langle (A \circ B)\mathbf{x}, \mathbf{x} \rangle \geq \lambda_1 \beta \|\mathbf{x}\|_2^2 > 0, \tag{13.5.9}
$$

so $A \circ B$ is positive definite.

(b) If B is positive definite, then

$$
b_{ii} = \langle B\mathbf{e}_i, \mathbf{e}_i \rangle > 0, \qquad i = 1, 2, \ldots, n,
$$

so the assertion follows from (a). $\qquad\square$

Example 13.5.10 Let $A, B \in \mathsf{M}_n$ be positive definite and consider the *Lyapunov equation*

$$
AX + XA = B. \tag{13.5.11}
$$

Since $\operatorname{spec} A \cap \operatorname{spec}(-A) = \varnothing$, Theorem 10.4.1 ensures that (13.5.11) has a unique solution X. We claim that X is positive definite. Let $A = U\Lambda U^*$ be a spectral decomposition, in which $\Lambda = \operatorname{diag}(\lambda_1, \lambda_2, \ldots, \lambda_n)$ and $U \in \mathsf{M}_n$ is unitary. Write (13.5.11) as

$$
U\Lambda U^* X + XU\Lambda U^* = B,
$$

which is equivalent to

$$
\Lambda(U^* XU) + (U^* XU)\Lambda = U^* BU. \tag{13.5.12}
$$

Use Theorem 13.5.3.g to write (13.5.12) as

$$
[\lambda_i + \lambda_j] \circ (U^* XU) = U^* BU,
$$

which is equivalent to

$$
U^* XU = [(\lambda_i + \lambda_j)^{-1}] \circ (U^* BU).
$$

Since B is positive definite, $U^* BU$ is positive definite (Theorem 13.1.23.b). The Hermitian matrix $[(\lambda_i + \lambda_j)^{-1}] \in \mathsf{M}_n$ has positive main diagonal entries and is positive semidefinite; see

P.13.33. Corollary 13.5.8 ensures that $[(\lambda_i + \lambda_j)^{-1}] \circ (U^*BU)$ is positive definite. The solution to (13.5.11) is

$$X = U\left([(\lambda_i + \lambda_j)^{-1}] \circ (U^*BU)\right) U^*.$$

Theorem 13.1.23.b tells us that X is positive definite.

13.6 Problems

P.13.1 Let $K \in \mathsf{M}_n$ be Hermitian. Show that $K = 0$ if and only if $\langle Kx, x \rangle = 0$ for all $x \in \mathbb{C}^n$. *Hint*: Consider $\langle K(x + e^{i\theta}y), x + e^{i\theta}y \rangle$.

P.13.2 Let $A \in \mathsf{M}_n$. (a) If A is Hermitian, show that $\langle Ax, x \rangle$ is real for all $x \in \mathbb{C}^n$. (b) If $\langle Ax, x \rangle$ is real for all $x \in \mathbb{C}^n$, show that A is Hermitian. *Hint*: Consider the Cartesian decomposition $A = H + iK$.

P.13.3 Let $\Lambda \in \mathsf{M}_n(\mathbb{R})$ be diagonal. Show that $\Lambda = P - Q$, in which $P, Q \in \mathsf{M}_n(\mathbb{R})$ are diagonal and positive semidefinite.

P.13.4 Let $A \in \mathsf{M}_n$ be Hermitian. Show that there are commuting positive semidefinite $B, C \in \mathsf{M}_n$ such that $A = B - C$.

P.13.5 Adopt the notation of Proposition 13.2.6. (a) Show that $\tau = \operatorname{tr} A^{1/2}$. (b) If $\operatorname{spec} A = \{\lambda, \mu\}$ and $\lambda \neq \mu$, use (9.8.4) to show that

$$A^{1/2} = \frac{\sqrt{\lambda} - \sqrt{\mu}}{\lambda - \mu} A + \frac{\lambda\sqrt{\mu} - \mu\sqrt{\lambda}}{\lambda - \mu} I = \frac{1}{\sqrt{\lambda} + \sqrt{\mu}}\left(A + \sqrt{\lambda\mu}I\right)$$

and verify that this is the same matrix as in (13.2.7). What happens if $\lambda = \mu$?

P.13.6 Let $A, B \in \mathsf{M}_n$ be positive semidefinite. Show that AB is positive semidefinite if and only if A commutes with B.

P.13.7 Let $A, B \in \mathsf{M}_n$ be positive semidefinite. (a) Show that $A = 0$ if and only if $\operatorname{tr} A = 0$. (b) Show that $A = B = 0$ if and only if $A + B = 0$.

P.13.8 Let $A, C \in \mathsf{M}_n$ be positive semidefinite and suppose that $C^2 = A$. Let $B = A^{1/2}$ and let p be a polynomial such that $B = p(A)$. Provide details for the following outline of a proof that $C = B$:

(a) B commutes with C.

(b) $(B - C)^*B(B - C) + (B - C)^*C(B - C) = (B^2 - C^2)(B - C) = 0.$

(c) $(B - C)B(B - C) = (B - C)C(B - C) = 0.$

(d) $B - C = 0.$ *Hint*: P.9.6.

P.13.9 Let $A, B \in \mathsf{M}_n$ be positive semidefinite and suppose that $A^2 = B^2$. (a) If A and B are invertible, show that $V = A^{-1}B$ is unitary. Why are the eigenvalues of V real and positive? Deduce that $V = I$ and $A = B$. (b) If $A = 0$, show that $B = 0$. *Hint*: P.9.6 or Theorem 13.1.10. (c) Suppose that A and B are not invertible and $A \neq 0$. Let $U = [U_1\ U_2] \in \mathsf{M}_n$ be unitary, in which the columns of U_2 are an orthonormal basis for $\operatorname{null} A$. Show that

$$U^*AU = \begin{bmatrix} U_1^*AU_1 & U_1^*AU_2 \\ U_2^*AU_1 & U_2^*AU_2 \end{bmatrix} = \begin{bmatrix} U_1^*AU_1 & 0 \\ 0 & 0 \end{bmatrix}$$

and let

$$U^*BU = \begin{bmatrix} U_1^*BU_1 & U_1^*BU_2 \\ U_2^*BU_1 & U_2^*BU_2 \end{bmatrix}.$$

Why is $B^2U_2 = 0$? Why is $BU_2 = 0$? Why is $(U^*AU)^2 = (U^*BU)^2$? Deduce from (a) that $A = B$. (d) State a theorem that summarizes what you have proved in (a)–(c).

P.13.10 Let $A, B \in M_n$ and suppose that A is positive semidefinite. (a) If B is Hermitian, show that AB has real eigenvalues. (b) If B is positive semidefinite, show that AB has real nonnegative eigenvalues. *Hint*: Consider the eigenvalues of $AB = A^{1/2}(A^{1/2}B)$ and $A^{1/2}BA^{1/2}$.

P.13.11 Let $A \in M_n$ be Hermitian and let $\mathbf{x} \in \mathbb{C}^n$ be a unit vector. Is it possible for $A\mathbf{x}$ to be nonzero and orthogonal to $\mathbf{x}$? Is that possible if A is positive semidefinite? Discuss.

P.13.12 If $A \in M_n$ is positive definite, show that $(A^{-1})^{1/2} = (A^{1/2})^{-1}$.

P.13.13 Let $A = [a_{ij}] \in M_2$ be Hermitian.

(a) Show that A is positive semidefinite if and only if $\operatorname{tr} A \geq 0$ and $\det A \geq 0$.

(b) Show that A is positive definite if and only if $\operatorname{tr} A \geq 0$ and $\det A > 0$.

(c) Show that A is positive definite if and only if $a_{11} > 0$ and $\det A > 0$.

(d) If $a_{11} \geq 0$ and $\det A \geq 0$, show that A need not be positive semidefinite.

P.13.14 Give an example of a Hermitian $A \in M_3$ for which $\operatorname{tr} A \geq 0$, $\det A \geq 0$, and A is not positive semidefinite.

P.13.15 Prove the Cauchy–Schwarz inequality in $\mathbb{F}^n$ by computing $\det A^*A$, in which $A = [\mathbf{x}\ \mathbf{y}]$ and $\mathbf{x}, \mathbf{y} \in \mathbb{F}^n$.

P.13.16 Let $A, B \in M_n$ be positive definite. Although AB need not be positive definite, show that $\operatorname{tr} AB > 0$ and $\det AB > 0$.

P.13.17 Let $A \in M_n$ be normal. Show that A is positive semidefinite if and only if $\operatorname{tr} AB$ is real and nonnegative for every positive semidefinite $B \in M_n$.

P.13.18 Consider

$$A = \begin{bmatrix} 10 & 0 \\ 0 & 1 \end{bmatrix}, \quad B = \begin{bmatrix} 1 & -1 \\ -1 & 2 \end{bmatrix}, \quad \text{and} \quad C = \begin{bmatrix} 3 & 5 \\ 5 & 10 \end{bmatrix}.$$

Show that A, B, and C are positive definite, but $\operatorname{tr} ABC < 0$. What can you say about $\det ABC$?

P.13.19 Let $A, B \in M_n$ and suppose that A is positive semidefinite. Show that $AB = BA$ if and only if $A^{1/2}B = BA^{1/2}$.

P.13.20 Review Example 13.3.6. (a) Prove by induction or otherwise that the $n \times n$ min matrix has a Cholesky factorization LL^*, in which $L \in M_n$ is lower triangular and has every entry on and below the diagonal equal to 1. (b) Show that the inverse of L^* is the upper bidiagonal matrix with entries 1 on the diagonal, and entries -1 in the superdiagonal. (c) Conclude that the inverse of the $n \times n$ min matrix is the tridiagonal matrix with entries -1 in the superdiagonal and subdiagonal, and entries 2 on the main diagonal except for the (n, n) entry, which is 1.

P.13.21 Consider the $n \times n$ tridiagonal matrix with entries 2 on the main diagonal and entries -1 in the superdiagonal and subdiagonal. This matrix arises in numerical solution schemes for differential equations. Use the preceding problem to show that it is positive definite.

P.13.22 Let $A = [a_{ij}] \in M_n$ be positive definite and let $k \in \{1, 2, \ldots, n - 1\}$. Show in the following two ways that the $k \times k$ leading principle submatrix of A is positive definite. (a) Use the Cholesky factorization of A. (b) Consider $\langle Ax, x \rangle$, in which the last $n - k$ entries of $\mathbf{x}$ are zero.

P.13.23 Let $A \in M_n$ be positive definite, let $k \in \{1, 2, \ldots, n - 1\}$, and let A_k be the $k \times k$ leading principal submatrix of A. Suppose that $A_k = L_k L_k^*$ is the Cholesky factorization of A_k. (a) Show that the Cholesky factorization of A_{k+1} has the form

$$A_{k+1} = \begin{bmatrix} A_k & \mathbf{a} \\ \mathbf{a}^* & \alpha \end{bmatrix} = \begin{bmatrix} L_k & \mathbf{0}_{k \times 1} \\ \mathbf{x}^* & \lambda \end{bmatrix} \begin{bmatrix} L_k^* & \mathbf{x} \\ \mathbf{0}_{1 \times k} & \lambda \end{bmatrix}, \quad \mathbf{x} \in \mathbb{C}^k, \lambda > 0.$$

(b) Why does the linear system $L_k \mathbf{x} = \mathbf{a}$ have a unique solution? (c) Why does the equation $\lambda^2 = \alpha - \mathbf{x}^* \mathbf{x}$ have a unique positive solution λ? *Hint*: Consider $\alpha - \mathbf{a}^* A_k^{-1} \mathbf{a}$. (d) Describe an inductive algorithm to compute the Cholesky factorization of A. (e) Use your algorithm to compute the Cholesky factorization of the 4×4 min matrix.

P.13.24 Let $A = [a_{ij}] \in M_n$ be positive definite. Let $A = LL^*$ be its Cholesky factorization, in which $L = [\ell_{ij}]$. Partition $L^* = [\mathbf{x}_1 \ \mathbf{x}_2 \ \ldots \ \mathbf{x}_n]$ according to its columns. Show that $a_{ii} = \|\mathbf{x}_i\|_2^2 \geq \ell_{ii}^2$ and deduce that $\det A \leq a_{11} a_{22} \cdots a_{nn}$, with equality if and only if A is diagonal. This is one version of *Hadamard's Inequality*.

P.13.25 Partition $A = [\mathbf{a}_1 \ \mathbf{a}_2 \ \ldots \ \mathbf{a}_n] \in M_n$ according to its columns. Apply the result in the preceding problem to A^*A and prove that $|\det A| \leq \|\mathbf{a}_1\|_2 \|\mathbf{a}_2\|_2 \cdots \|\mathbf{a}_n\|_2$, with equality if and only if either A has orthogonal columns or at least one of its columns is zero. This is another version of *Hadamard's Inequality*; see (6.7.2).

P.13.26 Let $A, B \in M_n$ be Hermitian and suppose that A is invertible. If there is an invertible $S \in M_n$ such that S^*AS and S^*BS are both diagonal, show that $A^{-1}B$ is diagonalizable and has real eigenvalues. This necessary condition for simultaneous diagonalization of two Hermitian matrices is known to be sufficient as well; see [HJ13, Thm. 4.5.17].

P.13.27 Let $A = \begin{bmatrix} 0 & 1 \\ 1 & 0 \end{bmatrix}$. Show that there is no invertible $S \in M_2$ such that S^*AS and S^*BS are both diagonal if (a) $B = \begin{bmatrix} 1 & 0 \\ 0 & -1 \end{bmatrix}$, or (b) if $B = \begin{bmatrix} 1 & 1 \\ 1 & 0 \end{bmatrix}$.

P.13.28 Let $A \in M_n$ and suppose that $A + A^*$ is positive definite. Show that there is an invertible $S \in M_n$ such that S^*AS is diagonal.

P.13.29 Show that a square complex matrix is the product of two positive definite matrices if and only if it is diagonalizable and has positive eigenvalues.

P.13.30 Let $A, B \in M_n$ be Hermitian and suppose that A is positive definite. Show that $A + B$ is positive definite if and only if spec $A^{-1}B \subseteq (-1, \infty)$.

P.13.31 Let $A \in M_n$ be positive semidefinite and partitioned as in (13.1.13). (a) If A_{11} is not invertible, prove that A is not invertible. (b) If A_{11} is invertible, must A be invertible?

P.13.32 Let $\mathcal{V}$ be an $\mathbb{F}$-inner product space and let $\mathbf{u}_1, \mathbf{u}_2, \ldots, \mathbf{u}_n \in \mathcal{V}$. (a) Show that the Gram matrix $G = [\langle \mathbf{u}_j, \mathbf{u}_i \rangle] \in M_n(\mathbb{F})$ is positive semidefinite. (b) Show that G is positive definite if and only if $\mathbf{u}_1, \mathbf{u}_2, \ldots, \mathbf{u}_n$ are linearly independent. (c) Show that the matrix $[(i + j - 1)^{-1}]$ in P.4.23 is positive definite.

P.13.33 Let $\lambda_1, \lambda_2, \ldots, \lambda_n$ be real and positive. (a) Show that

$$\frac{1}{\lambda_i + \lambda_j} = \int_0^\infty e^{-\lambda_i t} e^{-\lambda_j t}\, dt, \qquad i, j = 1, 2, \ldots, n.$$

(b) Show that the *Cauchy matrix* $[(\lambda_i + \lambda_j)^{-1}] \in M_n$ is positive semidefinite.

P.13.34 Let $A = [a_{ij}] \in M_3$ be positive semidefinite and partitioned as in (13.1.13). Suppose that

$$A_{11} = \begin{bmatrix} 1 & e^{i\theta} \\ e^{-i\theta} & 1 \end{bmatrix}, \qquad \theta \in \mathbb{R}.$$

If $a_{13} = \frac{1}{2}$, what is a_{23}?

P.13.35 Let $A = [a_{ij}] \in M_n$ be positive semidefinite and partitioned as in (13.1.13). Show that there is a $Y \in M_{k \times (n-k)}$ such that $A_{12} = Y A_{22}$. Deduce that $\operatorname{rank} A_{12} \leq \operatorname{rank} A_{22}$.

P.13.36 Let $A = [a_{ij}] \in M_n$ be partitioned as in (13.1.13). (a) Show that

$$\operatorname{rank} A \leq \operatorname{rank} \begin{bmatrix} A_{11} & A_{12} \end{bmatrix} + \operatorname{rank} \begin{bmatrix} A_{12}^* & A_{22} \end{bmatrix}.$$

(b) If A is positive semidefinite, deduce that $\operatorname{rank} A \leq \operatorname{rank} A_{11} + \operatorname{rank} A_{22}$. (c) Show by example that the inequality in (b) need not be correct if A is Hermitian but not positive semidefinite.

P.13.37 Let $A, B \in M_n$ be normal. Use Definition 12.1.1 and Theorem 13.1.10 to show that $\operatorname{rank} AB = \operatorname{rank} BA$. See P.12.38 for a different proof.

P.13.38 Let $H \in M_n$ be Hermitian and suppose that $\operatorname{spec} H \subseteq [-1, 1]$. (a) Show that $I - H^2$ is positive semidefinite and let $U = H + i(I - H^2)^{1/2}$. (b) Show that U is unitary and $H = \frac{1}{2}(U + U^*)$.

P.13.39 Let $A \in M_n$. Use the Cartesian decomposition and the preceding problem to show that A is a linear combination of at most four unitary matrices. See P.15.40 for a related result.

P.13.40 Prove (a)–(f) in Theorem 13.5.3.

P.13.41 Prove (a)–(c) in Theorem 13.5.4.

P.13.42 Let $A = [a_{ij}] \in M_n$ be normal and let $A = U \Lambda U^*$ be a spectral decomposition, in which $\Lambda = \operatorname{diag}(\lambda_1, \lambda_2, \ldots, \lambda_n)$ and $U = [u_{ij}] \in M_n$ is unitary. Let $\mathbf{a} = [a_{11}\ a_{22}\ \cdots\ a_{nn}]^\mathsf{T}$ and $\lambda = [\lambda_1\ \lambda_2\ \cdots\ \lambda_n]^\mathsf{T}$. (a) Show that

$$\mathbf{a} = (U \circ \overline{U}) \lambda. \tag{13.6.1}$$

(b) Show that $U \circ \overline{U}$ is a Markov matrix. (c) Is $(U \circ \overline{U})^\mathsf{T}$ a Markov matrix?

P.13.43 Let $A \in M_3$ be normal and suppose that 1, i, and $2 + 2i$ are its eigenvalues. (a) Use (13.6.1) to draw a region in the complex plane that must contain the diagonal entries of A. (b) Could $2i$ be a diagonal entry of A? How about -1 or $1 + i$? (c) Why must every main diagonal entry of A have nonnegative real part? (d) How large could the real part of a diagonal entry be?

P.13.44 Let $A = [a_{ij}] \in M_n$ be normal and let $\operatorname{spec} A = \{\lambda_1, \lambda_2, \ldots, \lambda_n\}$. Use (13.6.1) to show that $\min_{1 \leq i \leq n} \operatorname{Re} \lambda_i \leq \min_{1 \leq i \leq n} \operatorname{Re} a_{ii}$. Illustrate this inequality with a figure that plots the eigenvalues and main diagonal entries in the complex plane. What can you say about $\max \operatorname{Re} a_{ii}$ and $\max \operatorname{Re} \lambda_i$?

P.13.45 Let $A = [a_{ij}] \in \mathsf{M}_n$ be Hermitian and let $\lambda_1 \leq \lambda_2 \leq \cdots \leq \lambda_n$ be its eigenvalues. Use (13.6.1) to show that $\lambda_1 \leq \min_{1 \leq i \leq n} a_{ii}$. What can you say about the maximum diagonal entry and λ_n?

P.13.46 Let $A, B \in \mathsf{M}_n$ be positive semidefinite. Let γ be the largest main diagonal entry of B and let λ_n be the largest eigenvalue of A. Show that

$$\langle (A \circ B)\mathbf{x}, \mathbf{x} \rangle \leq \gamma \lambda_n \|\mathbf{x}\|_2^2, \qquad \mathbf{x} \in \mathbb{C}^n. \tag{13.6.2}$$

Hint: Proceed as in the proof of Corollary 13.5.8.

P.13.47 Let $A, B \in \mathsf{M}_n$ be positive definite. (a) Show that there is a unique $X \in \mathsf{M}_n$ such that $H(AX) = B$; see Definition 12.6.6. (b) Show that X is positive definite.

P.13.48 Let $k > 2$ and let $A_1, A_2, \ldots, A_k \in \mathsf{M}_n$ be positive semidefinite. (a) Show that $A_1 \circ A_2 \circ \cdots \circ A_k$ is positive semidefinite. (b) If every main diagonal entry of A_i is positive for all $i = 1, 2, \ldots, k$ and if A_j is positive definite for some $j \in \{1, 2, \ldots, k\}$, show that $A_1 \circ A_2 \circ \cdots \circ A_k$ is positive semidefinite.

P.13.49 Let $A = [a_{ij}] \in \mathsf{M}_n$ be positive semidefinite. Show that $B = [e^{a_{ij}}] \in \mathsf{M}_n$ is positive semidefinite. *Hint*: See P.A.13 and represent B as a sum of positive real multiples of Hadamard products.

P.13.50 Let $A, B \in \mathsf{M}_n$ be positive semidefinite. Show that $\mathrm{rank}(A \circ B) \leq (\mathrm{rank}\, A)(\mathrm{rank}\, B)$. *Hint*: Only $\mathrm{rank}\, A$ summands are needed to represent A in (13.5.6).

P.13.51 If $A \in \mathsf{M}_m$ and $B \in \mathsf{M}_n$ are positive semidefinite, show that $A \otimes B$ is positive semidefinite.

P.13.52 Let $A = [a_{ij}] \in \mathsf{M}_3$ and $B = [b_{ij}] \in \mathsf{M}_3$. Write down the 3×3 principal submatrix of $A \otimes B \in \mathsf{M}_9$ whose entries lie in the intersections of its first, fifth, and ninth rows with its first, fifth, and ninth columns. Do you recognize this matrix?

P.13.53 Let $A = [a_{ij}] \in \mathsf{M}_n$ and $B = [b_{ij}] \in \mathsf{M}_n$. (a) Show that $A \circ B \in \mathsf{M}_n$ is a principal submatrix of $A \otimes B \in \mathsf{M}_{n^2}$. (b) Deduce Theorem 13.5.5 from (a) and P.13.51.

13.7 Notes

See [HJ13, Sect. 3.5] for various ways in which a matrix can be factored as a product of triangular and permutation matrices. For example, each $A \in \mathsf{M}_n$ can be factored as $A = PLR$, in which L is lower triangular, R is upper triangular, and P is a permutation matrix.

See [HJ13, Sect. 4.5] for a discussion of how Theorem 13.4.3 is used to analyze small oscillations of a mechanical system about a stable equilibrium.

In 1899, the French mathematician Jacques Hadamard published a paper about power series

$$f(z) = \sum_{k=1}^{\infty} a_k z^k \qquad \text{and} \qquad g(z) = \sum_{k=1}^{\infty} b_k z^k,$$

in which he studied the product

$$(f \circ g)(z) = \sum_{k=1}^{\infty} a_k b_k z^k.$$

Because of that paper, Hadamard's name has been linked with term-by-term (or entrywise) products of various kinds. The first systematic study of entrywise products of matrices was published in 1911 by Isaai Schur. For a historical survey of the Hadamard product, see [HJ94, Sect. 5.0]. For a detailed treatment of Hadamard products and applications, see [HJ94, Ch. 5 and Sect. 6.3]; see also [HJ13, Sect. 7.5].

The eigenvalues of the tridiagonal matrix in P.13.20 are known to be $4\sin^2((2k+1)\pi/(4n+2))$, for $k = 0, 1, \ldots, n-1$, so the eigenvalues of the $n \times n$ min matrix are $\frac{1}{4}\csc^2((2k+1)\pi/(4n+2))$, for $k = 0, 1, \ldots, n-1$. For a continuous analog of the Cholesky factorization of the min matrix, see P.14.27.

13.8 Some Important Concepts

- Positive definite and semidefinite matrices.

- Characterization as Gram matrices and as normal matrices with real nonnegative eigenvalues.

- Column inclusion.

- Diagonal dominance and positive definiteness (Theorem 13.1.24).

- Unique positive semidefinite square root of a positive semidefinite matrix.

- Cholesky factorization of a positive semidefinite matrix.

- Simultaneous diagonalization (by *congruence) of a positive definite matrix and a Hermitian matrix (Theorem 13.4.3).

- Hadamard product and the Schur product theorem.

The Singular Value and Polar Decompositions

This chapter is about two closely related factorizations of a complex matrix. Both are matrix analogs of the polar form of a complex number. If $z = |z|e^{i\theta}$, then $e^{-i\theta}$ is a complex number with modulus 1 such that $ze^{-i\theta} \geq 0$. For an $m \times n$ complex matrix A, the polar decomposition theorem says that there is a unitary matrix U such that either AU or UA is positive semidefinite. The singular value decomposition theorem says that there are unitary matrices U and V such that $UAV = [\sigma_{ij}]$ has $\sigma_{ij} = 0$ if $i \neq j$ and each $\sigma_{ii} \geq 0$. Information provided by these factorizations finds important applications in data analysis, image compression, least squares solutions of linear systems, approximation problems, and error analysis for solutions of linear systems.

14.1 The Singular Value Decomposition

Schur's triangularization theorem uses one unitary matrix to reduce a square matrix to upper triangular form by similarity. The singular value decomposition uses two unitary matrices to reduce a matrix to a diagonal form. The singular value decomposition has many applications to theoretical and computational problems.

If $A \in M_{m \times n}$ and rank $A = r \geq 1$, Theorem 9.7.2 says that the positive semidefinite matrices $A^*A \in M_n$ and $AA^* \in M_m$ have the same nonzero eigenvalues. They are positive, and there are r of them (Theorem 13.1.10). The positive square roots of these eigenvalues are given a special name.

Definition 14.1.1 Let $A \in M_{m \times n}$ and let $q = \min\{m, n\}$. If rank $A = r \geq 1$, let $\sigma_1 \geq \sigma_2 \geq \cdots \geq \sigma_r > 0$ be the decreasingly ordered positive eigenvalues of $(A^*A)^{1/2}$. The *singular values* of A are

$$\sigma_1, \sigma_2, \ldots, \sigma_r, \quad \text{and} \quad \sigma_{r+1} = \sigma_{r+2} = \cdots = \sigma_q = 0.$$

If $A = 0$, then the *singular values* of A are $\sigma_1 = \sigma_2 = \cdots = \sigma_q = 0$.

The singular values of $A \in M_n$ are the eigenvalues of $(A^*A)^{1/2}$, which are the same as the eigenvalues of $(AA^*)^{1/2}$.

Example 14.1.2 Consider

$$A = \begin{bmatrix} 0 & 20 & 30 \\ 0 & 0 & 20 \\ 0 & 0 & 0 \end{bmatrix} \quad \text{and} \quad A^* = \begin{bmatrix} 0 & 0 & 0 \\ 20 & 0 & 0 \\ 30 & 20 & 0 \end{bmatrix}.$$

Then spec $A = \{0\}$, rank $A = 2$,

$$A^*A = \begin{bmatrix} 0 & 0 & 0 \\ 0 & 400 & 600 \\ 0 & 600 & 1300 \end{bmatrix}, \quad (A^*A)^{1/2} = \begin{bmatrix} 0 & 0 & 0 \\ 0 & 16 & 12 \\ 0 & 12 & 34 \end{bmatrix},$$

and $\operatorname{spec}(A^*A)^{1/2} = \{40, 10, 0\}$. The singular values of A are $40, 10, 0$ and its eigenvalues are $0, 0, 0$.

Theorem 14.1.3 *Let* $A \in \mathsf{M}_{m \times n}$, *let* $r = \operatorname{rank} A$ *and* $q = \min\{m, n\}$, *let* $\sigma_1 \geq \sigma_2 \geq \cdots \geq \sigma_q$ *be the singular values of* A, *and let* $c \in \mathbb{C}$.

(a) $\sigma_1^2, \sigma_2^2, \ldots, \sigma_r^2$ *are the positive eigenvalues of* A^*A *and* AA^*.

(b) $\sum_{i=1}^{q} \sigma_i^2 = \operatorname{tr} A^*A = \operatorname{tr} AA^* = \|A\|_F^2$.

(c) A, A^*, A^T, *and* $\overline{A}$ *have the same singular values.*

(d) *The singular values of* cA *are* $|c|\sigma_1, |c|\sigma_2, \ldots, |c|\sigma_q$.

Proof (a) Theorem 13.2.2 ensures that $\sigma_1^2, \sigma_2^2, \ldots, \sigma_r^2$ are the positive eigenvalues of $A^*A = ((A^*A)^{1/2})^2$. Theorem 9.7.2 says that they are the positive eigenvalues of AA^*.

(b) $\operatorname{tr} A^*A = \|A\|_F^2$ is a definition (see Example 4.5.5), and $\operatorname{tr} A^*A$ is the sum of the eigenvalues of A^*A.

(c) The positive semidefinite matrices A^*A, AA^*, $A^\mathsf{T}\overline{A} = \overline{A^*A}$, and $\overline{A}A^\mathsf{T} = \overline{AA^*}$ have the same nonzero eigenvalues; see Theorem 9.2.6.

(d) The eigenvalues of $(cA^*)(cA) = |c|^2 A^*A$ are $|c|^2\sigma_1, |c|^2\sigma_2, \ldots, |c|^2\sigma_q$; see Corollary 10.1.4. $\qquad\square$

The symbol Σ in the following theorem is conventional notation for the *singular value matrix*; in this context it has nothing to do with summation.

Theorem 14.1.4 (Singular Value Decomposition) *Let* $A \in \mathsf{M}_{m \times n}(\mathbb{F})$ *be nonzero and let* $r = \operatorname{rank} A$. *Let* $\sigma_1 \geq \sigma_2 \geq \cdots \geq \sigma_r > 0$ *be the positive singular values of* A *and define*

$$\Sigma_r = \begin{bmatrix} \sigma_1 & & 0 \\ & \ddots & \\ 0 & & \sigma_r \end{bmatrix} \in \mathsf{M}_r(\mathbb{R}).$$

Then there are unitary matrices $V \in \mathsf{M}_m(\mathbb{F})$ *and* $W \in \mathsf{M}_n(\mathbb{F})$ *such that*

$$A = V \Sigma W^*,$$

in which

$$\Sigma = \begin{bmatrix} \Sigma_r & 0_{r \times (n-r)} \\ 0_{(m-r) \times r} & 0_{(m-r) \times (n-r)} \end{bmatrix} \in \mathsf{M}_{m \times n}(\mathbb{R}) \tag{14.1.5}$$

is the same size as A. *If* $m = n$, *then* $V, W \in \mathsf{M}_n(\mathbb{F})$ *and* $\Sigma = \Sigma_r \oplus 0_{n-r}$.

Proof Suppose that $m \geq n$. Let $A^*A = W \Lambda W^*$ be a spectral decomposition, in which

$$\Lambda = \operatorname{diag}(\sigma_1^2, \sigma_2^2, \ldots, \sigma_r^2, 0, \ldots, 0) = \Sigma_r^2 \oplus 0_{n-r} \in \mathsf{M}_n.$$

Theorem 12.2.2 ensures that $W \in M_n(\mathbb{F})$ may be taken to be real if A is real. Let $D = \Sigma_r \oplus I_{n-r}$ and let

$$B = AWD^{-1} \in M_{m \times n}(\mathbb{F}). \tag{14.1.6}$$

Compute

$$
\begin{aligned}
B^*B = D^{-1}W^*A^*AWD^{-1} &= D^{-1}W^*(W\Lambda W^*)WD^{-1} \\
&= D^{-1}\Lambda D^{-1} \\
&= (\Sigma_r^{-1} \oplus I_{n-r})(\Sigma_r^2 \oplus 0_{n-r})(\Sigma_r^{-1} \oplus I_{n-r}) \\
&= I_r \oplus 0_{n-r}.
\end{aligned}
$$

Partition $B = [B_1 \; B_2]$, in which $B_1 \in M_{m \times r}(\mathbb{F})$ and $B_2 \in M_{m \times (n-r)}(\mathbb{F})$. The preceding identity becomes

$$B^*B = \begin{bmatrix} B_1^* \\ B_2^* \end{bmatrix} [B_1 \; B_2] = \begin{bmatrix} B_1^*B_1 & B_1^*B_2 \\ B_2^*B_1 & B_2^*B_2 \end{bmatrix} = \begin{bmatrix} I_r & 0 \\ 0 & 0_{n-r} \end{bmatrix}.$$

The identity $B_1^*B_1 = I_r$ says that B_1 has orthonormal columns; the identity $B_2^*B_2 = 0_{n-r}$ implies that $B_2 = 0$ (see Theorem 13.1.10.c). Theorem 6.2.17 ensures that there is a $B' \in M_{m \times (m-r)}(\mathbb{F})$ such that $V = [B_1 \; B'] \in M_m(\mathbb{F})$ is unitary. Then

$$
\begin{aligned}
BD &= [B_1 \; 0](\Sigma_r \oplus I_{n-r}) \\
&= [B_1 \; 0](\Sigma_r \oplus 0_{n-r}) \\
&= [B_1 \; B'](\Sigma_r \oplus 0_{n-r}) \\
&= V\Sigma.
\end{aligned}
$$

It follows from (14.1.6) that $A = BDW^* = V\Sigma W^*$.

If $m < n$, apply the preceding construction to A^*; see P.14.7. $\qquad\square$

Definition 14.1.7 A *singular value decomposition* of $A \in M_{m \times n}(\mathbb{F})$ is a factorization of the form $A = V\Sigma W^*$, in which $V \in M_m(\mathbb{F})$ and $W \in M_n(\mathbb{F})$ are unitary and $\Sigma \in M_{m \times n}(\mathbb{R})$ is the matrix (14.1.5).

The matrices $\Sigma^T\Sigma = \Sigma_r^2 \oplus 0_{n-r} \in M_n$ and $\Sigma\Sigma^T = \Sigma_r^2 \oplus 0_{m-r} \in M_m$ are diagonal. Thus,

$$A^*AW = W(\Sigma^T\Sigma)W^*W = W(\Sigma^T\Sigma) = W(\Sigma_r^2 \oplus 0_{n-r})$$

and

$$AA^*V = (V\Sigma\Sigma^TV^*)V = V(\Sigma\Sigma^T) = V(\Sigma_r^2 \oplus 0_{m-r}),$$

so the columns of $W = [\mathbf{w}_1 \; \mathbf{w}_2 \; \ldots \; \mathbf{w}_n] \in M_n$ are eigenvectors of A^*A and the columns of $V = [\mathbf{v}_1 \; \mathbf{v}_2 \; \ldots \; \mathbf{v}_m] \in M_m$ are eigenvectors AA^*. We have

$$A^*A\mathbf{w}_i = \begin{cases} \sigma_i^2\mathbf{w}_i & \text{for } i = 1, 2, \ldots, r, \\ 0 & \text{for } i = r+1, r+2, \ldots, n, \end{cases}$$

and

$$AA^*\mathbf{v}_i = \begin{cases} \sigma_i^2\mathbf{v}_i & \text{for } i = 1, 2, \ldots, r, \\ 0 & \text{for } i = r+1, r+2, \ldots, m. \end{cases}$$

The columns of V and W are linked by the identities

$$AW = V\Sigma W^*W = V\Sigma \quad \text{and} \quad A^*V = W\Sigma^T V^*V = W\Sigma^T,$$

that is,

$$A\mathbf{w}_i = \sigma_i \mathbf{v}_i \quad \text{and} \quad A^*\mathbf{v}_i = \sigma_i \mathbf{w}_i \quad \text{for } i = 1, 2, \ldots, r.$$

Definition 14.1.8 The columns of V are *left singular vectors of A*; the columns of W are *right singular vectors of A*.

Example 14.1.9 Here are two singular value decompositions of a rank-2 real 3×2 matrix A:

$$A = \begin{bmatrix} 4 & 0 \\ -5 & -3 \\ 2 & 6 \end{bmatrix} = \underbrace{\begin{bmatrix} \frac{1}{3} & -\frac{2}{3} & \frac{2}{3} \\ -\frac{2}{3} & \frac{1}{3} & \frac{2}{3} \\ \frac{2}{3} & \frac{2}{3} & \frac{1}{3} \end{bmatrix}}_{V} \underbrace{\begin{bmatrix} 6\sqrt{2} & 0 \\ 0 & 3\sqrt{2} \\ 0 & 0 \end{bmatrix}}_{\Sigma} \underbrace{\begin{bmatrix} \frac{1}{\sqrt{2}} & \frac{1}{\sqrt{2}} \\ -\frac{1}{\sqrt{2}} & \frac{1}{\sqrt{2}} \end{bmatrix}}_{W^*}$$

$$= \begin{bmatrix} -\frac{1}{3} & -\frac{2}{3} & \frac{2}{3} \\ \frac{2}{3} & \frac{1}{3} & \frac{2}{3} \\ -\frac{2}{3} & \frac{2}{3} & \frac{1}{3} \end{bmatrix} \begin{bmatrix} 6\sqrt{2} & 0 \\ 0 & 3\sqrt{2} \\ 0 & 0 \end{bmatrix} \begin{bmatrix} -\frac{1}{\sqrt{2}} & -\frac{1}{\sqrt{2}} \\ -\frac{1}{\sqrt{2}} & \frac{1}{\sqrt{2}} \end{bmatrix}.$$

In the second factorization, we changed the signs of the entries in the first columns of V and W, but that did not change the value of the product. The singular value matrix Σ in a singular value decomposition of A is always uniquely determined by A. The two unitary factors are never uniquely determined.

Example 14.1.10 Because the last row of Σ in the preceding example is zero, the third column of V plays no role in representing the entries of A. Consequently, there is a more compact representation

$$A = \begin{bmatrix} 4 & 0 \\ -5 & -3 \\ 2 & 6 \end{bmatrix} = \underbrace{\begin{bmatrix} \frac{1}{3} & -\frac{2}{3} \\ -\frac{2}{3} & \frac{1}{3} \\ \frac{2}{3} & \frac{2}{3} \end{bmatrix}}_{X} \underbrace{\begin{bmatrix} 6\sqrt{2} & 0 \\ 0 & 3\sqrt{2} \end{bmatrix}}_{\Sigma_2} \begin{bmatrix} \frac{1}{\sqrt{2}} & \frac{1}{\sqrt{2}} \\ -\frac{1}{\sqrt{2}} & \frac{1}{\sqrt{2}} \end{bmatrix},$$

in which $X \in \mathsf{M}_{3\times 2}$ has orthonormal columns and Σ_2 is 2×2; see Theorem 14.2.3.

Example 14.1.11 Let $\mathbf{x} \in \mathsf{M}_{n\times 1}$ be nonzero and let $V \in \mathsf{M}_n$ be a unitary matrix whose first column is $\mathbf{x}/\|\mathbf{x}\|_2$ (Corollary 6.4.10.b). Let $\Sigma = [\|\mathbf{x}\|_2 \; 0 \; \ldots \; 0]^T \in \mathsf{M}_{n\times 1}$ and let $W = [1] \in \mathsf{M}_1$. Then $\mathbf{x} = V\Sigma W^*$ is a singular value decomposition. The last $n - 1$ columns of V play no role in representing the entries of $\mathbf{x}$.

The terminology for multiplicities of singular values is parallel to that of eigenvalues.

Definition 14.1.12 Let $A \in \mathsf{M}_{m\times n}$ and suppose that $\operatorname{rank} A = r$. The *multiplicity* of a positive singular value σ_i of A is the multiplicity of σ_i^2 as an eigenvalue of A^*A (or of AA^*).

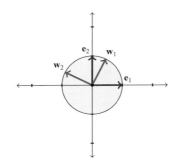

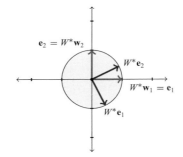

(a) The unit disk, the standard basis $\mathbf{e}_1, \mathbf{e}_2$, and the right singular vectors $\mathbf{w}_1, \mathbf{w}_2$ of A.

(b) Action of W^* (rotation by $\tan^{-1}(-2) \approx -63.4°$) on the vectors in (a).

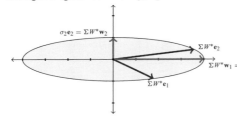

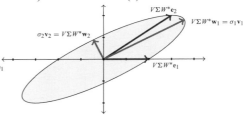

(c) Action of $\Sigma = \operatorname{diag}(4, 1)$ on the vectors in (b).

(d) Action of V (rotation by $\tan^{-1}(\frac{1}{2}) \approx 26.6°$) on the vectors in (c).

Figure 14.1 Singular value decomposition analysis of the matrix A in (14.1.14). It transforms the unit disk into an ellipse whose semi-axes have lengths equal to the singular values of A.

The multiplicity of a zero singular value of A (if any) is $\min\{m, n\} - r$. Singular values with multiplicity 1 are *simple*. If every singular value of A is simple, its singular values are *distinct*.

Example 14.1.13 Consider

$$A = \begin{bmatrix} 2 & 3 \\ 0 & 2 \end{bmatrix}, \tag{14.1.14}$$

which has a singular value decomposition

$$\begin{bmatrix} 2 & 3 \\ 0 & 2 \end{bmatrix} = \underbrace{\begin{bmatrix} \frac{2}{\sqrt{5}} & -\frac{1}{\sqrt{5}} \\ \frac{1}{\sqrt{5}} & \frac{2}{\sqrt{5}} \end{bmatrix}}_{V} \underbrace{\begin{bmatrix} 4 & 0 \\ 0 & 1 \end{bmatrix}}_{\Sigma} \underbrace{\begin{bmatrix} \frac{1}{\sqrt{5}} & \frac{2}{\sqrt{5}} \\ -\frac{2}{\sqrt{5}} & \frac{1}{\sqrt{5}} \end{bmatrix}}_{W^*}.$$

Let $V = [\mathbf{v}_1 \ \mathbf{v}_2]$ and $W = [\mathbf{w}_1 \ \mathbf{w}_2]$. Figure 14.1 illustrates how the three factors in the singular value decomposition of A act on the unit disk.

14.2 The Compact Singular Value Decomposition

If the rank of $A \in \mathsf{M}_{m \times n}$ is much smaller than m and n, or if m and n are substantially different, much of the information encoded in the two unitary factors of a singular value decomposition of A is irrelevant to the task of representing the entries of A; see Example 14.1.11. In this section, we investigate how to organize the singular value decomposition in a more compact way that preserves all the information needed to compute every entry of A. In Section 15.1, we investigate a broader question that arises in image compression: how to truncate the singular value decomposition in a way that permits us to compute an approximation to every entry of A, to any desired accuracy.

Let $A = V\Sigma W^*$ be a singular value decomposition and let $\Sigma_r = \mathrm{diag}(\sigma_1, \sigma_2, \ldots, \sigma_r)$ be the $r \times r$ upper-left corner of Σ. Partition

$$V = [V_r \ V'] \in \mathsf{M}_m \quad \text{and} \quad W = [W_r \ W'] \in \mathsf{M}_n,$$

in which

$$V_r \in \mathsf{M}_{m \times r}, \quad V' \in \mathsf{M}_{m \times (m-r)}, \quad W_r \in \mathsf{M}_{n \times r}, \quad \text{and} \quad W' \in \mathsf{M}_{n \times (n-r)}.$$

Then

$$
\begin{aligned}
A &= V\Sigma W^* \\
&= [V_r \ V'] \begin{bmatrix} \Sigma_r & 0_{r \times (n-r)} \\ 0_{(m-r) \times r} & 0_{(m-r) \times (n-r)} \end{bmatrix} \begin{bmatrix} W_r^* \\ W'^* \end{bmatrix} \\
&= [V_r \ V'] \begin{bmatrix} \Sigma_r W_r^* \\ 0_{(m-r) \times r} \end{bmatrix} \quad\quad\quad\quad\quad (14.2.1) \\
&= V_r \Sigma_r W_r^* + V' 0_{(m-r) \times r} \\
&= V_r \Sigma_r W_r^*. \quad\quad\quad\quad\quad\quad\quad\quad (14.2.2)
\end{aligned}
$$

Unless $r = m = n$, the last $m - r$ columns of V and/or the last $n - r$ columns of W are absent in the representation (14.2.2) of A.

Theorem 14.2.3 (Compact Singular Value Decomposition) *Let $A \in \mathsf{M}_{m \times n}(\mathbb{F})$ be nonzero, let $r = \mathrm{rank}\, A$, and let $A = V\Sigma W^*$ be a singular value decomposition.*

(a) *Partition $V = [V_r \ V'] \in \mathsf{M}_m(\mathbb{F})$ and $W = [W_r \ W'] \in \mathsf{M}_n(\mathbb{F})$, in which $V_r \in \mathsf{M}_{m \times r}(\mathbb{F})$ and $W_r \in \mathsf{M}_{n \times r}(\mathbb{F})$, and let $\Sigma_r = \mathrm{diag}(\sigma_1, \sigma_2, \ldots, \sigma_r)$. Then*

$$A = V_r \Sigma_r W_r^*. \quad\quad\quad\quad\quad (14.2.4)$$

(b) *Conversely, if $r \geq 1$, $X \in \mathsf{M}_{m \times r}(\mathbb{F})$ and $Y \in \mathsf{M}_{n \times r}(\mathbb{F})$ have orthonormal columns, $\Sigma_r = \mathrm{diag}(\sigma_1, \sigma_2, \ldots, \sigma_r)$, and $\sigma_1 \geq \sigma_2 \geq \cdots \geq \sigma_r > 0$, then $\sigma_1, \sigma_2, \ldots, \sigma_r$ are the positive singular values of $B = X\Sigma_r Y^* \in \mathsf{M}_{m \times n}(\mathbb{F})$.*

Proof The identity (14.2.4) is exhibited in (14.2.2). For the converse, observe that the positive eigenvalues of

$$BB^* = X\Sigma_r Y^* Y \Sigma_r X^* = X\Sigma_r I_r \Sigma_r X^* = (X)(\Sigma_r^2 X^*)$$

are the same as the positive eigenvalues of $(\Sigma_r^2 X^*)X = \Sigma_r^2 (X^*X) = \Sigma_r^2$; see Theorem 9.7.2. $\quad\square$

Definition 14.2.5 Let $A \in \mathsf{M}_{m \times n}(\mathbb{F})$ be nonzero and let $r = \mathrm{rank}\, A$. A factorization $A = X\Sigma_r Y^*$ is a *compact singular value decomposition* if $X \in \mathsf{M}_{m \times r}(\mathbb{F})$ and $Y \in \mathsf{M}_{n \times r}(\mathbb{F})$ have orthonormal columns, $\Sigma_r = \mathrm{diag}(\sigma_1, \sigma_2, \ldots, \sigma_r)$, and $\sigma_1 \geq \sigma_2 \geq \cdots \geq \sigma_r > 0$.

Theorem 14.2.3 ensures that every $A \in \mathsf{M}_{m \times n}(\mathbb{F})$ with rank $r \geq 1$ has a compact singular value decomposition in which the matrices X and Y are, respectively, the first r columns of the unitary matrices V and W in some singular value decomposition of A. Other choices of X and Y are always possible, so a compact singular value decomposition of A is not unique. Fortunately, all of the compact singular value decompositions of A can be obtained from any one of them; we return to this matter in Theorem 14.2.15.

Example 14.2.6 See Example 14.1.10 for a compact singular value decomposition of a 3×2 matrix with rank 2.

Example 14.2.7 If $\mathbf{x} \in \mathsf{M}_{n \times 1}$ is a nonzero vector, then a compact singular value decomposition is $\mathbf{x} = (\mathbf{x}/\|\mathbf{x}\|) [\|\mathbf{x}\|] [1]$, which represents $\mathbf{x}$ as a positive scalar multiple of a unit vector.

Example 14.2.8 If $A = X\Sigma_r Y^* \in \mathsf{M}_{m \times n}$ is a compact singular value decomposition, then $Y\Sigma_r X^*$ is a compact singular value decomposition of $A^* \in \mathsf{M}_{n \times m}$. Thus, $\mathbf{x}^* = [1][\|\mathbf{x}\|]$ $(\mathbf{x}^*/\|\mathbf{x}\|)$ is a compact singular value decomposition of $\mathbf{x}^*$.

Theorem 14.2.9 *Let $A \in \mathsf{M}_{m \times n}(\mathbb{F})$, let $r = \operatorname{rank} A \geq 1$, and let $A = X\Sigma_r Y^*$ be a compact singular value decomposition (14.2.4). Then*

$$\operatorname{col} A = \operatorname{col} X = (\operatorname{null} A^*)^{\perp} \tag{14.2.10}$$

and

$$\operatorname{null} A = (\operatorname{col} Y)^{\perp} = (\operatorname{col} A^*)^{\perp}. \tag{14.2.11}$$

Proof The representation $A = X\Sigma_r Y^*$ ensures that $\operatorname{col} A \subseteq \operatorname{col} X$. The reverse containment follows from

$$X = X\Sigma_r I_r \Sigma_r^{-1} = X\Sigma_r Y^* Y \Sigma_r^{-1} = AY\Sigma_r^{-1}.$$

This shows that $\operatorname{col} A = \operatorname{col} X$; the identity $\operatorname{col} A = (\operatorname{null} A^*)^{\perp}$ is (7.2.4). The identities (14.2.11) follow in the same way from applying (14.2.10) to A^* and taking orthogonal complements. $\qquad \square$

The positive definite diagonal factor Σ_r in (14.2.4) is uniquely determined by A; its diagonal entries are the positive singular values of A in decreasing order. However, the factors X and Y in (14.2.4) are never uniquely determined. For example, we can replace X by XD and Y by YD for any diagonal unitary matrix $D \in \mathsf{M}_r$. If some singular value has multiplicity greater than 1, then other types of nonuniqueness are possible. In Theorem 15.7.1, in Lemma 15.5.1, and in other applications, we need to be able to describe precisely how any two compact singular value decompositions of a given matrix are related. The following lemma is a first step toward that description.

Lemma 14.2.12 *Let $A, B, U, V \in \mathsf{M}_n$. Suppose that U and V are unitary, and suppose that A and B are positive definite. If*

$$UA = BV, \tag{14.2.13}$$

then $U = V$.

Proof The identity (14.2.13) implies that

$$AU^* = (UA)^* = (BV)^* = V^*B,$$

and hence

$$VA = BU. \tag{14.2.14}$$

Then (14.2.13) and (14.2.14) imply that

$$UA^2 = BVA = B^2U.$$

Theorem 0.8.1 ensures that

$$Up(A^2) = p(B^2)U$$

for any polynomial p. Now invoke Corollary 13.2.5. Choose p such that $A = p(A^2)$ and $B = p(B^2)$, and deduce that

$$UA = BU.$$

But $UA = BV$, so $BV = BU$. Since B is invertible, we conclude that $V = U$. $\qquad\square$

Theorem 14.2.15 *Let $A \in \mathsf{M}_{m \times n}(\mathbb{F})$ be nonzero, let $r = \operatorname{rank} A$, and let $A = X_1 \Sigma_r Y_1^* = X_2 \Sigma_r Y_2^*$ be compact singular value decompositions.*

(a) *There is a unitary $U \in \mathsf{M}_r(\mathbb{F})$ such that $X_1 = X_2 U$, $Y_1 = Y_2 U$, and $U\Sigma_r = \Sigma_r U$.*

(b) *If A has d distinct positive singular values $s_1 > s_2 > \cdots > s_d > 0$ with respective multiplicities $n_1, n_2, \ldots, n_d$, then $\Sigma_r = s_1 I_{n_1} \oplus s_2 I_{n_2} \oplus \cdots \oplus s_d I_{n_d}$ and the unitary matrix U in (a) has the form*

$$U = U_1 \oplus U_2 \oplus \cdots \oplus U_d, \quad U_{n_i} \in \mathsf{M}_{n_i}, \quad i = 1, 2, \ldots, d,$$

in which each summand U_i is unitary.

Proof (a) Theorem 14.2.9 tells us that

$$\operatorname{col} X_1 = \operatorname{col} X_2 = \operatorname{col} A,$$

and Theorem 7.3.9 ensures that there is a unitary $U \in \mathsf{M}_r$ such that $X_1 = X_2 U$. Similar reasoning shows that there is a unitary $V \in \mathsf{M}_r$ such that $Y_1 = Y_2 V$. Then

$$X_2 \Sigma_r Y_2^* = X_1 \Sigma_r Y_1^* = X_2 U \Sigma_r V^* Y_2^*,$$

so

$$X_2 (\Sigma_r - U\Sigma_r V^*) Y_2^* = 0.$$

Consequently,

$$0 = X_2^* 0 Y_2 = X_2^* X_2 (\Sigma_r - U\Sigma_r V^*) Y_2^* Y_2 = \Sigma_r - U\Sigma_r V^*,$$

which is equivalent to $U\Sigma_r = \Sigma_r V$. It follows from Lemma 14.2.12 that $U = V$, and hence U commutes with Σ_r.

(b) Since U commutes with Σ_r, a direct sum of distinct scalar matrices, the asserted block diagonal form of U follows from Lemma 3.3.21 and Theorem 6.2.7.b. $\qquad\square$

The geometric content of the preceding theorem is that, while the individual left and right singular vectors of A are never uniquely determined, the subspace spanned by the right (respectively, left) singular vectors associated with each distinct singular value of A is uniquely determined. This observation is actually a familiar fact expressed in a new vocabulary. The subspace of right (respectively, left) singular vectors of A associated with a singular value σ_i is the eigenspace of A^*A (respectively, AA^*) associated with the eigenvalue σ_i^2.

14.3 The Polar Decomposition

In this section, we explore matrix analogs of the polar form

$$z = re^{i\theta} = e^{i\theta}r, \quad r > 0, \quad \theta \in [0, 2\pi),$$

of a nonzero complex number z, in which $e^{i\theta}$ is a complex number with modulus 1. Since unitary matrices are natural analogs of complex numbers with modulus 1, we investigate matrix factorizations of the form $A = PU$ or $A = UQ$, in which P and Q are positive semidefinite and U is unitary. We find that such factorizations are possible (Theorem 14.3.9), and the commutativity of the factors is equivalent to the normality of A.

The matrix $(A^*A)^{1/2}$ appears in the polar decomposition, so the following definition is a convenience.

Definition 14.3.1 The *modulus* of $A \in M_{m \times n}(\mathbb{F})$ is the positive semidefinite matrix $|A| = (A^*A)^{1/2} \in M_n(\mathbb{F})$.

Example 14.3.2 If $A \in M_{m \times n}(\mathbb{F})$, then $|A^*| = (AA^*)^{1/2} \in M_m(\mathbb{F})$.

Example 14.3.3 The modulus of a 1×1 matrix $A = [a]$ is $|A| = [\bar{a}a]^{1/2} = [|a|]$, in which $|a|$ is the modulus of the complex number a.

Example 14.3.4 The modulus of the zero matrix in $M_{m \times n}$ is 0_n.

Example 14.3.5 The modulus of an $n \times 1$ matrix $A = \mathbf{x} \in \mathbb{F}^n$ is $|A| = (\mathbf{x}^*\mathbf{x})^{1/2} = \|\mathbf{x}\|_2$, its Euclidean norm.

Example 14.3.6 If $\mathbf{x} \in \mathbb{F}^n$ is nonzero, then the modulus of the $1 \times n$ matrix $A = \mathbf{x}^*$ is $|A| = (\mathbf{x}\mathbf{x}^*)^{1/2} = \|\mathbf{x}\|_2^{-1}\mathbf{x}\mathbf{x}^* \in M_n$. To verify this, compute

$$(\|\mathbf{x}\|_2^{-1}\mathbf{x}\mathbf{x}^*)^2 = \|\mathbf{x}\|_2^{-2}\mathbf{x}\mathbf{x}^*\mathbf{x}\mathbf{x}^* = \|\mathbf{x}\|_2^{-2}\|\mathbf{x}\|_2^2\mathbf{x}\mathbf{x}^* = \mathbf{x}\mathbf{x}^*.$$

Example 14.3.7 The moduli of

$$A = \begin{bmatrix} 0 & 1 \\ 0 & 0 \end{bmatrix} \quad \text{and} \quad B = \begin{bmatrix} 0 & 0 \\ 1 & 0 \end{bmatrix}$$

are

$$|A| = \begin{bmatrix} 0 & 0 \\ 0 & 1 \end{bmatrix} \quad \text{and} \quad |B| = \begin{bmatrix} 1 & 0 \\ 0 & 0 \end{bmatrix}.$$

Observe that

$$|A| \, |B| = \begin{bmatrix} 0 & 0 \\ 0 & 0 \end{bmatrix} \neq \begin{bmatrix} 1 & 0 \\ 0 & 0 \end{bmatrix} = |AB|.$$

The modulus of a matrix can be expressed as a product of factors extracted from its singular value decomposition.

Lemma 14.3.8 *Let $A \in M_{m \times n}(\mathbb{F})$ be nonzero, let* $\operatorname{rank} A = r \geq 1$, *and let* $A = V \Sigma W^*$ *be a singular value decomposition. Let $\sigma_1 \geq \sigma_2 \geq \cdots \geq \sigma_r > 0$ be the positive singular values of A and let $\Sigma_r = \operatorname{diag}(\sigma_1, \sigma_2, \ldots, \sigma_r)$. For any integer $k \geq r$, define $\Sigma_k = \Sigma_r \oplus 0_{k-r} \in M_k(\mathbb{R})$. Then*

$$|A| = W \Sigma_n W^* \in M_n(\mathbb{F}) \quad and \quad |A^*| = V \Sigma_m V^* \in M_m(\mathbb{F}).$$

Proof The matrix $\Sigma \in M_{m \times n}(\mathbb{R})$ is defined in (14.1.5). A computation reveals that

$$\Sigma^T \Sigma = \Sigma_r^2 \oplus 0_{n-r} = \Sigma_n^2 \quad and \quad \Sigma \Sigma^T = \Sigma_r^2 \oplus 0_{m-r} = \Sigma_m^2.$$

Now use (13.2.3) to compute

$$\begin{aligned} |A| &= (A^*A)^{1/2} = (W \Sigma^T V^* V \Sigma W^*)^{1/2} \\ &= (W \Sigma^T \Sigma W^*)^{1/2} = (W \Sigma_n^2 W^*)^{1/2} = W \Sigma_n W^* \end{aligned}$$

and

$$\begin{aligned} |A^*| &= (AA^*)^{1/2} = (V \Sigma W^* W \Sigma^T V^*)^{1/2} \\ &= (V \Sigma \Sigma^T V^*)^{1/2} = (V \Sigma_m^2 V^*)^{1/2} = V \Sigma_m V^*. \end{aligned} \qquad \square$$

Theorem 14.3.9 (Polar Decomposition) *Let $A \in M_{m \times n}(\mathbb{F})$ and suppose that* $\operatorname{rank} A = r \geq 1$. *Let $A = V \Sigma W^*$ be a singular value decomposition.*

(a) *If $m \geq n$, let $U = V_n W^*$, in which $V = [V_n \ V']$ and $V_n \in M_{m \times n}(\mathbb{F})$.*

(b) *If $m \leq n$, let $U = V W_m^*$, in which $W = [W_m \ W']$ and $W_m \in M_{n \times m}(\mathbb{F})$.*

(c) *If $m = n$, let $U = V W^*$.*

Then $U \in M_{m \times n}(\mathbb{F})$ has orthonormal columns (if $m \geq n$) or orthonormal rows (if $m \leq n$) and

$$A = U|A| = |A^*|U. \tag{14.3.10}$$

Proof Let $A = V \Sigma W^*$ be a singular value decomposition with unitary $V \in M_m(\mathbb{F})$ and $W \in M_n(\mathbb{F})$. Suppose that $m \geq n$. Partition

$$\Sigma = \begin{bmatrix} \Sigma_n \\ 0_{(m-n) \times n} \end{bmatrix},$$

in which $\Sigma_n \in M_n(\mathbb{R})$ is diagonal and positive semidefinite. Partition $V = [V_n \ V']$, in which $V_n \in M_{m \times n}$ and $V' \in M_{m \times (m-n)}$. Let $U = V_n W^*$. Then U has orthonormal columns,

$$\begin{aligned} A &= [V_n \ V'] \begin{bmatrix} \Sigma_n \\ 0_{(m-n) \times n} \end{bmatrix} W^* \\ &= V_n \Sigma_n W^* = (V_n W^*)(W \Sigma_n W^*) = U|A|, \end{aligned}$$

and

$$A = V_n \Sigma_n W^* = (V_n \Sigma_n V_n^*)(V_n W^*)$$

$$= [V_n \; V'] \begin{bmatrix} \Sigma_n & 0 \\ 0 & 0_{m-n} \end{bmatrix} [V_n \; V']^* U$$

$$= (V \Sigma_m V^*) U = |A^*| U.$$

If $m \le n$, apply the preceding results to A^*.   □

Definition 14.3.11 Let $A \in \mathsf{M}_{m \times n}$. A factorization $A = U|A|$ as in the preceding theorem is a *right polar decomposition*; $A = |A^*|U$ is a *left polar decomposition* of A.

Example 14.3.12 Consider

$$A = \begin{bmatrix} 2 & 3 \\ 0 & 2 \end{bmatrix}, \tag{14.3.13}$$

which has right and left polar decompositions

$$\begin{bmatrix} 2 & 3 \\ 0 & 2 \end{bmatrix} = \underbrace{\begin{bmatrix} \frac{4}{5} & \frac{3}{5} \\ -\frac{3}{5} & \frac{4}{5} \end{bmatrix}}_{U} \underbrace{\begin{bmatrix} \frac{8}{5} & \frac{6}{5} \\ \frac{6}{5} & \frac{17}{5} \end{bmatrix}}_{|A|} = \underbrace{\begin{bmatrix} \frac{17}{5} & \frac{6}{5} \\ \frac{6}{5} & \frac{8}{5} \end{bmatrix}}_{|A^*|} \underbrace{\begin{bmatrix} \frac{4}{5} & \frac{3}{5} \\ -\frac{3}{5} & \frac{4}{5} \end{bmatrix}}_{U}.$$

Figure 14.2 illustrates the actions of $|A|$ and U on the unit circle.

Eigenvectors for $|A| = (A^*A)^{1/2}$ are $\mathbf{w}_1 = \frac{1}{\sqrt{5}}[2 \; 1]^\mathsf{T}$ and $\mathbf{w}_2 = \frac{1}{\sqrt{5}}[-1 \; 2]^\mathsf{T}$. Eigenvectors of $|A^*| = (AA^*)^{1/2}$ are $\mathbf{v}_1 = \frac{1}{\sqrt{5}}[1 \; 2]^\mathsf{T}$ and $\mathbf{v}_2 = \frac{1}{\sqrt{5}}[-2 \; 1]^\mathsf{T}$. Notice that $\mathbf{v}_1$ and $\mathbf{v}_2$ are left singular vectors of A; $\mathbf{w}_1$ and $\mathbf{w}_2$ are right singular vectors of A. The eigenvalues of $|A|$ and $|A^*|$ (4 and 1) are the singular values of A.

We next investigate the uniqueness of the factors in left and right polar decompositions.

Example 14.3.14 The factorizations

$$A = \begin{bmatrix} 0 & 1 \\ 0 & 0 \end{bmatrix} = \underbrace{\begin{bmatrix} 0 & 1 \\ e^{i\theta} & 0 \end{bmatrix}}_{U} \underbrace{\begin{bmatrix} 0 & 0 \\ 0 & 1 \end{bmatrix}}_{|A|} = \underbrace{\begin{bmatrix} 1 & 0 \\ 0 & 0 \end{bmatrix}}_{|A^*|} \underbrace{\begin{bmatrix} 0 & 1 \\ e^{i\theta} & 0 \end{bmatrix}}_{U}$$

are valid for every $\theta \in [0, 2\pi)$. The unitary factor in a polar decomposition of a square matrix need not be unique if the matrix is not invertible.

Theorem 14.3.15 *Let* $A \in \mathsf{M}_{m \times n}$.

(a) *If* $m \ge n$ *and* $A = UB$, *in which* B *is positive semidefinite and* $U \in \mathsf{M}_{m \times n}$ *has orthonormal columns, then* $B = |A|$.

(b) *If* $\operatorname{rank} A = n$, $U \in \mathsf{M}_{m \times n}$, *and* $A = U|A|$, *then* $U = A|A|^{-1}$ *has orthonormal columns.*

(c) *If* $m \le n$ *and* $A = CU$, *in which* C *is positive semidefinite and* $U \in \mathsf{M}_{m \times n}$ *has orthonormal rows, then* $C = |A^*|$.

(d) *If* $\operatorname{rank} A = m$, $U \in \mathsf{M}_{m \times n}$, *and* $A = |A^*|U$, *then* $U = |A^*|^{-1}A$ *has orthonormal rows.*

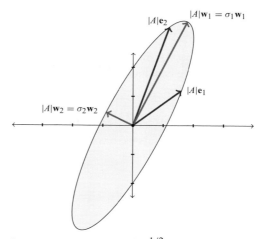

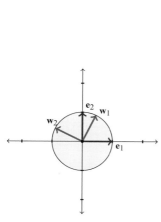

(a) The unit disk, the standard basis $\mathbf{e}_1, \mathbf{e}_2$, and the eigenvectors $\mathbf{w}_1, \mathbf{w}_2$ of $|A|$.

(b) Action of $|A| = (A^*A)^{1/2}$ on the vectors in (a). Notice that the angle between $\mathbf{e}_1$ and $|A|\mathbf{e}_1$ (and between $\mathbf{e}_2$ and $|A|\mathbf{e}_2$) is less than $\frac{\pi}{2}$.

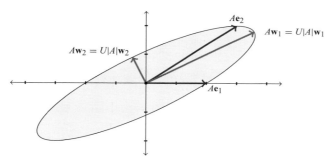

(c) Action of U (rotation by $\tan^{-1}(-\frac{3}{4}) \approx -36.9°$) on the vectors in (b).

Figure 14.2 Right polar decomposition of the matrix A in (14.3.13).

(e) *Let $m = n$ and let $U, V \in \mathsf{M}_n$ be unitary.*

   (i) *If $A = U|A|$, then $A = |A^*|U$.*

   (ii) *If $A = |A^*|V$, then $A = V|A|$.*

   (iii) *If A is invertible and $A = U|A| = |A^*|V$, then*

$$U = V = |A^*|^{-1}A = A|A|^{-1}.$$

Proof (a) If $A = UB$, then $A^*A = BU^*UB = B^2$, so Theorem 13.2.2 ensures that $|A| = (A^*A)^{1/2} = (B^2)^{1/2} = B$.

(b) If $\operatorname{rank} A = n$, then Theorems 13.1.10 and 13.2.2 tell us that A^*A and $|A|$ are invertible. Consequently, $U = A|A|^{-1}$ and $U^*U = |A|^{-1}A^*A|A|^{-1} = |A|^{-1}|A|^2|A|^{-1} = I_n$.

(c) Apply (a) to A^*.

(d) Apply (b) to A^*.

(e.i) Since $A = U|A| = (U|A|U^*)U$, it follows from (c) that $U|A|U^* = |A^*|$.

(e.ii) Since $A = |A^*|V = V(V^*|A^*|V)$, it follows from (a) that $V^*|A^*|V = |A|$.

(e.iii) If $A = U|A|$, then it follows from (e.i) that $A = |A^*|U$. Since $A = |A^*|V$, and $|A^*|$ is invertible, we conclude that $U = V$. One could also invoke Lemma 14.2.12 to obtain this conclusion. $\qquad\square$

Example 14.3.16 What are right and left polar decompositions of a nonzero $\mathbf{x} \in M_{n \times 1}$? The preceding theorem tells us to compute

$$|\mathbf{x}| = (\mathbf{x}^*\mathbf{x})^{1/2} = \|\mathbf{x}\|_2$$

and

$$U = \mathbf{x}|\mathbf{x}|^{-1} = \|\mathbf{x}\|_2^{-1}\mathbf{x}.$$

Then

$$\mathbf{x} = U|\mathbf{x}| = (\|\mathbf{x}\|_2^{-1}\mathbf{x})\|\mathbf{x}\|_2 \qquad (14.3.17)$$

is the right polar decomposition. Now compute

$$|\mathbf{x}^*| = (\mathbf{x}\mathbf{x}^*)^{1/2} = \|\mathbf{x}\|_2^{-1}\mathbf{x}\mathbf{x}^*,$$

and use the matrix U in (14.3.17) to obtain a left polar decomposition

$$\mathbf{x} = |\mathbf{x}^*|U = (\|\mathbf{x}\|_2^{-1}\mathbf{x}\mathbf{x}^*)(\mathbf{x}/\|\mathbf{x}\|_2).$$

Example 14.3.18 Let

$$A = \begin{bmatrix} 1 & 1 \\ 0 & 1 \end{bmatrix}, \qquad (14.3.19)$$

and compute

$$AA^* = \begin{bmatrix} 2 & 1 \\ 1 & 1 \end{bmatrix} \quad \text{and} \quad A^*A = \begin{bmatrix} 1 & 1 \\ 1 & 2 \end{bmatrix}.$$

Proposition 13.2.6 tells us that

$$|A| = (A^*A)^{1/2} = \frac{1}{\sqrt{5}}\begin{bmatrix} 2 & 1 \\ 1 & 3 \end{bmatrix}$$

and

$$|A^*| = (AA^*)^{1/2} = \frac{1}{\sqrt{5}}\begin{bmatrix} 3 & 1 \\ 1 & 2 \end{bmatrix}.$$

The (unique) unitary factor in the polar decompositions of A is

$$U = |A^*|^{-1}A = \frac{1}{\sqrt{5}}\begin{bmatrix} 2 & -1 \\ -1 & 3 \end{bmatrix}\begin{bmatrix} 1 & 1 \\ 0 & 1 \end{bmatrix} = \frac{1}{\sqrt{5}}\begin{bmatrix} 2 & 1 \\ -1 & 2 \end{bmatrix}. \qquad (14.3.20)$$

Then

$$|A^*|U = \frac{1}{5}\begin{bmatrix} 3 & 1 \\ 1 & 2 \end{bmatrix}\begin{bmatrix} 2 & 1 \\ -1 & 2 \end{bmatrix} = \begin{bmatrix} 1 & 1 \\ 0 & 1 \end{bmatrix} = A$$

and

$$U|A| = \frac{1}{5}\begin{bmatrix} 2 & 1 \\ -1 & 2 \end{bmatrix}\begin{bmatrix} 2 & 1 \\ 1 & 3 \end{bmatrix} = \begin{bmatrix} 1 & 1 \\ 0 & 1 \end{bmatrix} = A.$$

A normal matrix has a polar decomposition with a special property.

Theorem 14.3.21 *Let $A \in M_n$ and suppose that $A = U|A|$, in which $U \in M_n$ is unitary. Then A is normal if and only if $U|A| = |A|U$.*

Proof If $U|A| = |A|U$, compute

$$A^*A = |A|^2 = |A|^2UU^* = |A|U|A|U^* = U|A|\,|A|U^* = AA^*,$$

which shows that A is normal. Conversely, if A is normal, then $A^*A = AA^*$, so $|A^*| = (AA^*)^{1/2} = (A^*A)^{1/2} = |A|$. Theorem 14.3.9 says that

$$A = U|A| = (U|A|U^*)U.$$

It follows from Theorem 14.3.15.c that $U|A|U^* = |A^*| = |A|$, so $U|A| = |A|U$. $\square$

14.4 Problems

P.14.1 Let $A \in M_{m \times n}$ and suppose that rank $A = n$. Let $P \in M_m$ be the orthogonal projection onto col A. Use the polar decomposition to show that $A(A^*A)^{-1}A^*$ is well defined and equals P.

P.14.2 Let $a, b \in \mathbb{C}$ and let $A \in M_2$ be the matrix (12.1.7). (a) Show that the eigenvalues of A are $a \pm ib$. (b) Show that the singular values of A are $(|a|^2 + |b|^2 \pm 2|\operatorname{Im} \bar{a}b|)^{1/2}$. (c) Verify that $\|A\|_F^2 = \sigma_1^2 + \sigma_2^2$ and $|\det A| = \sigma_1\sigma_2$. (d) If a and b are real, show that A is a real scalar multiple of a real orthogonal matrix.

P.14.3 Let $A \in M_m$ and $B \in M_n$. Let $A = V\Sigma_A W^*$ and $B = X\Sigma_B Y^*$ be singular value decompositions. Find a singular value decomposition of $A \oplus B$.

P.14.4 Give 2×2 examples to show that a matrix can have: (a) equal singular values but distinct eigenvalues, or (b) equal eigenvalues but distinct singular values.

P.14.5 Let $A \in M_n$ be normal. If A has distinct singular values, show that it has distinct eigenvalues. What can you say about the converse?

P.14.6 Let $A = [a_{ij}] \in M_n$ have rank $r \geq 1$ and let $A = V\Sigma W^*$ be a singular value decomposition in which $\Sigma = \operatorname{diag}(\sigma_1, \sigma_2, \ldots, \sigma_n)$, $V = [v_{ij}]$, and $W = [w_{ij}]$ are in M_n. Show that each entry of A can be represented as $a_{ij} = \sum_{k=1}^r \sigma_k v_{ik}\overline{w_{jk}}$.

P.14.7 Let $A \in M_n$. Show that the adjoint of a singular value decomposition of A is a singular value decomposition of the adjoint of A.

P.14.8 (a) Describe an algorithm, similar to the one employed in the proof of Theorem 14.1.4, that begins with a spectral decomposition $AA^* = V\Sigma^2 V^*$ and produces a singular value decomposition of A. (b) Can any matrix V of orthonormal eigenvectors of AA^* be a factor in a singular value decomposition $A = V\Sigma W^*$? (c) Can a singular value decomposition of A be obtained by finding a spectral decomposition $A^*A = W\Sigma^2 W^*$, a spectral decomposition $AA^* = V\Sigma^2 V^*$, and setting $A = V\Sigma W^*$? Discuss. *Hint*: Consider $A = I$.

P.14.9   Let $A \in M_{m \times n}$ with $m \geq n$. Show that there are orthonormal vectors $x_1, x_2, \ldots, x_n \in \mathbb{C}^n$ such that $Ax_1, Ax_2, \ldots, Ax_n$ are orthogonal.

P.14.10   The proof of Theorem 14.2.3 shows how to construct a compact singular value decomposition from a given singular value decomposition. Describe how to construct a singular value decomposition from a given compact singular value decomposition.

P.14.11   Let $A \in M_{m \times n}$ have rank $r \geq 1$ and let $A = V_1 \Sigma W_1^* = V_2 \Sigma W_2^*$ be singular value decompositions. Suppose that A has d distinct positive singular values $s_1 > s_2 > \cdots > s_d > 0$ and that $\Sigma_r = s_1 I_{n_1} \oplus s_2 I_{n_2} \oplus \cdots \oplus s_d I_{n_d}$. Show that

$$V_1 = V_2(U_1 \oplus U_2 \oplus \cdots \oplus U_d \oplus Z_1) = V_2(U \oplus Z_1), \text{ and}$$
$$W_1 = W_2(U_1 \oplus U_2 \oplus \cdots \oplus U_d \oplus Z_2) = W_2(U \oplus Z_2),$$

in which $U_1 \in M_{n_1}, U_2 \in M_{n_2}, \ldots, U_d \in M_{n_d}, Z_1 \in M_{m-r}$, and $Z_2 \in M_{n-r}$. The summand Z_1 is absent if $r = m$; Z_2 is absent if $r = n$.

P.14.12   Show that Lemma 14.2.12 is false if the hypothesis "A and B are positive definite" is replaced by "A and B are Hermitian and invertible."

P.14.13   Show that Lemma 14.2.12 is false if the hypothesis "U and V are unitary" is replaced by "U and V are invertible."

P.14.14   Let $A \in M_n$, $m \geq n$, and $B \in M_{m \times n}$. If $\text{rank } B = n$, use a singular value decomposition of B to show that there is a unitary $V \in M_m$ and an invertible $S \in M_n$ such that $BAB^* = V(SAS^* \oplus 0_{m-n})V^*$.

P.14.15   Let $A \in M_n$ and let $A = V\Sigma W^*$ be a singular value decomposition, in which $\Sigma = \text{diag}(\sigma_1, \sigma_2, \ldots, \sigma_n)$. The eigenvalues of the $n \times n$ Hermitian matrix A^*A are $\sigma_1^2, \sigma_2^2, \ldots, \sigma_n^2$. (a) Use P.9.10 to show that the eigenvalues of the $2n \times 2n$ Hermitian matrix

$$B = \begin{bmatrix} 0 & A \\ A^* & 0 \end{bmatrix}$$

are $\pm\sigma_1, \pm\sigma_2, \ldots, \pm\sigma_n$. (b) Compute the block matrix $C = (V \oplus W)^* B(V \oplus W)$. (c) Show that C is permutation similar to

$$\begin{bmatrix} 0 & \sigma_1 \\ \sigma_1 & 0 \end{bmatrix} \oplus \begin{bmatrix} 0 & \sigma_2 \\ \sigma_2 & 0 \end{bmatrix} \oplus \cdots \oplus \begin{bmatrix} 0 & \sigma_n \\ \sigma_n & 0 \end{bmatrix}.$$

Deduce that the eigenvalues of B are $\pm\sigma_1, \pm\sigma_2, \ldots, \pm\sigma_n$.

P.14.16   Let $A \in M_2$ be nonzero and let $s = (\|A\|_F^2 + 2|\det A|)^{1/2}$. Show that the positive semidefinite factors in left and right polar decompositions of A are

$$|A^*| = \frac{1}{s}(AA^* + |\det A|I) \quad \text{and} \quad |A| = \frac{1}{s}(A^*A + |\det A|I).$$

P.14.17   Let $A, B \in M_n$. Show that $AA^* = BB^*$ if and only if there is a unitary $U \in M_n$ such that $A = BU$.

P.14.18   Suppose that $A \in M_n$ is not invertible, let x be a unit vector such that $A^*x = 0$, let $A = |A^*|U$ be a polar decomposition, and consider the Householder matrix $U_x =$

$I - 2\mathbf{x}\mathbf{x}^*$. Show that (a) $PU_\mathbf{x} = P$, (b) $U_\mathbf{x}U$ is unitary, (c) $A = P(U_\mathbf{x}U)$ is a polar decomposition, and (d) $U \neq U_\mathbf{x}U$.

P.14.19 Let $A \in M_n$. Prove that the following are equivalent:

(a) A is invertible.

(b) The unitary factor in a left polar decomposition $A = |A^*|U$ is unique.

(c) The unitary factor in a right polar decomposition $A = U|A|$ is unique.

P.14.20 Let $a, b \in \mathbb{C}$ be nonzero scalars, let $\theta \in \mathbb{R}$, and let

$$A = \begin{bmatrix} 0 & a & 0 \\ 0 & 0 & b \\ 0 & 0 & 0 \end{bmatrix}, \quad Q = \begin{bmatrix} 0 & 0 & 0 \\ 0 & |a| & 0 \\ 0 & 0 & |b| \end{bmatrix}, \quad U_\theta = \begin{bmatrix} 0 & \frac{a}{|a|} & 0 \\ 0 & 0 & \frac{b}{|b|} \\ e^{i\theta} & 0 & 0 \end{bmatrix}.$$

(a) Show that Q is positive semidefinite, U_θ is unitary, and $A = U_\theta Q$ for all $\theta \in \mathbb{R}$. (b) Find a positive semidefinite P such that $A = PU_\theta$ for all $\theta \in \mathbb{R}$. (c) Are there any values of a and b for which A is normal?

P.14.21 Let $A \in M_n$. Give three proofs that $A(A^*A)^{1/2} = (AA^*)^{1/2}A$ using (a) the singular value decomposition, (b) right and left polar decomposition, and (c) the fact that $(A^*A)^{1/2} = p(A^*A)$ for some polynomial p.

P.14.22 Let $W, S \in M_n$. Suppose that W is unitary, S is invertible (but not necessarily unitary), and SWS^* is unitary. Show that W and SWS^* are unitarily similar. *Hint*: Represent $S = PU$ as a polar decomposition and use Theorem 14.3.15.

P.14.23 Let $A, B \in M_n$. We know that AB and BA have the same eigenvalues (Theorem 9.7.2), but they need not have the same singular values. Let

$$A = \begin{bmatrix} 0 & 0 \\ 1 & 0 \end{bmatrix}, \quad B = \begin{bmatrix} 0 & 0 \\ 0 & 2 \end{bmatrix}.$$

What are the singular values of AB and BA? Which matrix is not normal?

P.14.24 Let $A, B \in M_n$ and suppose that $AB = 0$. (a) Give an example to show that $BA \neq 0$ is possible. Which of the matrices in your example is not normal? (b) If A and B are Hermitian, prove that $BA = 0$. (c) If A and B are normal, use the polar decomposition to show that $|A| |B| = 0$. Deduce that $BA = 0$. (d) If A and B are normal, use P.13.37 to show that $BA = 0$.

P.14.25 Let $A, B \in M_n$ be normal. Use the polar decomposition and Theorem 15.3.13 to show that AB and BA have the same singular values. Where does this argument fail if either A or B is not normal? *Hint*: Why are the singular values of AB the same as those of $|A| |B|$ and $(|A| |B|)^*$?

P.14.26 Let $A \in M_{m \times n}(\mathbb{F})$ with $m \geq n$. Let $A = U|A|$ be a polar decomposition. Show that $\|A\mathbf{x}\|_2 = \||A|\mathbf{x}\|_2$ for all $\mathbf{x} \in \mathbb{F}^n$.

P.14.27 Let T be the Volterra operator on $C[0, 1]$; see P.5.11 and P.8.27. (a) Show that

$$(TT^*f)(t) = \int_0^1 \min(t, s)f(s)\, ds \tag{14.4.1}$$

and

$$(T^*Tf)(t) = \int_0^1 \min(1 - t, 1 - s)f(s)\, ds.$$

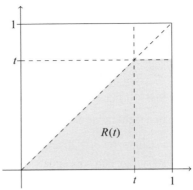

Figure 14.3 The region $R(t)$ in P.14.27.

Hint: Express $(TT^*f)(t)$ as a double integral over the region $R(t)$ in Figure 14.3. (b) Deduce that

$$\langle T^*f, T^*f\rangle = \int_0^1 \int_0^1 \min(t, s)f(s)\overline{f(t)}\, ds\, dt \geq 0 \quad \text{for all } f \in C[0, 1].$$

(c) Show that $(\pi^{-2}(n+\tfrac{1}{2})^{-2}, \sin(\pi(n+\tfrac{1}{2})t))$, $n = 0, 1, 2\ldots$, are eigenpairs of TT^*. For a discrete analog of (14.4.1), see P.13.20.

P.14.28 Let $A \in \mathsf{M}_m$ and $B \in \mathsf{M}_n$, and let $A = V\Sigma_A W^*$ and $B = X\Sigma_B Y^*$ be singular value decompositions. Show $A \otimes B$ has a singular value decomposition of the form $A \otimes B = (V \otimes W)D(V \otimes W)^*$. What is D? What are the singular values of $A \otimes B$?

14.5 Notes

For an explicit formula for a 2×2 unitary factor to accompany the positive semidefinite factors in P.14.16, see [HJ13, 7.3.P28].

14.6 Some Important Concepts

- Singular values.

- Singular value decomposition.

- Compact singular value decomposition.

- Uniqueness of the factors in a compact singular value decomposition (Theorem 14.2.15).

- Modulus of a matrix.

- Left and right polar decompositions.

- Uniqueness of the factors in polar decompositions (Theorem 14.3.15).

- Polar factors of a normal matrix.

Singular Values and the Spectral Norm

In this chapter we investigate applications and consequences of the singular value decomposition. For example, it provides a systematic way to approximate a matrix by another matrix of lower rank. It also permits us to define a generalized inverse for matrices that are not invertible (and need not even be square). It has a pleasant special form for complex symmetric matrices.

The largest singular value is especially important; it turns out to be a norm (the spectral norm) on matrices. We use the spectral norm to say something useful about how the solution of a linear system can change if the system is perturbed, and how the eigenvalues of a matrix can change if it is perturbed.

15.1 Singular Values and Approximations

With the help of the outer product representation in (3.1.19), a compact singular value decomposition of a rank-r matrix $A \in \mathsf{M}_{m \times n}$ can be written as

$$A = V_r \Sigma_r W_r^* = \sum_{i=1}^{r} \sigma_i \mathbf{v}_i \mathbf{w}_i^*, \tag{15.1.1}$$

in which $V_r = [\mathbf{v}_1 \ \mathbf{v}_2 \ \ldots \ \mathbf{v}_r] \in \mathsf{M}_{m \times r}$ and $W_r = [\mathbf{w}_1 \ \mathbf{w}_2 \ \ldots \ \mathbf{w}_r] \in \mathsf{M}_{n \times r}$. The summands are mutually orthogonal with respect to the Frobenius inner product

$$\begin{aligned}
\langle \sigma_i \mathbf{v}_i \mathbf{w}_i^*, \sigma_j \mathbf{v}_j \mathbf{w}_j^* \rangle_F &= \mathrm{tr}(\sigma_i \sigma_j \mathbf{w}_j \mathbf{v}_j^* \mathbf{v}_i \mathbf{w}_i^*) \\
&= \sigma_i \sigma_j \delta_{ij} \, \mathrm{tr} \, \mathbf{w}_j \mathbf{w}_i^* = \sigma_i \sigma_j \delta_{ij} \, \mathrm{tr} \, \mathbf{w}_i^* \mathbf{w}_j \\
&= \sigma_i \sigma_j \delta_{ij}, \quad i,j = 1, 2, \ldots, r,
\end{aligned} \tag{15.1.2}$$

and monotone decreasing in the Frobenius norm:

$$\|\sigma_i \mathbf{v}_i \mathbf{w}_i^*\|_F^2 = \sigma_i^2 \geq \sigma_{i+1}^2 = \|\sigma_{i+1} \mathbf{v}_{i+1} \mathbf{w}_{i+1}^*\|_F^2.$$

Orthogonality facilitates computation of the Frobenius norm of sums such as (15.1.1):

$$\|A\|_F^2 = \left\| \sum_{i=1}^{r} \sigma_i \mathbf{v}_i \mathbf{w}_i^* \right\|_F^2 = \sum_{i=1}^{r} \|\sigma_i \mathbf{v}_i \mathbf{w}_i^*\|_F^2 = \sum_{i=1}^{r} \sigma_i^2. \tag{15.1.3}$$

Monotonicity suggests the possibility of approximating A by a matrix of the form

$$A_{(k)} = \sum_{i=1}^{k} \sigma_i \mathbf{v}_i \mathbf{w}_i^*, \quad k = 1, 2, \ldots, r, \tag{15.1.4}$$

obtained by truncating the sum (15.1.1). The orthogonality relations (15.1.2) also facilitate computation of the norm of the residual in this approximation:

$$\|A - A_{(k)}\|_F^2 = \left\| \sum_{i=k+1}^{r} \sigma_i \mathbf{v}_i \mathbf{w}_i^* \right\|_F^2 = \sum_{i=k+1}^{r} \sigma_i^2 \|\mathbf{v}_i \mathbf{w}_i^*\|_F^2 = \sum_{i=k+1}^{r} \sigma_i^2. \tag{15.1.5}$$

If m, n, and r are all large, and if the singular values of A decrease rapidly, then $A_{(k)}$ can be close to A in norm for modest values of k. In order to store or transmit A itself, mn numbers

Original image, Rank: 800   Data used: 2.50%, Rank: 10

Data used: 5.00%, Rank: 20   Data used: 10.00%, Rank: 40

Data used: 15.00%, Rank: 60   Data used: 25.00%, Rank: 100

Figure 15.1 The original Mona Lisa image is represented by an 800×800 matrix A of full rank. The approximations $A_{(k)} = \sum_{i=1}^{k} \sigma_i \mathbf{v}_i \mathbf{w}_i^\mathsf{T}$ are displayed for $k = 10, 20, 40, 60, 100$.

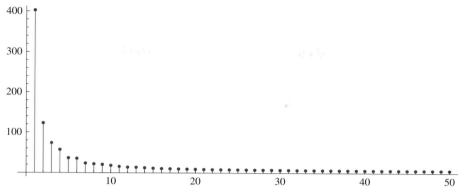

Figure 15.2 Plot of the first 50 singular values of the Mona Lisa image.

are required; $r(m+n+1)$ numbers are required to construct A from its compact singular value decomposition; $k(m+n+1)$ numbers are required to construct the approximation $A_{(k)}$ from the representation (15.1.4).

With the help of the outer product representation (3.1.19) again, we can write the truncated sum (15.1.4) as a *truncated singular value decomposition*

$$A_{(k)} = V_k \Sigma_k W_k^*, \quad k = 1, 2, \ldots, r,$$

in which the columns of $V_k \in \mathsf{M}_{m \times k}$ and $W_k \in \mathsf{M}_{n \times k}$ are, respectively, the first k columns of V_r and W_r, respectively, and $\Sigma_k = \mathrm{diag}(\sigma_1, \sigma_2, \ldots, \sigma_k)$.

Example 15.1.6 Figure 15.1 illustrates how truncated singular value decompositions can be used to construct approximations to a familiar image. The Mona Lisa is represented as a real 800×800 matrix of digitized gray scale values. Using only 2.5% of the original data, a recognizable image is recovered; with 15% of the original data an excellent image appears. Figure 15.2 shows that the singular values decay rapidly.

15.2 The Spectral Norm

Imagine that some physical or economic process is modeled as a linear transformation $\mathbf{x} \mapsto A\mathbf{x}$, in which $A \in \mathsf{M}_{m \times n}$. We want to know if a likely range of inputs $\mathbf{x}$ could result in outputs $A\mathbf{x}$ that exceed specified limits. Could a certain level of engine power and rate of climb result in vibrations severe enough to tear the wings off a new type of aircraft? Could a proposed program of future bond purchases by the Federal Reserve result in housing price inflation that exceeds 5%? The singular value decomposition can help us bound the size of the outputs as a scalar multiple of the size of the inputs.

Let $A \in \mathsf{M}_{m \times n}$ have rank r and let $A = V \Sigma W^*$ be a singular value decomposition. The transformation $\mathbf{x} \mapsto A\mathbf{x}$ is a composition of a unitary transformation

$$\mathbf{x} \mapsto W^* \mathbf{x},$$

scaling along the coordinate axes

$$W^* \mathbf{x} \mapsto \Sigma(W^* \mathbf{x}),$$

and another unitary transformation

$$\Sigma W^* \mathbf{x} \mapsto V(\Sigma W^* \mathbf{x}).$$

The two unitary transformations are isometries, so any difference between $\|\mathbf{x}\|_2$ and $\|A\mathbf{x}\|_2$ is due to the singular value matrix Σ. This qualitative observation is motivation for some careful analysis.

Definition 15.2.1 Let $X \in \mathsf{M}_{m \times n}(\mathbb{F})$, let $q = \min\{m, n\}$, and let $\sigma_1(X) \geq \sigma_2(X) \geq \cdots \geq \sigma_q(X) \geq 0$ be the singular values of X. Define $\sigma_{\max}(X) = \sigma_1(X)$.

Theorem 15.2.2 *Let $A \in \mathsf{M}_{m \times n}(\mathbb{F})$, let $q = \min\{m, n\}$, and let $A = V\Sigma W^*$ be a singular value decomposition, in which $V = [\mathbf{v}_1 \ \mathbf{v}_2 \ \ldots \ \mathbf{v}_m] \in \mathsf{M}_m(\mathbb{F})$ and $W = [\mathbf{w}_1 \ \mathbf{w}_2 \ \ldots \ \mathbf{w}_n] \in \mathsf{M}_n(\mathbb{F})$ are unitary.*

(a) $\sigma_{\max}(I_n) = 1$ *and* $\|I_n\|_F = \sqrt{n}$.

(b) $\|A\mathbf{x}\|_2 \leq \sigma_{\max}(A)\|\mathbf{x}\|_2$ *for all $\mathbf{x} \in \mathbb{F}^n$, with equality for $\mathbf{x} = \mathbf{w}_1$.*

(c) $\sigma_{\max}(A) = \max_{\|\mathbf{x}\|_2=1} \|A\mathbf{x}\|_2 = \|A\mathbf{w}_1\|_2$.

(d) $\sigma_{\max}(A) \leq \|A\|_F$, *with equality if and only if* $\operatorname{rank} A = 1$.

(e) $\|A\mathbf{x}\|_2 \leq \|A\|_F \|\mathbf{x}\|_2$ *for all $\mathbf{x} \in \mathbb{F}^n$.*

Proof Let $\operatorname{rank} A = r$. Because all the assertions about A are vacuous if $A = 0$, we may assume that $r \geq 1$.

(a) The largest eigenvalue of $I_n^* I_n$ is 1 and $\operatorname{tr} I_n^* I_n = n$.

(b) Compute

$$\|A\mathbf{x}\|_2^2 = \|V\Sigma W^*\mathbf{x}\|_2^2 = \|\Sigma W^*\mathbf{x}\|_2^2 = \sum_{i=1}^r \sigma_i^2 |\langle \mathbf{x}, \mathbf{w}_i \rangle|^2 \tag{15.2.3}$$

$$\leq \sigma_{\max}^2(A) \sum_{i=1}^r |\langle \mathbf{x}, \mathbf{w}_i \rangle|^2 \leq \sigma_{\max}^2(A) \sum_{i=1}^n |\langle \mathbf{x}, \mathbf{w}_i \rangle|^2 \tag{15.2.4}$$

$$= \sigma_{\max}^2(A) \|W\mathbf{x}\|_2^2 = \sigma_{\max}^2(A)\|\mathbf{x}\|_2^2. \tag{15.2.5}$$

(c) This follows from (b).

(d) $\sigma_{\max}^2(A) \leq \sigma_1^2 + \sigma_2^2 + \cdots + \sigma_r^2 = \|A\|_F^2$ with equality if and only if $\sigma_2^2 = \cdots = \sigma_r^2 = 0$, which occurs if and only if $r = 1$.

(e) This follows from (b) and (d). $\qquad \square$

In addition, the function $\sigma_{\max} : \mathsf{M}_{m \times n} \to [0, \infty)$ has the following properties.

Theorem 15.2.6 *Let $A, B \in \mathsf{M}_{m \times n}(\mathbb{F})$.*

(a) $\sigma_{\max}(A)$ *is real and nonnegative.*

(b) $\sigma_{\max}(A) = 0$ *if and only if $A = 0$.*

(c) $\sigma_{\max}(cA) = |c|\sigma_{\max}(A)$ *for all $c \in \mathbb{F}$.*

(d) $\sigma_{\max}(A + B) \leq \sigma_{\max}(A) + \sigma_{\max}(B)$.

Proof (a) Every singular value is real and nonnegative by definition.

(b) Since singular values are decreasingly ordered, $\sigma_{\max}(A) = 0$ if and only if all $\sigma_i(A) = 0$, which occurs if and only if $\|A\|_F = 0$.

(c) The largest eigenvalue of $(cA)^*(cA) = |c|^2 A^* A$ is $|c|^2 \sigma_1(A)^2$.

(d) Use Theorem 15.2.2.c to compute

$$
\begin{aligned}
\sigma_{\max}(A + B) = \max_{\|\mathbf{x}\|_2=1} \|(A + B)\mathbf{x}\|_2 &= \max_{\|\mathbf{x}\|_2=1} \|A\mathbf{x} + B\mathbf{x}\|_2 \\
&\leq \max_{\|\mathbf{x}\|_2=1} \left(\|A\mathbf{x}\|_2 + \|B\mathbf{x}\|_2 \right) \\
&\leq \max_{\|\mathbf{x}\|_2=1} \|A\mathbf{x}\|_2 + \max_{\|\mathbf{x}\|_2=1} \|B\mathbf{x}\|_2 = \sigma_{\max}(A) + \sigma_{\max}(B). \qquad \square
\end{aligned}
$$

The preceding theorem says that $\sigma_{\max}(\cdot) : \mathsf{M}_{m \times n}(\mathbb{F}) \to [0, \infty)$ satisfies the four axioms in Definition 4.6.1. Therefore, this function is a *norm* on the $\mathbb{F}$-vector space $\mathsf{M}_{m \times n}(\mathbb{F})$.

Definition 15.2.7 The function $\| \cdot \|_2 : \mathsf{M}_{m \times n} \to [0, \infty)$ defined by $\|A\|_2 = \sigma_{\max}(A)$ is the *spectral norm*.

Which is better, the spectral norm or the Frobenius norm? It depends. If rank $A \geq 2$, then $\|A\|_2 < \|A\|_F$ (Theorem 15.2.2.d), so the spectral norm typically gives tighter upper bounds in an estimate. However, the Frobenius norm of a matrix is likely to be easier to compute than its spectral norm: $\|A\|_F^2$ can be computed via arithmetic operations (see Example 4.5.5), but $\|A\|_2^2$, the largest eigenvalue of A^*A, is typically not so easy to compute.

Example 15.2.8 Let $\mathbf{x} \in \mathsf{M}_{n \times 1}$. Then $(\mathbf{x}^* \mathbf{x})^{1/2} = \|\mathbf{x}\|_2 = \sigma_{\max}(\mathbf{x})$. The spectral norm of the $n \times 1$ matrix $\mathbf{x}$ equals the Euclidean norm of the vector $\mathbf{x} \in \mathbb{C}^n$.

Example 15.2.9 Consider

$$
A = \begin{bmatrix} 1 & 0 \\ 0 & 0 \end{bmatrix}, \quad B = \begin{bmatrix} 0 & 0 \\ 0 & 1 \end{bmatrix}, \quad A + B = \begin{bmatrix} 1 & 0 \\ 0 & 1 \end{bmatrix}, \quad \text{and} \quad A - B = \begin{bmatrix} 1 & 0 \\ 0 & -1 \end{bmatrix}.
$$

Then

$$
\|A + B\|_2^2 + \|A - B\|_2^2 = 2
$$

and

$$
2\|A\|_2^2 + 2\|B\|_2^2 = 4.
$$

Therefore, the spectral norm is not derived from an inner product; see Theorem 4.5.9.e.

The following theorem says that the Frobenius and spectral norms are both *submultiplicative*.

Theorem 15.2.10 *Let* $A \in \mathsf{M}_{m \times n}$ *and* $B \in \mathsf{M}_{n \times k}$.

(a) $\|AB\|_F \leq \|A\|_F \|B\|_F$.

(b) $\|AB\|_2 \leq \|A\|_2 \|B\|_2$.

Proof (a) Partition $B = [\mathbf{b}_1 \ \mathbf{b}_2 \ \ldots \ \mathbf{b}_k]$ according to its columns. The basic idea in the following computation is that the square of the Frobenius norm of a matrix is not only the sum of the squares of the moduli of its entries, it is also the sum of the squares of the Euclidean norms of its columns (or its rows). With the help of the upper bound in Theorem 15.2.2.e, compute

$$\|AB\|_F^2 = \|[A\mathbf{b}_1 \ A\mathbf{b}_2 \ \ldots \ A\mathbf{b}_k]\|_F^2$$
$$= \|A\mathbf{b}_1\|_2^2 + \|A\mathbf{b}_2\|_2^2 + \cdots + \|A\mathbf{b}_k\|_2^2$$
$$\leq \|A\|_F^2\|\mathbf{b}_1\|_2^2 + \|A\|_F^2\|\mathbf{b}_2\|_2^2 + \cdots + \|A\|_F^2\|\mathbf{b}_k\|_2^2$$
$$= \|A\|_F^2\big(\|\mathbf{b}_1\|_2^2 + \|\mathbf{b}_2\|_2^2 + \cdots + \|\mathbf{b}_k\|_2^2\big)$$
$$= \|A\|_F^2\|B\|_F^2.$$

(b) For any $\mathbf{x} \in \mathbb{C}^k$, use Theorem 15.2.2.b to compute

$$\|AB\mathbf{x}\|_2 \leq \|A\|_2\|B\mathbf{x}\|_2 \leq \|A\|_2\|B\|_2\|\mathbf{x}\|_2.$$

Theorem 15.2.2.c ensures that

$$\|AB\|_2 = \max_{\|\mathbf{x}\|_2=1} \|AB\mathbf{x}\|_2 \leq \max_{\|\mathbf{x}\|_2=1} \big(\|A\|_2\|B\|_2\|\mathbf{x}\|_2\big) = \|A\|_2\|B\|_2. \qquad \square$$

15.3 Singular Values and Eigenvalues

Many interesting inequalities involve singular values and eigenvalues; we discuss just a few of them. In this section, it is convenient to index the eigenvalues of $A \in \mathsf{M}_n$ in such a way that $|\lambda_1(A)| = \rho(A)$ (the spectral radius) and their moduli are decreasingly ordered:

$$|\lambda_1(A)| \geq |\lambda_2(A)| \geq \cdots \geq |\lambda_n(A)|.$$

For brevity, we sometimes write $|\lambda_1| \geq |\lambda_2| \geq \cdots \geq |\lambda_n|$.

If A is invertible, then all of its eigenvalues are nonzero. Corollary 10.1.6 ensures that the eigenvalues of A^{-1} are the reciprocals of the eigenvalues of A. Therefore, the decreasingly ordered moduli of the eigenvalues of A^{-1} are

$$|\lambda_n^{-1}(A)| \geq |\lambda_{n-1}^{-1}(A)| \geq \cdots \geq |\lambda_1^{-1}(A)|.$$

Consequently, $\rho(A^{-1}) = |\lambda_n^{-1}(A)|$. There is an analogous identity for singular values.

Definition 15.3.1 Let $\sigma_1(A) \geq \sigma_2(A) \geq \cdots \geq \sigma_n(A) \geq 0$ be the singular values of $A \in \mathsf{M}_n$ and define $\sigma_{\min}(A) = \sigma_n(A)$.

Theorem 15.3.2 *Let $A \in \mathsf{M}_n$.*

(a) *If A is invertible, then $\|A^{-1}\|_2 = 1/\sigma_{\min}(A)$.*

(b) $|\lambda_1(A)\lambda_2(A) \cdots \lambda_n(A)| = |\det A| = \sigma_1(A)\sigma_2(A) \cdots \sigma_n(A).$

(c) $\rho(A) \leq \|A\|_2.$

(d) $\sigma_{\min}(A) \leq |\lambda_n(A)|.$

(e) $\lambda_n(A) = 0$ *if and only if $\sigma_{\min}(A) = 0$.*

Proof (a) If A is invertible and $A = V\Sigma W^*$ is a singular value decomposition, then $A^{-1} = W\Sigma^{-1}V^*$ and the largest diagonal entry in Σ^{-1} is $1/\sigma_{\min}(A)$. This is the spectral norm of A^{-1}.

(b) Corollary 10.1.3 ensures that $\lambda_1\lambda_2\cdots\lambda_n = \det A$. Let $A = V\Sigma W^*$ be a singular value decomposition and compute

$$|\det A| = |\det V\Sigma W^*| = |(\det V)(\det \Sigma)(\overline{\det W})|$$
$$= |\det V|\,(\det \Sigma)\,|\det W| = \sigma_1\sigma_2\cdots\sigma_n.$$

(c) Let $(\lambda_1, \mathbf{x})$ be an eigenpair of A with $\|\mathbf{x}\|_2 = 1$ and use Theorem 15.2.2.b to compute

$$\rho(A) = |\lambda_1| = \|\lambda_1\mathbf{x}\|_2 = \|A\mathbf{x}\|_2 \leq \|A\|_2\|\mathbf{x}\|_2 = \|A\|_2. \tag{15.3.3}$$

(d) If A is invertible, apply (15.3.3) to A^{-1}:

$$\rho(A^{-1}) = \frac{1}{|\lambda_n(A)|} = |\lambda_1(A^{-1})| \leq \sigma_{\max}(A^{-1}) = \frac{1}{\sigma_{\min}(A)} = \|A^{-1}\|_2,$$

so $\sigma_{\min} \leq |\lambda_n|$. If A is not invertible, then $\det A = 0$, so $\sigma_{\min} = \lambda_n = 0$.

(e) If $\lambda_n = 0$, then (d) implies that $\sigma_{\min} = 0$. If $\sigma_{\min} = 0$, then $\operatorname{rank} A < n$, so the dimension theorem implies that $\dim \operatorname{null} A \geq 1$. Consequently, A has an eigenvector associated with the eigenvalue zero. $\qquad\square$

The singular values and eigenvalues of a matrix are typically different, but the special cases of normal and positive semidefinite matrices are noteworthy.

Theorem 15.3.4 *Let $A = [a_{ij}] \in \mathsf{M}_n$.*

(a) $|\lambda_1(A)|^2 + |\lambda_2(A)|^2 + \cdots + |\lambda_n(A)|^2 \leq \sigma_1^2(A) + \sigma_2^2(A) + \cdots + \sigma_n^2(A)$.

(b) A *is normal if and only if* $|\lambda_i(A)| = \sigma_i(A)$ *for each* $i = 1, 2, \ldots, n$.

(c) A *is positive semidefinite if and only if* $\lambda_i(A) = \sigma_i(A)$ *for each* $i = 1, 2, \ldots, n$.

Proof (a) Schur's inequality (12.3.2) says that $\sum_{i=1}^{n} |\lambda_i|^2 \leq \operatorname{tr} A^*A$, and $\operatorname{tr} A^*A = \sum_{i=1}^{n} \sigma_i^2$.

(b) If $|\lambda_i| = \sigma_i$ for each $i = 1, 2, \ldots, n$, then

$$|\lambda_1|^2 + |\lambda_2|^2 + \cdots + |\lambda_n|^2 = \sigma_1^2 + \sigma_2^2 + \cdots + \sigma_n^2 = \operatorname{tr} A^*A,$$

so Theorem 12.3.1 ensures that A is normal. Conversely, if A is normal, the spectral theorem says that $A = U\Lambda U^*$, in which $\Lambda = \operatorname{diag}(\lambda_1, \lambda_2 \ldots, \lambda_n)$ and U is unitary. Then $A^*A = U\overline{\Lambda}U^*U\Lambda U^* = U\operatorname{diag}(|\lambda_1|^2, |\lambda_2|^2, \ldots, |\lambda_n|^2)U^*$, so the squared singular values of A are $|\lambda_1|^2, |\lambda_2|^2, \ldots, |\lambda_n|^2$.

(c) It follows from (b) that $\lambda_i = \sigma_i$ for each $i = 1, 2, \ldots, n$ if and only if A is normal and has real nonnegative eigenvalues. This is the case if and only if A is Hermitian and has nonnegative eigenvalues (Theorem 12.6.1), that is, if and only if A is positive semidefinite (Theorem 13.1.2). $\qquad\square$

Example 15.3.5 Since the singular values of a normal matrix are the moduli of its eigenvalues, the spectral norm of a normal matrix A equals its spectral radius:

$$\|A\|_2 = \rho(A) \text{ if } A \in \mathsf{M}_n \text{ is normal.} \tag{15.3.6}$$

In particular, if $A = \operatorname{diag}(\lambda_1, \lambda_2, \ldots, \lambda_n)$, then $\|A\|_2 = \max_{1 \leq i \leq n} |\lambda_i| = \rho(A)$.

Example 15.3.7 Since $\|A\|_2 \geq \rho(A)$ for every $A \in \mathsf{M}_n$ (Theorem 15.3.2.c), it follows from (15.3.6) that A cannot be normal if $\|A\|_2 > \rho(A)$. For example,

$$B = \begin{bmatrix} 2 & 3 \\ 0 & 2 \end{bmatrix} \qquad (15.3.8)$$

has singular values 4 and 1, so $\|B\|_2 > \rho(B)$ and hence B is not normal. However, if $\|A\|_2 = \rho(A)$, it does not follow that A is normal. For example, consider the non-normal matrix

$$C = [4] \oplus \begin{bmatrix} 2 & 3 \\ 0 & 2 \end{bmatrix}. \qquad (15.3.9)$$

We have spec $C = \{4, 2\}$, spec $C^*C = \{16, 1\}$, and $\|C\|_2 = \rho(C)$.

The singular values of a matrix can be any nonnegative real numbers, but the singular values of unitary matrices are severely constrained.

Theorem 15.3.10 *A square matrix A is a scalar multiple of a unitary matrix if and only if $\sigma_{\max}(A) = \sigma_{\min}(A)$.*

Proof If $A = cU$ for some scalar c and a unitary matrix U, then $AA^* = |c|^2 UU^* = |c|^2 I$, so every singular value of A is equal to $|c|$. Conversely, let $A = V\Sigma W^*$ be a singular value decomposition of $A \in \mathsf{M}_n$. The hypothesis is that $\Sigma = sI$ for some $s \geq 0$. Then $A = sVW^*$ is a scalar multiple of the unitary matrix VW^*. $\qquad\square$

Similarity preserves eigenvalues, but it need not preserve singular values. The following example illustrates this point.

Example 15.3.11 The matrices

$$A = \begin{bmatrix} 1 & 1 \\ 0 & 1 \end{bmatrix} \quad \text{and} \quad B = \begin{bmatrix} 1 & 2 \\ 0 & 1 \end{bmatrix} \qquad (15.3.12)$$

are similar via diag$(2, 1)$. A computation reveals that

$$A^*A = \begin{bmatrix} 1 & 1 \\ 1 & 2 \end{bmatrix} \quad \text{and} \quad B^*B = \begin{bmatrix} 1 & 2 \\ 2 & 5 \end{bmatrix}.$$

Since tr $A^*A \neq$ tr B^*B, the eigenvalues of A^*A and B^*B are not the same. Consequently, A and B do not have the same singular values.

The following theorem identifies a broad class of transformations that preserve singular values. Certain transformations in this class also preserve eigenvalues.

Theorem 15.3.13 *Let $A \in \mathsf{M}_{m \times n}$, and let $U \in \mathsf{M}_m$ and $V \in \mathsf{M}_n$ be unitary.*

(a) *A and UAV have the same singular values. In particular, these two matrices have the same spectral and Frobenius norms.*

(b) *If $m = n$, then A and UAU^* have the same eigenvalues and the same singular values.*

Proof (a) The matrices $(UAV)^*(UAV) = V^*A^*AV$ and A^*A are unitarily similar, so they have the same eigenvalues. These eigenvalues are the squared singular values of UAV and of A. The spectral and Frobenius norms of a matrix are determined by its singular values, so UAV and A have the same spectral and Frobenius norms.

(b) The matrices A and UAU^* are unitarily similar, so they have the same eigenvalues. Part (a) ensures that they have the same singular values. $\qquad\square$

How is the sum of the eigenvalues of a square matrix related to the sum of its singular values?

Theorem 15.3.14 *If $A \in M_n$, then $|\operatorname{tr} A| \leq \operatorname{tr}|A|$, that is,*

$$\left| \sum_{i=1}^{n} \lambda_i(A) \right| \leq \sum_{i=1}^{n} \sigma_i(A). \tag{15.3.15}$$

Proof Let $A = [a_{ij}] = V\Sigma W^*$ be a singular value decomposition, in which $V = [v_{ij}]$ and $W = [w_{ij}]$ are unitary, and $\Sigma = \operatorname{diag}(\sigma_1, \sigma_2, \ldots, \sigma_n)$. The assertion that $\operatorname{tr}|A| = \sum_{i=1}^{n} \sigma_i$ follows from Definition 14.1.1 and Definition 14.3.1. In the following argument, we use the fact that the columns of V and W are unit vectors. A calculation reveals that

$$a_{ii} = \sum_{k=1}^{n} \sigma_k v_{ik} \overline{w_{ik}}, \quad i = 1, 2, \ldots, n,$$

and

$$\operatorname{tr} A = \sum_{i=1}^{n} a_{ii} = \sum_{i=1}^{n} \sum_{k=1}^{n} \sigma_k v_{ik} \overline{w_{ik}} = \sum_{k=1}^{n} \sigma_k \sum_{i=1}^{n} v_{ik} \overline{w_{ik}}.$$

Now use the triangle and Cauchy–Schwarz inequalities to compute

$$|\operatorname{tr} A| = \left| \sum_{k=1}^{n} \sigma_k \sum_{i=1}^{n} v_{ik} \overline{w_{ik}} \right| \leq \sum_{k=1}^{n} \sigma_k \left| \sum_{i=1}^{n} v_{ik} \overline{w_{ik}} \right|$$

$$\leq \sum_{k=1}^{n} \sigma_k \left(\sum_{i=1}^{n} |v_{ik}|^2 \right)^{1/2} \left(\sum_{i=1}^{n} |\overline{w_{ik}}|^2 \right)^{1/2}$$

$$= \sum_{k=1}^{n} \sigma_k(A). \qquad\square$$

Example 15.3.16 For

$$A = \begin{bmatrix} 2 & 3 \\ 0 & 2 \end{bmatrix},$$

we have $\operatorname{tr} A = 4$ and $\sigma_1(A) + \sigma_2(A) = 4 + 1 = 5$.

We now use the invariance principles in Theorem 15.3.13 to modify (15.3.15) in two different ways.

Theorem 15.3.17 *Let $A = [a_{ij}] \in M_n$. Then*

$$|\lambda_1(A)| + |\lambda_2(A)| + \cdots + |\lambda_n(A)| \le \sigma_1(A) + \sigma_2(A) + \cdots + \sigma_n(A) \tag{15.3.18}$$

and

$$|a_{11}| + |a_{22}| + \cdots + |a_{nn}| \le \sigma_1(A) + \sigma_2(A) + \cdots + \sigma_n(A). \tag{15.3.19}$$

Proof Since $\operatorname{tr} A = \sum_{i=1}^{n} \lambda_i = \sum_{i=1}^{n} a_{ii}$, the inequalities (15.3.18) and (15.3.19) follow from the triangle inequality:

$$\left| \sum_{i=1}^{n} \lambda_i \right| \le \sum_{i=1}^{n} |\lambda_i| \quad \text{and} \quad \left| \sum_{i=1}^{n} a_{ii} \right| \le \sum_{i=1}^{n} |a_{ii}|.$$

To obtain the second inequality in (15.3.18), let $U \in M_n$ be unitary and such that $U^*AU = T$ is upper triangular and has diagonal entries $\lambda_1, \lambda_2, \ldots, \lambda_n$; this is Theorem 10.1.1. Then let $\theta_1, \theta_2, \ldots, \theta_n$ be real numbers such that

$$\lambda_1 = e^{i\theta_1} |\lambda_1|, \qquad \lambda_2 = e^{i\theta_2} |\lambda_2|, \ldots, \qquad \lambda_n = e^{i\theta_n} |\lambda_n|.$$

Let $D = \operatorname{diag}(e^{-i\theta_1}, e^{-i\theta_2}, \ldots, e^{-i\theta_n})$, which is unitary. The matrix DT has main diagonal entries $|\lambda_1|, |\lambda_2|, \ldots, |\lambda_n|$. Because $DT = DU^*AU$, Theorem 15.3.13.a ensures that DT and A have the same singular values. Now apply (15.3.15) to DT:

$$\sum_{i=1}^{n} |\lambda_i| = \operatorname{tr} DT = |\operatorname{tr} DT| \le \sum_{k=1}^{n} \sigma_k(DT) = \sum_{k=1}^{n} \sigma_k(A).$$

Now let $\theta_1, \theta_2, \ldots, \theta_n$ be real numbers such that

$$a_{11} = e^{i\theta_1} |a_{11}|, \qquad a_{22} = e^{i\theta_2} |a_{22}|, \ldots, \qquad a_{nn} = e^{i\theta_n} |a_{nn}|.$$

Let $E = \operatorname{diag}(e^{-i\theta_1}, e^{-i\theta_2}, \ldots, e^{-i\theta_n})$, which is unitary. Then EA has main diagonal entries $|a_{11}|, |a_{22}|, \ldots, |a_{nn}|$ and the same singular values as A, so an application of (15.3.15) to ET gives the desired inequality:

$$\sum_{i=1}^{n} |a_{ii}| = \operatorname{tr} EA = |\operatorname{tr} EA| \le \sum_{k=1}^{n} \sigma_k(EA) = \sum_{k=1}^{n} \sigma_k(A). \qquad \square$$

15.4 An Upper Bound for the Spectral Norm

Let $A \in M_n$ and let $A = H + iK$ be its Cartesian decomposition; see Definition 12.6.6. The triangle inequality and (15.3.6) ensure that

$$\|A\|_2 = \|H + iK\|_2 \le \|H\|_2 + \|K\|_2 = \rho(H) + \rho(K). \tag{15.4.1}$$

Example 15.4.2 Consider

$$A = \begin{bmatrix} 4 & 6 \\ 0 & 4 \end{bmatrix}, \qquad H = \begin{bmatrix} 4 & 3 \\ 3 & 4 \end{bmatrix}, \quad \text{and} \quad K = \begin{bmatrix} 0 & -3i \\ 3i & 0 \end{bmatrix}.$$

Then $\operatorname{spec} A^*A = \{4, 64\}$, $\operatorname{spec} H = \{1, 7\}$, and $\operatorname{spec} K = \{-3, 3\}$, so

$$8 = \|A\|_2 \le \rho(H) + \rho(K) = 7 + 3 = 10.$$

We may be able to do better by first shifting the Cartesian components of A and then invoking the triangle inequality. For any $z \in \mathbb{C}$, the matrices $H - zI$ and $iK + zI$ are normal (Theorem 12.1.9.a). Consequently,

$$\|A\|_2 = \|A - zI + zI\|_2 = \|H - zI + iK + zI\|_2$$
$$\leq \|H - zI\|_2 + \|iK + zI\|_2 = \rho(H - zI) + \rho(iK + zI). \tag{15.4.3}$$

The proof of the following theorem examines a particular choice of the shift parameter z in (15.4.3).

Theorem 15.4.4 *Let $A = H + iK \in \mathsf{M}_n$, in which H, K are Hermitian, $\operatorname{spec} H \subseteq [a, b]$, and $\operatorname{spec} K \subseteq [c, d]$. Then*

$$\|A\|_2 \leq \frac{1}{2}\sqrt{(b - a)^2 + (c + d)^2} + \frac{1}{2}\sqrt{(a + b)^2 + (d - c)^2}. \tag{15.4.5}$$

Proof Let $\alpha = \frac{1}{2}(a + b)$ and $\beta = \frac{1}{2}(c + d)$; these are the midpoints of the intervals $[a, b]$ and $[c, d]$. Then $H - \alpha I$ and $K - \beta I$ are Hermitian, so they have real eigenvalues. Compute

$$\|H - \alpha I\|_2 = \rho(H - \alpha I) = \max_{\lambda \in \operatorname{spec} H} |\lambda - \alpha| \leq \max\{b - \alpha, \alpha - a\} = \frac{b - a}{2}$$

and

$$\|K - \beta I\|_2 = \rho(K - \beta I) = \max_{\lambda \in \operatorname{spec} K} |\lambda - \beta| \leq \max\{d - \beta, \beta - c\} = \frac{d - c}{2}.$$

Let $z = \alpha - i\beta$ and consider

$$A = H - zI + iK + zI = (H - \alpha I + i\beta I) + (iK - i\beta I + \alpha I).$$

Since $H - zI$ and $iK + zI$ are normal,

$$\|H - \alpha I + i\beta I\|_2 = \max_{\lambda \in \operatorname{spec} H} |\lambda - \alpha + i\beta| = \max_{\lambda \in \operatorname{spec} H} \sqrt{|\lambda - \alpha|^2 + \beta^2}$$
$$= \sqrt{\max_{\lambda \in \operatorname{spec} H} |\lambda - \alpha|^2 + \beta^2} = \sqrt{\|H - \alpha I\|_2^2 + \beta^2}$$

and

$$\|iK - i\beta I + \alpha I\|_2 = \max_{\lambda \in \operatorname{spec} K} |i\lambda - i\beta + \alpha| = \max_{\lambda \in \operatorname{spec} K} \sqrt{|i\lambda - i\beta|^2 + \alpha^2}$$
$$= \sqrt{\max_{\lambda \in \operatorname{spec} K} |\lambda - \beta|^2 + \alpha^2} = \sqrt{\|K - \beta I\|_2^2 + \alpha^2}.$$

Thus,

$$\|H + iK\|_2 = \|(H - \alpha I + i\beta I) + (iK - i\beta I + \alpha I)\|_2$$
$$\leq \|(H - \alpha I) + i\beta I\|_2 + \|(iK - i\beta I) + \alpha I\|_2$$
$$= \sqrt{\|H - \alpha I\|_2^2 + \beta^2} + \sqrt{\|K - \beta I\|_2^2 + \alpha^2}$$
$$\leq \frac{1}{2}\sqrt{(b - a)^2 + (c + d)^2} + \frac{1}{2}\sqrt{(a + b)^2 + (d - c)^2}. \qquad \square$$

Example 15.4.6 Consider the matrix A in Example 15.4.2, for which $a = 1$, $b = 7$, $c = -3$, and $d = 3$. The upper bound (15.4.5) is

$$8 = \|A\|_2 \leq \sqrt{3^2 + 0^2} + \sqrt{4^2 + 3^2} = 3 + 5 = 8.$$

This is an improvement over the estimate obtained in Example 15.4.2.

15.5 The Pseudoinverse

With the help of the singular value decomposition, we can define a function of a matrix $A \in M_{m \times n}$ that has many properties of an inverse. The following lemma is the first step in defining that function.

Lemma 15.5.1 *Let* $A \in M_{m \times n}(\mathbb{F})$, *let* $r = \operatorname{rank} A \geq 1$, *and let*

$$A = X_1 \Sigma_r Y_1^* = X_2 \Sigma_r Y_2^*$$

be compact singular value decompositions. Then

$$Y_1 \Sigma_r^{-1} X_1^* = Y_2 \Sigma_r^{-1} X_2^*.$$

Proof Theorem 14.2.15 ensures that there is a unitary $U \in M_r$ such that $X_1 = X_2 U$, $Y_1 = Y_2 U$, and $U \Sigma_r = \Sigma_r U$. Then

$$\Sigma_r^{-1} U = \Sigma_r^{-1} (U \Sigma_r) \Sigma_r^{-1} = \Sigma_r^{-1} (\Sigma_r U) \Sigma_r^{-1} = U \Sigma_r^{-1}.$$

It follows that

$$Y_1 \Sigma_r^{-1} X_1^* = Y_2 U \Sigma_r^{-1} U^* X_2^* = Y_2 \Sigma_r^{-1} U U^* X_2^* = Y_2 \Sigma_r^{-1} X_2^*. \qquad \square$$

Definition 15.5.2 Let $A \in M_{m \times n}(\mathbb{F})$. If $r = \operatorname{rank} A \geq 1$ and $A = X \Sigma_r Y^*$ is a compact singular value decomposition, then

$$A^\dagger = Y \Sigma_r^{-1} X^* \in M_{n \times m}(\mathbb{F})$$

is the *pseudoinverse* (or *Moore–Penrose pseudoinverse*) of A. If $A = 0_{m \times n}$, define $A^\dagger = 0_{n \times m}$.

Although there are compact singular value decompositions

$$A = X_1 \Sigma_r Y_1^* = X_2 \Sigma_r Y_2^*$$

in which $X_1 \neq X_2$ or $Y_1 \neq Y_2$, Lemma 15.5.1 ensures that $Y_1 \Sigma_r^{-1} X_1^* = Y_2 \Sigma_r^{-1} X_2^*$. Therefore, $A^\dagger$ is a well-defined function of A and it is correct to speak of *the* pseudoinverse of A.

Example 15.5.3 A compact singular value decomposition of a nonzero $\mathbf{x} \in M_{n \times 1}$ is $\mathbf{x} = (\mathbf{x}/\|\mathbf{x}\|)\,[\|\mathbf{x}\|]\,[1]$; see Example 14.2.7. Therefore,

$$\mathbf{x}^\dagger = [1]\,[\|\mathbf{x}\|^{-1}]\,(\mathbf{x}/\|\mathbf{x}\|)^* = \mathbf{x}^*/\|\mathbf{x}\|^2.$$

Example 15.5.4 The compact singular value decomposition

$$A = \begin{bmatrix} 0 & 1 \\ 0 & 0 \end{bmatrix} = \begin{bmatrix} 1 \\ 0 \end{bmatrix} [1] \begin{bmatrix} 0 & 1 \end{bmatrix} \tag{15.5.5}$$

permits us to compute

$$A^\dagger = \begin{bmatrix} 0 \\ 1 \end{bmatrix} [1] \begin{bmatrix} 1 & 0 \end{bmatrix} = \begin{bmatrix} 0 & 0 \\ 1 & 0 \end{bmatrix}.$$

Example 15.5.6 If $A \in \mathsf{M}_{m \times n}$ and $\operatorname{rank} A = n$, then $m \geq n$ and A has a singular value decomposition

$$A = V \Sigma W^* = [V_n \ V'] \begin{bmatrix} \Sigma_n \\ 0 \end{bmatrix} W^* = V_n \Sigma_n W^*,$$

in which Σ_n is invertible. Then $A^\dagger = W \Sigma_n^{-1} V_n^*$ and $A^* = W \Sigma_n V_n^*$, so

$$A^* A = W \Sigma_n V_n^* V_n \Sigma_n W^* = W \Sigma_n^2 W^*.$$

Thus,

$$(A^* A)^{-1} = W \Sigma_n^{-2} W^*$$

and

$$A^\dagger = (A^* A)^{-1} A^*.$$

Theorem 15.5.7 *Let $A \in \mathsf{M}_{m \times n}(\mathbb{F})$ be nonzero.*

(a) $\operatorname{col} A^\dagger = \operatorname{col} A^*$ *and* $\operatorname{null} A^\dagger = \operatorname{null} A^*$.

(b) $AA^\dagger \in \mathsf{M}_m(\mathbb{F})$ *is the orthogonal projection of $\mathbb{F}^m$ onto $\operatorname{col} A$ and $I - AA^\dagger$ is the orthogonal projection of $\mathbb{F}^m$ onto $\operatorname{null} A^*$.*

(c) $A^\dagger A \in \mathsf{M}_n(\mathbb{F})$ *is the orthogonal projection of $\mathbb{F}^n$ onto $\operatorname{col} A^*$ and $I - A^\dagger A$ is the orthogonal projection of $\mathbb{F}^n$ onto $\operatorname{null} A$.*

(d) $AA^\dagger A = A$.

(e) $A^\dagger AA^\dagger = A^\dagger$.

(f) *If $m = n$ and A is invertible, then $A^\dagger = A^{-1}$.*

(g) $(A^\dagger)^\dagger = A$.

Proof Let $r = \operatorname{rank} A$ and let $A = X \Sigma_r Y^*$ be a compact singular value decomposition. Then $\Sigma_r \in \mathsf{M}_r$ is invertible, and X and Y have orthonormal columns, so they have full column rank.

(a) $A^* = Y \Sigma_r X^*$, so $\operatorname{col} A^* = \operatorname{col} Y = \operatorname{col} A^\dagger$ and $\operatorname{null} A^* = (\operatorname{col} X)^\perp = \operatorname{null} A^\dagger$.

(b) Compute $AA^\dagger = X \Sigma_r Y^* Y \Sigma_r^{-1} X^* = X \Sigma_r \Sigma_r^{-1} X^* = XX^*$. Then invoke (14.2.10) and Example 7.3.5.

(c) Compute $A^\dagger A = Y \Sigma_r^{-1} X^* X \Sigma_r Y^* = Y \Sigma_r^{-1} \Sigma_r Y^* = YY^*$. Then invoke (14.2.11) and Example 7.3.5.

(d) This follows from (b).

(e) This follows from (c) and (a).

(f) If $m = n$ and A is invertible, then $\operatorname{rank} A = n$, $\Sigma_n \in \mathsf{M}_n$ is invertible, and $X, Y \in \mathsf{M}_n$ are unitary. Consequently, $A^{-1} = (X\Sigma_n Y^*)^{-1} = Y\Sigma_n^{-1}X^* = A^\dagger$.

(g) $(Y\Sigma_r^{-1}X^*)^\dagger = X\Sigma_r Y^* = A$. □

In Section 7.2 we studied consistent linear systems of the form

$$A\mathbf{x} = \mathbf{y}, \qquad A \in \mathsf{M}_{m \times n}(\mathbb{F}), \quad \mathbf{y} \in \mathbb{F}^m. \tag{15.5.8}$$

If (15.5.8) has more than one solution, we may wish to find a solution that has minimum Euclidean norm; see Theorem 7.2.5. The following theorem shows that the pseudoinverse identifies the minimum norm solution, which is the only solution in $\operatorname{col} A^*$.

Theorem 15.5.9 *Let* $A \in \mathsf{M}_{m \times n}(\mathbb{F})$ *and* $\mathbf{y} \in \mathbb{F}^n$. *Suppose that* $A\mathbf{x} = \mathbf{y}$ *is consistent and let* $\mathbf{x}_0 = A^\dagger \mathbf{y}$.

(a) $\mathbf{x}_0 \in \operatorname{col} A^*$ *and* $A\mathbf{x}_0 = \mathbf{y}$.

(b) *If* $\mathbf{x} \in \mathbb{F}^n$ *and* $A\mathbf{x} = \mathbf{y}$, *then* $\|\mathbf{x}\|_2 \geq \|\mathbf{x}_0\|_2$, *with equality if and only if* $\mathbf{x} \in \operatorname{col} A^*$.

(c) *If* $\mathbf{x} \in \operatorname{col} A^*$ *and* $A\mathbf{x} = \mathbf{y}$, *then* $\mathbf{x} = \mathbf{x}_0$.

Proof (a) Since there is an $\mathbf{x} \in \mathbb{F}^n$ such that $A\mathbf{x} = \mathbf{y}$, parts (a) and (d) of Theorem 15.5.7 ensure that $\mathbf{x}_0 \in \operatorname{col} A^*$ and

$$A\mathbf{x}_0 = AA^\dagger \mathbf{y} = AA^\dagger A\mathbf{x} = A\mathbf{x} = \mathbf{y}.$$

(b) Compute

$$\begin{aligned}
\|\mathbf{x}\|_2^2 &= \|A^\dagger A\mathbf{x} + (I - A^\dagger A)\mathbf{x}\|_2^2 \\
&= \|A^\dagger A\mathbf{x}\|_2^2 + \|(I - A^\dagger A)\mathbf{x}\|_2^2 \\
&= \|A^\dagger \mathbf{y}\|_2^2 + \|(I - A^\dagger A)\mathbf{x}\|_2^2 \\
&= \|\mathbf{x}_0\|_2^2 + \|(I - A^\dagger A)\mathbf{x}\|_2^2 \geq \|\mathbf{x}_0\|_2^2.
\end{aligned}$$

The preceding inequality is an equality if and only if $(I - A^\dagger A)\mathbf{x} = \mathbf{0}$, which occurs if and only if $\mathbf{x} \in \operatorname{col} A^*$ (Theorem 15.5.7.c).

(c) If $\mathbf{x} \in \operatorname{col} A^*$, then $\mathbf{x} = A^\dagger A\mathbf{x}$. If, in addition, $A\mathbf{x} = \mathbf{y}$, then

$$\|\mathbf{x} - \mathbf{x}_0\|_2 = \|A^\dagger A\mathbf{x} - A^\dagger \mathbf{y}\|_2 = \|A^\dagger A\mathbf{x} - A^\dagger A\mathbf{x}\|_2 = 0. \qquad \square$$

The pseudoinverse can also help us analyze least squares problems. If a system of the form (15.5.8) is inconsistent, then there is no $\mathbf{x}$ such that $A\mathbf{x} = \mathbf{y}$. However, we may wish to find an $\mathbf{x}_0$ such that $\|A\mathbf{x}_0 - \mathbf{y}\|_2 \leq \|A\mathbf{x} - \mathbf{y}\|_2$ for all $\mathbf{x} \in \mathbb{F}^n$; see Section 7.5. Among all such vectors, we may also wish to identify one that has minimum Euclidean norm. The pseudoinverse identifies a vector that has both of these minimality properties.

Theorem 15.5.10 *Let* $A \in \mathsf{M}_{m \times n}(\mathbb{F})$, *let* $\mathbf{y} \in \mathbb{F}^m$, *and let* $\mathbf{x}_0 = A^\dagger \mathbf{y}$.

(a) *For every* $\mathbf{x} \in \mathbb{F}^n$,

$$\|A\mathbf{x}_0 - \mathbf{y}\|_2 \leq \|A\mathbf{x} - \mathbf{y}\|_2,$$

with equality if and only if $\mathbf{x} - \mathbf{x}_0 \in \operatorname{null} A$.

(b) *If* $\mathbf{x} \in \mathbb{F}^n$ *and* $\|A\mathbf{x} - \mathbf{y}\|_2 = \|A\mathbf{x}_0 - \mathbf{y}\|_2$, *then* $\|\mathbf{x}\|_2 \geq \|\mathbf{x}_0\|_2$, *with equality if and only if* $\mathbf{x} = \mathbf{x}_0$.

Proof (a) Use the Pythagorean theorem and the facts that $A\mathbf{x} - AA^{\dagger}\mathbf{y} \in \operatorname{col} A$ and $(I - AA^{\dagger})\mathbf{y} \in (\operatorname{col} A)^{\perp}$ to compute

$$\|A\mathbf{x} - \mathbf{y}\|^2 = \|A\mathbf{x} - AA^{\dagger}\mathbf{y} + AA^{\dagger}\mathbf{y} - \mathbf{y}\|_2^2$$
$$= \|A\mathbf{x} - AA^{\dagger}\mathbf{y}\|_2^2 + \|AA^{\dagger}\mathbf{y} - \mathbf{y}\|_2^2$$
$$= \|A(\mathbf{x} - \mathbf{x}_0)\|_2^2 + \|A\mathbf{x}_0 - \mathbf{y}\|_2^2 \geq \|A\mathbf{x}_0 - \mathbf{y}\|_2^2,$$

with equality if and only if $A(\mathbf{x} - \mathbf{x}_0) = \mathbf{0}$.

(b) If $\|A\mathbf{x} - \mathbf{y}\|_2 = \|A\mathbf{x}_0 - \mathbf{y}\|_2$, then (a) ensures that $\mathbf{x} = \mathbf{x}_0 + \mathbf{w}$, in which $\mathbf{x}_0 = A^{\dagger}\mathbf{y} \in \operatorname{col} A^*$ and $\mathbf{w} \in \operatorname{null} A = (\operatorname{col} A^*)^{\perp}$. Therefore,

$$\|\mathbf{v}\|_2^2 = \|\mathbf{x}_0 + \mathbf{w}\|_2^2 = \|\mathbf{x}_0\|_2^2 + \|\mathbf{w}\|_2^2 \geq \|\mathbf{x}_0\|_2^2,$$

with equality if and only if $\mathbf{w} = \mathbf{0}$ if and only if $\mathbf{x} = \mathbf{x}_0$. $\qquad\square$

Here is a summary of the preceding theorems.

(a) $\mathbf{x}_0 = A^{\dagger}\mathbf{y}$ is a least squares solution of the linear system $A\mathbf{x} = \mathbf{y}$.

(b) Among all the least squares solutions, $\mathbf{x}_0$ is the only one with smallest norm.

(c) If $A\mathbf{x} = \mathbf{y}$ is consistent, then $\mathbf{x}_0$ is a solution.

(d) Among all the solutions, $\mathbf{x}_0$ is the only one of smallest norm and it is the only one in $\operatorname{col} A^*$.

15.6 The Spectral Condition Number

Definition 15.6.1 Let $A \in \mathsf{M}_n$. If A is invertible, its *spectral condition number* is

$$\kappa_2(A) = \|A\|_2 \|A^{-1}\|_2.$$

If A is not invertible, its spectral condition number is undefined.

Theorem 15.3.2.a and the definition of the spectral norm permit us to express the spectral condition number of A as a ratio of singular values:

$$\kappa_2(A) = \frac{\sigma_{\max}(A)}{\sigma_{\min}(A)}. \qquad (15.6.2)$$

Theorem 15.6.3 *Let* $A, B \in \mathsf{M}_n$ *be invertible.*

(a) $\kappa_2(AB) \leq \kappa_2(A)\kappa_2(B)$.

(b) $\kappa_2(A) \geq 1$, *with equality if and only if* A *is a nonzero scalar multiple of a unitary matrix.*

(c) *If* c *is a nonzero scalar, then* $\kappa_2(cA) = \kappa_2(A)$.

(d) $\kappa_2(A) = \kappa_2(A^*) = \kappa_2(\overline{A}) = \kappa_2(A^{\mathsf{T}}) = \kappa_2(A^{-1})$.

(e) $\kappa_2(A^*A) = \kappa_2(AA^*) = \kappa_2(A)^2$.

(f) If $U, V \in \mathbf{M}_n$ are unitary, then $\kappa_2(UAV) = \kappa_2(A)$.

Proof (a) Use Theorem 15.2.10.b to compute

$$\kappa_2(AB) = \|AB\|_2 \|B^{-1}A^{-1}\|_2 \leq \|A\|_2 \|B\|_2 \|B^{-1}\|_2 \|A^{-1}\|_2 = \kappa_2(A)\kappa_2(B).$$

(b) It follows from (a) and Theorem 15.2.2.a that

$$1 = \|I_n\|_2 = \|AA^{-1}\|_2 \leq \|A\|_2 \|A^{-1}\|_2 = \kappa_2(A),$$

with equality if and only if $\sigma_{\max}(A) = \sigma_{\min}(A)$ if and only if A is a nonzero scalar multiple of a unitary matrix (Theorem 15.3.10).

(c) Compute

$$\kappa_2(cA) = \|cA\|_2 \|(cA)^{-1}\|_2 = \|cA\|_2 \|c^{-1}A^{-1}\|_2$$
$$= |c|\|A\|_2 |c^{-1}| \|A^{-1}\|_2 = \|A\|_2 \|A^{-1}\|_2 = \kappa_2(A).$$

(d) A, A^*, $\overline{A}$, and A^{T} all have the same singular values; see Theorem 14.1.3. Also, $\kappa_2(A^{-1}) = \|A^{-1}\|_2 \|A\|_2 = \kappa_2(A)$.

(e) $\kappa_2(A^*A) = \sigma_{\max}(A^*A)/\sigma_{\min}(A^*A) = \sigma_{\max}^2(A)/\sigma_{\min}^2(A) = \kappa_2(A)^2 = \kappa_2(A^*)^2 = \kappa_2(AA^*)$.

(f) The singular values of A and UAV are the same, so their spectral condition numbers are the same. □

The spectral condition number of an invertible matrix plays a key role in understanding how the solution of a linear system and computation of eigenvalues can be affected by errors in the data or in the computation.

Let $A \in \mathbf{M}_n(\mathbb{F})$ be invertible, let $\mathbf{y} \in \mathbb{F}^n$ be nonzero, and let $\mathbf{x} \in \mathbb{F}^n$ be the unique (necessarily nonzero) solution of the linear system $A\mathbf{x} = \mathbf{y}$. Then

$$\|\mathbf{y}\|_2 = \|A\mathbf{x}\|_2 \leq \|A\|_2 \|\mathbf{x}\|_2,$$

which we rewrite as

$$\frac{1}{\|\mathbf{x}\|_2} \leq \frac{\|A\|_2}{\|\mathbf{y}\|_2}. \tag{15.6.4}$$

Now let $\Delta\mathbf{y} \in \mathbb{F}^n$ and let $\Delta\mathbf{x} \in \mathbb{F}^n$ be the unique solution of the linear system $A(\Delta\mathbf{x}) = \Delta\mathbf{y}$. Linearity ensures that

$$A(\mathbf{x} + \Delta\mathbf{x}) = \mathbf{y} + \Delta\mathbf{y}. \tag{15.6.5}$$

We have $\Delta\mathbf{x} = A^{-1}(\Delta\mathbf{y})$, and therefore

$$\|\Delta\mathbf{x}\|_2 = \|A^{-1}(\Delta\mathbf{y})\|_2 \leq \|A^{-1}\|_2 \|\Delta\mathbf{y}\|_2. \tag{15.6.6}$$

Now combine (15.6.4) and (15.6.6) to obtain

$$\frac{\|\Delta\mathbf{x}\|_2}{\|\mathbf{x}\|_2} \leq \frac{\|A\|_2 \|A^{-1}\|_2 \|\Delta\mathbf{y}\|_2}{\|\mathbf{y}\|_2} = \kappa_2(A)\frac{\|\Delta\mathbf{y}\|_2}{\|\mathbf{y}\|_2}, \tag{15.6.7}$$

which provides an upper bound

$$\frac{\|\Delta \mathbf{x}\|_2}{\|\mathbf{x}\|_2} \leq \kappa_2(A)\frac{\|\Delta \mathbf{y}\|_2}{\|\mathbf{y}\|_2} \tag{15.6.8}$$

on the relative error $\|\Delta \mathbf{x}\|_2/\|\mathbf{x}\|_2$ involved when solving the perturbed linear system (15.6.5).

If $\kappa_2(A)$ is not large, the relative error in the solution of $A\mathbf{x} = \mathbf{y}$ cannot be much worse than the relative error in the data. In this case, A is *well conditioned*. For example, if A is a nonzero scalar multiple of a unitary matrix, then $\kappa_2(A) = 1$ and A is *perfectly conditioned*. In this case, the relative error in the solution is at most the relative error in the data. If $\kappa_2(A)$ is large, however, the relative error in the solution could be much larger than the relative error in the data. In this case, A is *ill conditioned*.

Example 15.6.9 Consider $A\mathbf{x} = \mathbf{y}$, in which

$$A = \begin{bmatrix} 1 & 2 \\ 2 & 4.001 \end{bmatrix} \quad \text{and} \quad \mathbf{y} = \begin{bmatrix} 4 \\ 8.001 \end{bmatrix}.$$

It has the unique solution $\mathbf{x} = [2\ 1]^\mathsf{T}$. Let $\Delta \mathbf{y} = [0.001 \ -0.002]^\mathsf{T}$. The system $A(\Delta \mathbf{x}) = \Delta \mathbf{y}$ has the unique solution $\Delta \mathbf{x} = [8.001 \ -4.000]^\mathsf{T}$. The relative errors in the data and in the solution are

$$\frac{\|\Delta \mathbf{y}\|_2}{\|\mathbf{y}\|_2} = 2.5001 \times 10^{-4} \quad \text{and} \quad \frac{\|\Delta \mathbf{x}\|_2}{\|\mathbf{x}\|_2} = 4.0004.$$

The spectral condition number of A is $\kappa_2(A) = 2.5008 \times 10^4$ and the upper bound in (15.6.8) is

$$\frac{\|\Delta \mathbf{x}\|_2}{\|\mathbf{x}\|_2} = 4.0004 \leq 6.2523 = \kappa_2(A)\frac{\|\Delta \mathbf{y}\|_2}{\|\mathbf{y}\|_2}.$$

Because A is ill conditioned, a relatively small change in the data can (and in this case does) cause a large relative change in the solution.

Conditioning issues were one reason we expressed reservations in Section 7.5 about finding a least squares solution of an inconsistent linear system $A\mathbf{x} = \mathbf{y}$ by solving the normal equations

$$A^*A\mathbf{x} = A^*\mathbf{y}. \tag{15.6.10}$$

If A is square and invertible (but not a scalar multiple of a unitary matrix), the matrix A^*A in the normal equations (15.6.10) is more ill conditioned than A because $\kappa_2(A) > 1$ and $\kappa_2(A^*A) = \kappa_2(A)^2$. On the other hand, if one uses a QR factorization of A, then Theorem 15.6.3.f ensures that the equivalent linear system $R\mathbf{x} = Q^*\mathbf{y}$ (see (7.5.9)) is as well conditioned as A:

$$\kappa_2(R) = \kappa_2(Q^*A) = \kappa_2(A).$$

Interpretation of results from eigenvalue computations must take into account uncertainties in the input data (perhaps it arises from physical measurements) as well as roundoff errors due to finite precision arithmetic. One way to model uncertainty in the results is to imagine that the computation has been done with perfect precision, but for a slightly different matrix. The following discussion adopts that point of view, and gives a framework to assess the quality of computed eigenvalues.

Lemma 15.6.11 *Let $A \in \mathsf{M}_n$. If $\|A\|_2 < 1$, then $I + A$ is invertible.*

Proof Theorem 15.3.2.c ensures that every $\lambda \in \mathrm{spec}\, A$ satisfies $|\lambda| \leq \|A\|_2 < 1$. Therefore, $-1 \notin \mathrm{spec}\, A$ and $0 \notin \mathrm{spec}(I + A)$, so $I + A$ is invertible. □

The following theorem relies on the preceding lemma and on the fact that the spectral norm of a diagonal matrix is the largest of the moduli of its diagonal entries; see Example 15.3.5.

Theorem 15.6.12 (Bauer–Fike) *Let $A \in \mathsf{M}_n$ be diagonalizable and let $A = S\Lambda S^{-1}$, in which $S \in \mathsf{M}_n$ is invertible and $\Lambda = \mathrm{diag}(\lambda_1, \lambda_2, \ldots, \lambda_n)$. Let $\Delta A \in \mathsf{M}_n$. If $\lambda \in \mathrm{spec}(A + \Delta A)$, then there is some $i \in \{1, 2, \ldots, n\}$ such that*

$$|\lambda_i - \lambda| \leq \kappa_2(S)\|\Delta A\|_2. \qquad (15.6.13)$$

Proof If λ is an eigenvalue of A, then $\lambda = \lambda_i$ for some $i \in \{1, 2, \ldots, n\}$ and (15.6.13) is satisfied. Now assume that λ is not an eigenvalue of A, so $\Lambda - \lambda I$ is invertible. Since λ is an eigenvalue of $A + \Delta A$, we know that $A + \Delta A - \lambda I$ is not invertible. Therefore,

$$
\begin{aligned}
I + (\Lambda - \lambda I)^{-1} S^{-1} \Delta A S &= (\Lambda - \lambda I)^{-1}(\Lambda - \lambda I + S^{-1}\Delta A S) \\
&= (\Lambda - \lambda I)^{-1}(S^{-1}AS - \lambda S^{-1}S + S^{-1}\Delta A S) \\
&= (\Lambda - \lambda I)^{-1} S^{-1}(A + \Delta A - \lambda I)S
\end{aligned}
$$

is not invertible. It follows from the preceding lemma and Theorem 15.2.10.b that

$$
\begin{aligned}
1 &\leq \|(\Lambda - \lambda I)^{-1} S^{-1} \Delta A S\|_2 \leq \|(\Lambda - \lambda I)^{-1}\|_2 \|S^{-1}\Delta A S\|_2 \\
&= \max_{1 \leq i \leq n} |\lambda_i - \lambda|^{-1} \|S^{-1}\Delta A S\|_2 = \frac{\|S^{-1}\Delta A S\|_2}{\min_{1 \leq i \leq n} |\lambda_i - \lambda|} \\
&\leq \frac{\|S^{-1}\|_2 \|\Delta A\|_2 \|S\|_2}{\min_{1 \leq i \leq n} |\lambda_i - \lambda|} = \kappa_2(S) \frac{\|\Delta A\|_2}{\min_{1 \leq i \leq n} |\lambda_i - \lambda|}.
\end{aligned}
$$

This inequality implies (15.6.13). □

If A is Hermitian or normal, then the matrix of eigenvectors S may be chosen to be unitary. In this case, $\kappa_2(S) = 1$ and the error in the computed eigenvalue cannot be greater than the spectral norm of ΔA. However, if our best choice of S has a large spectral condition number, we should be skeptical about the quality of a computed eigenvalue. For example, the latter situation could occur if A has distinct eigenvalues and two of its eigenvectors are nearly collinear.

Example 15.6.14 The matrix

$$
A = \begin{bmatrix} 1 & 1000 \\ 0 & 2 \end{bmatrix}
$$

has distinct eigenvalues, so it is diagonalizable. Let

$$\Delta A = \begin{bmatrix} 0 & 0 \\ 0.01 & 0 \end{bmatrix}.$$

Then $\|\Delta A\|_2 = 0.01$,

$$A + \Delta A = \begin{bmatrix} 1 & 1000 \\ 0.01 & 2 \end{bmatrix}, \quad \text{and} \quad \text{spec}(A + \Delta A) \approx \{-1.702, 4.702\}.$$

Even though ΔA has small norm, neither eigenvalue of $A + \Delta A$ is a good approximation to an eigenvalue of A. A matrix of eigenvectors of A is

$$S = \begin{bmatrix} 1 & 1 \\ 0 & 0.001 \end{bmatrix},$$

for which $\kappa_2(S) \approx 2000$. For the eigenvalue $\lambda \approx 4.702$ of $A + \Delta A$, the bound in (15.6.13) is

$$2.702 \approx |2 - \lambda| \le \kappa_2(S) \|\Delta A\|_2 \approx 20,$$

so our computations are consistent with the Bauer–Fike bound.

15.7 Complex Symmetric Matrices

If $A \in M_n$ and $A = A^T$, then A is *complex symmetric*. Real symmetric matrices are normal, but non-real symmetric matrices need not be normal; they need not even be diagonalizable.

Theorem 15.7.1 (Autonne) *Let $A \in M_n$ be nonzero and symmetric, let* rank $A = r$, *let $\sigma_1 \ge \sigma_2 \ge \cdots \ge \sigma_r > 0$ be the positive singular values of A, and let $\Sigma_r = \text{diag}(\sigma_1, \sigma_2, \ldots, \sigma_r)$.*

(a) *There is a unitary $U \in M_n$ such that*

$$A = U \Sigma U^T \quad \text{and} \quad \Sigma = \Sigma_r \oplus 0_{n-r}. \tag{15.7.2}$$

(b) *There is a $U_r \in M_{n \times r}$ with orthonormal columns such that*

$$A = U_r \Sigma_r U_r^T. \tag{15.7.3}$$

(c) *If $U_r, V_r \in M_{n \times r}$ have orthonormal columns and $A = U_r \Sigma_r U_r^* = V_r \Sigma_r V_r^*$, then there is a real orthogonal $Q \in M_r$ such that $V_r = U_r Q$ and Q commutes with Σ_r.*

Proof (a) Let $A = V \Sigma W^*$ be a singular value decomposition, let

$$B = W^T A W = W^T (V \Sigma W^*) W = (W^T V) \Sigma,$$

and let $X = W^T V$. Then B is symmetric, X is unitary, and $B = X \Sigma$ is a polar decomposition. Since $B^* B = \Sigma X^* X \Sigma = \Sigma^2$ is real and $B = B^T$,

$$B^* B = \overline{B^* B} = B^T \overline{B} = B B^*,$$

that is, B is normal. Partition

$$X = \begin{bmatrix} X_{11} & X_{12} \\ X_{21} & X_{22} \end{bmatrix},$$

in which $X_{11} \in M_r$. The normality of B and Theorem 14.3.21 ensure that $\Sigma = \Sigma_r \oplus 0_{n-r}$ commutes with X, and Lemma 3.3.21.b says that $X_{12} = 0$ and $X_{21} = 0$. Therefore,

$$B = X\Sigma = X_{11}\Sigma_r \oplus X_{22}0_{n-r} = X_{11}\Sigma_r \oplus 0_{n-r},$$

in which X_{11} is unitary (Theorem 6.2.7.b) and commutes with Σ_r. The symmetry of B tells us that

$$X_{11}\Sigma_r \oplus 0_{n-r} = B = B^\mathsf{T} = \Sigma_r X_{11}^\mathsf{T} \oplus 0_{n-r}.$$

Since $X_{11}\Sigma_r$ and $\Sigma_r X_{11}^\mathsf{T}$ are right and left polar decompositions of an invertible matrix, Theorem 14.3.15.e.iii says that $X_{11} = X_{11}^\mathsf{T}$, that is, X_{11} is unitary and symmetric. Lemma 12.6.10 ensures that there is a symmetric unitary $Y \in M_r$ such that $X_{11} = Y^2$ and Y is a polynomial in X_{11}. Since X_{11} commutes with Σ_r, it follows that Y commutes with Σ_r (Theorem 0.8.1). Therefore,

$$\begin{aligned} B &= X_{11}\Sigma_r \oplus 0_{n-r} = Y^2\Sigma_r \oplus 0_{n-r} = Y\Sigma_r Y \oplus 0_{n-r} \\ &= (Y \oplus I_r)(\Sigma_r \oplus 0_{n-r})(Y \oplus I_r) \\ &= Z\Sigma_r Z, \end{aligned}$$

in which $Z = Y \oplus I_{n-r}$ is unitary and symmetric. Consequently,

$$\begin{aligned} A &= \overline{W}BW^* = \overline{W}Z\Sigma_r ZW^* = (\overline{W}Z)\Sigma_r(\overline{W}Z^\mathsf{T})^\mathsf{T} \\ &= (\overline{W}Z)\Sigma_r(\overline{W}Z)^\mathsf{T} \\ &= U\Sigma U^\mathsf{T}, \end{aligned}$$

in which $U = \overline{W}Z$ is unitary.

(b) Partition Σ and U in (a) as $\Sigma = \Sigma_r \oplus 0_{n-r}$ and $U = [U_r \ U']$, in which $U_r \in M_{n \times r}$. Then $A = U\Sigma U^\mathsf{T} = U_r\Sigma_r U_r^\mathsf{T}$ is a compact singular value decomposition.

(c) The hypothesis is that U_r and V_r have orthonormal columns and

$$A = U_r\Sigma_r\overline{U}_r^* = V_r\Sigma_r\overline{V}_r^*.$$

Theorem 14.2.15.a ensures that there is a unitary $Q \in M_r$ that commutes with Σ_r and is such that $V_r = U_r Q$ and $\overline{V}_r = \overline{U}_r Q$. The identities

$$Q = U_r^* V_r \quad \text{and} \quad Q = \overline{U}_r^* \overline{V}_r$$

imply that $Q = \overline{Q}$, so Q is real orthogonal. $\qquad\square$

Example 15.7.4 Consider the 1×1 symmetric matrix $[-1]$. Then $[-1] = [i][1][i]$ and $[i]$ is unitary. There is no real number c such that $[-1] = [c][1][c] = [c^2]$. Thus, if A is real symmetric, there need not be a real orthogonal U such that $A = U\Sigma U^\mathsf{T}$ is a singular value decomposition. However, Autonne's theorem says that this factorization is always possible with a complex unitary U.

Example 15.7.5 Here is a singular value decomposition of the type guaranteed by Autonne's theorem:

$$A = \begin{bmatrix} 1 & i \\ i & -1 \end{bmatrix} = \underbrace{\begin{bmatrix} \frac{1}{\sqrt{2}} & -\frac{1}{\sqrt{2}} \\ \frac{i}{\sqrt{2}} & \frac{i}{\sqrt{2}} \end{bmatrix}}_{U} \underbrace{\begin{bmatrix} 2 & 0 \\ 0 & 0 \end{bmatrix}}_{\Sigma} \underbrace{\begin{bmatrix} \frac{1}{\sqrt{2}} & \frac{i}{\sqrt{2}} \\ -\frac{1}{\sqrt{2}} & \frac{i}{\sqrt{2}} \end{bmatrix}}_{U^\mathsf{T}}.$$

Notice that $A^2 = 0$, so both eigenvalues of A are zero (see Theorem 8.3.3). If A were diagonalizable, it would be similar to the zero matrix, which it is not. Consequently, A is a complex symmetric matrix that is not diagonalizable.

15.8 Idempotent Matrices

Orthogonal projections and the non-Hermitian matrix

$$A = \begin{bmatrix} 1 & 2 \\ 0 & 0 \end{bmatrix}$$

are idempotent. In the following theorem, we use the singular value decomposition to find a standard form to which any idempotent matrix is unitarily similar.

Theorem 15.8.1 *Suppose that $A \in \mathsf{M}_n(\mathbb{F})$ is idempotent and has rank $r \geq 1$. If A has any singular values that are greater than 1, denote them by $\sigma_1 \geq \sigma_2 \geq \cdots \geq \sigma_k > 1$. Then there is a unitary $U \in \mathsf{M}_n(\mathbb{F})$ such that*

$$U^*AU = \left(\begin{bmatrix} 1 & \sqrt{\sigma_1^2 - 1} \\ 0 & 0 \end{bmatrix} \oplus \cdots \oplus \begin{bmatrix} 1 & \sqrt{\sigma_k^2 - 1} \\ 0 & 0 \end{bmatrix} \right) \oplus I_{r-k} \oplus 0_{n-r-k}. \qquad (15.8.2)$$

One or more of the direct summands may be absent. The first is present only if $k \geq 1$; the second is present only if $r > k$; the third is present only if $n > r + k$.

Proof Example 10.4.13 shows that A is unitarily similar to

$$\begin{bmatrix} I_r & X \\ 0 & 0_{n-r} \end{bmatrix}, \quad X \in \mathsf{M}_{r \times (n-r)}. \qquad (15.8.3)$$

If $X = 0$, then A is unitarily similar to $I_r \oplus 0_{n-r}$ and A is an orthogonal projection.

Suppose that rank $X = p \geq 1$ and let $\tau_1 \geq \tau_2 \geq \cdots \geq \tau_p > 0$ be the singular values of X. Since A is unitarily similar to (15.8.3), AA^* is unitarily similar to

$$\begin{bmatrix} I_r & X \\ 0 & 0_{n-r} \end{bmatrix} \begin{bmatrix} I_r & 0 \\ X^* & 0_{n-r} \end{bmatrix} = (I_r + XX^*) \oplus 0_{n-r}.$$

Therefore, the eigenvalues of AA^* are $1 + \tau_1^2, 1 + \tau_2^2, \ldots, 1 + \tau_p^2$, together with 1 (with multiplicity $r - p$) and 0 (with multiplicity $n - r$). Since A has k singular values greater than 1, it follows that $p = k$. Moreover,

$$\sigma_i^2 = \tau_i^2 + 1 \quad \text{and} \quad \tau_i = \sqrt{\sigma_i^2 - 1}, \qquad i = 1, 2, \ldots, k.$$

Let $X = V\Sigma W^*$ be a singular value decomposition, in which $V \in \mathsf{M}_r$ and $W \in \mathsf{M}_{n-r}$ are unitary. Let $\Sigma_k = \operatorname{diag}(\tau_1, \tau_2, \ldots, \tau_k)$ be the $k \times k$ upper-left corner of Σ. Consider the unitary similarity

$$
\begin{bmatrix} V & 0 \\ 0 & W \end{bmatrix}^* \begin{bmatrix} I_r & X \\ 0 & 0_{n-r} \end{bmatrix} \begin{bmatrix} V & 0 \\ 0 & W \end{bmatrix}
$$

$$
= \begin{bmatrix} V^*V & V^*XW \\ 0 & 0_{n-r} \end{bmatrix} = \begin{bmatrix} I_r & \Sigma \\ 0 & 0_{n-r} \end{bmatrix}
$$

$$
= \begin{bmatrix} I_p & 0 & \Sigma_k & 0 \\ 0 & I_{r-k} & 0 & 0 \\ 0_k & 0 & 0_k & 0 \\ 0 & 0 & 0 & 0_{n-r-k} \end{bmatrix}. \tag{15.8.4}
$$

Then (15.8.4) is permutation similar to

$$
\begin{bmatrix} I_k & \Sigma_k & 0 & 0 \\ 0_k & 0_k & 0 & 0 \\ 0 & 0 & I_{r-k} & 0 \\ 0 & 0 & 0 & 0_{n-r-k} \end{bmatrix} = \begin{bmatrix} I_k & \Sigma_k \\ 0_k & 0_k \end{bmatrix} \oplus I_{r-k} \oplus 0_{n-r-k}. \tag{15.8.5}
$$

The 2×2 block matrix in (15.8.5) is permutation similar to

$$
\begin{bmatrix} 1 & \tau_1 \\ 0 & 0 \end{bmatrix} \oplus \cdots \oplus \begin{bmatrix} 1 & \tau_k \\ 0 & 0 \end{bmatrix} = \begin{bmatrix} 1 & \sqrt{\sigma_1^2 - 1} \\ 0 & 0 \end{bmatrix} \oplus \cdots \oplus \begin{bmatrix} 1 & \sqrt{\sigma_k^2 - 1} \\ 0 & 0 \end{bmatrix};
$$

see (6.3.8). Therefore, A is unitarily similar to the direct sum (15.8.2). If A is real, Corollary 12.6.12 ensures that it is real orthogonally similar to (15.8.2). □

15.9 Problems

P.15.1 Let $A \in \mathsf{M}_2$. Show that

$$
\|A\|_2 = \frac{1}{2} \left(\sqrt{\|A\|_F^2 + 2|\det A|} + \sqrt{\|A\|_F^2 - 2|\det A|} \right).
$$

Use this formula to find the spectral norms of A and B in (15.3.12).

P.15.2 Let $A, B \in \mathsf{M}_n$. (a) Show that $\rho(AB) = \rho(BA) \leq \min\{\|AB\|_2, \|BA\|_2\}$. (b) Give an example with $n = 2$ in which $\|AB\|_2 \neq \|BA\|_2$.

P.15.3 Let $A, B \in \mathsf{M}_n$ and suppose that AB is normal. Show that $\|AB\|_2 \leq \|BA\|_2$. *Hint:* Theorem 15.3.2 and (15.3.6).

P.15.4 Let $A = [\mathbf{a}_1 \ \mathbf{a}_2 \ \ldots \ \mathbf{a}_n] \in \mathsf{M}_{m \times n}$. Show that $\|\mathbf{a}_i\|_2 \leq \|A\|_2$ for each $i = 1, 2, \ldots, n$.

P.15.5 Consider

$$
A = \begin{bmatrix} 1 & i \\ 0 & 1 \end{bmatrix}.
$$

Compute the Cartesian decomposition $A = H + iK$. Compare $\|A\|_2$ with $\|H\|_2 + \|K\|_2$ and the upper bound in Theorem 15.4.4.

P.15.6 Let $A, B \in \mathsf{M}_n$. Show that $\|AB\|_F \leq \|A\|_2 \|B\|_F$.

P.15.7 Let $A, B \in M_{m \times n}$. Matrix analysts say that A and B are *unitarily equivalent* if there is a unitary $U \in M_m$ and a unitary $V \in M_n$ such that $A = UBV$. This relation arose in Theorem 15.3.13. Show that unitary equivalence is an equivalence relation.

P.15.8 Let $A \in M_{m \times n}$. Use Theorem 14.1.3, (15.1.3), and Definition 15.2.1 to show that $\|A\|_F = \|A^*\|_F$ and $\|A\|_2 = \|A^*\|_2$.

P.15.9 Let $A, B \in M_{m \times n}$. (a) Prove that $\|A \circ B\|_F \leq \|A\|_F \|B\|_F$. This is an analog of Theorem 15.2.10.a. (b) Verify the inequality in (a) for the matrices in Example 13.5.2.

P.15.10 Let $A, B \in M_n$. Suppose that A is positive definite and B is symmetric. Use Theorem 15.7.1 to show that there is an invertible $S \in M_n$ and a diagonal $\Sigma \in M_n$ with nonnegative diagonal entries such that $A = SIS^*$ and $B = S\Sigma S^\mathsf{T}$. *Hint:* $A^{-1/2}B\overline{A}^{-1/2} = U\Sigma U^\mathsf{T}$ is symmetric. Consider $S = A^{1/2}U$.

P.15.11 Let $A \in M_n$, let $\mathbf{x} \in \mathbb{C}^n$, and consider the complex quadratic form $q_A(\mathbf{x}) = \mathbf{x}^\mathsf{T} A\mathbf{x}$. (a) Show that $q_A(\mathbf{x}) = q_{\frac{1}{2}(A + A^\mathsf{T})}(\mathbf{x})$, so we may assume that A is symmetric. (b) If A is symmetric and has singular values $\sigma_1, \sigma_2, \ldots, \sigma_n$, show that there is a change of variables $\mathbf{x} \mapsto U\mathbf{x} = \mathbf{y} = [y_i]$ such that U is unitary and $q_A(\mathbf{y}) = \sigma_1 y_1^2 + \sigma_2 y_2^2 + \cdots + \sigma_n y_n^2$.

P.15.12 If $A \in M_n$ is symmetric and rank $A = r$, show that it has a full-rank factorization of the form $A = BB^\mathsf{T}$ with $B \in M_{n \times r}$.

P.15.13 Let $A \in M_n$ be symmetric and have distinct singular values. Suppose that $A = U\Sigma U^\mathsf{T}$ and $A = V\Sigma V^\mathsf{T}$ are singular value decompositions of A. Show that $U = VD$, in which $D = \operatorname{diag}(d_1, d_2, \ldots, d_n)$ and $d_i = \pm 1$ for each $i = 1, 2, \ldots, n-1$. If A is invertible, then $d_n = \pm 1$. If A is not invertible, show that d_n can be any complex number with modulus 1.

P.15.14 Let $A \in M_n$. Show that A is symmetric if and only if it has polar decompositions $A = |A|^\mathsf{T} U = U|A|$, in which the unitary factor U is symmetric.

P.15.15 Let $A \in M_{m \times n}$ and let c be a nonzero scalar. Show that $(A^\mathsf{T})^\dagger = (A^\dagger)^\mathsf{T}$, $(A^*)^\dagger = (A^\dagger)^*$, $(\overline{A})^\dagger = \overline{A^\dagger}$, and $(cA)^\dagger = c^{-1}A^\dagger$.

P.15.16 Let $A \in M_{m \times n}(\mathbb{F})$, let $B, C \in M_{n \times m}(\mathbb{F})$, and consider the *Penrose identities*

$$(AB)^* = AB, \quad (BA)^* = BA, \quad ABA = A, \quad \text{and} \quad BAB = B. \tag{15.9.1}$$

(a) Verify that $B = A^\dagger$ satisfies these identities. (b) If $(AC)^* = AC$, $(CA)^* = CA$, $ACA = A$, and $CAC = C$, show that $B = C$ by justifying every step in the following calculation:

$$B = B(AB)^* = BB^*A^* = B(AB)^*(AC)^* =$$
$$= BAC = (BA)^*(CA)^*C = A^*C^*C = (CA)^*C = C.$$

(c) Conclude that $B = A^\dagger$ if and only if B satisfies the identities (15.9.1).

P.15.17 Verify that

$$A^\dagger = \frac{1}{10}\begin{bmatrix} -2 & -1 & 0 & 1 & 2 \\ 6 & 4 & 2 & 0 & -2 \end{bmatrix}$$

is the pseudoinverse of the matrix A in Example 7.5.7 and show that $A^\dagger \mathbf{y}$ gives the parameters of the least squares line in that example.

P.15.18 Let $A \in M_{m \times n}(\mathbb{F})$ have rank $r \geq 1$ and let $A = RS$ be a full-rank factorization; see Theorem 3.2.15. Show that R^*AS^* is invertible. Prove that

$$A^\dagger = S^*(R^*AS^*)^{-1}R^* \tag{15.9.2}$$

by verifying that $S^*(R^*AS^*)^{-1}R^*$ satisfies the identities (15.9.1). If A has low rank, this can be an efficient way to calculate the pseudoinverse.

P.15.19 Suppose that $A \in M_n$ is invertible. Use (15.9.1) and (15.9.2) to calculate $A^\dagger$.

P.15.20 Let $\mathbf{x} \in \mathbb{F}^m$ and $\mathbf{y} \in \mathbb{F}^n$ be nonzero and let $A = \mathbf{xy}^*$. Use (15.9.2) to show that $A^\dagger = \|\mathbf{x}\|_2^{-2}\|\mathbf{y}\|_2^{-2}A^*$. What is $A^\dagger$ if A is an all-ones matrix?

P.15.21 Let $\Lambda = \mathrm{diag}(\lambda_1, \lambda_2, \ldots, \lambda_n) \in M_n$. Show that $\Lambda^\dagger = [\lambda_1]^\dagger \oplus [\lambda_2]^\dagger \oplus \cdots \oplus [\lambda_n]^\dagger$.

P.15.22 If $A \in M_n$ is normal and $A = U\Lambda U^*$ is a spectral decomposition, show that $A^\dagger = U\Lambda^\dagger U^*$.

P.15.23 Let

$$A = \begin{bmatrix} 1 & 0 \\ 0 & 2 \end{bmatrix} \quad \text{and} \quad B = \begin{bmatrix} 1 & 1 \\ 1 & 1 \end{bmatrix}.$$

Show that $(AB)^\dagger \neq B^\dagger A^\dagger$.

P.15.24 Let $A \in M_n$ be invertible. Find a noninvertible matrix $B \in M_n$ such that $\|A - B\|_F = \|A - B\|_2 = \sigma_{\min}(A)$.

P.15.25 If $A \in M_n$ is symmetric, prove that $A^\dagger$ is symmetric.

P.15.26 Let $A \in M_n$ be idempotent. (a) Show that $I - A$ is idempotent. (b) Show that A and $I - A$ have the same singular values that are greater than 1 (if any). (c) If $A \neq 0$ and $A \neq I$, show that the largest singular value of A is equal to the largest singular value of $I - A$.

P.15.27 Let $A, B \in M_n$ be idempotent. Show that A and B are unitarily similar if and only if they have the same singular values.

P.15.28 Show that the idempotent matrix in Example 7.3.16 is unitarily similar to $\begin{bmatrix} 1 & 3 \\ 0 & 0 \end{bmatrix}$, which is a matrix of the form (7.7.2).

P.15.29 Let $A \in M_n$ have rank $r \geq 1$, let $\sigma_1 \geq \sigma_2 \geq \cdots \geq \sigma_r > 0$ be the positive singular values of A, and suppose that A in nilpotent of index two, that is, $A^2 = 0$. (a) Prove that A is unitarily similar to

$$\begin{bmatrix} 0 & \sigma_1 \\ 0 & 0 \end{bmatrix} \oplus \cdots \oplus \begin{bmatrix} 0 & \sigma_r \\ 0 & 0 \end{bmatrix} \oplus 0_{n-2r}. \tag{15.9.3}$$

Hint: Adapt the proof of Theorem 15.8.1. (b) If A is real, show that it is real orthogonally similar to (15.9.3).

P.15.30 Let $A, B \in M_n$ and suppose that $A^2 = 0 = B^2$. Show that A and B are unitarily similar if and only if they have the same singular values.

P.15.31 (a) Show that

$$C = \begin{bmatrix} 1 & \tau \\ 0 & -1 \end{bmatrix}, \quad \tau > 0,$$

satisfies $C^2 = I$. (b) Compute the singular values of C, show that they are reciprocals, and explain why one of them is greater than 1.

P.15.32 Let $A \in M_n$ and suppose that A is an involution. Let $\sigma_1 \geq \sigma_2 \geq \cdots \geq \sigma_k > 1$ be the singular values of A that are greater than 1, if any. (a) Show that $\operatorname{spec} A \subseteq \{1, -1\}$. (b) Show that the singular values of A are $\sigma_1, \sigma_1^{-1}, \sigma_2, \sigma_2^{-1}, \ldots, \sigma_k, \sigma_k^{-1}$, together with 1 (with multiplicity $n - 2k$). *Hint*: $(AA^*)^{-1} = A^*A$.

P.15.33 Let $A \in M_n$. (a) Show that A is idempotent if and only if $2A - I$ is an involution. (b) Show that A is an involution if and only if $\frac{1}{2}(A + I)$ is idempotent.

P.15.34 Let $A \in M_n(\mathbb{F})$, let p be the multiplicity of 1 as an eigenvalue of A, and let k be the number of singular values of A that are greater than 1. Show that A is an involution if and only if there is a unitary $U \in M_n(\mathbb{F})$ and positive real numbers $\tau_1, \tau_2, \ldots, \tau_k$ such that

$$U^*AU = \begin{bmatrix} 1 & \tau_1 \\ 0 & -1 \end{bmatrix} \oplus \cdots \oplus \begin{bmatrix} 1 & \tau_k \\ 0 & -1 \end{bmatrix} \oplus I_{p-k} \oplus (-I_{n-p-k}). \qquad (15.9.4)$$

Hint: Use the preceding problem and (15.8.2).

P.15.35 Let $A, B \in M_n$ and suppose that $A^2 = I = B^2$. Show that A and B are unitarily similar if and only if they have the same singular values, and $+1$ is an eigenvalue with the same multiplicity for each of them.

P.15.36 Let $A \in M_{m \times n}$. Show that $\|A\|_2 \leq 1$ if and only if $I_n - A^*A$ is positive semidefinite. A matrix that satisfies either of these conditions is a *contraction*.

P.15.37 Let $A \in M_n$ be normal and let $A = U\Lambda U^*$ be a spectral decomposition. (a) Show that $H(A) = UH(\Lambda)U^*$; see Definition 12.6.6. (b) If A is a contraction, show that $H(A) = I$ if and only if $A = I$.

P.15.38 Let $U, V \in M_n$ be unitary and let $C = \frac{1}{2}(U + V)$. (a) Show that C is a contraction. (b) Show that C is unitary if and only if $U = V$.

P.15.39 Let $C \in M_n$ be a contraction. Show that there are unitary $U, V \in M_n$ such that $C = \frac{1}{2}(U + V)$. *Hint*: If $0 \leq \sigma \leq 1$ and $s_\pm = \sigma \pm i\sqrt{1 - \sigma^2}$, then $\sigma = \frac{1}{2}(s_+ + s_-)$ and $|s_\pm| = 1$.

P.15.40 Deduce from the preceding problem that every square matrix is a linear combination of at most two unitary matrices. See P.13.39 for a related result.

P.15.41 Let $A, B \in M_n$ and let p be a polynomial. (a) Show that $Bp(AB) = p(BA)B$. (b) If A is a contraction, use (a) to show that

$$A^*(I - AA^*)^{1/2} = (I - A^*A)^{1/2}A^*. \qquad (15.9.5)$$

(c) Use (a) to solve P.14.21.

P.15.42 Let $A \in M_n$. Use the singular value decomposition to prove (15.9.5).

P.15.43 Let $A \in M_n$. Show that A is a contraction if and only if for some $m \geq n$, there is a block matrix of the form

$$U = \begin{bmatrix} A & B \\ C & D \end{bmatrix} \in M_m$$

that is unitary. Thus, every contraction is a principal submatrix of some unitary matrix. *Hint*: Start with $B = (I - AA^*)^{1/2}$.

P.15.44 Let $A \in M_n$ be idempotent. Show that the following are equivalent:

(a) A is a Householder matrix.

(b) $\operatorname{rank}(A - I) = 1$ and A is unitary.

(c) $\operatorname{rank}(A - I) = 1$ and A is normal.

(d) $\operatorname{rank}(A - I) = 1$ and A is a contraction.

P.15.45 Show that the spectral norm of a nonzero idempotent matrix can be any number in the real interval $[1, \infty)$, but it cannot be any number in the real interval $[0, 1)$.

P.15.46 Under the same assumptions used to derive the upper bound (15.6.8), prove the lower bound

$$\frac{1}{\kappa_2(A)} \frac{\|\Delta \mathbf{y}\|_2}{\|\mathbf{y}\|_2} \leq \frac{\|\Delta \mathbf{x}\|_2}{\|\mathbf{x}\|_2}$$

on the relative error in a solution of the perturbed linear system (15.6.5).

P.15.47 Define the *Frobenius condition number* of an invertible matrix $A \in M_n$ by $\kappa_F(A) = \|A\|_F \|A^{-1}\|_F$. Let $B \in M_n$ be invertible. Prove the following:

(a) $\kappa_F(AB) \leq \kappa_F(A)\kappa_F(B)$.

(b) $\kappa_F(A) \geq \sqrt{n}$.

(c) $\kappa_F(cA) = \kappa_F(A)$ for any nonzero scalar c.

(d) $\kappa_F(A) = \kappa_F(A^*) = \kappa_F(\overline{A}) = \kappa_F(A^\mathsf{T})$.

P.15.48 Prove the upper bound

$$\frac{\|\Delta \mathbf{x}\|_2}{\|\mathbf{x}\|_2} \leq \kappa_F(A) \frac{\|\Delta \mathbf{y}\|_2}{\|\mathbf{y}\|_2} \tag{15.9.6}$$

for the relative error in the linear system analyzed in Section 15.6. If A is unitary, compare the bounds (15.6.8) and (15.9.6) and discuss.

P.15.49 Let $A = [A_{ij}] \in M_{2n}$ be a 2×2 block matrix with each $A_{ij} \in M_n$. Let $M = [\|A_{ij}\|_2] \in M_2$. Let $\mathbf{x} \in \mathbb{C}^{2n}$ be a unit vector, partitioned as $\mathbf{x} = [\mathbf{x}_1^\mathsf{T}\ \mathbf{x}_2^\mathsf{T}]^\mathsf{T}$, in which $\mathbf{x}_1, \mathbf{x}_2 \in \mathbb{C}^n$. Provide details for the computation

$$\|A\mathbf{x}\|_2 = \left(\|A_{11}\mathbf{x}_1 + A_{12}\mathbf{x}_2\|_2^2 + \| \cdots \|_2^2 \right)^{1/2}$$

$$\leq \left(\left(\|A_{11}\|_2 \|\mathbf{x}_1\|_2 + \|A_{12}\|_2 \|\mathbf{x}_2\|_2 \right)^2 + \left(\cdots \right)^2 \right)^{1/2}$$

$$= \left\| M \begin{bmatrix} \|\mathbf{x}_1\|_2 \\ \|\mathbf{x}_2\|_2 \end{bmatrix} \right\|_2 \leq \max\{\|M\mathbf{y}\|_2 : \mathbf{y} \in \mathbb{C}^2, \|\mathbf{y}\|_2 = 1\}.$$

Conclude that $\|A\|_2 \leq \|M\|_2$.

P.15.50 Let $A = [A_{ij}] \in M_{kn}$ be a $k \times k$ block matrix with each $A_{ij} \in M_n$. Let $M = [\|A_{ij}\|_2] \in M_k$. Show that $\|A\|_2 \leq \|M\|_2$.

P.15.51 Let $A = [a_{ij}] \in M_n$ and let $M = [|a_{ij}|]$. Show that $\|A\|_2 \leq \|M\|_2$.

P.15.52 Compute the spectral norms of the matrices

$$A = \begin{bmatrix} 1 & 1 \\ -1 & 1 \end{bmatrix} \quad \text{and} \quad B = \begin{bmatrix} 1 & 1 \\ 0 & 1 \end{bmatrix}.$$

Conclude that replacing an entry of a matrix by zero can increase its spectral norm. What can you say in this regard about the Frobenius norm?

P.15.53 Let $A \in M_{n \times r}$ and $B \in M_{r \times n}$. Suppose that $\operatorname{rank} A = \operatorname{rank} B = r \geq 1$. Show that $A^\dagger = (A^*A)^{-1}A^*$ and $B^\dagger = B^*(BB^*)^{-1}$.

P.15.54 Let $A \in \mathsf{M}_{m \times n}$ be nonzero and suppose that $A = XY$ is a full-rank factorization. Show that $A^\dagger = Y^\dagger X^\dagger$.

15.10 Notes

Error bounds for eigenvalues or solutions of linear systems have been studied by numerical analysts since the 1950s. For more information and references to an extensive literature, see [GVL13].

The structure of the example (15.3.9) is not an accident. If $A \in \mathsf{M}_n$ is not a scalar matrix and $\rho(A) = \|A\|_2$, then A is unitarily similar to a matrix of the form $\rho(A)I \oplus B$, in which $\rho(B) < \rho(A)$ and $\|B\|_2 \leq \rho(A)$. For a proof, see [HJ13, Prob. 27, Sect. 1.5].

In functional analysis, *unitary equivalence* (see P.15.7) refers to the notion we have called *unitary similarity*.

15.11 Some Important Concepts

- Approximate a matrix with a truncated singular value decomposition.

- The largest singular value is a submultiplicative norm: the spectral norm.

- Unitary similarity, unitary equivalence, eigenvalues, and singular values (Theorem 15.3.13).

- Pseudoinverse of a matrix; minimum norm and least squares solutions of a linear system.

- Spectral condition number.

- Error bounds for solutions of linear systems.

- Error bounds for eigenvalue computations (Bauer–Fike theorem).

- Singular value decomposition of a complex symmetric matrix (Autonne theorem).

- Unitary similarity of idempotent matrices (Theorem 15.8.1 and P.15.27).

16   Interlacing and Inertia

If a Hermitian matrix of size n is perturbed by adding a rank-1 Hermitian matrix to it, or by bordering it to obtain a Hermitian matrix of size $n + 1$, the eigenvalues of the respective matrices are related by a pattern known as interlacing. We use subspace intersections (Theorem 2.2.10 and Corollary 2.2.17) to study eigenvalue interlacing and the related inequalities of Weyl (Theorem 16.7.1). We discuss applications of eigenvalue interlacing, including Sylvester's principal minor criterion for positive definiteness, singular value interlacing between a matrix and a submatrix, and majorization inequalities between the eigenvalues and diagonal entries of a Hermitian matrix.

Sylvester's inertia theorem (1852) says that although *congruence can change all the nonzero eigenvalues of a Hermitian matrix, it cannot change the number of positive eigenvalues or the number of negative ones. We use subspace intersections again to prove Sylvester's theorem, and then use the polar decomposition to prove a generalization of the inertia theorem to normal matrices.

16.1 The Rayleigh Quotient

The eigenvalues of a Hermitian $A = [a_{ij}] \in \mathsf{M}_n$ are real, so we can arrange them in increasing order

$$\lambda_1(A) \leq \lambda_2(A) \leq \cdots \leq \lambda_n(A).$$

We abbreviate $\lambda_i(A)$ to λ_i if only one matrix is under discussion. The foundation upon which many eigenvalue results stand is that if $\mathbf{x} \in \mathbb{C}^n$ is nonzero, then the ratio

$$\frac{\langle A\mathbf{x}, \mathbf{x} \rangle}{\langle \mathbf{x}, \mathbf{x} \rangle} = \frac{\mathbf{x}^* A \mathbf{x}}{\mathbf{x}^* \mathbf{x}}$$

is a lower bound for λ_n and an upper bound for λ_1. This principle was discovered by John William Strutt, the third Baron Rayleigh and 1904 Nobel laureate for his discovery of argon.

Theorem 16.1.1 (Rayleigh) *Let $A \in \mathsf{M}_n$ be Hermitian, with eigenvalues $\lambda_1 \leq \lambda_2 \leq \cdots \leq \lambda_n$ and corresponding orthonormal eigenvectors $\mathbf{u}_1, \mathbf{u}_2, \ldots, \mathbf{u}_n \in \mathbb{C}^n$. Let p, q be integers such that $1 \leq p \leq q \leq n$. If $\mathbf{x} \in \operatorname{span}\{\mathbf{u}_p, \mathbf{u}_{p+1}, \ldots, \mathbf{u}_q\}$ is a unit vector, then*

$$\lambda_p \leq \langle A\mathbf{x}, \mathbf{x} \rangle \leq \lambda_q. \tag{16.1.2}$$

Proof The list $\mathbf{u}_p, \mathbf{u}_{p+1}, \ldots, \mathbf{u}_q$ is an orthonormal basis for $\operatorname{span}\{\mathbf{u}_p, \mathbf{u}_{p+1}, \ldots, \mathbf{u}_q\}$. Theorem 5.2.5 ensures that

$$\mathbf{x} = \sum_{i=p}^{q} c_i \mathbf{u}_i,$$

in which each $c_i = \langle \mathbf{x}, \mathbf{u}_i \rangle$ and

$$\sum_{i=p}^{q} |c_i|^2 = \|\mathbf{x}\|_2^2 = 1. \tag{16.1.3}$$

The ordering of the eigenvalues and (16.1.3) ensure that

$$\lambda_p = \sum_{i=p}^{q} \lambda_p |c_i|^2 \le \sum_{i=p}^{q} \lambda_i |c_i|^2 \le \sum_{i=p}^{q} \lambda_q |c_i|^2 = \lambda_q. \tag{16.1.4}$$

Now compute

$$\langle A\mathbf{x}, \mathbf{x} \rangle = \left\langle \sum_{i=p}^{q} c_i A\mathbf{u}_i, \mathbf{x} \right\rangle = \left\langle \sum_{i=p}^{q} c_i \lambda_i \mathbf{u}_i, \mathbf{x} \right\rangle$$

$$= \sum_{i=p}^{q} \lambda_i c_i \langle \mathbf{u}_i, \mathbf{x} \rangle = \sum_{i=p}^{q} \lambda_i c_i \overline{c_i} = \sum_{i=p}^{q} \lambda_i |c_i|^2.$$

The central term in (16.1.4) is $\langle A\mathbf{x}, \mathbf{x} \rangle$, which verifies (16.1.2). $\qquad \square$

The upper bound in (16.1.2) is attained by $\mathbf{x} = \mathbf{u}_q$ and the lower bound is attained by $\mathbf{x} = \mathbf{u}_p$. Therefore,

$$\min_{\substack{\mathbf{x} \in \mathcal{U} \\ \|\mathbf{x}\|_2 = 1}} \langle A\mathbf{x}, \mathbf{x} \rangle = \lambda_p \quad \text{and} \quad \max_{\substack{\mathbf{x} \in \mathcal{U} \\ \|\mathbf{x}\|_2 = 1}} \langle A\mathbf{x}, \mathbf{x} \rangle = \lambda_q, \tag{16.1.5}$$

in which $\mathcal{U} = \text{span}\{\mathbf{u}_p, \mathbf{u}_{p+1}, \ldots, \mathbf{u}_q\}$. If $p = 1$ and $q = n$, then $\mathcal{U} = \mathbb{C}^n$ and (16.1.5) becomes

$$\min_{\|\mathbf{x}\|_2 = 1} \langle A\mathbf{x}, \mathbf{x} \rangle = \lambda_1 \quad \text{and} \quad \max_{\|\mathbf{x}\|_2 = 1} \langle A\mathbf{x}, \mathbf{x} \rangle = \lambda_n.$$

Consequently,

$$\lambda_1 \le \langle A\mathbf{x}, \mathbf{x} \rangle \le \lambda_n \tag{16.1.6}$$

for any unit vector $\mathbf{x}$. For example, if $\mathbf{x} = \mathbf{e}_i$, then $\langle A\mathbf{e}_i, \mathbf{e}_i \rangle = a_{ii}$ is a diagonal entry of A and (16.1.6) tells us that

$$\lambda_1 \le a_{ii} \le \lambda_n, \quad i = 1, 2, \ldots, n. \tag{16.1.7}$$

Example 16.1.8 Let

$$A = \begin{bmatrix} 6 & 3 & 0 & 3 \\ 3 & -2 & 5 & 2 \\ 0 & 5 & -2 & 3 \\ 3 & 2 & 3 & 4 \end{bmatrix}.$$

The inequalities (16.1.7) tell us that $\lambda_1 \le -2$ and $\lambda_4 \ge 6$. Additional bounds can be obtained from (16.1.6) with any unit vector $\mathbf{x}$. For example, with $\mathbf{x} = [\frac{1}{2} \ \frac{1}{2} \ \frac{1}{2} \ \frac{1}{2}]^\mathsf{T} \in \mathbb{R}^4$ we obtain $\langle A\mathbf{x}, \mathbf{x} \rangle = 9.5$. With $\mathbf{x} = [\frac{1}{2} \ -\frac{1}{2} \ \frac{1}{2} \ -\frac{1}{2}]^\mathsf{T}$ we obtain $\langle A\mathbf{x}, \mathbf{x} \rangle = -4.5$, so $\lambda_1 \le -4.5$ and $\lambda_4 \ge 9.5$.

16.2 Eigenvalue Interlacing for Sums of Hermitian Matrices

Example 16.2.1 Consider the Hermitian matrices

$$A = \begin{bmatrix} 6 & 3 & 0 & 3 \\ 3 & -2 & 5 & 2 \\ 0 & 5 & -2 & 3 \\ 3 & 2 & 3 & 4 \end{bmatrix} \quad \text{and} \quad E = \begin{bmatrix} 4 & -2 & 0 & 2 \\ -2 & 1 & 0 & -1 \\ 0 & 0 & 0 & 0 \\ 2 & -1 & 0 & 1 \end{bmatrix}.$$

Rounded to two decimal places,

$$\text{spec } A = \{-7.46, 0.05, 3.22, 10.29\},$$
$$\text{spec } E = \{0, 0, 0, 6\},$$
$$\text{spec}(A + E) = \{-6.86, 0.74, 4.58, 13.54\}, \quad \text{and}$$
$$\text{spec}(A - E) = \{-8.92, -1.98, 2.21, 8.70\}.$$

The increasingly ordered eigenvalues of A and $A \pm E$ satisfy

$$\lambda_1(A) < \lambda_1(A + E) < \lambda_2(A) < \lambda_2(A + E) < \lambda_3(A) < \lambda_3(A + E)$$

and

$$\lambda_1(A - E) < \lambda_1(A) < \lambda_2(A - E) < \lambda_2(A) < \lambda_3(A - E) < \lambda_3(A).$$

Adding the positive semidefinite matrix E to A increases each of its eigenvalues. Subtracting E from A decreases each of its eigenvalues. The eigenvalues of $A \pm E$ interlace the eigenvalues of A. Figure 16.1 illustrates the eigenvalue interlacings in this example.

The interlacing patterns in the preceding example are not accidental.

Theorem 16.2.2 *Let $A, E \in \mathsf{M}_n$ be Hermitian and suppose that E is positive semidefinite and has rank 1. The increasingly ordered eigenvalues of A and $A \pm E$ satisfy*

$$\lambda_i(A) \leq \lambda_i(A + E) \leq \lambda_{i+1}(A) \leq \lambda_n(A + E), \quad i = 1, 2, \ldots, n - 1, \qquad (16.2.3)$$

and

$$\lambda_1(A - E) \leq \lambda_i(A) \leq \lambda_{i+1}(A - E) \leq \lambda_{i+1}(A), \quad i = 1, 2, \ldots, n - 1. \qquad (16.2.4)$$

Proof Let $\mathbf{u}_1, \mathbf{u}_2, \ldots, \mathbf{u}_n$ and $\mathbf{v}_1, \mathbf{v}_2, \ldots, \mathbf{v}_n$ be orthonormal eigenvectors corresponding to the increasingly ordered eigenvalues of A and $A + E$, respectively. Fix $i \in \{1, 2, \ldots, n - 1\}$.

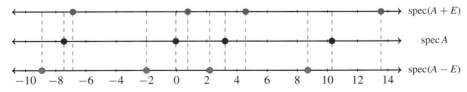

Figure 16.1 Eigenvalues of A and $A \pm E$ in Example 16.2.1; rank $E = 1$.

Let

$$\mathcal{U} = \mathrm{span}\{\mathbf{u}_1, \mathbf{u}_2, \ldots, \mathbf{u}_{i+1}\},$$
$$\mathcal{V} = \mathrm{span}\{\mathbf{v}_i, \mathbf{v}_{i+1}, \ldots, \mathbf{v}_n\}, \quad \text{and}$$
$$\mathcal{W} = \mathrm{null}\, E.$$

Then $\dim \mathcal{U} = i + 1$, $\dim \mathcal{V} = n - i + 1$, and $\dim \mathcal{W} = n - \mathrm{rank}\, E = n - 1$. Because

$$\dim \mathcal{U} + \dim \mathcal{V} = n + 2,$$

Corollary 2.2.17.b ensures that $\dim(\mathcal{U} \cap \mathcal{V}) \geq 2$. Since

$$\dim(\mathcal{U} \cap \mathcal{V}) + \dim \mathcal{W} \geq 2 + (n - 1) = n + 1,$$

Corollary 2.2.17.a ensures that there is a unit vector $\mathbf{x} \in \mathcal{U} \cap \mathcal{V} \cap \mathcal{W}$. Use Theorem 16.1.1 to compute

$$
\begin{aligned}
\lambda_i(A + E) &\leq \langle (A + E)\mathbf{x}, \mathbf{x} \rangle & &\text{(since } \mathbf{x} \in \mathcal{V}) \\
&= \langle A\mathbf{x}, \mathbf{x} \rangle + \langle E\mathbf{x}, \mathbf{x} \rangle & & \\
&= \langle A\mathbf{x}, \mathbf{x} \rangle & &\text{(since } \mathbf{x} \in \mathcal{W}) \\
&\leq \lambda_{i+1}(A) & &\text{(since } \mathbf{x} \in \mathcal{U}).
\end{aligned}
$$

Now let

$$\mathcal{U} = \mathrm{span}\{\mathbf{u}_i, \mathbf{u}_{i+1}, \ldots, \mathbf{u}_n\} \quad \text{and} \quad \mathcal{V} = \mathrm{span}\{\mathbf{v}_1, \mathbf{v}_2, \ldots, \mathbf{v}_i\}.$$

Then $\dim \mathcal{U} = n - i + 1$ and $\dim \mathcal{V} = i$, so

$$\dim \mathcal{U} + \dim \mathcal{V} = n + 1.$$

Corollary 2.2.17.a ensures that there is a unit vector $\mathbf{x} \in \mathcal{U} \cap \mathcal{V}$. Use Theorem 16.1.1, and the fact that E is positive semidefinite to compute

$$
\begin{aligned}
\lambda_i(A) &\leq \langle A\mathbf{x}, \mathbf{x} \rangle & &\text{(since } \mathbf{x} \in \mathcal{U}) \\
&\leq \langle A\mathbf{x}, \mathbf{x} \rangle + \langle E\mathbf{x}, \mathbf{x} \rangle & &\text{(since } \langle E\mathbf{x}, \mathbf{x} \rangle \geq 0) \\
&= \langle (A + E)\mathbf{x}, \mathbf{x} \rangle & & \\
&\leq \lambda_i(A + E) & &\text{(since } \mathbf{x} \in \mathcal{V}).
\end{aligned}
$$

We have now proved the interlacing inequalities (16.2.3). Replace A with $A - E$ and replace $A + E$ with A in (16.2.3) to obtain (16.2.4). $\qquad\square$

The preceding theorem provides bounds on the eigenvalues of a Hermitian matrix $A \in \mathsf{M}_n$ that is perturbed by adding or subtracting a rank-1 positive semidefinite Hermitian matrix E. No matter how large the entries of E are, the eigenvalues $\lambda_2(A \pm E), \lambda_3(A \pm E), \ldots, \lambda_{n-1}(A \pm E)$ are confined to intervals between a pair of adjacent eigenvalues of A. Since any Hermitian matrix H is a real linear combination of rank-1 Hermitian matrices, we can apply Theorem 16.2.2 repeatedly to find bounds for the eigenvalues of $A + H$. For example, the following corollary considers the case in which H has at most one positive eigenvalue and at most one negative eigenvalue. We make use of it in our discussion of bordered matrices in the next section.

Corollary 16.2.5 *Let $A, H \in \mathsf{M}_n$ be Hermitian with* rank $H \leq 2$. *If H has at most one positive eigenvalue and at most one negative eigenvalue, then the increasingly ordered eigenvalues of A and $A + H$ satisfy*

$$\lambda_1(A + H) \leq \lambda_2(A) \tag{16.2.6}$$
$$\lambda_{i-1}(A) \leq \lambda_i(A + H) \leq \lambda_{i+1}(A), \quad i = 2, 3, \ldots, n - 1, \tag{16.2.7}$$
$$\lambda_{n-1}(A) \leq \lambda_n(A + H). \tag{16.2.8}$$

Proof Let γ_1 and γ_n be the least and greatest eigenvalues of H (we know that $\gamma_1 \leq 0$ and $\gamma_n \geq 0$), and let $\mathbf{v}_1$ and $\mathbf{v}_n$ be corresponding orthonormal eigenvectors. The spectral theorem ensures that $H = -E_1 + E_2$, in which $E_1 = -\gamma_1 \mathbf{v}_1 \mathbf{v}_1^*$ and $E_2 = \gamma_n \mathbf{v}_n \mathbf{v}_n^*$ have rank 1 and are positive semidefinite. For $i \in \{2, 3, \ldots, n - 1\}$, compute

$$\lambda_{i-1}(A) \leq \lambda_i(A - E_1) \qquad \text{(by (16.2.4))}$$
$$\leq \lambda_i(A - E_1 + E_2) \qquad \text{(by (16.2.3))}$$
$$= \lambda_i(A + H)$$
$$\leq \lambda_{i+1}(A - E_1) \qquad \text{(by (16.2.3))}$$
$$\leq \lambda_{i+1}(A) \qquad \text{(by (16.2.4))}.$$

This proves (16.2.7). Now compute

$$\lambda_1(A - E_1 + E_2) \leq \lambda_1(A + E_2) \leq \lambda_2(A) \quad \text{(by (16.2.4) and (16.2.3))}.$$

Invoke (16.2.4) and (16.2.3) again to obtain

$$\lambda_{n-1}(A) \leq \lambda_n(A - E_1) \leq \lambda_n(A - E_1 + E_2) = \lambda_n(A + H).$$

This proves (16.2.6) and (16.2.8). $\qquad\qquad\qquad\qquad\qquad\qquad\qquad\qquad\square$

Example 16.2.9 Let

$$A = \begin{bmatrix} 1 & 1 & 1 & 0 & 1 \\ 1 & 3 & 0 & 0 & 0 \\ 1 & 0 & 5 & 0 & 0 \\ 0 & 0 & 0 & 7 & 0 \\ 1 & 0 & 0 & 0 & 9 \end{bmatrix} \quad \text{and} \quad H = \begin{bmatrix} -1 & -1 & 0 & 0 & 0 \\ -1 & -1 & 0 & 0 & 0 \\ 0 & 0 & 1 & 1 & 1 \\ 0 & 0 & 1 & 1 & 1 \\ 0 & 0 & 1 & 1 & 1 \end{bmatrix}.$$

Rounded to two decimal places,

$$\operatorname{spec} H = \{-2, 0, 0, 0, 3\},$$
$$\operatorname{spec} A = \{0.30, 3.32, 5.25, 7.00, 9.13\}, \quad \text{and}$$
$$\operatorname{spec}(A + H) = \{-0.24, 2.00, 5.63, 7.72, 10.90\}.$$

Then

$$\lambda_1(A + H) < \lambda_2(A),$$
$$\lambda_1(A) < \lambda_2(A + H) < \lambda_3(A),$$
$$\lambda_2(A) < \lambda_3(A + H) < \lambda_4(A), \quad \text{and}$$
$$\lambda_3(A) < \lambda_4(A + H),$$

which is consistent with the inequalities in the preceding corollary; see Figure 16.2.

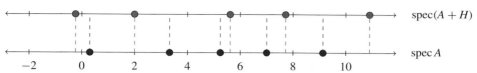

Figure 16.2 Eigenvalues of A and $A + H$ in Example 16.2.9; rank $H = 2$; H has one positive and one negative eigenvalue.

16.3 Eigenvalue Interlacing for Bordered Hermitian Matrices

We now study another context in which eigenvalue interlacing occurs.

Example 16.3.1 Let

$$A = \begin{bmatrix} 4 & -2 & 1 & 3 & 1 \\ -2 & 6 & 4 & -1 & -3 \\ 1 & 4 & 4 & -1 & -2 \\ 3 & -1 & -1 & 2 & -2 \\ 1 & -3 & -2 & -2 & -2 \end{bmatrix} = \begin{bmatrix} B & \mathbf{y} \\ \mathbf{y}^* & c \end{bmatrix},$$

in which

$$B = \begin{bmatrix} 4 & -2 & 1 & 3 \\ -2 & 6 & 4 & -1 \\ 1 & 4 & 4 & -1 \\ 3 & -1 & -1 & 2 \end{bmatrix}, \quad \mathbf{y} = \begin{bmatrix} 1 \\ -3 \\ -2 \\ -2 \end{bmatrix}, \quad \text{and} \quad c = -2.$$

Rounded to two decimal places,

$$\operatorname{spec} A = \{-4.38, -0.60, 2.15, 6.15, 10.68\} \quad \text{and}$$
$$\operatorname{spec} B = \{-1.18, 1.38, 5.91, 9.88\}.$$

The increasingly ordered eigenvalues of A and B satisfy

$$\lambda_1(A) < \lambda_1(B) < \lambda_2(A) < \lambda_2(B) < \lambda_3(A) < \lambda_3(B) < \lambda_4(A) < \lambda_4(B) < \lambda_5(A);$$

see the first two lines in Figure 16.3.

To understand the interlacing pattern in the preceding example, we begin with a special case.

Lemma 16.3.2 *Let $\mathbf{y} \in \mathbb{C}^n$ and $c \in \mathbb{R}$. Then*

$$H = \begin{bmatrix} 0_n & \mathbf{y} \\ \mathbf{y}^* & c \end{bmatrix} \in \mathsf{M}_{n+1}$$

is Hermitian, rank $H \le 2$, *and H has at most one negative eigenvalue and at most one positive eigenvalue.*

Proof If $\mathbf{y} = \mathbf{0}$, then $H = 0_n \oplus [c]$, so rank $H \le 1$ and $\operatorname{spec} H = \{0, c\}$.

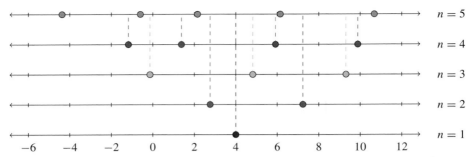

Figure 16.3 Spectra of the $n \times n$ leading principal submatrices of A in Example 16.3.1 for $n = 1, 2, 3, 4, 5$; $n = 5$ corresponds to A and $n = 4$ to B.

Now assume that $\mathbf{y} \neq \mathbf{0}$. The Cauchy expansion (3.4.13) for the determinant of a bordered matrix tells us that

$$p_H(z) = \det \begin{bmatrix} zI_n & -\mathbf{y} \\ -\mathbf{y}^* & z - c \end{bmatrix} = (z - c - \mathbf{y}^*(zI_n)^{-1}\mathbf{y}) \det(zI_n)$$

$$= \left(z - c - \frac{\|\mathbf{y}\|_2^2}{z} \right) z^n = (z^2 - cz - \|\mathbf{y}\|_2^2)z^{n-1}.$$

The zeros of $p_H(z)$ are 0 (with multiplicity $n - 1$) and $(c \pm (c^2 + 4\|\mathbf{y}\|_2^2)^{1/2})/2$. In the latter pair, one is positive

$$\frac{1}{2}\left(c + \sqrt{c^2 + 4\|\mathbf{y}\|_2^2} \right) > \frac{1}{2}(c + |c|) \geq 0,$$

and the other is negative

$$\frac{1}{2}\left(c - \sqrt{c^2 + 4\|\mathbf{y}\|_2^2} \right) < \frac{1}{2}(c - |c|) \leq 0. \qquad \square$$

Theorem 16.3.3 *Let $B \in M_n$ be Hermitian, $\mathbf{y} \in \mathbb{C}^n$, $c \in \mathbb{R}$, and*

$$A = \begin{bmatrix} B & \mathbf{y} \\ \mathbf{y}^* & c \end{bmatrix} \in M_{n+1}.$$

Then A is Hermitian and the increasingly ordered eigenvalues of A and B satisfy

$$\lambda_i(A) \leq \lambda_i(B) \leq \lambda_{i+1}(A), \quad i = 1, 2, \ldots, n. \tag{16.3.4}$$

Proof Add μ to each of the three terms in (16.3.4) and obtain the equivalent inequalities

$$\lambda_i(A) + \mu \leq \lambda_i(B) + \mu \leq \lambda_{i+1}(A) + \mu, \quad i = 1, 2, \ldots, n. \tag{16.3.5}$$

Since $\lambda_i(A) + \mu = \lambda_i(A + \mu I)$ and $\lambda_i(B) + \mu = \lambda_i(B + \mu I)$, the inequalities (16.3.5) are equivalent to

$$\lambda_i(A + \mu I) \leq \lambda_i(B + \mu I) \leq \lambda_{i+1}(A + \mu I), \quad i = 1, 2, \ldots, n.$$

With suitable choices of μ, we can ensure that $B + \mu I$ is either positive definite or negative definite. Therefore, there is no loss of generality if we assume that B is either positive definite or negative definite in the asserted inequalities (16.3.4).

Let

$$D = \begin{bmatrix} B & \mathbf{0} \\ \mathbf{0}^{\mathsf{T}} & 0 \end{bmatrix} \quad \text{and} \quad H = \begin{bmatrix} 0_n & \mathbf{y} \\ \mathbf{y}^* & c \end{bmatrix}.$$

Then $A = D + H$. Lemma 16.3.2 ensures that H has at most one positive eigenvalue and at most one negative eigenvalue. Corollary 16.2.5 tells us that

$$\lambda_1(A) \le \lambda_2(D), \tag{16.3.6}$$
$$\lambda_{i-1}(D) \le \lambda_i(A) \le \lambda_{i+1}(D), \quad i = 2, 3, \ldots, n, \quad \text{and} \tag{16.3.7}$$
$$\lambda_n(D) \le \lambda_{n+1}(A). \tag{16.3.8}$$

Since $D = B \oplus [0]$, we have

$$\operatorname{spec} D = \{\lambda_1(B), \lambda_2(B), \ldots, \lambda_n(B), 0\}.$$

First assume that B is positive definite. Then the $n + 1$ increasingly ordered eigenvalues of D are

$$0 \le \lambda_1(B) \le \lambda_2(B) \le \cdots \le \lambda_n(B),$$

that is, $\lambda_1(D) = 0$ and $\lambda_{i+1}(D) = \lambda_i(B)$ for each $i = 1, 2, \ldots, n$. The inequality (16.3.6) is

$$\lambda_1(A) \le \lambda_2(D) = \lambda_1(B)$$

and the right-hand inequality in (16.3.7) is

$$\lambda_i(A) \le \lambda_{i+1}(D) = \lambda_i(B), \quad i = 2, 3, \ldots, n.$$

This proves the left-hand inequality in (16.3.4).

Now assume that B is negative definite. In this case, the $n + 1$ increasingly ordered eigenvalues of D are

$$\lambda_1(B) \le \lambda_2(B) \le \cdots \le \lambda_n(B) \le 0,$$

so $\lambda_i(D) = \lambda_i(B)$ for each $i = 1, 2, \ldots, n$ and $\lambda_{n+1}(D) = 0$. The inequality (16.3.8) is

$$\lambda_n(B) = \lambda_n(D) \le \lambda_{n+1}(A).$$

The left-hand inequality in (16.3.6) is

$$\lambda_{i-1}(B) = \lambda_{i-1}(D) \le \lambda_i(A), \quad i = 2, 3, \ldots, n,$$

that is,

$$\lambda_i(B) \le \lambda_{i+1}(A), \quad i = 1, 2, \ldots, n - 1. \tag{16.3.9}$$

This proves the right-hand inequality in (16.3.4). $\qquad \square$

Our first corollary of the preceding theorem is an interlacing theorem for singular values. In the proof, we have to deal with two conflicting conventions: singular values are routinely indexed in decreasing order, while eigenvalues are indexed in increasing order.

Corollary 16.3.10 *Let $n \geq 2$ and let $A \in \mathsf{M}_n$. Delete a row or column of A and denote the resulting matrix by B, which is either in $\mathsf{M}_{(n-1)\times n}$ or in $\mathsf{M}_{n\times(n-1)}$. The decreasingly ordered singular values of A and B satisfy*

$$\sigma_i(A) \geq \sigma_i(B) \geq \sigma_{i+1}(A), \qquad i = 1, 2, \ldots, n-1. \tag{16.3.11}$$

Proof Suppose that a row is deleted and denote that row by $\mathbf{y}^* \in \mathsf{M}_{1\times n}$. Let P be a permutation matrix such that

$$PA = \begin{bmatrix} B \\ \mathbf{y}^* \end{bmatrix}.$$

Theorem 15.3.13.a ensures that A and PA have the same singular values, the squares of which are the eigenvalues of the bordered matrix

$$P(AA^*)P^\mathsf{T} = (PA)(PA)^* = \begin{bmatrix} BB^* & B\mathbf{y} \\ \mathbf{y}^*B^* & \mathbf{y}^*\mathbf{y} \end{bmatrix}.$$

Arranged in increasing order, the eigenvalues of AA^* are

$$\sigma_n^2(A) \leq \sigma_{n-1}^2(A) \leq \cdots \leq \sigma_1^2(A),$$

that is

$$\lambda_i(AA^*) = \sigma_{n-i+1}^2(A), \quad i = 1, 2, \ldots, n.$$

The increasingly ordered eigenvalues of BB^* are

$$\sigma_{n-1}^2(B) \leq \sigma_{n-2}^2(B) \leq \cdots \leq \sigma_1^2(B),$$

that is,

$$\lambda_i(BB^*) = \sigma_{n-i}^2(B), \quad i = 1, 2, \ldots, n-1.$$

Now invoke (16.3.4), which says that the increasingly ordered eigenvalues of BB^* interlace the increasingly ordered eigenvalues of AA^*, that is,

$$\sigma_{n-i+1}^2(A) \leq \sigma_{n-i}^2(B) \leq \sigma_{n-i}^2(A), \qquad i = 1, 2, \ldots, n-1.$$

Change variables in the indices to obtain

$$\sigma_i^2(A) \geq \sigma_i^2(B) \geq \sigma_{i+1}^2(A), \qquad i = 1, 2, \ldots, n-1,$$

which is equivalent to (16.3.11).

If a column of A is deleted, apply the preceding result to A^*. $\qquad\square$

Example 16.3.12 Consider

$$A = \begin{bmatrix} 1 & 2 & 3 & 4 \\ 8 & 7 & 6 & 5 \\ 10 & 11 & 9 & 8 \\ 12 & 13 & 15 & 14 \end{bmatrix} \quad \text{and} \quad B = \begin{bmatrix} 1 & 2 & 3 & 4 \\ 8 & 7 & 6 & 5 \\ 12 & 13 & 15 & 14 \end{bmatrix}.$$

Rounded to two decimal places, the singular values of A and B are $\{35.86, 4.13, 1.00, 0.48\}$ and $\{30.42, 3.51, 0.51\}$, respectively.

Our second corollary generalizes Theorem 16.3.3 to address leading principal submatrices of any size.

Corollary 16.3.13 *Let $B \in \mathsf{M}_n$ and $C \in \mathsf{M}_m$ be Hermitian, and let $Y \in \mathsf{M}_{n \times m}$. Then*

$$A_m = \begin{bmatrix} B & Y \\ Y^* & C \end{bmatrix} \in \mathsf{M}_{n+m}$$

is Hermitian and the increasingly ordered eigenvalues of A_m and B satisfy

$$\lambda_i(A_m) \leq \lambda_i(B) \leq \lambda_{i+m}(A_m), \quad i = 1, 2, \ldots, n. \tag{16.3.14}$$

Proof We proceed by induction on m. The preceding theorem establishes the base case $m = 1$. For the induction step, suppose that $m \geq 1$ and the inequalities (16.3.14) are valid. Write A_{m+1} as a bordered matrix

$$A_{m+1} = \begin{bmatrix} A_m & \mathbf{y} \\ \mathbf{y}^* & c \end{bmatrix} \in \mathsf{M}_{n+m+1}, \qquad A_m = \begin{bmatrix} B & Y \\ Y^* & C \end{bmatrix} \in \mathsf{M}_{n+m},$$

in which $\mathbf{y}$ is the last column of Y and c is the (m, m) entry of C. Theorem 16.3.3 ensures that the increasingly ordered eigenvalues of A_m and A_{m+1} satisfy

$$\lambda_i(A_{m+1}) \leq \lambda_i(A_m) \leq \lambda_{i+1}(A_{m+1}), \qquad i = 1, 2, \ldots, n. \tag{16.3.15}$$

The induction hypothesis and (16.3.15) ensure that

$$\lambda_i(A_{m+1}) \leq \lambda_i(A_m) \leq \lambda_i(B), \qquad i = 1, 2, \ldots, n, \tag{16.3.16}$$

as well as

$$\lambda_i(B) \leq \lambda_{i+m}(A_m) \leq \lambda_{i+m+1}(A_{m+1}), \qquad i = 1, 2, \ldots, n. \tag{16.3.17}$$

Now combine (16.3.16) and (16.3.17), and conclude that

$$\lambda_i(A_{m+1}) \leq \lambda_i(B) \leq \lambda_{i+m+1}(A_{m+1}), \qquad i = 1, 2, \ldots, n.$$

This completes the induction. □

Example 16.3.18 Figure 16.3 illustrates the interlacing inequalities (16.3.14) for the 5×5 matrix A in Example 16.3.1 and $m = 5, 4, 3, 2$, and 1.

16.4 Sylvester's Criterion

Suppose that $A \in \mathsf{M}_n$ is positive definite and partition it as

$$A = \begin{bmatrix} A_{11} & A_{12} \\ A_{12}^* & A_{22} \end{bmatrix},$$

in which $A_{11} \in \mathsf{M}_k$. If $\mathbf{y} \in \mathbb{C}^k$ is nonzero and $\mathbf{x} = [\mathbf{y}^\mathsf{T} \ \mathbf{0}^\mathsf{T}]^\mathsf{T} \in \mathbb{C}^n$, then $\mathbf{x} \neq \mathbf{0}$ and

$$0 < \langle A\mathbf{x}, \mathbf{x} \rangle = \langle A_{11}\mathbf{y}, \mathbf{y} \rangle.$$

We conclude that each leading $k \times k$ principal submatrix of A is positive definite, and consequently it has positive determinant (Theorem 13.1.9.b).

Definition 16.4.1 The determinant of a principal submatrix of $A \in \mathsf{M}_n$ is a *principal minor* of A. The determinant of a leading principal submatrix of $A \in \mathsf{M}_n$ is a *leading principal minor* of A.

Example 16.4.2 Consider the matrix A in Example 16.3.1. Its leading principal minors are 4, 20, -6, -95, and 370. The principal minor obtained by deleting the first two rows and columns of A is -46. The principal minor obtained by deleting the third row and column of A is -104.

Theorem 16.4.3 (Sylvester's Criterion) *A Hermitian matrix is positive definite if and only if all of its leading principal minors are positive.*

Proof Let $A \in \mathsf{M}_n$ be Hermitian. The preceding discussion shows that if A is positive definite, then its leading principal minors are positive. To prove the converse, we proceed by induction on n. For the base case $n = 1$, the hypothesis is that $\det A = a_{11} > 0$. Since $\operatorname{spec} A = \{a_{11}\}$, Theorem 13.1.8 ensures that A is positive definite.

For the induction step, assume that an $n \times n$ Hermitian matrix is positive definite if all of its leading principal minors are positive. Let $A \in \mathsf{M}_{n+1}$ be Hermitian and suppose that all of its leading principal minors are positive. Partition A as

$$A = \begin{bmatrix} B & \mathbf{y} \\ \mathbf{y}^* & a \end{bmatrix}, \quad B \in \mathsf{M}_n.$$

Then $\det A > 0$ and each leading principal minor of B is positive. The induction hypothesis ensures that B is positive definite, so its eigenvalues are positive. Theorem 16.3.3 tells us that the eigenvalues of B interlace those of A, that is,

$$\lambda_1(A) \le \lambda_1(B) \le \lambda_2(A) \le \cdots \le \lambda_n(B) \le \lambda_{n+1}(A).$$

Since $\lambda_1(B) > 0$, it follows that $\lambda_2(A), \lambda_3(A), \ldots, \lambda_{n+1}(A)$ are positive. We also know that

$$\lambda_1(A)\big(\lambda_2(A)\lambda_3(A) \cdots \lambda_{n+1}(A)\big) = \det A > 0,$$

so $\lambda_1(A) > 0$ and A is positive definite. This completes the induction. $\square$

Example 16.4.4 Consider the Hermitian matrix

$$A = \begin{bmatrix} 2 & 2 & 2 \\ 2 & 3 & 3 \\ 2 & 3 & 4 \end{bmatrix},$$

which is not diagonally dominant. Its leading principal minors are 2, 2, and 2, so Sylvester's criterion ensures that it is positive definite.

16.5 Diagonal Entries and Eigenvalues of Hermitian Matrices

If $A \in \mathsf{M}_n$ has real diagonal entries, there is a permutation matrix $P \in \mathsf{M}_n$ such that the diagonal entries of PAP^T are in increasing order; see the discussion following Definition 6.3.3. The matrices A and PAP^T are similar, so they have the same eigenvalues.

Definition 16.5.1 The *increasingly ordered diagonal entries* of a Hermitian matrix $A = [a_{ij}] \in M_n$ are the real numbers $a_1 \leq a_2 \leq \cdots \leq a_n$ defined by

$$a_1 = \min\{a_{11}, a_{22}, \ldots, a_{nn}\}$$

$$a_j = \min\{a_{11}, a_{22}, \ldots, a_{nn}\} \setminus \{a_1, a_2, \ldots, a_{j-1}\}, \quad j = 2, 3, \ldots, n.$$

Theorem 16.5.2 (Schur) *Let $A = [a_{ij}] \in M_n$ be Hermitian. Let $\lambda_1 \leq \lambda_2 \leq \cdots \leq \lambda_n$ be its increasingly ordered eigenvalues and let $a_1 \leq a_2 \leq \cdots \leq a_n$ be its increasingly ordered diagonal entries. Then*

$$\sum_{i=1}^{k} \lambda_i \leq \sum_{i=1}^{k} a_i, \quad k = 1, 2, \ldots, n, \tag{16.5.3}$$

with equality for $k = n$.

Proof A permutation similarity of A does not change its eigenvalues, so we may assume that $a_{ii} = a_i$, that is, $a_{11} \leq a_{22} \leq \cdots \leq a_{nn}$. Let $k \in \{1, 2, \ldots, n\}$ and partition

$$A = \begin{bmatrix} A_{11} & A_{12} \\ A_{12}^* & A_{22} \end{bmatrix},$$

in which $A_{11} \in M_k$. The inequalities (16.3.14) ensure that the increasingly ordered eigenvalues of A and its principal submatrix A_{11} satisfy

$$\lambda_i(A) \leq \lambda_i(A_{11}), \quad i = 1, 2, \ldots, k.$$

Therefore,

$$\sum_{i=1}^{k} \lambda_i(A) \leq \sum_{i=1}^{k} \lambda_i(A_{11}) = \operatorname{tr} A_{11} = \sum_{i=1}^{k} a_{ii} = \sum_{i=1}^{k} a_i,$$

which proves (16.5.3). If $k = n$, then both sides of (16.5.3) equal $\operatorname{tr} A$. $\square$

Inequalities of the form (16.5.3) are *majorization inequalities*. See P.16.27 for a different, but equivalent, version of the majorization inequalities.

Example 16.5.4 For the matrix A in Example 16.3.1, the quantities in the inequalities (16.5.3) are

k	1	2	3	4	5
$\sum_{i=1}^{k} \lambda_i$	-4.38	-4.98	-2.83	3.32	14
$\sum_{i=1}^{k} a_i$	-2	0	4	8	14

16.6 *Congruence and Inertia of Hermitian Matrices

In Section 13.4 we encountered the transformation $B \longmapsto SBS^*$ in our discussion of simultaneous diagonalization of two quadratic forms. We now study the properties of this transformation, with initial emphasis on its effect on the eigenvalues of a Hermitian matrix.

Definition 16.6.1 Let $A, B \in \mathsf{M}_n$. Then A is *congruent* to B if there is an invertible $S \in \mathsf{M}_n$ such that $A = SBS^*$.

Example 16.6.2 If two matrices are unitarily similar, then they are *congruent. In particular, real orthogonal similarity and permutation similarity are *congruences.

Like similarity, *congruence is an equivalence relation; see P.16.18.

Suppose that $A \in \mathsf{M}_n$ is Hermitian and $S \in \mathsf{M}_n$ is invertible. Then SAS^* is Hermitian and $\operatorname{rank} A = \operatorname{rank} SAS^*$ (Theorem 3.2.9). Therefore, if A is not invertible, the multiplicities of 0 as an eigenvalue of A and SAS^* are the same (Theorem 9.4.12). What other properties do A and SAS^* have in common? As a first step toward an answer, we introduce a type of diagonal matrix to which each Hermitian matrix is *congruent.

Definition 16.6.3 A *real inertia matrix* is a diagonal matrix whose entries are in $\{-1, 0, 1\}$.

Theorem 16.6.4 *Let $A \in \mathsf{M}_n$ be Hermitian. Then A is *congruent to a real inertia matrix D, in which the number of $+1$ diagonal entries in D is equal to the number of positive eigenvalues of A. The number of -1 diagonal entries in D is equal to the number of negative eigenvalues of A.*

Proof Let $A = U \Lambda U^*$ be a spectral decomposition, in which U is unitary,

$$\Lambda = \operatorname{diag}(\lambda_1, \lambda_2, \ldots, \lambda_n),$$

and the eigenvalues $\lambda_1, \lambda_2, \ldots, \lambda_n$ are in any desired order. Let $G = \operatorname{diag}(g_1, g_2, \ldots, g_n)$, in which

$$g_i = \begin{cases} |\lambda_i|^{-1/2} & \text{if } \lambda_i \neq 0, \\ 1 & \text{if } \lambda_i = 0. \end{cases}$$

Then

$$g_i \lambda_i g_i = \begin{cases} \frac{\lambda_i}{|\lambda_i|} = \pm 1 & \text{if } \lambda_i \neq 0, \\ 0 & \text{if } \lambda_i = 0. \end{cases}$$

If $S = GU^*$, then

$$SAS^* = GU^*(U \Lambda U^*)UG = G \Lambda G \qquad (16.6.5)$$

is a real inertia matrix that is *congruent to A and has the asserted numbers of positive, negative, and zero diagonal entries. $\qquad \square$

Lemma 16.6.6 *If two real inertia matrices of the same size have the same number of $+1$ diagonal entries and the same number of -1 diagonal entries, then they are *congruent.*

Proof Let $D_1, D_2 \in \mathsf{M}_n$ be real inertia matrices. The hypotheses ensure that D_1 and D_2 have the same eigenvalues, and Theorem 12.7.1.d tells us that they are permutation similar. Since permutation similarity is a *congruence, D_1 and D_2 are *congruent. $\qquad \square$

Example 16.6.7 The real inertia matrices

$$D_1 = \begin{bmatrix} 1 & 0 & 0 \\ 0 & -1 & 0 \\ 0 & 0 & 1 \end{bmatrix} \quad \text{and} \quad D_2 = \begin{bmatrix} -1 & 0 & 0 \\ 0 & 1 & 0 \\ 0 & 0 & 1 \end{bmatrix}$$

are *congruent via the permutation matrix

$$P = \begin{bmatrix} 0 & 1 & 0 \\ 1 & 0 & 0 \\ 0 & 0 & 1 \end{bmatrix}.$$

The following result is the key to a proof of the converse of Lemma 16.6.6.

Lemma 16.6.8 *Let* $A \in \mathsf{M}_n$ *be Hermitian, let* k *be a positive integer, and let* $\mathcal{U}$ *be a* k*-dimensional subspace of* $\mathbb{C}^n$. *If* $\langle A\mathbf{x}, \mathbf{x} \rangle > 0$ *for every nonzero* $\mathbf{x} \in \mathcal{U}$, *then* A *has at least* k *positive eigenvalues.*

Proof Let $\lambda_1 \leq \lambda_2 \leq \cdots \leq \lambda_n$ be the increasingly ordered eigenvalues of A, with corresponding orthonormal eigenvectors $\mathbf{u}_1, \mathbf{u}_2, \ldots, \mathbf{u}_n \in \mathbb{C}^n$, and let

$$\mathcal{V} = \text{span}\{\mathbf{u}_1, \mathbf{u}_2, \ldots, \mathbf{u}_{n-k+1}\}.$$

Then

$$\dim \mathcal{U} + \dim \mathcal{V} = k + (n - k + 1) = n + 1,$$

so Corollary 2.2.17.a ensures that there is a nonzero vector $\mathbf{x} \in \mathcal{U} \cap \mathcal{V}$. Then $\langle A\mathbf{x}, \mathbf{x} \rangle > 0$ because $\mathbf{x} \in \mathcal{U}$. Since $\mathbf{x} \in \mathcal{V}$, Theorem 16.1.1 tells us that $\langle A\mathbf{x}, \mathbf{x} \rangle \leq \lambda_{n-k+1}$. Therefore,

$$0 < \langle A\mathbf{x}, \mathbf{x} \rangle \leq \lambda_{n-k+1} \leq \lambda_{n-k+2} \leq \cdots \leq \lambda_n,$$

that is, the k largest eigenvalues of A are positive. $\qquad\square$

The following theorem is the main result of this section.

Theorem 16.6.9 (Sylvester's Inertia Theorem) *Two Hermitian matrices of the same size are* *congruent if and only if they have the same number of positive eigenvalues and the same number of negative eigenvalues.*

Proof Let $A, B \in \mathsf{M}_n$ be Hermitian. If A and B have the same number of positive eigenvalues and the same number of negative eigenvalues, then they have the same numbers of zero eigenvalues. Theorem 16.6.4 ensures that A and B are *congruent to real inertia matrices that have the same eigenvalues, and Lemma 16.6.6 ensures that those real inertia matrices are *congruent. It follows from the transitivity of *congruence that A and B are *congruent.

For the converse, suppose that $A = S^*BS$ and $S \in \mathsf{M}_n$ is invertible. Since A and B are diagonalizable and $\text{rank}\, A = \text{rank}\, B$, they have the same number of zero eigenvalues and the same number of nonzero eigenvalues (Theorem 9.4.12). If they have the same number of positive eigenvalues, then they also have the same number of negative eigenvalues.

If A has no positive eigenvalues, then every eigenvalue of A is negative or zero. Therefore, $-A$ is positive semidefinite (Theorem 13.1.2), in which case $-B = S^*(-A)S$ is also positive semidefinite (Theorem 13.1.23.b). It follows (Theorem 13.1.2 again) that every eigenvalue of $-B$ is nonnegative. Therefore, B has no positive eigenvalues.

Let $\lambda_1 \leq \lambda_2 \leq \cdots \leq \lambda_n$ be the increasingly ordered eigenvalues of A, with corresponding orthonormal eigenvectors $\mathbf{u}_1, \mathbf{u}_2, \ldots, \mathbf{u}_n \in \mathbb{C}^n$. Suppose that A has exactly k positive eigenvalues for some $k \in \{1, 2, \ldots, n\}$. Then

$$0 < \lambda_{n-k+1} \leq \lambda_{n-k+2} \leq \cdots \leq \lambda_n$$

and $\mathcal{U} = \mathrm{span}\{\mathbf{u}_{n-k+1}, \mathbf{u}_{n-k+2}, \ldots, \mathbf{u}_n\}$ has dimension k. If $\mathbf{x} \in \mathcal{U}$ is nonzero, then $\langle A\mathbf{x}, \mathbf{x} \rangle \geq \lambda_{n-k+1} \|\mathbf{x}\|_2^2 > 0$ (Theorem 16.1.1). The set

$$S\mathcal{U} = \{S\mathbf{x} : \mathbf{x} \in \mathcal{U}\}$$

is a subspace of $\mathbb{C}^n$ (see Example 1.3.13). The vectors $S\mathbf{u}_{n-k+1}, S\mathbf{u}_{n-k+2}, \ldots, S\mathbf{u}_n$ span $S\mathcal{U}$, and Example 3.2.11 shows that they are linearly independent. Therefore, $\dim S\mathcal{U} = k$. If $\mathbf{y} \in S\mathcal{U}$ is nonzero, then there is a nonzero $\mathbf{x} \in \mathcal{U}$ such that $\mathbf{y} = S\mathbf{x}$. Then

$$\langle B\mathbf{y}, \mathbf{y} \rangle = (S\mathbf{x})^*B(S\mathbf{x}) = \mathbf{x}^*S^*BS\mathbf{x} = \langle A\mathbf{x}, \mathbf{x} \rangle > 0.$$

Lemma 16.6.8 ensures that B has at least k positive eigenvalues, that is, B has at least as many positive eigenvalues as A. Now interchange the roles of A and B in the preceding argument, and conclude that A has at least as many positive eigenvalues as B. Therefore, A and B have the same number of positive eigenvalues. $\qquad\qquad\square$

Example 16.6.10 Consider the matrices

$$A = \begin{bmatrix} -2 & -4 & -6 \\ -4 & -5 & 0 \\ -6 & 0 & 34 \end{bmatrix} \quad \text{and} \quad L = \begin{bmatrix} 1 & 0 & 0 \\ -2 & 1 & 0 \\ 5 & -4 & 1 \end{bmatrix}.$$

Then

$$LAL^* = \begin{bmatrix} -2 & 0 & 0 \\ 0 & 3 & 0 \\ 0 & 0 & 4 \end{bmatrix},$$

so A is *congruent to the real inertia matrix $\mathrm{diag}(-1, 1, 1)$. Theorem 16.6.9 tells us that A has two positive eigenvalues and one negative eigenvalue. Rounded to two decimal places, the eigenvalues of A are -8.07, 0.09, and 34.98. See P.16.21 for a discussion of how L is determined.

16.7 Weyl's Inequalities

Interlacing inequalities such as those in Theorem 16.2.2 are special cases of inequalities for the eigenvalues of two Hermitian matrices and their sum.

Theorem 16.7.1 (Weyl) *Let $A, B \in \mathsf{M}_n$ be Hermitian. The increasingly ordered eigenvalues of A, B, and $A + B$ satisfy*

$$\lambda_i(A + B) \leq \lambda_{i+j}(A) + \lambda_{n-j}(B), \quad j = 0, 1, \ldots, n - i, \tag{16.7.2}$$

and

$$\lambda_{i-j+1}(A) + \lambda_j(B) \leq \lambda_i(A+B), \quad j = 1, 2, \ldots, i, \quad (16.7.3)$$

for each $i = 1, 2, \ldots, n$.

Proof Let

$$\mathbf{u}_1, \mathbf{u}_2, \ldots, \mathbf{u}_n, \qquad \mathbf{v}_1, \mathbf{v}_2, \ldots, \mathbf{v}_n, \quad \text{and} \quad \mathbf{w}_1, \mathbf{w}_2, \ldots, \mathbf{w}_n$$

be orthonormal eigenvectors of A, $A + B$, and B, respectively, corresponding to their increasingly ordered eigenvalues.

Let $i \in \{1, 2, \ldots, n\}$ and $j \in \{0, 1, \ldots, n - i\}$. Let $\mathcal{U} = \text{span}\{\mathbf{u}_1, \mathbf{u}_2, \ldots, \mathbf{u}_{i+j}\}$, $\mathcal{V} = \text{span}\{\mathbf{v}_i, \mathbf{v}_{i+1}, \ldots, \mathbf{v}_n\}$, and $\mathcal{W} = \text{span}\{\mathbf{w}_1, \mathbf{w}_2, \ldots, \mathbf{w}_{n-j}\}$. Then

$$\dim \mathcal{U} + \dim \mathcal{V} = (i+j) + (n-i+1) = n+j+1,$$

so Corollary 2.2.17.b ensures that $\dim(\mathcal{U} \cap \mathcal{V}) \geq j+1$. Since

$$\dim(\mathcal{U} \cap \mathcal{V}) + \dim \mathcal{W} \geq (j+1) + (n-j) = n+1,$$

Corollary 2.2.17.a tells us that there is a unit vector $\mathbf{x} \in \mathcal{U} \cap \mathcal{V} \cap \mathcal{W}$. Now compute

$$
\begin{aligned}
\lambda_i(A+B) &\leq \langle (A+B)\mathbf{x}, \mathbf{x} \rangle & \text{(since } \mathbf{x} \in \mathcal{V}) \\
&= \langle A\mathbf{x}, \mathbf{x} \rangle + \langle B\mathbf{x}, \mathbf{x} \rangle \\
&\leq \lambda_{i+j}(A) + \langle B\mathbf{x}, \mathbf{x} \rangle & \text{(since } \mathbf{x} \in \mathcal{U}) \\
&\leq \lambda_{i+j}(A) + \lambda_{n-j}(B) & \text{(since } \mathbf{x} \in \mathcal{W}).
\end{aligned}
$$

This verifies the inequalities (16.7.2).

Now consider the Hermitian matrices $-A$, $-(A+B)$, and $-B$. The increasingly ordered eigenvalues of $-A$ are

$$-\lambda_n(A) \leq -\lambda_{n-1}(A) \leq \cdots \leq -\lambda_1(A),$$

that is, $\lambda_i(-A) = -\lambda_{n-i+1}(A)$ for each $i = 1, 2, \ldots, n$. The inequalities (16.7.2) involve the increasingly ordered eigenvalues

$$
\begin{aligned}
\lambda_i(-(A+B)) &= -\lambda_{n-i+1}(A+B), \\
\lambda_{i+j}(-A) &= -\lambda_{n-i-j+1}(A), \quad \text{and} \\
\lambda_{n-j}(-B) &= -\lambda_{j+1}(B).
\end{aligned}
$$

We have

$$-\lambda_{n-i+1}(A+B) \leq -\lambda_{n-i-j+1}(A) - \lambda_{j+1}(B), \quad j = 0, 1, \ldots, n-i,$$

that is,

$$\lambda_{n-i+1}(A+B) \geq \lambda_{n-i-j+1}(A) + \lambda_{j+1}(B), \quad j = 0, 1, \ldots, n-i.$$

After a change of variables in the indices we obtain

$$\lambda_i(A+B) \geq \lambda_{i-j+1}(A) + \lambda_j(B), \quad j = 1, 2, \ldots, i.$$

This verifies the inequalities (16.7.3).   □

Example 16.7.4 Weyl's inequalities imply the interlacing inequalities (16.2.3). Take $B = E$ and observe that $\lambda_i(B) = 0$ for $i = 1, 2, \ldots, n - 1$. If we invoke (16.7.2) and take $j = 1$, we obtain

$$\lambda_i(A + E) \le \lambda_{i+1}(A) + \lambda_{n-1}(E) = \lambda_{i+1}(A), \quad i = 1, 2, \ldots, n - 1. \tag{16.7.5}$$

If we invoke (16.7.3) and take $j = 1$, we obtain

$$\lambda_i(A + E) \ge \lambda_i(A) + \lambda_1(E) = \lambda_i(A). \tag{16.7.6}$$

Example 16.7.7 Weyl's inequalities provide some new information about the eigenvalues of A and $A + E$, in which E is positive semidefinite and has rank 1. If we invoke (16.7.2), take $j = 0$, and observe that $\lambda_n(E) = \|E\|_2$, we obtain

$$\lambda_i(A + E) \le \lambda_i(A) + \|E\|_2, \quad i = 1, 2, \ldots, n. \tag{16.7.8}$$

The bounds (16.7.5), (16.7.6), and (16.7.8) tell us that

$$0 \le \lambda_i(A + E) - \lambda_i(A) \le \min\left\{\|E\|_2, \lambda_{i+1}(A) - \lambda_i(A)\right\}, \quad i = 1, 2, \ldots, n - 1.$$

Each eigenvalue $\lambda_i(A + E)$ differs from $\lambda_i(A)$ by no more than $\|E\|_2$ for each $i = 1, 2, \ldots, n$. This gives valuable information if E has small norm. However, no matter how large the norm of E is, $\lambda_i(A + E) \le \lambda_{i+1}(A)$ for each $i = 1, 2, \ldots, n - 1$.

16.8 *Congruence and Inertia of Normal Matrices

A version of Sylvester's inertia theorem is valid for normal matrices, and the generalization gives some insight into the role of the numbers of positive and negative eigenvalues in the theory of *congruence for Hermitian matrices. We begin with a natural generalization of real inertia matrices.

Definition 16.8.1 An *inertia matrix* is a complex diagonal matrix whose nonzero diagonal entries (if any) have modulus 1.

If an inertia matrix is nonzero, it is permutation similar to the direct sum of a diagonal unitary matrix and a zero matrix. An inertia matrix with real entries is a real inertia matrix (Definition 16.6.3).

Theorem 16.8.2 *Let $A \in \mathsf{M}_n$ be normal. Then A is *congruent to an inertia matrix D, in which for each $\theta \in [0, 2\pi)$, the number of diagonal entries $e^{i\theta}$ in D is equal to the number of eigenvalues of A that lie on the ray $\{\rho e^{i\theta} : \rho > 0\}$ in the complex plane.*

Proof Let $A = U\Lambda U^*$ be a spectral decomposition, in which U is unitary and $\Lambda = \operatorname{diag}(\lambda_1, \lambda_2, \ldots, \lambda_n)$. Each nonzero eigenvalue has the polar form $\lambda_j = |\lambda_j| e^{i\theta_j}$ for a unique $\theta_j \in [0, 2\pi)$. Let $G = \operatorname{diag}(g_1, g_2, \ldots, g_n)$, in which

$$g_j = \begin{cases} |\lambda_j|^{-1/2} & \text{if } \lambda_j \ne 0, \\ 1 & \text{if } \lambda_j = 0. \end{cases}$$

Then

$$g_j \lambda_j g_j = \begin{cases} \frac{\lambda_j}{|\lambda_j|} = e^{i\theta_j} & \text{if } \lambda_j \neq 0, \\ 0 & \text{if } \lambda_j = 0. \end{cases}$$

If $S = GU^*$, then

$$SAS^* = GU^*(U\Lambda U^*)UG = G\Lambda G \tag{16.8.3}$$

is an inertia matrix that is *congruent to A and has the asserted number of diagonal entries $e^{i\theta}$. $\qquad\qquad\square$

Figure 16.4 indicates the locations of the nonzero eigenvalues of a normal matrix A; the points on the unit circle indicate the nonzero eigenvalues of an inertia matrix that is *congruent to A.

The preceding theorem tells us that each normal matrix A is *congruent to an inertia matrix D that is determined by the eigenvalues of A; this is an analog of Theorem 16.6.4. We must now show that D is uniquely determined, up to permutation of its diagonal entries. The following lemma is the key to our proof of this result. It is motivated by the observation that a nonzero inertia matrix is permutation similar to the direct sum of a diagonal unitary matrix and a zero matrix (which may be absent). To make productive use of this observation, we need to understand the relationship between two *congruent unitary matrices.

Lemma 16.8.4 *Two unitary matrices are *congruent if and only if they are similar.*

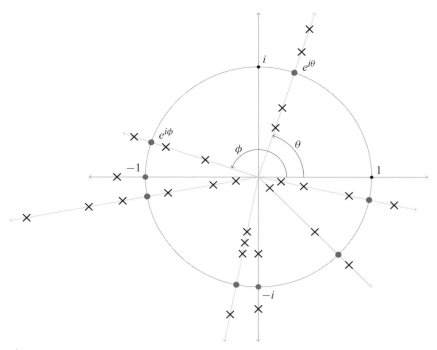

Figure 16.4 Eigenvalues ($\times$) of a normal matrix A on rays in $\mathbb{C}$. Nonzero eigenvalues ($\bullet$) of an inertia matrix D that is *congruent to A. The respective multiplicities of 1, $e^{i\theta}$, i, $e^{i\phi}$, -1, and $-i$ as eigenvalues of D are 0, 4, 0, 3, 1, and 2.

Proof Let $V, W \in M_n$ be unitary. If they are similar, they have the same eigenvalues, so Theorem 12.7.1.b ensures that they are unitarily similar. Unitary similarity is a *congruence.

Conversely, if V and W are *congruent, then there is an invertible $S \in M_n$ such that $V = SWS^*$. Let $S = PU$ be a left polar decomposition, in which P is positive definite and U is unitary. Then $V = SWS^* = PUWU^*P$, so

$$P^{-1}V = (UWU^*)P, \tag{16.8.5}$$

in which P and P^{-1} are positive definite and V and UWU^* are unitary. Let $A = P^{-1}V$. The left-hand side of (16.8.5) is a left polar decomposition of the invertible matrix A, while the right-hand side of (16.8.5) is a right polar decomposition of A. Theorem 14.3.15.e.iii ensures that $V = UWU^*$, that is, V and W are (unitarily) similar. □

Theorem 16.8.6 *Two inertia matrices are *congruent if and only if they are permutation similar.*

Proof Let $D_1, D_2 \in M_n$ be inertia matrices. If they are permutation similar, they are *congruent since a permutation similarity is a *congruence.

Conversely, suppose that D_1 and D_2 are *congruent and let $r = \operatorname{rank} D_1 = \operatorname{rank} D_2$ (Theorem 3.2.9). If $r = 0$, then $D_1 = D_2 = 0$. If $r = n$, then the preceding lemma ensures that D_1 and D_2 are similar. Theorem 12.7.1.d tells us that they are permutation similar. Suppose that $1 \leq r < n$. After permutation similarities (*congruences), if necessary, we may assume that $D_1 = V_1 \oplus 0_{n-r}$ and $D_2 = V_2 \oplus 0_{n-r}$, in which $V_1, V_2 \in M_r$ are diagonal unitary matrices. Let $S \in M_n$ be invertible and such that $SD_1S^* = D_2$. Partition $S = [S_{ij}]$ conformally with D_1 and D_2, so $S_{11} \in M_r$. Then

$$\begin{bmatrix} V_2 & 0 \\ 0 & 0 \end{bmatrix} = D_2 = SD_1S^* = \begin{bmatrix} S_{11} & S_{12} \\ S_{21} & S_{22} \end{bmatrix} \begin{bmatrix} V_1 & 0 \\ 0 & 0 \end{bmatrix} \begin{bmatrix} S_{11}^* & S_{21}^* \\ S_{12}^* & S_{22}^* \end{bmatrix}$$

$$= \begin{bmatrix} S_{11}V_1S_{11}^* & S_{11}V_1S_{21}^* \\ S_{21}V_1S_{11}^* & S_{21}V_1S_{21}^* \end{bmatrix},$$

which tells us that $S_{11}V_1S_{11}^* = V_2$. Since V_2 is invertible, we conclude that S_{11} is invertible and V_1 is *congruent to V_2. The preceding lemma ensures that V_1 and V_2 are similar, from which it follows that D_1 and D_2 are similar. Theorem 12.7.1.d tells us that they are permutation similar. □

Theorem 16.8.7 (Inertia Theorem for Normal Matrices) *Two normal matrices of the same size are *congruent if and only if, for each $\theta \in [0, 2\pi)$, they have the same number of eigenvalues on the ray $\{\rho e^{i\theta} : \rho > 0\}$ in the complex plane.*

Proof Let $A, B \in M_n$ be normal. Let D_1 and D_2 be inertia matrices that are *congruent to A and B, respectively, and are constructed according to the recipe in Theorem 16.8.2.

If A is *congruent to B, the transitivity of *congruence ensures that D_1 is *congruent to D_2. Theorem 16.8.6 tells us that D_1 and D_2 are permutation similar, so they have the same eigenvalues (diagonal entries) with the same multiplicities. The construction in Theorem 16.8.2 ensures that A and B have the same number of eigenvalues on the ray $\{\rho e^{i\theta} : \rho > 0\}$ for each $\theta \in [0, 2\pi)$.

Conversely, the hypotheses on the eigenvalues of A and B ensure that the diagonal entries of D_1 can be obtained by permuting the diagonal entries of D_2, so D_1 and D_2 are permutation similar and *congruent. The transitivity of *congruence again ensures that A and B are *congruent.   □

Theorem 16.6.9 is a corollary of the inertia theorem for normal matrices.

Corollary 16.8.8 *Two Hermitian matrices of the same size are *congruent if and only if they have the same number of positive eigenvalues and the same number of negative eigenvalues.*

Proof If a Hermitian matrix has any nonzero eigenvalues, they are real and hence are on the rays $\{\rho e^{i\theta} : \rho > 0\}$ with $\theta = 0$ or $\theta = \pi$.   □

A square matrix is Hermitian if and only if it is *congruent to a real inertia matrix. Although every normal matrix is *congruent to an inertia matrix, a matrix can be *congruent to an inertia matrix without being normal; see P.16.36. Moreover, not every matrix is *congruent to an inertia matrix; see P.16.40.

16.9 Problems

P.16.1 Let $A \in \mathsf{M}_n$ be Hermitian and let $\lambda_1 \leq \lambda_2 \leq \cdots \leq \lambda_n$ be its increasingly ordered eigenvalues. Show that

$$\min_{\substack{\mathbf{x} \in \mathbb{C}^n \\ \mathbf{x} \neq \mathbf{0}}} \frac{\langle A\mathbf{x}, \mathbf{x} \rangle}{\langle \mathbf{x}, \mathbf{x} \rangle} = \lambda_1 \quad \text{and} \quad \max_{\substack{\mathbf{x} \in \mathbb{C}^n \\ \mathbf{x} \neq \mathbf{0}}} \frac{\langle A\mathbf{x}, \mathbf{x} \rangle}{\langle \mathbf{x}, \mathbf{x} \rangle} = \lambda_n.$$

P.16.2 Let $A \in \mathsf{M}_n$ be Hermitian, with increasingly ordered eigenvalues $\lambda_1 \leq \lambda_2 \leq \cdots \leq \lambda_n$ and corresponding orthonormal eigenvectors $\mathbf{u}_1, \mathbf{u}_2, \ldots, \mathbf{u}_n \in \mathbb{C}^n$. Let $i_1, i_2, \ldots, i_k$ be integers such that $1 \leq i_1 < i_2 < \cdots < i_k \leq n$. If $\mathbf{x} \in \text{span}\{\mathbf{u}_{i_1}, \mathbf{u}_{i_2}, \ldots, \mathbf{u}_{i_k}\}$ is a unit vector, show that $\lambda_{i_1} \leq \langle A\mathbf{x}, \mathbf{x} \rangle \leq \lambda_{i_k}$.

P.16.3 Adopt the notation of Theorem 16.1.1. Show that

$$\{\langle A\mathbf{x}, \mathbf{x} \rangle : \mathbf{x} \in \mathbb{C}^n, \|\mathbf{x}\|_2 = 1\} = [\lambda_1, \lambda_n] \subseteq \mathbb{R}.$$

If A is real symmetric, show that

$$\{\langle A\mathbf{x}, \mathbf{x} \rangle : \mathbf{x} \in \mathbb{R}^n, \|\mathbf{x}\|_2 = 1\} = [\lambda_1, \lambda_n] \subseteq \mathbb{R}.$$

P.16.4 Suppose that the eigenvalues $\lambda_1, \lambda_2, \ldots, \lambda_n$ of a normal matrix $A \in \mathsf{M}_n$ are the vertices of a convex polygon P in the complex plane; see Figure 16.5 for an illustration with $n = 5$. Show that

$$\{\langle A\mathbf{x}, \mathbf{x} \rangle : \mathbf{x} \in \mathbb{C}^n \text{ and } \|\mathbf{x}\|_2 = 1\} = P.$$

This set is the *numerical range* (also known as the *field of values*) of A.

P.16.5 Let $A = J_2$. Sketch the sets

$$\{\langle A\mathbf{x}, \mathbf{x} \rangle : \mathbf{x} \in \mathbb{R}^n \text{ and } \|\mathbf{x}\|_2 = 1\} \quad \text{and} \quad \{\langle A\mathbf{x}, \mathbf{x} \rangle : \mathbf{x} \in \mathbb{C}^n \text{ and } \|\mathbf{x}\|_2 = 1\}.$$

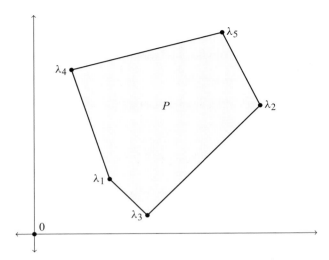

Figure 16.5 Numerical range of a 5×5 normal matrix.

P.16.6 Let $A \in M_n$. (a) Show that $\operatorname{rank} A = 1$ if and only if there are nonzero $\mathbf{u}, \mathbf{v} \in \mathbb{C}^n$ such that $A = \mathbf{u}\mathbf{v}^*$. (b) Show that A is Hermitian and $\operatorname{rank} A = 1$ if and only if there is a nonzero $\mathbf{u} \in \mathbb{C}^n$ such that either $A = \mathbf{u}\mathbf{u}^*$ or $A = -\mathbf{u}\mathbf{u}^*$.

P.16.7 Adopt the notation in Theorem 16.1.1. If $\mathbf{x} \in \mathcal{U}$, show that (a) $\langle A\mathbf{x}, \mathbf{x} \rangle = \lambda_p$ if and only if $A\mathbf{x} = \lambda_p \mathbf{x}$, and (b) $\langle A\mathbf{x}, \mathbf{x} \rangle = \lambda_q$ if and only if $A\mathbf{x} = \lambda_q \mathbf{x}$.

P.16.8 Let $A = [a_{ij}] \in M_n$ be Hermitian and have eigenvalues $\lambda_1 \leq \lambda_2 \leq \cdots \leq \lambda_n$. If $i \in \{1, 2, \ldots, n\}$ is such that $a_{ii} = \lambda_1$ or $a_{ii} = \lambda_n$, show that $a_{ij} = a_{ji} = 0$ for all $j \neq i$. *Hint*: Suppose that A is positive definite and $a_{11} = \lambda_n(A)$ (otherwise, consider a permutation similarity of $\pm A + \mu I$). Partition $A = \begin{bmatrix} \lambda_n(A) & \mathbf{x}^* \\ \mathbf{x} & B \end{bmatrix}$. Use Theorem 12.3.1 and (16.3.4) to show that $\|A\|_F^2 = \lambda_n(A)^2 + 2\|\mathbf{x}\|_2^2 + \|B\|_F^2 \geq \lambda_n(A)^2 + 2\|\mathbf{x}\|_2^2 + \sum_{i=1}^{n-1} \lambda_i(A)^2 = \|A\|_F^2 + 2\|\mathbf{x}\|_2^2$.

P.16.9 Let $A = [a_{ij}] \in M_4$ be Hermitian and suppose that $\operatorname{spec} A = \{1, 2, 3, 4\}$. If $a_{11} = 1$, $a_{22} = 2$, and $a_{44} = 4$, what are the other entries? Why?

P.16.10 Adopt the notation in Theorem 16.2.2. If $\lambda \in \operatorname{spec} A$ has multiplicity 2 or greater, show that $\lambda \in \operatorname{spec}(A + E)$.

P.16.11 Give an example of a 2×2 Hermitian matrix that has nonnegative leading principal minors but is not positive semidefinite. Which of its principal minors is negative?

P.16.12 Suppose that $A \in M_n$ is Hermitian and let $B \in M_{n-1}$ be its $(n-1) \times (n-1)$ leading principal submatrix. If B is positive semidefinite and $\operatorname{rank} A = \operatorname{rank} B$, show that A is positive semidefinite.

P.16.13 Use Sylvester's criterion to show that

$$A = \begin{bmatrix} 2 & 2 & 3 \\ 2 & 5 & 5 \\ 3 & 5 & 7 \end{bmatrix}$$

is positive definite.

P.16.14 Show that

$$A = \begin{bmatrix} 2 & 0 & 0 & 0 & 1 & 1 \\ 0 & 2 & 0 & 0 & 1 & 1 \\ 0 & 0 & 3 & 1 & 1 & 1 \\ 0 & 0 & 1 & 3 & 1 & 1 \\ 1 & 1 & 1 & 1 & 5 & 1 \\ 1 & 1 & 1 & 1 & 1 & 5 \end{bmatrix}$$

is positive definite.

P.16.15 Suppose that $0, 0, \ldots, 0, 1$ are the eigenvalues of a Hermitian $A = [a_{ij}] \in M_n$. Show that $0 \le a_{ii} \le 1$ for each $i = 1, 2, \ldots, n$.

P.16.16 If $10, 25, 26, 39, 50$ are the eigenvalues of a Hermitian $A \in M_5$, show that A cannot have main diagonal entries $14, 16, 30, 44, 46$.

P.16.17 Deduce the inequalities (16.1.7) from the majorization inequalities (16.5.3).

P.16.18 Show that $*$congruence is an equivalence relation on M_n.

P.16.19 Deduce from Theorem 16.6.9 that two real inertia matrices are $*$congruent if and only if they are similar.

P.16.20 Let $D = \mathrm{diag}(1, -1) \in M_2$. Is D $*$congruent to $-D$? Is I_2 $*$congruent to $-I_2$? Why?

P.16.21 (a) In Example 16.6.10, show that

$$L = \begin{bmatrix} 1 & 0 & 0 \\ 0 & 1 & 0 \\ 0 & -4 & 1 \end{bmatrix} \begin{bmatrix} 1 & 0 & 0 \\ -2 & 1 & 0 \\ -3 & 0 & 1 \end{bmatrix}.$$

(b) What are the elementary row operations that reduce A to upper triangular form, and how are they related to L? (c) Since LA is upper triangular, why must LAL^* be diagonal? (d) To determine the signs of the eigenvalues of A, why does it suffice to examine only the main diagonal entries of LA?

P.16.22 If $H \in M_n$ is Hermitian, define

$$i_+(H) = \text{the number of positive eigenvalues of } H,$$
$$i_-(H) = \text{the number of negative eigenvalues of } H, \quad \text{and}$$
$$i_0(H) = \text{the number of zero eigenvalues of } H.$$

Let $A = [A_{ij}] \in M_n$ be a Hermitian 2×2 block matrix, in which $A_{11} \in M_k$ is invertible. Define

$$S = \begin{bmatrix} I_k & 0 \\ -A_{12}A_{11}^{-1} & I_{n-k} \end{bmatrix} \in M_n.$$

(a) Show that $SAS^* = A_{11} \oplus A/A_{11}$, in which $A/A_{11} = A_{22} - A_{12}^* A_{11}^{-1} A_{12}$ is the Schur complement of A_{11} in A.

(b) Prove *Haynsworth's inertia theorem*:

$$i_+(A) = i_+(A_{11}) + i_+(A/A_{11}),$$
$$i_-(A) = i_-(A_{11}) + i_-(A/A_{11}), \quad \text{and}$$
$$i_0(A) = i_0(A/A_{11}).$$

(c) If A is positive definite, show that A/A_{11} is positive definite.

P.16.23 Let $A, B \in \mathsf{M}_n$ be Hermitian. (a) Use Weyl's inequalities to show that the increasingly ordered eigenvalues of A and $A + B$ satisfy

$$\lambda_i(A) + \lambda_1(B) \le \lambda_i(A + B) \le \lambda_i(A) + \lambda_n(B), \quad i = 1, 2, \ldots, n.$$

(b) Show that $|\lambda_i(A + B) - \lambda_i(A)| \le \|B\|_2$ for each $i = 1, 2, \ldots, n$. (c) Compare the bounds in (b) with the Bauer–Fike bounds in Theorem 15.6.12. Discuss.

P.16.24 Let $A, B \in \mathsf{M}_n$ be Hermitian. Suppose that B has exactly p positive eigenvalues and exactly q negative eigenvalues, in which $p, q \in \{0, 1, \ldots, n\}$ and $p + q \le n$. Use Weyl's inequalities to show that the increasingly ordered eigenvalues of A and $A + B$ satisfy

$$\lambda_{i-q}(A) \le \lambda_i(A + B), \quad i = q + 1, q + 2, \ldots, n,$$

and

$$\lambda_i(A + B) \le \lambda_{i+p}(A), \quad i = 1, 2, \ldots, n - p.$$

P.16.25 Let $A, B \in \mathsf{M}_n$ be Hermitian and suppose that B is positive semidefinite. Show that the increasingly ordered eigenvalues of A and $A + B$ satisfy

$$\lambda_i(A) \le \lambda_i(A + B), \quad i = 1, 2, \ldots, n.$$

This is the *monotonicity theorem*. What can you say if B is positive definite?

P.16.26 Consider the non-Hermitian matrices

$$A = \begin{bmatrix} 0 & 0 \\ 1 & 0 \end{bmatrix} \quad \text{and} \quad B = \begin{bmatrix} 0 & 1 \\ 0 & 0 \end{bmatrix}.$$

Show that the eigenvalues of A, B, and $A + B$ do not satisfy Weyl's inequalities.

P.16.27 Adopt the notation of Theorem 16.5.2. Show that the "bottom-up" majorization inequalities (16.5.3) are equivalent to the "top-down" inequalities

$$a_n + a_{n-1} + \cdots + a_{n-k+1} \le \lambda_n + \lambda_{n-1} + \cdots + \lambda_{n-k+1}, \quad k = 1, 2, \ldots, n,$$

with equality for $k = n$.

P.16.28 Let $A \in \mathsf{M}_n$, let $\sigma_1 \ge \sigma_2 \ge \cdots \ge \sigma_n$ be its singular values, and let $c_1 \ge c_2 \ge \cdots \ge c_n$ be the decreasingly ordered Euclidean norms of the columns of A. (a) Show that

$$c_n^2 + c_{n-1}^2 + \cdots + c_{n-k+1}^2 \ge \sigma_n^2 + \sigma_{n-1}^2 + \cdots + \sigma_{n-k+1}^2, \tag{16.9.1}$$

$k = 1, 2, \ldots, n$, with equality for $k = n$. (b) If A has some columns or rows with small Euclidean norm, why must it have some small singular values? (c) If A is normal, what does (16.9.1) tell you about its eigenvalues?

P.16.29 Let $\lambda, c \in \mathbb{R}$, $\mathbf{y} \in \mathbb{C}^n$, and

$$A = \begin{bmatrix} \lambda I_n & \mathbf{y} \\ \mathbf{y}^* & c \end{bmatrix} \in \mathsf{M}_{n+1}.$$

(a) Use Theorem 16.3.3 to show that λ is an eigenvalue of A with multiplicity at least $n - 1$. (b) Use the Cauchy expansion (3.4.13) to compute p_A and determine the eigenvalues of A.

P.16.30 Adopt the notation of Definition 10.3.10. (a) Show that the companion matrix of $f = z^n + c_{n-1}z^{n-1} + \cdots + c_1 z + c_0$ can be written as

$$C_f = \begin{bmatrix} \mathbf{0}^\mathsf{T} & -c_0 \\ I_{n-1} & \mathbf{y} \end{bmatrix} \in \mathsf{M}_n, \qquad \mathbf{y} = \begin{bmatrix} -c_1 \\ -c_2 \\ \vdots \\ -c_{n-1} \end{bmatrix}.$$

(b) Show that

$$C_f^* C_f = \begin{bmatrix} I_{n-1} & \mathbf{y} \\ \mathbf{y}^* & c \end{bmatrix}, \qquad c = |c_0|^2 + |c_1|^2 + \cdots + |c_n|^2.$$

(c) Show that the singular values $\sigma_1 \geq \sigma_2 \geq \cdots \geq \sigma_n$ of C_f are

$$\sigma_1 = \frac{1}{2}\left(c + 1 + \sqrt{(c+1)^2 - 4|c_0|^2} \right),$$

$$\sigma_2 = \sigma_3 = \cdots = \sigma_{n-1} = 1, \quad \text{and}$$

$$\sigma_n = \frac{1}{2}\left(c + 1 - \sqrt{(c+1)^2 - 4|c_0|^2} \right).$$

P.16.31 Let $A \in \mathsf{M}_{n+1}$ be Hermitian and let $i \in \{1, 2, \ldots, n+1\}$. Let B denote the $n \times n$ matrix obtained by deleting the ith row and ith column of A. Show that the eigenvalues of A and B satisfy the interlacing inequalities (16.3.4).

P.16.32 Let $A \in \mathsf{M}_n$ and let B denote the matrix obtained by deleting either (a) two rows; (b) two columns; or (c) a row and a column from A. What is the analog of (16.3.11) in this case?

P.16.33 Let $A = [a_{ij}] \in \mathsf{M}_n(\mathbb{R})$ be tridiagonal and suppose that $a_{i,i+1}a_{i+1,i} > 0$ for each $i = 1, 2, \ldots, n-1$. Show that: (a) There is a real diagonal matrix D with positive diagonal entries such that DAD^{-1} is symmetric. (b) A is diagonalizable. (c) $\operatorname{spec} A \subset \mathbb{R}$. (d) Each eigenvalue of A has geometric multiplicity 1. *Hint*: Why is $\operatorname{rank}(A - \lambda I) \geq n - 1$ for every $\lambda \in \mathbb{C}$? (e) A has distinct eigenvalues. (f) If B is obtained from A by deleting any one row and the corresponding column, then the eigenvalues of B are real, distinct, and interlace the eigenvalues of A.

P.16.34 Let $A, B \in \mathsf{M}_n$ be positive semidefinite. Use (13.6.2) and (16.1.7) to show that $\|A \circ B\|_2 \leq \|A\|_2 \|B\|_2$.

P.16.35 Let $A, B \in \mathsf{M}_n$. (a) Prove that $\|A \circ B\|_2 \leq \|A\|_2 \|B\|_2$. This is an analog of Theorem 15.2.10.b. *Hint*: Use P.16.53, (16.3.11), and P.14.28 (b) Verify the inequality in (a) for the matrices in Example 13.5.2.

P.16.36 A normal matrix is *congruent to an inertia matrix, but a matrix can be *congruent to an inertia matrix without being normal. Let

$$S = \begin{bmatrix} 1 & 1 \\ 0 & -1 \end{bmatrix} \quad \text{and} \quad D = \begin{bmatrix} 1 & 0 \\ 0 & i \end{bmatrix}.$$

Compute $A = SDS^*$ and show that A is not normal.

P.16.37 Consider the matrix A defined in the preceding problem. Compute $A^{-*}A$ and show that it is similar to a unitary matrix. Why is this not surprising?

P.16.38 If $A \in \mathsf{M}_n$ is invertible, let $B = A^{-*}A$. (a) Show that B is similar to B^{-*}. (b) What can you say about the Jordan canonical form of B? (c) Let A be the matrix defined in P.16.40. Verify that B is similar to B^{-*}.

P.16.39 Deduce from Theorem 16.8.7 that two unitary matrices are *congruent if and only if they are similar.

P.16.40 Suppose that $A \in \mathsf{M}_n$ is invertible and *congruent to an inertia matrix D. (a) Show that $A^{-*}A$ is similar to the unitary matrix D^2. (b) Show that

$$A = \begin{bmatrix} 0 & 1 \\ 2 & 0 \end{bmatrix}$$

is not *congruent to an inertia matrix.

16.10 Notes

Some authors use a *decreasing* order $\lambda_1 \geq \lambda_2 \geq \cdots \geq \lambda_n$ for the eigenvalues of a Hermitian matrix. This convention has the advantage that the eigenvalues of a positive semidefinite matrix are its singular values, with the same indexing. It has the disadvantage that the indices and eigenvalues are not ordered in the same way as real numbers.

P.16.11 shows that a Hermitian matrix whose leading principal submatrices have nonnegative determinants need not be positive semidefinite. However, if *every* principal submatrix of a Hermitian matrix has nonnegative determinant, then the matrix is positive semidefinite; see [HJ13, Thm. 7.2.5.a].

The inequalities (16.2.3), (16.3.4), and (16.5.3) characterize the structures associated with them. For example, if real numbers λ_i and μ_i satisfy

$$\lambda_i \leq \mu_i \leq \lambda_{i+1} \leq \mu_n, \quad i = 1, 2, \ldots, n-1,$$

then there is a Hermitian $A \in \mathsf{M}_n$ and a positive semidefinite rank-1 matrix $E \in \mathsf{M}_n$ such that $\lambda_1, \lambda_2, \ldots, \lambda_n$ are the eigenvalues of A and $\mu_1, \mu_2, \ldots, \mu_n$ are the eigenvalues of $A + E$. If

$$\mu_1 \leq \lambda_1 \leq \mu_2 \leq \cdots \leq \lambda_n \leq \mu_{n+1},$$

then there is a Hermitian $A \in \mathsf{M}_n$ such that $\lambda_1, \lambda_2, \ldots, \lambda_n$ are the eigenvalues of A and $\mu_1, \mu_2, \ldots, \mu_{n+1}$ are the eigenvalues of a Hermitian matrix that borders it. If

$$\mu_1 + \mu_2 + \cdots + \mu_k \leq \lambda_1 + \lambda_2 + \cdots + \lambda_k, \quad k = 1, 2, \ldots, n,$$

with equality for $k = n$, then there is a Hermitian $A \in \mathsf{M}_n$ whose eigenvalues and main diagonal entries are $\lambda_1, \lambda_2, \ldots, \lambda_n$ and $\mu_1, \mu_2, \ldots, \mu_n$, respectively. Proofs of these and other inverse eigenvalue problems are in [HJ13, Sect. 4.3].

Theorem 16.6.9 was proved for real symmetric matrices in 1852 by James Joseph Sylvester, who wrote "my view of the physical meaning of quantity of matter inclines me, upon the ground of analogy, to give [it] the name of the Law of Inertia for Quadratic Forms." Sylvester's theorem was generalized to normal matrices (Theorem 16.8.7) in 2001, and to all square complex matrices in 2006. For more information about *congruence, see [HJ13, Sect. 4.5].

Matrices $A, B \in \mathsf{M}_n$ are *congruent* if there is an invertible $S \in \mathsf{M}_n$ such that $A = SBS^\mathsf{T}$. Congruence is an equivalence relation on M_n, but many of its properties are quite different from those of *congruence. For example, two symmetric matrices of the same size are congruent if and only if they have the same rank. Although $-A$ need not be *congruent to A, $-A$ is always congruent to A. However, A^T is always both congruent and *congruent to A. See [HJ13, Sect. 4.5] for more information about congruence.

Equality occurs in the Weyl inequalities (16.7.2) if and only if there is a nonzero vector $\mathbf{x}$ such that $A\mathbf{x} = \lambda_{i+j}(A)\mathbf{x}$, $B\mathbf{x} = \lambda_{n-j}(B)\mathbf{x}$, and $(A + B)\mathbf{x} = \lambda_i(A + B)\mathbf{x}$. For a discussion of this and other cases of equality for the inequalities in this chapter, see [HJ13, Sect. 4.3].

For more information about the numerical range see [HJ94, Ch. 1].

16.11 Some Important Concepts

- The Rayleigh quotient and eigenvalues of a Hermitian matrix.

- Eigenvalue interlacing for a Hermitian matrix and a rank-1 additive perturbation.

- Weyl's inequalities for eigenvalues of a sum of two Hermitian matrices.

- Eigenvalue interlacing for a bordered Hermitian matrix.

- Eigenvalue interlacing for a principal submatrix of a Hermitian matrix.

- Singular value interlacing if a row or column is deleted from a matrix.

- Sylvester's principal minor criterion for a positive definite matrix.

- Majorization inequalities for the eigenvalues and diagonal entries of a Hermitian matrix.

- Sylvester's inertia theorem about *congruence of Hermitian matrices and its generalization to normal matrices.

APPENDIX A
Complex Numbers

A.1 The Complex Number System

A *complex number* is an expression of the form $a + bi$, in which $a, b \in \mathbb{R}$ and i is a symbol with the property $i^2 = -1$. The set of all complex numbers is denoted by $\mathbb{C}$. The real numbers a and b are the *real* and *imaginary parts*, respectively, of the complex number $z = a + bi$. We write $\operatorname{Re} z = a$ and $\operatorname{Im} z = b$ ($\operatorname{Im} z$ is a real number). The number $i = 0 + 1i$ is the *imaginary unit*.

We identify a complex number $z = a + bi$ with the ordered pair $(a, b) \in \mathbb{R}^2$. In this context, we refer to $\mathbb{C}$ as the *complex plane*; see Figure A.1. We regard $\mathbb{R}$ as a subset of $\mathbb{C}$ by identifying the real number a with the complex number $a + 0i = (a, 0)$. If b is a real number, we identify the *purely imaginary* number bi with $(0, b)$. The number $0 = 0 + 0i$ is simultaneously real and imaginary; it is sometimes called the *origin*.

Definition A.1.1 Let $z = a + bi$ and $w = c + di$ be complex numbers. The *sum* $z + w$ and *product* zw of z and w are defined by

$$z + w = (a + c) + (b + d)i, \tag{A.1.2}$$

and

$$zw = (ac - bd) + (ad + bc)i. \tag{A.1.3}$$

If we write $z = (a, b)$ and $w = (c, d)$, then (A.1.2) is vector addition in $\mathbb{R}^2$; see Figure A.2. Therefore, $\mathbb{C}$ can be thought of as $\mathbb{R}^2$, endowed with the additional operation (A.1.3) of complex multiplication.

Example A.1.4 Let $z = 1 + 2i$ and $w = 3 + 4i$. Then $z + w = 4 + 6i$ and $zw = -5 + 10i$.

Although the definition of complex multiplication may appear unmotivated, there are good algebraic and geometric reasons to define the product of complex numbers in this manner. If we manipulate the imaginary unit i in the same manner as we manipulate real numbers, we find that

$$
\begin{aligned}
zw &= (a + bi)(c + di) \\
&= ac + bic + adi + bidi \\
&= ac + bdi^2 + i(ad + bc) \\
&= (ac - bd) + i(ad + bc)
\end{aligned}
$$

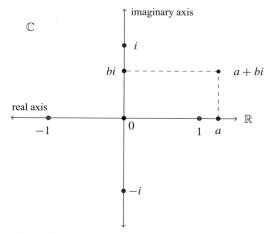

Figure A.1 The complex number $a + bi$ is identified with the point whose Cartesian coordinates are (a, b).

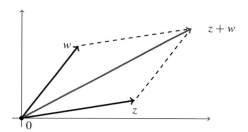

Figure A.2 Addition in the complex plane corresponds to vector addition in $\mathbb{R}^2$.

since $i^2 = -1$. Thus, the definition (A.1.3) is a consequence of the assumption that complex arithmetic obeys the same algebraic rules as real arithmetic.

We cannot assume without proof that complex arithmetic is commutative, associative, and distributive. The computation

$$(a + bi)(c + di) = (ac - bd) + (ad + bc)i = (ca - db) + (da + cb)i = (c + di)(a + bi)$$

shows that complex multiplication is *commutative*: $zw = wz$ for all $z, w \in \mathbb{C}$. To verify directly that the complex numbers satisfy the associative and distributive rules requires more work. Fortunately, a linear algebraic device saves us much effort and provides valuable geometric insight.

Consider

$$\mathfrak{C} = \left\{ \begin{bmatrix} a & -b \\ b & a \end{bmatrix} : a, b \in \mathbb{R} \right\} \subseteq \mathsf{M}_2(\mathbb{R})$$

and observe that the sum

$$\begin{bmatrix} a & -b \\ b & a \end{bmatrix} + \begin{bmatrix} c & -d \\ d & c \end{bmatrix} = \begin{bmatrix} a + c & -(b + d) \\ b + d & a + c \end{bmatrix} \tag{A.1.5}$$

and product

$$\begin{bmatrix} a & -b \\ b & a \end{bmatrix} \begin{bmatrix} c & -d \\ d & c \end{bmatrix} = \begin{bmatrix} ac - bd & -(ad + bc) \\ ad + bc & ac - bd \end{bmatrix} \tag{A.1.6}$$

of two matrices from $\mathfrak{C}$ also belongs to $\mathfrak{C}$. Thus, $\mathfrak{C}$ is closed under matrix addition and matrix multiplication.

For each complex number $z = a + bi$, define $M_z \in \mathfrak{C}$ by

$$M_z = \begin{bmatrix} a & -b \\ b & a \end{bmatrix}. \tag{A.1.7}$$

Then (A.1.5) and (A.1.6) are equivalent to

$$M_z + M_w = M_{z+w}, \qquad M_z M_w = M_{zw}, \tag{A.1.8}$$

in which $z = a + bi$ and $w = c + di$. In particular,

$$M_1 = \begin{bmatrix} 1 & 0 \\ 0 & 1 \end{bmatrix}, \qquad M_i = \begin{bmatrix} 0 & -1 \\ 1 & 0 \end{bmatrix},$$

and $(M_i)^2 = M_{-1}$, which reflects the fact that $i^2 = -1$.

The map $z \mapsto M_z$ is a one-to-one correspondence between $\mathbb{C}$ and $\mathfrak{C}$ that respects the operations of complex addition and multiplication. Anything that we can prove about addition and multiplication in $\mathfrak{C}$ translates immediately into a corresponding statement about $\mathbb{C}$ and vice-versa. For example, we may conclude that complex arithmetic is *associative* and *distributive* because matrix arithmetic is associative and distributive. For a direct verification of associativity of multiplication, see P.A.8.

Every nonzero complex number $z = a + bi$ has a multiplicative inverse, denoted by z^{-1}, which has the property that $zz^{-1} = z^{-1}z = 1$. The existence of multiplicative inverses is not obvious from the definition (A.1.3), although (A.1.8) helps make this clear. The inverse of z should be the complex number corresponding to

$$(M_z)^{-1} = \frac{1}{\det M_z} \begin{bmatrix} a & b \\ -b & a \end{bmatrix} = \begin{bmatrix} \frac{a}{a^2+b^2} & \frac{b}{a^2+b^2} \\ \frac{-b}{a^2+b^2} & \frac{a}{a^2+b^2} \end{bmatrix}.$$

This leads to

$$z^{-1} = \frac{a}{a^2 + b^2} - \frac{b}{a^2 + b^2}i, \tag{A.1.9}$$

which can be verified by direct computation.

Example A.1.10 If $z = 1 + 2i$, then $z^{-1} = \frac{1}{5} - \frac{2}{5}i$.

One can think of $\mathfrak{C}$ as a copy of the complex numbers system that is contained in $\mathsf{M}_2(\mathbb{R})$. Complex addition and multiplication correspond to matrix addition and matrix multiplication, respectively, in $\mathfrak{C}$.

The linear transformation on $\mathbb{R}^2$ induced by M_z acts as follows:

$$\begin{bmatrix} a & -b \\ b & a \end{bmatrix} \begin{bmatrix} c \\ d \end{bmatrix} = \begin{bmatrix} ac - bd \\ ad + bc \end{bmatrix}.$$

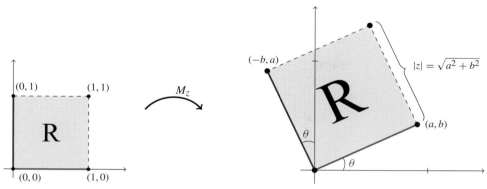

Figure A.3 Multiplication by the complex number $z = a + bi$; it scales by a factor of $|z| = \sqrt{a^2 + b^2}$ and induces a rotation of the complex plane about the origin through an angle of $\theta = \tan^{-1} \frac{b}{a}$ if $a \neq 0$.

Thus, multiplication by z maps the ordered pair (c, d), which represents the complex number $w = c + di$, to $(ac - bd, ad + bc)$, which represents the product zw. Figure A.3 illustrates how multiplication by $z = a + bi$ rotates the complex plane about the origin through an angle of $\theta = \tan^{-1} \frac{b}{a}$ if $a \neq 0$; it scales by a factor of $\sqrt{a^2 + b^2}$. Here and throughout the book, our convention is that the square root of a positive real number is a positive real number.

A.2 Modulus, Argument, and Conjugation

Definition A.2.1 Let $z = a + bi$ be a complex number. The *modulus* of z is the nonnegative real number defined by $|z| = \sqrt{a^2 + b^2}$. The *argument* of z, denoted arg z, is defined as follows:

(a) If $\operatorname{Re} z \neq 0$, then arg z is any angle θ that satisfies $\tan \theta = \frac{b}{a}$ (see Figure A.4).

(b) If $\operatorname{Re} z = 0$ and $\operatorname{Im} z > 0$, then arg z is any angle θ that differs from $\frac{\pi}{2}$ by an integer multiple of 2π.

(c) If $\operatorname{Re} z = 0$ and $\operatorname{Im} z < 0$, then arg z is any angle θ that differs from $-\frac{\pi}{2}$ by an integer multiple of 2π.

The argument of $z = 0$ is undefined.

Example A.2.2 If $z = 1 + 2i$, then $|z| = \sqrt{5}$ and $\theta = \tan^{-1} 2 = 1.107\,148\,7 \ldots$.

The argument of a nonzero complex number is determined only up to an additive multiple of 2π. For example, $\theta = -\frac{\pi}{2}$, $\theta = \frac{3\pi}{2}$, and $\theta = -\frac{5\pi}{2}$ all represent the argument of $z = -i$. The modulus of z is sometimes called the *absolute value* or *norm* of z. Since $|a + 0i| = \sqrt{a^2}$, the modulus function is an extension to $\mathbb{C}$ of the absolute value function on $\mathbb{R}$. We have $|z| = 0$ if and only if $z = 0$.

Definition A.2.3 The *conjugate* of $z = a + bi \in \mathbb{C}$ is $\bar{z} = a - bi$.

As a map from the complex plane to itself, $z \mapsto \bar{z}$ is reflection across the real axis; see Figure A.5. The map $z \mapsto -z$ is inversion with respect to the origin; see Figure A.6. The map

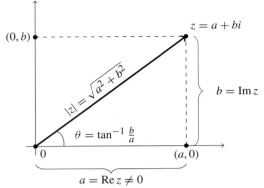

Figure A.4 The argument and modulus of a complex number that is not purely imaginary.

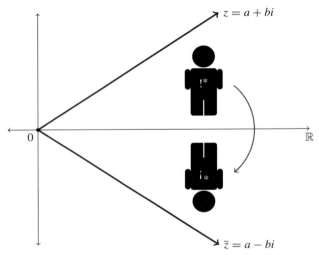

Figure A.5 Complex conjugation reflects the complex plane across the real axis.

$z \mapsto -\bar{z}$ is reflection across the imaginary axis; see Figure A.7. Multiplication by $\bar{z}$ stretches by $|z|$ and rotates about the origin through an angle of $-\arg z$.

We can express the real and imaginary parts of $z = a + bi$ using complex conjugation:

$$a = \operatorname{Re} z = \frac{1}{2}(z + \bar{z}) \quad \text{and} \quad b = \operatorname{Im} z = \frac{1}{2i}(z - \bar{z}). \tag{A.2.4}$$

Moreover, z is real if and only if $\bar{z} = z$, and z is purely imaginary if and only if $\bar{z} = -z$.

Conjugation is *additive* and *multiplicative*, that is,

$$\overline{z + w} = \bar{z} + \bar{w}$$

and

$$\overline{zw} = \bar{z}\,\bar{w}, \tag{A.2.5}$$

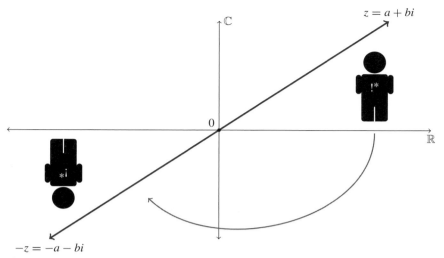

Figure A.6 The map $z \mapsto -z$ is inversion with respect to the origin.

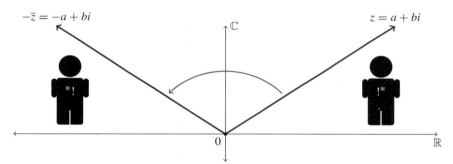

Figure A.7 The map $z \mapsto -\bar{z}$ is reflection across the imaginary axis.

respectively. To verify (A.2.5), let $z = a + bi$, let $w = c + di$, and compute

$$\overline{zw} = \overline{(ac - bd) + i(ad + bc)} = (ac - bd) - i(ad + bc)$$
$$= \big(ac - (-b)(-d)\big) + i\big(a(-d) + b(-c)\big)$$
$$= (a - bi)(c - di)$$
$$= \bar{z}\,\bar{w}.$$

Another proof of (A.2.5) follows from (A.1.8) and the fact that $M_{\bar{z}} = M_z^{\mathsf{T}}$. Since

$$M_{\overline{zw}} = (M_{zw})^{\mathsf{T}} = (M_z M_w)^{\mathsf{T}} = M_w^{\mathsf{T}} M_z^{\mathsf{T}} = M_{\bar{w}} M_{\bar{z}} = M_{\bar{z}} M_{\bar{w}},$$

we conclude that $\overline{zw} = \bar{z}\,\bar{w}$

An important identity relates the modulus to conjugation:

$$|z|^2 = z\bar{z}. \qquad (A.2.6)$$

This can be verified by direct computation:

$$z\bar{z} = (a + bi)(a - bi) = a^2 + b^2 + i(ab - ba) = a^2 + b^2 = |z|^2.$$

For a geometric explanation of (A.2.6), observe that multiplication by z and $\bar{z}$ rotate by $\arg z$ and $-\arg z$, respectively, while both scale by a factor of $|z|$. Multiplication by $z\bar{z}$ therefore induces no net rotation and scales by a factor of $|z|^2$.

We can use (A.2.6) to simplify quotients of complex numbers. If $w \neq 0$, write

$$\frac{z}{w} = \frac{z\bar{w}}{w\bar{w}} = \frac{z\bar{w}}{|w|^2},$$

in which the denominator on the right-hand side is real and positive. We also obtain from (A.2.6) another formula for the multiplicative inverse of a complex number:

$$z^{-1} = \frac{\bar{z}}{|z|^2}, \qquad z \neq 0.$$

Substituting $z = a + bi$ into the preceding yields our original formula (A.1.9).

Like its real counterpart, the complex absolute value is *multiplicative*, that is,

$$|zw| = |z||w| \tag{A.2.7}$$

for all $z, w \in \mathbb{C}$. This is more subtle than it appears, for (A.2.7) is equivalent to the algebraic identity

$$\underbrace{(ac - bd)^2 + (ad + bc)^2}_{|zw|^2} = \underbrace{(a^2 + b^2)}_{|z|^2} \underbrace{(c^2 + d^2)}_{|w|^2}.$$

One can verify (A.2.7) by using (A.2.6):

$$|zw|^2 = (zw)(\overline{zw}) = (z\bar{z})(w\bar{w}) = |z|^2|w|^2.$$

We can also establish (A.2.7) using determinants and the fact that $\det M_z = a^2 + b^2 = |z|^2$:

$$|zw|^2 = \det M_{zw} = \det(M_z M_w) = (\det M_z)(\det M_w) = |z|^2|w|^2.$$

An important consequence of (A.2.7) is that $zw = 0$ if and only if $z = 0$ or $w = 0$. If $zw = 0$, then $0 = |zw| = |z||w|$, so at least one of $|z|$ or $|w|$ is zero, which implies that $z = 0$ or $w = 0$.

Observe that

$$|\operatorname{Re} z| \leq |z| \quad \text{and} \quad |\operatorname{Im} z| \leq |z|, \tag{A.2.8}$$

since $|z|$ is the length of the hypotenuse of a right triangle with sides of length $|\operatorname{Re} z|$ and $|\operatorname{Im} z|$; see Figure A.4. To derive (A.2.8) algebraically, observe that $|a| \leq \sqrt{a^2 + b^2}$ and $|b| \leq \sqrt{a^2 + b^2}$ for all $a, b \in \mathbb{R}$. More generally, we have the complex *triangle inequality*,

$$|z + w| \leq |z| + |w|, \tag{A.2.9}$$

with equality if and only if one of z or w is a real nonnegative multiple of the other.

If we think of z and w as vectors in $\mathbb{R}^2$, then the triangle inequality asserts that the sum of the lengths of two sides of a triangle is always greater than or equal to the length of the other side; see Figure A.2.

It is useful to give an algebraic proof of the triangle inequality since it demonstrates some techniques to deal with complex numbers:

$$\left(|z| + |w|\right)^2 - |z + w|^2 = \left(|z|^2 + 2|z||w| + |w|^2\right) - (z + w)(\bar{z} + \bar{w})$$
$$= |z|^2 + 2|z||w| + |w|^2 - |z|^2 - |w|^2 - z\bar{w} - \bar{z}w$$
$$= 2|z||\bar{w}| - (z\bar{w} + \bar{z}w)$$
$$= 2|z\bar{w}| - 2\operatorname{Re} z\bar{w}$$
$$= 2(|z\bar{w}| - \operatorname{Re} z\bar{w})$$
$$\geq 0.$$

We invoke (A.2.8) for the final equality. We also see that $|z+w| = |z|+|w|$ if and only if $|z\bar{w}| = \operatorname{Re} z\bar{w}$, that is, (A.2.9) is an equality if and only if $z\bar{w} = r$ is real and nonnegative. If $r = 0$, then either $z = 0$ or $w = 0$. If $r > 0$, then $z|w|^2 = z\bar{w}w = rw$, so $w = z|w|^2 r^{-1}$. Thus, (A.2.9) is an equality if and only if one of z or w is a real nonnegative multiple of the other. Geometrically this means that z and w both lie on the same ray from the origin in the complex plane.

A.3 Polar Form of a Complex Number

Since complex multiplication is equivalent to rotating and scaling in the plane, polar coordinates might yield some insights. For $z = a + bi$, let $a = r\cos\theta$ and $b = r\sin\theta$, in which $r = |z|$ and $\theta = \arg z$. Thus,

$$a + bi = r(\cos\theta + i\sin\theta),$$

in which $r \geq 0$; see Figure A.8. The complex number $\cos\theta + i\sin\theta$ lies on the *unit circle* $\{z \in \mathbb{C} : |z| = 1\}$ since

$$|\cos\theta + i\sin\theta| = \sqrt{\cos^2\theta + \sin^2\theta} = 1.$$

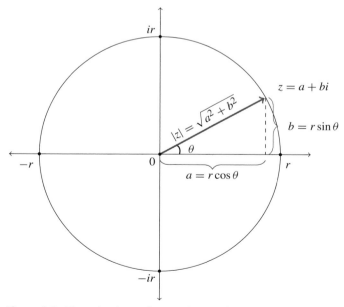

Figure A.8 The polar form of a complex number.

Writing z in its *polar form* $z = r(\cos \theta + i \sin \theta)$ separates the two distinct transformations induced by multiplication by z:

- Multiplication by $r \geq 0$ scales by a factor of r.

- Multiplication by $\cos \theta + i \sin \theta$ rotates about the origin through an angle of θ.

This suggests a general rule for complex multiplication: to multiply two complex numbers, add the arguments and multiply the lengths. A short computation shows why this is true. The product of $z = r_1(\cos \theta_1 + i \sin \theta_1)$ and $w = r_2(\cos \theta_2 + i \sin \theta_2)$ is

$$zw = r_1 r_2 (\cos \theta_1 + i \sin \theta_1)(\cos \theta_2 + i \sin \theta_2)$$
$$= r_1 r_2 [(\cos \theta_1 \cos \theta_2 - \sin \theta_1 \sin \theta_2) + i(\cos \theta_1 \sin \theta_2 + \sin \theta_1 \cos \theta_2)]$$
$$= r_1 r_2 [\cos(\theta_1 + \theta_2) + i \sin(\theta_1 + \theta_2)].$$

Thus, $|zw| = r_1 r_2 = |z||w|$ and $\arg zw = \theta_1 + \theta_2 = \arg z + \arg w$.

If we let $z = r(\cos \theta + i \sin \theta)$ with $r > 0$ and apply the preceding rule repeatedly, we obtain *de Moivre's formula*

$$z^n = r^n(\cos n\theta + i \sin n\theta), \tag{A.3.1}$$

which is valid for all $n \in \mathbb{Z}$.

Example A.3.2 Set $r = 1$ and $n = 2$ in (A.3.1) to obtain

$$(\cos \theta + i \sin \theta)^2 = \cos 2\theta + i \sin 2\theta.$$

Expand the left-hand side of the preceding equation to obtain

$$(\cos^2 \theta - \sin^2 \theta) + (2 \sin \theta \cos \theta)i = \cos 2\theta + i \sin 2\theta. \tag{A.3.3}$$

Compare the real and imaginary parts of (A.3.3) to obtain the *double angle formulas*

$$\cos 2\theta = \cos^2 \theta - \sin^2 \theta \quad \text{and} \quad \sin 2\theta = 2 \sin \theta \cos \theta.$$

Definition A.3.4 For a positive integer n, the solutions to the equation $z^n = 1$ are the *nth roots of unity*.

Example A.3.5 Compute the cube roots of unity using de Moivre's formula. If $z^3 = 1$, then (A.3.1) ensures that $r^3 = 1$ and $\cos 3\theta + i \sin 3\theta = 1$. Since $r \geq 0$, it follows that $r = 1$. The restrictions on θ tell us that $3\theta = 2\pi k$ for some integer k. Since $\theta = \frac{2\pi k}{3}$, there are only three distinct angles $\theta \in [0, 2\pi)$ that can arise this way, namely 0, $\frac{2\pi}{3}$, and $\frac{4\pi}{3}$. Therefore, the cube roots of unity are

$$\cos 0 + i \sin 0 = 1,$$
$$\cos \tfrac{2\pi}{3} + i \sin \tfrac{2\pi}{3} = -\tfrac{1}{2} + i \tfrac{\sqrt{3}}{2}, \quad \text{and}$$
$$\cos \tfrac{4\pi}{3} + i \sin \tfrac{4\pi}{3} = -\tfrac{1}{2} - i \tfrac{\sqrt{3}}{2}.$$

These three points form the vertices of an equilateral triangle; see Figure A.9.

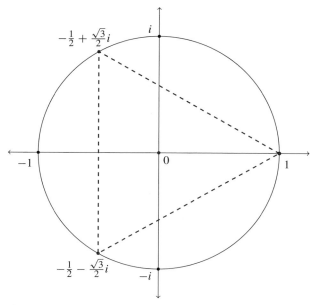

Figure A.9 The cube roots of unity are the vertices of an equilateral triangle inscribed in the unit circle.

In general, for a nonzero $z = r(\cos\theta + i\sin\theta)$ there are precisely n distinct nth roots of z, given by the formula

$$r^{1/n}\left[\cos\left(\frac{\theta + 2\pi k}{n}\right) + i\sin\left(\frac{\theta + 2\pi k}{n}\right)\right], \qquad k = 0, 1, \ldots, n - 1,$$

in which $r^{1/n}$ denotes the positive nth root of the positive real number r.

Definition A.3.6 Let $z = x + iy$, in which x and y are real. The *complex exponential* function e^z is defined by

$$e^z = e^{x+iy} = e^x(\cos y + i\sin y). \tag{A.3.7}$$

It can be shown that the complex exponential function satisfies the addition formula $e^{z+w} = e^z e^w$ for all complex z, w. A special case of (A.3.7) is *Euler's formula*

$$e^{iy} = \cos y + i\sin y, \qquad y \in \mathbb{R}. \tag{A.3.8}$$

Although the definition (A.3.7) may appear mysterious, it can be justified through the use of power series; see P.A.13. Since cosine is an even function and sine is odd, it follows from (A.3.8) that

$$\cos y = \frac{1}{2}(e^{iy} + e^{-iy}), \qquad \sin y = \frac{1}{2i}(e^{iy} - e^{-iy}).$$

The complex exponential permits us to express the polar form of a complex number as

$$z = r(\cos\theta + i\sin\theta) = re^{i\theta}.$$

Thus, de Moivre's formula can be stated as

$$z^n = r^n e^{in\theta}.$$

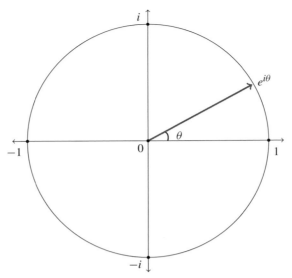

Figure A.10 The unit circle in the complex plane.

Example A.3.9 One can use Euler's formula to deduce many trigonometric identities. For example, equate real and imaginary parts in

$$\cos(\theta_1 + \theta_2) + i\sin(\theta_1 + \theta_2) = e^{i(\theta_1 + \theta_2)} = e^{i\theta_1} e^{i\theta_2}$$
$$= (\cos\theta_1 + i\sin\theta_1)(\cos\theta_2 + i\sin\theta_2)$$
$$= (\cos\theta_1 \cos\theta_2 - \sin\theta_1 \sin\theta_2)$$
$$+ i(\cos\theta_1 \sin\theta_2 + \sin\theta_1 \cos\theta_2)$$

to obtain the addition formulas for sine and cosine.

Euler's formula (A.3.8) implies that

$$|e^{i\theta}|^2 = (\cos\theta + i\sin\theta)(\cos\theta - i\sin\theta) = \cos^2\theta + \sin^2\theta = 1.$$

Thus, the set $\{e^{i\theta} : -\pi < \theta \le \pi\}$ is the unit circle in the complex plane; see Figure A.10.

A.4 Problems

P.A.1 Show that:

(a) $(1 + i)^4 = -4$

(b) $(1 - i)^{-1} - (1 + i)^{-1} = i$

(c) $(i - 1)^{-4} = -1/4$

(d) $10(1 + 3i)^{-1} = 1 - 3i$

(e) $(\sqrt{3} + i)^3 = 8i$

P.A.2 Evaluate the following expressions. Write your answer in the form $a + bi$, in which a and b are real.

(a) $(1 + i)(2 + 3i)$

(b) $\dfrac{2 + 3i}{1 + i}$

(c) $\left(\dfrac{(2 + i)^2}{4 - 3i}\right)^2$

(d) $e^{i\alpha}\overline{e^{i\beta}}$, in which α and β are real

(e) $\overline{(1 - i)\overline{(2 + 2i)}}$

(f) $|2 - i|^3$

P.A.3 Let $z = 1 + i$. Plot and label the following points in the complex plane: $z, -z, \bar{z}, -\bar{z},$ $1/z, 1/\bar{z},$ and $-1/\bar{z}$.

P.A.4 Write $z = \sqrt{3} - i$ in polar form $z = re^{i\theta}$.

P.A.5 Write the following in polar form: (a) $1 + i$, (b) $(1 + i)^2$, (c) $(1 + i)^3$.

P.A.6 What are the square roots of $1 + i$? Draw a picture that shows their locations.

P.A.7 What are the third roots of $z = 8i$? Draw a picture that shows their locations.

P.A.8 Verify that complex multiplication is associative, that is, show that

$$[(a + bi)(c + di)](e + fi) = (a + bi)[(c + di)(e + fi)]$$

for all $a, b, c, d, e, f \in \mathbb{R}$.

P.A.9 Show that $|z| = 1$ if and only if $z \neq 0$ and $1/z = \bar{z}$.

P.A.10 If $z \neq 0$, show that $\overline{1/z} = 1/\bar{z}$.

P.A.11 Show that $\operatorname{Re} z^{-1} > 0$ if and only if $\operatorname{Re} z > 0$. What about $\operatorname{Im} z^{-1} > 0$?

P.A.12 Show that $|z + w|^2 - |z - w|^2 = 4\operatorname{Re}(z\bar{w})$.

P.A.13 Derive Euler's formula (A.3.8) by substituting $z = iy$ into the power series expansion

$$e^z = \sum_{n=0}^{\infty} \frac{z^n}{n!}$$

for the exponential function. *Hint*:

$$\cos z = \sum_{n=0}^{\infty} (-1)^n \frac{1}{(2n)!} z^{2n} \quad \text{and} \quad \sin z = \sum_{n=0}^{\infty} (-1)^n \frac{1}{(2n + 1)!} z^{2n+1}.$$

P.A.14 Let a, b, c be distinct complex numbers. Prove that a, b, c are the vertices of an equilateral triangle if and only if $a^2 + b^2 + c^2 = ab + bc + ca$. *Hint*: Consider the relationship between $b - a$, $c - b$, and $a - c$.

P.A.15 Suppose that $|a| = |b| = |c| \neq 0$ and $a + b + c = 0$. What can you conclude about a, b, c?

P.A.16 Prove that

$$1 + 2\sum_{n=1}^{N} \cos nx = \frac{\sin((N + \frac{1}{2})x)}{\sin \frac{x}{2}}$$

by using the formula

$$1 + z + \cdots + z^{n-1} = \frac{1 - z^n}{1 - z}$$

for the sum of a finite geometric series in which $z \neq 1$; see P.0.8.

References

[Bha05] Rajendra Bhatia, *Fourier Series*, Classroom Resource Materials Series, Mathematical Association of America, Washington, DC, 2005, Reprint of the 1993 edition [Hindustan Book Agency, New Delhi].

[ĆJ15] Branko Ćurgus and Robert I. Jewett, Somewhat stochastic matrices, *Amer. Math. Monthly* **122** (2015), no. 1, 36–42.

[Dav63] Philip J. Davis, *Interpolation and Approximation*, Blaisdell Publishing Co., Ginn and Co., New York, Toronto, London, 1963.

[DF04] David S. Dummit and Richard M. Foote, *Abstract Algebra*, 3rd edn., John Wiley & Sons, Inc., Hoboken, NJ, 2004.

[Dur12] Peter Duren, *Invitation to Classical Analysis*, Pure and Applied Undergraduate Texts, vol. 17, American Mathematical Society, Providence, RI, 2012.

[GVL13] Gene H. Golub and Charles F. Van Loan, *Matrix Computations*, 4th edn., Johns Hopkins Studies in the Mathematical Sciences, Johns Hopkins University Press, Baltimore, MD, 2013.

[HJ94] Roger A. Horn and Charles R. Johnson, *Topics in Matrix Analysis*, Cambridge University Press, Cambridge, 1994, Corrected reprint of the 1991 original.

[HJ13] Roger A. Horn and Charles R. Johnson, *Matrix Analysis*, 2nd edn., Cambridge University Press, Cambridge, 2013.

[JS96] Charles R. Johnson and Erik A. Schreiner, The relationship between AB and BA, *Amer. Math. Monthly* **103** (1996), no. 7, 578–582.

[Var04] Richard S. Varga, *Geršgorin and his Circles*, Springer Series in Computational Mathematics, vol. 36, Springer-Verlag, Berlin, 2004.

Index